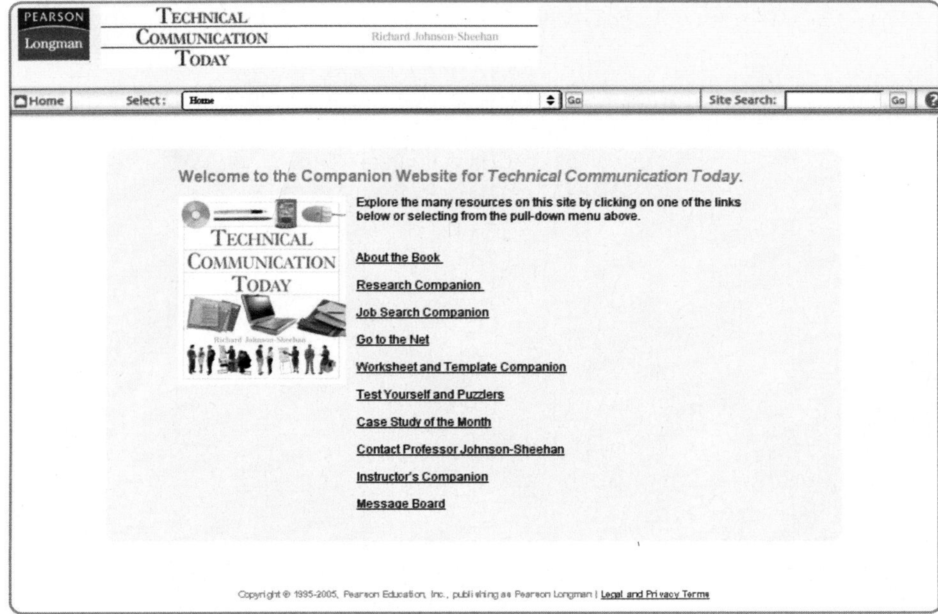

In addition to the **Go to the Net** resources available on Johnsonweb, you will also find:

Worksheet and Template Companion—Help yourself analyze readers, invent ideas, design documents, and revise your work by downloading the worksheets and templates provided in this section of the site.

Test Yourself and Puzzlers—Improve your retention of key concepts and practice important skills by using the ready-made quizzes and essay questions tailored for each chapter.

Case Study of the Month—Read a featured case study and discuss your thoughts with your instructor. Appropriate responses will be posted, and for those interested in revisiting case studies of the previous months, you may access them through the archives.

Research Companion—Start your research through this invaluable asset, which links you to other powerful search engines, online libraries, reference tools, and strategies for developing effective research methodologies.

Job Search Companion—Begin your search for employment opportunities now. Go to this one-stop spot for finding and landing a job. Here, you will also find links to sites that offer a variety of tools and templates for developing winning resumes and letters of application; excellent advice on effective interviewing; online job search engines; and additional advice about finding jobs in technical disciplines.

Contact Professor Johnson-Sheehan—Keep in touch with Richard Johnson-Sheehan by using this space for correspondence via blog, listserv, and e-mail dedicated exclusively to the book.

Technical Communication Today

Richard Johnson-Sheehan
Purdue University

New York San Francisco Boston
London Toronto Sydney Tokyo Singapore Madrid
Mexico City Munich Paris Cape Town Hong Kong Montreal

To Tracey, Emily, and Collin

Longman Publishers
Publisher: Joseph Opiela
Development Manager: Janet Lanphier
Senior Marketing Manager: Melanie Craig
Senior Supplements Editor: Donna Campion
Media Supplements Editor: Jenna Egan
Managing Editor: Valerie Zaborski
Production Managers: Joseph Vella and Doug Bell
Project Coordination: Elm Street Publishing Services, Inc.
Cover Design Manager: John Callahan
Manufacturing Manager: Mary Fischer
Printer and Binder: R. R. Donnelley and Sons, Inc.—Crawfordsville
Cover Printer: The Lehigh Press

Dorling-Kindersley, UK
DK Design Director: Stuart Jackman
Project Manager; DK Designs: Nigel Duffield
Senior Project Art Editor: Anthony Limerick
Senior Designer: Lee Redmond
Designer: Yvonne Cornes
DTP: David McDonald
Picture Researchers: Rachael Swann, Cecilia Mackay
Cover Design: Stuart Jackman

For permission to use copyrighted material, grateful acknowledgment is made to the copyright holders on pp. A-42–A-43, which are hereby made part of this copyright page.

Library of Congress Cataloging-in-Publication Data
Johnson-Sheehan, Richard.
Technical communication today / Richard D. Johnson-Sheehan.
p. cm.
Includes bibliographical references and index.
ISBN 0-321-11764-6
1. Communication of technical information. I. Title.
T10.5.J65 2005
601'.4—dc22

2004018221

Please visit our website at http://www.ablongman.com/johnsonweb.

ISBN 0-321-11764-6

3 4 5 6 7 8 9 10—DOC—07 06

Contents

Part 4: Genres of Technical Communication 445

Preface

As we move further into the 21st century, the technical workplace is dramatically redefining its relationship to the written and spoken word. Right before our eyes, the advent of the computer is reshaping our understanding of text in ways that we are only now beginning to understand. It's an exciting and challenging time.

Technical Communication Today addresses these shifts in workplace communication. Today, people use their computers to help them think, research, compose, design, and edit. Computers are no longer merely helpful tools that replace the typewriter in the writing process. Instead, for most people, they are essential *thinking tools* that powerfully influence how technical documents and presentations are developed, produced, designed, and delivered. Moreover, with e-mail, instant messaging, and the Internet, computers are important communication portals through which people join the ongoing electronic conversation around them. By centralizing the computer in the writing process, this book shows students and professionals how to take full advantage of this important workplace tool.

This book was written to help students and professionals negotiate the computer-centered technical workplace. Traditionally, books like this one addressed engineers and engineering students. As the technical workplace has expanded during the Information Age, though, people working in other fields find themselves needing to communicate technical information. To meet this need, this book addresses a broader spectrum of people, including those who need to communicate in the natural sciences, social sciences, computer science, medicine, public relations, law, business—and engineering.

Guiding Themes

In times of change, like ours, it is important to develop new communication strategies while retaining proven approaches to writing and speaking. In this book, I have tried to incorporate the newest technology-driven changes in workplace communication. You will find in-depth instruction on using e-mail, working in virtual teams, making websites, using digital audio and visual technology, negotiating electronic ethics, using instant messaging (IM), working with wireless networks, and using personal digital assistants (PDAs), among many other Information Age strategies and tools. But, you will also find that the book is based on a solid core of rhetorical principles that have been around for at least two and a half millennia. In fact, these core principles hold up surprisingly well in this Information Age, perhaps becoming even more relevant as we return to a more visual and oral culture.

It has been my intent from the beginning to write a book that teaches students core principles of rhetoric, while showing them how to use computers in a rapidly evolving, information-based society.

Computers as Thinking Tools

The foremost theme of this book is that computers are now integral and indispensable in technical communication. This premise may seem obvious to many readers; yet the majority of technical communication textbooks do not successfully integrate computers into their discussions of workplace communication. Instead, most of these books have simply grafted computer use onto a writing process that was originally developed for the pen and typewriter. These old warhorse textbooks march students through the stages of prewriting, drafting, revising, and proofreading, with little recognition that these stages are mostly suited to the kinds of writing that existed before computers. As a result, they often limit computers to their word-processing abilities. They do not adequately show students how to fully use their computers to succeed in a networked technical workplace.

An important step toward a new understanding of the writing process involves centering the computer in the technical workplace and in student learning. We need to recognize that students use the computer as a thinking tool from beginning to end, inventing their ideas and composing text at the same time. In this book, the writing process has been redefined around the computer as a communication medium. As a result, the writing process described here is far more in line with the kinds of computer-centered activities in the technical workplace.

Visual-Spatial Reading, Thinking, and Composing

This book also reflects an ongoing evolution in technical communication from *literal-linear* texts toward *visual-spatial* documents and presentations. The invention of the computer, like the invention of the printing press, is bringing about radical changes in our society. One of these changes is that people are thinking more visually and spatially. They rarely read documents linearly word for word from front to back. Instead, they conceptualize texts as spaces of information that need to be visually navigated. They rely on visual and spatial strategies to "raid" documents and presentations for the information they need. Here is perhaps one of the more profound outgrowths of the Information Age. People see documents as spaces where information is stored and flows. Visual-spatial reading, thinking, and composing means interacting with text in three dimensions, much like navigating a building.

This book addresses this evolution toward visual-spatial in thinking four ways:

- First, it teaches writers and speakers how to harness their visual-spatial abilities. This book shows them how to use visual-spatial techniques to research, invent, draft, design, and edit their work.
- Second, it shows how to compose visual-spatial documents and presentations that take advantage of readers' inclination to navigate texts. It teaches students how to write and speak visually, while designing highly navigable documents and presentations.
- Third, the book shows how to compose visual-spatial documents like hypertexts, websites, and multimedia presentations. Writing in these environments is becoming increasingly important as companies move their communications and documentation on-line.
- Finally, it practices what it preaches by presenting information in a visual-spatial way that will be more accessible to today's students. Clearly, students learn differently now than they did even a couple of decades ago. This book reflects their ability to think visually and spatially.

This visual-spatial turn is an important intellectual shift in our culture—one that we do not fully understand at the moment. We do know, however, that communicating visually and spatially involves more than adding headings and charts to documents or using PowerPoint to enhance oral presentations. Instead, we must recognize that the advent of the computer, which is a visual-spatial medium, is revolutionizing how we conceptualize the world and how we communicate. Increasingly, people are thinking visually and spatially in addition to literally and linearly. This book tries to incorporate this important change.

The Activity of Technical Communication

Finally, an important theme of this book is its stress on the *activity* of technical communication. In this computer-centered age, people learn by doing, not by passively listening or reading. So, this book emphasizes the activity of producing effective documents and presentations. Each chapter follows a process approach that mirrors how professionals communicate in the technical workplace. Meanwhile, it shows students how to pay close attention to the evolving workplace contexts in which communication happens.

Perhaps this theme comes about through my experiences with students and my observations of people using books like this one. As someone who has consulted and taught technical communication for nearly two decades, I began to realize that people, especially young people, learn differently than they did before. Today, students rarely read the textbook. Instead, they *raid* their textbooks for the specific information they need to complete a task. They use their textbooks like they use websites. They ask questions of the text and then look for the answers.

We might condemn our students as lazy or undisciplined (as some mistakenly do), but those of us who also work in the technical workplace know that *everyone* reads like this now. People read with an activity in mind. They use their books as mentors that teach them step by step how to accomplish a task. My assumption is that students will not read this book page by page. Instead, they will have this book open next to their computers, reading/raiding it *as they are composing*. This book has been written and designed with this kind of reading in mind.

Review of the Content

Let us turn to a review of the content of the book. *Technical Communication Today* is divided into four parts:

Part 1: Elements of Technical Communication

In this part of the book, the basic features of technical communication are defined and discussed. Chapter 1 introduces students to the study of technical communication, showing them its importance to their careers. In Chapter 2, students learn about the electronic writing process, studying how computers have changed our approach to communicating technical information. Chapter 3 provides some strategies for using computers to do detailed reader and context analysis when preparing to write a document or make a presentation. In this chapter, students also learn how to handle the needs and conventions of international and cross-cultural readers. In Chapter 4, students learn how to negotiate the complex ethical issues involved in the technical workplace, exploring strategies for resolving these thorny ethical dilemmas.

Part 2: Communicating in the Technical Workplace

This part describes the electronic writing process in greater detail, showing how to use computers as tools for researching, organizing, drafting, improving style, designing documents, creating graphics, and revising and editing. Chapter 5 stresses that the challenge in technical communication is often not generating new information; rather, it is effectively managing the information that already exists. This chapter shows students how to use their computers to "triangulate" the many sources of information available to them. In Chapter 6, students learn how to organize information into common genres and conventions used in the technical workplace. Chapter 7 teaches students how to use plain and persuasive style to strengthen the clarity and effectiveness of technical prose. Chapter 8 describes the design process for documents and interfaces, showing how computers are used to create visually effective documents and websites. In Chapter 9, students learn how to create and use graphics that are commonly found in technical documents and presentations. Chapter 10 provides tools and strategies for revising and editing to improve documents and presentations. Chapter 11 discusses how to present technical information orally, using computers to organize and deliver information effectively.

Part 3: Working in the Wired Workplace

Here is perhaps the most innovative part of this book. It shows students how technical professionals work through the Internet at the workplace. Chapter 12 discusses the importance and use of e-mail and instant messaging, offering strategies for using these communication tools effectively while avoiding typical problems. In Chapter 13, students learn how to work in companies that are increasingly moving to virtual teaming, in which the Internet plays a central role. Chapter 14 discusses the development and design of websites and other multimedia hypertexts in the technical workplace. Chapter 15 offers a basic discussion of conducting an electronic job search, including creating a portfolio, designing a resume, and crafting an effective letter of application. In this chapter, students learn how to take full advantage of their computers as career development tools.

Part 4: Genres of Technical Communication

Finally, in Part 4 the book offers detailed discussions of the major genres in technical communication, paying close attention to how the advent of computers is changing these genres to suit the Information Age. Chapter 16 discusses how to write effective letters and memos, showing how the invention of e-mail has significantly changed how these documents are written and interpreted. Chapter 17 describes how to write technical definitions in a rapidly evolving workplace. Chapter 18 shows how to describe technical products and services, demonstrating the importance of specifications to an increasingly globalized economy. Chapter 19 discusses writing instructions and procedures, stressing the fact that these documents are increasingly being placed online and using multimedia elements. Chapter 20 offers an advanced discussion of writing proposals, taking full advantage of the computer's ability to make documents more persuasive. Chapter 21 shows how to write activity reports and make briefings in a workplace that is becoming increasingly paperless. Chapter 22 discusses the writing of analytical reports, demonstrating how research has shifted to take full advantage of the networked computer.

In Part 4, the chapters are admittedly repetitious in some ways. They are designed to lead students through a consistent electronic writing process that reflects the activities in the technical workplace.

Features of Chapters

Each chapter includes regular features that students and instructors will find helpful and engaging.

Chapter Objectives and Chapter Review

At the beginning of each chapter, the *Chapter Objectives* are listed, so students know exactly the learning objectives for the chapter. Then, at the end of each chapter, a *Chapter Review* summarizes the main points to help students review what they have learned. These bookend features should help instructors and students identify the most important concepts and issues covered in each chapter. They will also help students identify the key points that they should pay attention to as they use the chapter.

Help Boxes

Each chapter includes a *Help* box that describes how a particular computer application or computer-related strategy can be used to improve the production of documents and presentations or communications in the workplace. The *Help* boxes describe applications that are common to most computer software packages. In many cases, they offer helpful tips and strategies to help students enhance and streamline their research, drafting, editing, and communicating with computers.

Go to the Net

At the bottom of pages, students will find helpful *Go to the Net* notes. These notes refer students to the book's Companion Website. At this website, students will find a wealth of additional materials as examples. They will also find worksheets to download and links to other helpful websites.

At Work Boxes

In the *At Work* boxes in each chapter, professionals who work in technical workplaces answer questions about technical communication. These boxes offer helpful tips and strategies for improving technical communication, but they also offer students a glimpse into the challenges faced in real technical workplaces.

At a Glance Boxes

The *At a Glance* boxes in each chapter summarize key content for quick student review of major concepts. They are designed to make basic information highly accessible to the scanning reader. Moreover, they provide access points at which readers can enter the text and read further on important subjects.

Links

Links are provided in the text to alert students that additional information is available on a related topic elsewhere in the book. This textbook is designed to be used like a hypertext. These links allow students to move spatially through the text to find the information they need on demand.

Take Note

The *Take Note* comments in each chapter often provide quick asides to the readers, offering tips that may help them conceptualize difficult ideas and concepts. These notes are supplemental to the main text. They can be skipped by readers who feel comfortable with the content of the chapter. They are provided for readers who want more information on the subject.

Annotated Sample Documents

In each chapter, sample documents and texts are provided with annotations. The samples show students good examples of documents that have worked in the technical workplace. The annotations point out their major features. At the bottom of the page, *Go to the Net* notes will help students find more examples on the book's Companion Website.

Exercises and Projects

Every chapter ends with a number of *Exercises and Projects.* The "Individual or Team Projects" can be completed either individually or in small teams. The "Collaborative Project" is a larger project designed for larger teams. The *Exercises and Projects* are designed to challenge students with realistic situations, allowing them to put the information in the chapter into practice.

Case Studies

The *Case Studies* at the end of each chapter present situations typical of those students are likely to encounter in their jobs, usually involving an ethical dilemma. Students are encouraged to consider the personal, social, political, ethical, economic, and environmental issues that shape decisions in the technical workplace. These case studies are especially helpful for running class discussions, whether on-line or in the classroom.

Supplements to the Book

Included with this book are three important tools that instructors and students will find especially helpful: the Companion Website, a manual for grammar and style, and an *Instructor's Manual.*

Companion Website

The Companion Website (http://www.ablongman.com/johnsonweb) includes a wealth of sample documents, materials, and links that greatly expand on the content of the book. The website is designed to be used side by side with the book. Students are able to look at more samples, download worksheets, and learn from the excellent communication-related websites available on the Internet. I will be an active presence on the website, offering answers to common questions and providing a case study of the month to challenge students further.

Appendixes

At the end of the book, three appendixes have been included for quick reference by students. *Appendix A,* a grammar and punctuation guide, offers remedies for common grammar problems in technical communication and discussses the proper use of punctuation. *Appendix B* is a guide for those who speak English as a second language. *Appendix C,* a documentation guide, demonstrates the use of APA, CBE, and MLA citation styles.

Instructor's Manual

The *Instructor's Manual* offers teaching strategies for each chapter while providing prompts for class discussion and strategies for improving student writing and presenting. The *Instructor's Manual* will be available on-line, offering additional materials for downloading, such as slides and PowerPoint lectures. It will also offer additional ideas for assignments and projects.

Acknowledgments

Ironically, this is a book I swore I would never write. As a technical writer, consultant, and editor, I had always been skeptical of technical communication books like this one. They never seemed to get things just right, and they were always one or two steps behind the kinds of written and oral communication that were happening in the technical workplace (the "real world," as my students refer to it). Now I have a new respect for the brave souls who try to put these kinds of books together.

Many people collaborated on this book both formally and informally. The editors, Joseph Opiela, Janet Lanphier, and Judith Fifer, were essential in the writing of the book. Joe was responsible for the original concept of the book, and Janet and Judy made me practice what I preach. I would like to thank my colleagues, Professors Scott Sanders and Charles Paine. Our collaborations led to many of the concepts and strategies discussed in this book. Also, I would like to thank Paul Lynch, who wrote the *Instructor's Manual,* did much of the development work for the Companion Website, and secured many of the permissions.

Technical Communication Today represents the collaborative effort of many people. The reviewers of the book, especially, deserve mention. Their comments and suggestions were very helpful toward shaping and refining all facets of this book. My cordial thanks go to the following people: Valerie Balester, Texas A&M University; Brian Ballentine, Case Western Reserve University; Jeanelle Barrett, Tarleton State University; Gaby Bedetti, Eastern Kentucky University; Ann M. Blakeslee, Eastern Michigan University; Carol Brown, South Puget Sound Community College; Patricia Cearley, South Plains College; Frankie Greco Chadwick, University of Arkansas, Little Rock; Jim Clemmer, Austin Peay State University; Susan Codone, Mercer University;

Rocky Colavito, Northwestern State University; Paul Dombrowski, University of Central Florida; Teresa Dunn, Clemson University; Alexander Friedlander, Drexel University; Norman Gayford, Genesee Community College; Mark Gellis, Kettering University; Jeff Grabill, Michigan State University; Monika E. Gross, Bowie State University; Linda Harris, University of Maryland, Baltimore; Dawn Hayden, Thomas Nelson Community College; Rebecca Hettich, University of the Incarnate Word; Glenda Hudson, California State University, Bakersfield; Traci Irani, University of Florida; Susan Karberg, Purdue University; Marsha Kruger, University of Nebraska, Omaha; Jamie Larsen, North Carolina State University; Martha Levine, Southwest Missouri State University; Chris McKitterick, University of Kansas; Miriam Y. Miller, University of New Orleans; Joseph Moxley, University of South Florida; Mary Ellen Muesing, University of North Carolina, Charlotte; Sushil K. Oswal, University of Hartford; Elizabeth Pass, James Madison University; Seamus Reilly, Parkland College; Patrick M. Scanlon, Rochester Institute of Technology; Jennifer Scheidt, Palo Alto College; Kristi Siegel, Mount Mary College; Rick Simmons, Louisiana Tech; Ronald E. Smith, University of North Alabama; Clay Spinuzzi, University of Texas, Austin; Lee S. Tesdell, Minnesota State University, Mankato; Thomas L. Warren, Oklahoma State University; Judith A. Wootten, Kent State University, Salem.

Most important, I would like to thank my wife, Tracey, and my children, Emily and Collin, for being patient while I wrote this book. This book is for them.

RICHARD JOHNSON-SHEEHAN

PART 1

Elements of Technical Communication

CHAPTER 1

Communicating in the Workplace

CHAPTER CONTENTS

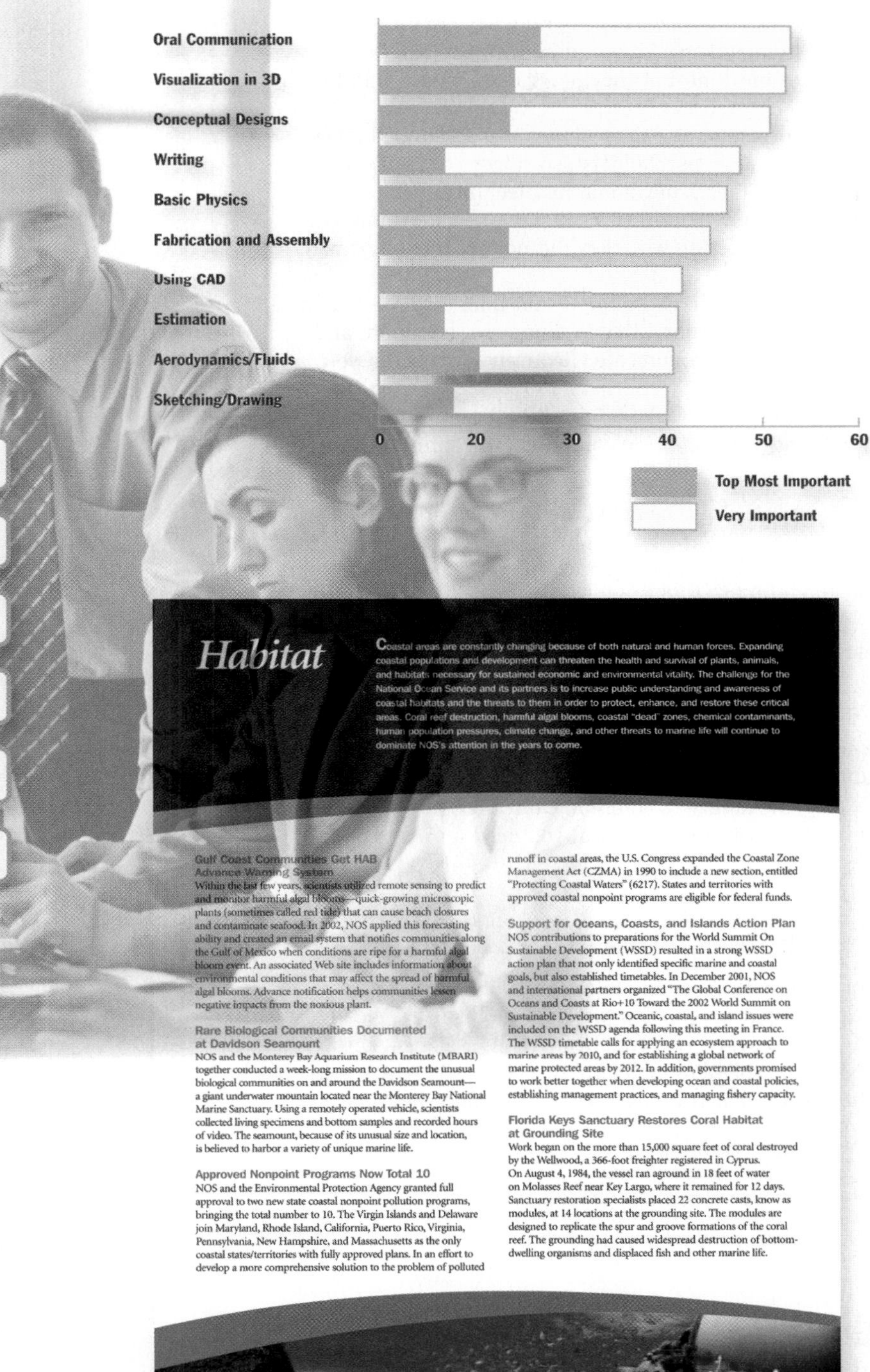

Habitat

Coastal areas are constantly changing because of both natural and human forces. Expanding coastal populations and development can threaten the health and survival of plants, animals, and habitats necessary for sustained economic and environmental vitality. The challenge for the National Ocean Service and its partners is to increase public understanding and awareness of coastal habitats and the threats to them in order to protect, enhance, and restore these critical areas. Coral reef destruction, harmful algal blooms, coastal "dead" zones, chemical contaminants, human population pressures, climate change, and other threats to marine life will continue to dominate NOS's attention in the years to come.

Gulf Coast Communities Get HAB Advance Warning System

Within the last few years, scientists utilized remote sensing to predict and monitor harmful algal blooms—quick-growing microscopic plants (sometimes called red tide) that can cause beach closures and contaminate seafood. In 2002, NOS applied this forecasting ability and created an email system that notifies communities along the Gulf of Mexico when conditions are ripe for a harmful algal bloom event. An associated Web site includes information about environmental conditions that may affect the spread of harmful algal blooms. Advance notification helps communities lessen negative impacts from the noxious plant.

Rare Biological Communities Documented at Davidson Seamount

NOS and the Monterey Bay Aquarium Research Institute (MBARI) together conducted a week-long mission to document the unusual biological communities on and around the Davidson Seamount—a giant underwater mountain located near the Monterey Bay National Marine Sanctuary. Using a remotely operated vehicle, scientists collected living specimens and bottom samples and recorded hours of video. The seamount, because of its unusual size and location, is believed to harbor a variety of unique marine life.

Approved Nonpoint Programs Now Total 10

NOS and the Environmental Protection Agency granted full approval to two new state coastal nonpoint pollution programs, bringing the total number to 10. The Virgin Islands and Delaware join Maryland, Rhode Island, California, Puerto Rico, Virginia, Pennsylvania, New Hampshire, and Massachusetts as the only coastal states/territories with fully approved plans. In an effort to develop a more comprehensive solution to the problem of polluted runoff in coastal areas, the U.S. Congress expanded the Coastal Zone Management Act (CZMA) in 1990 to include a new section, entitled "Protecting Coastal Waters" (6217). States and territories with approved coastal nonpoint programs are eligible for federal funds.

Support for Oceans, Coasts, and Islands Action Plan

NOS contributions to preparations for the World Summit On Sustainable Development (WSSD) resulted in a strong WSSD action plan that not only identified specific marine and coastal goals, but also established timetables. In December 2001, NOS and international partners organized "The Global Conference on Oceans and Coasts at Rio+10 Toward the 2002 World Summit on Sustainable Development." Oceanic, coastal, and island issues were included on the WSSD agenda following this meeting in France. The WSSD timetable calls for applying an ecosystem approach to marine areas by 2010, and for establishing a global network of marine protected areas by 2012. In addition, governments promised to work better together when developing ocean and coastal policies, establishing management practices, and managing fishery capacity.

Florida Keys Sanctuary Restores Coral Habitat at Grounding Site

Work began on the more than 15,000 square feet of coral destroyed by the Wellwood, a 366-foot freighter registered in Cyprus. On August 4, 1984, the vessel ran aground in 18 feet of water on Molasses Reef near Key Largo, where it remained for 12 days. Sanctuary restoration specialists placed 22 concrete casts, know as modules, at 14 locations at the grounding site. The modules are designed to replicate the spur and groove formations of the coral reef. The grounding had caused widespread destruction of bottom-dwelling organisms and displaced fish and other marine life.

CHAPTER OBJECTIVES

In this chapter, you will learn:

- The importance of communication in today's technical workplace.
- To define technical communication as a process of managing information in ways that allow people to take action.
- To recognize technical communication as a blend of action, words, and images.
- To distinguish technical communication from other kinds of communication, especially expository writing.
- That technical communication is interactive and adaptable, reader centered, collaborative, and highly visual.
- The importance of ethical, legal, political, international, and cross-cultural factors in technical communication.
- The importance of effective written and spoken communication to your career.

When entering the technical workplace, new college graduates are often surprised by the amount of writing and speaking required. Of course, they knew technical communication would be important, but they never realized it would be so crucial to their success.

Effective communication is the cornerstone of the technical workplace, whether you are an engineer, scientist, doctor, nurse, psychologist, social worker, anthropologist, architect, technical writer, or any other professional in a technical field. People who are able to write and speak effectively tend to succeed. People who cannot communicate well often find themselves wondering why they did not get the job or why they were passed over for promotions.

Today, effective technical communication is more important than ever. We live in an age in which whole industries are built around the development, retention, and application of information. The ability to communicate effectively is crucial if you plan to survive and succeed.

Computers are the central nervous system of the information-based workplace. Throughout your career, you will find yourself using computers to communicate with others:

- Your desktop and laptop computers will be networked to local area networks (LANs) and the Internet.
- Your personal digital assistant (PDA) will keep track of your calendar for you.
- You will use e-mail, websites, and blogs to communicate with people inside and outside your office.
- You will use your wireless phone to talk to people around the world.
- Wireless networks will allow you to use your computer, PDA, and other communication tools to keep in touch with your colleagues and coworkers.

In essence, computers are communication tools. And, as they have become more centralized in our lives, the ability to communicate with computers has become an essential part of our careers. Today, the ability to communicate effectively and efficiently with computers might be the most important skill you need for a successful career in technology.

How Important Is Technical Communication?

Surveys regularly show that oral and written communication skills are among the most important in the technical workplace. For example, in a 2003 survey, members of the American Institute of Aeronautics and Astronautics (AIAA) were asked to evaluate their educational preparation for their jobs (AIAA, 2003). The AIAA members ranked major items in their workplace training needs in the following order:

Ranked Importance of Workplace Training

1. Oral communication
2. Visualization in three dimensions
3. Technical writing
4. Understanding processes of fabrication and assembly

To see the AIAA survey, go to
www.ablongman.com/johnsonweb/1.1

5. Using CAD, CAM, and solid modeling

6. Estimating solutions to complex problems without using computer models

7. Sketching and drawing

The membership of the AIAA is made up mostly of engineers, so it is interesting that two out of three of the top skills they listed stress the importance of technical communication. In fact, 34 percent of the surveyed AIAA members ranked oral communication "top most important," while 54 percent ranked it "very important." Meanwhile, writing was ranked "top most important" by 14 percent of respondents, while 61 percent ranked it "very important" (Figure 1.1).

This survey's results are very much in line with surveys in other technical fields. An article in an engineering journal, *Professional Issues in Engineering Education and Practice*, reported that:

> Within 2–3 years of graduation, engineers spend about 30 percent of their time on the job writing; engineers in middle management spend 50–70 percent; and engineers in senior management spend 70 percent or more—up to 95 percent (Silyn-Roberts, 1998).

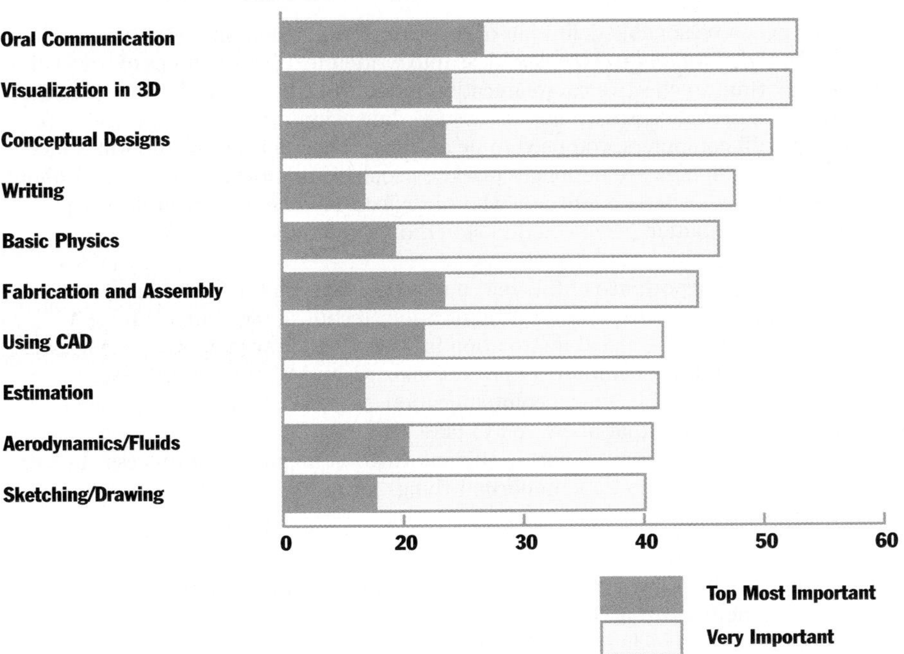

Source: AIAA, 2003.

Figure 1.1: Professionals in technical fields regularly put communication skills at the top of their lists of important workplace skills.

To see other surveys concerning workplace writing, go to **www.ablongman.com/johnsonweb/1.2**

Meanwhile, in a 1996 survey of recent graduates, researchers at North Carolina State University found that across all disciplines, professionals spent an average of 31 percent of their time writing at work (Miller, Larsen & Gaitens, 1996). In some technical fields, professionals said they were spending close to half their time communicating with others.

These kinds of results highlight an important paradox in the technical workplace. Obviously, technical communication is very important; yet managers regularly report that college graduates do not have adequate communication skills. To make up for these deficiencies, companies need to pay millions and perhaps billions each year to train and retrain their employees how to write and speak effectively in the technical workplace.

Fortunately, you can learn how to write and speak effectively in the technical workplace. The ability to communicate effectively is not something people are born with. With a little effort, anyone can learn to write and speak well.

The purpose of this book is to help you acquire the communication skills you need for success in the technical workplace. You will find throughout your career that people who write and speak clearly and persuasively are much more likely to succeed than those who don't. Right now, you have a golden opportunity to develop these important technical communication skills. They will help you land the job you want, and they will help you succeed.

What Is Technical Communication?

Not long ago, a reasonable definition of technical communication might have said something about *translating* complex ideas into written text. This concept of translation is still important to effective communication in the workplace. But the centralization of computers in the workplace has added new dimensions and responsibilities.

Today, with computers, you need to be an *information manager* in the workplace. The experts in an area are no longer just the people who know a great amount about a subject. Rather, experts are people who can effectively manage and utilize large amounts of information generated and stored on computers.

Technical communication is a process of managing technical information in ways that allow people to take action.

To reflect the importance of this information management, this book will use the following definition for technical communication: Technical communication is a process of managing technical information in ways that allow people to take action. The key words in this definition are *process, manage*, and *action*. In this book, you will learn the *process* of technical communication, so you can *manage* large amounts of information in ways that allow you to take *action*.

Computers are an essential part of this information management process. As communication tools, they do three important things for us:

- They store large quantities of data and information, allowing us to emphasize using information, not storing information.
- They provide easy access to these large quantities of information (e.g., data bases, the Internet, CD-ROMs), requiring us to be selective about what kinds of information to include in a document.
- They allow us to communicate and interact easily with others, often crossing corporate, national, and international boundaries.

For other definitions of technical communication, go to **www.ablongman.com/johnsonweb/1.3**

In this Information Age, managing information means we are primarily concerned with how information *flows*. In other words, technical communication is about guiding the flow of information into useful areas, so people, including ourselves, can use that information.

Technical Communication: Actions, Words, Images

If you are a college student, much of your writing experience until now has probably been *expository* in nature. The root word for expository is "exposition," meaning you have been taught to *display* or *exhibit* information. Primarily, you have been taught how to demonstrate that you have acquired and retained knowledge on specific subjects. Your readers have mostly been teachers and professors.

Technical communication is different from expository writing, because it puts more focus on taking action with words and images. It focuses more on achieving a specific *purpose* with language. It also puts a much heavier emphasis on anticipating the needs of the readers and communicating information clearly and persuasively. Moreover, issues of ethics, legality, politics, and culture in technical communication are often much more tangible, because technology has such an immediate impact on people. Figure 1.2 shows some of the characteristics of technical communication that can set it apart from expository writing.

Fortunately, the skills you learned in expository writing are all adaptable to technical communication. The main difference is that technical communication puts a much greater emphasis on achieving a specific purpose with words and images.

The Qualities of Technical Communication

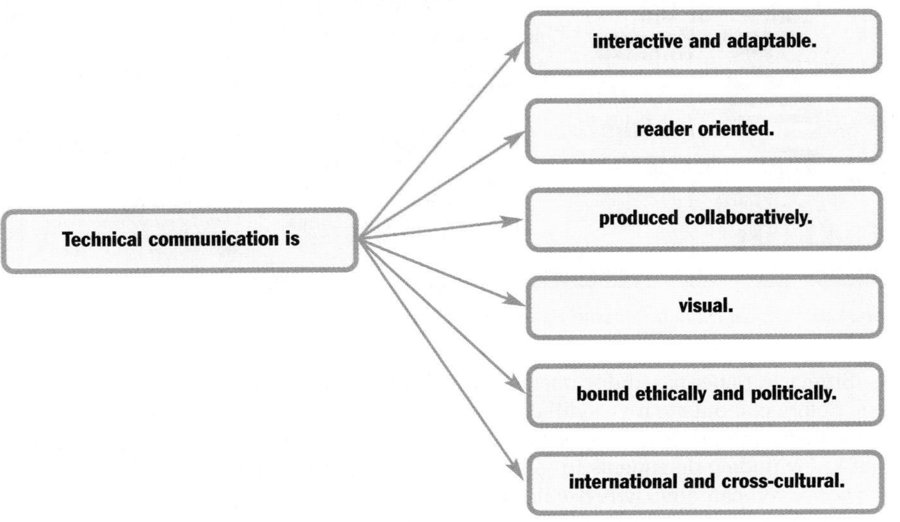

Figure 1.2: Technical communication puts much more emphasis on managing information and taking action than do most other forms of writing.

GO TO THE NET

To find other resources on technical communication, go to **www.ablongman.com/johnsonweb/1.4**

Technical Communication Is Interactive and Adaptable

One of the most significant changes brought about by computers is the amount of *interactivity* among people in the technical workplace. In the computer-networked workplace, people are constantly communicating with each other and sharing their ideas.

As a result of this interactivity, it is possible for you to quickly adapt documents and presentations to fit the specific needs of many different kinds of readers and situations. Websites are an especially interactive form of technical communication (Figure 1.3). Using a website, people can find the information that is most helpful to them. And, if they cannot find the information they need on the website, they can send an e-mail to get the answers they need.

Sample Webpage

Figure 1.3: Websites are highly interactive, allowing readers to follow a variety of different paths to find information.

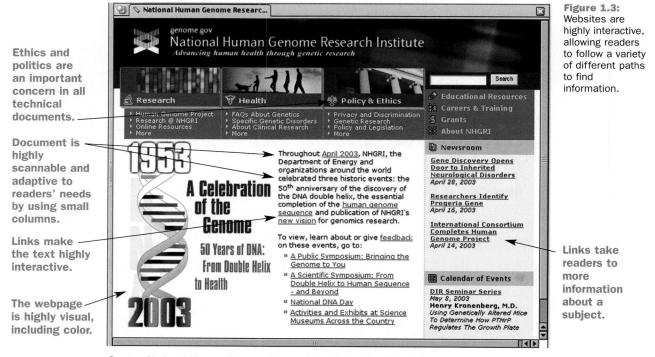

Source: National Human Genome Research Institute, http://www.genome.gov.

Similarly, paper-based documents can be adapted to the changing needs of readers. Before computers, it was difficult to adjust and revise paper-based documents. Once they were printed, documents were hard to change. Today, with computers, you can easily update documents to reflect changes in your company's products and services. Or, you can quickly revise documents to address unexpected changes in the workplace. With little effort, you can adapt documents to readers whose needs may be different than you expected or anticipated.

A Document as a Meeting Space

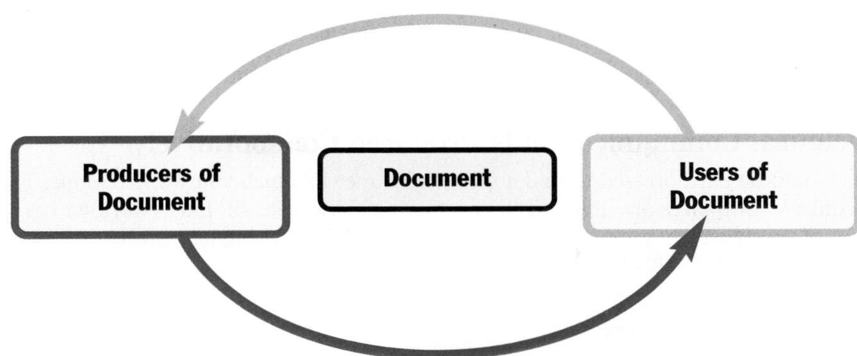

Figure 1.4: Today, producers and users of documents interact when needed, allowing the writer to adjust the text to suit the needs of specific users and contexts.

As illustrated in Figure 1.4, the relationship between the producers of documents and the users has become highly interactive. The document itself, whether it is a website, a manual, a report, a proposal, or an oral presentation, is simply the meeting space between its producers and users. That space can be changed to suit the changing situations in which the document is needed.

The highly interactive nature of communication means you will be continually adapting your documents to the changing needs of the readers and the situations in which they will use your materials.

Technical Communication Is Reader Centered

In technical communication, readers play a much more significant role than in other kinds of writing. When writing a typical college essay, you are trying to express *your* ideas and opinions. Technical communication turns this situation around. It concentrates on what the readers "need to know" to take action, not only what you, as the writer, want to tell them.

Because it is reader centered, effective technical communication tends to be highly pragmatic. Effective technical communication is—

- **Efficient**—providing only information that readers need to know.
- **Easy to understand**—employing an organization and style that are easy for readers to follow and comprehend.
- **Accessible**—providing visual cues like headings, lists, and graphics to help readers quickly access important information.
- **Action oriented**—highlighting how the information can be or should be used.
- **Adaptable**—allowing documents to be freely changed to suit the specific needs of different readers.

Of course, your needs as the writer are important too. Your *need to tell* and the readers' *need to know* should be aligned. When the needs of both writers and readers are met, you will have produced an effective document.

LINK To learn about adapting texts to readers and contexts, go to Chapter 3, page 42.

Technical Communication Is Produced Collaboratively

You would be hard pressed to find a technical career in which you worked alone. Technical workplaces are highly collaborative, meaning you will likely work with a team of specialists on almost every project (Figure 1.5). Writing and presenting with a team are crucial skills in any technical workplace.

Working with a Team

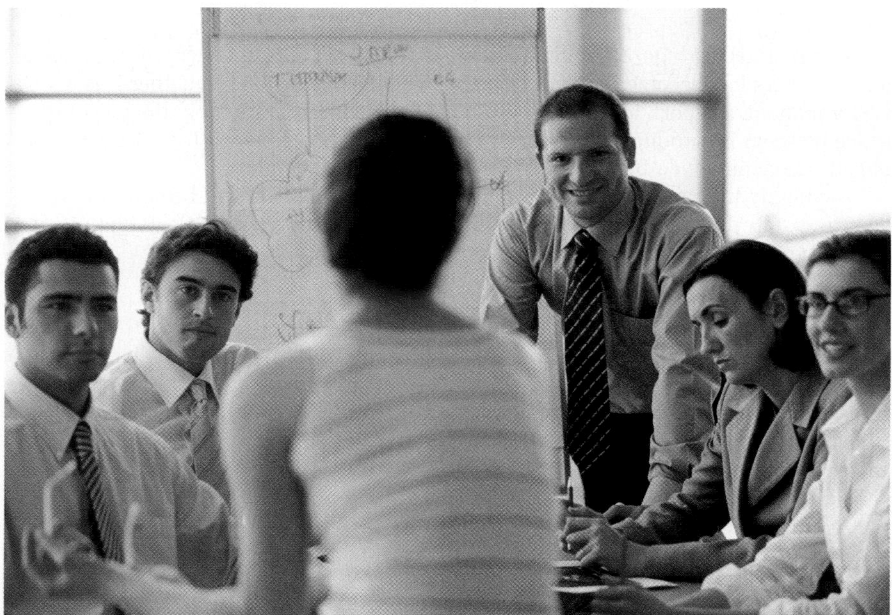

Figure 1.5: Working with a team can be fun and rewarding. Teams take advantage of the strengths and knowledge of different people to succeed.

Computers have only heightened the collaborative nature of the technical workplace. Today, because documents can be shared through e-mail or the Internet, it is common for many people to be working on a document at the same time. In some cases, your team might be adjusting and updating information on an ongoing basis.

Learning to work with teams, especially on documents, is an important skill that you need to develop if you are going to succeed in the technical workplace. Yes, working with a team is often difficult and sometimes frustrating. But, it can also be rewarding, helping you become more creative and productive.

LINK For more information on working in teams, see Chapter 13, page 352.

To learn more about collaborative writing, go to **www.ablongman.com/johnsonweb/1.5**

While in school, people are often given the false impression that they can successfully work alone. After all, writing essays and taking tests is usually an individual effort. However, in the workplace, you will almost always work in a team with other people. Knowing how to communicate and negotiate with those other people is a strength that will be an important part of your professional career.

Technical Communication Is Visual

Increasingly, technical documents take full advantage of the computer's design capabilities. People don't usually read technical documents for enjoyment or personal satisfaction. They read technical documents so they can take action—do something.

LINK For more information on visual design, see Chapter 8, page 192.

By making texts highly visual, you can help readers quickly locate the information they need. Visual cues, like headings, lists, diagrams, and margin comments, are common in technical documents (Figure 1.6). Graphics also play an important role in technical communication. By using charts, graphs, drawings, and pictures, you can clarify and strengthen your arguments in any technical document. Today's readers grow quickly impatient with large blocks of text. They prefer graphics that reinforce the text and help them quickly gain access to important information.

LINK To learn about using graphics in documents, turn to Chapter 9, page 234.

One important change in our computer-centered society is that people think more visually. Your readers grew up with television and computers, so they are used to receiving information through visual media. Therefore, technical documents, whether they are manuals, reports, or websites, need to present information visually.

Technical Communication Has Ethical, Legal, and Political Dimensions

In the increasingly complex technical workplace, issues involving ethics, laws, and politics are always present. With the increased freedom allowed by computers, we now have an increased ability to intentionally and unintentionally violate ethical and legal standards. Moreover, computers have created new micro- and macropolitical challenges that need to be negotiated in the workplace.

Employees at all levels need to know how to negotiate ethical, legal, and political issues. As management structures become flatter—meaning there are fewer layers of management—employees are being asked to take on more decision-making responsibilities than ever. In most corporations, fewer checks and balances exist, meaning all employees need to be able to sort out the ethical, legal, and political aspects of a decision for themselves.

- All technical documents involve ethical issues of rights, justice, and fairness. When you are communicating with others, you need to know how to analyze the ethical issues at stake.
- Legal issues are especially important in technical communication. Issues of product liability and copyright play an important role in how documents are written and how important information is conveyed.

For more on the importance of visual design in technical communication, go to **www.ablongman.com/johnsonweb/1.6**

Figure 1.6: Visual design is an essential part of technical communication.

An easy-to-read title identifies the document's subject.

Habitat

Coastal areas are constantly changing because of both natural and human forces. Expanding coastal populations and development can threaten the health and survival of plants, animals, and habitats necessary for sustained economic and environmental vitality. The challenge for the National Ocean Service and its partners is to increase public understanding and awareness of coastal habitats and the threats to them in order to protect, enhance, and restore these critical areas. Coral reef destruction, harmful algal blooms, coastal "dead" zones, chemical contaminants, human population pressures, climate change, and other threats to marine life will continue to dominate NOS's attention in the years to come.

Headings make the text highly scannable.

The two-column format makes the text easy to scan.

Visuals add color and emotion.

Gulf Coast Communities Get HAB Advance Warning System

Within the last few years, scientists utilized remote sensing to predict and monitor harmful algal blooms—quick-growing microscopic plants (sometimes called red tide) that can cause beach closures and contaminate seafood. In 2002, NOS applied this forecasting ability and created an email system that notifies communities along the Gulf of Mexico when conditions are ripe for a harmful algal bloom event. An associated Web site includes information about environmental conditions that may affect the spread of harmful algal blooms. Advance notification helps communities lessen negative impacts from the noxious plant.

Rare Biological Communities Documented at Davidson Seamount

NOS and the Monterey Bay Aquarium Research Institute (MBARI) together conducted a week-long mission to document the unusual biological communities on and around the Davidson Seamount—a giant underwater mountain located near the Monterey Bay National Marine Sanctuary. Using a remotely operated vehicle, scientists collected living specimens and bottom samples and recorded hours of video. The seamount, because of its unusual size and location, is believed to harbor a variety of unique marine life.

Approved Nonpoint Programs Now Total 10

NOS and the Environmental Protection Agency granted full approval to two new state coastal nonpoint pollution programs, bringing the total number to 10. The Virgin Islands and Delaware join Maryland, Rhode Island, California, Puerto Rico, Virginia, Pennsylvania, New Hampshire, and Massachusetts as the only coastal states/territories with fully approved plans. In an effort to develop a more comprehensive solution to the problem of polluted runoff in coastal areas, the U.S. Congress expanded the Coastal Zone Management Act (CZMA) in 1990 to include a new section, entitled "Protecting Coastal Waters" (6217). States and territories with approved coastal nonpoint programs are eligible for federal funds.

Support for Oceans, Coasts, and Islands Action Plan

NOS contributions to preparations for the World Summit On Sustainable Development (WSSD) resulted in a strong WSSD action plan that not only identified specific marine and coastal goals, but also established timetables. In December 2001, NOS and international partners organized "The Global Conference on Oceans and Coasts at Rio+10 Toward the 2002 World Summit on Sustainable Development." Oceanic, coastal, and island issues were included on the WSSD agenda following this meeting in France. The WSSD timetable calls for applying an ecosystem approach to marine areas by 2010, and for establishing a global network of marine protected areas by 2012. In addition, governments promised to work better together when developing ocean and coastal policies, establishing management practices, and managing fishery capacity.

Florida Keys Sanctuary Restores Coral Habitat at Grounding Site

Work began on the more than 15,000 square feet of coral destroyed by the Wellwood, a 366-foot freighter registered in Cyprus. On August 4, 1984, the vessel ran aground in 18 feet of water on Molasses Reef near Key Largo, where it remained for 12 days. Sanctuary restoration specialists placed 22 concrete casts, know as modules, at 14 locations at the grounding site. The modules are designed to replicate the spur and groove formations of the coral reef. The grounding had caused widespread destruction of bottom-dwelling organisms and displaced fish and other marine life.

Source: National Ocean Service, 2002.

- Political issues involve negotiating among the interests and needs of many stakeholders. The politics of the technical workplace can be complex, influencing how you communicate and how your messages are received.

LINK Ethical, legal, and political issues are discussed in Chapter 4, starting on page 70.

To communicate effectively in the technical workplace, you need to be aware of the ethical, legal, and political issues that shape your writing and speaking.

Technical Communication Is International and Cross-Cultural

Computers have also increased the international nature of the technical workplace. Today, it is common for professionals to be regularly communicating with people around the world.

The growth of international trade means you will find yourself working with people who speak other languages and have other customs. They will also hold different expectations about technical documents and presentations.

Always keep in mind that communication practices that North Americans might consider "normal" or "commonsense" can be strange or even offensive to people from other cultures. Even simple things can make a difference. For example, in many cultures the "OK" hand signal in a document would be highly offensive (just as you might find the use of the middle finger to be offensive). Awareness of these kinds of cultural differences can make your documents much more effective, while avoiding embarrassment.

Cross-cultural communication is also essential. Our society is becoming more diverse than ever. As a result, readers will interpret your documents from a variety of perspectives. Your readers may all be North Americans, but they will be reading your document from cultural perspectives that are different from yours. You need to be aware of these perspectives, so you can adjust your message to suit their needs.

AT A GLANCE

Qualities of Technical Communication

Technical communication is:

- interactive and adaptable.
- reader centered.
- produced collaboratively.
- visual.
- influenced by ethics, laws, and politics.
- international and cross-cultural.

LINK To learn about communicating internationally and cross-culturally, go to Chapter 3, page 60.

Your Career and Technical Communication

It is typical for students in technical disciplines to wonder why they need to learn how to write and speak effectively. But, if you ask professionals who work in technical fields, you will quickly find that communication is one of their most valued skills (Silyn-Roberts, 1998). After all, in technical workplaces, almost every activity involves writing or speech. So, writing and speaking clearly are often the difference between success and failure.

To learn more about ethics, go to **www.ablongman.com/johnsonweb/1.7**

HELP

E-Literacy and Overcoming Cyberphobia

If you don't have much experience with computers, now is a good time to begin strengthening your skills in this area. Most people who fear computers simply haven't had much opportunity to work with them. If you are one of those people, a little experience will quickly bring you up to speed.

Here are some tips for gaining more experience with computers and, if needed, overcoming any fears about them.

Dive right in—The best way to learn how to use computers is to use them. User manuals can be helpful, but nothing is better than hands-on experience. To help you get started, you might try out the tutorials that come with the computer and various software packages. These tutorials can be found under the "Help" menu on your computer's desktop.

Learn by playing—Children learn how to use computers very quickly. Why? Because they start out by playing with the technology. They mess around with computers to see what they can do. Similarly, give yourself time to play with the technology. Don't be too goal oriented right away.

Learn by exploring—Go surfing on the web to find things that interest you. Your favorite foods, hobbies, people, movies, news programs, and anything else are all discussed on the web. Before long, you will be having so much fun surfing, you will have completely forgotten that you are using a computer.

Allow yourself to fail . . . and learn—When things go wrong (and they will), don't become frustrated or defeated. Mistakes happen, even to the most experienced computer users. You simply need to see each "failure" as a learning opportunity.

View the computer as a tool—In the end, computers are tools that help you do things. Like any tool (hammer, car, microscope, bicycle pump), they take some time to learn and master. Remember, the computer is there to help you be more effective. You're in control (even though it might not seem like it sometimes).

Your campus or workplace more than likely has classes available that can help you learn about computers. Many of these classes are free or inexpensive. These classes will help you get comfortable with computers.

Take a look at the help wanted advertisements on Internet job sites or in your local paper. You will notice that many technical jobs stress communication skills. Managers in these disciplines consistently rate effective communication as a "very important" skill for employees. Employers have little patience with employees who cannot write and speak effectively.

Need help fighting cyberphobia? Go to
www.ablongman.com/johnsonweb/1.8

Effective technical communication will be essential to your career in a variety of ways:

- Your managers will immediately notice whether you can communicate clearly and efficiently, because your writing and speaking skills are the first outward signs of your abilities.
- Your performance evaluations will often be based on the documents (memos, reports, proposals, websites, presentations, etc.) that you wrote or helped produce.
- Your colleagues will judge your abilities by the effectiveness of your documents or presentations.
- Your company's clients will judge the quality of your entire company by the quality of your documents and the presentations you make to them.

Here is your golden opportunity to master these important writing and speaking skills. If you are reading this book, we can safely assume that you are in a class on technical communication or looking to improve your skills in the technical workplace. This book will give you the tools you need for success.

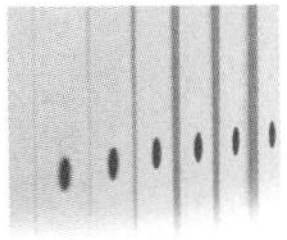

CHAPTER REVIEW

- Computers, the Internet, and instant forms of communication have had an enormous impact on communication in the technical workplace.
- Technical communication is defined as a process of managing technical information in ways that allow people to take action.
- Technical communication is a blend of action, words, and images. Readers expect technical documents to use writing, visuals, and design to communicate effectively.
- Technical communication is interactive, adaptable, user oriented, and is often produced collaboratively.
- Technical communication has ethical, political, international, and cross-cultural dimensions that must be considered.
- Effective written and spoken communication will be vital to your career.

GO TO THE NET

For more information about writing and careers, go to **www.ablongman.com/johnsonweb/1.9**

EXERCISES AND PROJECTS

Individual or Team Projects

1. Locate a document that is used in a technical workplace through a search engine like Google.com, Altavista.com, or Yahoo.com. To find documents, type in keywords like "report," "proposal," "instructions," and "presentation." Links to sample documents are also available at www.ablongman.com/johnsonweb/1.10.

   What characteristics make the document you found a form of technical communication? Develop a two-minute presentation for your class in which you highlight these characteristics of the document. Compare and contrast the document with academic essays you have written for your other classes.

2. Using a search engine on the Internet, locate a professional who works in your chosen field. Using e-mail, ask that person what kinds of documents or presentations he or she needs to produce. Ask how much time he or she devotes to communication on the job. Ask whether he or she has some advice about how to gain and improve communication skills that will be needed in your career. Write a memo to your instructor in which you summarize your findings.

3. Make a list of your goals for this class. What are some skills you would like to learn? What are some experiences you would like to have? In a memo to your instructor, briefly summarize these goals and discuss how you are going to reach them.

4. In a memo or e-mail, write a short biographical description of someone in your class. Tell your readers what you found interesting about this person, concentrating on information others in your class would likely want to know. Then, turn your memo into a one-minute presentation in which you introduce your classmate to the rest of your class.

Links to sample technical documents are available at
www.ablongman.com/johnsonweb/1.10

Collaborative Project: Writing a Course Mission Statement

As you begin this semester, it is a good idea for your class to develop a common understanding of the course objectives and outcomes. Companies develop mission statements to help focus their efforts and keep their employees striving toward common ends. Corporate mission statements are typically general and nonspecific, but they set an agenda or tone for how the company will do business internally and with its clients.

Your task in this assignment is to work with a group of others to develop a "Course Mission Statement" in which you lay out your expectations for the course, your instructor, and yourselves. To write the mission statement, follow these steps:

1. Go to www.ablongman.com/johnsonweb/1.11 to find links to sample mission statements. Or, you can use an Internet search engine to find your own examples of mission statements. Just type "mission statement" in Lycos.com, Webcrawler.com, or Google.com.

2. In class, with your group, identify the common characteristics of these mission statements. Pay special attention to their content, organization, and style. Make note of their common features.

3. With your group, write your own Course Mission Statement. Be sure to include goals you would like the course to meet. You might also want to develop an "ethics statement" that talks about your approach to ethical issues with assignments, course readings, and attendance.

4. Compare your Course Mission Statement with other groups' mission statements. Note places where your statement is similar and different from their statements.

When your Course Mission Statement is complete, it should provide a one-paragraph description of what you are trying to achieve in your class.

For sites that show examples of mission statements, go to **www.ablongman.com/johnsonweb/1.11**

CHAPTER

2

The Technical Writing Process Today

CHAPTER CONTENTS

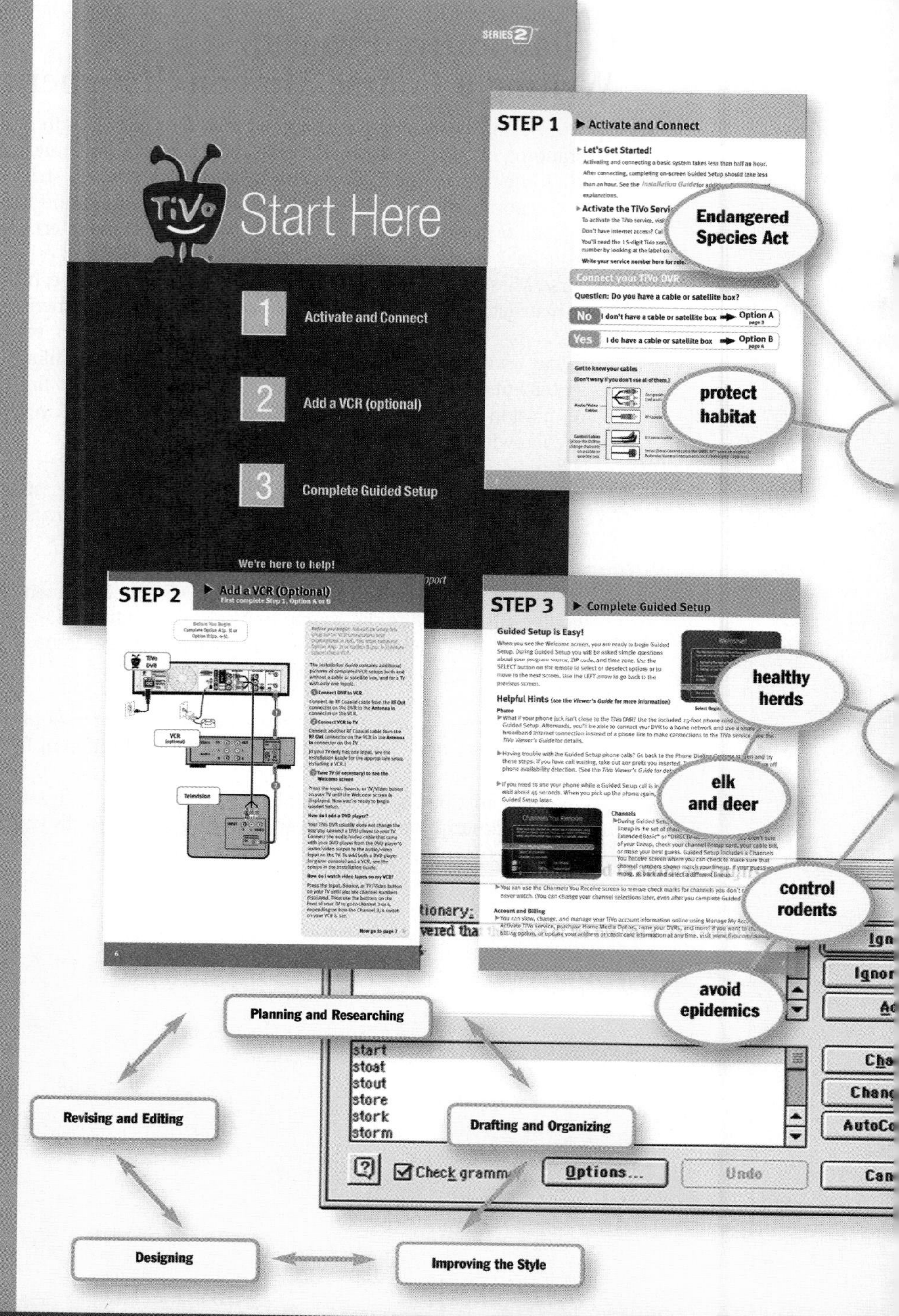

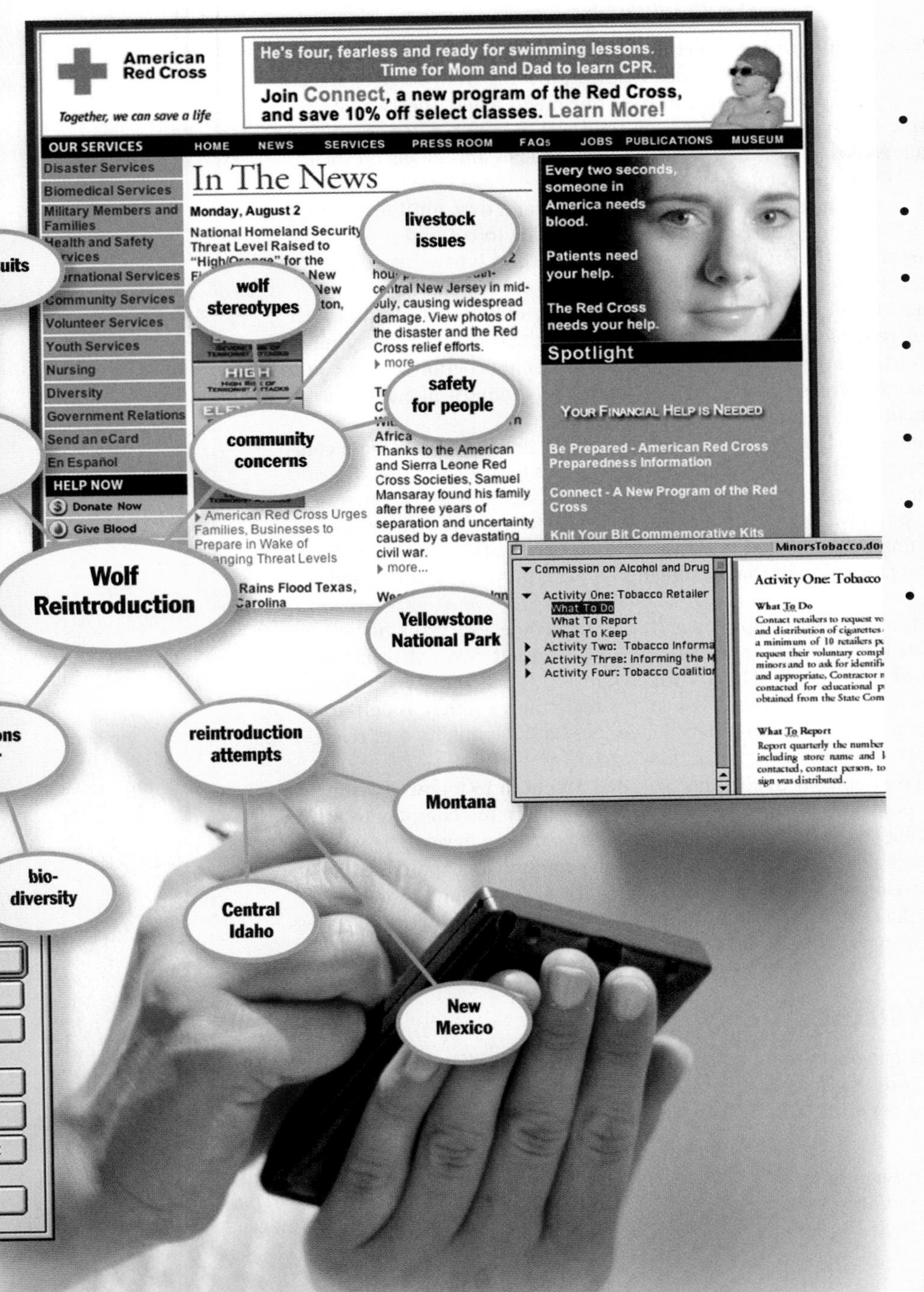

CHAPTER OBJECTIVES

In this chapter, you will learn:

- How to use a writing process that takes advantage of your computer's capabilities.
- Strategies for defining a document's *rhetorical situation.*
- The importance of defining your purpose.
- Techniques for organizing and drafting your ideas with common genres.
- How to use your computer to overcome writer's block.
- The importance of style and design in informative and persuasive documents.
- The basics of revising and editing your work.

Not long ago, most people wrote at their desk with a pen and a blank sheet of paper. So, their writing process was mostly linear, moving from stage to stage. First, they would develop a detailed outline of their document. Then, they would use a variety of prewriting strategies to put words on the sheet of paper. After prewriting, they might have walked over to a library to track down articles and books on their subject. Eventually, they would compose a rough draft.

This rough draft would then be revised (rewritten) over and over until it reached a desired quality level. Some people even used scissors and glue to cut and paste paragraphs while revising the draft. Finally, they would turn to their typewriter to produce the final draft, with white correction fluid at their elbow.

If you are like most people today, you probably compose directly on the computer screen. Much of your research is conducted on the Internet, and you can find printed materials through your library's website. Your word-processing program allows you to revise and edit as you are writing. It also makes the text look almost like a finished product. The document can be printed out at any point.

Writing with pen, paper, and typewriter was a bit more tedious than using a computer, but this approach had one advantage. People needed to work through a *writing process* with defined stages. Each stage (prewriting, outlining, drafting, rewriting, proofreading, and typing) was handled more or less separately, giving writers several opportunities to rethink, reshape, and improve the text. The final document often reflected this deeper consideration of the document's subject, purpose, readers, and context (Elbow, 1981; Perl, 1994).

One problem with writing on computers is that some people, especially novice writers, limit their writing process to one drafting stage. They sit down at their computer and hammer out a document. The printlike words on the screen give the false impression that the document is "finished" when it is really just a rough, rough draft. As a result, documents written on computers sometimes lack the quality and refinement of documents written with pen, paper, and typewriter.

The purpose of this chapter is to convince you that you need to follow a writing process that goes through stages. Most technical documents are too complex to be handled in one drafting stage. Consequently, you need to work on developing your writing process, allowing you take full advantage of the computer's capabilities.

By consciously developing a writing process that suits your individual work habits, you will find that you write better and more efficiently. A good writing process will help you "work smarter, not harder" when you are writing technical documents.

Technical Writing Today

Of course, we all have our own way of doing things. Writing is no different. Each of us will develop a writing process that suits his or her individual interests, needs, and work habits. Whatever your personal habits, though, it helps to see writing as a series of stages.

- Planning and Researching
- Organizing and Drafting
- Improving the Style
- Designing
- Revising and Editing

For more discussions of the writing process, go to **www.ablongman.com/johnsonweb/2.1**

The boundaries between each of these stages are blurred, and you will find yourself working back and forth among them (Figure 2.1). While drafting, for example, you might discover that you need to do more research. While you are editing your document, you may decide that you need to draft an additional section. For the most part, though, these stages will lead you from the beginning of a writing project to the end.

The Writing Process in Electronic Environments

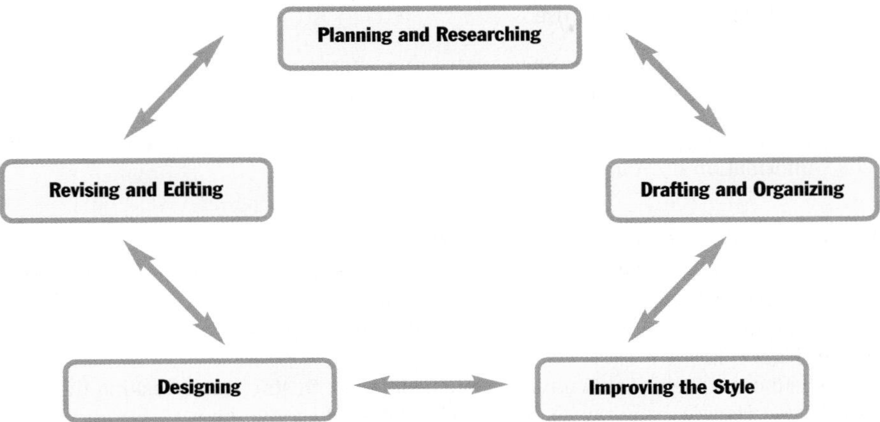

Figure 2.1: Effective writers in technical workplaces follow a process that includes stages.

Another thing to keep in mind is that writing technical documents is not something you will do alone. At each of these stages, your colleagues and supervisors will be involved, helping you shape your ideas, organize the information, design the document, and handle some of the editing. Writing in technical workplaces is regularly a team effort.

LINK For more information on writing in teams, see Chapter 13, page 352.

When you are working with a team of writers on a document, following a defined writing process is especially important. On a calendar, you and your team can map out when each of these stages will be completed, resulting in a finished document.

TAKE NOTE Without a defined process, the document will often not be written until the last minute. Inevitably, this kind of mad dash to the deadline leads to poor thinking and poor writing.

GO TO THE NET

To learn more about working in teams, go to **www.ablongman.com/johnsonweb/2.2**

Stage 1: Planning and Researching

Before you begin drafting the document on your computer, you should give yourself time to plan and do research. Even if time is short, don't just sit down and put words on the screen. After all, half-baked ideas may do more harm than good if the document only confuses the readers.

When preparing and researching, you should spend some time doing three activities:

Defining the rhetorical situation—Identify your document's subject, purpose, readers, and context of use.

Defining your purpose—Spend extra time sharpening your purpose into a one-sentence statement that will guide your research and drafting of the document.

Researching your subject—Use electronic, print, and empirical sources, collect information on your subject.

Defining the Rhetorical Situation

A good first step is to define the *rhetorical situation* that will shape the content, organization, style, and design of your document. Understanding the rhetorical situation means gaining a firm grasp of your document's subject, purpose, readers, and context of use (Figure 2.2).

To define the rhetorical situation, some people like to start out by asking the *Five-W and How Questions:* who, what, why, where, when, and how.

Once you have informally answered the Five-W and How questions, you are ready to fully define the rhetorical situation. The Five-W and How questions should give you an overall sense of your document's rhetorical situation. Now, you can use that information to more firmly define your document's subject, purpose, readers, and context of use.

Spend some time taking notes on the following four issues:

Subject—What is the document about? What is it *not* about? What kinds of information will my readers need to make a decision or complete a task? What is the scope of the project?

Purpose—Why is this document needed? What does it need to achieve or prove?

Readers—Who are the readers of this document? What are their needs and interests?

Context of use—Where and when will this document be used? What physical, economic, political, and ethical constraints will shape this text?

The Five-W and How Questions

- Who are the readers of this document?
- What am I writing about?
- When will the document be used?
- Where will the document be used?
- Why am I writing this document or presenting this material?
- How will the document be used?

LINK For more information about defining your readers and context of use, see Chapter 3, page 42.

To find worksheets to help you analyze the rhetorical situation, go to **www.ablongman.com/johnsonweb/2.3**

As you begin the writing process, it is helpful to consider these four categories separately. In some cases, especially with larger documents, it is helpful to write down your answers, so you can keep yourself focused and on track.

TAKE NOTE With team projects, it is essential that you begin with a solid understanding of the rhetorical situation. Otherwise, your team will waste hours or days with false starts and dead ends.

Defining the Rhetorical Situation

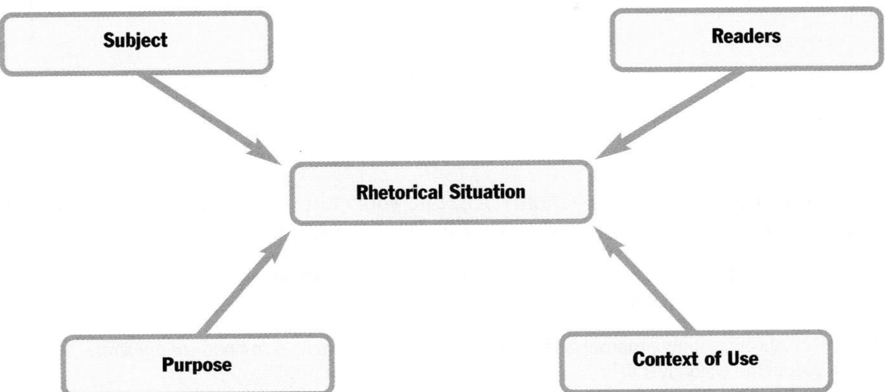

Figure 2.2: The rhetorical situation is a blending of the document's subject, purpose, readers, and context of use.

Defining the rhetorical situation may seem like an added step that is keeping you from writing. Actually, knowing your document's rhetorical situation will save you time and effort, because you will avoid dead ends and unnecessary obstacles. Also, you won't find yourself staring at the screen, wondering what to write.

Defining Your Purpose

The purpose of your document deserves some special consideration. Among the four elements of the rhetorical situation, your purpose is probably the most important. It is what you want to do—and what you want the document to achieve.

Your purpose statement is like a compass for the document. Once you have clearly defined your purpose for yourself and your readers, you can use that purpose statement to guide your decisions about the content, organization, style, and design of your document. Also, a clearly defined purpose statement saves you time, because it will help you avoid following tangents and dead ends as you are researching and drafting.

When defining your purpose, try to express exactly what you want your document to achieve. Sometimes it helps to think of an action verb and then build your purpose statement around it. Here are some useful action verbs that you might use:

Want to know more about rhetoric? Go to **www.ablongman.com/johnsonweb/2.4**

Informative documents	Persuasive documents
to inform	to persuade
to describe	to convince
to define	to influence
to review	to recommend
to notify	to change
to instruct	to advocate
to advise	to urge
to announce	to defend
to explain	to justify
to demonstrate	to support

Once you have chosen an action verb, state your purpose in one sentence. It might help to finish the phrase "The purpose of my document is to . . . . "

> The purpose of my report is to review the successes and failures of wolf reintroduction programs in the western United States.

> The purpose of my proposal is to recommend significant changes to flood control strategies in the Ohio River Valley.

Sometimes it is very hard to hammer your purpose statement down into one sentence, but the effort is worth it. After all, by expressing your purpose in one sentence, you are ensuring that your document is focused. Your clear sense of purpose will help you write the document, saving you time and effort.

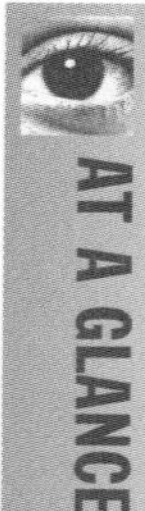

Two Types of Purpose in Technical Documents

- To inform—to give readers information they need to understand a subject or make a decision.
- To persuade—to attempt to convince readers that they should hold a particular opinion or take a specific action.

Researching Your Subject

Once you have defined your purpose, you can begin researching your subject to collect the content you need to write the document. While researching, you need to gather information from a variety of sources, including the Internet, print documents, and empirical methods (e.g., experiments, surveys, observations, interviews). Methods for doing these kinds of research are discussed in Chapter 5.

LINK For more information on research methods, go to Chapter 5, starting on page 105.

Computers have significantly changed the way we do research in technical workplaces. Before computers, finding enough information was usually the problem. Today, there is almost too much information available on any given subject. So, it is important that you learn how to *manage* information, sorting through all the texts, scraps, junk, and distortions to uncover what you need.

For more help writing a purpose statement, go to **www.ablongman.com/johnsonweb/2.5**

As the writer, you need to effectively manage the flow of information for your readers. Your document should give them the information they need to know to make a decision or take action, but that is all. Any excess information should be left out of the document.

Stage 2: Organizing and Drafting

Most people view drafting as the hardest part of the writing process. But with a thorough understanding of the rhetorical situation and solid research, you should find the organizing and drafting stage much easier.

While organizing and drafting, you are essentially doing two things at the same time:

Organizing the content—Using common genres to shape your ideas into documents that will be familiar to the readers.

Drafting the content—Generating and composing the content of your document by including facts, data, reasoning, and examples.

As you write, you will move back and forth between organizing and drafting your document.

Organizing the Content

During the planning and researching stage of the writing process, you developed a clear sense of purpose and collected some information on your subject. Now, it is time to consider what organization would best achieve your purpose and present this information to the readers.

GENRES Technical documents generally follow patterns of organization called *genres.* These genres arrange information in ways that the readers will find predictable and easily accessible.

A genre is a familiar pattern that readers will expect the document to follow.

A genre is a familiar pattern that readers will expect the document to follow. For example, a *report* is a genre that would usually include sections on methodology, results, discussion, and recommendations. A set of *instructions*, quite differently, would lead the readers step by step through a process (Figure 2.3). Both of these genres achieve different purposes, so they organize information differently.

Chapters 16 through 23 of this book provide detailed discussions of the most common genres in technical workplaces. In most situations, you will know which genre you need because your supervisor or instructor will ask you to write a "report," a "proposal," or a "letter." But, if you are uncertain which genre suits your needs, pay attention to your document's purpose. Then, find the genre that best suits the purpose you are trying to achieve.

TAKE NOTE Genres are not formulas to be followed mechanically. Rather, view them as maps that help you arrive at a particular destination. Just as a map usually offers a variety of different ways to go somewhere (some paths are more efficient than others), a genre offers a pattern for writing a document. You can modify the genre to fit your or your readers' specific needs.

For helpful tools for doing research online, go to **www.ablongman.com/johnsonweb/2.6**

Sample of Genre: Instructions

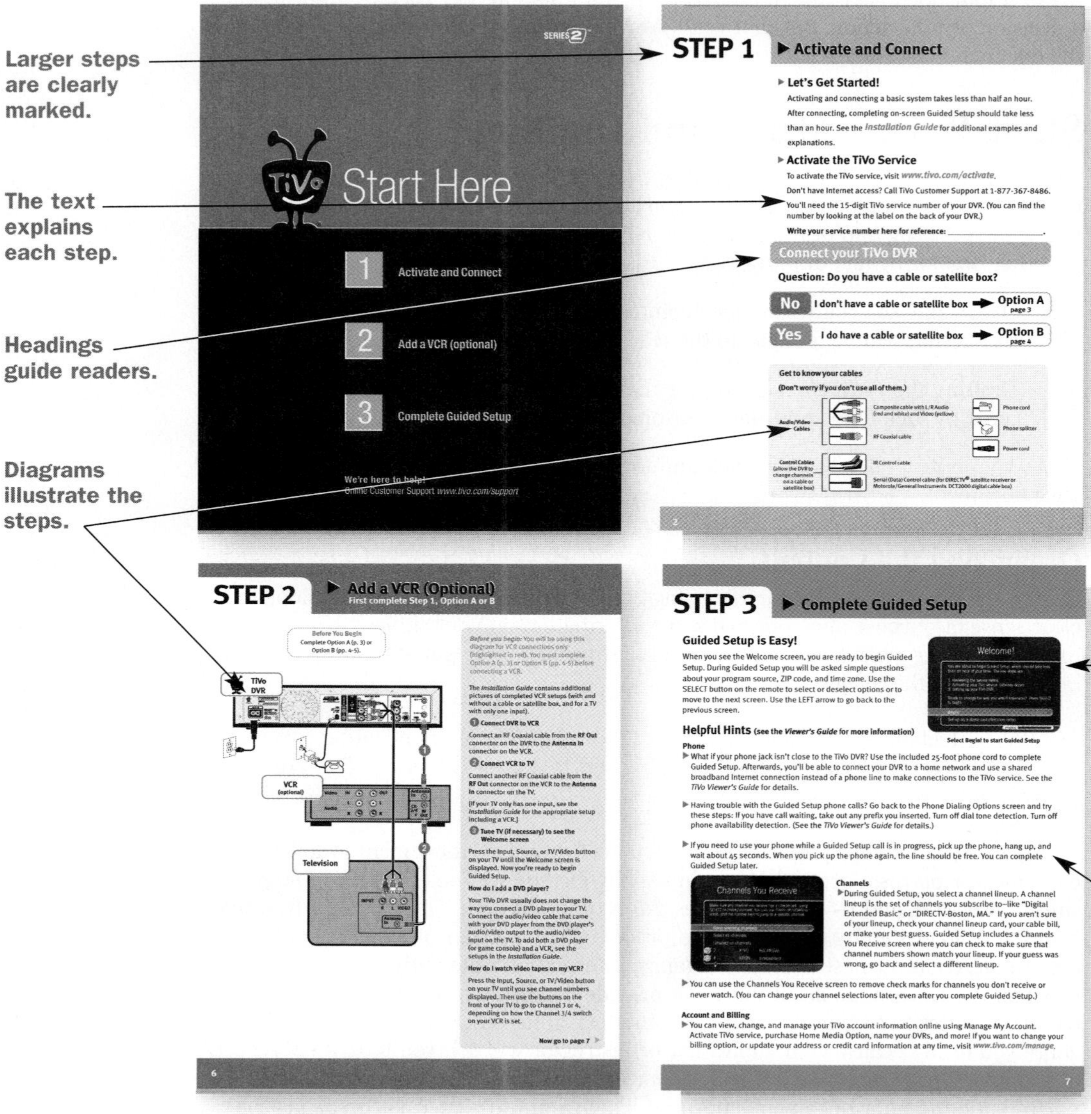

Larger steps are clearly marked.

The text explains each step.

Headings guide readers.

Diagrams illustrate the steps.

SERIES2

TiVo Start Here

1 Activate and Connect

2 Add a VCR (optional)

3 Complete Guided Setup

We're here to help!
Online Customer Support www.tivo.com/support

STEP 1 ▶ **Activate and Connect**

▶ **Let's Get Started!**

Activating and connecting a basic system takes less than half an hour. After connecting, completing on-screen Guided Setup should take less than an hour. See the *Installation Guide* for additional examples and explanations.

▶ **Activate the TiVo Service**

To activate the TiVo service, visit *www.tivo.com/activate*.

Don't have Internet access? Call TiVo Customer Support at 1-877-367-8486.

You'll need the 15-digit TiVo service number of your DVR. (You can find the number by looking at the label on the back of your DVR.)

Write your service number here for reference: ______________.

Connect your TiVo DVR

Question: Do you have a cable or satellite box?

No I don't have a cable or satellite box ➡ **Option A** page 3

Yes I do have a cable or satellite box ➡ **Option B** page 4

Get to know your cables
(Don't worry if you don't use all of them.)

Audio/Video Cables — Composite cable with L/R Audio (red and white) and Video (yellow); RF Coaxial cable

Control Cables (allow the DVR to change channels on a cable or satellite box) — IR Control cable; Serial (Data) Control cable (for DIRECTV® satellite receiver or Motorola/General Instruments DCT2000 digital cable box)

Phone cord; Phone splitter; Power cord

2

Figure 2.3: A genre follows a pattern that the readers will find familiar. Readers would immediately recognize this document as a set of instructions and be able to use it.

STEP 2 ▶ **Add a VCR (Optional)**
First complete Step 1, Option A or B

Before You Begin
Complete Option A (p. 3) or Option B (pp. 4–5).

TiVo DVR

VCR (optional)

Television

The *Installation Guide* contains additional pictures of completed VCR setups (with and without a cable or satellite box, and for a TV with only one input).

❶ **Connect DVR to VCR**

❷ **Connect VCR to TV**

❸ **Tune TV (if necessary) to see the Welcome screen**

How do I add a DVD player?

How do I watch video tapes on my VCR?

Now go to page 7

6

STEP 3 ▶ **Complete Guided Setup**

Guided Setup is Easy!

When you see the Welcome screen, you are ready to begin Guided Setup. During Guided Setup you will be asked simple questions about your program source, ZIP code, and time zone. Use the SELECT button on the remote to select or deselect options or to move to the next screen. Use the LEFT arrow to go back to the previous screen.

Welcome!

Select Begin! to start Guided Setup

Helpful Hints (see the *Viewer's Guide* for more information)

Phone

▶ What if your phone jack isn't close to the TiVo DVR? Use the included 25-foot phone cord to complete Guided Setup. Afterwards, you'll be able to connect your DVR to a home network and use a shared broadband Internet connection instead of a phone line to make connections to the TiVo service. See the *TiVo Viewer's Guide* for details.

▶ Having trouble with the Guided Setup phone calls? Go back to the Phone Dialing Options screen and try these steps: If you have call waiting, take out any prefix you inserted. Turn off dial tone detection. Turn off phone availability detection. (See the *TiVo Viewer's Guide* for details.)

▶ If you need to use your phone while a Guided Setup call is in progress, pick up the phone, hang up, and wait about 45 seconds. When you pick up the phone again, the line should be free. You can complete Guided Setup later.

Channels You Receive

Channels

▶ During Guided Setup, you select a channel lineup. A channel lineup is the set of channels you subscribe to–like "Digital Extended Basic" or "DIRECTV-Boston, MA." If you aren't sure of your lineup, check your channel lineup card, your cable bill, or make your best guess. Guided Setup includes a Channels You Receive screen where you can check to make sure that channel numbers shown match your lineup. If your guess was wrong, go back and select a different lineup.

▶ You can use the Channels You Receive screen to remove check marks for channels you don't receive or never watch. (You can change your channel selections later, even after you complete Guided Setup.)

Account and Billing

▶ You can view, change, and manage your TiVo account information online using Manage My Account. Activate TiVo service, purchase Home Media Option, name your DVRs, and more! If you want to change your billing option, or update your address or credit card information at any time, visit *www.tivo.com/manage*.

7

Screenshots are used to illustrate results of steps.

Additional notes help readers adjust to their specific needs.

Source: TiVo.

To find websites that discuss various genres, go to **www.ablongman.com/johnsonweb/2.7**

In the end, genres are simply common patterns of writing used by professional communities. They are not set in stone. Rather, each community modifies them to suit its specific needs. A report written by an anthropologist, for example, is different from a report written by an electrical engineer. Nevertheless, both reports will have similar features and follow similar patterns.

AT A GLANCE

Genres in Technical Communication

Here are the genres covered later in this book:

- Letters and Memos—Chapter 16
- Technical Definitions—Chapter 17
- Technical Descriptions—Chapter 18
- Instructions—Chapter 19
- Proposals—Chapter 20
- Activity Reports—Chapter 21
- Analytical Reports—Chapter 22

INTRODUCTIONS, BODIES, AND CONCLUSIONS All genres share a few organizational features in common. In general, all documents have a beginning (introduction), a middle (body), and an end (conclusion). All three of these parts play a different role in any given document.

> **Introduction** sets a context for the document by telling readers your subject, purpose, and main point. It might also offer background information on the subject, stress the importance of the subject, and forecast the structure of the document's body.
>
> **Body** presents the content of the document. The body includes facts, data, reasoning, and examples that help the document achieve its purpose and prove its main point.
>
> **Conclusion** reestablishes the context for the document by restating the main point, stressing the importance of the subject to the readers, and looking to the future.

A well-written technical document begins by offering contextual information up front in the introduction (Figure 2.4). Basically, the introduction tells readers the who, what, where, when, why, and how. Then, the body provides the content (the need-to-know information) that readers want. The conclusion reframes the discussion by circling back to the contextual information stated in the introduction.

TAKE NOTE It might help to think of your document as a building. The introduction and conclusion are part of the framework that defines the overall shape of the building. The body includes the rooms and things inside the building.

LINK For more information on writing introductions and conclusions, see Chapter 6, page 134.

To find websites that discuss writing introductions and conclusions, go to **www.ablongman.com/johnsonweb/2.8**

Basic Organization of a Document

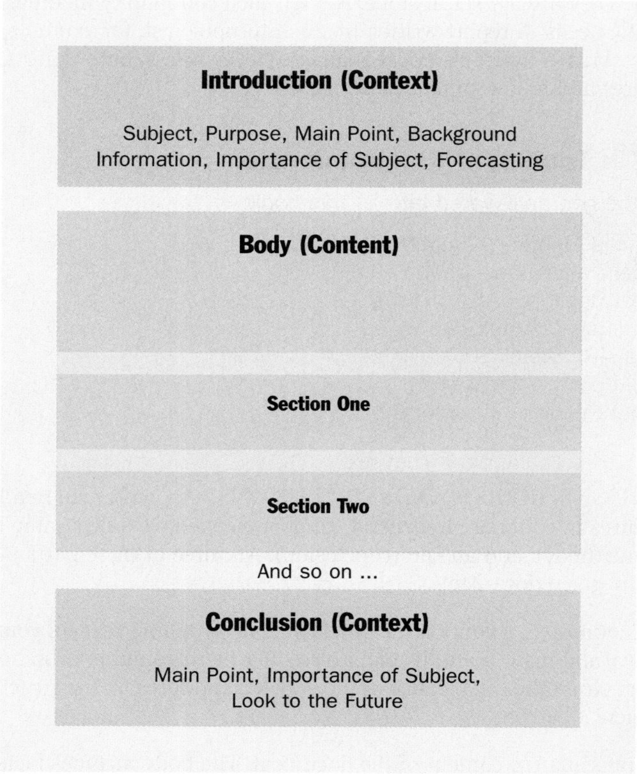

Introduction (Context)
Subject, Purpose, Main Point, Background Information, Importance of Subject, Forecasting
Body (Content)
Section One
Section Two
And so on ...
Conclusion (Context)
Main Point, Importance of Subject, Look to the Future

Figure 2.4: Technical documents should have a beginning, middle, and end. The introduction sets a context. The body includes the content. The conclusion resets the context.

Drafting the Content

Once you have a good sense about how your document will be organized, you can start filling it with words and images. It is important that you recognize that the purpose of drafting is *not* to create the final version of your document. The purpose of drafting is to put your ideas on the screen so you can work with them.

Let the genre you are following roughly guide your drafting of the document. Then, during the revision and editing phase, you can transform your words and images into the final version.

Sometimes the hardest part of drafting is just putting words on the screen. After all, you probably already have a wealth of important ideas in your head. You have facts and data stored on your computer or written in your notes. But you just can't turn your thoughts into sentences and paragraphs. Well, don't—at least for now.

Several *invention techniques* can help you put words on the screen. Three of the best techniques for technical communication are freewriting, logical mapping, and outlining/boxing.

FREEWRITING The aim of freewriting is to get your ideas out on the screen, so you can work with them. While freewriting, simply put your fingers on the keyboard and start typing.

Don't worry about the constraints of writing, like sentences, paragraphs, grammatical correctness, or citations. Just keep typing. Eventually, you will find that you have filled one or more screens with words, sentences, and fragments of sentences (Figure 2.5).

TAKE NOTE Some people even turn the computer screen off as they freewrite. They just write whatever comes to mind. Then, when they turn the screen on again, there are a couple of pages of raw text to start them out.

Freewriting

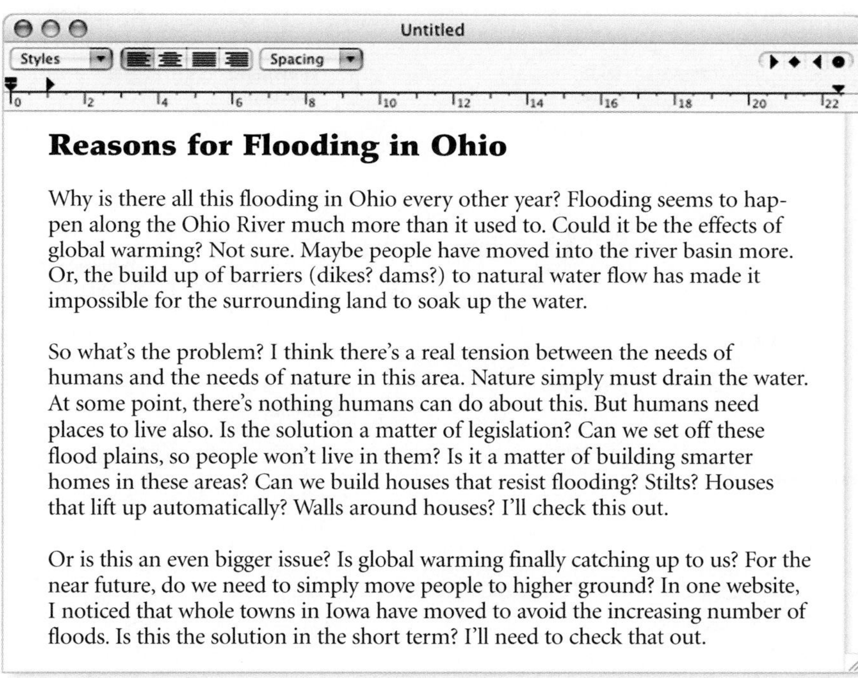

Untitled
Styles
Spacing
Reasons for Flooding in Ohio
Why is there all this flooding in Ohio every other year? Flooding seems to happen along the Ohio River much more than it used to. Could it be the effects of global warming? Not sure. Maybe people have moved into the river basin more. Or, the build up of barriers (dikes? dams?) to natural water flow has made it impossible for the surrounding land to soak up the water.
So what's the problem? I think there's a real tension between the needs of humans and the needs of nature in this area. Nature simply must drain the water. At some point, there's nothing humans can do about this. But humans need places to live also. Is the solution a matter of legislation? Can we set off these flood plains, so people won't live in them? Is it a matter of building smarter homes in these areas? Can we build houses that resist flooding? Stilts? Houses that lift up automatically? Walls around houses? I'll check this out.
Or is this an even bigger issue? Is global warming finally catching up to us? For the near future, do we need to simply move people to higher ground? In one website, I noticed that whole towns in Iowa have moved to avoid the increasing number of floods. Is this the solution in the short term? I'll need to check that out.

Figure 2.5: While freewriting, just get your ideas on the screen. Simply writing ideas down will help you locate important ideas and directions for research.

With some raw text on the screen, you can go back and identify the important topics, concepts, and themes in your document. You may or may not end up using some of the words and sentences in your freewriting draft. However, you will find that freewriting helps you put your ideas into words. It helps you fight through writer's block.

LOGICAL MAPPING Logical mapping is a highly visual way to invent your ideas. Essentially, mapping allows you to find the logical relationships among the ideas that will form the content of your document.

To map the content of your document, start by putting your subject in the middle of the screen. Put a circle or box around it. Then, start typing your other ideas around the subject, and put circles or boxes around them (Figure 2.6).

Logical Map

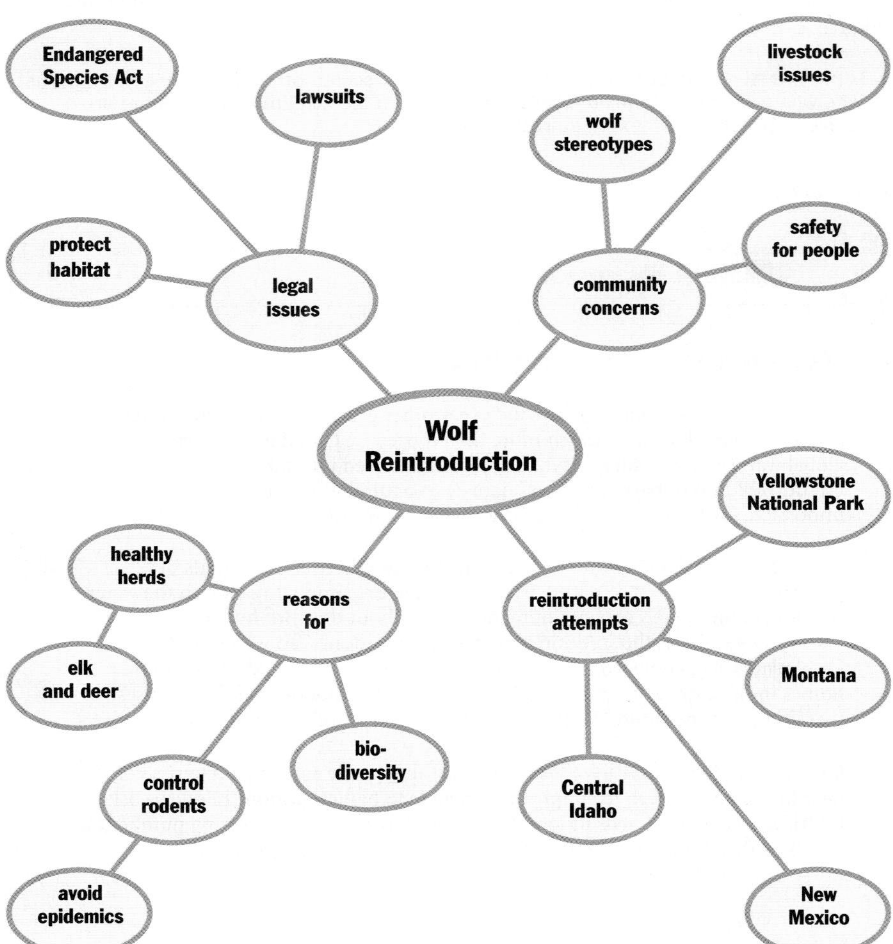

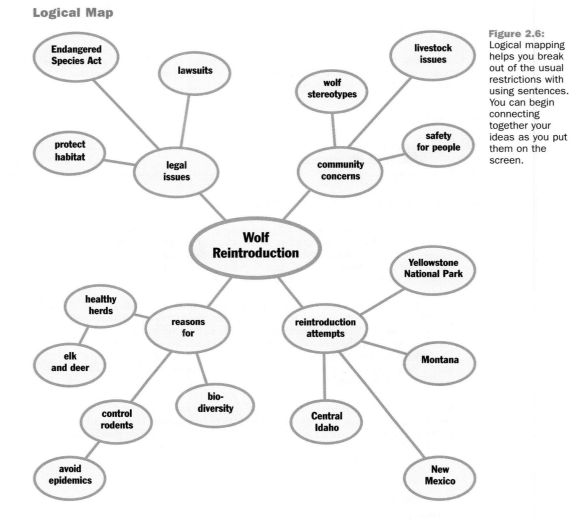

Figure 2.6: Logical mapping helps you break out of the usual restrictions with using sentences. You can begin connecting together your ideas as you put them on the screen.

Eventually, you will fill a screen with words and phrases related to the subject. At this point, start connecting related ideas by drawing lines among them. As you draw lines, you will begin to see the major topics, concepts, or themes that will be important parts of the document you are writing.

Software programs such as Inspiration, Visio, MindManager, and IHMC Concept Mapping Software can help you do logical mapping on screen. Otherwise, use the

GO TO THE NET

To find links to logical mapping software, go to **www.ablongman.com/johnsonweb/2.9**

Draw function of your word processor to create "text boxes" and draw lines among them. With a little practice, you will find that you can create logical maps on the screen with little effort.

Writing theorists have argued that logical mapping draws ideas from the right side of your brain. They also suggest that it taps into your "visual thinking" abilities. No matter how it works, the advantage to logical mapping is that it allows you to put your ideas onto the screen or sheet of paper. You can then visualize the argument you are going to make.

OUTLINING OR BOXING Outlining has changed somewhat since computers came on the scene. Most word-processing programs will allow you to draft in *Outline* mode or *Document Map* mode (Figure 2.7).

Outlining or Document Mapping

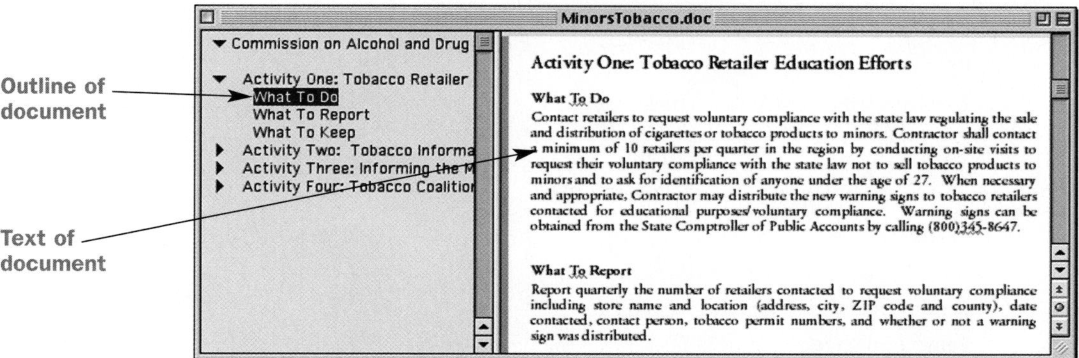

Figure 2.7: In Document Map mode, the computer automatically outlines your document on the left. This feature allows you to keep the whole structure of the document in mind as you work on individual parts.

Outlines can be used throughout the drafting process. Sometimes it helps to sketch out an outline before you start drafting. That way, you can see how the document will be structured. When revising, you might find it helpful to re-outline your document or print out the computer-generated outline in Outline mode. This outline will help you find places where need-to-know information should be added or unnecessary information taken out.

LINK For more information on using electronic outlining, see Chapter 6, page 138.

Boxing is less formal. As you plan your document, draw boxes on the screen or a piece of paper that will contain the ideas or topics in your document (Figure 2.8). Then, type your ideas into the boxes. If you want to make multiple levels in your text, simply create boxes within boxes.

TAKE NOTE The Table function in your word-processing software will allow you to make boxes easily. You can start out with a few boxes. If you need more, you can always create more.

These invention tools don't work for you? To learn about other invention techniques, go to **www.ablongman.com/johnsonweb/2.10**

Boxing

Introduction: Report on Flooding in Ohio

Purpose Statement: This report will provide strategies for managing flooding in the Ohio River Valley.

Main Point: Solving flooding means restoring wetlands and slowing development.

Importance of Subject: If we don't do something now, it will only become worse as the effects of global warming are felt.

Section One: The Problem

Development in Ohio River Valley

Increased water due to global warming

Additional dams and retaining walls

Section Two: The Plan

Restore wetlands

Limit development along rivers

Create holding reservoirs for water

Remove some retaining walls

Conclusions: We're Running Out of Time

We need to restore wetlands and lessen development on river

Advantages of these recommendations

The future

Figure 2.8: Boxing is like outlining. Each of the cells can be filled with your ideas. Then, you organize these ideas into a more structured document.

Both outlining and boxing have their advantages and disadvantages. Outlining allows you to put all your ideas down on paper in a structured way. The problem with outlining is that you might not be ready to produce such a detailed version of your document. Boxing gives you more flexibility, because you are only filling spaces with ideas. The drawback to boxing is that you still need to turn those filled boxes into sentences and paragraphs.

HELP

Using Your Computer to Overcome Writer's Block

Have you ever stared at a blank computer screen and had it stare back? Many people have trouble starting the writing process. Here are 10 quick tips for overcoming writer's block:

Just write something—Don't wait to be inspired. Those moments are rare. Instead, just go ahead and write, even if you are putting garbage on the screen. The act of writing will get you going.

Write something every day—Writer's block is directly linked to anxiety, which is often linked to an impending deadline or due date. If you write for just a half hour every day (that's all), you will usually finish all your documents with time to spare.

Lower your standards—Stop trying to get it right on the first try. Instead, put your ideas on the screen without worrying about whether they are intelligent or grammatically correct. Then, spend more time during the revision phase turning those ideas into a document that embodies the quality you expect.

Talk it out—Professional writers can often be found talking to themselves or friends about their work. Sometimes it helps to just say what you want to write. Then, after you have rehearsed the document a few times, write it down.

Use a notebook or your PDA—Jot down notes in a notebook or your personal digital assistant (PDA) as they come to mind (Figure A). Then, when you are ready to write, you will find a cache of ideas to draw from.

Using Your PDA to Fight Procrastination

Figure A: In a notebook or a personal digital assistant, keep records of your ideas. Good ideas can come to you at any time. Make sure you keep track of them.

Having trouble with writer's block? For more help, go to **www.ablongman.com/johnsonweb/2.11**

Use both sides of your brain—The right side of your brain is more visual than your left. Use logical mapping, freewriting, and boxing to tap into that visual creativity. These invention techniques will help you put your ideas on the screen. Then, the left side of your brain can organize them into sentences and paragraphs.

Surf and dump—Surf the Internet, looking for items related to the subject of your document. Cut and paste, or "dump," the information you find into your document. Then start writing. Remember to cite your sources, though, and don't plagiarize the works of others.

"What I really mean is . . ."—Whenever you are blocked, finish the sentence "What I really mean is . . ." You will find that simply finishing this sentence will help you get past the temporary block.

Write an e-mail—Start writing your document as an e-mail to a friend or yourself. E-mail is often a more familiar writing environment, allowing you to relax as you write.

Stop procrastinating—Procrastination is the usual culprit behind writer's block. The pressure of a deadline can cause your brain to freeze. So, start each project early, and write a little every day. Your writer's block will evaporate.

Stage 3: Improving the Style

All documents have a style, whether it is chosen or not. If properly chosen, good style can make your documents easier to read and more persuasive. But, if you don't pay attention to style, your writing might seem chaotic and difficult to read.

Mistakenly, some people believe that considerations of style are not important in their technical documents. After all, they say, the content of the document is the most important thing. Style is just window dressing.

Actually, your style reflects the values, beliefs, and relationships you want with the readers. In a word, good style is about quality. It is about your and your company's commitment to excellence and achievement. As a result, when writers do not pay attention to style, technical documents tend to use erratic styles that annoy their readers. Have you ever read a text that just didn't feel right? More than likely, you were reacting to its erratic style.

TAKE NOTE Technical documents that seem difficult to understand are usually just poorly written on a stylistic level. Science and technology are not naturally obscure or ambiguous. Unfortunately, lack of attention to style has fostered an image that science and technology are too difficult and complex for the average person to understand.

Good style is a choice you can and should make. In Chapter 7 of this book, you will learn about two kinds of style that are widely used in technical documents: plain style and persuasive style.

Plain style—This style stresses clarity and accuracy. By simply paying attention to where words appear in a sentence and in paragraphs, you can make your ideas clearer and easier to understand.

Persuasive style—Using persuasion strategies, you can motivate readers by appealing to their values and emotions. You can use similes and analogies to add a visual quality to your work. You can use metaphors to change your readers' perspective on issues. Meanwhile, you can use tone and pace to add energy and color to your work.

Most technical documents should be written in the plain style to make the information as clear and concrete as possible. Periodically, though, you will need to turn to the persuasive style to help guide or change readers' minds.

LINK For more about improving style, go to Chapter 7, page 165.

Your goal should always be to make information as clear and concrete as possible. When persuasion is needed, you will want to energize your words.

AT WORK

Tim Raine
RESEARCHER/WRITER, RAINE TITLE SEARCH, LLC.
Raine Title Search, in Anamosa, Iowa, researches the history of property for sale.

Do writers in the technical workplace *really* follow a writing process as they write?

Yes. Experienced writers may not consciously consider each step required to complete daily writing tasks. Like any skill, when you learn the fundamentals of technical writing and then practice what you learn on a daily basis, the writing process becomes a habit that you don't think about. The more you write and the more you pay attention to the process you follow as you write, the easier it becomes to recognize and produce high-quality writing.

Learning technical writing is like learning any workplace skill. When you're acquiring a new skill, you pay great attention to the process. As you grow more and more familiar with the skill, you tend to stop thinking so much about the process and pay more attention to achieving your goals. You don't forget the process—it just becomes a habit that you can rely on.

However, good writers also know their own writing process and routinely evaluate it. Thinking critically about how you write allows you to streamline your process to handle the task at hand as efficiently as possible.

Knowing your writing process also gives you a frame of reference for estimating the time and effort required to complete specific writing tasks. The ability to make the most of your time and to estimate what it takes to produce high-quality writing is extremely useful in any work environment.

Want to improve your writing style? Go to **www.ablongman.com/johnsonweb/2.12**

Stage 4: Designing

Before computers, designing a document was a difficult task. Typewriters usually limited people to underlining headings or using all-capital letters. Centering a title on a page required some calculations and the skillful use of the backspace button. Charts and graphics took hours to create.

With computers, you can lay out a page in minutes and create graphics with a few clicks of a button. So, design is not only possible, it is expected. You should make document design a regular part of your writing process.

LINK For more ideas about designing documents and screens, go to Chapter 8, page 198.

Moreover, few people read documents word for word anymore. Right now, you are probably scanning this chapter, looking for the highlights. If so, you are like most readers of technical documents. You are a "raider" for information as well as a reader. So keep in mind this saying: Readers of technical documents are "raiders" for information.

Your readers, like you, want the important parts to be highlighted. They prefer documents that use graphics and document layout to make the information more accessible, interesting, and attractive (Figure 2.9).

LINK To learn about creating and using graphics, turn to Chapter 9, page 234.

Chapter 8 discusses document design techniques and strategies. Chapter 9 discusses the creation and use of graphics in documents. As you draft and revise your document, keep looking for places where you can use highlighting to help your readers locate information. Look for places where graphics might support or reinforce the written text. And, think about how the layout of the document can improve its readability and your readers' interest in it.

Stage 5: Revising and Editing

When you have finished drafting and designing the document, you are about two-thirds finished. In technical communication, it is important to leave plenty of time for revising, editing, and proofreading the document. Clarity and accuracy are essential if your readers are going to understand what you are trying to tell them.

In Chapter 10, you will learn about four levels of revising and editing:

Level 1: Revising—Once you have drafted the document, it is time to reconsider your subject and purpose while considering the needs of your readers and the places where they will use the document. During revision, you are trying to "re-vision" the document to see if it meets your original goals.

Level 2: Substantive editing—Look at the content, organization, and design of the document. The content of your document should be complete and organized in a way that is familiar to your readers. Using the genre of the document, locate areas in the document where information has been misplaced. Then, make sure your document has an effective beginning (introduction), middle (body), and end (conclusion). Your goal is to boil down your document's content to its core, relying exclusively on need-to-know information.

If you want more information on document design, go to **www.ablongman.com/johnsonweb/2.13**

Document Design Is Very Important

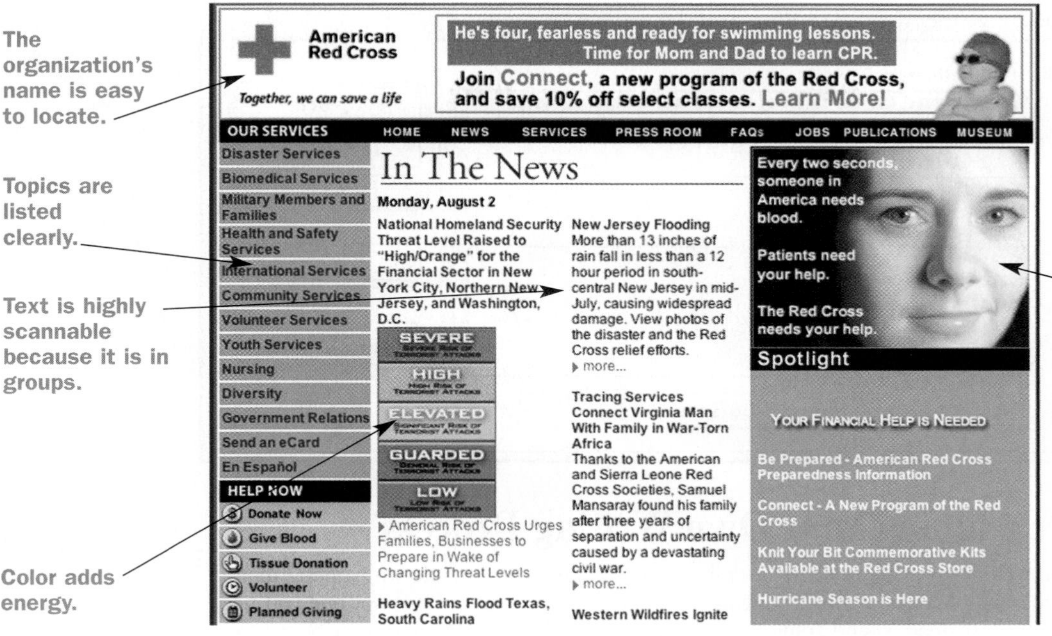

Pictures add a human quality.

Figure 2.9: Because readers are raiders of information, you want the design of your document to be visually accessible.

Source: American Red Cross, http://www.redcross.org.

Level 3: Copyediting—Effective writers often spend a significant amount of time copyediting the sentences, paragraphs, and graphics in their documents. Copyediting makes the document easier to read and more persuasive. It also ensures that graphics are accurate and support the written text.

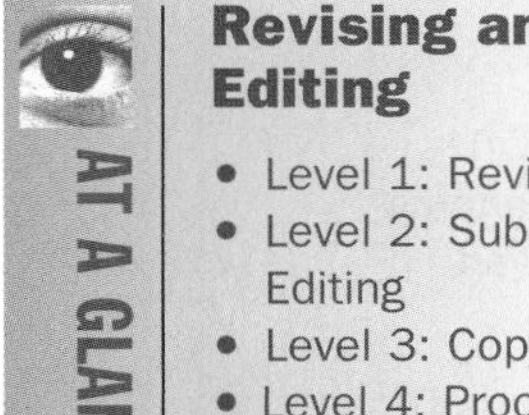

AT A GLANCE
Revising and Editing
• Level 1: Revising
• Level 2: Substantive Editing
• Level 3: Copyediting
• Level 4: Proofreading

Level 4: Proofreading—Finally, don't forget to carefully proofread your document. In the technical workplace, quality is taken very seriously. So, when readers encounter a document that has spelling mistakes, grammar problems, typos, spelling errors, and usage problems, they will question the quality of the information. Minimally, the spelling and grammar checker on your computer can catch errors as you proofread (Figure 2.10). But, you should also carefully read the document to catch mistakes that slip through.

LINK To learn more about revising and editing, go to Chapter 10, page 264.

At the end of a difficult project, it is sometimes tempting to simply throw up your hands and send in the document. You can even trick yourself into believing that readers don't really care about the quality of your document. They do. It is important to spend time editing your work to meet the readers' high expectations.

Want to know more about editing? Go to **www.ablongman.com/johnsonweb/2.14**

Spelling and Grammar Checker

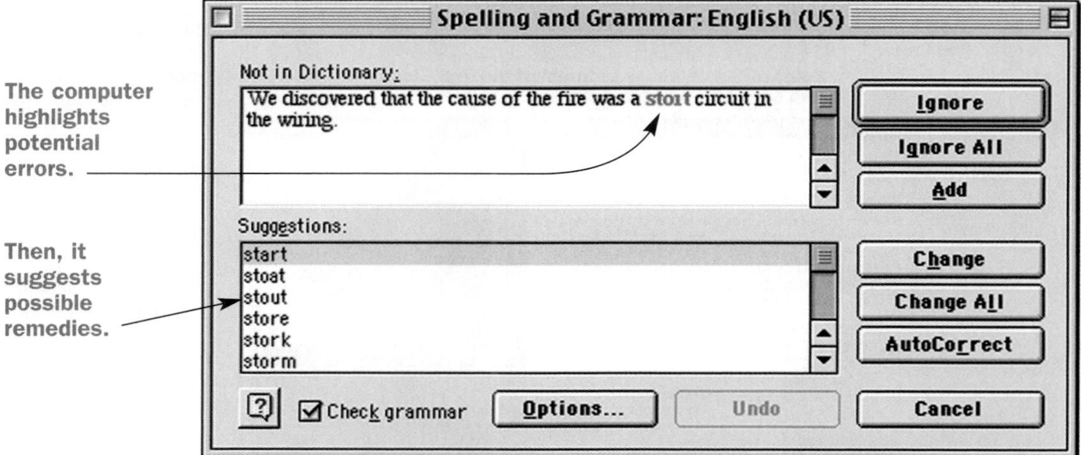

Figure 2.10: The spelling and grammar checker in your word processor software can help locate typos and misspellings. It is less helpful for locating grammatical errors.

Developing Your Own Writing Process

The writing process described in this chapter is just one possible way among many of producing effective technical documents. This process is useful but not one you need to adopt. Instead, spend some time thinking about how you go about writing documents. Then, adapt your writing process to the needs of technical documents.

In the end, it is important for you to consciously construct your writing process for yourself. Without a conscious writing process, you will find writing more difficult, less effective, and less satisfying. Once you have consciously constructed a reliable writing process for yourself, you will find writing technical documents much easier.

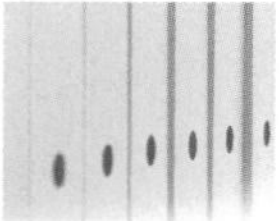

CHAPTER REVIEW

- By consciously developing a writing process, you will learn how to write more efficiently. In other words, you will "work smarter, not harder."
- A useful computer-based writing process includes the following stages: planning and researching, organizing and drafting, improving the style, designing, and revising and editing.
- The key to effective technical writing is to define your rhetorical situation, putting special emphasis on the purpose of your document.
- Technical writing genres are helpful for organizing information into patterns that your readers will expect.
- When drafting your text, you might use freewriting, logical mapping, or boxing/outlining to put your ideas on the screen.

GO TO THE NET

Some good advice about developing a writing process is available at **www.ablongman.com/johnsonweb/2.15**

- You can use the plain or persuasive style to clarify and energize your message.
- Document design should be used to enhance readability, support the written text, and make the document more attractive.
- Revising and editing a document is important to improve clarity and accuracy while ensuring that the content provides only need-to-know information.

EXERCISES AND PROJECTS

Individual or Team Projects

1. Interview a professional in your field or a professor in your major about his or her writing process. Ask about smaller documents like memos, letters, and procedures. Then, ask about larger documents. How does this person's process follow any or all of the stages described in this chapter? Here are some questions you might ask:
   - How much planning and research do you do before drafting?
   - Do you follow any genres or patterns in your documents?
   - How much revising do you do?
   - How do you use style to clarify or amplify your message?
   - When do you start thinking about the visuals and design of the document?

   Write a memo to your instructor in which you discuss the results of your interview. Compare and contrast the results of your interview with the writing process described here.

2. Write an e-mail to your instructor in which you describe your own writing process. Each person has a unique approach to writing. What are your unique approaches to writing? How is your writing process similar to the one described in this chapter? How is your writing process different? How can you adapt your writing process to fit the technical workplace?

   Compare your writing process with that of others in your class. How does your writing process differ from theirs? Why do you think each of you follows a different writing process?

3. More than likely, you are working on a document right now. Describe the rhetorical situation in which this document will be used. In other words, (1) define the subject, (2) state your purpose, (3) describe the readers, and (4) describe the document's context of use. In a memo to your instructor, write a description of the rhetorical situation. Also, tell your instructor how this document's rhetorical situation will require you to adapt information to the document's subject, purpose, readers, or context of use.

Collaborative Project

Imagine you and your group are writing a report on sleep deprivation on campus and its effects. Using the Internet, run searches on sleep deprivation to collect information. You can use search engines like Google.com, Altavista.com, Lycos.com, or Yahoo.com. Once you have gathered some information, have each member of your group do one of the following forms of drafting:

Freewriting—Write a screen or two worth of text in which you describe everything you know or have found about the effects of sleep deprivation.

Logical mapping—Put "sleep deprivation" in the middle of your screen or a sheet of paper. Then, use logical mapping to write down all the ideas, topics, and facts that you might include in the report. Connect related ideas.

Boxing or outlining—Use boxing or outlining to start developing an organization for your report. When boxing, draw large boxes that define topics. Then, fill those boxes with the ideas, facts, and arguments that will be covered under each topic in the report.

Present the results of your research and drafting to your class. Show them how your freewriting, logical mapping, and boxing/outlining might shape your writing process.

For more information on sleep deprivation, go to **www.ablongman.com/johnsonweb/2.16**

CASE STUDY

The Procrastinator

Elizabeth Brown manages a team of engineers on a robotics project. A month ago, Jane Ganders, the vice president of research and development, asked each team in the company to provide a progress report discussing the status of their project. The buzz around the company says that finances are growing tight, and management is looking for weak or bloated projects that can be cut to save money. The report is due a month from now.

Elizabeth and her team gathered information for the report, including measurements, data, and costs. Now, she needs each member of her team to write part of the report, describing his or her contribution to the project and listing any successes and shortcomings.

Her problem is Bill Hands. He is a good engineer, but he is a notorious procrastinator when it comes to writing. Elizabeth often notices that he frantically slaps together his progress reports at the last minute. Meanwhile, his proposals and reports are usually shoddy because he always waits until the last moment to write them. He says "I write best under pressure." In reality, though, Elizabeth knows he writes very poorly under pressure.

But this report to the vice president is crucial, because it might determine whether Elizabeth's team keeps working on this project. A shoddy report could mean a few members of the team would be laid off. So, Elizabeth can't allow Bill or anyone else on the team to sabotage the report because they procrastinated.

If you were Elizabeth, how might you use the techniques you learned in this chapter to ensure that Bill does not procrastinate on his part of the report? How can you use the writing process to ensure that the report is finished at a high quality level?

GO TO THE NET

To learn about ways to overcome procrastination, go to **www.ablongman.com/johnsonweb/2.17**

CHAPTER

3

Readers and Contexts of Use

CHAPTER CONTENTS

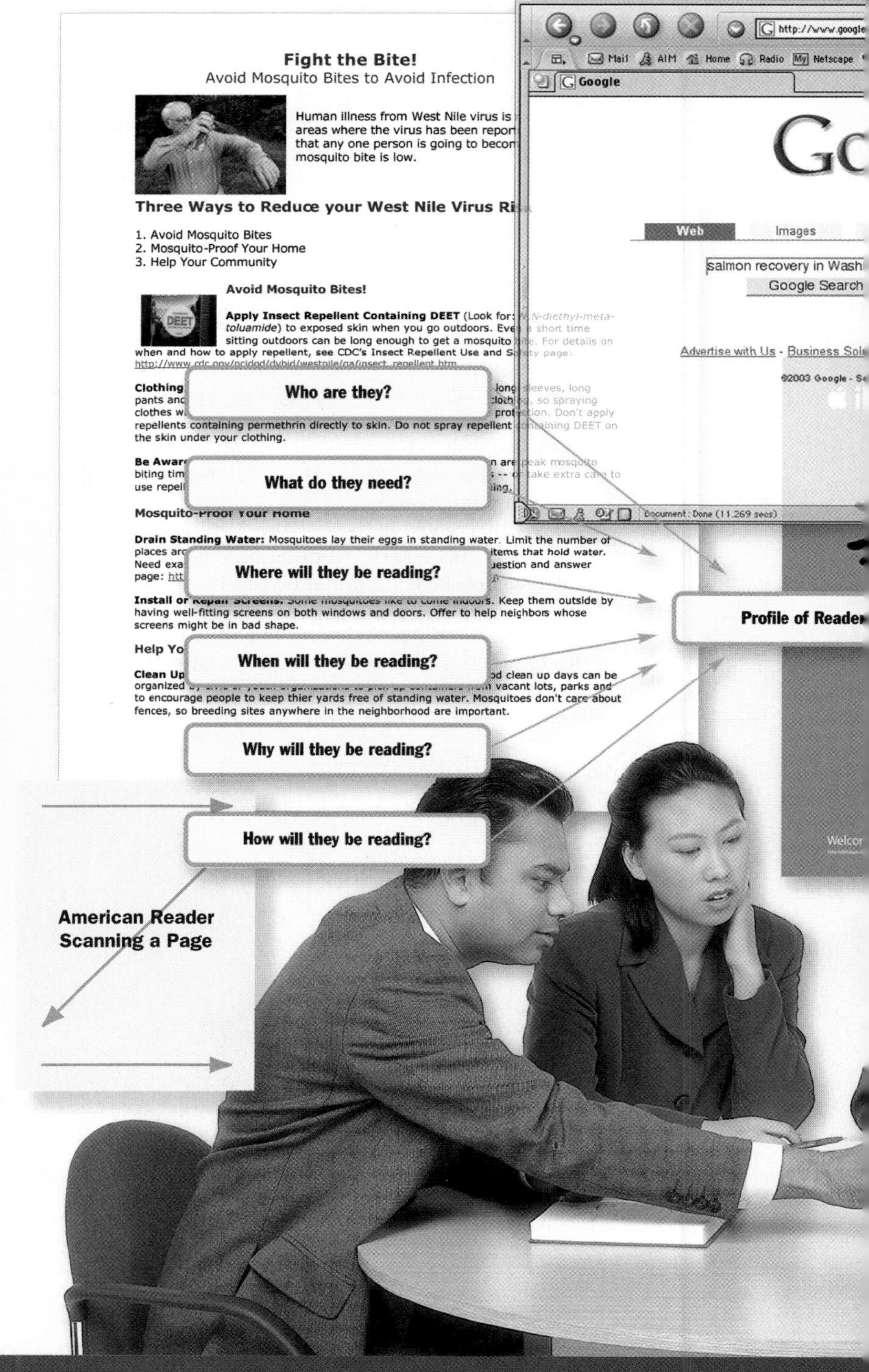

CHAPTER OBJECTIVES

In this chapter, you will learn:

- How to use the computer as a reader analysis tool.
- How to develop a comprehensive profile of a document's readers.
- How to sort your readers into primary, secondary, tertiary, or gatekeeper audiences.
- Techniques for identifying readers' needs, values, and attitudes about you and your document.
- How to analyze the physical, economic, political, and ethical contexts of use that influence how readers will interpret your text.
- How to anticipate the needs of international and cross-cultural readers.

An important advantage we have today is the ability to use the computer as a ***reader analysis tool.*** A quick search on the Internet will usually turn up information on just about anyone. So, as you prepare to write a document, you can develop a helpful profile of the people to whom you are writing. Or, you can build a profile of the sorts of people who might use your document. These profiles help you to tailor your text exactly to your readers' needs.

On the other hand, anticipating the broad network of people who may gain access to your text is challenging. The intended reader of your document might send it to others in his or her company. With a few clicks of the mouse, your document can be shared with a limitless number of people—technical experts, consultants, reporters, lawyers, auditors, etc.

In the Information Age, it is rare to find a document that is read by only one person. Instead, your document may have many readers, who will all have some influence on its reception, interpretation, and results.

Another challenge is the increasing significance of international communication through electronic networks. In technical fields, you ***will*** find yourself regularly communicating with people who speak other languages, have different customs, and hold different expectations. Computers have broken down many of the geographical barriers that once separated people and cultures. It is now common to communicate with people around the world on a daily basis.

Readers and contexts of use have always been important considerations when writing or presenting ideas. The use of computers has only heightened that importance, because documents can be more easily tailored to the needs of specific readers. You need to learn how to use the computer as a reader analysis tool, so you can communicate more effectively with the people who might use your text.

Profiling Your Readers

In technical communication, there is no such thing as a "general reader." Rather, each document is designed to suit the needs of specific types of readers. For this reason, early in the writing process, you should profile the types of people who might be interested in your document (Figure 3.1).

Reader profiles are sketches of your readers' tendencies, abilities, experience, needs, values, and attitudes.

Reader profiles are sketches of your readers' tendencies, abilities, experience, needs, values, and attitudes. To build a profile, begin asking yourself the Five-W and How questions about your readers.

Who might read this document?

What information do they need?

Where will they read the document?

When will they read the document?

Why will they be reading it?

How will they be reading it?

For websites that offer other ways to use the Five-W and How questions, go to **www.ablongman.com/johnsonweb/3.1**

Developing a Reader Profile

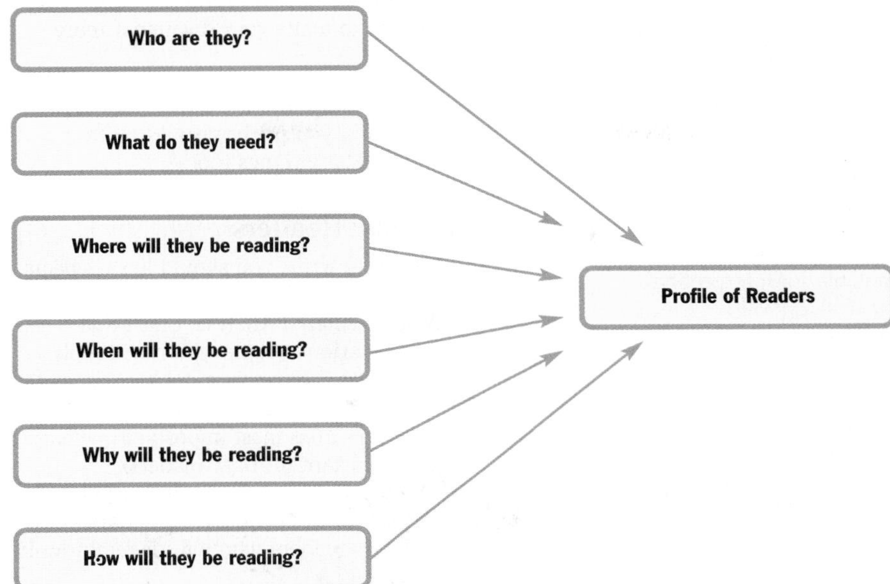

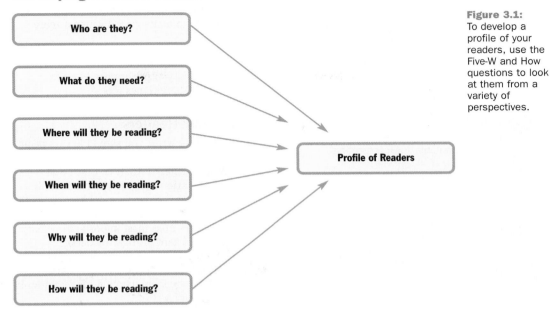

Figure 3.1: To develop a profile of your readers, use the Five-W and How questions to look at them from a variety of perspectives.

As you answer these questions, keep in mind the following guidelines about your readers and how they prefer to read.

Guideline One: Readers are wholly responsible for interpreting your text—Documents are only pages or screens with symbols on them. Since you will not be available to explain your document's meaning, it is up to your readers to make meaning from those symbols. What you "meant" does not matter if the readers create a different meaning than you expected.

Guideline Two: Readers are "raiders" for information—People rarely read technical documents for pleasure. Instead, most readers are simply raiding your document for information. They are scanning the document, looking for the information they need.

Guideline Three: Readers want only "need-to-know" information—Readers want only the information they *need* to make a decision. Any additional material beyond need-to-know information only makes the information they need harder to find.

Guideline Four: Readers prefer concise texts—The shorter, the better. In most cases, the longer the document is, the less likely that people are going to read it. Your readers prefer documents that get to the point and highlight the important information for them.

What else do readers want? For answers, go to **www.ablongman.com/johnsonweb/3.2**

Guideline Five: Readers prefer documents with graphics and effective page design—We live in a visual culture. Large blocks of text intimidate most readers. So, include graphics and use page design to make your document more readable.

Think about how you are reading this book right now. You are likely raiding for need-to-know information. You prefer conciseness and you prefer visual text that makes information easy to find. Your readers prefer these things too.

AT A GLANCE

About Your Readers

- Readers are wholly responsible for interpreting your text.
- Readers are "raiders" for information.
- Readers want only "need-to-know" information.
- Readers prefer concise texts.
- Readers prefer documents with graphics and effective page design.

Identifying Your Readers

When you sit down to write, you should always begin by identifying the readers of your document. Figure 3.2 shows a Writer-Centered Analysis Chart that will help you locate the various people who might look over your text (Mathes & Stevenson, 1976). You, as the writer, are in the center ring. Each ring in the chart identifies your readers from most important (primary readers) to least important (tertiary readers).

To use the Writer-Centered Chart, begin filling in the names and titles of the primary, secondary, tertiary, and gatekeeper readers who will or might look over your work.

Writer-Centered Analysis Chart

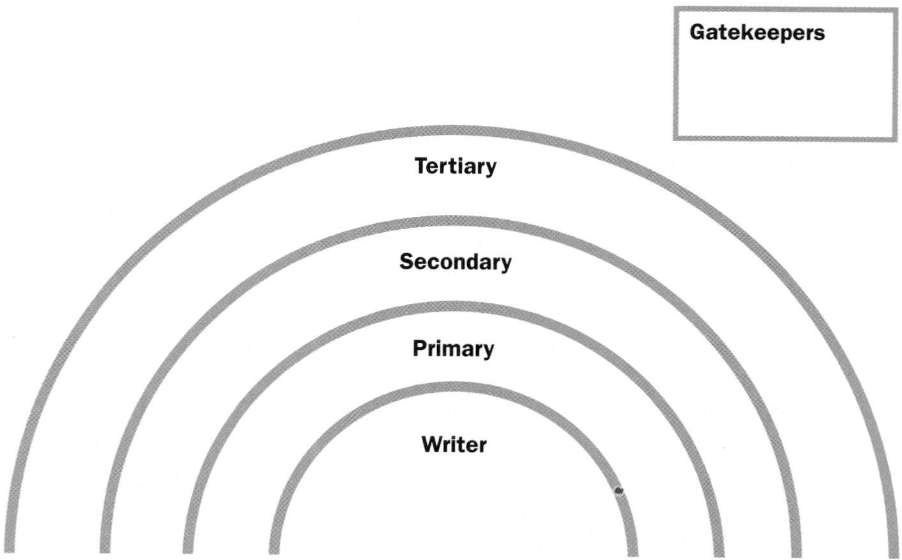

Figure 3.2: A Writer-Centered Analysis Chart starts with you in the center and identifies the various people who may be interested in your document.

For a downloadable version of the Writer-Centered Analysis Chart, go to **www.ablongman.com/johnsonweb/3.3**

PRIMARY READERS The primary readers are the people to whom your document is addressed. They are usually *action takers* because the information you are providing them will allow them to do something or make a decision. Usually your document will have only one or two primary readers, or types of primary readers.

SECONDARY READERS The secondary readers are people who *advise* the primary readers. Usually, they are experts in the field, or they have special knowledge that the primary readers require to make a decision. They might be engineers, technicians, lawyers, scientists, doctors, accountants, and others to whom the primary readers will turn for advice.

TERTIARY READERS The tertiary readers include anyone else who may have an interest in your document's information. They are often *evaluators* of you, your team, or your company. These readers might include local reporters, lawyers, auditors, historians, politicians, community activists, environmentalists, or perhaps your company's competitors. Even if you never expect your document to fall into these readers' hands, you should keep them in mind to avoid saying anything that could put you or your company at risk. Figure 3.3, for example, shows a memo in which the tertiary readers were not kept in mind.

GATEKEEPERS The gatekeepers are people who will need to look over your document before it is sent to the primary readers. Your most common gatekeeper is your immediate supervisor. In some cases, though, your company's lawyers, accountants, and others may need to sign off on the document before it is sent out.

AT A GLANCE

Types of Readers

- Primary Readers: Action Takers
- Secondary Readers: Advisors
- Tertiary Readers: Evaluators
- Gatekeepers: Supervisors

Each of these four types of readers will look for different kinds of information. The primary readers are the most important, so their needs come first. Nevertheless, a well-written document also anticipates the needs of the secondary, tertiary, and gatekeeper readers.

Ultimately, you want the primary readers to take action. But, the secondary readers (advisors) and tertiary readers (evaluators) may have direct or indirect influence over the primary readers' decision to do something.

Determining Readers' Needs, Values, and Attitudes

Now that you have identified the various readers of your document, you can start developing a more thorough profile of them. Specifically, you need to make some strategic assumptions about their needs, values, and attitudes.

TAKE NOTE Try not to fall into the "common sense" trap. Different communities have different ideas about what is common sense. Keep in mind that your views about what is common sense may differ significantly from the views of others.

A Memo that Does Not Consider Tertiary Readers

Primary readers of the document are listed here.

The first paragraph is clear about the purpose of memo.

The main point of the memo is stated up front.

This list summarizes the current situation.

October 16, 2003

TO: Gen. Dick Myers
Paul Wolfowitz
Gen. Pete Pace
Doug Feith

FROM: Donald Rumsfeld

SUBJECT: Global War on Terrorism

The questions I posed to combatant commanders this week were: Are we winning or losing the Global War on Terror? Is DoD changing fast enough to deal with the new 21st century security environment? Can a big institution change fast enough? Is the USG changing fast enough?

DoD has been organized, trained and equipped to fight big armies, navies and air forces. It is not possible to change DoD fast enough to successfully fight the global war on terror; an alternative might be to try to fashion a new institution, either within DoD or elsewhere—one that seamlessly focuses the capabilities of several departments and agencies on this key problem.

With respect to global terrorism, the record since September 11th seems to be:

— We are having mixed results with Al Qaida, although we have put considerable pressure on them—nonetheless, a great many remain at large.

— USG has made reasonable progress in capturing or killing the top 55 Iraqis.

— USG has made somewhat slower progress tracking down the Taliban—Omar, Hekmatyar, etc.

— With respect to the Ansar Al-Islam, we are just getting started.

Have we fashioned the right mix of rewards, amnesty, protection and confidence in the US?

Does DoD need to think through new ways to organize, train, equip and focus to deal with the global war on terror?

Source: The Smoking Gun, http://www.thesmokinggun.com.

Figure 3.3: In this memo, U.S. Secretary of Defense Donald Rumsfeld was strikingly candid about the situation in Iraq at the time. His memo was leaked to the press (a tertiary reader), creating a minor controversy because the memo did not reflect what the U.S. Administration had been saying about progress in Iraq. You can see the original memo at www.ablongman.com/johnsonweb/3.4.

GO TO THE NET

Want to see other controversial documents? Go to **www.ablongman.com/johnsonweb/3.4**

Figure 3.3: (*continued.*)

Are the changes we have and are making too modest and incremental? My impression is that we have not yet made truly bold moves, although we have made many sensible, logical moves in the right direction, but are they enough?

Today, we lack metrics to know if we are winning or losing the global war on terror. Are we capturing, killing or deterring and dissuading more terrorists every day than the madrassas and the radical clerics are recruiting, training and deploying against us?

Author returns to the main point of memo.

Does the US need to fashion a broad, integrated plan to stop the next generation of terrorists? The US is putting relatively little effort into a long-range plan, but we are putting a great deal of effort into trying to stop terrorists. The cost-benefit ratio is against us! Our cost is billions against the terrorists' costs of millions.

This list frames questions that the primary readers should answer.

- Do we need a new organization?
- How do we stop those who are financing the radical madrassa schools?
- Is our current situation such that "the harder we work, the behinder we get"?

It is pretty clear that the coalition can win in Afghanistan and Iraq in one way or another, but it will be a long, hard slog.

Does CIA need a new finding?

Should we create a private foundation to entice radical madrassas to a more moderate course?

What else should we be considering?

The final action item is stated for the primary readers.

Please be prepared to discuss this at our meeting on Saturday or Monday.

Thanks.

DHR:dh
101503-58

Please Respond by ____________________

What is the common sense trap?
To find out, go to
www.ablongman.com/johnsonweb/3.5

Sometimes, writers mistakenly assume that their readers have the same needs, values, and attitudes as they do. In reality, readers often have very different characteristics than the writers of a document. As you begin considering your readers for the document, think about some of the following issues:

- Readers' familiarity with the subject.
- Readers' professional experience.
- Readers' educational level.
- Readers' reading and comprehension level.
- Readers' skill level.

TAKE NOTE Making assumptions based on the readers' age or gender can be risky. You can make some serious mistakes assuming that people from a particular generation or gender behave in predictable ways.

With these reader characteristics in mind, you can begin viewing your document from their perspective. To help you deepen this perspective, see Figure 3.4, a Reader Analysis Chart that will help you identify your readers' needs, values, and attitudes toward your text.

Reader Analysis Chart

Readers	Needs	Values	Attitudes
Primary			
Secondary			
Tertiary			
Goalkeepers			

Figure 3.4: To better understand your readers, fill this Reader Analysis Chart with notes about their characteristics.

To use the Reader Analysis Chart, consider each level of readers separately, filling in what you know about their needs, values, and attitudes.

NEEDS Try to identify what your readers need from your document. What do your primary readers need to make a decision or take action? What do the secondary readers need if they are going to make positive recommendations to the primary readers? What are the tertiary and gatekeeper readers looking for in your document?

For a downloadable version of the Reader Analysis Chart, go to **www.ablongman.com/johnsonweb/3.6**

VALUES Different readers value different things. Managers may value efficiency and consistency. Experts may value accuracy and feasibility. For some readers, profit and market share are of highest value. Perhaps other readers put a premium on environmental concerns or social ethics. By anticipating your readers' values, you can determine what they will find important.

ATTITUDES Readers will adopt various attitudes toward your document. Are your readers excited, upset, wary, positive, hopeful, careful, concerned, skeptical, or gladdened by what you are telling them? In your document, you can use any positive attitudes to your advantage, or you can counteract any negative attitudes.

TAKE NOTE One thing to keep in mind is that primary, secondary, and tertiary readers may have very different attitudes about the same document. A manager may be very interested in your project, but her technical advisors may be skeptical about what you have to say. To be effective, you need to address the contradictory attitudes of both of these kinds of readers.

As you fill out the Reader Analysis Chart, you will be making guesses about your readers. If you aren't sure about your readers' needs, values, or attitudes, simply put a question mark (?) in those spaces. Your question marks signal that you need to do a little more research on your readers.

To fill in any gaps in your understanding of the readers, you might interview people who are Subject Matter Experts (SMEs) at your company. Or, you can hire consultants who are SMEs in the area about which you are writing. These experts may be able to give you insights into your readers' likely characteristics.

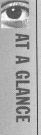

The Internet is a great place to find information on your readers. Most companies have a website where they place biographical information about their employees. Many people also have personal websites that you can look over. You can learn a great amount about readers' needs, values, and attitudes by paying attention to what they say about themselves.

AT A GLANCE

Determining How Readers Make Decisions

- *Needs*—information the readers need to take action or make a decision.
- *Values*—issues, goals, or beliefs that the readers feel are important.
- *Attitudes*—the readers' emotional response to you, your project, or your company.

TAKE NOTE Avoid invading someone's privacy. In general, it is fine to use information freely available on the Internet. But, you should avoid personal information services that allow you to go beyond simple biographical information.

LINK For more information on using search engines, see Chapter 5, page 106.

Otherwise, you can make strategic assumptions about your readers. Most managers are concerned about efficiency and budgets. Most engineers are interested in how things work. Most nurses and doctors are concerned about how their actions will affect their patients. Most environmentalists are concerned about how decisions will affect wildlife and habitats. You can learn more about your readers by simply viewing the situation from their perspective.

To find websites that help you search for information about people, go to **www.ablongman.com/johnsonweb/3.7**

Profiling Readers with Search Engines

Search engines on the Internet are great tools for collecting information on just about any topic, including your readers and the contexts in which they will use your document.

The most popular search engines are Google, Altavista, Yahoo, Northern Light, Lycos, HotBot, Excite, and AllTheWeb. Each of these search engines will allow you to type in keywords, a phrase, or even a whole question (Figure A). The search engine will find any websites with those words in them.

You can learn a great amount about your readers by typing in a few keywords. For example, let's say you are writing a proposal to the Salmon Recovery Program in Washington state. You can type in the phrase:

Salmon Recovery Program in Washington

The search engine will pull up thousands of pages that refer to this subject (Google found over 100,000 pages). The engine will usually rank them for you, trying to present you with the most relevant pages first.

An Internet Search Engine

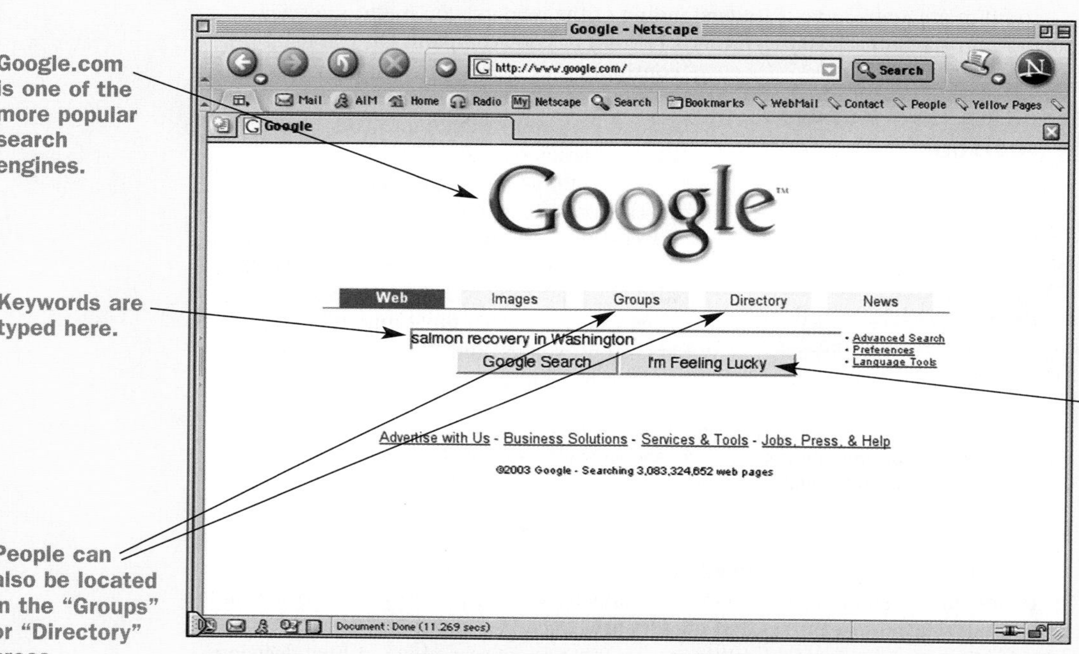

Figure A: Type in some keywords, and the search engine will look for information on that subject.

Source: Google, http://www.google.com.

GO TO THE NET

Links to several search engines are available at **www.ablongman.com/johnsonweb/3.8**

Of course, you can't read that many pages in your lifetime. You need to sharpen your search with some helpful symbols and strategies. For example, perhaps you notice that Alice Guthrie is the director of the Salmon Recovery Program. So, you refine your search with plus and minus signs.

The + sign—Putting a + in front of words tells the search engine to find only pages that have those words in them.

> Salmon Recovery Program + Washington + Alice + Guthrie

Here, the search engine will pull up only pages with "Alice" and "Guthrie" in them.

The – sign—Putting a – in front of words will eliminate pages with information you don't want.

> Salmon Recovery Program + Washington + Alice + Guthrie – restaurant

With this minus sign, you can eliminate any pages that refer to Arlo Guthrie's song, "Alice's Restaurant."

Quotation marks—If you put a phrase in quotes, the search engine will look for that exact phrase:

> "Salmon Recovery Program" + Washington + Alice + Guthrie – restaurant

Wildcard symbols—Some search engines also have symbols for "wildcards." These symbols are helpful when you know most of a phrase, but not all of it.

> "Salmon ? Program" + Washington + Alice + Guthrie – restaurant

Different search engines use different wildcard symbols. Commonly used symbols include ?, *, and %.

It is amazing how much you can learn about your readers (and other topics) with search engines. These helpful tools will allow you to tailor your documents to the specific needs of your readers.

Understanding Contexts of Use

How people interpret your document depends on the *contexts of use* in which they are reading or using it. Many outside factors may influence their decision-making process.

Identifying the Context of Use

Perhaps the most obvious concern is the *physical context* in which the document will be read. Will your readers be in their office? On the factory floor? In the emergency room, trying to save someone's life? Are they sitting on their living room floor, trying desperately to figure out how to assemble their child's new bike before a birthday party? Each of these different physical contexts will alter the way your readers interpret your document.

Want more search engine tips? Go to **www.ablongman.com/johnsonweb/3.9**
Want to find someone? Go to **www.ablongman.com/johnsonweb/3.10**

But context of use goes beyond your readers' physical context. The context of use may also include economic, ethical, and political issues that will shape how your readers interpret your document. To help you sort out these various contexts, you can use a Context Analysis Chart like the one shown in Figure 3.5.

Context Analysis Chart

	Physical Context	Economic Context	Political Context	Ethical Context
Primary Readers				
Readers' Company				
Readers' Industry				

Figure 3.5: Each reader is influenced by physical, economic, political, and ethical concerns. A Context Analysis Chart anticipates these concerns for the primary readers, their company, and their industry.

Here is how the Context Analysis Chart is used. Fill in what you know about the physical, economic, political, and ethical issues that might influence the primary readers, their company, and their industry.

PHYSICAL CONTEXT Imagine the places where the primary readers will use your document. If they are on a factory floor, they will read the text differently than they might if they are in an office. How can you write and design the document to anticipate where they will read it?

AT A GLANCE

Context of Use

- *Physical Context*—the places where the readers will use your document.
- *Economic Context*—the money-related issues that will restrain the kinds of actions possible.
- *Political Context*—the micropolitical and macropolitical trends that will guide your readers.
- *Ethical Context*—the personal, social, and environmental issues that shape the readers' responses.

ECONOMIC CONTEXT Identify the money-related issues that influence the readers' decision-making process. What are the costs and benefits of your ideas? Will accepting your ideas change the financial situation of your readers, their company, or their industry? Is the industry in a recession or an expansion?

POLITICAL CONTEXT All readers are influenced by political issues. On a micropolitical level, your readers may be concerned about how a decision will affect their relationships with their supervisor or colleagues.

For a downloadable version of the Context Analysis Chart, go to **www.ablongman.com/johnsonweb/3.11**

Meanwhile, macropolitical trends can also influence their decisions. Keep in mind the larger political trends at the local, state, federal, and international levels that might influence how readers interpret your ideas.

ETHICAL CONTEXT All decisions involve ethics at some level. How do your ideas affect the rights, values, and well-being of others? Does your document involve any social or environmental issues that might be of concern to your readers? Will any laws or rules be bent or broken if they take action? Try to anticipate the ethical factors that may shape your readers' decisions.

LINK For more help on identifying ethical issues, see Chapter 4, page 72.

As with the Reader Analysis Chart, if you don't know specific information about your readers' physical, economic, political, and ethical contexts, just put a question mark (?) in that space. You can then turn to the Internet for answers, or you can interview people who may have the answers you need.

Using Your Reader and Context Analyses

Filling out the Reader and Context Analysis charts may seem like a lot of work, especially if the document you are writing is small or not too important. At a minimum, you should do some reader analysis for any document, identifying the important characteristics of the primary, secondary, tertiary, and gatekeeper audiences. You should also think about the physical, economic, political, and ethical issues that will shape their responses. After all, even the most insignificant workplace document can have unforeseen positive or negative consequences.

For important documents, though, you should always conduct a thorough analysis of your readers and the document's contexts of use. As you fill out the charts, you will begin to understand your readers at a much deeper level. You will better understand how they approach the subject of your document and what kinds of information they need.

In your charts, circle or highlight important terms, concepts, and phrases. As you draft your document, your analysis of readers and contexts of use will help you—

- make strategic decisions about what information to include in your document. Readers don't need all the information you have available. They need only enough to make a decision or take action.
- organize your document to highlight the information that is most important to your readers. Crucial information that readers need should be easy to locate.
- develop a persuasive style that will appeal to your readers. Your notes about their values and attitudes can be used to shape the document to their biases and beliefs. Your notes about contexts of use will help you anticipate the forces that will shape how they make decisions.
- design the document for the places where it will be used. Your understanding of the context in which the document will be used will help you use document design to give your text the maximum impact.

For important documents, this deeper understanding will give your writing extra potency and legitimacy. Readers will appreciate your efforts to see the situation from their point of view.

Want more help with reader analysis?
go to
www.ablongman.com/johnsonweb/3.12

Document Written to the General Public

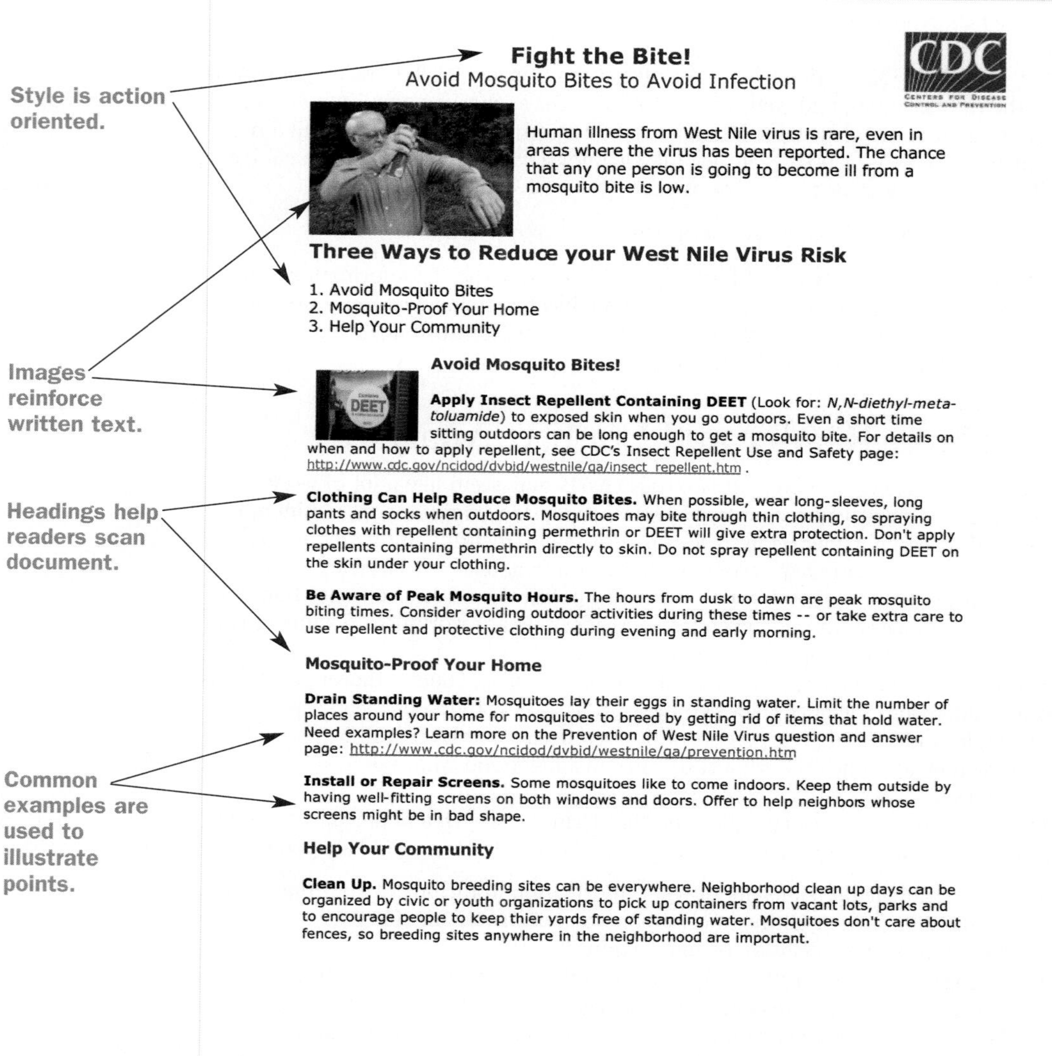

Style is action oriented.
Images reinforce written text.
Headings help readers scan document.
Common examples are used to illustrate points.

Fight the Bite!
Avoid Mosquito Bites to Avoid Infection

CDC
CENTERS FOR DISEASE CONTROL AND PREVENTION

Human illness from West Nile virus is rare, even in areas where the virus has been reported. The chance that any one person is going to become ill from a mosquito bite is low.

Three Ways to Reduce your West Nile Virus Risk

1. Avoid Mosquito Bites
2. Mosquito-Proof Your Home
3. Help Your Community

Avoid Mosquito Bites!

Apply Insect Repellent Containing DEET (Look for: *N,N-diethyl-meta-toluamide*) to exposed skin when you go outdoors. Even a short time sitting outdoors can be long enough to get a mosquito bite. For details on when and how to apply repellent, see CDC's Insect Repellent Use and Safety page: http://www.cdc.gov/ncidod/dvbid/westnile/qa/insect_repellent.htm .

Clothing Can Help Reduce Mosquito Bites. When possible, wear long-sleeves, long pants and socks when outdoors. Mosquitoes may bite through thin clothing, so spraying clothes with repellent containing permethrin or DEET will give extra protection. Don't apply repellents containing permethrin directly to skin. Do not spray repellent containing DEET on the skin under your clothing.

Be Aware of Peak Mosquito Hours. The hours from dusk to dawn are peak mosquito biting times. Consider avoiding outdoor activities during these times -- or take extra care to use repellent and protective clothing during evening and early morning.

Mosquito-Proof Your Home

Drain Standing Water: Mosquitoes lay their eggs in standing water. Limit the number of places around your home for mosquitoes to breed by getting rid of items that hold water. Need examples? Learn more on the Prevention of West Nile Virus question and answer page: http://www.cdc.gov/ncidod/dvbid/westnile/qa/prevention.htm

Install or Repair Screens. Some mosquitoes like to come indoors. Keep them outside by having well-fitting screens on both windows and doors. Offer to help neighbors whose screens might be in bad shape.

Help Your Community

Clean Up. Mosquito breeding sites can be everywhere. Neighborhood clean up days can be organized by civic or youth organizations to pick up containers from vacant lots, parks and to encourage people to keep thier yards free of standing water. Mosquitoes don't care about fences, so breeding sites anywhere in the neighborhood are important.

Source: Centers for Disease Control and Prevention (CDC), http://www.cdc.gov/ncidod/dvbid/westnile/index.htm.

Figure 3.6: This document on West Nile virus was written to the general public. It is action oriented and not very technical. The images also help reinforce the message, and the layout makes it highly scannable.

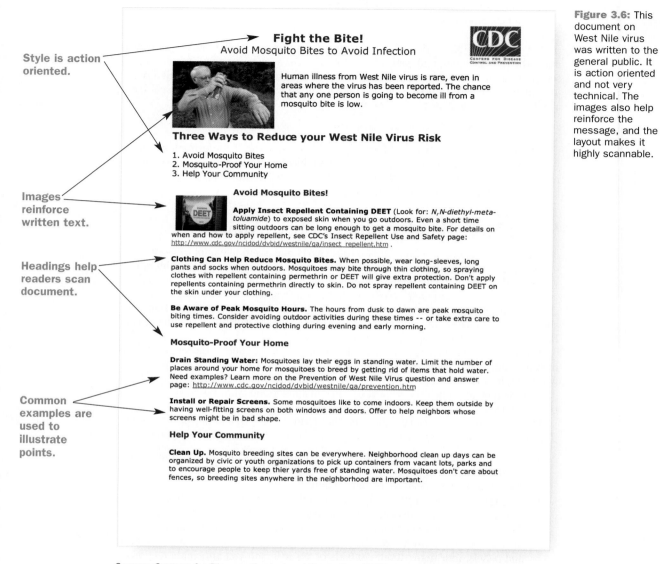

Figures 3.6 and 3.7 show documents from the same website about the same topic, the West Nile virus, that were written to two different types of readers. The first document, Figure 3.6, is written to the general public. Notice how it uses content, organi-

Figure 3.6: (*continued.*)

Report Dead Birds to Local Authorities. Dead birds may be a sign that West Nile virus is circulating between birds and the mosquitoes in an area. Over 110 species of birds are known to have been infected with West Nile virus, though not all infected birds will die.

By reporting dead birds to state and local health departments, the public plays an important role in monitoring West Nile virus. Because state and local agencies have different policies for collecting and testing birds check the Links to State and Local Government Sites page to find information about reporting dead birds in your area: http://www.cdc.gov/ncidod/dvbid/westnile/city_states.htm. This page contains mo re information about reporting dead birds and dealing with bird carcasses: http://www.cdc.gov/ncidod/dvbid/westnile/qa/wnv_birds.htm

Links provide more information.

Mosquito Control Programs. Check with local health authorities to see if there is an organized mosquito control program in your area. If no program exists, work with your local government officials to establish a program. The American Mosquito Control Association (www.mosquito.org) can provide advice, and their book Organization for Mosquito Control is a useful reference. More questions about mosquito control? A source for information about pesticides and repellents is the National Pesticide Information Center: http://npic.orst.edu/ , which also operates a toll-free information line: 1-800-858-7378 (check their Web site for hours).

Find out more about local prevention efforts. Find state and local West Nile virus information and contacts on the Links to State and Local Government Sites page.

zation, style, and design to appeal to this audience. The second document, Figure 3.7, is written to medical personnel. Notice how the content is far more complex and the style is less personal than in the Figure 3.6 document.

Document Written to Experts

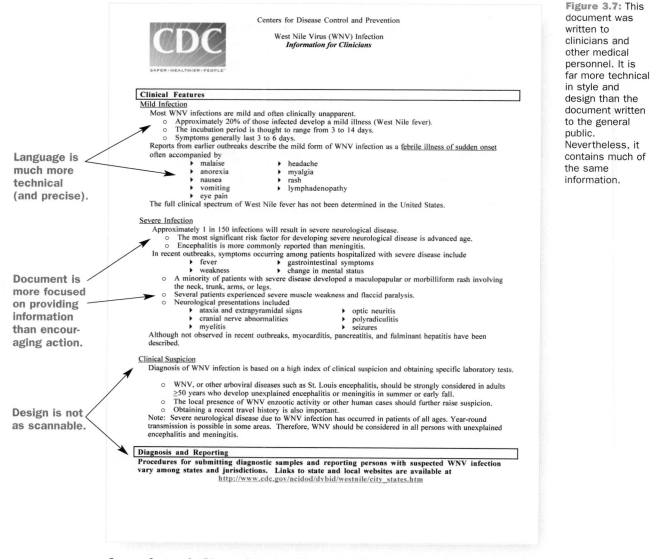

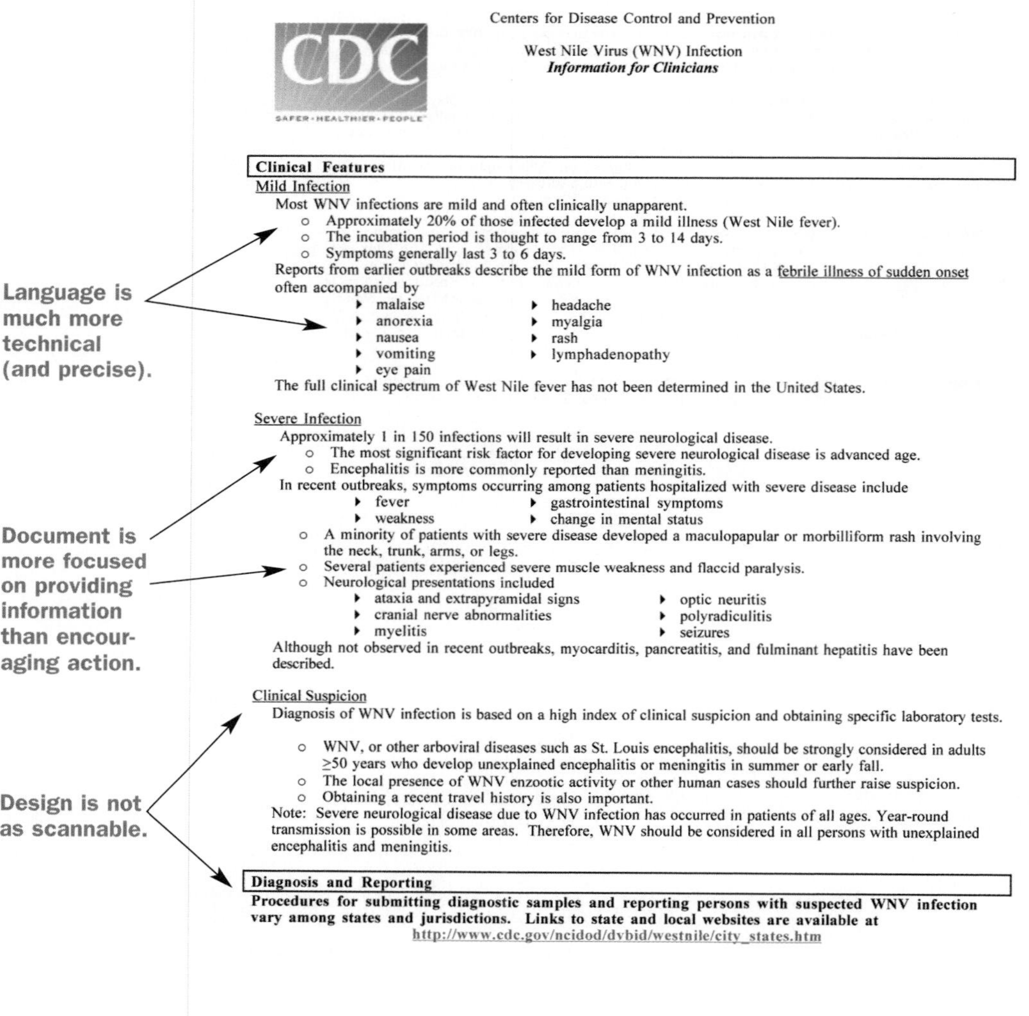

Centers for Disease Control and Prevention

CDC
SAFER · HEALTHIER · PEOPLE

West Nile Virus (WNV) Infection
Information for Clinicians

Clinical Features

Mild Infection

Most WNV infections are mild and often clinically unapparent.

- Approximately 20% of those infected develop a mild illness (West Nile fever).
- The incubation period is thought to range from 3 to 14 days.
- Symptoms generally last 3 to 6 days.

Reports from earlier outbreaks describe the mild form of WNV infection as a febrile illness of sudden onset often accompanied by

- malaise
- anorexia
- nausea
- vomiting
- eye pain
- headache
- myalgia
- rash
- lymphadenopathy

The full clinical spectrum of West Nile fever has not been determined in the United States.

Severe Infection

Approximately 1 in 150 infections will result in severe neurological disease.

- The most significant risk factor for developing severe neurological disease is advanced age.
- Encephalitis is more commonly reported than meningitis.

In recent outbreaks, symptoms occurring among patients hospitalized with severe disease include

- fever
- weakness
- gastrointestinal symptoms
- change in mental status

- A minority of patients with severe disease developed a maculopapular or morbilliform rash involving the neck, trunk, arms, or legs.
- Several patients experienced severe muscle weakness and flaccid paralysis.
- Neurological presentations included
  - ataxia and extrapyramidal signs
  - cranial nerve abnormalities
  - myelitis
  - optic neuritis
  - polyradiculitis
  - seizures

Although not observed in recent outbreaks, myocarditis, pancreatitis, and fulminant hepatitis have been described.

Clinical Suspicion

Diagnosis of WNV infection is based on a high index of clinical suspicion and obtaining specific laboratory tests.

- WNV, or other arboviral diseases such as St. Louis encephalitis, should be strongly considered in adults ≥50 years who develop unexplained encephalitis or meningitis in summer or early fall.
- The local presence of WNV enzootic activity or other human cases should further raise suspicion.
- Obtaining a recent travel history is also important.

Note: Severe neurological disease due to WNV infection has occurred in patients of all ages. Year-round transmission is possible in some areas. Therefore, WNV should be considered in all persons with unexplained encephalitis and meningitis.

Diagnosis and Reporting

Procedures for submitting diagnostic samples and reporting persons with suspected WNV infection vary among states and jurisdictions. Links to state and local websites are available at http://www.cdc.gov/ncidod/dvbid/westnile/city_states.htm

Language is much more technical (and precise).
Document is more focused on providing information than encouraging action.
Design is not as scannable.

Source: Centers for Disease Control and Prevention (CDC), http://www.cdc.gov/ncidod/dvbid/westnile/index.htm.

Figure 3.7: This document was written to clinicians and other medical personnel. It is far more technical in style and design than the document written to the general public. Nevertheless, it contains much of the same information.

Figure 3.7: *(continued.)*

Diagnosis and Reporting – *continued*

Diagnostic Testing

WNV testing for patients with encephalitis or meningitis can be obtained through local or state health departments.

- The most efficient diagnostic method is detection of IgM antibody to WNV in serum or cerebral spinal fluid (CSF) collected within 8 days of illness onset using the IgM antibody capture enzyme-linked immunosorbent assay (MAC-ELISA).
- Since IgM antibody does not cross the blood-brain barrier, IgM antibody in CSF strongly suggests central nervous system infection.
- Patients who have been recently vaccinated against or recently infected with related flaviviruses (e.g., yellow fever, Japanese encephalitis, dengue) may have positive WNV MAC-ELISA results.

Reporting Suspected WNV Infection

Refer to local and state health department reporting requirements:
www.cdc.gov/ncidod/dvbid/westnile/city_states.htm

- WNV encephalitis is on the list of designated nationally notifiable arboviral encephalitides.
- Aseptic meningitis is reportable in some jurisdictions.

The timely identification of persons with acute WNV or other arboviral infection may have significant public health implications and will likely augment the public health response to reduce the risk of additional human infections.

Laboratory Findings

Among patients in recent outbreaks

- Total leukocyte counts in peripheral blood were mostly normal or elevated, with lymphocytopenia and anemia also occurring.
- Hyponatremia was sometimes present, particularly among patients with encephalitis.
- Examination of the cerebrospinal fluid (CSF) showed pleocytosis, usually with a predominance of lymphocytes.
- Protein was universally elevated.
- Glucose was normal.
- Computed tomographic scans of the brain mostly did not show evidence of acute disease, but in about one-third of patients, magnetic resonance imaging showed enhancement of the leptomeninges, the periventricular areas, or both.

Treatment

Treatment is supportive, often involving hospitalization, intravenous fluids, respiratory support, and prevention of secondary infections for patients with severe disease.

- Ribavirin in high doses and interferon alpha-2b were found to have some activity against WNV in vitro, but no controlled studies have been completed on the use of these or other medications, including steroids, antiseizure drugs, or osmotic agents, in the management of WNV encephalitis.

For additional clinical information, please refer to Petersen LR and Marfin AA, "West Nile Virus: A Primer for the Clinician[Review]," Annals of Internal Medicine (August 6) 2002: 137:173-9.

For clinical and laboratory case definitions, see "Epidemic/Epizootic West Nile Virus in the United States: Revised Guidelines for Surveillance, Prevention, and Control, 2001," at
www.cdc.gov/ncidod/dvbid/westnile/surv&control.htm

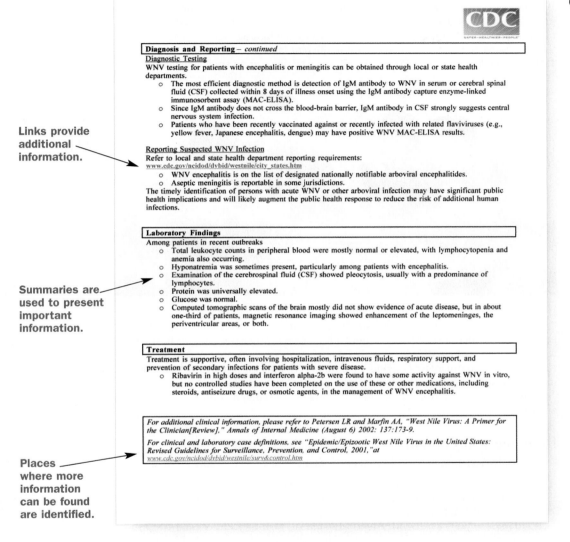

Links provide additional information.

Summaries are used to present important information.

Places where more information can be found are identified.

International and Cross-Cultural Communication

Information technology allows us to blur geographical and political boundaries. Whether you are writing a user manual for a toaster or describing a heart transplant procedure, it is possible your work will be read and used by people from different cultures (Figure 3.8). The use of computers, especially the Internet, has only heightened the need to communicate with people from different lands and backgrounds (Hoft, 1995; Reynolds & Valentine, 2004). It's all very exciting—and very challenging.

Figure 3.8: Your documents will likely be used by people around the world.

Even within North America, cross-cultural communication is important. Your company's product or service, even though it is offered exclusively in North America, may be used by people who speak only Spanish or a Native American language. In America, cultural differences among these audiences may influence how your document or presentation is received.

International and cross-cultural issues will affect the content, organization, style, and design of your document.

Differences in Content

Cultures have different expectations about content in technical documentation.

- North Americans and Europeans tend to see empirically tested evidence as the most credible form of information. They rank personal credibility second and appeals to emotion last.
- In Mexico and most South American countries, family and personal backgrounds are of greater importance. It is common for family-related issues to be mentioned in public relations, advertising, and documentation. Business relationships and meetings usually start with exchanges about families and personal interests.
- In Asian countries, the reputation of the writer or company is essential toward establishing the credibility of the information (Haneda & Shima, 1983). Interpersonal relationships and prior experiences can sometimes even trump empirical evidence in Asia.

Want to learn more about international readers? Go to **www.ablongman.com/johnsonweb/3.13**

- Also in Asia, contextual cues can be more important than content. In other words, *how* a Japanese person says something may be more important than *what* she is saying. For example, when speaking or writing in their own language, the Japanese rarely use the word "no." Instead, they rely on contextual cues to signal the refusal. As a result, when Japanese is translated into English, these "high-context" linguistic strategies are often misunderstood (Chaney & Martin, 2004).

Differences in Organization

The organization of a document often needs to be altered to suit an international audience. Organizational structures that Americans perceive to be "logical" or "common sense" are often seen as confusing and even rude in some cultures.

- In Arabic cultures, documents and meetings often start out with statements of appreciation and attempts to build common bonds among people. The American tendency to "get to the point" is often seen as rude.
- Also, in Arabic cultures, documents rely on repetition to make their points. To North Americans, this repetition might seem like the document is moving one step back for every two steps forward. To Arabs, American documentation seems incomplete because it lacks this repetition.
- Asians often prefer to start out with contextual information about nonbusiness issues. For example, it is common for Japanese writers to start out letters by saying something about the weather. To some Asians, American documents seem abrupt, because Americans tend to bluntly highlight goals and objectives up front.

Differences in Style

Beyond difficulties with translation, style is usually an important difference among cultures:

- Arabic style may seem overly ornamental to North American tastes. So, Arabic documents and presentations can seem colorful or even bombastic to non-Arabs. On the other hand, the American reliance on "plain language" can rub against the sensibilities of Arabs, who prefer a more ornate style in formal documents.
- In Mexico and much of South America, the American preference for informal style often suggests a lack of respect for the project, the product, or the readers. Mexicans especially value formality in business settings, so the use of first names and contractions in business prose can be offensive.
- Some Native Americans prefer the sense that everyone had input on the document. Therefore, a direct writing or presentation style will meet resistance because it will seem to be the opinion of only one person.
- In North America, women are more direct than women in other parts of the world, including Europe. This directness often works to their advantage in other countries, because they are viewed as confident and forward thinking. However, as writers and speakers, women should not be too surprised when people from other cultures resist their directness.

TAKE NOTE Go ahead and be yourself. Be direct. Just keep in mind that some of the resistance from your readers might be due more to gender expectations than resistance to your ideas.

For information sources on other countries, go to **www.ablongman.com/johnsonweb/3.14**

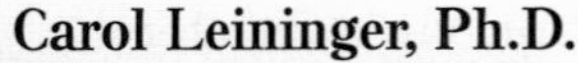

Carol Leininger, Ph.D.

COMMUNICATIONS MANAGER, F. HOFFMAN-LA ROCHE, SWISS HEADQUARTERS

F. Hoffman-La Roche is a pharmaceutical company that works closely with a partner company, Chugai Pharmaceuticals in Tokyo, Japan.

What are some strategies for communicating with people from another culture?

Physical distance may be the biggest hurdle in global communication. I believe that distance affects how people work together more than does language or culture. When people are working face to face, they tend to figure out issues by interacting with each other, even if they are not fluent in the same languages. There are awkward moments, but that's to be expected.

Mostly, solid *preparation* is the key to working across distances and languages. Communication needs to be structured and simple. Your readers may be reading your text in their second- or third-best language, so make their job easier by writing as simply and clearly as you can.

All the rules for good technical communication in the United States apply to international communication—only more so. What helps second-language or non-U.S. English speakers?

- State your objectives and purpose clearly.
- Use language consistently (i.e., the same terms for the same things).
- Do not attempt humor until a relationship has been established.
- Rank issues by importance.
- Handle only one message per e-mail or paragraph.
- Use headings and subheadings that convey a specific meaning.
- Minimize use of adjectives and adverbs.
- Minimize prepositional phrases.
- Highlight actions, deadlines, and dates.
- Spell everything correctly (always check the spelling).

Always be as polite as you can by your own cultural standards (e.g., formal language, politeness markers like "please" and "thank you," use of full names in greeting and salutation). Even if your cultural view of what is "polite" is different from that of your audience, your intention to be polite will be recognized by international readers as courtesy and civility.

Differences in Design

Even the design of documents is important when you are working with international and cross-cultural readers:

- Arabic and some Chinese scripts are read right to left, unlike English, which is read left to right (Figure 3.9). As a result, Arabic and some Chinese readers tend to scan pages and images quite differently than do Americans or Europeans.
- Some icons that show hand gestures, like the OK sign, a pointing finger, or a peace sign with the back of the hand facing outward can be highly offensive in

Want to learn more about Native American cultures? Go to **www.ablongman.com/johnsonweb/3.15**

some cultures. Imagine a document in which a hand with the middle finger extended is used to point at things. You get the picture.

- In some Asian cultures a white flower or white dress can symbolize death. As a result, a photograph using white flowers or white dresses can signal a funeral or mourning.
- Europeans find that American texts include too many graphics and use too much white space. Americans, meanwhile, often find that the small margins in European texts make them look crowded and cramped.
- Graphs and charts that seem to have obvious meanings to Americans can be baffling and confusing to readers from other cultures.
- In some Native American cultures, hand gestures during presentations should be limited and eye contact should be minimized. Ironically, this advice is exactly the opposite of what most public speaking trainers suggest.

Different Ways of Scanning a Page

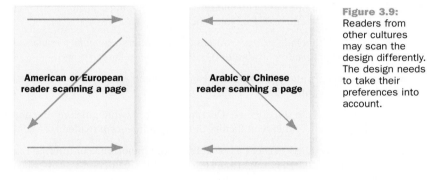

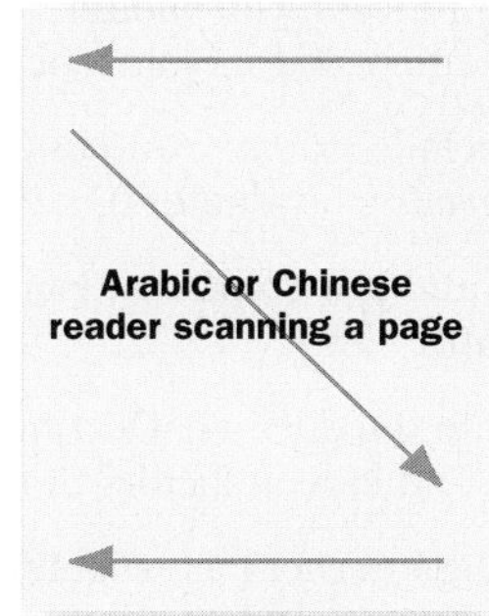

Figure 3.9: Readers from other cultures may scan the design differently. The design needs to take their preferences into account.

Adapting to International and Cross-Cultural Readers

With all these differences in content, organization, style, and design, how can you possibly write for international or cross-cultural readers? Fortunately, readers are becoming more familiar with cultural differences. So, mistakes that previously offended or baffled readers are now simply tolerated. Readers usually give international texts some leeway.

Nevertheless, a good rule of thumb is to listen carefully to your readers' expectations. Careful listening is a valued quality in all cultures, and you will learn a great amount by simply listening to what your readers expect the document to include and how it should look.

Likewise, another good rule of thumb is that politeness in one culture tends to translate well into other cultures. For example, words like "please" and "thank you" are universally seen as polite.

Use the Internet to do some research into your readers' cultural expectations for technical documents. On the Internet or at your workplace, you might also find some model texts from the readers' culture. Use them to help guide your decisions about content, organization, style, and design.

For more about using graphics in other cultures, go to **www.ablongman.com/johnsonweb/3.16**

TAKE NOTE International readers will usually be gratified that you took the time to learn their conventions, as long as it does not look like you are simply using stereotypes (e.g., the cliché of mentioning cherry blossoms in a letter to a Japanese reader).

You may look for coworkers who are from the target culture or who have lived in the places where you need to communicate. You can ask them about conventions that might make your document or presentation more effective. They can also help you avoid doing anything awkward or offensive.

Overall, when you are communicating to international readers or people from different cultures, be observant and listen to what they tell you. Do some research into their expectations, and be ready to learn from your mistakes.

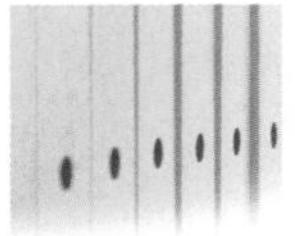

CHAPTER REVIEW

- You can use your computer as a reader analysis tool to better tailor your document's content, organization, style, and design to the needs of its readers.
- Early in the writing process, you should begin developing a profile of the types of people who may be interested in your document.
- Your readers will include *primary readers* (action takers), *secondary readers* (advisors), *tertiary readers* (evaluators), and *gatekeepers* (supervisors).
- In your documents and presentations, you should anticipate various readers' needs, values, and attitudes.
- You should anticipate the document's contexts of use, which include the physical, economic, political, and ethical factors that may influence a reader's ideas.
- The emergence of the Internet has heightened the importance of international and cross-cultural communication. You need to adjust the content, organization, style, and design of your text to be sensitive to cross-cultural needs.

EXERCISES AND PROJECTS

Individual or Team Projects

1. Choose two websites that are designed for very different types of readers. Write a memo to your instructor in which you compare and contrast the websites, showing how they approach their readers differently. How do they use content, organization, style, and design to meet the needs, values, and attitudes of their readers?

   Some pairs of websites you might consider include websites for cars (chevrolet.com versus honda.com), magazines (time.com versus outsidemag.com), or computers (dell.com versus apple.com). Look for websites for products that are similar but pursue different kinds of customers.

2. Consider the advertisement in Figure 3.10 and "reverse-engineer" its reader analysis. Using a Writer-Centered Chart and a Reader Analysis Chart, identify the primary, secondary, and tertiary readers of the text. Then, make guesses about the needs, values, and attitudes of these readers.

   Write a report to your instructor in which you use your charts to discuss the readers of this document. Then, show how the document anticipates these readers' needs.

To find websites that discuss politeness strategies, go to **www.ablongman.com/johnsonweb/3.17**

An Advertisement

iPod
Welcome to the digital music revolution. 7,500 songs in your pocket.
Works with Mac or PC. Over a million sold. The new iPod.

Figure 3.10: This advertisement is aimed at specific kinds of readers. Who are they?

Source: Apple, 2003.

3. For a document you are writing, conduct a thorough reader analysis. Start out by identifying the primary, secondary, tertiary, and gatekeeper readers. Then, identify these readers' needs, values, and attitudes. And finally, identify the physical, economic, political, and ethical issues that may influence how your readers interpret your document.

   Offer a presentation about your readers to your class. Discuss how various readers in various contexts will require you to adjust the content, organization, style, and design of your document.

4. Choose a country or culture that interests you. Then, find three texts written by someone from that country or culture. Write a memo to your instructor in which you discuss any similarities or differences between the texts and your own expectations for texts. Pay close attention to differences in content, organization, style, and design of these texts.

Collaborative Project

With a group of people from your class, create a website that explores the needs, values, and attitudes of people from a different country or culture. The website does not need to be complex. Rather, on the Internet, identify various websites that offer information on that country or culture. Then, organize those websites by content and create links to them. Specifically, pay attention to the ways in which this country's physical, economic, political, and ethical contexts shape the way its people live their lives.

When you are finished with the website, offer a presentation to your class in which your group discusses how this country or culture differs from your own. Answer the following question: If you were going to offer a product or service to the people of this country or culture, what considerations would you need to keep in mind? If you needed to write a proposal or a set of instructions to people from this country or culture, how might you need to adjust it to fit their unique qualities?

Installing a Medical Waste Incinerator

Duane Jackson knew this decision was going to be difficult. As the assistant city engineer for Dover City, he was frequently asked to study construction proposals sent to the city council. He would then write a report with a recommendation. So, when the proposal for constructing a medical waste incinerator crossed his desk, he knew there was going to be trouble.

Overall, the proposal from Valley Medical, Inc., looked solid. The incinerator would be within three miles of the two major hospitals and a biotech research facility. And, it would bring about 30 good jobs to the Blue Park neighborhood, an economically depressed part of town.

The problem was that people in Blue Park were going to be skeptical. Duane grew up in a neighborhood like Blue Park, primarily African-American and lower middle class. He knew that hazardous industries often put their operations in these kinds of neighborhoods because the people did not have the financial resources or political clout to fight them. In the past, companies had taken advantage of these neighborhoods' political weaknesses, leaving the area polluted and unhealthy.

Powerful interests were weighing in on this issue. Dover City's mayor wanted the incinerator badly because she wanted the economic boost the new business would provide. Certainly, the hospitals and research laboratory were enthusiastic, because a nearby incinerator would help them cut costs. The city councilor who represented Blue Park wanted the jobs, but not at the expense of his constituents' health. Environmental groups, health advocates, and neighborhood associations were cautious about the incinerator, but they seemed to be keeping an open mind.

Analyzing the Readers

After a few weeks of intense study, Duane's research convinced him that the incinerator was not a health hazard to the people of Blue Park. Similar incinerators built by Valley Medical had spotless records. Emissions would be minimal because advanced "scrubbers" would remove almost all the particles left over after incineration. The scrubbers were very advanced, almost completely removing the pollutants, like dioxin and mercury, emitted by other incinerators.

For more information about waste incinerators, go to **www.ablongman.com/johnsonweb/3.18**

Also, the company had a good plan for ensuring that medical waste would not sit around in trucks or containers waiting to be burned. The waste would be immediately incinerated on arrival.

Duane decided to write a report to the city council that recommended the incinerator be built. That decision was the easy part. Now he needed to write a report that would convince the skeptics.

After identifying the subject and purpose of the report, Duane decided to do a thorough analysis of his readers and the report's contexts of use. He began with a Writer-Centered Chart (Figure 3.11). He then used a Reader Analysis Chart to identify the various readers' needs, values, and attitudes (Figure 3.12). Finally, Duane filled out a Context Analysis Chart to identify the physical, economic, political, and ethical issues involved (Figure 3.13).

Duane's Writer-Centered Chart

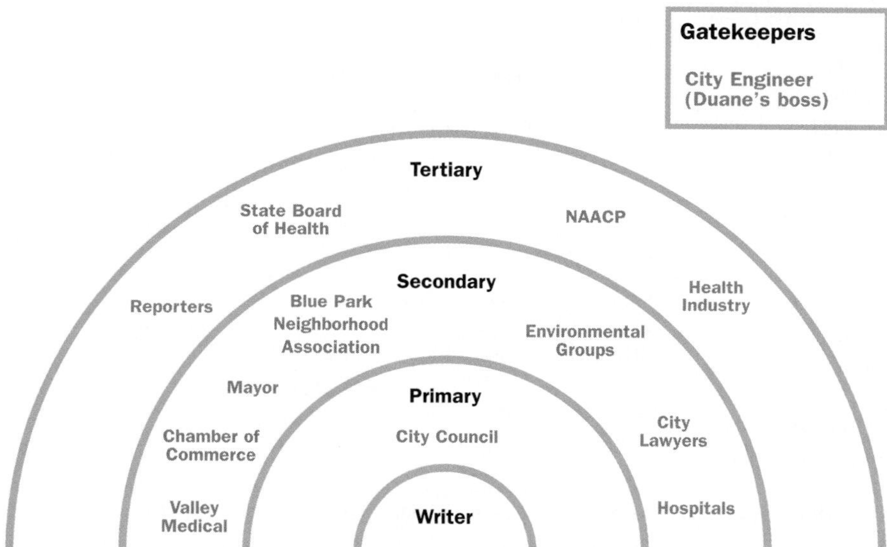

Figure 3.11: Duane's Writer-Centered Chart showed how many readers his report would have.

Duane's Reader Analysis Chart

	Needs	Values	Attitudes
Primary • City Council	Reliable information Environmental impact data Clear recommendation Impartial commentary	Citizen safety Economic development Fairness	Optimistic Cautious
Secondary • Valley Medical • Chamber of Commerce • Mayor • Neighborhood Association • Environmentalists • City Lawyers • Hospitals	Impartial commentary Specific facts about emissions Cultural and social considerations Valley Medical wants profits	Maintaining character of Blue Park Economic development Environmental safety Safe disposal of waste	Mayor and hospitals are positive and hopeful Neighborhood association and environmental groups are skeptical, perhaps resistant Valley Medical hopeful
Tertiary • Reporters • State Board of Health • NAACP • Health industry	Impartial decision Reassurance that race or poverty are not factors Basic facts about incinerator	Fairness and honesty Lack of bias Protection of people with little political power	Skeptical Open-minded
Gatekeepers • City Engineer	Reliable information Clear decision based on reliable data	Minimal trouble with mayor and council Low profile. Stay out of politics Honesty	Impartial to project Concerned that report may cause tensions

Figure 3.12
Duane filled in a Reader Analysis Chart, noting everything he knew about the various readers of his report. He noticed that most readers wanted impartial and reliable information. Some readers were positive; others were skeptical.

Duane's Context Analysis Chart

	Physical Context	Economic Context	Political Context	Ethical Context
Primary Readers • City Council	Initially at their office Further in city council meeting	Looking to improve city economics	Re-election is always an issue Mayor wants it BP city council would like it	Exploiting poor neighborhood? Racial issues? Environmental issues
Readers' Company • Dover City	City Hall Engineers' office City website	The jobs are a financial plus. Home values in neighborhood decline? Good for hospitals	Voters are wary Neighborhood and environmental groups draw attention Racial politics?	City liable if mistake? Environmental impact Infringement of people's rights? Legal issues
Readers' Industry • City government	Reports on Internet about project	Job creation Economic growth	Pressure on mayor Shows Dover City is positive about this kind of business	Public relations Don't want to seem exploitative

Figure 3.13: Duane's Context Analysis Chart revealed some interesting tensions between the economics and politics of the decision on the incinerator.

What Should Duane Do Now?

On the facts alone, Duane was convinced that the incinerator would be well placed in the Blue Park neighborhood. But his reader analysis charts showed that facts alone were not going to win over his primary audience, the members of the city council. They would have numerous other economic, political, and ethical issues to consider as well as the facts. In many ways, these factors were more important than the empirical evidence for the incinerator.

Indeed, one important thing Duane noticed was that the city council members, though they were the "action takers," were heavily influenced by the secondary readers. These secondary readers, the "advisors," would play a large role in the council's decision.

How can Duane adjust the content, organization, style, and design to better write his report? If you were Duane's readers, what kinds of information would you expect in this kind of report?

CHAPTER

4

Ethics in the Technical Workplace

CHAPTER CONTENTS

k, which is at present be...
of University laboratories, by
ed, through his contacts with y
contributions for this cause,
eration of industrial laboratories

ually stopped the sale of uranium
e has taken over. That she should
haps be understood on the ground
tary of State, von Weizsäcker, is
t in Berlin where some of the
repeated.

Yours very truly,

(Albert Einstein)

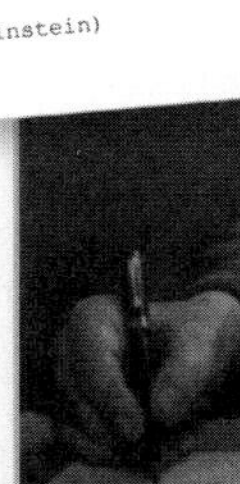

IEEE Code of Ethics

We, the members of the IEEE, in recognition of the importance of our technologies in affecting the quality of life throughout the world, and in accepting a personal obligation to our profession, its members and the communities we serve, do hereby commit ourselves to the highest ethical an[d] professional conduct and agree:

1. to accept responsibility in making engineering decisions consistent with th[e] safety, health and welfare of the public, and to disclose promptly factors that might endanger the public or the environment;
2. to avoid real or perceived conflicts of interest whenever possible, and to disclose them to affected parties when they do exist;
3. to be honest and realistic in stating claims or estimates based on available data;
4. to reject bribery in all its forms;
5. to improve the understanding of technology, its appropriate application, and potential consequences;
6. to maintain and improve our technical competence and to undertake technological tasks for others only if qualified by training or experience, or after full disclosure of pertinent limitations;
7. to seek, accept, and offer honest criticism of technical work, to acknowledge and correct errors, and to credit properly the contributions of others;
8. to treat fairly all persons regardless of such factors as race, religion, gender, disability, age, or national origin;
9. to avoid injuring others, their pr... nt by false or malicious action;
10. to assist colleagues and co-work... ment and to support them in following th...

Approved by the IEEE Board of Directors August 1990

Sir:

Some
municated t
ium may be
mediate fut
to call for
of the Admi
to your att

In th
through the
America - t
in a large
ities of ne
almost cert

This
and it is c
ful bombs c
type, ca
the whol
such bom
air.

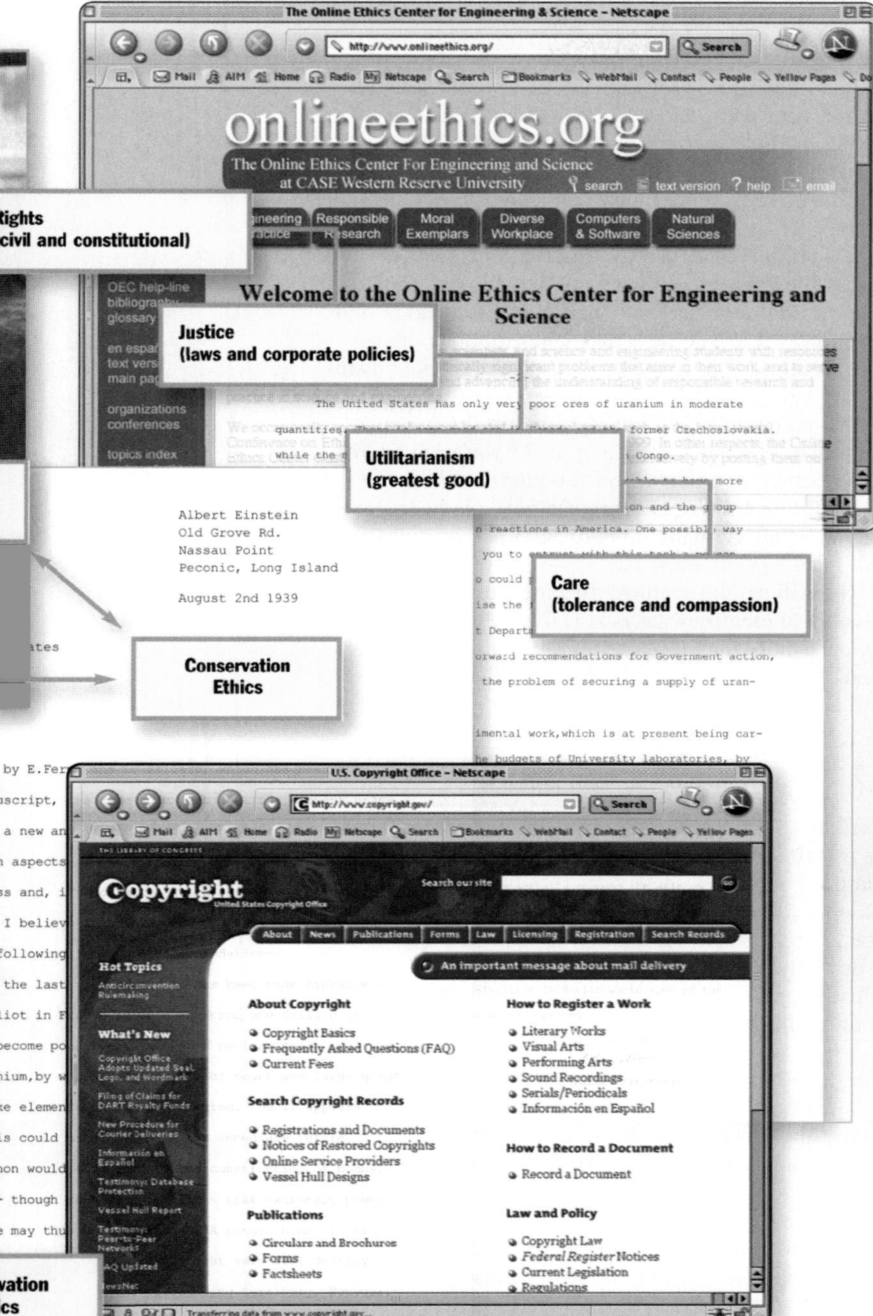

CHAPTER OBJECTIVES

In this chapter, you will learn:

- A working definition and understanding of ethics.
- To identify three ethical systems: personal, social, and conservation.
- To consider social ethics as issues of rights, justice, utility, and care.
- How copyright law affects technical communication.
- Strategies for resolving ethical conflicts in technical workplaces.
- How to balance the many issues involved in an ethical dilemma.
- About the new ethical challenges that face the computer-centered workplace.

Ethics have taken a beating recently. Enron, Tyco, WorldCom, Global Crossing, and PCINet have added their names to a growing list of corporate scandals. Meanwhile, the collapse of the "dot com" industry was largely due to massive amounts of money being thrown at companies built on questionable, perhaps even unethical, foundations. When the dot com bubble burst in the year 2000, many people lost their jobs, their retirement savings, and their confidence in the technology sector.

We could blame these scandals on unethical, greedy, unscrupulous people—and certainly they deserve blame—but the newness of the computerized workplace is probably also to blame. As we evolve into an electronic culture, the ethical boundaries are not as clear as they were only a few decades ago. Unfortunately, some people are willing to exploit these ethical gray areas for their own financial advantage, often hurting others.

For example, copyright law is a renewed ethical battleground in the Information Age, especially where these laws involve music. More than likely, you or your friends have MP3 players and many "free" songs downloaded off the Internet onto your computer's hard drive. Is downloading songs for free ethical? The music industry says no, and copyright law backs it up. Nevertheless, many users of MP3 players think sharing music is ethical and have decided to violate the law.

In the technical workplace, you will run into ethical dilemmas regularly. At these decision points, you need to be able to identify what is at stake and make an informed decision. Ethical behavior is more than a matter of personal virtue—it is good business. The scandals and failures of the early 21st century have quite clearly proven the importance of ethics in the technical workplace.

What Are Ethics?

People have different definitions for ethics. For some, ethics are about issues of morality. For others, ethics are a matter of law. Actually, ethics bring together many different ideas about appropriate behavior in a society.

Ethics are a system of moral, social, or cultural values that govern the conduct of an individual or community.

Ethics are a system of moral, social, or cultural values that govern the conduct of an individual or community. For many people, acting ethically simply means "doing the right thing." In fact, this phrase sums up ethics quite well. The hard part, of course, is figuring out what is the right thing to do. Ethical choices, after all, are not always straightforward.

Every decision you make has an ethical dimension, whether it is apparent or not. In most workplace situations, the ethical choice is apparent, so you do not pause to consider whether you are acting ethically. Occasionally, though, you will be presented with an *ethical dilemma* that needs more consideration. An ethical dilemma offers a choice among two or more unsatisfactory courses of action. At these decision points, it is helpful to ponder the ethics of each path, so you can make the best choice.

In technical workplaces, ethical dilemmas are not uncommon. Resources, time, and reputations are at stake, so you will feel pressure to overpromise, underdeliver, bend the rules, cook the numbers, or exaggerate results. Technical fields are also highly competitive, urging people to stretch a little further than they should. Ethical dilemmas force us into situations where all choices seem unsatisfactory.

Want to see some other definitions of ethics? Go to **www.ablongman.com/johnsonweb/4.1**

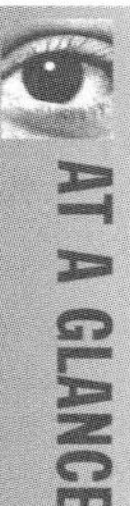

Definitions of Ethics and Ethical Dilemma

- Ethics—systems of moral, social, or cultural values that govern the conduct of an individual or community.
- Ethical Dilemma—a choice among two or more unsatisfactory courses of action.

Why do some people behave unethically? People rarely set out to do something unethical. Rather, they usually find themselves facing a tough decision in which moving forward means taking risks or treating others unfairly. In these situations, they may be tempted to act unethically due to a fear of failure, a desire to survive, pressure from others, or just a series of bad decisions. Small lies lead to bigger lies until the whole house of cards collapses on them.

Keep in mind, though, that ethics are not always about deception or fraud. A famous example involving Albert Einstein (Figure 4.1) proves this point. Figure 4.2 shows a letter from Einstein to President Franklin Roosevelt encouraging research into the development of the atom bomb. Throughout the rest of his life, Einstein, who was an ardent pacifist, was troubled by this letter. Five months before his death, he stated:

> I made one great mistake in my life . . . when I signed the letter to President Roosevelt recommending that atom bombs be made; but there was some justification—the danger that the Germans would make them. (Clark, p. 752)

In this quote, you see the ethical dilemma weighing on Einstein. He deeply regretted the atom bomb's development and use on Japan. However, he also recognized that his letter may have alerted Roosevelt to a very real danger. Historians have pointed out that Einstein's letter may have helped prevent the Nazis from creating an atom bomb themselves. Ethical dilemmas put people in these kinds of quandaries.

Einstein with Robert Oppenheimer

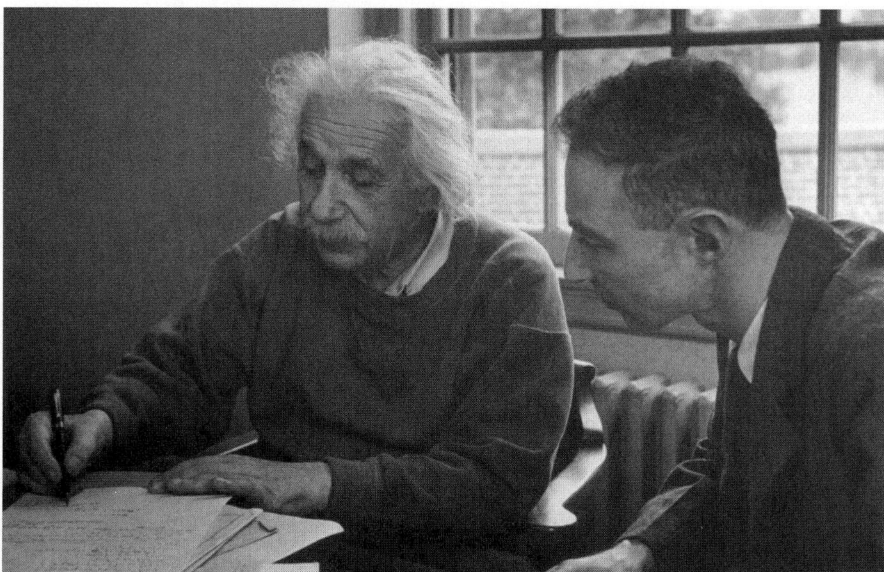

Figure 4.1: Einstein meets with Robert Oppenheimer, the leader of the United States' efforts to develop an atom bomb. Later, Einstein regretted his involvement, though minimal, with its development.

To learn more about Einstein, go to **www.ablongman.com/johnsonweb/4.2**

Einstein's Letter to Roosevelt about the Atom Bomb

Figure 4.2: In 1939, Einstein wrote this letter to President Franklin Roosevelt. The atomic bomb would have been built without Einstein's letter, but his prodding jump-started the United States nuclear program.

Albert Einstein
Old Grove Rd.
Nassau Point
Peconic, Long Island

August 2nd 1939

F.D. Roosevelt
President of the United States
White House
Washington, D.C.

Sir:

Some recent work by E.Fermi and L. Szilard, which has been communicated to me in manuscript, leads me to expect that the element uranium may be turned into a new and important source of energy in the immediate future. Certain aspects of the situation which has arisen seem to call for watchfulness and, if necessary, quick action on the part of the Administration. I believe therefore that it is my duty to bring to your attention the following facts and recommendations:

Here is Einstein's main point.

In the course of the last four months it has been made probable - through the work of Joliot in France as well as Fermi and Szilard in America - that it may become possible to set up a nuclear chain reaction in a large mass of uranium,by which vast amounts of power and large quantities of new radium-like elements would be generated. Now it appears almost certain that this could be achieved in the immediate future.

Einstein expresses the imminent problem.

This new phenomenon would also lead to the construction of bombs, and it is conceivable - though much less certain - that extremely powerful bombs of a new type may thus be constructed. A single bomb of this type, carried by boat and exploded in a port, might very well destroy the whole port together with some of the surrounding territory. However, such bombs might very well prove to be too heavy for transportation by air.

He points out the possible threat of these new kinds of weapons.

Source: Argonne National Laboratory, http://www.anl.gov/OPA/frontiers96arch/aetofdr.html.

Figure 4.2: (continued.)

-2-

The United States has only very poor ores of uranium in moderate quantities. There is some good ore in Canada and the former Czechoslovakia. while the most important source of uranium is Belgian Congo.

In view of the situation you may think it desirable to have more permanent contact maintained between the Administration and the group of physicists working on chain reactions in America. One possible way of achieving this might be for you to entrust with this task a person who has your confidence and who could perhaps serve in an inofficial capacity. His task might comprise the following:

Einstein offers a potential solution.

a) to approach Government Departments, keep them informed of the further development, and put forward recommendations for Government action, giving particular attention to the problem of securing a supply of uranium ore for the United States;

b) to speed up the experimental work,which is at present being carried on within the limits of the budgets of University laboratories, by providing funds, if such funds be required, through his contacts with y private persons who are willing to make contributions for this cause, and perhaps also by obtaining the co-operation of industrial laboratories which have the necessary equipment.

I understand that Germany has actually stopped the sale of uranium from the Czechoslovakian mines which she has taken over. That she should have taken such early action might perhaps be understood on the ground that the son of the German Under-Secretary of State, von Weizsäcker, is attached to the Kaiser-Wilhelm-Institut in Berlin where some of the American work on uranium is now being repeated.

He points out that the Nazis may already be working on nuclear technology, potentially a bomb.

Yours very truly,

A. Einstein

(Albert Einstein)

Where Do Ethics Come From?

How can you identify ethical issues and make appropriate choices? To begin, consider where values come from:

> **Personal ethics**—Values derived from family, culture, and faith.
>
> **Social ethics**—Values derived from constitutional, legal, utilitarian, and caring sources.
>
> **Conservation ethics**—Values required to protect and preserve the ecosystem in which we live.

These ethical systems intertwine, and sometimes they even conflict with each other (Figure 4.3).

Intertwined Ethical Systems

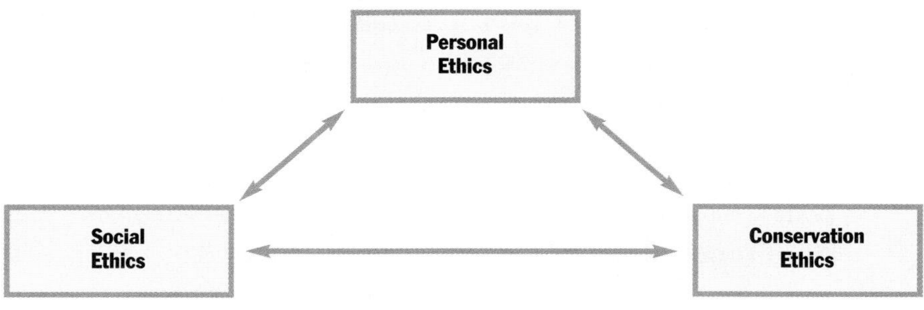

Figure 4.3: Ethics come from a variety of sources, which overlap. Your personal sense of ethics guides the majority of your daily decisions. Social and conservation ethics play significant roles in the technical workplace.

Personal Ethics

By this point in your life, you have developed a good sense of right and wrong. More than likely, your personal ethics derive from your family, your culture, and your faith. Your family, especially your parents, probably taught you some principles to live by. Meanwhile, your culture, including the people in your neighborhood or even the people you watch on television, have shaped how you make decisions. And, for many people, their faith gives them specific principles about how they should live their lives.

The basis for almost all personal ethics is the "Golden Rule," which has been championed by numerous philosophers and religious figures, including Buddha, Confucius, Jesus, Mohammed, Moses, and Socrates.

Golden Rule: Do unto others as you would have them do unto you.

The Golden Rule states that you should do unto others as you would have them do unto you. This simple rule offers a strong foundation for personal ethics. In fact, it probably already guides the majority of your daily ethical decisions. Each day, you have opportunities to lie, steal, cheat, vandalize, or hurt people. But you don't, because you know that hurting others ultimately hurts yourself. Further, your family, culture, and faith have taught you that respecting the needs of others means others will respect your needs.

For websites that discuss sources of the Golden Rule, go to **www.ablongman.com/johnsonweb/4.3**

In the technical workplace, a strong sense of personal values is essential, because these values offer a reliable touchstone for ethical behavior. A good exercise is to make a list of values that you hold dear. Perhaps some values that end up on your list might include honesty, integrity, respect, candor, loyalty, politeness, thoughtfulness, cautiousness, thriftiness, and caring. By articulating these values and following them, you will likely find yourself acting ethically in almost all situations.

Social Ethics

In technical workplaces, the most difficult ethical dilemmas are usually found in the social realm. Here is where ethical dilemmas occasionally go beyond or even against your personal values, asking you to think more globally about the consequences of your or your company's actions.

Ethics scholar Manuel Velasquez (2002) offers a helpful four-part categorization of social ethical situations:

> **Rights**—Rights are fundamental freedoms that are innate to humans or granted by a nation to its citizens. *Human rights,* like those mentioned in the *U.S. Declaration of Independence* (life, liberty, and the pursuit of happiness), are innate to humans and cannot be taken away. *Constitutional rights* (freedom of speech, right to bear arms, protection against double jeopardy) are the rights held in common by citizens of a nation.
>
> **Justice**—Justice involves fairness among equals. Justice takes its most obvious form in the laws that govern a society. Our laws are a formalized ethical system that is designed to ensure that people are treated equally and fairly. Similarly, *corporate policies* are the rules that ensure fairness within a company.
>
> **Utility**—Utility suggests that the interests of the majority should outweigh the interests of the few. Of paramount importance to utilitarianism is *the greatest good for the greatest number of people.*
>
> **Care**—Care suggests that tolerance and compassion take precedence over rigid, absolute rules. Ethics of care suggest that each situation should be judged on its own, putting heightened attention on concern for the welfare of people and preserving relationships. It also recognizes that some relationships, like those involving friends and family, will often lead to ethical choices that transcend rights, justice, and utility.

Legal issues, usually involving rights and justice, are especially important in technical communication, because the temptation to break the law to gain a competitive edge can be great. The law is a collection of rules of conduct that individuals and businesses are obligated to follow. In democracies, codes of laws are enacted by legislatures and interpreted by judges. Violating these laws can result in civil sanctions and/or criminal prosecution. Legal issues of copyright law, patent law, liability, privacy, and fraud (which are all discussed later in this chapter) are crucial concerns that affect how individuals and companies conduct themselves. You should be aware of the laws that apply to your discipline.

More information on Manuel Velasquez's ethical system is available at **www.ablongman.com/johnsonweb/4.4**

When facing an ethical dilemma or controversy involving ethics, you should first identify which of these four ethical categories applies to your situation. Ethical issues that involve human or constitutional rights are usually given the most gravity (Figure 4.4). Issues involving care are still important, but they have the least gravity. In other words, if an ethical decision involves human or constitutional rights, it will take on much more importance than a decision that involves issues of justice, utility, or care.

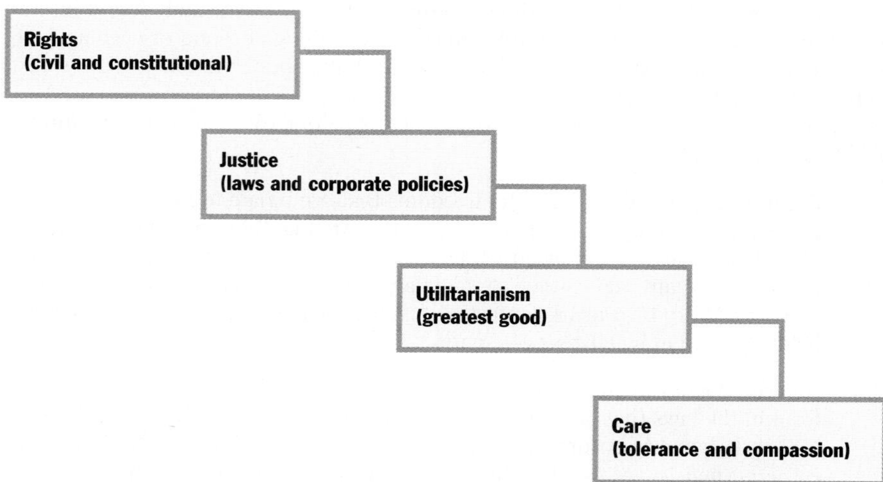

Figure 4.4: Social ethics can be ranked. Concerns about rights usually have more gravity than justice issues, and so on.

By sorting ethical dilemmas into these four categories, you can often decide which course of action is best. For example, consider the following case study:

> **Case Study:** Your company makes a popular action figure toy with many tiny accessories. Countless children enjoy the toy, especially all those tiny boots, hats, backpacks, weapons, etc. However, a few children have choked on some of the small pieces that go with the toy. How would the four levels of ethics govern how to handle this situation?
>
> **Answer:** In this case, a human right (life) has more weight than utility (thousands of children versus a few). So, the ethical choice would be to alter the toy or stop selling it.

Social ethical issues are rarely this clear cut, though. After all, rights, justice, utility, and care are open to interpretation and debate. A union organizer, for example, may see a company's resistance to unionizing as a violation of the right to free assembly provided by the U.S. Constitution. The company, on the other hand, may point to federal laws that allow it to curb union activities. This kind of debate over rights and justice happens all the time.

TAKE NOTE Keep in mind that readers from other cultures may have very different sets of values. If you are communicating with international readers, make sure you are aware of their ethical values.

Want to see the U.S. Constitution? Go to **www.ablongman.com/johnsonweb/4.5**

LINK For more information on working with international readers, see Chapter 3, Page 60.

Another problem is that people miscategorize their ethical issue.

Case Study: Your town's city council has decided to implement a no smoking policy that includes all public property and restaurants. Many smokers now find it impossible to have their "smoke break" around public buildings and in restaurants. They argue that their right to smoke is being violated. How might you resolve this ethical issue?

Answer: Actually, there is no such thing as a right to smoke. Smoking is a legal issue (a matter of justice) and perhaps a utility issue (the health interests of the nonsmoking majority versus the interests of a smoking minority). So, if the city chooses to ban smoking on public property or in restaurants, it can do so legally as long as it applies the law fairly to all. It is not violating anyone's human or constitutional rights.

Of course, defenders of smoking may point out that restricting smoking hurts businesses like bars and restaurants (a utility argument). Advocates of nonsmoking places might counter by pointing out that secondhand smoke may cause cancer in patrons and employees of these establishments (also a utility argument). The city council may take these arguments into consideration before passing a new law.

When making decisions about social ethical issues, it is important to first decide which ethical categories fit the ethical dilemma you are pondering. Then, decide which set of ethics has more significance, or gravity. In most cases:

- issues involving *rights* will have more gravity than issues involving *laws.*
- issues involving *laws* will have more gravity than issues involving *utility.*
- issues involving *utility* will have more gravity than issues involving *care.*

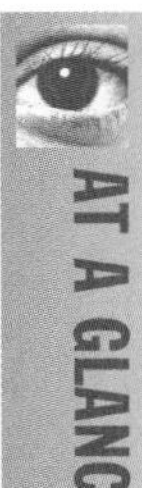

Four Categories of Social Ethics

- Rights—Civil rights and constitutional rights
- Justice—Laws and corporate policies
- Utility—Greatest good (majority rules)
- Care—Tolerance and compassion for others

There are, of course, exceptions. In some cases, utility may be used to argue against laws that are antiquated or unfair. For example, it was once legal to smoke just about anywhere, including the workplace (and the college classroom). By using utility arguments, opponents of smoking have successfully changed those laws. Today, smoking is evermore restricted in public.

Conservation Ethics

Increasingly, the ecosystem is becoming a source for ethical dilemmas. With issues of global warming, nuclear waste storage, toxic waste disposal, and overpopulation, among others, we must move beyond the idea that conservation is a "personal virtue." We are forced to realize that human health and survival are closely tied to the health and survival of the entire ecosystem in which we live. Conservation ethics involves issues of water conservation, chemical and nuclear production and waste, management of insects and weeds in agriculture, mining, energy production and use, land use, pollution, and other environmental issues.

For more information on conservation ethics, go to **www.ablongman.com/johnsonweb/4.6**

One of America's prominent naturalists, Aldo Leopold, suggested that humans need to develop a *land ethic.* He argued:

> All ethics so far evolved rest upon a single premise: that the individual is a member of a community of interdependent parts. . . . The land ethic simply enlarges the boundaries of the community to include soils, waters, plants, and animals, or collectively: the land. (p. 239)

In other words, your considerations of ethics should go beyond the impact on humans and their communities. The health and welfare of the ecosystem around you should also be carefully considered.

Technical fields need to be especially aware of conservation ethics, because we handle so many tools and products that can damage the ecosystem. Without careful concern for use and disposal of materials and wastes, we can do great harm to the environment.

Ultimately, conservation ethics are about *sustainability.* Can humans interact with their ecology in ways that are sustainable in the long term? Conservation ethics recognize that resources must be used. They simply ask that people use resources in sustainable ways. They ask us to pay attention to the impact our decisions have on the air, water, soil, plants, and animals on this planet.

Conservation ethics are becoming increasingly important. The 21st century has been characterized as the "Green Century," because humans have reached a point where we can no longer ignore the ecological damage caused by our decisions. For example, within this century, estimates suggest that human-caused climate change will raise global temperatures from 2 to 10 degrees. Such a rise would radically alter our ecosystem.

HELP

Copyright Law in Technical Communication

An interesting flashpoint today is copyright law. A copyright gives someone an exclusive legal right to reproduce, publish, or sell his or her literary, musical, or artistic works. Copyright law in the United States was established by Article 1 of the United States Constitution. The U.S. law that governs copyright protection is called "Title 17" of the United States Code.

Essentially, a copyright means creative work is someone's property. If others would like to duplicate that work, they need to ask permission and possibly pay the owner. Authors, musicians, and artists often sign over their copyrights to publishers, who pay them royalties for the right to duplicate their work.

New electronic media, however, have complicated copyright law. For example,

- When you purchase something, like a music CD, you have the right to duplicate it for your own personal use. What happens if you decide to copy a song off a CD and put it on your website for downloading? You might claim that you put the song on your website for your personal use, but now anyone else can download the song for free. Are you violating copyright law?
- According to Title 17, section 107, you can reproduce the work of others "for purposes such as criticism, comment, news reporting, teaching (including mul-

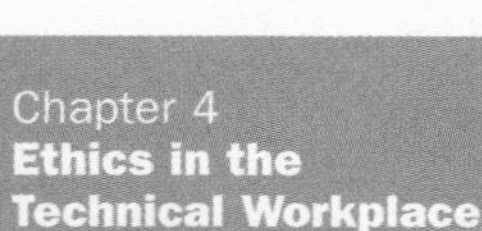

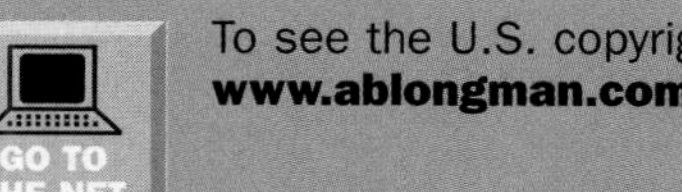

To see the U.S. copyright law, go to
www.ablongman.com/johnsonweb/4.7

tiple copies for classroom use), scholarship, or research."This is referred to as "fair use." So, is it illegal to scan whole chapters of books for "teaching purposes" and put them on a CD for fellow students or coworkers?

- New technology like webcasting (using digital cameras to broadcast over the Internet) allows people to produce creative works. If you decided to webcast your and your roommates' dorm room antics each evening, would you be protected by copyright law?
- Blogs, or web logs, are becoming a popular way to broadcast news and opinions. Are these materials copyrighted? Is it illegal to share words and images on blogs?

The answer to these questions is "yes," but the laws are still being worked out. It is illegal to allow others to download songs off your website. It would be illegal to scan large parts of a book, even if you claimed they were being used for educational purposes. Meanwhile, you can protect webcasting and blogs through the copyright laws.

The problem is the ease of duplication. Before computers, copyrights were easier to protect because expensive equipment like printing presses, sound studios, and heavy cameras were required to copy someone else's work. Today, anyone can easily duplicate the works of others with a scanner, CD/DVD recorder, or digital video recorder.

Ultimately, violating copyright is like stealing someone else's property. The fact that it is easier to steal today does not make it all right. Nevertheless, a few scholars have argued that copyright law is antiquated and that this kind of electronic sharing is how people will use text and music in the future.

Asking Permission

To avoid legal problems, it is best to follow copyright law as it is currently written. You need to ask permission if you would like to duplicate or take something from someone else's work. You can ask permission by writing a letter or e-mail to the publisher of the materials. Publishers can almost always be found on the Internet. On their websites, they will often include a procedure for obtaining permissions. Tell them exactly what you want to use and how it will be used.

In some cases, especially when you are a student, your use may fall under the "fair use clause" of the Copyright Act. Fair use allows people to copy works for purposes of "criticism, comment, news reporting, teaching (including multiple copies for classroom use), scholarship, or research" (17 U.S. Code sec. 107). If your use of the materials falls under these guidelines, you may have a *limited* right to use the materials without asking permission.

For example, fair use would likely allow you to use a song legally downloaded from the Internet as background music in a presentation for your class. However, it does not allow you to distribute that song freely to your friends, even if you claim you are doing so for educational purposes.

Copyrighting Your Work

What if you write a novel, take a picture, produce a movie, or create a song? How do you copyright it? The good news is that you already have. In the United States, a work is copyrighted as soon as it exists in written form. If you want, you can add the copyright symbol "©" to your work to signal that it is copyrighted. The copyright symbol, however, is no longer necessary to protect a work.

If you want to formally protect your work from copyright infringement (i.e., so you can sue someone who uses your work without your permission), you should register your copyright with the U.S. Copyright Office (Figure A). This step is not necessary to protect your work, but it makes settling who owns the material much easier.

The U.S. Copyright Office Website

Figure A: You can visit the U.S. Copyright Office website to learn more about copyright law or to protect your own work.

Source: United States Copyright Office, http://www.loc.gov/copyright.

Plagiarism

One type of copyright infringement is plagiarism. In Chapter 5 of this book, plagiarism is discussed in depth, but the subject is worth briefly mentioning here. Plagiarism is the use of someone else's text or ideas as your own without giving credit. Plagiarism is a violation of copyright law, but it is also a form of academic dishonesty that can have consequences for your education and career.

For example, cutting and pasting words and images off the Internet and "patchwriting" them into your documents is a form of plagiarism, unless those materials are properly cited. To avoid questions of plagiarism, make sure you cite your sources properly and, when needed, ask permission to use someone else's work.

Resolving Ethical Dilemmas

No doubt, you will be faced with numerous ethical dilemmas during your career. There is no formula or mechanism you can use to come out with the "right" answer. Rather, ethical dilemmas usually force us to choose among uncomfortable alternatives.

Doing the right thing can mean putting your reputation and your career on the line. It might mean putting the interests of people above profits. It also might mean putting the long-term interests of the environment above short-term solutions to waste disposal and use of resources.

Confronting an Ethical Dilemma

When faced with an ethical dilemma, start considering it from all three ethical perspectives: personal, social, and conservation (Figure 4.5).

> **Personal ethics**—How does my upbringing in a family, culture, and faith guide my decision? How can I do unto others as I would have them do unto me?
>
> **Social ethics**—What rights or laws are involved in my decision? What is best for the majority? How can I demonstrate caring by being tolerant and compassionate?
>
> **Conservation ethics**—How will my decision affect the ecosystem? Will my choice be ecologically sustainable in the long term?

Balancing the Different Issues in an Ethical Dilemma

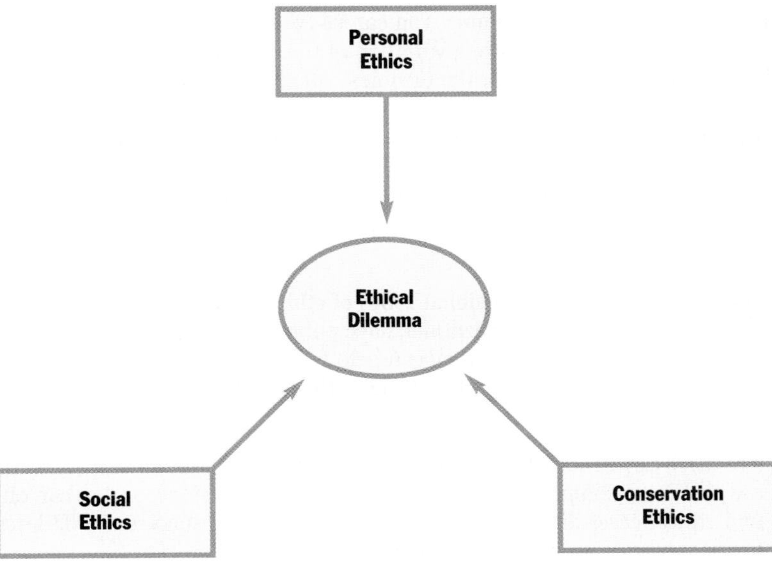

Figure 4.5: Resolving an ethical dilemma requires you to consider it from various ethical perspectives.

Do you want to try solving some more ethical dilemmas? Go to **www.ablongman.com/johnsonweb/4.8**

With most ethical dilemmas, you will find that ethical stances conflict. To resolve the dilemma, it helps to first locate the "ethical tension"—the point where two or more ethical stances are not compatible. For example:

- As a doctor who treats gunshot victims all the time, you may be opposed to gun ownership, but social ethics in the form of constitutional law make gun ownership a right. *Here, rights are coming into conflict with utility.*
- Someone offers you a draft of your competitor's proposal for an important project. With that information your company would almost certainly win the bid. However, your industry's code of ethics regarding proprietary information forbids you to look at it. *Here, justice is coming into conflict with utility.*
- Your company owns the rights to the timber in a forest, but an endangered species of owl lives there and its habitat, by law, should be protected. *Here, justice, rights, and conservation are in conflict.*
- A legal loophole allows your company to pump tons of pollution into the air, even though this pollution is obviously harming the health of the residents in a small town a few miles downwind. *Here, personal ethics and utilitarianism are in conflict with justice.*

Product liability is a place where these kinds of conflicts become especially important. If your company produces a product that harms people unintentionally, the company may still be found liable for damages. It is not enough to simply splash warnings all over your documentation. Even if warnings are provided, your company might be found negligent by the courts and ordered to pay damages.

LINK For more information on liability in technical documents, see Chapter 19, page 546.

Resolving an Ethical Dilemma

When faced with an ethical dilemma, you can use the following five questions to help you resolve it. These questions are a variation of the ones developed by Professor Sam Dragga in an article on ethics in technical communication (1996).

Do any laws or rules govern my decision?—In many cases, laws at the federal, state, and local levels will specify the appropriate action in an ethical case. You can look to your company's legal counsel for guidance in these matters. Otherwise, companies often have written rules or procedures that address ethical situations.

Do any corporate or professional codes of ethics offer guidance?—Most companies and professional organizations have published codes of ethics. They are usually rather abstract, but they can help frame ethical situations, so you can make clearer decisions. Figure 4.6 shows the code of ethics for the Institute of Electrical and Electronics Engineers (IEEE).

Are there any historical records to learn from?—Look for similar situations in the past. Your company may keep records of past decisions, or you can often find ethical cases discussed on the Internet. By noting successes or failures in the past, you can make a more informed decision.

To find websites discussing moral leaders, go to **www.ablongman.com/johnsonweb/4.9**

Sample Code of Ethics

Figure 4.6: The IEEE Code of Ethics. Just about every established field has a code of ethics you can turn to for guidance.

IEEE Code of Ethics

We, the members of the IEEE, in recognition of the importance of our technologies in affecting the quality of life throughout the world, and in accepting a personal obligation to our profession, its members and the communities we serve, do hereby commit ourselves to the highest ethical and professional conduct and agree:

1. to accept responsibility in making engineering decisions consistent with the safety, health and welfare of the public, and to disclose promptly factors that might endanger the public or the environment;
2. to avoid real or perceived conflicts of interest whenever possible, and to disclose them to affected parties when they do exist;
3. to be honest and realistic in stating claims or estimates based on available data;
4. to reject bribery in all its forms;
5. to improve the understanding of technology, its appropriate application, and potential consequences;
6. to maintain and improve our technical competence and to undertake technological tasks for others only if qualified by training or experience, or after full disclosure of pertinent limitations;
7. to seek, accept, and offer honest criticism of technical work, to acknowledge and correct errors, and to credit properly the contributions of others;
8. to treat fairly all persons regardless of such factors as race, religion, gender, disability, age, or national origin;
9. to avoid injuring others, their property, reputation, or employment by false or malicious action;
10. to assist colleagues and co-workers in their professional development and to support them in following this code of ethics.

Approved by the IEEE Board of Directors

August 1990

Source: Institute of Electrical and Electronics Engineers, 1990.

Want to see other codes of ethics? Go to **www.ablongman.com/johnsonweb/4.10**

> **AT A GLANCE**
> ### Resolving Ethical Dilemmas
> - Do any laws or rules govern my decision?
> - Do any corporate or professional codes of ethics offer guidance?
> - Are there any historical records to learn from?
> - What do my colleagues think?
> - What would moral leaders do?

What do my colleagues think?—Your co-workers, especially people who have been around for awhile, may have some insight into handling difficult ethical situations. First, they can help you assess the seriousness of the situation. Second, they may be able to help you sort out the impact on others. At a minimum, though, talking through the ethical dilemma may help you sort out the facts.

What would moral leaders do?—You can look for guidance from moral leaders that you respect. These might include spiritual leaders, civil rights advocates, business pioneers, or even your friends and relatives. In your situation, what would they do? Sometimes their convictions will help guide your own. Their stories may give you the confidence to do what is right.

Facing an ethical dilemma, you will probably need to make a judgment call. In the end, you want to make an informed decision. If you fully consider the personal, social, and conservation perspectives, you will likely make a good decision.

When You Disagree with the Company

Ethical conflicts between you and your company need to be handled carefully. If you suspect your company or your supervisors are acting unethically, there are a few paths you can take:

Persuasion through costs and benefits—After you have collected the facts, take some time to discuss the issue with your supervisors in terms of costs and benefits. Usually unethical practices are costly in the long term. Show them that the ethical choice will be beneficial over time.

LINK For more on persuasion strategies, see Chapter 7, page 180.

Seek legal advice—Your company likely has an attorney who can offer legal counsel on some issues. You may visit legal counsel to sort out the laws involved in the situation. If your company does not have legal counsel or you don't feel comfortable using it, you may need to look outside the company for legal help.

Mediation—Companies often offer access to mediators who can facilitate meetings between you and others. Mediators will not offer judgments on your ethical case, but they can help you and others identify the issues at stake and work toward solutions.

Memos to file—In some cases, you will be overruled by your supervisors. If you believe a mistake is being made, you may decide to write a *memo to file* in which you express your concerns. In the memo, write down all the facts and your concerns. Then, present the memo to your supervisors and keep a

For more strategies for handling unethical situations at work, go to **www.ablongman.com/johnsonweb/4.11**

copy for yourself. These memos are usually filed for future reference to show your doubts.

LINK For more information on writing memos, see Chapter 16, page 446.

Whistleblowing—In serious cases, especially where people's lives are at stake, you may even choose to be a whistleblower. Whistleblowing usually involves going to legal authorities, regulatory agencies, or the news media. Being a whistleblower is a serious decision. It will affect your career and your company. Federal laws exist that protect whistleblowers, but there is always a personal price to be paid.

Ethical situations should be carefully considered, but they should not be ignored. When faced with an ethical dilemma, it is tempting to walk away from it or pretend it isn't there. In any ethical situation, you should take some kind of action. Inaction on your part is both ethically wrong and might leave you or your company vulnerable to liability lawsuits. At a minimum, taking action will allow you to live with your conscience.

Websites exist that can help you make your decision by considering ethical case studies. For example, the Online Ethics Center for Engineering and Science at Case Western University offers many case studies that are discussed by ethics experts (Figure 4.7). Perhaps one of these cases is similar to the one you face, and you can use the wisdom of these experts to make the ethical decision.

On-Line Ethics

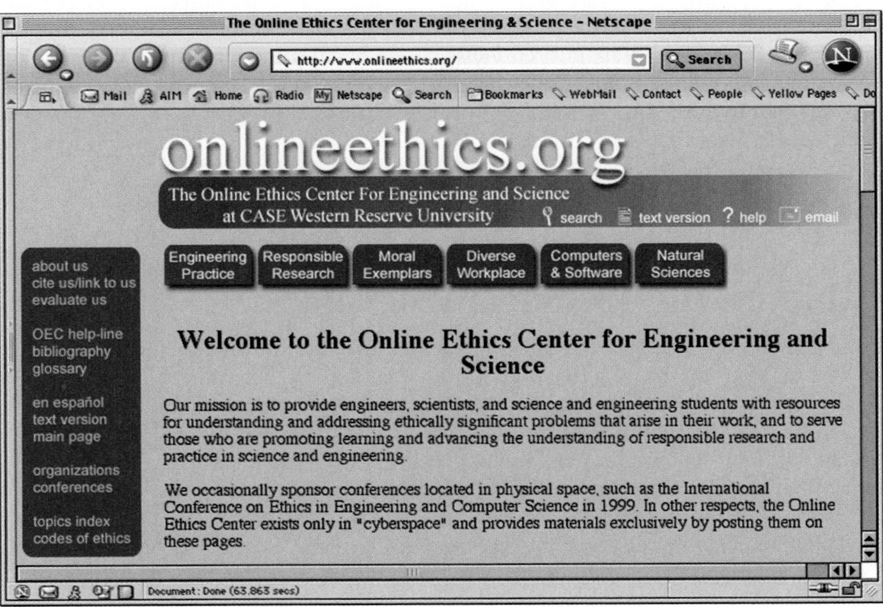

Figure 4.7: The Online Ethics Center for Engineering and Science at Case Western University is a great place to learn about ethics in scientific and technical disciplines.

Source: The Online Ethics Center for Engineering and Science, http://www.onlineethics.org.

Does whistleblowing really work? Go to **www.ablongman.com/johnsonweb/4.12**
To access the Online Ethics Center and similar sites, go to **www.ablongman.com/johnsonweb/4.13**

AT WORK

Caroline Whitbeck, Ph.D.

DIRECTOR OF THE ONLINE ETHICS CENTER FOR ENGINEERING & SCIENCE, CASE WESTERN UNIVERSITY

The Center for Engineering & Science in Cleveland, Ohio, is an academic division that researches issues involving technology.

Why should technical professionals learn about ethics?

The practice of a profession, such as the profession of engineering, is characterized by two elements: (1) the practice directly influences one or more major aspects of human well-being and (2) it requires mastery of a complex body of knowledge and specialized skills. To become a professional in a field requires both formal education and practical experience.

The responsibility to achieve certain ends characterizes the core of professional ethics. For example, engineers have a responsibility for the public safety, and research investigators have a responsibility for the integrity of research. Achieving ends requires judgment in the application of professional knowledge.

Following moral rules, such as "Do not offer or accept bribes," important as they are, does not demand the exercise of judgment that is required to fulfill responsibilities. Because the judgment needed to fulfill responsibilities requires professional knowledge, those without professional knowledge cannot judge whether a professional is making responsible judgments, that is, behaving both competently and with due concern. This is the reason why professions establish standards of responsible practice for their practitioners.

Of course, individual practitioners and sometimes the professions themselves may prove untrustworthy, but when they do, everyone loses. When professionals prove irresponsible, they may be monitored more closely. Monitoring may work to see if the rules are being followed, but it does not readily check on the trustworthiness of professional judgments.

If society loses trust in a profession, people avoid relying on the service of members of that profession.

Ethics in the Technical Workplace

Some legal and ethics scholars have speculated that the Information Age requires a new sense of ethics, or at least an updating of commonly held ethics. These scholars may be right. After all, our ethical systems, especially those involving forms of communication, are based on the printing press as the prominent technology. Laws and guidelines about copyright, plagiarism, privacy, information sharing, and proprietary information are all based on the idea that information is "owned" and shared on paper.

The fluid, shareable, changeable nature of electronic files and text brings many of these laws and guidelines into discussion. For example, consider the following case study:

Want to learn more about wired ethics? Go to **www.ablongman.com/johnsonweb/4.14**

Case Study: You and your coworkers are pulling together a training package by collecting information off the Internet. You find numerous sources of information on the websites of consultants and college professors. Most of it is well written, so you cut and paste some of the text directly into your materials. You also find some great pictures and drawings on the Internet to add to your presentation. At what point does your cutting and pasting of text become a violation of copyright law? How can you avoid any copyright problems?

In the past, these kinds of questions were easier to resolve, because text was almost exclusively paper based. Printed text is rather static, so determining who "owns" something is a bit easier. Today, the flexibility and speed of electronic media make these questions much more complicated.

At this time, many of our laws governing the use of information and text are evolving to suit new situations.

Copyright law—Today, copyright law is being strained by the electronic sharing of information, images, and music. In legal and illegal forms, copies of books, songs, and software are all available on the Internet. According to the law, these materials are owned by the people who wrote or produced them; however, how can these materials be protected when they can be shared with a few clicks of a mouse?

Trademarks—People or companies can claim a symbol, word, or phrase as their property by trademarking it. Usually, a trademark is signaled with a ™ symbol. For example, the Internet search engine, Google™, is a trademarked name. The trademark signals that the company is claiming this word for its use in the area of Internet search engines.

To gain further protection, a company might *register* its trademark with the U.S. Patent and Trademark Office, allowing it to use the symbol ® after the logo, word, or phrase. Once the item is registered, the trademark owner has exclusive rights to use that symbol, word, or phrase. For example, IBM's familiar blue symbol is its registered trademark, and it has the exclusive right to use it.

There are exceptions, though. The First Amendment of the U.S. Constitution, which protects free speech, has allowed trademarked items to be parodied or critiqued without permission of the trademark's owner.

Patents—Inventors of machines, processes, products, and other items can protect their inventions by patenting them. Obtaining a patent is very difficult because the mechanism being patented must be demonstrably unique. But once something is patented, the inventor is protected against others' use of his or her ideas to create other products.

Privacy—Whether you realize it or not, your movements online are regularly monitored through electronic networks. Websites will send or ask for *cookies* that identify your computer. These cookies can be used to build a profile of you. Meanwhile, at a workplace, your e-mail and phone conversations can be monitored by your supervisors. Privacy laws are only now being established to cover these issues.

For more information on trademark and patent laws, go to **www.ablongman.com/johnsonweb/4.15**

Information sharing—Through electronic networks, companies can build databases with information about their customers and employees that can be shared with other companies. Information sharing, especially involving medical information, is an important issue that will probably be resolved in the courts.

Proprietary information—As an employee of a company, you have access to proprietary information that you are expected not to share with others outside the company. In government-related work, you may even have a *security clearance* that determines what kinds of information you have access to. When you leave that company, you cannot take copies of documents, databases, or software with you. In some cases, you may even be asked to sign papers that prevent you from sharing your previous employer's secrets with your new employer.

Libel and slander—You or your company can be sued for printing falsehoods (libel) or speaking untruths (slander) that damage the reputation or livelihood of another person or company. With the broadcasting capabilities of the Internet, libel and slander have much greater reach. A website, for example, that libels another person or company, may be a target for legal retaliation. If you use e-mail to libel others, these messages could be used against you.

Fraud—The Internet is also opening whole new avenues for fraud. Con artists are finding new victims with classic fraud schemes. Websites, especially, can sometimes give the appearance of legitimacy to a fraudulent operation. Meanwhile, con artists use e-mail to find victims (had any e-mails from the widows of wealthy Nigerian dictators recently?).

Only the future can tell whether a new understanding of ethics will arise. Most of the changes will be subtle. Many will be worked out in the courts.

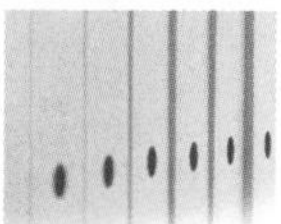

- Ethics are a system of moral, social, or cultural values that govern the conduct of an individual or community.
- Ethical dilemmas force us to choose among uncomfortable alternatives.
- When you are faced with an ethical dilemma, consider it from all three ethical perspectives: personal, social, and conservation.
- You can turn to sources like laws, professional codes of ethics, historical records, your colleagues, or moral leaders to help you make ethical choices.
- When you disagree with the company, use persuasion first to discuss costs and benefits. You may turn to legal avenues if persuasion won't work.
- Ethical guidelines are evolving to suit the new abilities of computers.
- Copyright law and plagiarism are two rapidly evolving areas of ethics in this computer-centered world.
- Privacy and information sharing are also becoming hot topics, because computer networks facilitate the collection of so much information.

GO TO THE NET

Want to learn more about Internet libel, slander, and fraud? Go to **www.ablongman.com/johnsonweb/4.16**

Individual or Team Projects

1. At www.ablongman.com/johnsonweb/4.10, you will find codes of ethics that apply to a number of career paths. If you are an engineer, you might consider looking to the IEEE Code of Ethics. If you are going into medicine, you might look to the American Medical Association (AMA). Write a memo to your instructor in which you summarize the codes of ethics, and highlight their important features. Discuss which specific ethical issues seem to pertain to your field.

2. Describe a real or fictional situation that involves a communications-related ethical dilemma. As you describe the situation, try to bring personal, social, and conservation ethics into conflict. At the end of the situation, leave the readers with a difficult question to answer.

   In a memo to your instructor, identify the ethical issues at stake in the situation you described and offer a solution to the ethical dilemma. Then, give your description to someone else in your class. He or she should write a memo to you and your instructor discussing the ethical issue at stake and offering a solution to the problem. Compare your original solution to your classmate's solution.

3. Visit the U.S. Copyright Office at http://www.loc.gov/copyright and write a brief synopsis of copyright law. What are some copyright issues that seem to be changing? What are some issues that will likely stay the same? Do you believe that the current copyright law will stand the test of time? Do you think an alternative to copyright law is available, specifically, an alternative that allows authors, musicians, and artists to be paid for their creative work?

4. Find examples of advertising that seem to stretch ethics by making unreasonable claims. Choose one of these examples and write a short report to your instructor in which you discuss why you find the advertisement unethical. Use the terminology from this chapter to show how the advertisement challenges your sense of personal ethics, your social ethics, or your conservation ethics.

5. After researching conservation ethics on the Internet (see www.ablongman.com/johnsonweb/4.6), develop a Conservation Code of Ethics for your campus. In your code, you might discuss issues of recycling, water usage, chemical usage, testing on animals, or release of pollution. If you were an administrator at your university, how might you go about putting your code of ethics into action?

Collaborative Project

The case study at the end of this chapter discusses a difficult case in which an engineer feels forced to do something unethical. Read and discuss this case with a group of others. Sort out the ethical issues involved by paying attention to the personal, social, and conservation factors that shape the ethical dilemma.

If you were this engineer, how would you react to this ethical dilemma? How would you turn your reaction into action with writing? Would you write a memo to Frank's supervisor? Would you write a letter to the state Environmental Protection

Agency (EPA)? Would you try to write a new policy regarding violations of pollution releases? Would a memo to file be enough? Would you blow the whistle?

Whichever path you choose, write a letter or memo to a specific reader (e.g., Frank, his supervisor, the state EPA, a newspaper journalist) that takes action. Summarize the situation for the readers of your document, and then suggest an appropriate course of action. Support your decision by highlighting the ethical issues involved and discussing the ramifications of inaction.

To visit the EPA and other environmental protection agencies, go to
www.ablongman.com/johnsonweb/4.17

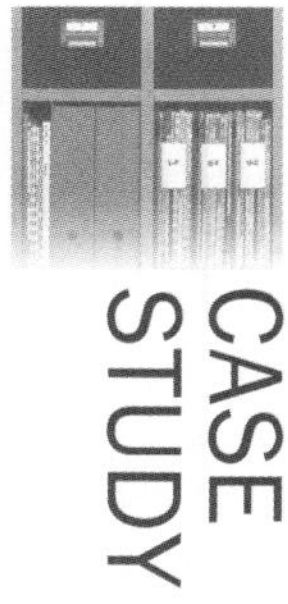

CASE STUDY

The Chemical Spill

As soon as she saw the data, Hanna Roberts knew her supervisor, Frank Gathers, the company's senior environmental engineer, was not going to be pleased. Hanna's company, IDC Chemical Products, was allowed by the state's Environmental Protection Agency (EPA) to release small levels of pollution into a local river, which eventually flowed into a popular lakeside resort area. Hanna's data showed that IDC had violated the allowable limits last month.

The violation was almost inevitable. Frank insisted they release as much pollution as legally allowed, thus saving the company \$300,000 a year in disposal costs. With his careful management of the waste, Frank had become the darling of the company, because he had found a way to keep disposal costs extraordinarily low.

Soon after she joined the company, however, Hanna suspected that one of the problems with Frank's approach was that the company would exceed the allowable limits whenever anything went wrong. The company's monthly reports to the state's EPA were supposed to highlight any violations, and the company would be charged a fine. More than two violations per year would trigger an environmental audit from the state government.

Hanna's data on the most recent spill showed that the company had clearly gone over the limits, even though the spill was a minor one. Being new to the company, she looked for records of spills in past reports, assuming that these accidents had happened before. She couldn't find any. When she asked coworkers about the accuracy of the reports, they said Frank was "careful" and they reminded her that the last person in her position was fired due to "personality conflicts" with Frank.

Hanna decided to include the spill data in her report anyway. She put it in Frank's mailbox. That afternoon, Frank visited her office, clearly angry. "These numbers are unacceptable," he said. "If we submit a report like this one to the state EPA, we'll have a bunch of wacko environmentalist regulators all over us. That spill was a minor one. We can adjust the numbers to show we are in compliance. Let me show you how."

Frank proceeded to "round off" a few numbers, so that the results showed that the company had complied with the law—just barely. Obviously, Frank had done this before. With a smile, he said, "Now, rewrite the report with these new results. And don't worry about those folks at the state EPA. The regulations in this state are too strict anyway, and they know it. We could drink the water coming out of this plant, even when there is an accidental spill."

Hanna knew that wasn't true. Frank was clearly in denial about the kinds of pollution released by the plant. But she needed this job. She and her kids didn't want to move. In fact, one of the reasons Hanna took the job was because her kids loved the resort and liked to swim in the lake. Now she was responsible for deciding whether that lake would be polluted.

So, she used Frank's altered numbers in her report. She signed the report and showed it to Frank. She sent it to the state EPA. That night, her conscience wouldn't let her sleep. What should Hanna do?

PART 2

Communicating in the Technical Workplace

CHAPTER

5

Researching and Managing Information

CHAPTER CONTENTS

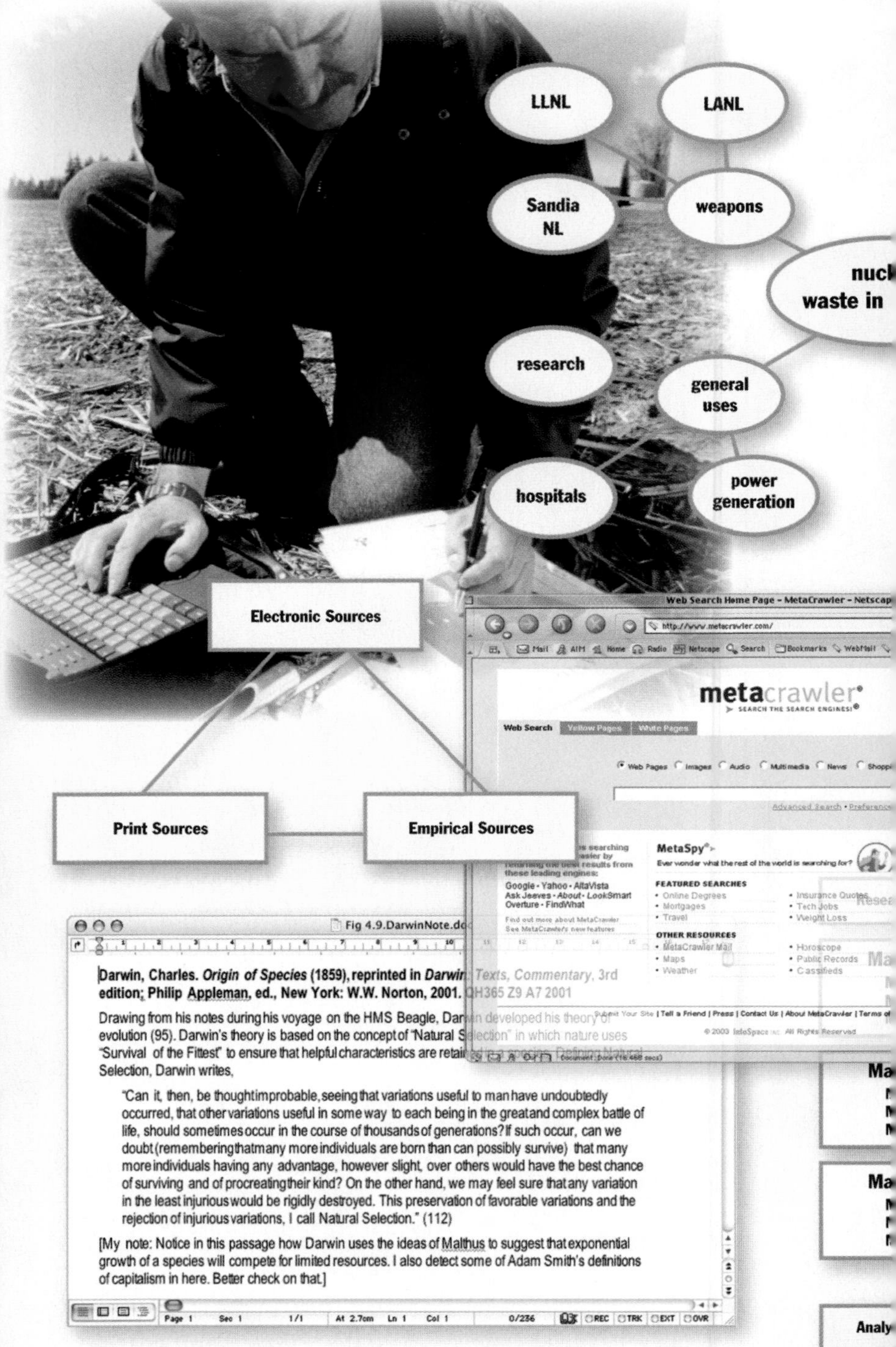

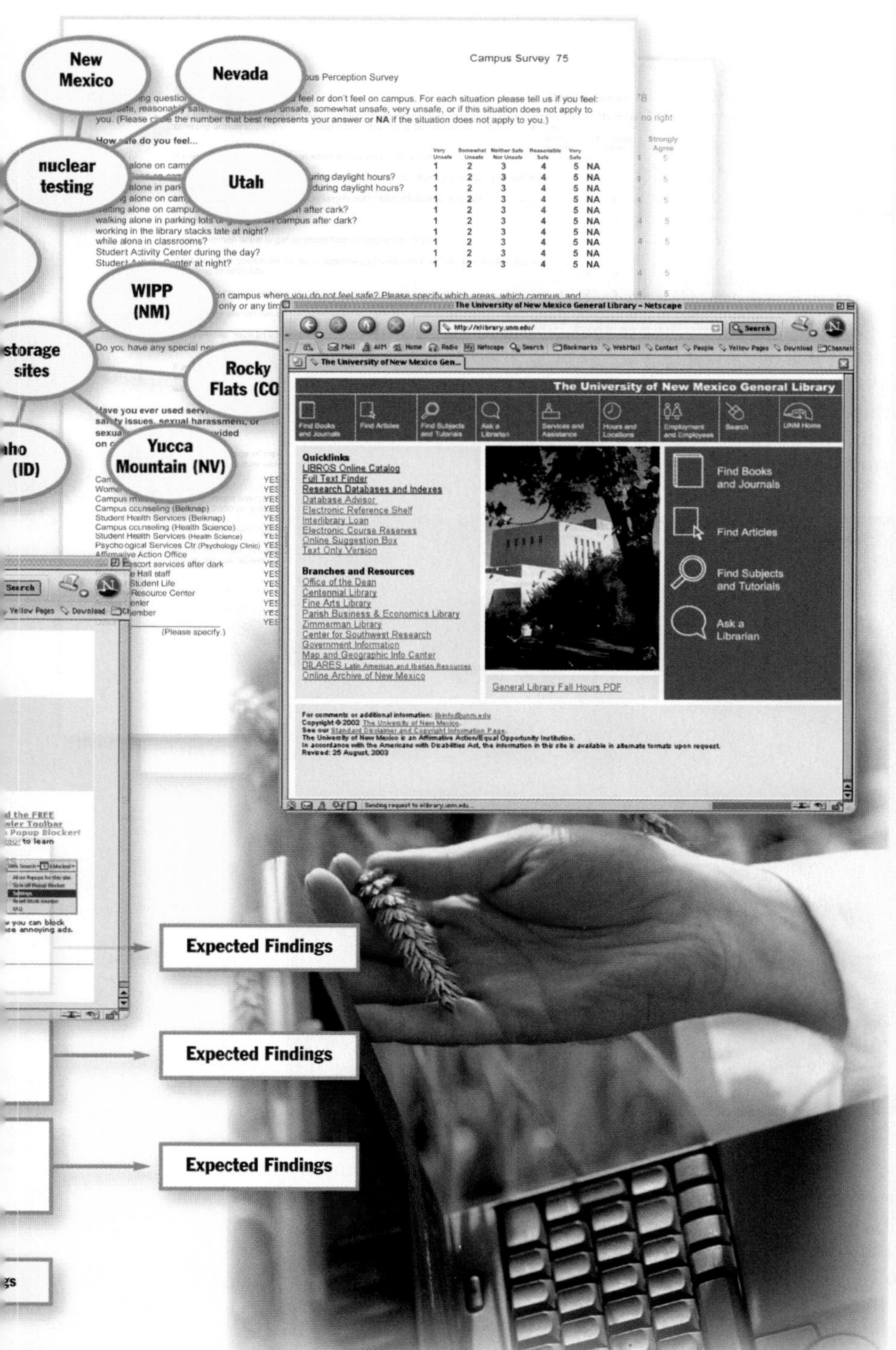

CHAPTER OBJECTIVES

In this chapter, you will learn:

- How to define your research subject using logical mapping to define its boundaries.
- How to formulate a research question or hypothesis that will guide your research.
- To develop a research methodology and revise it as needed.
- Methods for triangulating information to ensure reliability.
- To use the many available electronic research tools.
- To find electronic, print, and empirical sources for your research.
- How to take accurate notes and use your notes appropriately in paraphrasing, summarizing, and quoting.
- How to appraise the information you gather for reliability and bias while avoiding plagiarism.

Computers and computer networks have made research both easier and more challenging. Not long ago, finding *enough* information was the hard part of doing research. Today, with access to the Internet, you will find seemingly endless amounts of information available on any given topic. If you run a search on Google.com or Altavista.com, thousands of webpages might refer to your subject. Even a traditional search at your library will unearth more information than you could ever collect.

Information glut: more information than there is time available to collect, interpret, and synthesize that information.

The problem caused by this overwhelming universe of facts, data, and opinions is called an *information glut.* An information glut exists when more information than time is available to collect, interpret, and synthesize that information.

What should you do about this overwhelming access to information? You should view "research" as a form of *information management.* Research is now a process of shaping the flow of information, so you can locate and utilize the information you need. As an information manager, you need to learn how to evaluate, prioritize, interpret, and store that information so you can use it effectively.

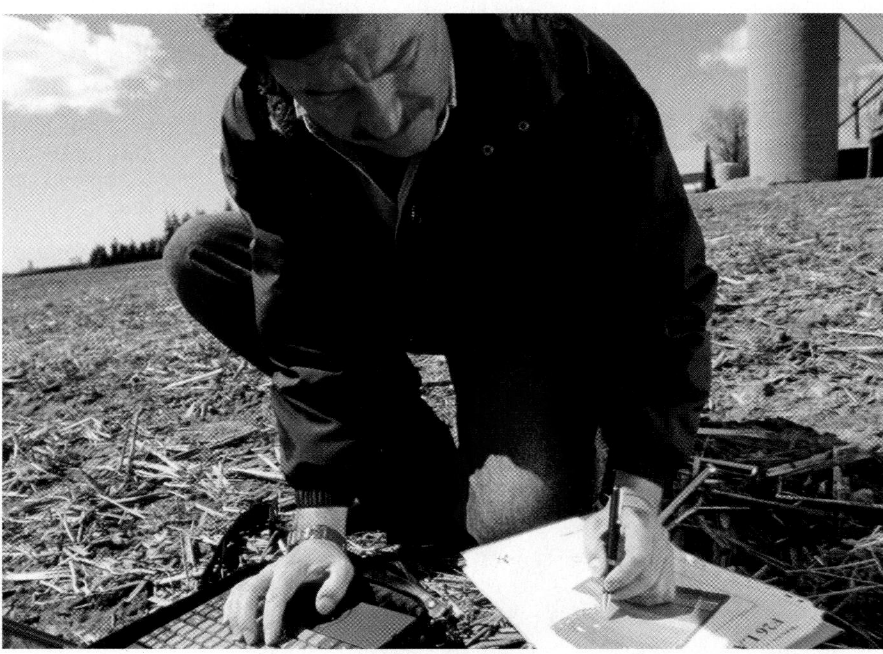

Figure 5.1: Empirical research is a critical part of working in technical disciplines.

Of course, in addition to collecting existing information, *primary research* (empirical research) is still important in the technical workplace. Primary research involves observing and/or directly experiencing the subject of your study. By conducting experiments, doing field studies, using surveys, and following other empirical methods, you can make your own observations and collect your own data. The most effective research usually blends these kinds of empirical observations with existing information available through computers and libraries.

GO TO THE NET

For websites that discuss the information glut, go to **www.ablongman.com/johnsonweb/5.1**

Beginning Your Research

In technical fields, research typically uses a combination of primary and secondary sources to gain a full understanding of a particular subject.

> **Primary sources**—Information collected from observations, experiments, surveys, interviews, ethnographies, testing.
>
> **Secondary sources**—Information drawn from academic journals, magazine articles, books, websites, CD-ROMs, and reference materials.

Most researchers begin their research by first locating the secondary sources available on their subject. Once they have a thorough understanding of their subject, they use primary research to expand on these existing materials.

Your research into primary and secondary sources should follow a process like the following:

1. Define the research subject.
2. Formulate a research question and hypothesis.
3. Develop a research methodology.
4. Triangulate electronic, print, and empirical sources of information.
5. Appraise collected information to determine reliability.

A good research process begins by clearly defining the research subject. Then, it follows a research methodology in which a variety of sources are located and appraised for reliability (Figure 5.2).

A Research Process

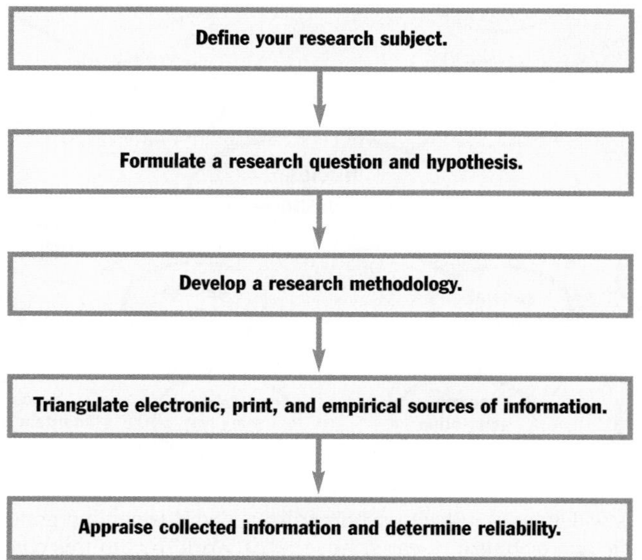

Figure 5.2: To ensure collection of reliable information, it helps to follow a predictable research process.

Defining Your Research Subject

Your first task is to define your research subject as clearly as possible. Therefore, it is a good idea to begin by identifying what you already know about the subject and highlighting areas where you need to do more research.

A good way to start is to first develop a *logical map* of your research subject (Figure 5.3). To create a logical map, write your subject in the middle of your screen or a piece of paper. Then, around that subject, begin noting everything you already know or believe about the subject. As you find relationships among these ideas, you can begin drawing lines to connect them together into clusters.

TAKE NOTE As you are making a logical map, you will notice that some ideas will lead to other, unexpected ideas—some seemingly unrelated to your subject. When this happens, just keep writing them down. Don't stop. These unexpected ideas are evidence that you are thinking creatively by tapping into your visual-spatial abilities. You may end up crossing many of these ideas out, but some may offer you new insights into the subject that will be helpful.

In places where you are not sure of yourself, simply jot down your thoughts and put question marks (?) after them.

Mapping takes advantage of your ability to think visually and spatially about an issue. It will help you define the boundaries of your research subject and also help you focus your research.

Using Mapping to Find the Boundaries of a Subject

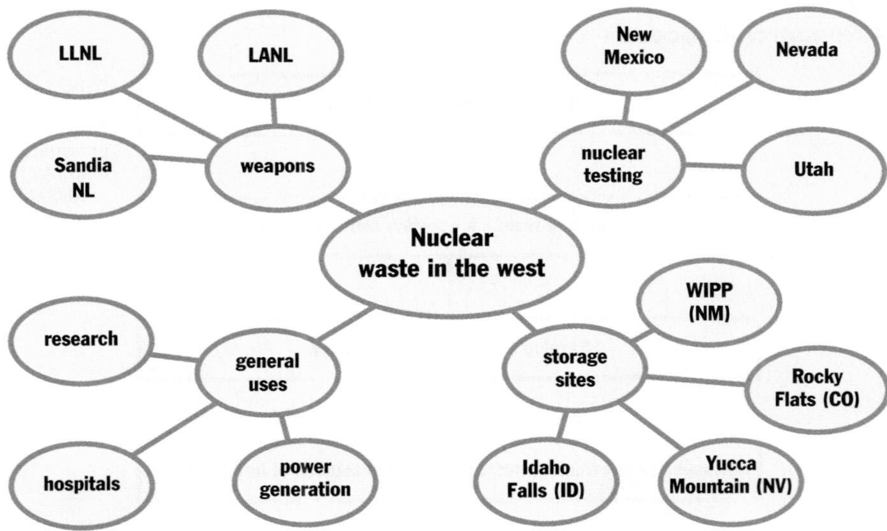

Figure 5.3: A logical map can help you generate ideas about your subject. It can also show you where you need to do research.

Mapping is widely used in technical disciplines, and it is gaining popularity in highly scientific and technical research (see the "At Work" box in this chapter). You might find it strange to begin your research by drawing circles and lines, but map-

GO TO THE NET

Want to learn more about being creative? Go to **www.ablongman.com/johnsonweb/5.2**

AT WORK

Greg Wilson

RESEARCHER, SYSTEMS ETHNOGRAPHY AND QUALITATIVE MODELING TEAM AT LOS ALAMOS NATIONAL LABORATORY

Los Alamos National Laboratory is a U.S. Government research center in Los Alamos, New Mexico.

Do researchers in the technical workplace *really* use logical mapping?

Mapping is a regular part of our research at Los Alamos National Laboratory (LANL). The mission of LANL is to address big science problems related to the U.S. nuclear stockpile, terrorism and weapons of mass destruction, defense, energy, environment, and infrastructure. These types of problems involve complex technical systems, political and social concerns, and interdisciplinary teams of experts.

We use mapping methods to create representations that help the teams of experts understand the important elements of a problem and how they relate to each other. I do textual research, interviewing, and ethnographies to identify the ways that different experts understand the problem. Then, I create maps so the experts can understand each other's perspectives and communicate about how to solve the problem. For example, if you have a biologist, epidemiologist, physician, emergency response planner, police chief, and political decision maker all trying to plan how a major city should respond in the event of a bioterrorist attack, each of them understands the problem in a different way. They talk about the problem in different ways, making efficient communication and problem solving difficult.

Using mapping as a centerpiece of research, I work with teams like this to build a common graphical representation of the problem. This representation also serves as a framework for identifying what data and information exist at each node in the graph, so that all relevant information is available to the team. Usually, a simple picture is the best way to begin to understand a complex problem.

ping will reveal relationships that you would not otherwise discover. It is a great way to tap into your creativity through your visual abilities.

Narrowing Your Research Subject

After defining your subject, you also look for ways to narrow and focus your research. Often, when people start the research process, they begin with a very broad subject (e.g., nuclear waste, raptors, lung cancer). Your logical map and a brief search on the Internet will soon show you that these kinds of subjects are too large for you to handle in the time available.

To help narrow your subject, you need to choose an *angle* on the subject. An angle is a specific direction that your research will follow. For example, nuclear waste may be too large a subject, but "the hazards of transporting nuclear waste in the Western United States" might be a good angle for your research. Research on raptors is probably too large a subject, but "the restoration of bald eagles along the Mississippi River" might be a manageable project.

Want to see other workplaces that use mapping? Go to **www.ablongman.com/johnsonweb/5.3**

General subject (too broad)	Angled research area (narrowed)
Nuclear Waste	Transportation of Nuclear Waste in Western States
Eagles	Bald Eagles on the Mississippi
Lung Cancer	Effects of Secondhand Smoke
Water Usage	Water Usage on the TTU Campus
Violence	Domestic Abuse in Rural Areas

By choosing an angle, you will help yourself narrow your research subject into a manageable size.

Formulating a Research Question or Hypothesis

Once you have narrowed your subject, you should then formulate a *research question* or *hypothesis.*

Research question or hypothesis: A question or statement that is tentatively forwarded to guide empirical or analytical research.

A research question or hypothesis is a question or statement that is tentatively forwarded to guide empirical or analytical research. Your research question or hypothesis does not need to be very specific when you begin your research. It simply needs to give your research a direction to follow.

Try to devise a research question that is as specific as possible:

Why do crows like to gather on our campus during the winter?

What are the effects of violent television on boys between the ages of 10 and 16?

Is solar power a viable energy source for South Dakota?

Your hypothesis is your best guess about an answer to the research question:

The campus is the best source of available food in the wintertime, because students leave food around. Crows naturally congregate because of the food.

Boys between the ages of 10 and 16 model what they see on violent television, causing them to be more violent than boys who do not watch violent television.

Solar power is a viable energy source in summer, but cloudiness in the winter makes it less economical than other forms of renewable energy.

As you move forward with your research, you will probably need to refine or sharpen your original research question or hypothesis. For now, though, ask the question that you would most like to answer. Then, to form your hypothesis, answer this question to the best of your knowledge. Your hypothesis should be your best guess for the moment.

Having trouble refining your hypothesis? Go to **www.ablongman.com/johnsonweb/5.4**

Developing a Research Methodology

With a research question or hypothesis formed, you are ready to start developing your *research methodology.* A methodology is a plan that describes how you are going to collect information, answer your research question, and test your hypothesis.

A research methodology is a step-by-step procedure that you will use to study the subject. As you and your research team consider how to study your subject, begin thinking about all the different ways you can collect information.

A research methodology is a step-by-step procedure that you will use to study the subject.

Mapping Out a Methodology

Logical mapping can help. Put the purpose of your research in the middle of your screen or a piece of paper. Ask "*How* are we going to achieve this purpose?" Then, answer this question by formulating the two to five major steps you will need to take in your research. Each of these major steps can then be broken down into minor steps (Figure 5.4).

Mapping Out a Methodology

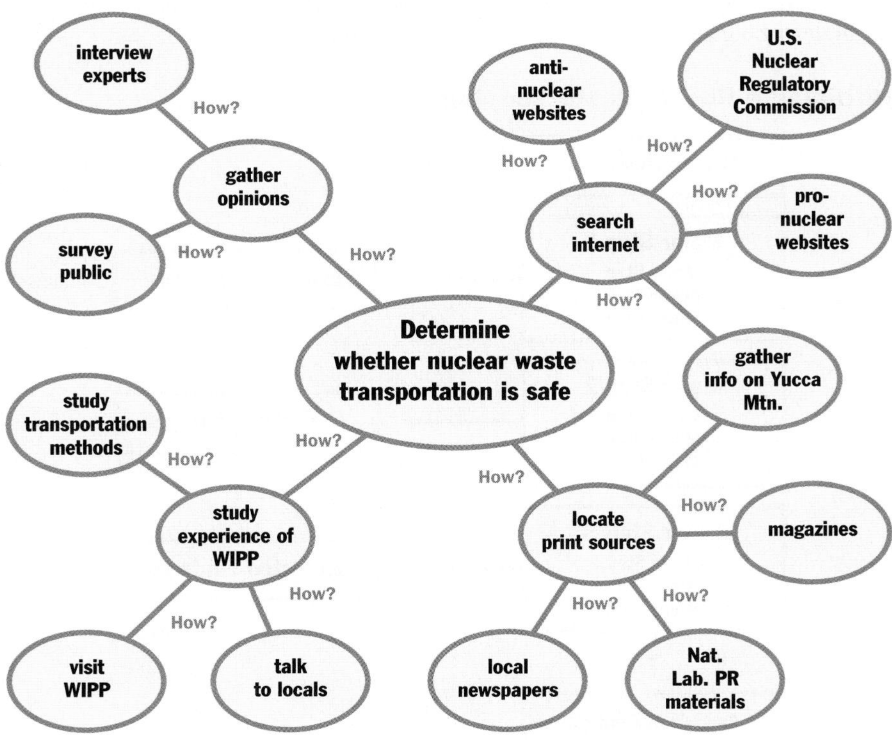

Figure 5.4: Logical mapping can help you rough out a methodology. Keep asking the How question as you consider the steps needed to complete your project.

Using the map in Figure 5.4, for example, a team of researchers might devise the following methodology for studying their research question:

Methodology for Researching Nuclear Waste Transportation:

- Collect information off the Internet from sources for and against nuclear waste storage and transportation.
- Track down news stories in the print media and collect any journal articles available on nuclear waste transportation.
- Interview experts and survey members of the general public.
- Study the Waste Isolation Pilot Plant (WIPP) in New Mexico to see if transportation to the site has been a problem.

Note that these researchers are drawing information from a variety of sources. They are planning to collect information from a range of electronic, print, and empirical sources.

Describing Your Methodology

You can keep mapping indefinitely. Eventually, though, you will notice that you have probably developed your methodology in sufficient depth. At this point, you can begin describing your methodology in outline form (Figure 5.5).

Outlining a Research Methodology

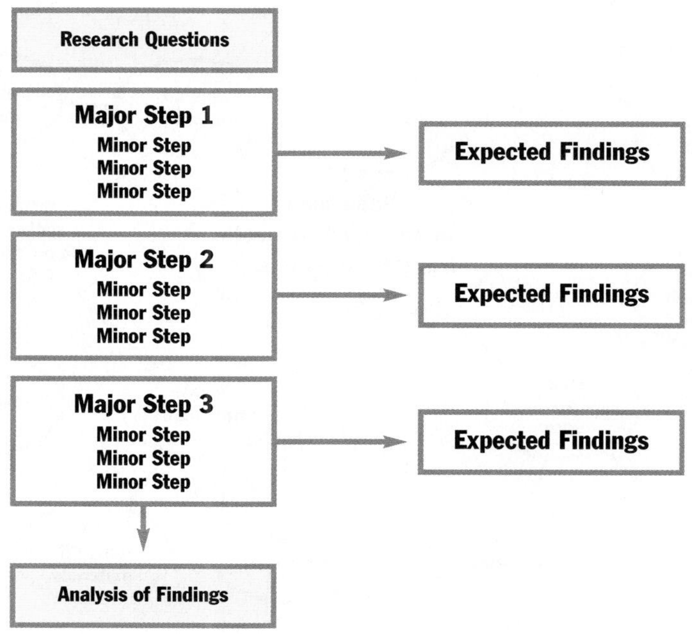

Figure 5.5: The major and minor steps in the research methodology should result in specific kinds of findings. At the end of the methodology, leave time for assessing your findings.

Need help developing a methodology? Go to **www.ablongman.com/johnsonweb/5.5**

Sometimes, as shown in Figure 5.5, it is also helpful to identify the kinds of information you expect to find in each step. By clearly stating your *expected findings* before you start collecting information, you will know if your research methodology is working the way you expected.

At the end of your methodology, add a step called "Analysis of Findings." If you collected numbers, you will need to do some statistical analysis. If you conducted interviews or tracked down information on the Internet, you will need to spend some time checking and verifying your sources.

Using and Revising Your Methodology

A good methodology is like a treasure map. You and your research team can use it as a guide to uncover answers to questions that intrigue you.

Almost certainly, you will deviate from your methodology while doing your research. Sometimes you will find unexpected information that takes you down an unexpected path. Sometimes information you expected to find is not available. In some cases, experiments and surveys provide unexpected results.

When you deviate from your methodology, note these changes in direction. A change in the methodology is not a failure. It is simply a recognition that research is not formulaic and can be unpredictable. Research is a process of discovery. Sometimes the most important discoveries are made when we deviate from the plan.

Triangulating Materials

To ensure that your methodology is reliable, you should draw information from a variety of sources. Specifically, you should always try to *triangulate* your materials by collecting information from electronic, print, and empirical sources (Figure 5.6). Triangulation allows you to compare and contrast sources, helping you determine which information is reliable and which information is not. Plus, triangulation gives your readers confidence in your research, because you will have collected information from a variety of sources.

The Research Triangle

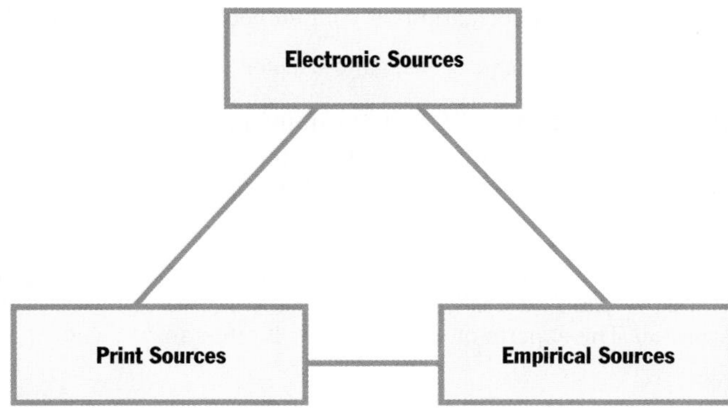

Figure 5.6: In any research project, try to draw information from electronic, print, and empirical sources.

GO TO THE NET

Tools for helping you do research electronically are available at **www.ablongman/johnsonweb/5.6**

Solid research draws from three kinds of information:

- **Electronic Sources:** Websites, CD-ROMs, listservs, television and radio, videos, blogs
- **Print Sources:** Books, journals, magazines and newspapers, government publications, reference materials, microform/microfiche
- **Empirical Sources:** experiments, surveys, interviews, field observations, ethnographies, case studies

By drawing information from all three kinds of sources, you will be able to verify the facts you find:

- If you find similar facts in all three kinds of sources, you can be reasonably confident that the information is reliable.
- If you find the information in two of three kinds of sources, the information is probably still reliable, though you should be less confident.
- If, however, you find the information in only one kind of source, it might not be reliable and needs further confirmation.

Something to remember is that "truth" or "facts" are more slippery than we want to admit. A source may claim it is providing you the truth, but until you can confirm that source's facts through triangulation, you should always treat any claims skeptically.

Also, keep in mind that there are always two sides to every issue. So, avoid restricting your research to only one side. If you look only for sources that confirm what you already believe, you will probably not gain a deeper understanding of the subject. After all, even when you absolutely disagree with someone else, his or her argument may give you additional insight into the issue you are researching. Keep an open mind.

Using Electronic Sources

Because electronic sources are so convenient, a good place to start collecting information is through your computer.

Websites—Websites are accessible through browsers like Mozilla, Netscape, Explorer, or Safari. Using search engines like MetaCrawler.com, Altavista.com, Google.com, Yahoo.com, and Lycos.com, among many others, you can run keyword searches to find information on your subject (Figures 5.7 and 5.8).

LINK To improve your use of search engines, go to Chapter 3, page 52.

CD-ROM—A compact disc (CD-ROM) can hold a library worth of text and images. Often available at your local or campus library, CD-ROMs are usually searchable through keywords or subjects. Encyclopedias and databases are available on CD-ROM.

Listservs—Listservs are ongoing e-mail discussions, usually among specialists in a field. Once you find a listserv on your subject, you can usually subscribe to the discussion. The experts on the listserv will often answer your questions.

For links to Internet search engines, go to **www.ablongman.com/johnsonweb/5.7**
For links to solid electronic sources of information, go to **www.ablongman.com/johnsonweb/5.8**

LINK To help find a listserv on your subject, go to Chapter 12, page 344.

Television and radio—You can often locate television and radio documentaries or news programs that address your subject. In some cases, copies of these materials will be available at your library. Or, increasingly, they are available for download off a television or radio station's website.

Videos—Increasingly, documentaries and training videos are available on digital videodiscs (DVDs) or videotapes (VHS). Your library or even video rental stores may have these kinds of materials available.

Blogs—Blogs are Internet sites where a commentator or group of commentators often "publish" raw information, opinions, and hearsay. Blogs can be a good source for keeping up with the cutting edge of a research area.

TAKE NOTE Electronic information, especially materials found on the Internet, can be short lived. So, if you find something useful, take careful notes. Make sure you write down the place where you found the information and the date on which you found it.

Internet Search Engine

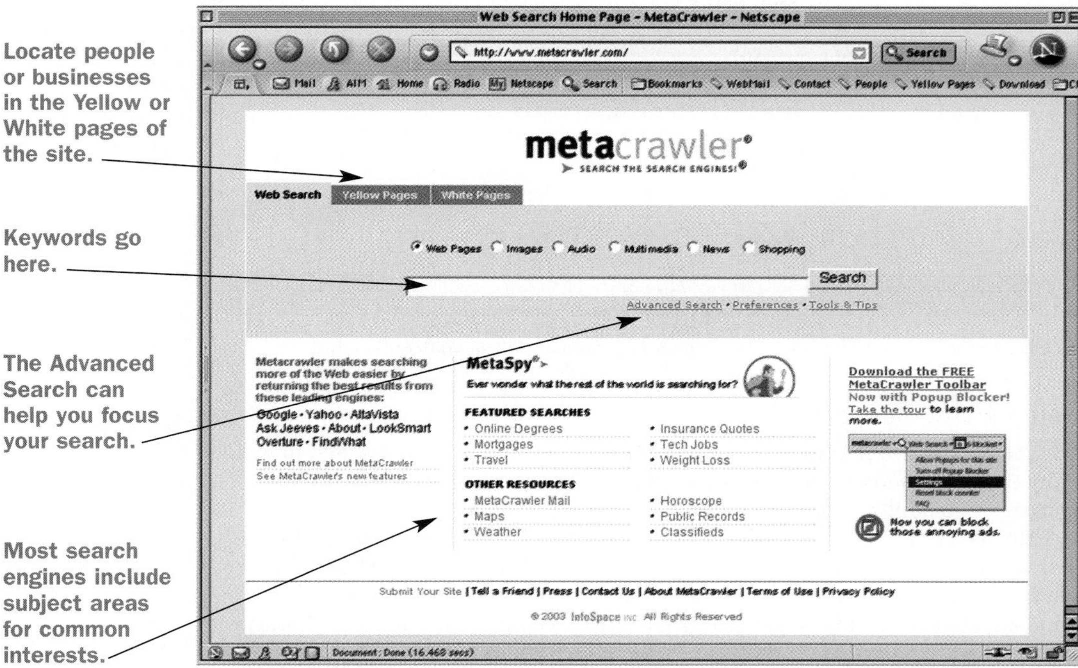

Figure 5.7: MetaCrawler is one of the more useful Internet search engines. By typing in some keywords, you can locate limitless amounts of information.

Source: Metacrawler, http://www.metacrawler.com.

Want to know more about blogs? Go to **www.ablongman.com/johnsonweb/5.9**

Advanced Search

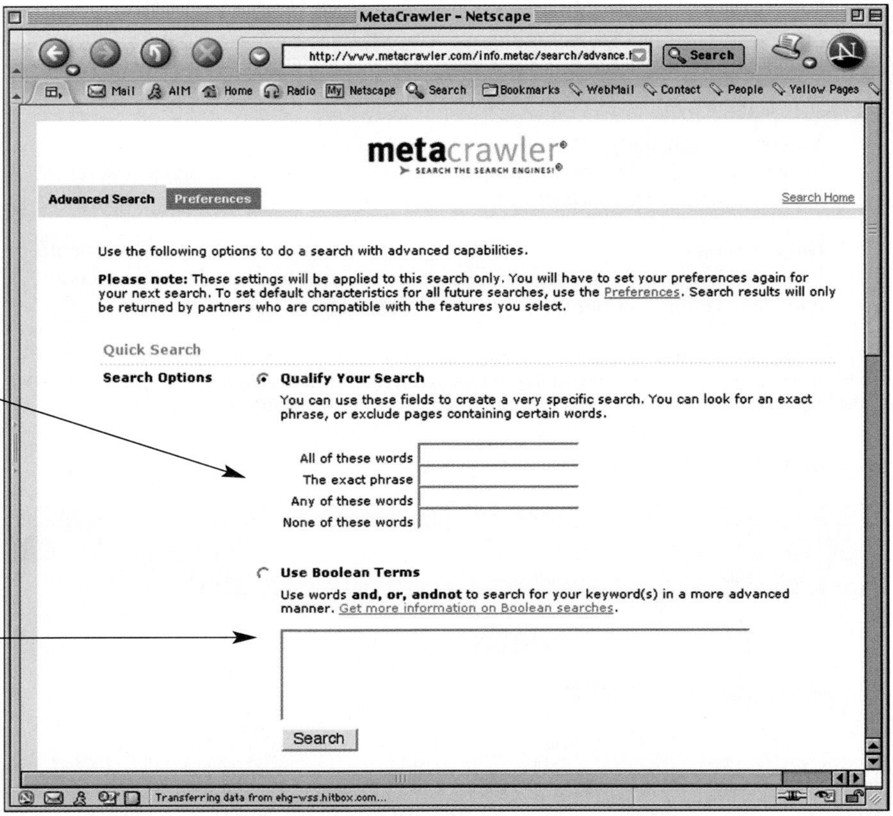

MetaCrawler - Netscape

http://www.metacrawler.com/info.metac/search/advance.h

metacrawler®
SEARCH THE SEARCH ENGINES!®

Advanced Search | Preferences | Search Home

Use the following options to do a search with advanced capabilities.

Please note: These settings will be applied to this search only. You will have to set your preferences again for your next search. To set default characteristics for all future searches, use the Preferences. Search results will only be returned by partners who are compatible with the features you select.

Quick Search

Search Options **Qualify Your Search**
You can use these fields to create a very specific search. You can look for an exact phrase, or exclude pages containing certain words.

All of these words
The exact phrase
Any of these words
None of these words

Use Boolean Terms
Use words **and, or, andnot** to search for your keyword(s) in a more advanced manner. Get more information on Boolean searches.

Search

Transferring data from ehg-wss.hitbox.com...

Here, you can be more exact about the terms you want searched.

Boolean terms (and, or, not) can be used to sharpen the search considerably.

Figure 5.8: The Advanced Search area of an Internet search engine can help you refine your search to locate the exact information you are looking for.

Source: Metacrawler, http://www.metacrawler.com/info.metac/search/advance.html.

Using Print Sources

With easy access to information through the Internet, you may be tempted to forego using the print sources available at your library. Researchers who neglect print sources are making a serious mistake.

Printed documents are still the most abundant and reliable sources of information. In the rush to use electronic sources, many people have forgotten that their nearby library is loaded with books and periodicals on almost any subject. These print sources can usually be located by using your computer to access the library's website (Figure 5.9).

TAKE NOTE Print sources of information are usually considered more reliable than electronic sources, but that is not always true. Authors and publishers of books, magazines, and newspapers can still be biased, inaccurate, or even deceptive. Moreover, print materials can become dated or obsolete. In all, though, the print sources you find at your

For access to online news and legal and business information, go to **www.ablongman.com/johnsonweb/5.10**

A Library's Search Engine

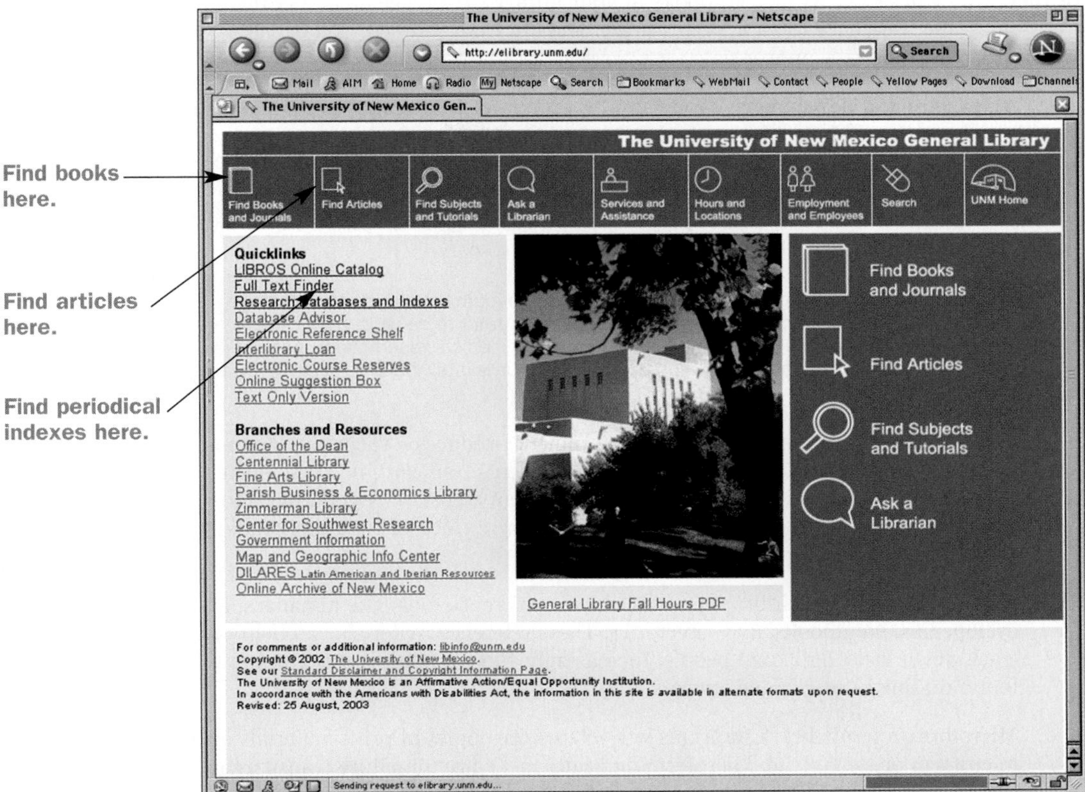

Figure 5.9: Your library likely has a website for finding print sources. This is the University of New Mexico's page for finding a variety of different kinds of materials. Your campus has trained librarians who are there to help you. Don't be afraid to ask.

Source: The University of New Mexico, http://elibrary.unm.edu.

library are usually reliable, because they have been chosen and examined by professional librarians.

Here are a few of the many kinds of print materials that you can use:

Books—Almost all libraries have electronic cataloging systems that allow you to use author name, subject title, and keywords to search for books on your subject. Once you have located a book on your subject, it is a good idea to look at the books shelved around it to find other useful materials.

Journals—Using a *periodical index,* you can search for journal articles on your subject. Journal articles are usually written by professors and scientists in a research field, so the articles can be rather detailed and hard to understand. Nevertheless, these articles offer some of the most exact research on any subject.

For *The Catalog of U.S. Government Publications,* go to **www.ablongman.com/johnsonweb/5.11**

Periodical indexes for journals are usually available on-line at your library's website, or they will be available as printed books in your library's reference area. Ask the librarian at the Reference Desk in your library. The librarian can help you decide which periodical index is appropriate for your search.

Magazines and newspapers—You can also search for magazine and newspaper articles on your subject by using the *Readers' Guide to Periodical Literature* or a newspaper index. The *Readers' Guide* and newspaper indexes are likely available on-line at your library or in print form. Recent editions of magazines or newspapers might be stored at your library. Older magazines and newspapers have usually been stored on microform or microfiche.

TAKE NOTE An important difference between articles in journals and articles in magazines and newspapers is who wrote them. Journal articles tend to be written by subject matter experts (SMEs) in the field (scientists, professors, medical researchers, engineers). Magazine and newspaper articles tend to be written by journalists who are reporting on the activities of these experts.

Government publications—The U.S. government produces a surprising amount of useful books, reports, maps, and other documents. You can find these documents through your library or through government websites. A good place to start is *The Catalog of U.S. Government Publications,* which offers a searchable listing of government publications and reports.

Reference materials—Libraries contain many reference tools like almanacs, encyclopedias, handbooks, and directories. These reference materials can help you track down facts, data, and people. Increasingly, these materials can also be found on-line in searchable formats.

Microform/microfiche—Librarians will often store copies of print materials on microform or microfiche. Microform and microfiche are miniature transparencies that can be read on projectors available at your library.

You will usually find that magazines and newspapers over a year old have been transferred to microform or microfiche to save space in the library. Also, delicate and older texts are available in this format to reduce the handling of the original documents.

Of course, finding print sources can be more work than finding electronic sources—at least, it requires a few more steps. However, the search for these materials is not difficult, because library websites offer useful search engines. The payoff is worth the effort, because print sources can be more reliable and stable than electronic sources.

For websites that discuss empirical research methods, go to **www.ablongman.com/johnsonweb/5.12**

Using Empirical Sources

You should also generate your own data and observations to support your research. Empirical studies can be *quantitative* or *qualitative*, depending on the kinds of information you are looking for. Quantitative research allows you to generate data that you can analyze statistically to find trends. Qualitative research allows you to observe patterns of behavior that cannot be readily boiled down into numbers.

Experiments—Each research field has its own experimental procedures. A controlled experiment allows you to test a hypothesis by generating data. From that data, you can confirm or dispute the hypothesis. Experiments should be *repeatable,* meaning their results can be replicated by another experimenter.

Surveys—You can ask a set of people to answer questions about your subject. Their answers can then be scored and analyzed for trends. Survey questions can be *closed-ended* or *open-ended.* Closed-ended questions ask respondents to choose among preselected answers. Open-ended questions allow them to write down their views in their own words. Figure 5.10 shows pages from a survey with both closed-ended and open-ended questions.

Interviews—You can ask experts to answer questions about your subject. On almost any given college campus, there are experts available on just about any subject. Your well-crafted questions can draw out very useful information and quotes.

Field observations—Researchers often carry field notebooks to record their observations of their research subjects (Figure 5.11). For example, an ornithologist might regularly note the birds she observes in her hikes around a lake. Her notebook would include her descriptions of birds and their activities.

Ethnographies—An ethnography is a systematic recording of your observations of a defined group or culture. Anthropologists use ethnographies to identify social or cultural trends and norms.

Case studies—Case studies typically offer in-depth observations of specific people or situations. For example, a case study might describe how a patient reacted to a new treatment regimen that manages diabetes.

When conducting empirical research, you should follow the *scientific method.* The concept of a scientific method was first conceived by Francis Bacon, a 17th-century English philosopher. Later in the 17th century, the London Royal Society, a club of scientists, gave the scientific method the form we recognize now.

For more information on conducting qualitative empirical studies, go to **www.ablongman.com/johnsonweb/5.13**

Pages from a Survey on Campus Safety

Introduction explains how to complete the survey.

Figure 5.10: A survey is a good way to generate data for your research. In this example, both closed-ended and open-ended questions are being used to solicit information.

Campus Survey 75

Campus Perception Survey

The following questions are about how safe you feel or don't feel on campus. For each situation please tell us if you feel: very safe, reasonably safe, neither safe nor unsafe, somewhat unsafe, very unsafe, or if this situation does not apply to you. (Please circle the number that best represents your answer or **NA** if the situation does not apply to you.)

How safe do you feel...

	Very Unsafe	Somewhat Unsafe	Neither Safe Nor Unsafe	Reasonable Safe	Very Safe	
walking alone on campus during daylight hours?	1	2	3	4	5	NA
waiting alone on campus for public transportation during daylight hours?	1	2	3	4	5	NA
walking alone in parking lots or garages on campus during daylight hours?	1	2	3	4	5	NA
walking alone on campus after dark?	1	2	3	4	5	NA
waiting alone on campus for public transportation after dark?	1	2	3	4	5	NA
walking alone in parking lots or garages on campus after dark?	1	2	3	4	5	NA
working in the library stacks late at night?	1	2	3	4	5	NA
while alone in classrooms?	1	2	3	4	5	NA
Student Activity Center during the day?	1	2	3	4	5	NA
Student Activity Center at night?	1	2	3	4	5	NA

Are there any specific areas on campus where you do not feel safe? Please specify which areas, which campus, and when; for example, evenings only or any time. ____________________

____

Do you have any special needs related to safety on campus? ____________________

These closed-ended questions yield numerical data.

Open-ended questions give participants an opportunity to elaborate on their answers.

Have you ever used services related to safety issues, sexual harassment, or sexual assault that are provided on campus by the following?

How satisfied were you with help from this source?

			Very Dissatisfied	Somewhat Dissatisfied	Neither Satisfied Nor Dissatisfied	Somewhat Satisfied	Very Satisfied
Campus Police	YES	NO	1	2	3	4	5
Women's Center	YES	NO	1	2	3	4	5
Campus ministry	YES	NO	1	2	3	4	5
Campus counseling (Belknap)	YES	NO	1	2	3	4	5
Student Health Services (Belknap)	YES	NO	1	2	3	4	5
Campus counseling (Health Science)	YES	NO	1	2	3	4	5
Student Health Services (Health Science)	YES	NO	1	2	3	4	5
Psychological Services Ctr (Psychology Clinic)	YES	NO	1	2	3	4	5
Affirmative Action Office	YES	NO	1	2	3	4	5
Security escort services after dark	YES	NO	1	2	3	4	5
Residence Hall staff	YES	NO	1	2	3	4	5
Office of Student Life	YES	NO	1	2	3	4	5
Disability Resource Center	YES	NO	1	2	3	4	5
Access Center	YES	NO	1	2	3	4	5
Faculty member	YES	NO	1	2	3	4	5
Other ____________ (Please specify.)	YES	NO	1	2	3	4	5

Source: Bledsoe & Sar, 2001.

Figure 5.10: *(continued.)*

The survey uses statements to measure the participant's reaction to specific situations or opinions.

Campus Survey 78

The following are some beliefs that may be held about the role of women and men in today's society. There are no right or wrong answers. (Please circle the response that best describes your opinion.)

	Strongly Disagree		Somewhat Agree		Strongly Agree
A man's got to show the woman who's boss right from the start.	1	2	3	4	5
Women are usually sweet until they've caught a man, but then they let their true self show.	1	2	3	4	5
In a dating relationship a woman is largely out to take advantage of a man.	1	2	3	4	5
Men are out for only one thing.	1	2	3	4	5
A lot of women seem to get pleasure from putting a man down.	1	2	3	4	5
A woman who goes to the home or apartment of a man on their first date implies that she is willing to have sex.	1	2	3	4	5
Any female can get raped.	1	2	3	4	5
Any healthy woman can successfully resist a rapist if she really wants to.	1	2	3	4	5
Many women have an unconscious wish to be raped, and may then unconsciously set up a situation in which they are likely to be attacked.	1	2	3	4	5
If a woman gets drunk at a party and has intercourse with a man she's just met there, she should be considered "fair game" to other males at the party who also want to have sex with her whether she wants to or not.	1	2	3	4	5

What **percentage of women** who report a rape would you say are lying because they are angry and want to get back at the man they accuse? ______%

What **percentage of reported rapes** would you guess were merely invented by women who discovered they were pregnant and wanted to protect their own reputation? ______%

Did you attend any type of student orientation conducted by University of Louisville during Summer 2000 or at the beginning of this term? (Please circle your answer). **YESNO**

If your answer was "No, I did not attend orientation", please continue on next page.

	None		Some		A Lot
At the orientation you attended, how much information about violence against women issues did you receive? (Please circle your answer).	1	2	3	4	5

Figure 5.11: Empirical research requires you to observe your subject directly.

The Scientific Method:

1. Observe and describe a phenomenon.
2. Formulate a hypothesis or theory that explains the phenomenon.
3. Use the hypothesis or theory to make predictions.
4. Use observations and experiments to generate results that confirm or deny your predictions.
5. Modify the hypothesis or theory to account for your results.
6. Repeat steps 2 through 5 until results match your hypothesis or theory OR you abandon the hypothesis or theory.

The scientific method can be used with quantitative or qualitative forms of empirical research. Whether you are doing an experiment in a laboratory or making field observations, following the scientific method will help you focus and streamline your research. It should help you produce the kinds of results that will provide a solid empirical foundation for your work.

GO TO THE NET

To learn more about the scientific method, go to **www.ablongman.com/johnsonweb/5.14**

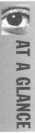

Triangulating Research

Solid research draws from three kinds of information:

- Electronic Sources—Internet, CD-ROMs, listservs, television and radio, videos, blogs
- Print Sources—books, journals, magazines and newspapers, government publications, reference materials, microform/microfiche
- Empirical Sources—experiments, surveys, interviews, field observations, ethnographies, case studies

Managing Information and Taking Notes

On almost any subject, you are going to find a wealth of information. At this point, you need to start thinking like an information manager. After all, only some of that information you collected will be important to your readers. Most of the information you find will not be needed by your readers to take action or make a decision (Figure 5.12).

Managing Information

As you decide what to include in the document you are writing, you need to distinguish between *need-to-know* information and *want-to-tell* information.

Need-to-know information: Material that the readers require to take action or make a decision.

Want-to-tell information: Material you would like to tell your readers but that is not necessary to take action or make a decision.

Need-to-know information includes material that your readers require to take action or make a decision. Want-to-tell information includes material that you would like to tell your readers, but that is not necessary for them to take action or make a decision.

After you have gone through all the effort to collect information, you will want to tell the readers about everything you found. But your readers don't need (or want) all that information. They only want information they need to take action or make an informed decision. Any extra want-to-tell information will just cloud their ability to understand your document.

TAKE NOTE You may find it discouraging that your final document uses only a quarter or even one-tenth of the information that you collected. Nevertheless, you need to put your readers' interests first. They will always prefer concise quality over bloated quantity.

Careful Note Taking

Reliable note taking is essential when you do research. If you are organized when you take notes, you will find the information you collected easy to use in the document you are writing.

Note-organizing software and database programs can help you keep track of the information you find. Many researchers write their notes exclusively on a laptop or their personal digital assistant (PDA). A pen and pad of paper is still a good way to keep track of information.

For some great ideas on note taking, go to **www.ablongman.com/johnsonweb/5.15**

Need-to-Know Versus Want-to-Tell Information

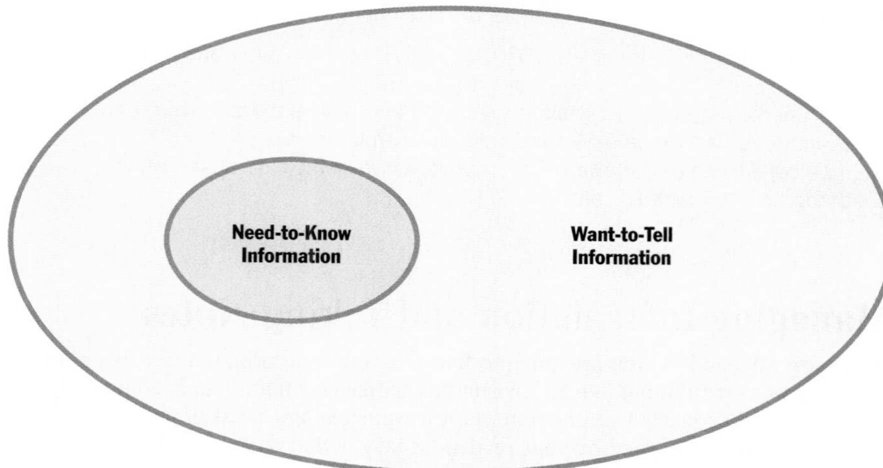

Figure 5.12: While researching, you will find much more information than your readers need. Give them only the information they need to make a decision.

TAKE NOTE Entering notes directly into a computer is a good idea. You can then use the "Find" function in your word processor to locate specific terms or information in your notes. Moreover, computers allow you to cut and paste quotes quickly. Handwritten notes are often hard to interpret later.

What is most important, though, is to have a workable system for taking notes. Here are some note-taking strategies you might consider using:

- Record each source separately.
- Take quotations.
- Paraphrase ideas.
- Summarize sources.
- Write commentary.

RECORD EACH SOURCE SEPARATELY Make sure you clearly identify the author, title of the work, and the place where you found the information (Figure 5.13). For information off the Internet, write down the webpage address (URL) and the date and time you found the information. For a print document, write down where the information was published and who published it. Also, record the library number of the document.

For large research projects, you might consider making a separate word-processing file for each of your authors or sources, like the one shown in Figure 5.13. That way, you can more easily keep your notes organized.

TAKE QUOTATIONS When an author makes an especially interesting statement or claim, you may want to take a direct quotation from the text. When taking a quotation, you need to ensure that you copied the exact wording of the author. If you are taking a quote from a website, you might avoid errors by using the "Copy" and

GO TO THE NET For links to note-taking software, go to **www.ablongman.com/johnsonweb/5.16**

Keeping Notes on Your Computer

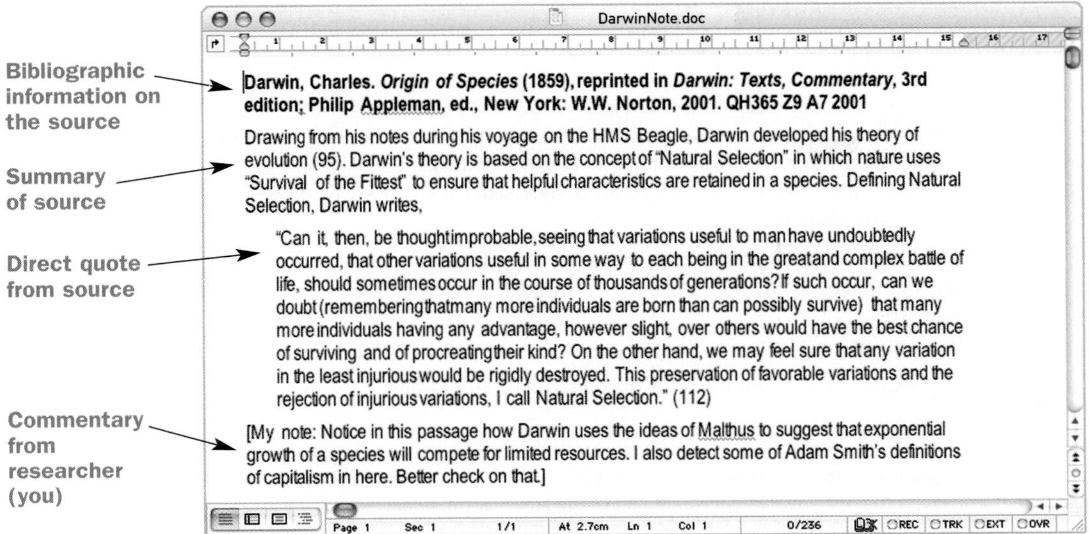

Figure 5.13: Most notes include a combination of summaries, paraphrases, direct quotes, and personal comments.

"Paste" functions of your computer to copy the statement directly from your source into your notes.

In your notes, you should put quotation marks around any material you copied word for word from a source.

> According to Louis Pakiser and Kaye Shedlock, scientists for the Earthquake Hazards Program at the U.S. Geological Survey, "the assumption of random occurrence with time may not be true" (1997, paragraph 3).

If the quoted material runs more than three lines in your text, you should set off the material by indenting it in the text.

> Louis Pakiser and Kaye Shedlock, scientists for the Earthquake Hazards Program at the U.S. Geological Survey, make the following point:
>
>> When plate movements build the strain in rocks to a critical level, like pulling a rubber band too tight, the rocks will suddenly break and slip to a new position. Scientists measure how much strain accumulates along a fault segment each year, how much time has passed since the last earthquake along the segment, and how much strain was released in the last earthquake. (1997, paragraph 4)
>
> If we apply this rubber band analogy to the earthquake risk here in California . . .

When you are quoting materials, you also need to include an in-text citation at the end of the quote. In these two examples, the in-text citation is the information in the parentheses.

LINK For more information on citing sources, go to Appendix C, page A-24.

Overall, you should use direct quotes sparingly in your technical writing. You might be tempted to use several quotes from a source, because the authors "said it right." If you use too many quotes, though, your writing will sound fragmented and patchy, because the quotes disrupt the flow of your text.

PARAPHRASE IDEAS One way to better incorporate someone else's ideas into your writing is to paraphrase them. When paraphrasing, you are presenting another person's ideas in your own words. You still need to give the original author credit for the ideas, but you do not need to use quotation marks around the text. To paraphrase something, you should:

- reorganize the information to highlight important points.
- use plain language, replacing jargon and technical terms with simpler words.
- include an in-text citation.

In the following example, a quote from an original document is paraphrased:

Original Quote

"But in many places, the assumption of random occurrence with time may not be true, because when strain is released along one part of the fault system, it may actually increase on another part. Four magnitude 6.8 or larger earthquakes and many magnitude 6–6.5 shocks occurred in the San Francisco Bay region during the 75 years between 1836 and 1911. For the next 68 years (until 1979), no earthquakes of magnitude 6 or larger occurred in the region. Beginning with a magnitude 6.0 shock in 1979, the earthquake activity in the region increased dramatically; between 1979 and 1989, there were four magnitude 6 or greater earthquakes, including the magnitude 7.1 Loma Prieta earthquake. This clustering of earthquakes leads scientists to estimate that the probability of a magnitude 6.8 or larger earthquake occurring during the next 30 years in the San Francisco Bay region is about 67 percent (twice as likely as not)."

In-text citation

Effective Paraphrase

Simpler language is used.

Pakiser and Shedlock (1997) report that large earthquakes are mostly predictable, because an earthquake in one place usually increases the likelihood of an earthquake somewhere nearby. They point out that the San Francisco area—known for earthquakes—has experienced long periods of minor earthquake activity (most notably from 1836 to 1911, when no earthquakes over magnitude 6 occurred). At other times in San Francisco, major earthquakes have arrived with more frequency, because large earthquakes tend to trigger other large earthquakes in the area.

Some of the more technical details have been removed to enhance understanding.

Improper Paraphrase

Much of the original wording is retained.

Pakiser and Shedlock (1997) report the assumption of random occurrence of earthquakes may not be accurate. Earthquakes along one part of a fault system may increase the frequency of earthquakes in another part. For example, the San Francisco Bay region experienced many large earthquakes between 1836 and 1911. For the next six decades until 1979, only smaller earthquakes (below magnitude 6) occurred in the area. Then, there

Language is still overly technical for the readers.

was a large rise in earthquakes between 1979 and 1989. Scientists estimate that the probability of an earthquake of a magnitude 6.8 or larger is 67 percent in the next 30 years in the Bay area.

The "effective" paraphrase shown here uses the ideas of the original quote, while reordering information to highlight important points and simplifying the language. The "improper" paraphrase above duplicates too much of the wording from the original source and does not effectively reorder information to highlight important points. In fact, this improper paraphrase is so close to the original, it could be considered plagiarism.

LINK For more information on plagiarism, see page 127 in this chapter.

In many ways, paraphrasing is superior to using direct quotes. A paraphrase allows you to simplify the language of a technical document, making the information easier to understand for the readers. Also, you can better blend the paraphrased information into your writing, because you are using your writing style, not the style of the source.

As a warning, when taking notes, make sure you are paraphrasing sources properly. Do not use the author's original words and phrases. Otherwise, when you are drafting your document, you may forget that you copied some of the wording from the original text. These duplications may leave you vulnerable to charges of plagiarism or copyright violation.

SUMMARIZE SOURCES When summarizing, your goal is to condense the ideas from your source into a brief passage. Summaries usually strip out many of the examples, details, data, and reasoning from the original text, leaving only the essential information that readers need to know. Like a paraphrase, summaries should be written in your own words. When you are summarizing a source for your notes:

- Read the source carefully to gain an overall understanding.
- Highlight or underline the main point and other key points.
- Condense key points into lists, where appropriate.
- Organize information from most important to least important.
- Use plain language to replace any technical terms or jargon in the original.
- Use in-text citations to identify important ideas from the source.

To demonstrate summarizing, consider the passage about predicting earthquakes shown in Figure 5.14. When summarizing this text, you would first need to identify the main point and key points in the text. The main point is that scientists are increasingly able to estimate the probability of an earthquake in a specific area in the near future.

Now, locate the other key points in the text. There are three other key points in this text: (1) the frequency of earthquakes in the past helps scientists predict them in the future, (2) earthquakes are not random events, and they tend to occur in clusters, and (3) measurements of the strain on the earth can help scientists measure the probability of a future earthquake.

For more tips on paraphrasing and summarizing, go to **www.ablongman.com/johnsonweb/5.17**

Original Text to Be Summarized

Figure 5.14: The original text contains many details that can be condensed in a summary.

Predicting Earthquakes
by Louis Pakiser and Kaye M. Shedlock
Earthquake Hazards Program, U.S. Geological Survey

The goal of earthquake prediction is to give warning of potentially damaging earthquakes early enough to allow appropriate response to the disaster, enabling people to minimize loss of life and property. The U.S. Geological Survey conducts and supports research on the likelihood of future earthquakes. This research includes field, laboratory, and theoretical investigations of earthquake mechanisms and fault zones. A primary goal of earthquake research is to increase the reliability of earthquake probability estimates. Ultimately, scientists would like to be able to specify a high probability for a specific earthquake on a particular fault within a particular year. Scientists estimate earthquake probabilities in two ways: by studying the history of large earthquakes in a specific area and the rate at which strain accumulates in the rock.

Scientists study the past frequency of large earthquakes in order to determine the future likelihood of similar large shocks. For example, if a region has experienced four magnitude 7 or larger earthquakes during 200 years of recorded history, and if these shocks occurred randomly in time, then scientists would assign a 50 percent probability (that is, just as likely to happen as not to happen) to the occurrence of another magnitude 7 or larger quake in the region during the next 50 years.

But in many places, the assumption of random occurrence with time may not be true, because when strain is released along one part of the fault system, it may actually increase on another part. Four magnitude 6.8 or larger earthquakes and many magnitude 6–6.5 shocks occurred in the San Francisco Bay region during the 75 years between 1836 and 1911. For the next 68 years (until 1979), no

Here is the main point.

These are the key points.

Source: Pakiser and Shedlock, U.S. Geological Survey, http://pubs.usgs.gov/gip/earthq1/predict.html.

Figure 5.14: (*continued.*)

These are the key points.

earthquakes of magnitude 6 or larger occurred in the region. Beginning with a magnitude 6.0 shock in 1979, the earthquake activity in the region increased dramatically; between 1979 and 1989, there were four magnitude 6 or greater earthquakes, including the magnitude 7.1 Loma Prieta earthquake. This clustering of earthquakes leads scientists to estimate that the probability of a magnitude 6.8 or larger earthquake occurring during the next 30 years in the San Francisco Bay region is about 67 percent (twice as likely as not).

Another way to estimate the likelihood of future earthquakes is to study how fast strain accumulates. When plate movements build the strain in rocks to a critical level, like pulling a rubber band too tight, the rocks will suddenly break and slip to a new position. Scientists measure how much strain accumulates along a fault segment each year, how much time has passed since the last earthquake along the segment, and how much strain was released in the last earthquake. This information is then used to calculate the time required for the accumulating strain to build to the level that results in an earthquake. This simple model is complicated by the fact that such detailed information about faults is rare. In the United States, only the San Andreas fault system has adequate records for using this prediction method.

Many of these details should be removed in the summary.

In the summary shown in Figure 5.15, pay attention to the highlighting of the main point and the listing of the other key points. Here, the details in the original text have been stripped away, leaving only a condensed version. As shown here, the summary uses the writer's own words, not the words from the original source.

Summary of Original Text

Summary of "Predicting Earthquakes" by Louis Pakiser and Kaye Shedlock (1997). http://pubs.usgs.gov/gip/earthq1/predict.html. Retrieved March 10, 2004.

The main point is expressed up front. →

The goal of earthquake prediction is to anticipate earthquakes that may cause major damage in a region. According to geologists Louis Pakiser and Kaye Shedlock from the U.S. Geological Survey, scientists are increasingly able to predict the likelihood of an earthquake in a region in the near future (1997). Scientists use two important methods to predict large earthquakes:

- They study the frequency of earthquakes in the past, especially the recent past, because large earthquakes tend to trigger other large earthquakes.
- They measure the strain on the earth to determine the buildup of pressure along fault lines. These measurements can be used to predict when the strain will be released as an earthquake.

The important points are put into list form.

As Pakiser and Shedlock point out, earthquakes are not random events. They tend to occur in "clusters" over periods of several years. By paying attention to these clusters of earthquakes, scientists can make rather reliable predictions about the probability of a future large quake.

The summary uses plain language. →

Figure 5.15: A summary highlights important points and puts the text in plain language.

WRITE COMMENTARY In your notes, you might offer your own commentary to help interpret your sources. Your commentary might help you remember why you collected the information and how you thought it could be used. To avoid plagiarism, it is important to visually distinguish your commentary from summaries, paraphrases, and quotations drawn from other sources. You might put brackets around your comments or use italic or bold type to set them off from your other notes.

Documenting Sources

As you draft your text, you will need to *document* your sources. Documentation involves (1) naming each source with an *in-text citation* and (2) recording your sources in the *References* list at the end of the document. Documenting your sources offers a few advantages:

- Supports your claims by referring to the research of others.
- Helps build your credibility with readers by showing them the support for your ideas.

- Reinforces the thoroughness of your research methodology.
- Allows your readers to explore your sources for more information.

When should you document your sources? Any ideas, text, or images that you draw from another text need to be properly acknowledged. If you are in doubt about whether you need to cite someone else's work, you should go ahead and do it. Citing sources will help you avoid any questions about the integrity and soundness of your work.

In Appendix C at the end of this book, you will find a full discussion of three documentation systems (APA, CBE, and MLA) that are used in technical fields. Each of these systems works differently.

The most common documentation style for technical fields is offered by the American Psychological Association (APA). The APA style, published in the *Publication Manual of the American Psychological Association*, is preferred in technical fields because it puts emphasis on the year of publication. As an example, let us briefly look at the APA style for in-text citations and full references.

APA IN-TEXT CITATIONS In the APA style, in-text citations can include the author's name, publication year, and the page number where the information was found.

> One important study showed that physicians were regularly misusing antibiotics to treat viruses (Reynolds, 2003).
>
> According to Reynolds (2003), physicians are regularly misusing antibiotics to treat viruses.
>
> According to Reynolds, "doctors are creating larger problems by mistakenly treating viruses with antibiotics" (2003, p. 743).

These in-text citations are intended to refer the readers back to the list of full references at the end of the document.

APA FULL REFERENCES The full references at the end of the document provide readers with the complete citation for each source.

> Jaspers, F. (2001). *Einstein online.* Retrieved March 9, 2003, from http://www.einsteinonlinetoo.com.
>
> Pauling, L., & Wilson, E. B. (1935). *Introduction to quantum mechanics.* New York, NY: Dover Publications.

LINK For a full discussion of documentation, including models for documenting references, turn to Appendix C, page A-24.

As you take notes, you should keep track of the information needed to properly cite your sources. That way, when you draft the document and create a references list, you will have this important information available. It's very difficult to locate the sources of your information *after* you finish drafting the document.

Need help citing sources? Go to **www.ablongman.com/johnsonweb/5.18**

Avoiding Junk Science on the Internet

The Internet has been a boon for the "junk science" industry. Junk science is really not science at all. It is a public relations tool that corporations, lawyers, and special-interest groups use to confuse the public or cast doubt on the findings of legitimate scientists. Junk science is the selective use of data, scientific style, or scientific-like methods to advance a political or economic agenda.

The tobacco industry has been the most flagrant user of junk science. For decades, it employed scientists to confuse and cast doubt on findings that smoking causes cancer and other health problems. The tobacco industry's deceptions were only exposed when their own scientists, most notably chemist Jeffrey Wigand, turned over key documents to the government and media.

One of the most infamous uses of junk science was the response to biologist Rachel Carson's book *Silent Spring*. In the book, Carson exposed DDT, an insecticide, as a dangerous cause of a variety of illnesses. Designed to kill insects, DDT was widely used in the United States, killing many forms of wildlife, especially birds. Humans also became dangerously ill. Whole neighborhoods and wildlife areas would be sprayed with DDT, leaving a path of dead birds, sickened animals, and poisoned people.

Instead of immediately pulling DDT off the market, Carson's opponents used junk science tactics to attack her credibility and her book. John Stauber and Sheldon Rampton in their book, *Toxic Sludge Is Good for You,* show how public relations agents did everything possible to undermine Carson and her findings.

The attacks were vicious, and Carson's reputation was compromised. But Carson was right. DDT is a very dangerous chemical. Eventually, it was banned in the United States as the scientific evidence mounted against it. Unfortunately, the junk science attacks on her reputation did great damage and allowed DDT to stay on the market longer than it should have.

Junk science can be used on all sides of a debate to support various causes. You will find active use of junk science by chemical companies, extreme environmental groups, anti-environmental groups, the diet industry, the oil industry, the tobacco industry, and proponents and opponents of genetically modified foods.

TAKE NOTE Ironically, some special-interest groups will use the term "junk science" to attack the legitimate work of real scientists. For example, global warming research is often attacked as junk science, even though these scientists are highly reputable and their studies are broadly correlated.

How do you tell real science from junk when you are doing research on the Internet? It's actually very difficult, but here are some pointers:

Follow the money—If the scientists behind a study receive all of their funding from a corporation or group that benefits from their results, there is a good chance these scientists are being paid to generate specific results. Their results could be biased.

Check the reputations—Some "experts" on scientific issues are not really scientists at all. They are public relations consultants whose real job is to spin

Want to learn more about junk science? Go to **www.ablongman.com/johnsonweb/5.19**

science for the media, casting doubt on the work of reliable scientists. So, look into the education and work experience of these "experts" to determine whether they are scientists.

Check the source—Reliable science is typically published in *peer-reviewed journals.* If the only "scientific" evidence you find for an argument is from a company or special-interest group, it might be junk science. Look for impartial sources to confirm that information.

Check the science—Reliable science follows the scientific method. The results should be repeatable or publicly available. If the methodology looks flimsy or the data are not available, the results are probably unreliable.

If it sounds too extreme, it probably is—Special-interest groups like to scare people by exaggerating the results of scientific studies. Be skeptical of studies that seem to contradict common sense (e.g., global warming is good for the planet) or seem too far-fetched (e.g., AIDS was a CIA experiment gone awry).

Junk science is a serious problem on the Internet. As you are researching an issue, you need to be skeptical and crosscheck your sources. Always rely on a broad base of sources to confirm your information.

Appraising Your Information

All information is not created equal. In fact, some information is downright wrong or misleading. Keep in mind that even the most respected authorities usually have agendas that they are pursuing with their research. Even the most objective experiment will include some tinge of bias.

To avoid misleading information and researcher biases, you need to appraise the information you have collected and develop an overall sense of what the truth might be (Figure 5.16). Here are some questions you might use to appraise your information:

- Is the source reliable?
- How biased is the source?
- Am I biased?
- Is the source up to date?
- Can the information be verified?

Is the Source Reliable?

Usually, the most reliable sources of information are sources that have limited personal, political, or financial stakes in the subject. For example, claims about the safety of pesticides from a company that sells pesticides need to be carefully verified. Meanwhile, a study on pesticides by a university professor should be less biased, because the professor is not selling the product.

Of course, keep in mind that professors might receive significant funding from companies that support their research. Even the seemingly least biased sources are receiving funding from somewhere. Those funding sources may bias results.

To ensure your sources are reliable, you should always do some checking on their authors. Use an Internet search engine like Altavista.com or Google.com to check out the authors, company, or organization that produced the materials. If the researchers have a good reputation, the information is probably reliable. If you can find little or no information about the researchers, company, or organization, you should be skeptical about their research.

LINK For more information on finding people on the Internet, go to Chapter 3, page 52.

Questions for Appraising Your Information

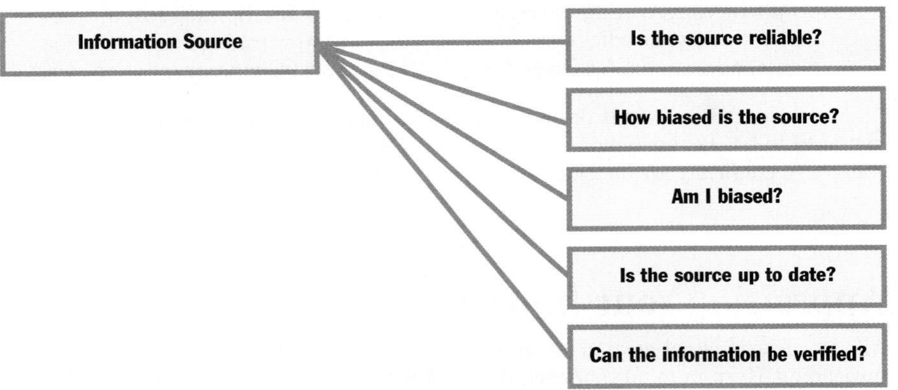

Figure 5.16: Challenge your sources by asking questions about their biases and validity.

How Biased Is the Source?

It is safe to say that all sources of information have some bias. There is no such thing as a completely objective source. So, you need to assess the amount of bias in your source. For example, facts from creation science websites that are used to dispute Darwinian evolution are usually biased toward theories that reinforce the biblical creation story. Their information is still useable in some situations—you might even accept it as true—but you need to recognize the inherent bias in such material.

Even the most reliable sources have some bias. Researchers, after all, very much want their hypotheses to be true, so irregularities in their results might be overlooked. Bias is a natural part of research. A researcher's interest in the subject alone indicates that he or she has some bias. You need to determine how biased the researchers are toward a specific outcome or theory.

When you are assessing bias, consider how much the researchers want their results to be true. If the researchers indicate they were open to a range of answers at the beginning of their research, then the bias of the material is probably minimal. If only one answer was acceptable to the researchers (e.g., smoking does not cause lung cancer), then the material should be considered heavily biased.

For help evaluating electronic sources, go to **www.ablongman.com/johnsonweb/5.20**

Am I Biased?

As a researcher, you need to carefully examine your own biases. We all go into a research project with our own beliefs and expectations of what we will find. Our own biases can cause us to overlook evidence that contradicts our beliefs or expectations. For example, our beliefs about gender, race, sexuality, poverty, or religion, among other social issues, can strongly influence the way we conduct research and interpret our findings. These influences cannot be completely avoided, but they can be identified and taken into consideration.

To keep your own biases in check, consider your research subject from an alternative or opposing perspective. At a minimum, considering alternative views will only strengthen your confidence in your research. But, in some cases, you may actually gain a new perspective that can help you further your research.

Is the Source Up to Date?

Depending on the field of study, results from prior research can become obsolete rather quickly. For instance, three-year-old research on skin cancer might already be considered outdated. On the other hand, climate measurements that are over 100 years old are still useable today.

Try to find the most recent sources on your subject. Reliable sources will usually offer a *literature review* that traces research on the subject back at least a few years. These literature reviews will show you how quickly the field is changing, while allowing you to judge whether the information you have located is current.

Can the Information Be Verified?

You should be able to find more than one independent source that verifies the information you find. If you find the same information from a few different independent sources, chances are good that the information is reliable. If you find the information in only one or two places, it is probably less reliable.

Triangulation is the key to verifying information. If you can find the information in diverse electronic and print sources, it is probably information you can trust. You might also use empirical methods to confirm or challenge the results of others.

Avoiding Plagiarism

One thing to watch out for is plagiarism in your own work, whether it is intentional or unintentional.

Plagiarism is the use of words, images, or ideas of others without acknowledgment or permission.

Plagiarism is the use of words, images, or ideas of others without acknowledgment or permission. In most cases, plagiarism is unintentional. While researching, a person might cut and paste information off websites or duplicate passages from a book. Later, he or she might use the exact text, forgetting that the information was copied directly from a source.

In rare cases, plagiarism is intentional and therefore a form of academic dishonesty. In these cases, teachers and colleges will often punish plagiarizers by having them fail the course, putting them on academic probation, or even expelling them from college. Intentional plagiarism is a serious form of dishonesty.

To avoid plagiarizing, keep careful track of your sources and acknowledge where you found your information.

Keep track of sources—Whenever you are drawing information from a source, carefully note where that information came from. If you are cutting and pasting information from an on-line source, make sure you put quotation marks around that material and clearly identify where you found it.

Acknowledge your sources—Any words, sentences, images, data, or unique ideas that you take from another source should be properly cited. If you are taking a direct quote from a source, use quotation marks to set it off from your writing. If you are paraphrasing the work of others, make sure you cite them with an in-text citation and put a full-text citation in a references list.

Ask permission—If you want to include others' images or large blocks of text in your work, write them an e-mail to ask permission. Downloading pictures and graphics off the Internet is really easy. But those images are usually someone's property. If you are using them for educational purposes, you can probably include them without asking permission. But, if you are using them for any other reason, you likely need to obtain permission from their owner.

LINK For more information on obtaining permission, go to Chapter 4, page 80.

You do not need to cite sources that offer information that is "common knowledge." If you find the same information in a few different sources, you probably do not need to document that information. But, if you have any doubts, you might cite the sources anyway to avoid any plagiarism problems.

Unfortunately, cases of plagiarism are on the rise. One of the downsides of on-line texts, such as websites, is the ease of plagiarism. Some students have learned techniques of "patchwriting," in which they cut and paste text from the Internet and then revise it into a document. This kind of writing is highly vulnerable to charges of plagiarism, so it should be avoided.

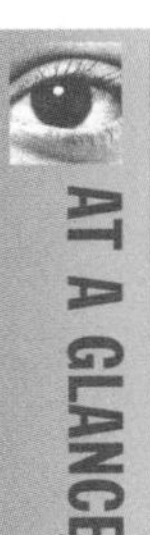

Assessing Your Information

- Is the source reliable?
- How biased is the source?
- Am I biased?
- Is the source up to date?
- Can the information be verified?

In the end, plagiarism mostly harms the person doing it. Plagiarism is kind of like running stoplights. People get away with it for only so long. Then, when they are caught, the penalties can be severe. Moreover, whether intentional or unintentional, plagiarizing reinforces some lazy habits. Before long, people who plagiarize find it difficult to do their own work, because they did not learn proper research skills. Your best approach is to avoid plagiarism in the first place.

To learn more about plagiarism, go to **www.ablongman/johnsonweb/5.21**

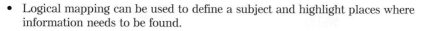

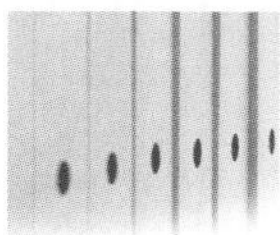

CHAPTER REVIEW

- Research today involves collecting information from diverse sources that are available in many media, including the Internet.
- Effectively managing existing information is often as important as creating new information.
- Logical mapping can be used to define a subject and highlight places where information needs to be found.
- A research methodology is a planned, step-by-step procedure that you will use to study the subject. Your research methodology could be revised as needed as your research moves forward.
- Triangulation is a process of using electronic, print, and empirical sources to obtain and evaluate your findings and conclusions.
- After collecting information, you should carefully assess whether it is biased or outdated. Also, be aware of your own biases.
- Careful note taking is essential for research. You should keep close track of your sources and use your research carefully in summarizing, paraphrasing, and quoting sources.

EXERCISES AND PROJECTS

Individual or Team Projects

1. Think of a technical subject that interests you. Then, collect information from electronic and print sources. Write a progress report to your instructor in which you highlight themes in the materials you find. Discuss any gaps in the information that might be filled with more searching or empirical study. Some possible topics might include the following:

   Wildlife on campus
   Surveillance in America
   Hybrid motor cars
   The problems with running red lights on or near campus
   Safety on campus at night
   The effects of acid rain in Canada
   Migration of humpback whales

2. On the Internet, find information on a subject that you think is junk science or influenced by junk science. Pay close attention to the reputations of the researchers and their results. Can you find any information to back up their claims? Pay special attention to where they receive their funding for the research. When you are finished searching the web, make a report to your class on your findings. Show your audience how junk science influences the debate on your subject.

Here are a few possible topics:

Evolution versus creation science
Genetically engineered foods
Managing forests to prevent fires
Cell phones and cancer
Experimentation on animals
Herbicides and insecticides
Diets and dietary supplements
Global warming
Smoking and secondhand smoke
Air and water pollution
Transporting nuclear waste
Welfare abuse

3. Survey your class on a campus issue that interests you. Write five questions and let your classmates select among answers like "strongly agree," "agree," "disagree," and "strongly disagree." Then, tabulate the results of the survey. Write a memo to your instructor in which you discuss the trends you found in your findings. In your memorandum, also point out places where your methodology might be challenged by someone who doubts your findings. Discuss how you might strengthen your survey if you wanted to do a larger study on this subject.

Collaborative Project

With a group, develop a methodology for studying substance abuse on campus (alcohol abuse or abuse of prescription drugs or illegal drugs). First, use logical mapping to identify what you already know or believe about substance abuse on your campus. Second, formulate a research question that your research will answer. Third, use logical mapping to sketch out a methodology that would help you generate results to answer your research question.

Your methodology should use triangulation to gather information from a broad range of sources. In other words, you should plan to gather information from electronic, print, and empirical sources.

Finally, write up your methodology, showing the step-by-step procedures you would use to study substance abuse on campus. Your methodology should be written in a way that makes it repeatable. It should also clearly identify the kinds of results you expect your research to generate.

Give your methodology to your instructor. At this point, your instructor may ask you to continue your research, following your methodology. As you do your research, note places where you changed your methodology or found information you did not expect.

For websites that discuss substance abuse on college campuses, go to **www.ablongman.com/johnsonweb/5.22**

CASE STUDY

Bye Bye Birdies

George Benks is a genetic engineering student at Missouri Tech University. During the summer of his junior year, he landed a great job as a field research assistant working with Professor Chad Henkle, an agronomist at the university. Professor Henkle and a team of other professors were field-testing a corn hybrid that had a built-in genetic insecticide. The inserted genes removed the need to spray insecticides on the fields, because insects were killed when they ate the leaves of the plant.

The experimental plots that George observed were showing great results. The new genetically modified corn plants seemed to be protected against the harmful pests. Insects that chewed on the plants were dying within 24 hours. George carefully recorded his observations in his field book.

Professor Henkle and the whole research team were very enthusiastic about George's and the other research assistants' observations. Confirmation of the results meant they would likely receive a large research grant from AgriMonz, an agribusiness company. Moreover, a patent on this new genetically modified corn hybrid might even make Professor Henkle and the others wealthy if AgriMonz purchased the rights to produce the corn.

One problem emerged, though. George began to find more than normal amounts of dead sparrows in the fields he was observing. He guessed that the sparrows were eating the poisoned insects and then dying themselves. He didn't find many dead birds, but enough to cause him some concern.

While eating lunch with Professor Henkle, George casually mentioned the dead sparrows. Professor Henkle told him that there are always "residual effects" with these kinds of new hybrids. Besides, sparrows aren't protected in any way. There are plenty of them, Henkle said, so the loss of some sparrows is not a problem.

George asked if he should mention the dead birds in his final report to the research team. Professor Henkle shrugged and said, "Go ahead, if you like. Personally, I don't think you should, because those observations are not within the boundaries of the experiment. We simply want to know whether the hybrid kills the bugs. The birds aren't important."

George knew that a few dead birds in experimental fields were probably not that important. But if the hybrid were used widely, the impact on the birds could be significant. Moreover, sparrows weren't the only kinds of birds that ate the poisoned insects. They were the ones George was finding dead.

What should George do at this point? Would you tell one of the other researchers? If you were in George's place, would you mention the dead birds in your report to the whole research team? If so, how would you do it? What do you think the research team should do if the hybrid leads to the death of sparrows and perhaps other kinds of birds?

Want to learn more about genetic engineering? Go to **www.ablongman.com/johnsonweb/5.23**

CHAPTER

6

Organizing and Drafting

CHAPTER CONTENTS

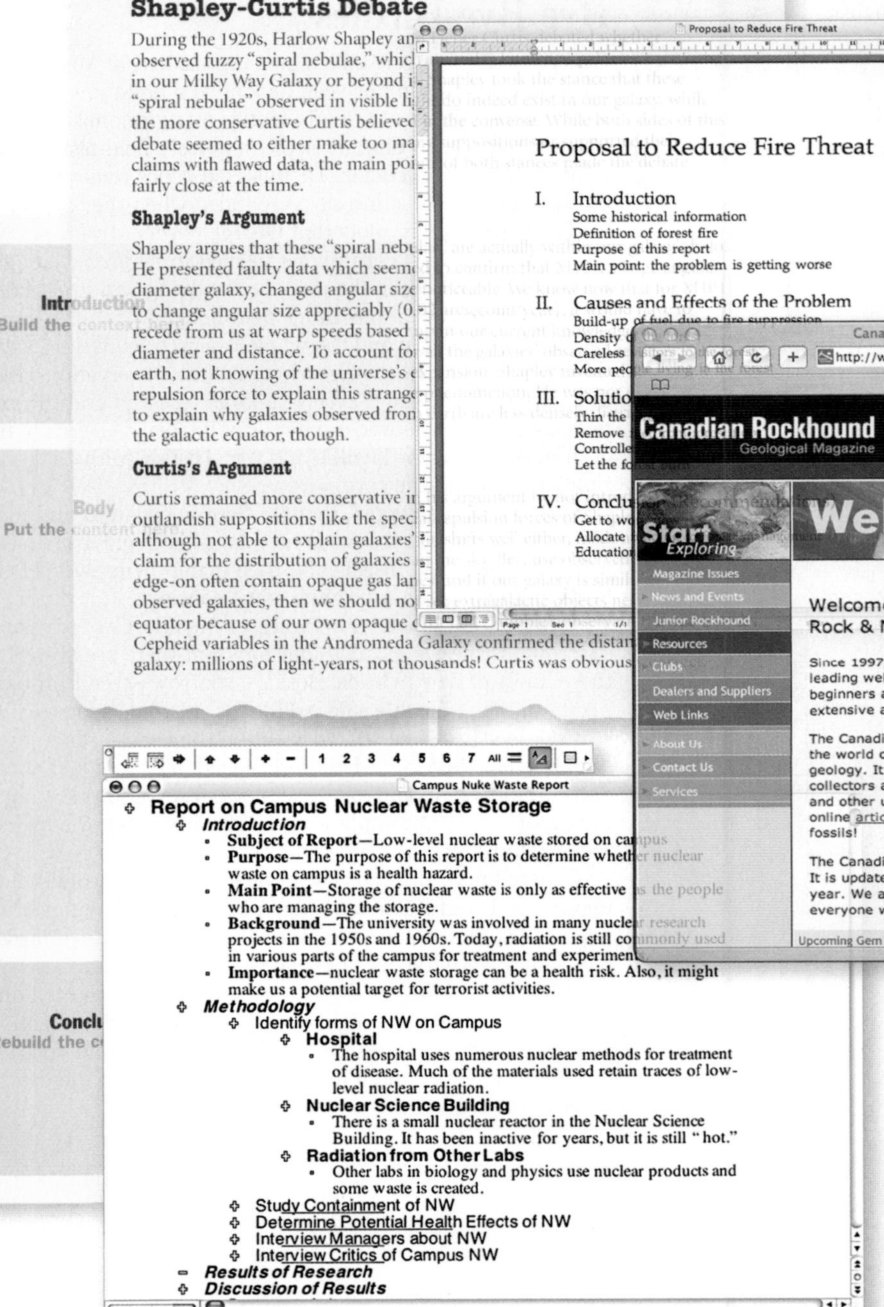

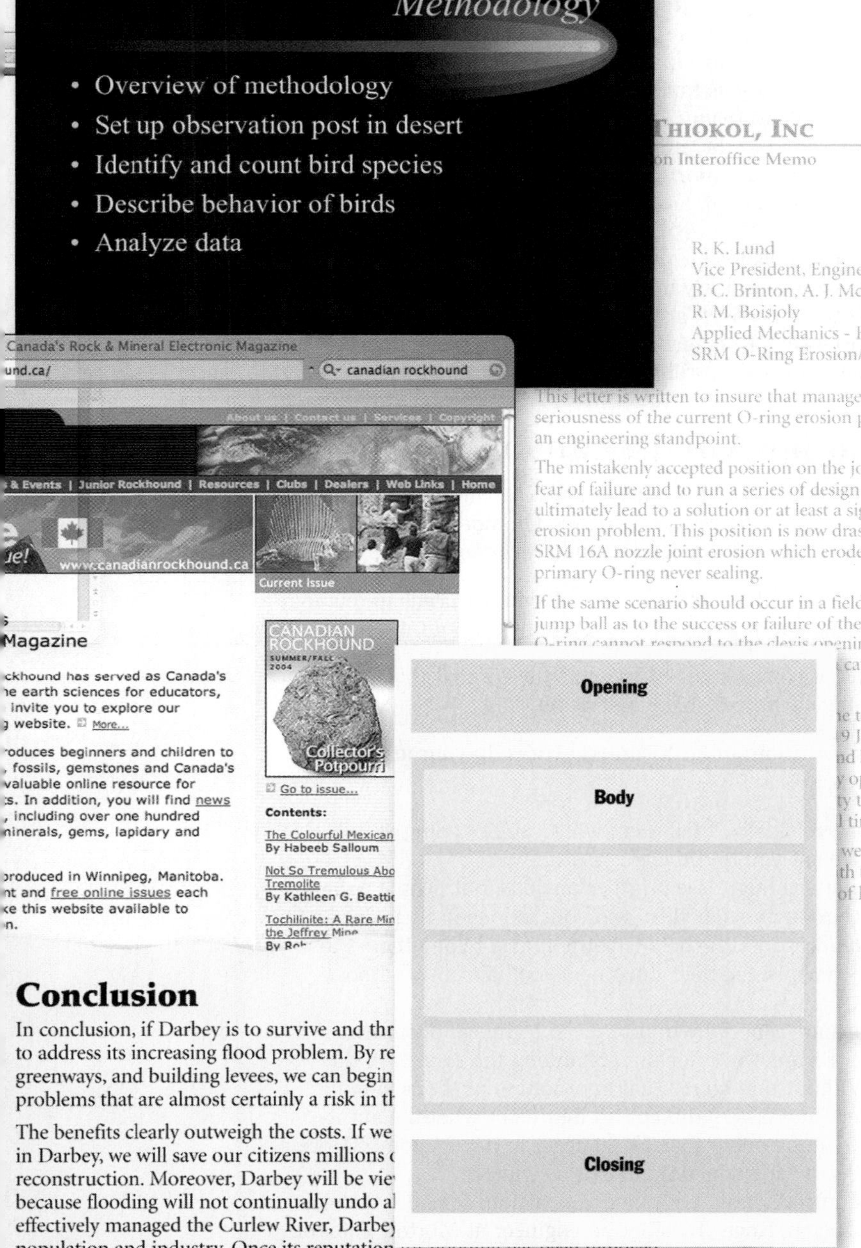

Methodology
Overview of methodology
Set up observation post in desert
Identify and count bird species
Describe behavior of birds
Analyze data
Canada's Rock & Mineral Electronic Magazine
www.canadianrockhound.ca
Current Issue
CANADIAN ROCKHOUND
Collector's Potpourri
Opening
Body
Closing
Conclusion
We appreciate your time and consideration. If you have any questions or
would like to meet with us about this report, please contact the task force
leader, Mary Subbock, at 555-0912 or e-mail her at msubbock@cdarbey.gov.

CHAPTER OBJECTIVES

In this chapter, you will learn:

- A basic organizational pattern that any document can follow.
- How genres offer patterns for organizing documents.
- How to outline the organization of a document.
- How to use presentation software to organize your information.
- How to organize and draft a document's introduction.
- Patterns of arrangement for organizing and drafting sections of a document's body.
- Basic moves used in a document's conclusion.

While doing research or working on a project, you will pull together many ideas, facts, and observations, drawn from a variety of sources. As you turn to draft your document, you will need to organize that information into patterns that are familiar to your readers. Your challenge is to arrange the information in ways that make it accessible and understandable to them.

Your readers, after all, are interested in the information you have gathered. But, they need you to organize and present that information in a predictable and usable way. Otherwise, they won't be able to take full advantage of your thoughts and research on the subject.

Fortunately, computers are great tools for helping us manage large amounts of information. They are also able to help us quickly move that information around and arrange it to suit our and our readers' needs. It's still up to you, though, to give that information shape, to make it accessible to readers. You still need to make important decisions about where and when information will appear in the document.

Basic Organization for Any Document

Despite their differences, almost all technical documents have one important thing in common. They should all have a beginning, middle, and end. Or, more specifically, all technical documents should have an *introduction, body,* and *conclusion.*

Introduction (Beginning)—The introduction of your document needs to tell readers what you are writing about and why you are writing about it.

Body (Middle)—The body of your document presents the content that your readers need to know to take action or make a decision.

Conclusion (End)—The conclusion of the document wraps up your argument by restating your main point(s).

To use the familiar speechwriters' advice, "Tell them what you are going to tell them. Tell them. Then, tell them what you told them."

This beginning-middle-end pattern might seem rather obvious, but people regularly forget to include distinct introductions, bodies, and conclusions in their documents. All too often, they toss readers into the details without first telling them the subject and purpose of their documents. Or, their documents end abruptly without summing up the major points.

Introductions and conclusions are especially important in technical documents, because they provide a *context*, or framework, for understanding the *content* in the body of the text (Figure 6.1). Without that contextual information at the beginning and end of the document, readers find it very difficult to figure out what the author is telling them. Have you ever read a document that didn't seem to have a point? More than likely, it was lacking an effective introduction and/or conclusion.

To see an example of a good introduction, body, and conclusion, consider the classic memo in Figure 6.2. In this memo, Roger Boisjoly, an engineer at Morton Thiokol, warns a vice president of the company that the Space Shuttle *Challenger* is in danger

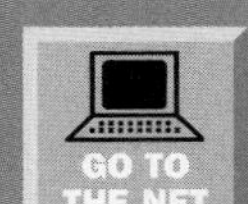

Why do documents need introductions and conclusions? For answers, see **www.ablongman.com/johnsonweb/6.1**

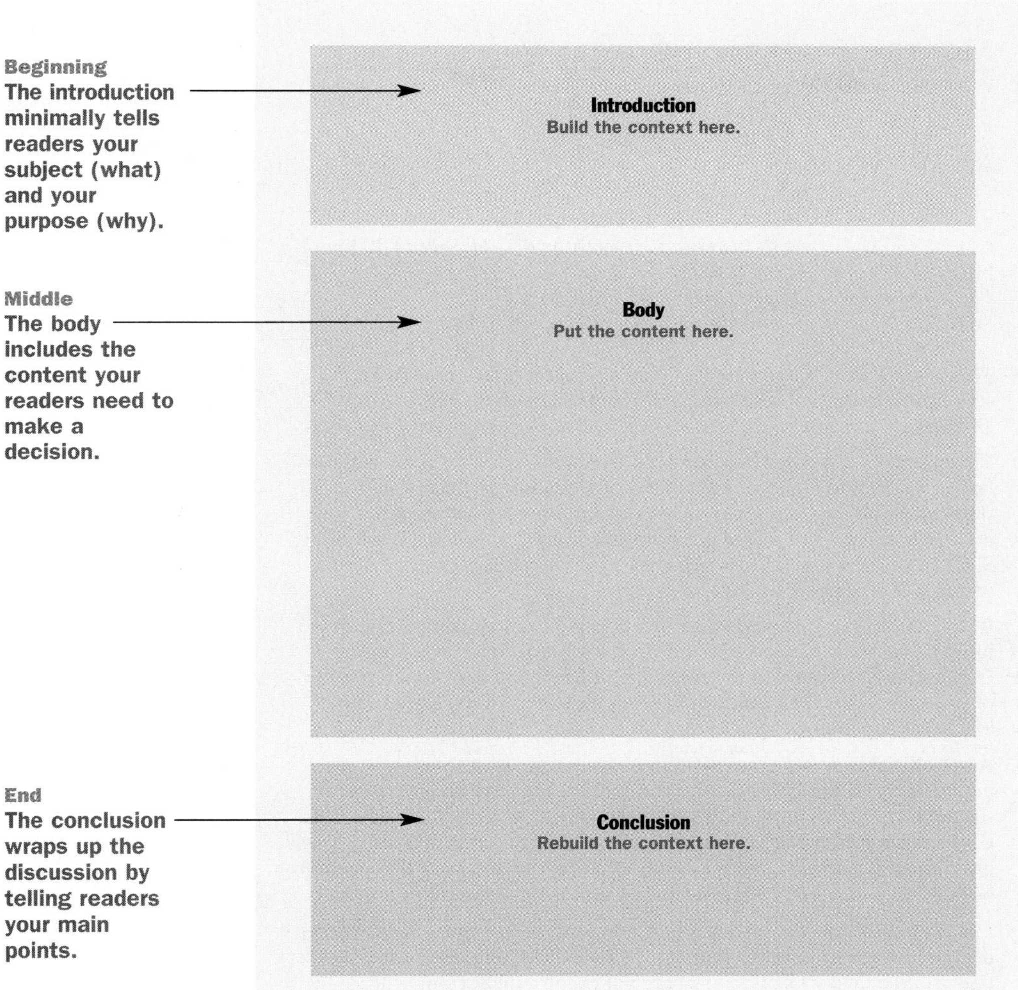
Standard Organization of a Document
Beginning
The introduction minimally tells readers your subject (what) and your purpose (why).
Introduction
Build the context here.
Middle
The body includes the content your readers need to make a decision.
Body
Put the content here.
End
The conclusion wraps up the discussion by telling readers your main points.
Conclusion
Rebuild the context here.

Figure 6.1: An effective technical text will have a beginning (introduction), middle (body), and end (conclusion).

of blowing up. This memo is often referred to as the "smoking gun" memo, which demonstrated that NASA and Morton Thiokol were ignoring erosion problems with the shuttle's O-rings. These O-rings kept explosive exhaust gases from destroying the rocket boosters—and the shuttle.

The problem described in this memo is clear, and Boisjoly does his best to stress the importance of the problem in the introduction and conclusion. Unfortunately, his and other engineers' warnings were not heeded by higher-ups.

The "Smoking Gun" *Challenger* Memo

Figure 6.2: Roger Boisjoly's classic memo warning of imminent disaster with the *Challenger* shuttle.

MORTON THIOKOL, INC

Wasatch Division Interoffice Memo

July 31, 1985

2870:FY86:073

TO:	R. K. Lund Vice President, Engineering
CC:	B. C. Brinton, A. J. McDonald, L. H. Sayer, J. R. Kapp
FROM:	R. M. Boisjoly Applied Mechanics - Ext. 3525
SUBJECT:	SRM O-Ring Erosion/Potential Failure Criticality

The introduction contains the purpose and main point.

This letter is written to insure that management is fully aware of the seriousness of the current O-ring erosion problem in the SRM joints from an engineering standpoint.

The body supports the argument.

The mistakenly accepted position on the joint problem was to fly without fear of failure and to run a series of design evaluations which would ultimately lead to a solution or at least a significant reduction of the erosion problem. This position is now drastically changed as a result of the SRM 16A nozzle joint erosion which eroded a secondary O-ring with the primary O-ring never sealing.

If the same scenario should occur in a field joint (and it could), then it is a jump ball as to the success or failure of the joint because the secondary O-ring cannot respond to the clevis opening rate and may not be capable of pressurization. The result would be a catastrophe of the highest order—loss of human life.

An unofficial team (a memo defining the team and its purpose was never published) with leader was formed on July 19, 1985 and was tasked with solving the problem for both the short and long term. This unofficial team is essentially nonexistent at this time. In my opinion, the team must be officially given the responsibility and the authority to execute the work that needs to be done on a non-interference basis (full time assignment until completed.)

The conclusion restates the main point.

It is my honest and very real fear that if we do not take immediate action to dedicate a team to solve the problem with the field joint having the number one priority, then we stand in jeopardy of losing a flight along with all the launch pad facilities.

R. M. Boisjoly

Concurred by: J. R. Kapp, Manager

Applied Mechanics

Source: Report of the Presidential Commission on the Space Shuttle Challenger *Accident, 1986.*

Using Genres to Organize Information

Once you move beyond the introduction, body, and conclusion pattern, technical documents can follow a variety of different organization patterns. A report, for example, is very different from a set of instructions. A proposal shows little resemblance to a technical specification. These documents follow completely different patterns of organization, or *genres.*

A genre is a predictable pattern for organizing information to achieve specific purposes.

A genre is a predictable pattern for organizing information to achieve specific purposes. For instance, a report will tend to include the following sections:

- introduction.
- methods.
- results.
- discussion.
- conclusion.

A set of instructions, in contrast, will tend to include these sections:

- introduction.
- list of parts/tools.
- steps.
- conclusion.

Each genre achieves a different purpose, requiring a different organizational pattern. Once you know your purpose, you should be able to easily figure out what kind of document you are being asked to write and how it should be organized (Figure 6.3).

Choosing a Genre

Which genre fits my writing task?

Purpose	Genre	Type of Document
I need to tell others about a decision or event.	Correspondence (Chapters 12 and 16)	E-mail, Letter, or Memo
I need to define a concept or set of terms.	Definition (Chapter 17)	Technical Definition, Glossary
I need to describe an item, product, or service.	Description (Chapter 18)	Technical Description, White Paper
I need to explain how to do something.	Procedure (Chapter 19)	Instructions, Specifications, Procedures
I need to make a suggestion or propose a new project.	Proposal (Chapter 20)	Research Proposal, Planning Proposal, Implementation Proposal
I need to present information or make a recommendation.	Report (Chapters 21 and 22)	Research Report, Feasibility Report, Recommendation Report, Lab Report, Progress Report
I need to provide on-line information to others.	Hypertext (Chapter 14)	Website, Multimedia Document, CD-ROM
I need to tell the public about an experience, product, or service.	Narrative (Chapter 23)	Article for Website, Magazine, Newsletter, or Newspaper
I need a job.	Resume (Chapter 15)	Portfolio, Resume, Scannable Resume

Figure 6.3: Once you know the purpose of your document, you can decide which genre fits your need. Each genre provides a pattern for you to follow as you organize and draft your document.

Want to learn more about genres? Go to **www.ablongman.com/johnsonweb/6.2**

Always keep in mind, though, that genres are patterns, not formulas. There are countless ways to organize reports and instructions, depending on the needs of your readers and the context in which the document will be used. However, you will find that reports and instructions tend to include the elements listed above. You can use these genres as starting places from which to begin organizing and drafting your documents.

Chapters 16 through 22 discuss each of these genres separately in depth. When you begin writing one of these documents, you should turn to the chapter that describes it.

Outlining the Document

Once you have identified the genre of your document, you can begin developing a rough outline of its organization.

Outlining may seem a bit old-fashioned, but it is very helpful when you are trying to sort out your ideas, especially when you are writing a large technical document. In the workplace, most people sketch out a rough outline to help them organize their ideas. In their outline, they type the document's main headings on the screen (Figure 6.4). Then, they list the contents of each section separately. The outline will usually change as new ideas, evidence, or issues emerge.

If you are unsure how to outline your document, let the genre you are following guide your thought process. The genre should give you a good sense of the larger sections of the document. From there, you can begin filling in the smaller topics in each section.

A Rough Outline

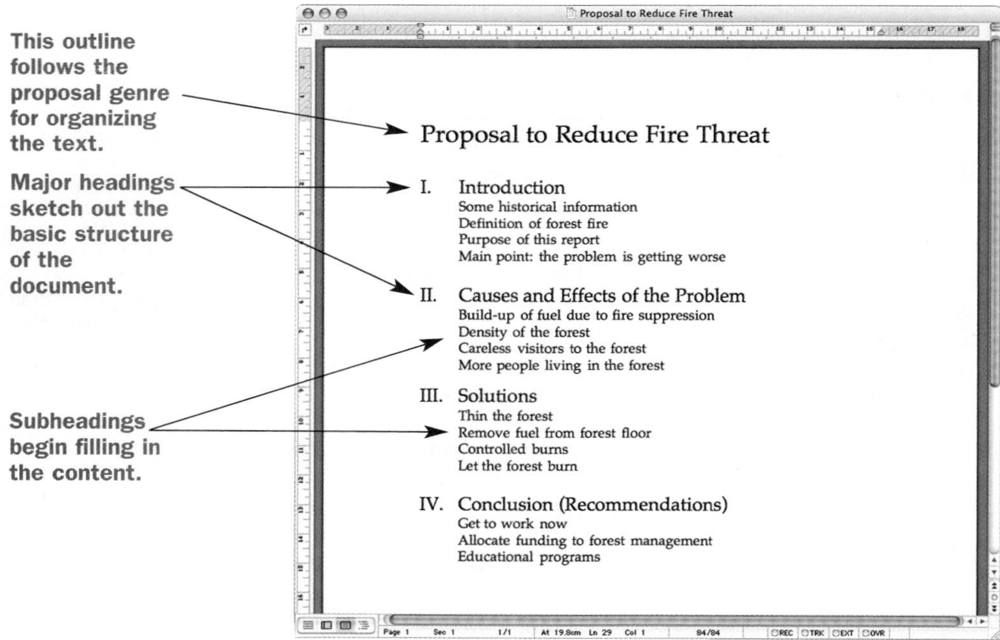

Figure 6.4: An outline doesn't need to be formal, and it should always be open to change. Here is a rough outline with some guesses about what kinds of topics will be discussed in the proposal.

For websites that offer outlining strategies, go to **www.ablongman.com/johnsonweb/6.3**

When outlining, you might also consider using the *Outline View* in your word processor. In Outline View, you can begin arranging your information by listing the headings for the document's major sections and subsections (Figure 6.5). Then, you can decide if—

- smaller sections should be merged or incorporated into larger sections.
- larger sections should be divided into separate sections.
- any need-to-know information is missing in sections.
- there is too much want-to-tell information.

While you are revising your document, Outline View also helps you quickly move information around. You can "collapse" a whole part of a document under a heading. Then, you can cut and paste to move that part somewhere else in the document.

Outline View

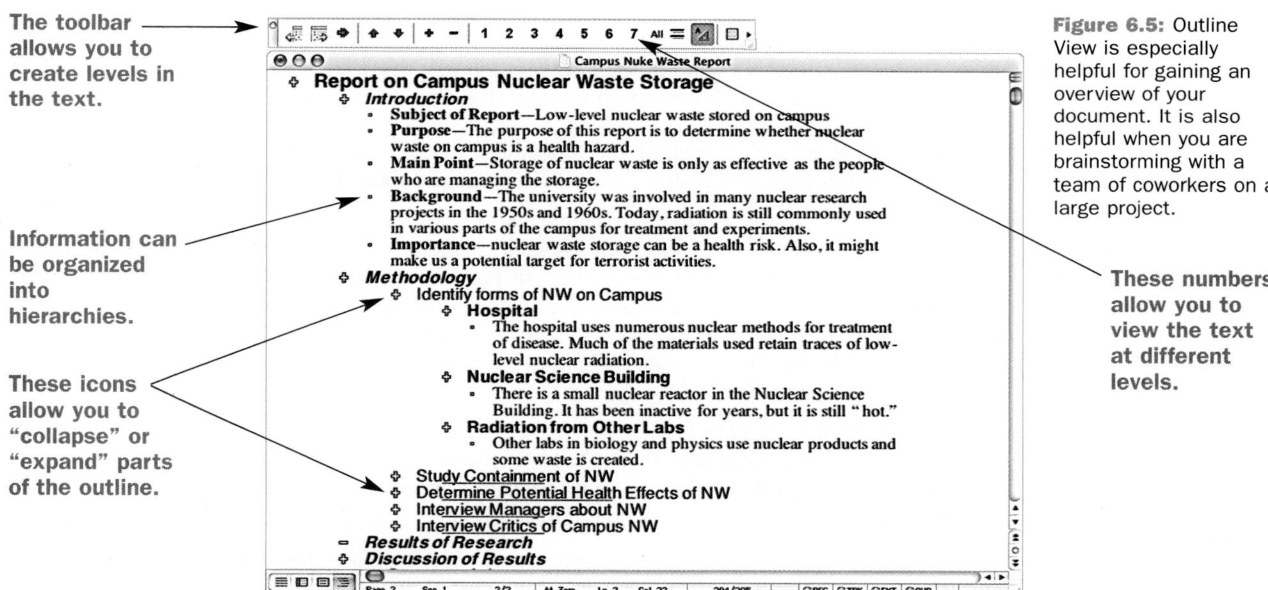

Figure 6.5: Outline View is especially helpful for gaining an overview of your document. It is also helpful when you are brainstorming with a team of coworkers on a large project.

Outline View is a great tool when you are working with a team. When you and your team are brainstorming about a project or document, just toss your ideas into Outline View. Then, when you're finished brainstorming, you can start organizing all that information into a more structured outline. At the end of the meeting, you can print out the outline or e-mail it to the whole group to guide the drafting of the document.

LINK For more information on working with a team, see Chapter 13, page 352.

Overall, an outline should be as flexible as the document itself. A computer-generated outline can be a helpful tool for planning, drafting, and revising your work.

For more help using Outline View, go to **www.ablongman.com/johnsonweb/6.4**

Organizing with Presentation Software

Not everyone likes using outlines, but as documents get larger, it becomes increasingly difficult to organize all that information without some kind of outline. One clever way to organize your information is to use presentation software, like Corel Presentations or MS PowerPoint, to frame and arrange your ideas. Even if you don't need to present your materials orally, using the presentation software will help you structure and clarify your thoughts.

Let's say you need to write a report. You have collected a large amount of information from the Internet, empirical studies, interviews, and trips to the library. You are almost overwhelmed by the amount of information you need to put into your report. Presentation software will force you to boil all that information down into headings and bullet points. Before too long, you will have created a very effective outline with which to start drafting.

To organize your materials with presentation software, follow these steps:

1. Create a title page and write a title.
2. State your document's main point as a subtitle on the title page.
3. Create an introduction page that states your subject, purpose, and the importance of the subject.
4. Create a new page for each section of your document, and give it a unique title.
5. Create a conclusion page that restates your main point and looks to the future.

Then, working section by section, fill in your information using only bulleted lists. For each slide, use a new title and limit yourself to two to five bullets (Figure A). If one of your slides has only one bullet of information, merge its contents with another neighboring slide. If a slide has more than five bullets, divide the slide into two separate slides with unique titles.

As you fill in your information, you will find that the presentation software forces you to think about

- whether information is need-to-know content.
- how smaller pieces of information fit into larger sections.
- where you are missing information that your readers will need to know.
- how you can express the information you gathered in concise ways.

When you have finished filling in the slides, you will be able to review the structure of your whole document. Now, you can sharpen and clarify the materials by editing your slides. You might move some slides around, consolidate a few, and expand others. You should also pay attention to the titles of the slides, because they will eventually become the headings in the written document.

When you are done filling and editing the slides, you will have created a very helpful outline for writing your document. Each of the slides will probably become a section of your written document. Each of the bullet points will probably need a paragraph or two of coverage.

Ultimately, you are using the presentation software as an outlining tool. But the presentation software is much more flexible and visual than a written outline. Try out this technique. You will find it very effective.

Want more advice on using presentation software? Go to **www.ablongman.com/johnsonweb/6.5**

Organizing Content on a Presentation Slide

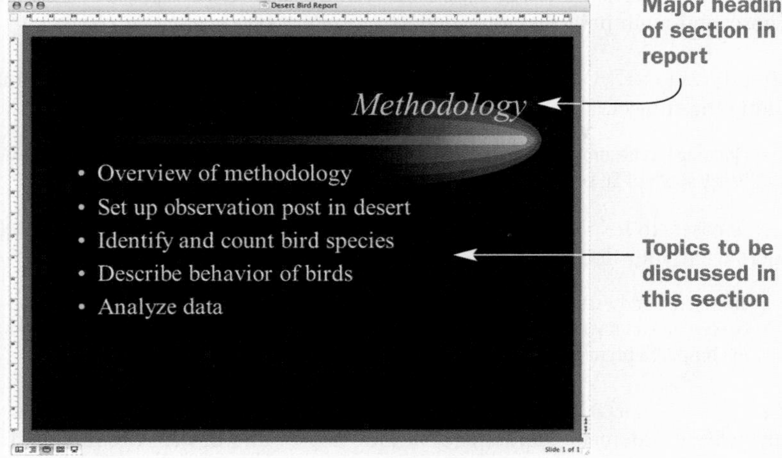

Figure A: Presentation software can be a very helpful tool for organizing your information, even if you are not going to present the material orally.

Organizing and Drafting the Introduction

As you begin organizing and drafting your document, put yourself in your readers' place. When you begin to read something, what information do you want to know up front? More than likely, you want answers to the following kinds of questions:

What is this document about?

Why did someone write this to me?

What is the main point?

Is this information important?

How is this document organized?

When an introduction answers these questions up front, readers are better able to understand the rest of the document.

Six Opening Moves in an Introduction

These kinds of questions translate into six opening "moves" made in an introduction:

- Define your subject.
- State your purpose.
- State your main point.
- Stress the importance of the subject.
- Provide background information.
- Forecast the content.

As you begin drafting your document on your computer, it is helpful to first put your responses to these opening moves on the screen. Later, you can shape your

To learn more about how people read introductions, go to **www.ablongman.com/johnsonweb/6.6**

responses into a smooth, coherent introduction. It's not critical that you write the introduction right now. However, you should have a clear idea about the subject, purpose, and main point before you start writing anything else.

MOVE 1: DEFINE YOUR SUBJECT Tell readers what your document is about by defining the subject.

> Flooding has become a recurring problem in Darbey, our small town nestled in the Curlew Valley south of St. Louis.

In some cases, to help define the boundaries of your subject, you might also tell readers what your document is *not* going to cover.

> This memo addresses the issue of smoking outside the main hospital building. We are not concerned about your smoking habits at home. We are only concerned about smoking on the hospital's premises.

MOVE 2: STATE YOUR PURPOSE Tell readers what you are trying to achieve. Your purpose statement should be clear and easy to find in the introduction. It should plainly tell your readers what the document will do.

> This proposal offers some strategies for managing flooding in the Darbey area.

You should be able to articulate your purpose in one sentence. Otherwise, your purpose may not be clear to your readers—and perhaps not even to yourself.

LINK For more information on crafting a purpose, go to Chapter 2, page 18.

Six Moves in an Introduction

- Move 1: Define your aubject.
- Move 2: State your purpose.
- Move 3: State your main point.
- Move 4: Stress the importance of the subject.
- Move 5: Provide background information.
- Move 6: Forecast the content.

MOVE 3: STATE YOUR MAIN POINT Tell your readers the key idea or main point that you would like them to take away from the document. Usually your main point is your overall decision, conclusion, or solution that you would like the readers to accept.

> The only long-term way to control flooding around Darbey is to purchase and restore the wetlands around the Curlew River, while enhancing some of the existing flood control mechanisms like levees and diversion ditches.

Are you giving away the ending by telling the readers your main point up front? Yes. But then, technical documents are not mystery stories. Just tell readers your main point in the introduction. That way, as they read the document, they can see how you came to that decision.

MOVE 4: STRESS THE IMPORTANCE OF THE SUBJECT Make sure you give your readers a reason to care about your subject. You need to answer their "so what?" questions if you want them to pay attention and continue reading.

> If development continues to expand between the town and the river, the flooding around Darbey will only continue to worsen, potentially causing millions of dollars in damage.

For sample introductions, see **www.ablongman.com/johnsonweb/6.7**

AT WORK

Timothy Myers
DIRECTOR OF CONSULTING SERVICES AT TRIPOINT SYSTEMS, CHICAGO, ILLINOIS
Tripoint Systems is a software development company.

Is it true that writers should not write the introduction first?

Never start at the beginning, and never end at the end. Start writing where you feel comfortable and weave your web of words from there. There is no law saying you should write the first paragraph of the document first. Truthfully, there ought to be a law *against* it. After all, the introduction is by far the most difficult piece to write.

Think about it: The opening paragraph is where you either grab your readers and pull them into your thoughts, or it is where they yawn, stretch, and look for something else to read.

Of course, take the time to create an organizational outline. Write down your purpose and main point. These are essential tasks to creating any strong document. You must know what you are writing about before banging out words, and you must have an understanding of the overall structure of the document. But don't feel obligated to write the first word your reader will see as *your* first word.

A document is really just a puzzle. Start by creating the pieces of the document puzzle. Then, fit these pieces into place once you are ready. If that means writing the introduction last, so be it. Your document will be better for it.

MOVE 5: PROVIDE BACKGROUND INFORMATION Typically, background information includes material that readers already know or won't find controversial. This material could be historical, or it could stress a connection with the readers.

> As we mentioned in our presentation to the City Council last month, Darbey has been dealing with flooding since it was founded. Previously, the downtown was flooded three times (1901, 1922, and 1954). In recent years, Darbey has experienced flooding with much more frequency. The downtown was flooded in 1995, 1998, 2000, and 2003.

MOVE 6: FORECAST THE CONTENT Forecasting describes the structure of the document for your readers by identifying the major topics it will cover.

> In this proposal, we will first identify the causes of Darbey's flooding problems. Then, we will offer some solutions for managing future flooding. And finally, we will discuss the costs and benefits of implementing our solutions.

Forecasting helps readers visualize the organization of the rest of the document. It gives them a map to anticipate the topics the document will cover.

Drafting with the Six Moves

In an introduction, these moves can be made in just about any order. Figure 6.6, for example, shows how the introductory moves can be used in different arrangements.

For more help with writing introductions, go to **www.ablongman.com/johnsonweb/6.8**

Two Versions of an Introduction

Version One

Solving the Flooding Problems in Darbey, Missouri: A Recommendation Report

Subject → Flooding has become a recurring problem in Darbey, our small community set in the Curlew Valley south of St. Louis. As we mentioned in our Background information → presentation to the Darbey City Council last month, the town has been dealing with flooding since it was founded. Unfortunately, the problem seems to be growing worse. Previously, the downtown was flooded three times (1901, 1922, and 1954). In recent years, though, Darbey has experienced flooding with much more frequency. The downtown has flooded in 1995, 1998, 2000, and 2003.

Purpose → This proposal offers some strategies for managing flooding in the Darbey Main point → area. We argue that the only long-term way to control flooding around Darbey is to purchase and restore the wetlands around the Curlew River, while enhancing some of the existing flood control mechanisms like levees and diversion ditches. Otherwise, if development continues to expand between the Importance → town and the river, the flooding around Darbey will only continue to worsen, potentially causing millions of dollars in damage.

Forecasting → In this proposal, we will first identify the causes of the Darbey's flooding problems. Then, we will offer some solutions for managing future flooding. And finally, we will discuss the costs and benefits of implementing our solutions.

Version Two

Solving the Flooding Problems in Darbey, Missouri: A Recommendation Report

Purpose and subject → This proposal offers strategies for managing flooding in the Darbey area. The only long-term way to control flooding around Darbey is to purchase and Main point → restore the wetlands around the Curlew River, while enhancing some of the existing flood control mechanisms like levees and diversion ditches. Importance → Otherwise, if development expands between the town and the river, the flooding around Darbey will only continue to worsen, potentially causing millions of dollars in damage.

We cannot ignore this problem. Flooding has become a recurring problem in Background information → Darbey, our small community set in the Curlew Valley south of St. Louis. As we mentioned in our presentation to the Darbey City Council last month, this town has been dealing with flooding since it was founded. Previously, the downtown was flooded three times (1901, 1922, and 1954). In recent years, however, Darbey has experienced flooding with much more frequency. The downtown has flooded in 1995, 1998, 2000, and 2003.

Forecasting → In this proposal, we will first identify the causes of the Darbey's flooding problems. Then, we will offer some solutions for managing future flooding. And finally, we will discuss the costs and benefits of implementing our solutions.

Figure 6.6: As shown in these two sample introductions, the six introductory moves can be arranged in just about any order. Some arrangements, however, are more effective than others, depending on the subject and purpose of the document.

Information that goes beyond these six moves should not be put in the introduction. It should be saved for the body of the document. After all, any information in the introduction that goes beyond the six moves will only make it more difficult for the readers to figure out the subject, purpose, and main point of your document.

As discussed more thoroughly in Chapter 14, homepages in websites are also introductions. For example, the homepage shown in Figure 6.7 also makes the typical moves found in an introduction.

A Homepage

Figure 6.7: Readers expect a homepage to make the same moves as an introduction in a print document.

Source: Canadian Rockhound, http://www.canadianrockhound.ca.

Organizing and Drafting the Body

The body of the document is where you are going to provide the *content* that readers need to know. Here is where you will give them the information they need (facts, details, examples, and reasoning) to understand your subject and/or take action.

Carving the Body into Sections

With the exception of small documents (e.g., small memos or letters), the bodies of technical documents are typically carved into *sections.* In many ways, sections are like miniature documents, needing their own beginning, middle, and end. They typically include an *opening, body,* and *closing* (Figure 6.8).

A Section with an Opening, Body, and Closing

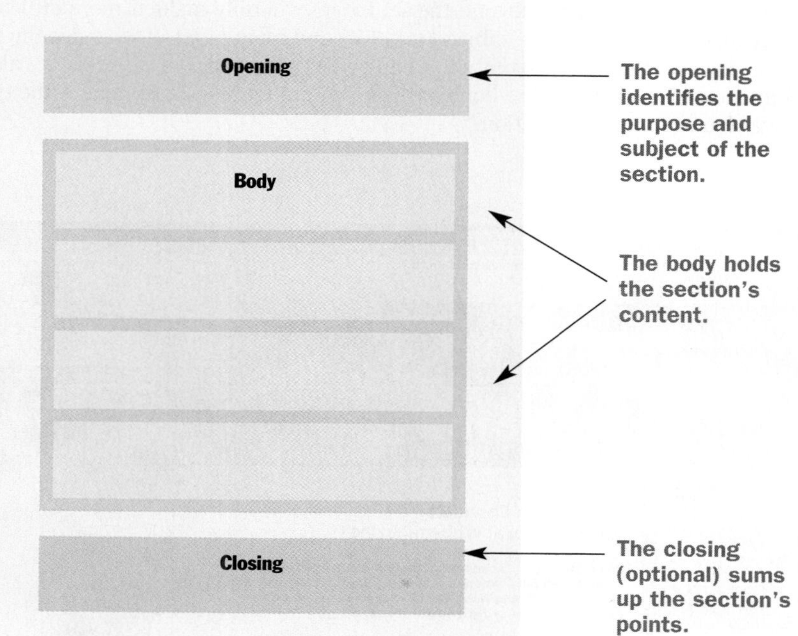

Opening
Body
Closing
The opening identifies the purpose and subject of the section.
The body holds the section's content.
The closing (optional) sums up the section's points.

Figure 6.8: A section includes an opening and body. The closing is optional, but it can be helpful to sum up the section's point or points.

OPENING An opening is usually a sentence or small paragraph that identifies the subject and purpose of the section. The opening includes a claim or claims that the rest of the section will support.

> **Results of Our Study**
>
> The results of our study allow us to draw two conclusions about the causes of flooding in Darbey. First, Darbey's flooding is mostly due to the recent construction of new levees by towns farther upriver. Second, development around the river is taking away some of the wetlands that have protected Darbey from flooding in the past. In this section, we will discuss each of these causes in depth.

BODY The body of a section is where you will offer support for the claim you made in the opening. The body of a section can run anywhere from one paragraph to many paragraphs, depending on the purpose of the section. For example, if you are discussing the results of a research study, your Results section may require three or more paragraphs in the body—one paragraph per major result.

CLOSING (OPTIONAL) Larger sections sometimes need a closing paragraph or sentence to wrap up the discussion. A closing usually restates the claim you made in the opening of the section. It might also look forward to the next section.

For sample sections that you can model, go to **www.ablongman.com/johnsonweb/6.9**

In sum, these two causes will likely only become more significant over time. As upriver towns grow in population, there will be more pressure than ever to build more levees to protect them. Meanwhile, if development continues in the available wetlands around the Curlew River, Darbey will find some of its last defenses against flooding have disappeared.

Overall, a section should be able to work as a stand-alone unit in the document. By having its own beginning, middle, and end, a well-written section feels like a miniature document that makes a specific point.

Patterns of Arrangement

When writing each section, you can usually follow a *pattern of arrangement* to organize your ideas. These patterns are based on logical principles, and they can help you organize your information so that your views will be presented in a reasoned way. Major patterns of arrangement are—

- Cause and Effect
- Comparison and Contrast
- Better and Worse
- Costs and Benefits
- If . . . Then
- Either . . . Or
- Chronological Order
- Problem/Needs/Solution
- Example.

You will find that each section in your document likely follows one of these patterns of arrangement. One section, for example, may discuss the *causes and effects* of a problem. A later section might use a discussion of the *costs and benefits* of doing something about the problem. So, as you are organizing and drafting each section, decide which pattern of arrangement best fits your needs.

Then, use these patterns to guide your drafting of each section. These patterns are not formulas, but they can provide you with helpful structures for presenting your ideas.

CAUSE AND EFFECT In a sense, all events are subject to *causes* and *effects*. For example, if a bridge suddenly collapsed, investigators would immediately try to determine the causes for the collapse.

In 2002, the I-40 bridge over the Arkansas River collapsed for a few different reasons. The actual collapse occurred when a runaway barge on the river rammed into one of the bridge's supports. But other causes were evident. The bridge was already weakened by erosion around the pilings, deteriorated concrete, and attrition due to recent seismic activity.

A cause can also have various effects.

Leaving a wound untreated can be dangerous. The wound may become contaminated with dirt and germs, thus requiring more healing time. In some cases, the wound may grow infected or even gangrenous, requiring much more treatment at a hospital. Infections can be life-threatening.

As you discuss causes and effects, show how effects are the results of specific causes (Figure 6.9).

Want to see examples of cause-and-effect sections? Go to **www.ablongman.com/johnsonweb/6.10**

A Section Using Cause and Effect

Buckinghamshire Flooding Facts

Flooding—Cause and Effect

In Buckinghamshire, as with other inland counties, there are three main causes of flooding, which are river (also known as riparian), flash, and groundwater. In all three cases the ability of the ground to absorb rainfall, like a sponge, plays a major part, and this will vary according to weather conditions.

River flooding occurs when rivers cannot cope with the amount of water draining into them off the land. In the winter months this is usually because the ground has become saturated and can no longer absorb water properly, so when rainfall is heavy and/or prolonged, runoff reaches the rivers faster and eventually they overtop their banks, as happened along the River Thames in January 2003.

Flash flooding can happen anywhere at any time, although it is more likely in the summer months when the ground is hard and dry. Sudden downpours, such as those associated with thunderstorms, cannot soak in fast enough and the water runs off quickly into drains, ditches, and culverts that cannot cope with the volume. Sometimes there is so much rain that a hillside may have the soil washed off it—often into nearby properties.

Groundwater flooding is rare, occurring only when the underground water table rises to an unusually high level. After the very heavy autumn and winter rainfall of 2000/01, the water table in the Chilterns was at its highest level for decades. Springs and wells that had been dry for, in some cases, 50 years or more began to flow again and did so throughout the summer, causing property flooding in many areas. Groundwater again rose significantly following the very wet winter of 2002/03, though fortunately the drier spring weather meant that levels did not reach a critical point.

Whatever the cause, the effects are much the same. If floodwater enters your property it will ruin carpets, furniture, household goods, decor, and electrics. Often floodwater will be mixed with raw sewage, as drains overflow. Even when the flood recedes, your home will take a long time to dry out and may smell for weeks. Plaster may have to be stripped off walls. If you are fully insured there should be no problem in getting the necessary repairs paid for, but the whole process will probably take months, and the experience is severely depressing. People who have been flooded will often say that it is far worse than being burgled.

Opening paragraph lists the causes of the problem.

Body paragraphs discuss each cause separately.

Closing paragraph discusses the effects of the problem.

Source: Buckinghamshire County Council, http://www.buckscc.gov.uk/emergency_planning/ff_cause_and_effect_2003.htm.

Figure 6.9: A section that uses the cause and effect pattern of arrangement.

COMPARISON AND CONTRAST You can *compare and contrast* just about anything. When comparing and contrasting two things, first identify all the features that make them similar. Then, contrast them by noting the features that make them different (Figure 6.10).

A Section Using Comparison and Contrast

Figure 6.10: Comparison and contrast puts two things side by side to show their similarities and differences.

Opening states the claim for the section to follow.

Comparisons are made by noting similarities.

Contrasts are made by noting differences.

Similar Plans of Hemyock and Bodiam Castles

The plan of Hemyock Castle has similarities with Bodiam Castle in Kent, built some five years later.

Both Hemyock and Bodiam are typical of small late medieval castles: a rectangular site with high, round corner towers and central interval towers, connected by a high curtain wall; all topped with crenellations; surrounded by a water-filled moat. Both had massive fortified gatehouses. The two castles were roughly the same size. Even the detail of Hemyock's NE Tower appears similar to Bodiam's Well Tower.

There were obvious differences:

- Hemyock Castle was built around the family's old manor house, whereas Bodiam Castle was built on a new site.
- Judging by the remains, Bodiam Castle appears to have been more lavish.
- Bodiam's interval towers were rectangular rather than round.
- Bodiam has a huge moat.

Presumably, Hemyock Castle was more functional. Much of the accommodations would have been provided by the old manor house, so the defensive outer walls could be simpler. This would also have allowed the use of stronger, more functional round towers. The rectangular towers at Bodiam were required to provide comfortable accommodations. Further accommodation was built into Bodiam's outer walls.

Bodiam had a huge, if easily drained, moat and complex entrance causeways.

Hemyock's moat was more functional but not easily drained. Hemyock's western entrance (now lost) may have included a short defensive causeway.

The massive gatehouse at Hemyock's eastern entrance was protected by a drawbridge across the moat and by outer bastions (now the "Guard Houses" holiday cottages). In recent centuries there have been great changes around Hemyock's eastern entrance, including diversion of the river (St. Margaret's Brook) and extension of St. Mary's Church. It is just possible that there was a more complex series of defenses and water obstacles.

Source: Hemyock Castle, http://www.hemyockcastle.co.uk/bodiam.htm.

By comparing and contrasting two similar things, you can give your readers a deeper understanding of both.

BETTER AND WORSE In technical workplaces, you are often faced with moments in which you need to choose among different paths. In these cases, you may need to play the advantages off the disadvantages.

The "better" is discussed in terms of advantages.

Automating our assembly line with robotic workstations has clear advantages. With proper maintenance, robots can work around the clock, every day of the week. They don't take vacations, and they don't require benefits. Moreover, after an initial up-front investment, they are less expensive per unit than human labor.

The "worse" is shown in a lesser light.

Our alternative to automation is to stay with human labor. Increasingly, we will become less profitable, because our competitors are moving their operations to offshore facilities, where labor is much cheaper and environmental laws are routinely ignored. Meanwhile, the increasing costs of health care and an aging workforce will eventually force us to close some of our manufacturing plants in North America.

COSTS AND BENEFITS By directly weighing the *costs and benefits*, you can show readers that the price is ultimately worth the benefits of moving forward with a project.

Economic and Other Benefits to the State

The benefits of a biomass-to-ethanol production industry for California's economy are potentially greater than the cost of state support for such an industry. The economic analysis estimates statewide economic benefits of $1 billion over a 20-year period, assuming state government incentives totaling $500 million for a 200-million-gallon-per-year ethanol industry.

The economic benefits of an in-state ethanol industry result from feedstock handling and processing activities, ethanol plant construction and operation, and product marketing. All contribute income to California's economy, due primarily to employment.

Important environmental benefits also stand to be realized by a California biomass-to-ethanol industry, although these benefits are difficult to quantify in monetary terms. Nevertheless, environmental benefits would be real and should be considered in public policymaking regarding development of such an industry.

Source: California Energy Commission, 2001, p. x.

Stressing the benefits to your readers is always a good way to reason with them. By putting the costs in contrast to these benefits, you can show how the advantages ultimately outweigh the price.

IF . . . THEN Perhaps the most common way to reason is using *if . . . then* sentences and paragraphs. Essentially, you are saying "If you believe in X, then you should do Y" or perhaps "If X happens, then Y is likely to happen also" (Figure 6.11).

When using if . . . then arguments, you are leveraging something the readers already believe or consider possible to convince them that they should believe or do something further.

EITHER . . . OR When using *either . . . or* statements, you are offering the readers a choice or showing them two sides of the issue. You are saying "Either you believe X, or you believe Y" or perhaps "Either X will happen, or Y will happen" (Figure 6.12).

To see other documents that argue costs and benefits, go to **www.ablongman.com/johnsonweb/6.11**

A Section Using If . . . Then

Figure 6.11: If . . . then patterns are used to build arguments on uncertainties.

The opening sets a premise with two if . . . then statements.

The body of the section explores the implications of those if . . . then statements.

The closing poses if . . . then statements that are more speculative.

Effects of Climate Change on Forests

The projected 2°C (3.6°F) warming could shift the ideal range for many North American forest species by about 300 km (200 mi.) to the north. If the climate changes slowly enough, warmer temperatures may enable the trees to colonize north into areas that are currently too cold, at about the same rate as southern areas became too hot and dry for the species to survive. If the earth warms 2°C (3.6°F) in 100 years, however, the species would have to migrate about 2 miles every year.

Trees whose seeds are spread by birds may be able to spread at that rate. But neither trees whose seeds are carried by the wind, nor nut-bearing trees such as oaks, are likely to spread by more than a few hundred feet per year. Poor soils may also limit the rate at which tree species can spread north. Thus, the range over which a particular species is found may tend to be squeezed as southern areas become inhospitably hot. The net result is that some forests may tend to have a less diverse mix of tree species.

Several other impacts associated with changing climate further complicate the picture. On the positive side, CO^2 has a beneficial fertilization effect on plants, and also enables plants to use water more efficiently. These effects might enable some species to resist the adverse effects of warmer temperatures or drier soils. On the negative side, forest fires are likely to become more frequent and severe if soils become drier. Changes in pest populations could further increase the stress on forests. Managed forests may tend to be less vulnerable than unmanaged forests, because the managers will be able to shift to tree species appropriate for the warmer climate.

Perhaps the most important complicating factor is uncertainty (see U.S. Climate in the Future Climate section) whether particular regions will become wetter or drier. If climate becomes wetter, then forests are likely to expand toward rangelands and other areas that are dry today; if climate becomes drier, then forests will retreat away from those areas. Because of these fundamental uncertainties, existing studies of the impact of climate change have ambiguous results.

Source: U.S. Environmental Protection Agency, http://yosemite.epa.gov/oar/globalwarming.nsf/content/ImpactsForests.html.

Statements using either . . . or patterns suggest that there is no middle ground, so you should use this strategy only when appropriate. When used appropriately, an either/or argument can prompt your readers to make a decision. When used inappropriately, these arguments can invite the readers to make a choice you didn't expect—or perhaps to make no choice at all.

A Section Using Either . . . Or

Figure 6.12: The either . . . or pattern is helpful when only one belief or action is the right one.

The opening sets up two sides with either . . . or statements.

The remainder of the section discusses the two sides.

Here, the section ends by saying which side won the debate.

Shapley-Curtis Debate

During the 1920s, Harlow Shapley and Heber Curtis debated whether observed fuzzy "spiral nebulae," which are today known as galaxies, exist either in our Milky Way Galaxy or beyond it. Shapley took the stance that these "spiral nebulae" observed in visible light do indeed exist in our galaxy, while the more conservative Curtis believed in the converse. While both sides of this debate seemed to either make too many suppositions or supported their claims with flawed data, the main points of both stances made the debate fairly close at the time.

Shapley's Argument

Shapley argues that these "spiral nebulae" are actually within our galaxy's halo. He presented faulty data which seemed to confirm that M101, a large angular-diameter galaxy, changed angular size noticeably. We know now that for M101 to change angular size appreciably (0.02 arcsecond/year), it would have to recede from us at warp speeds based upon our current knowledge of its diameter and distance. To account for all the galaxies' observed recession from earth, not knowing of the universe's expansion, Shapley invented a special repulsion force to explain this strange phenomenon. He was not cogently able to explain why galaxies observed from earth are less densely distributed along the galactic equator, though.

Curtis's Argument

Curtis remained more conservative in his argument by not introducing outlandish suppositions like the special repulsion forces of Shapley. Curtis, although not able to explain galaxies' redshifts well either, made a convincing claim for the distribution of galaxies in the sky. Because observed galaxies seen edge-on often contain opaque gas lanes, and if our galaxy is similar to those observed galaxies, then we should not see extragalactic objects near our equator because of our own opaque dust lane. Hubble's observation of Cepheid variables in the Andromeda Galaxy confirmed the distance to a galaxy: millions of light-years, not thousands! Curtis was obviously correct.

Source: A. Aversa, 2003.

CHRONOLOGICAL ORDER Time offers its own logic because events happen in chronological order. You can arrange information logically according to the sequence of events.

> Three things happen inside the bronchial tubes and airways in the lungs of people with asthma. The first change is inflammation: The tubes become red, irritated, and swollen. This inflamed tissue "weeps," producing thick mucus. If the inflammation persists, it can lead to permanent thickening in the airways.
>
> Next comes constriction: The muscles around the bronchial tubes tighten, causing the airways to narrow. This is called *bronchospasm* or *bronchoconstriction*.

GO TO THE NET

Want to see other models that use chronological order? Go to **www.ablongman.com/johnsonweb/6.12**

Finally, there's hyperreactivity. The chronically inflamed and constricted airways become highly reactive to so-called triggers: things like allergens (animal dander, dust mites, molds, pollens), irritants (tobacco smoke, strong odors, car and factory emissions), and infections (flu, the common cold). These triggers result in progressively more inflammation and constriction.

Source: G. Shapiro, 1998.

PROBLEM/NEEDS/SOLUTION The *problem/needs/solution* pattern is a common organizational scheme in technical documents, because technical work often involves solving problems. When using this pattern, you should start by identifying the problem that needs to be solved. Then, state what is needed to solve the problem. Finally, end the section by stating the solution (Figure 6.13).

The three-part structure leads the readers logically from the problem to the solution.

A Section Describing Problems, Needs, and Solutions

Figure 6.13: The problem/needs/solution pattern is a good way to lead people logically toward taking action.

North Creek Water Cleanup Plan (TMDL)

Opening paragraph identifies the problem.

North Creek does not meet state standards for swimming and wading because there is too much bacteria in the water. Also, the federal government has determined that Chinook salmon are threatened, and other salmon species face continuing pressure from urban development.

In the 1960s, much of the watershed was home to small ranches and hobby farms. Over the past 40 years, much of the land has been redeveloped with a trend towards more urban, commercial, and suburban residential development. The basin's hydrology, how water is stored and managed throughout the basin, has also changed.

Why Are These Waters Polluted?

Pollution in the North Creek watershed comes from thousands of sources that may not have clearly identifiable emission points; this category of pollution is called "nonpoint" pollution. These nonpoint sources can contribute a variety of pollutants that may come from failing septic systems, livestock and pet wastes, at-home car washing, lawn and garden care, leaky machinery, and other daily activities. Some of these nonpoint sources create fecal coliform bacterial pollution that indicate the presence of fecal wastes from warm-blooded animals. Ecology has confirmed that high levels of fecal coliform bacteria exist in North Creek. For this reason, Total Maximum Daily Loads

The section's body elaborates on the problem.

Continued on following page

Source: Washington State Department of Ecology, http://www.ecy.wa.gov/programs/wq/tmdl/watershed/north_creek/solution.html.

Figure 6.13: *(continued.)*

for fecal coliform bacteria were subsequently established at multiple locations through each watershed.

Although wildlife can also contribute bacteria, such sources are not defined as pollution; however, when such natural sources combine with nonpoint pollution, the result can cause the kind of problems found in North Creek.

North Creek became polluted because of the way we do things, not the activities themselves. For example, having dogs, cats, horses, and other animals as part of our life is not a problem; rather, it is the way that we care for these animals. Similarly, roads and parking lots are a necessity of our modern society, but the way we build roads, neighborhoods, and shopping centers is causing our local streams and creeks to be polluted. There are solutions that can be undertaken by local governments, businesses, organizations, and citizens to solve the problem.

What Can You Do to Improve North Creek?

These questions are designed to highlight what is needed.

If cleaning up local waters is important to you, think about what you can do on your own first. Do you always pick up after your pet? Can you use organic fertilizer? Do you wash your car on your lawn or take it to a car wash? Can you reduce the amount of stormwater runoff from your property? Can you develop a farm plan to ensure your horse's manure is not reaching local streams? Do you practice good on-site septic system maintenance? The draft Action Plan includes information about current and future activities to clean up local waters.

There are many things residents can do now to reduce pollution reaching water bodies and to improve water quality. Here are some ways that you can help:

Here are the recommended solutions to the problem.

- Be responsible for proper septic tank maintenance or repair. If you have questions about your on-site septic system (exactly where it is located, how to maintain it), you can call the Snohomish Health District for technical assistance.
- Can you reduce stormwater leaving your property? To have a free survey of your property for ways to reduce potential water quality problems and improve stormwater management, contact Craig Young, your North Creek Basin steward.
- Keep pet and other animal wastes out of your local streams. Pick up after your pet and work with your community, association, or local government to get a pet waste collection station installed where it is needed most.

Want to see people using examples in their documents? See **www.ablongman.com/johnsonweb/6.13**

Figure 6.13: (*continued.*)

- Use landscaping methods that eliminate or reduce fertilizers and pesticides. If fertilizers are needed, organic products break down more slowly and help prevent big flushes of pollution when we have heavy rains; they also improve soil structure.
- Join local volunteers in planting trees and performing other activities that help local streams. Snohomish County, the Stilly/Snohomish Task Force and North Creek Streamkeepers, and the City of Bothell water quality volunteer program improve water quality by helping with stream restoration activities. They provide help or other opportunities to plant trees on your property or at other needed locations to help water quality (you can also volunteer to plant trees in other areas that need help too).
- Get involved in your local government's programs. Folks interested in sampling their local waters can contact Snohomish County, North Creek Streamkeepers, or your local city to explore the availability of volunteer monitoring opportunities. If you live outside of a city, be a Salmon Watcher or Watershed Keeper, or get involved in other individual or group activities to improve local waters; these are coordinated by Snohomish County staff.

The section closes by restating the main point and looking to the future.

How else can you be involved? The solution to polluting our local waters is to do some things a little differently. In this way, we can still live a normal 21st-century lifestyle, have animals as a close part of our lives, and have clean water. Citizen involvement in deciding what needs to be done is essential to making our water bodies safe places for people and fish, and you may be part of helping to design future watershed activities that haven't even been thought of yet! Check out the Related Links page for more ideas.

EXAMPLE Using an *example* is a good way to support your claims. Your readers might struggle with facts and data, but an example can help put all those technical details into a broader picture.

> To fool predators, most butterflies have evolved colors and patterns that allow them to survive. For example, the delicious Red-spotted Purple is often not eaten by birds, because it looks like a Pipevine Swallowtail, a far less appetizing insect. Other butterflies, like anglewings, look like tree bark or leaves when seen from below. If a predator comes near, though, anglewings can surprise it with a sudden burst of color from the topside of their wings (see Pyle, p. 23).

In this passage, the two examples support the claim made in the first sentence. To signal an example, you can use transitions like *for example, for instance, to illustrate, in fact, of course, specifically,* and *such as.*

Organizing and Drafting the Conclusion

Conclusions are very important, and yet they are often forgotten or poorly written. Here, at the end of your document, is where you are going to drive home your main

For sample conclusions that you can model, see **www.ablongman.com/johnsonweb/6.14**

point. An effective conclusion rounds out the discussion by bringing readers back to the subject, purpose, and main point of your document. Conclusions should be concise and to the point.

As a general guideline, new information should not appear in the conclusion, because it will only redirect your discussion at this crucial point. Instead, you should concentrate on summarizing your main points.

Again, put yourself in your readers' place. As a document concludes, what do you want to know? More than likely, you want answers to these questions:

What is the main point?

Why is this information important to me?

Where do we go from here?

Five Closing Moves in a Conclusion

These concluding questions translate into five moves typically found in a conclusion:

- Make an obvious transition.
- Restate your main point.
- Restress the importance of the subject.
- Look to the future.
- Say thank you and offer contact information.

These five concluding moves will help round off your document, reestablishing the context that you created in the introduction. As with the introduction, sometimes it is helpful to type in these five moves separately on your computer screen. Then, later, you can craft them into a smooth, coherent conclusion.

MOVE 1: MAKE AN OBVIOUS TRANSITION After reading the body of your document, your readers need to wake up a bit. By using a heading such as "Final Points" or a transitional phrase such as "To sum up," you will signal to the readers that you are going to tell them your main points. Your readers will sit up and begin reading closely again. Here are some transitions that will wake up your readers:

In conclusion,	Put briefly,	Overall,
To sum up,	In brief,	As a whole,
Let us sum up,	Finally,	In the end,
In summary,	To finish up,	On the whole,
In closing,	Ultimately,	

The readers' heightened attention will last only a little while, perhaps for a paragraph or a page. If your conclusion runs more than a page or two, your readers will lose interest and even become annoyed by its length.

LINK For more information on using transitions, see Chapter 7, page 174.

MOVE 2: RESTATE YOUR MAIN POINT You have probably already stated your main point a couple of times in the document. In the conclusion, you need to restate this point one more time to drive it home. After all, your readers now have all the facts, so they should be ready to make a final decision.

If Darbey is to survive and thrive, we need to take action now to address its increasing flood problem. By restoring wetlands, developing greenways, and building levees, we can begin preparing for the flooding problems that are almost certainly a risk in the future.

If you need to be blunt about your main point, be blunt. Your main point should be absolutely obvious to readers as they finish the document. If you need to highlight your main point, you may need to say something direct, such as, "Our main point is . . .". Make sure you drive your point home.

MOVE 3: RESTRESS THE IMPORTANCE OF THE SUBJECT Sometimes readers need to be reminded why the subject of your document is important to them.

If we can reduce or eliminate flooding in Darbey, we will save our citizens millions of dollars in lost revenues and reconstruction. Moreover, Darbey will be viewed as a place with a future, because flooding will not continually undo all our hard work.

Don't try to scare the readers at this point. Stay positive. Stress the benefits of accepting your ideas or agreeing to your recommendations. Scaring your readers will take the shine off your document at this crucial moment.

MOVE 4: LOOK TO THE FUTURE Looking to the future is a good way to end any document (Figure 6.14). A sentence or paragraph that looks to the future will leave your readers with a positive image.

Sample Conclusion

Conclusion

In conclusion, if Darbey is to survive and thrive, we need to take action now to address its increasing flood problem. By restoring wetlands, developing greenways, and building levees, we can begin preparing for the flooding problems that are almost certainly a risk in the future.

The benefits clearly outweigh the costs. If we can reduce or eliminate flooding in Darbey, we will save our citizens millions of dollars in lost revenues and reconstruction. Moreover, Darbey will be viewed as a place with a future, because flooding will not continually undo all our hard work. When we have effectively managed the Curlew River, Darbey will likely see steady growth in population and industry. Once its reputation for flooding has been removed, people and businesses will likely move to this area for its riverside charm and outdoor activities. The town would experience a true revival.

We appreciate your time and consideration. If you have any questions or would like to meet with us about this report, please contact the task force leader, Mary Subbock, at 555-0912 or e-mail her at msubbock@cdarbey.gov.

Figure 6.14: This conclusion demonstrates all five closing moves that might be found in a document.

For more help with writing conclusions, go to **www.ablongman.com/johnsonweb/6.15**

When we have effectively managed the Curlew River, Darbey will likely see steady growth in population and industry. Once its reputation for flooding has been removed, people and businesses will likely move to this area for its riverside charm and outdoor activities. The town would experience a true revival.

MOVE 5: SAY THANK YOU AND OFFER CONTACT INFORMATION You might end your document by saying thank you and offering contact information.

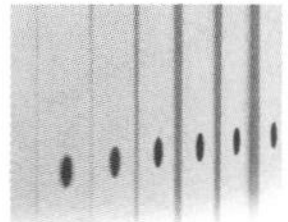

AT A GLANCE

Five Moves in a Conclusion

- Move 1: Make an obvious transition.
- Move 2: Restate your main point.
- Move 3: Restress the importance of the subject.
- Move 4: Look to the future.
- Move 5: Say thank you and offer contact information.

We appreciate your time and consideration. If you have any questions or would like to meet with us about this report, please contact the task force leader, Mary Subbock, at 555-0912 or e-mail her at msubbock@cdarbey.gov.

This kind of thank you statement leaves readers with a positive feeling and invites them to contact you if they need more information.

CHAPTER REVIEW

- The beginning of a document (introduction) builds a context. The middle (body) provides the content. And, the end (conclusion) rebuilds the context.
- A genre is a predictable pattern for organizing information to achieve specific purposes.
- Outlining may seem old-fashioned, but it is a very effective way to sketch out the organization of a document.
- Presentation software can help you organize complicated information.
- Introductions usually include up to six opening moves: (1) define your subject, (2) state your purpose, (3) state your main point, (4) stress the importance of the subject, (5) provide background information, and (6) forecast the content.
- The body of larger documents is usually carved into sections. Each section has an opening, a body, and perhaps a closing.
- Sections usually follow patterns of arrangement to organize information.
- Conclusions usually include up to five closing moves: (1) make an obvious transition, (2) restate your main point, (3) restress the importance of the subject, (4) look to the future, and (5) say thank you and offer contact information.

Sample documents for these exercises can be found at **www.ablongman.com/johnsonweb/6.16**

EXERCISES AND PROJECTS

Individual and Team Projects

1. Find a document that interests you at www.ablongman.com/johnsonweb/6.16 (or you can find one of your own). Write a memo to your instructor in which you critique the organization of the document. Does it have a clear beginning, middle, and end? Does the introduction make some or all of the opening six moves? Is the body divided into sections? Does the conclusion make some or all of the closing five moves? Explain to your instructor how the document's organization might be improved.

2. Find an introduction that you think is ineffective. Then, rewrite the introduction, so that it includes a clear subject, purpose, and main point. Also, stress the importance of the document's subject, provide some background information, and forecast the structure of the document's body. Sample introductions are available at www.ablongman.com/johnsonweb/6.16.

3. On the Internet, find a homepage that makes some or all of the six opening moves discussed in this chapter. Write a memo to your instructor in which you compare and contrast how homepages use the opening moves similarly and differently than introductions in paper-based documents.

4. On the Internet or in a print document, find a section that uses one of the following organizational patterns:

   - Cause and Effect
   - Comparison and Contrast
   - Better and Worse
   - Costs and Benefits
   - If . . . Then
   - Either . . . Or
   - Problem/Needs/Solution
   - Example.

   Prepare a brief presentation to your class in which you show how the text you found follows the pattern.

   You can find sample documents at www.ablongman.com/johnsonweb/6.16 that will use some or all of these organizational patterns.

Collaborative Project

Around campus or at www.ablongman.com/johnsonweb/6.16, find a large document, perhaps a report or proposal. With presentation software, outline the document by creating a slide for each section or subsection. Use the headings of the sections as titles on your slides. Then, on each slide use bulleted lists to highlight the important points in each section.

When you have finished outlining, look over the presentation. Are there any places where too much information was offered? Are there places where more information was needed? Where would you rearrange information to make it more effective?

Present your findings to your class. Using the presentation software, show the class how the document you studied is organized. Then, discuss some improvements you might make to the organization to highlight important information.

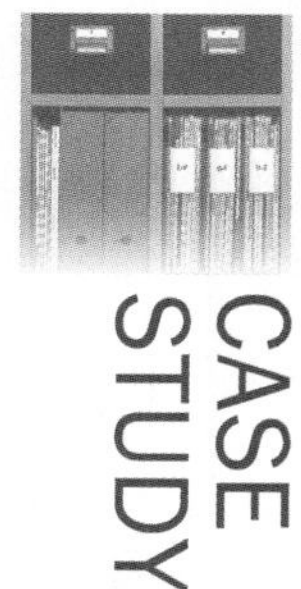

CASE STUDY

The Bad News

In most ways, the project had been a failure. Lisa Franklin was on a chemical engineering team developing a polymer that would protect lightweight tents against desert elements like extreme sun, sandstorms, and chewing insects. Hikers and campers were interested in tents with this kind of protection. But the company Lisa worked for, Outdoor Solutions, wanted to begin selling tents to the military. Bulk sales to the U.S. Army and Marines would improve the company's bottom line, as well as open new opportunities to sell other products.

Despite a promising start, Lisa and the other researchers were having little success developing a stable polymer that would provide the desired protection. The best polymers they had developed were highly flammable. Tent materials covered with these polymers went up like an inferno when exposed to flame. The nonflammable polymers, meanwhile, seemed to break down within three months of use in desert conditions. Lisa's boss, Jim Franklin, was convinced they were close to a breakthrough, but they just couldn't find the right combination to create a nonflammable polymer that would last.

Unfortunately, Jim had been giving the company president the impression that the polymer was already a success. So, the president had secured a million-dollar loan to retool a factory to produce new lightweight tents coated with the polymer. Renovation was due to begin in a month.

Before starting the factory retooling, the president of the company asked for a final update on the polymer. So, Jim wrote a progress report. Then, he gave the report to Lisa for final revisions, because he was going out of town on a business trip. He said, "Do whatever you want to revise the report. I'm so frustrated with this project, I don't want to look at this report anymore. When you're done, send the report to the company president. I don't need to see it again."

In the report, Lisa noticed, Jim's facts were all true, but the organization of the report hid the fact that they had not developed a workable polymer. For example, when Jim mentioned that their best polymers were highly flammable, he did so at the end of a long paragraph in the middle of the report. It was highly unlikely that the president would be reading closely enough to see that important fact.

Lisa knew Jim was trying to hide the research team's failure to develop a workable polymer. She too didn't want to admit that they had failed. But she also thought it was important that the company president understand that they had not developed a successful polymer. After all, there was a lot of money on the line.

If you were in Lisa's place, how would you handle this situation?

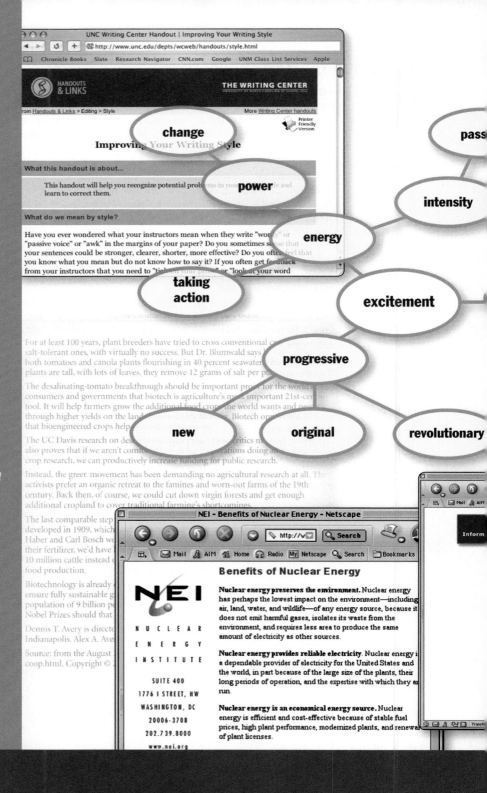

CHAPTER

7

Using Plain and Persuasive Style

CHAPTER CONTENTS

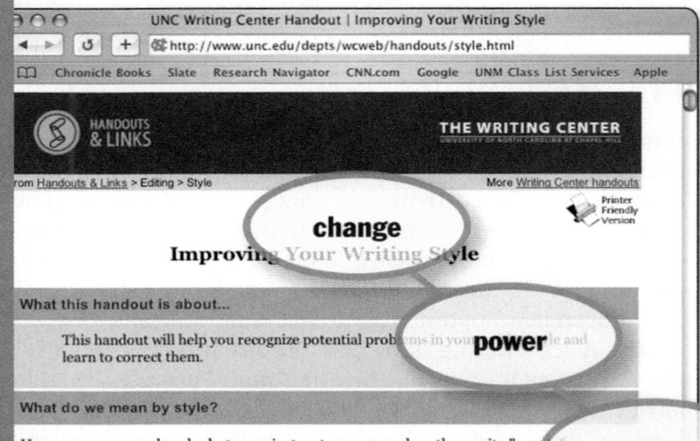

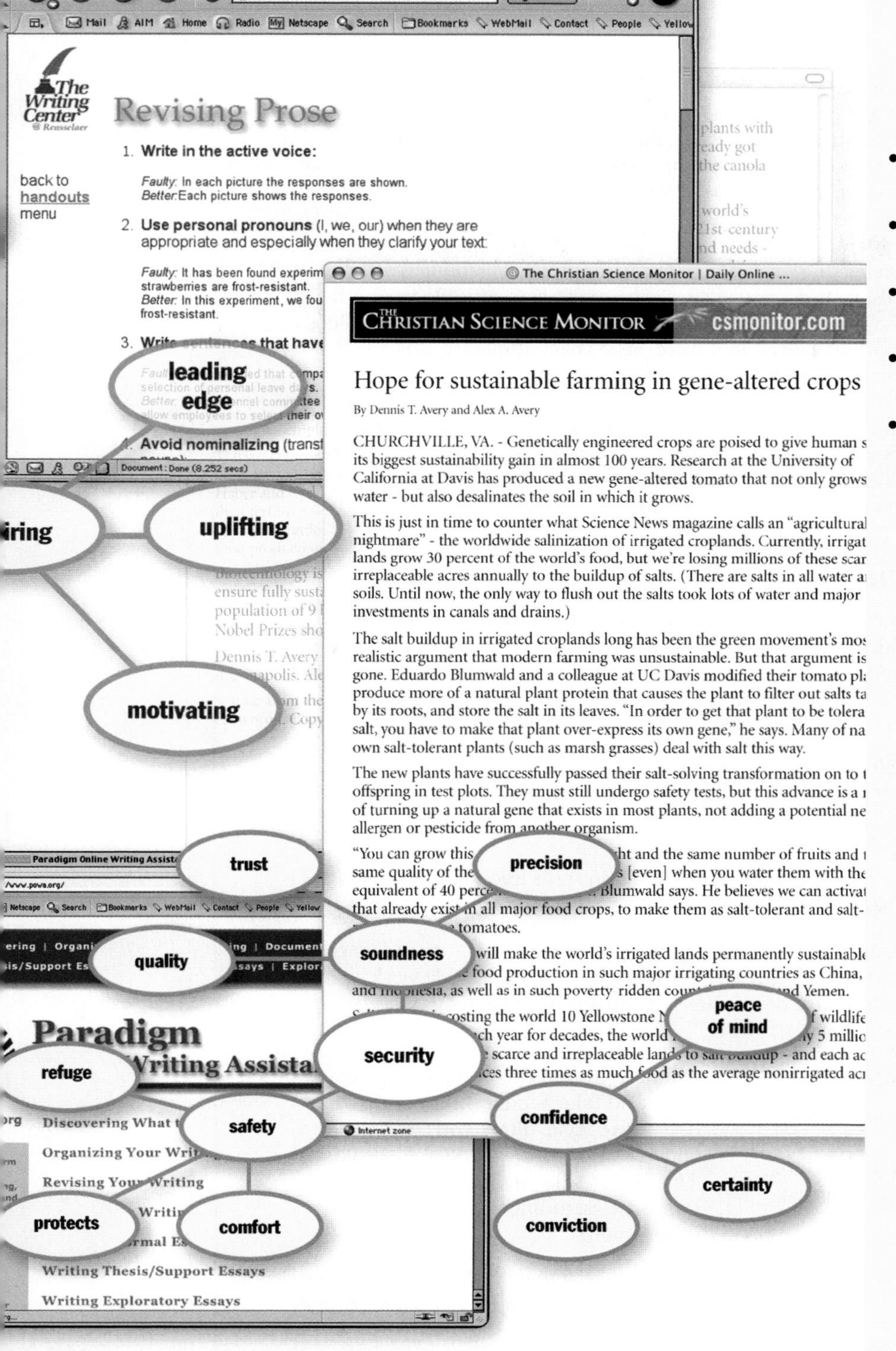

CHAPTER OBJECTIVES

In this chapter, you will learn:

- The importance of style in technical communication.
- The differences between plain and persuasive style.
- Strategies for writing plain sentences and paragraphs.
- How to tell when it's appropriate to use persuasive style.
- How to use techniques of persuasive style, including tone, similes, analogies, metaphors, and pace.

Some people believe that style is not important in technical communication. After all, they say, only the content matters. The rest is just window dressing.

Quite the opposite is true. In many cases, the style of your document says as much about your subject as the content. For example, as an engineer, if you have an exciting idea for a new product, you want to reflect that excitement in your proposal. As a doctor, if you are concerned about a specific problem at your hospital, you want your report's style to convey your concern.

An important guideline to remember is that *how you say something is often what you say.* In other words, your document's style expresses your attitude toward the work. It reflects your character by embodying the values, beliefs, and relationships you want to share with the readers. In a word, style is about quality. It is about your and your company's commitment to excellence.

One of the great advantages of computers is the ability to move text around to improve style. By learning some rather simple techniques, you can make your writing clearer and more persuasive. By putting words and sentences in the proper places, you can make your documents more interesting and easier to understand. In this chapter, you will learn how to use these easy-to-learn stylistic techniques.

What Is Style?

In technical communication, style is not embellishment or ornamentation. Style is not artificial flavoring added to a document at the last minute to make the content more "interesting." A few added adjectives or exclamation marks won't perk up the bland parts or make the text more colorful.

Good style goes beyond these kinds of superficial cosmetic changes. It involves

- choosing the right words and phrases.
- structuring sentences and paragraphs for clarity.
- using an appropriate tone.
- adding a visual sense to the text.

Historically, rhetoricians have classified style into three levels: plain style, persuasive style, and grand style.

Plain style—Plain style stresses clear wording and simple prose. It is most often used to instruct, teach, or present information. Plain style works best in documents like technical descriptions, instructions, and activity reports.

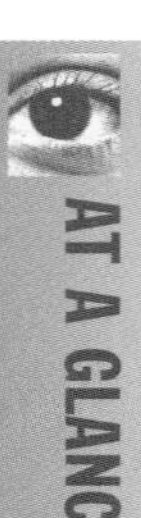

Two Common Styles in Technical Communication

- Plain Style—stresses clear wording and simple prose.
- Persuasive Style—is used to influence people to accept your ideas and take action.

Persuasive style—There are times when you will need to influence people to accept your ideas and take action. In these situations, persuasive style allows you to add energy and vision to your writing and speaking. This style works best with proposals, letters, articles, public presentations, and some reports.

Grand style—Grand style stresses eloquence. For example, Martin Luther King, Jr., and

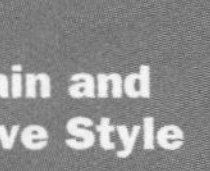

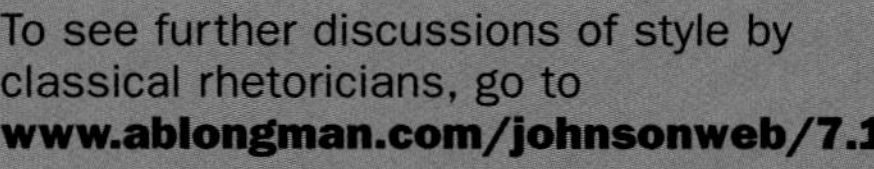
To see further discussions of style by classical rhetoricians, go to **www.ablongman.com/johnsonweb/7.1**

President John F. Kennedy often used the grand style to move their listeners to do what was right, even if they were reluctant to do it.

In this chapter, we will concentrate solely on plain and persuasive style, because they are most common in technical documents. Grand style is rarely used in technical communication, because it often sounds too ornate or formal for the workplace.

Writing Plain Sentences

In the past, you have probably been advised to "write clearly" or "write in concrete language" as though making up your mind to do so was all it took. In reality, writing plainly is a skill that requires practice and concentration. Fortunately, once a few basic guidelines have been mastered, plain style will soon become a natural strength in your writing.

Your computer can be your partner as you work toward plainer writing, because it allows you to quickly move words and phrases around. You can sharpen your sentences by putting subjects and verbs in the right places, while eliminating the phrasing problems that make your meaning unclear.

Basic Parts of a Sentence

To start, let us consider the parts of a basic sentence. A sentence in English typically has three main parts: a subject, a verb, and a comment.

Subject—What the sentence is about.

Verb—What the subject is doing.

Comment—What is being said about the subject.

English is a flexible language, allowing sentences to be organized in a variety of different ways. For example, consider these three variations of the same sentence:

Subject	Verb	Comment
The Institute	provided	the government with accurate crime statistics.
The government	was provided	with accurate crime statistics by the Institute.
Crime statistics	were provided	to the government by the Institute.

Notice that the *content* in these sentences has not changed, only the order of the parts. Nevertheless, the emphasis in each sentence changes when the subject is changed. The first sentence is *about* "The Institute." The second sentence is *about* "the government." The last sentence is *about* "crime statistics." By changing the subject of the sentence, you essentially shift its focus, drawing your readers' attention to different issues.

Eight Guidelines for Plain Sentences

This understanding of the different parts of a sentence is the basis for eight guidelines that can be used to write plainer sentences in technical documents.

For more information on how sentences are constructed, see **www.ablongman.com/johnsonweb/7.2**

GUIDELINE 1: THE SUBJECT OF THE SENTENCE SHOULD BE WHAT THE SENTENCE IS ABOUT. Confusion often creeps into texts when readers cannot easily identify the subjects of the sentences. For example, what is the subject of the following sentence?

1. Ten months after the Hartford Project began in which a team of our experts conducted close observations of management actions, our final conclusion is that the scarcity of monetary funds is at the basis of the inability of Hartford Industries to appropriate resources to essential projects that have the greatest necessity.

This sentence is difficult to read for a variety of reasons, but the most significant problem is the lack of a clear subject. What is this sentence about? The word *conclusion* is currently in the subject position, but the sentence might also be about "the experts," "the Hartford Project," or "the scarcity of monetary funds."

When you run into a sentence like this one, first decide what the sentence is about. Then, cut and paste to move that subject into the subject slot of the sentence. For example, when this sentence is restructured around "experts," readers will find it easier to understand:

The subject (underlined) is what this sentence is "about." →

1. Ten months after the Hartford Project began, our experts have concluded through close observations of management actions that the scarcity of monetary funds is at the basis of the inability of Hartford Industries to appropriate resources to essential projects that have the greatest necessity.

This sentence is still rather difficult to read. Nevertheless, it is easier to read because the noun in the subject slot ("experts") is what the sentence is about. We will return to this sentence later in this chapter.

GUIDELINE 2: THE SUBJECT SHOULD BE THE "DOER" IN THE SENTENCE. In your opinion, which revision of sentence 1 above is easier to read? Most people would point to sentence 1a, in which "experts" is in the subject slot. Why? In sentence 1a, "experts" (people) are actually doing something. In sentence 1b, "scarcity" is an inactive noun that is not doing anything. Whereas experts do things, scarcity is merely something that happens.

Readers tend to focus on who or what is doing something in a sentence. For example, which of the following sentences is easier to read?

2a. On Saturday morning, the paperwork was completed in a timely fashion by Jim.

2b. On Saturday morning, Jim completed the paperwork in a timely fashion.

Most people would say sentence 2b is easier to read because Jim, the subject of the sentence, is actually doing something. In the first sentence, the paperwork is merely sitting there. People make especially good subjects of sentences, because they are usually active.

GUIDELINE 3: THE VERB SHOULD STATE THE ACTION, OR WHAT THE DOER IS DOING. Once you have determined who or what is doing something, ask yourself what that person or thing is actually doing. Find the *action* and turn it into the verb of the sentence. For example, consider these sentences:

For some interesting discussions of plain style, go to **www.ablongman.com/johnsonweb/7.3**

In these sentences, it becomes harder and harder to figure out the action.

3a. The detective investigated the loss of the payroll.

3b. The detective conducted an investigation into the loss of the payroll.

3c. The detective is the person who is conducting an investigation of the loss of the payroll.

Sentence 3a is easier to understand because the action of the sentence is expressed in the verb. Sentences 3b and 3c are increasingly more difficult to understand, because the action, "investigate," is further removed from the verb of the sentence.

HELP

Using On-line Writing Centers

On-line writing centers, also called on-line writing labs (OWLs), are helpful websites that will help you improve your writing. They are typically written and supported by people at universities, often students. These sites offer you quickly accessible resources for improving your writing style, checking your grammar, and documenting sources (Figure A).

On-line Writing Centers

You can print out this document for your use.

Here are specific tips about improving style.

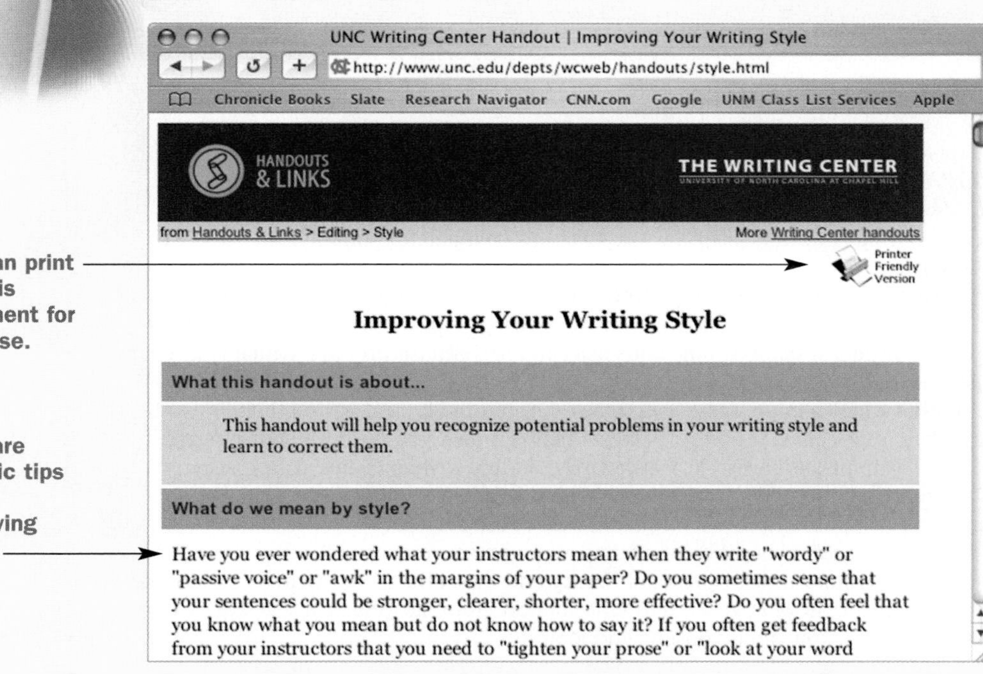

Figure A: On-line writing centers offer a wealth of information on improving writing style.

Source: The Writing Center, University of North Carolina at Chapel Hill, http://www.unc.edu/depts/wcweb.

Need more examples of revised sentences? Go to **www.ablongman.com/johnsonweb/7.4**

The Paradigm Online Writing Assistant

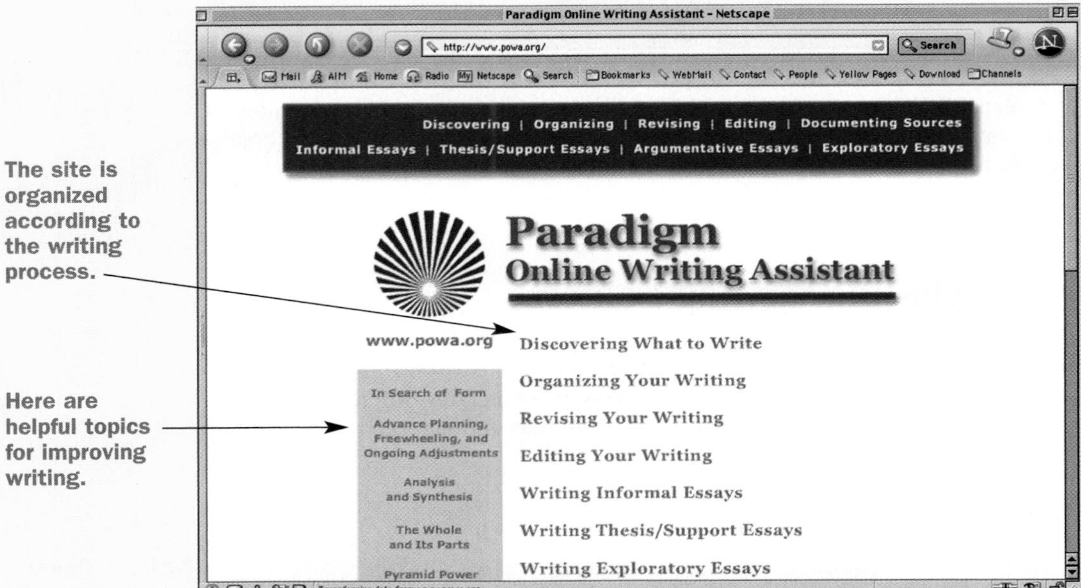

Figure B: The Paradigm Online Writing Assistant offers just about anything you might find in a writer's handbook. The sections on revising and editing are especially helpful for improving style.

Source: Paradigm Online Writing Assistant, http://www.powa.org.

Some of the better on-line writing centers include the ones at Purdue University (http://owl.english.purdue.edu), University of North Carolina (www.unc.edu/depts/wcweb), and Colorado State (http://writing.colostate.edu). The Paradigm Online Writing Assistant (http://www.powa.org) is a helpful nonuniversity resource (Figure B).

In many ways, on-line writing centers are like interactive versions of the writing handbooks you might have used in your first-year composition classes. Fortunately, they are a little less cumbersome and easier to use, because they are available at your fingertips.

In other ways, though, these websites go beyond the typical writing handbook, because they also include useful worksheets that can help you improve your writing style. For example, as shown in Figure C, the on-line writing center at Rensselaer Polytechnic Institute (RPI) offers some helpful handouts that you can print out for your own use (go to http://www.rpi.edu/web/writingcenter).

There are a variety of different on-line writing resources available. You don't even need to root around in your personal library and dust off that old writer's handbook from your freshman year. When in doubt about a grammar rule or if you want to improve your style, you can go on-line and look it up.

GO TO THE NET

To find on-line writing centers, go to **www.ablongman.com/johnsonweb/7.5**

Downloadable Worksheets on Style

Here are strategies for improving writing style.

These examples demonstrate "faulty" and "better" style.

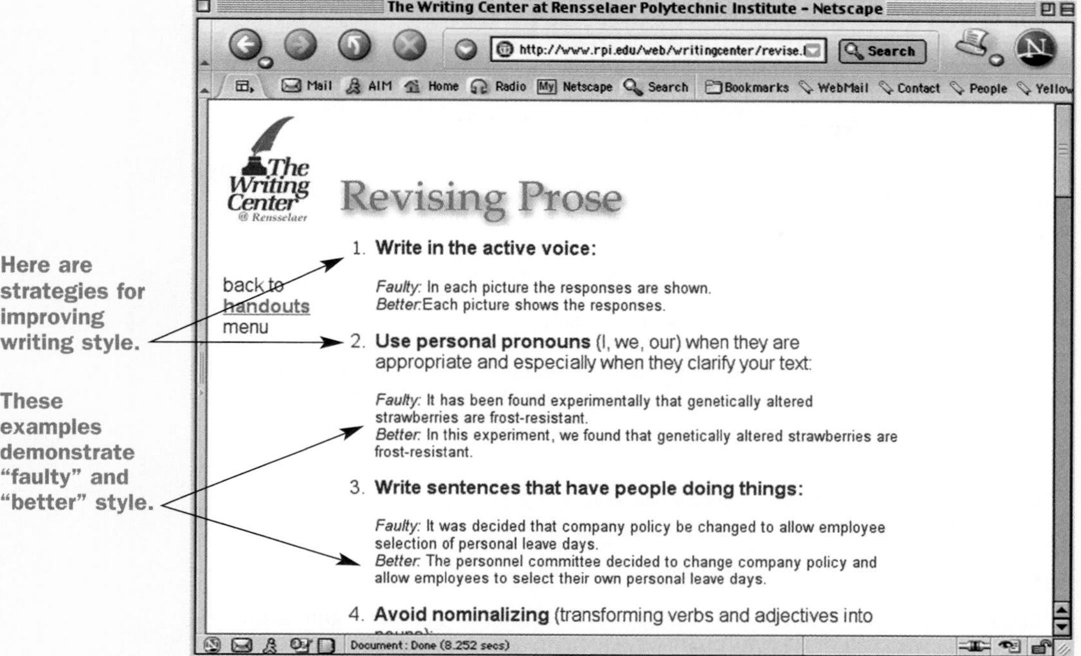
The Writing Center at Rensselaer Polytechnic Institute - Netscape
http://www.rpi.edu/web/writingcenter/revise.
Search
Mail
AIM
Home
Radio
Netscape
Search
Bookmarks
WebMail
Contact
People
The Writing Center @ Rensselaer
Revising Prose
back to handouts menu
1. Write in the active voice:
Faulty: In each picture the responses are shown.
Better: Each picture shows the responses.
2. Use personal pronouns (I, we, our) when they are appropriate and especially when they clarify your text:
Faulty: It has been found experimentally that genetically altered strawberries are frost-resistant.
Better: In this experiment, we found that genetically altered strawberries are frost-resistant.
3. Write sentences that have people doing things:
Faulty: It was decided that company policy be changed to allow employee selection of personal leave days.
Better: The personnel committee decided to change company policy and allow employees to select their own personal leave days.
4. Avoid nominalizing (transforming verbs and adjectives into
Document: Done (8.252 secs)

Source: The Writing Center at Rensselaer Polytechnic Institute, http://www.rpi.edu/web/writingcenter.

Figure C: Many on-line writing centers offer worksheets that can help you improve your style. In this screen, The Writing Center at Rensselaer Polytechnic Institute offers tips for revising prose.

GUIDELINE 4: THE SUBJECT OF THE SENTENCE SHOULD COME EARLY IN THE SENTENCE. Subconsciously, your readers start every sentence looking for the subject. The subject anchors the sentence, because it tells readers what the sentence is about. So, if the subject is buried somewhere in the middle of the sentence, readers will have greater difficulty finding it, and the sentence will be harder to understand.

Consider these two sentences:

4a. If deciduous and evergreen trees experience yet another year of drought like the one observed in 1997, the entire <u>Sandia Mountain ecosystem</u> will be heavily damaged.

It's easier to locate the subject of this sentence because it is up front. →

4b. <u>The entire Sandia Mountain ecosystem</u> will be heavily damaged if deciduous and evergreen trees experience yet another year of drought like the one observed in 1997.

The problem with sentence 4a is that it forces readers to hold all those details about trees, drought, and 1997 in short-term memory before it identifies the subject of the sentence. Readers almost feel a sense of relief when they come to the subject, because until that point they cannot figure out what the sentence is about.

GO TO THE NET

For websites that discuss bureaucratic language, go to **www.ablongman.com/johnsonweb/7.6**

Quite differently, sentence 4b tells readers what the sentence is about up front. With the subject early in the sentence, readers immediately know how to connect the comment with the subject.

TAKE NOTE Of course, introductory or transitional phrases do not always signal weak style. But when these phrases are used, they should be short and to the point. In most situations, longer introductory phrases should be moved to the end of the sentence.

GUIDELINE 5: ELIMINATE NOMINALIZATIONS. Nominalizations are perfectly good verbs and adjectives that have been turned into awkward nouns. For example, look at these sentences:

The first sentence is harder to read because the action is in a nominalization.

5a. Management has an expectation that the project will meet the deadline.

5b. Management expects the project to meet the deadline.

In sentence 5a, "expectation" is a nominalization. Here, a perfectly good verb is being used as a noun. Sentence 5b is not only shorter than sentence 5a, it also has more energy because the verb "expects" is now an action verb.

Consider these two sentences:

Here are a whole bunch of nominalizations.

6a. Our discussion about the matter allowed us to make a decision on the acquisition of the new x-ray machine.

See how reducing them improves the readability?

6b. We discussed the matter and decided to acquire the new x-ray machine.

Sentence 6a includes three nominalizations, "discussion," "decision," and "acquisition," making the sentence hard to understand. Sentence 6b is clearer, because the nominalizations "discussion" and "decision" have been turned into action verbs.

Why do writers use nominalizations in the first place? They use nominalizations for two reasons:

- First, humans generally think in terms of people, places, and things (nouns), so our first drafts are often filled with nominalizations, which are nouns. While revising, an effective writer will turn those first-draft nominalizations into action verbs.
- Second, some people mistakenly believe that using nominalizations makes their writing sound more formal or important. In reality, though, nominalizations only make sentences harder to read. The best way to sound important is to write sentences that readers can understand.

GUIDELINE 6: AVOID EXCESSIVE PREPOSITIONAL PHRASES. Prepositional phrases are necessary in writing, but they are often overused in ways that make text too long and too tedious. Prepositional phrases follow prepositions like *in, of, by, about, over,* and *under.* These phrases are used to modify nouns.

Prepositional phrases become problematic when used in excess. For example, sentence 7a is difficult to read because it chains together too many prepositional phrases (the prepositional phrases are underlined and the prepositions italicized). Sentence 7b shows the same sentence with fewer prepositional phrases:

For more practice eliminating prepositional phrases, go to **www.ablongman.com/johnsonweb/7.7**

The prepositional phrases have been reduced, clarifying the sentence's meaning.

7a. The decline *in* the number *of* businesses owned *by* locals *in* the town *of* Artesia is a demonstration *of* the increasing hardship faced *in* rural communities *in* the southwest.

7b. Artesia's declining number of locally owned businesses demonstrates the increased hardship faced by southwestern rural communities.

You should never feel obligated to eliminate all the prepositional phrases in a sentence. Rather, look for places where prepositional phrases are chained together into long sequences. Then, try to condense the sentence by turning some of the prepositional phrases into adjectives.

In sentence 7b, for example, the phrase "in the town of Artesia" was reduced to the adjective "Artesia's." The phrase "in rural communities in the southwest" was reduced to "southwestern rural communities." As a result, sentence 7b is much shorter and easier to read.

Figure 7.1 shows a website in which the authors write plainly. Notice how they have used minimal prepositional phrases.

Writing Plainly

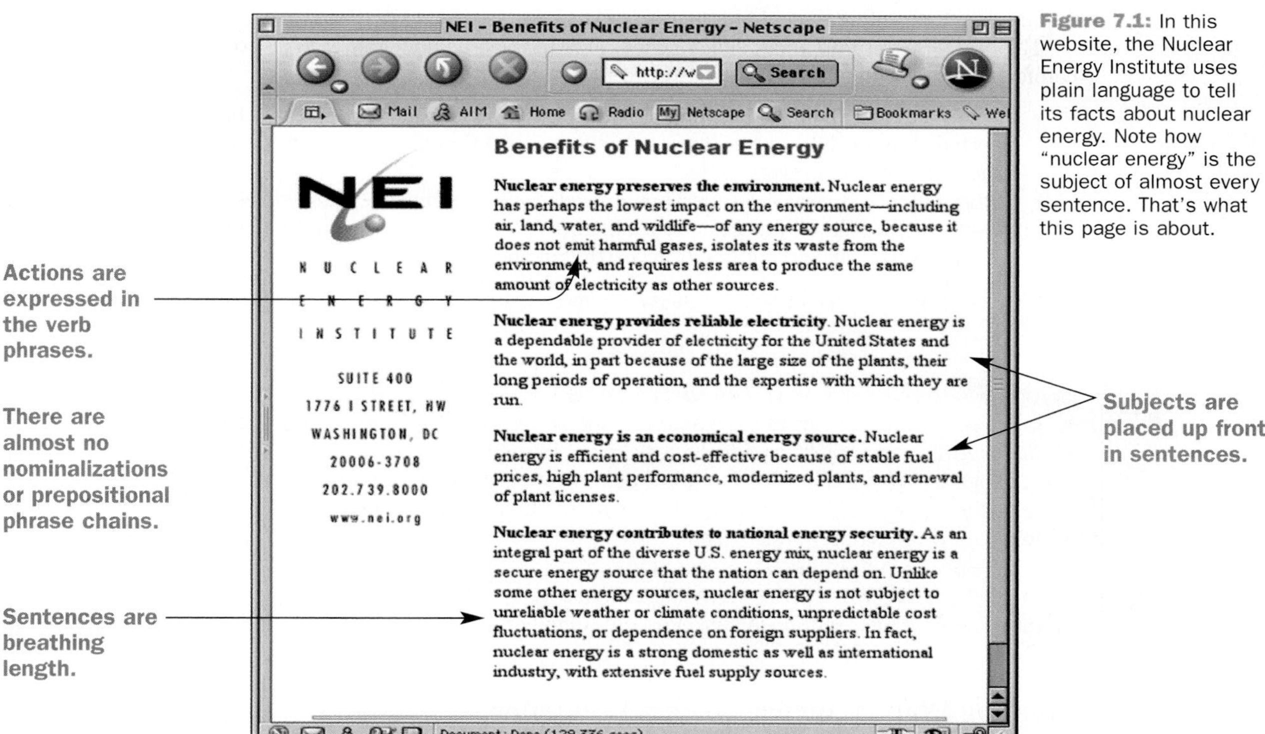

Figure 7.1: In this website, the Nuclear Energy Institute uses plain language to tell its facts about nuclear energy. Note how "nuclear energy" is the subject of almost every sentence. That's what this page is about.

Source: Nuclear Energy Institute, http://www.nei.org.

GUIDELINE 7: ELIMINATE REDUNDANCY IN SENTENCES. In our efforts to get our point across, we sometimes use redundant phrasing. For example, we might write "unruly mob" as though some mobs are orderly, or we might talk about "active participants" as though someone can participate without doing anything.

Sometimes buzzwords and jargon lead to redundancies like "We should collaborate together as a team" or "Empirical observations will provide a new understanding of the subject." In some cases, we might use a synonym to modify a synonym by saying something like "We are demanding important, significant changes."

Redundancies should be eliminated because they use two words to do the work of one word. As a result, readers need to work twice as hard to understand one basic idea.

GUIDELINE 8: WRITE SENTENCES THAT ARE "BREATHING LENGTH." You should be able to read a sentence out loud in one breath. At the end of the sentence, the period signals "take a breath." Of course, when reading silently, readers do not actually breathe when they see a period, but they do take a mental pause at the end of each sentence. If a sentence runs on and on, it forces the readers to mentally hold their breath. By the end of an especially long sentence, they are more concerned about when the sentence is going to end than what the sentence is saying.

The best way to think about sentence length is to imagine how long it takes to comfortably say a sentence out loud.

- If the written sentence is too long to say comfortably, it probably needs to be shortened or cut into two sentences. Avoid asphyxiating the readers with sentences that go on forever.
- If the sentence is very short, perhaps it needs to be combined with one of its neighbors to make it a more comfortable breathing length. You want to avoid making readers hyperventilate over a string of short sentences.

LINK For examples of different breathing-length sentences, see the discussion of pace in this chapter on page 185.

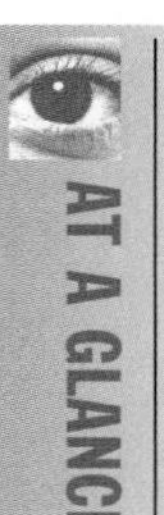

AT A GLANCE

Eight Guidelines for Plain Sentences

- Guideline 1: The subject of the sentence should be what the sentence is about.
- Guideline 2: The subject should be the "doer" in the sentence.
- Guideline 3: The verb should state the action, or what the doer is doing.
- Guideline 4: The subject of the sentence should come early in the sentence.
- Guideline 5: Eliminate nominalizations.
- Guideline 6: Avoid excessive prepositional phrases.
- Guideline 7: Eliminate redundancy in sentences.
- Guideline 8: Write sentences that are "breathing length."

Creating Plain Sentences with a Computer

Computers give us an amazing ability to manipulate sentences. So, take full advantage of your machine's capabilities. First, write out your draft as you normally would, not paying too much attention to the style. Then, as you revise, identify difficult sentences and follow these seven steps:

Want to learn other methods for improving style at the sentence level? Go to **www.ablongman.com/johnsonweb/7.8**

1. Identify who or what is doing something in the sentence.
2. Turn that who or what into the subject of the sentence.
3. Move the subject to an early place in the sentence.
4. Identify what the subject is doing, and move that action into the verb slot.
5. Eliminate prepositional phrases, where appropriate, by turning them into adjectives.
6. Eliminate unnecessary nominalizations and redundancies.
7. Shorten, lengthen, combine, or divide sentences to make them breathing length.

With these seven steps in mind, let us revisit sentence 1, the example of weak style earlier in this chapter:

Original

1. Ten months after the Hartford Project began in which a team of our experts conducted close observations of management actions, our final conclusion is that the scarcity of monetary funds is at the basis of the inability of Hartford Industries to appropriate resources to essential projects that have the greatest necessity.

Now, let's apply the seven-step method for revising sentences into the plain style. First, identify who or what is doing something in the sentence, make it the subject, and move it to an early place in the sentence.

The experts are doing something.

The action is now in the verb.

1a. After a 10-month study, our experts have concluded that the scarcity of monetary funds is at the basis of the inability of Hartford Industries to appropriate resources to essential projects that have the greatest necessity.

Now eliminate the prepositional phrases, nominalizations, and redundancies.

1b. After a 10-month study, our experts have concluded that Hartford Industries' budget shortfalls are limiting the company's support for priority projects.

In the revision, the "doers" (our experts) were moved into the subject position and then moved to an early place in the sentence. Then, the action of the sentence (concluded) was moved into the verb slot. Prepositional phrases like "after the Hartford Project" and "to appropriate resources to essential projects" were turned into adjectives. Nominalizations ("conclusion," "necessity") were turned into verbs or adjectives. And finally, the sentence was shortened to breathing length.

The resulting sentence still sends the same message—just more plainly.

Writing Plain Paragraphs

Some rather simple methods are available to help you write plainer paragraphs. As with writing plain sentences, the computer really helps, because you can quickly move sentences around in paragraphs to clarify your meaning.

The Elements of a Paragraph

Paragraphs tend to include four kinds of sentences: a *transition* sentence, a *topic* sentence, *support* sentences, and a *point* sentence (Figure 7.2). Each of these sentences plays a different role in the paragraph.

To learn more about paragraphing, go to
www.ablongman.com/johnsonweb/7.9

Types of Sentences in a Typical Paragraph

Transition Sentence (optional)
Topic Sentence (needed)
Support Sentences (needed)
examples
reasoning
facts
data
definitions
descriptions
Point Sentence (optional)

Figure 7.2: Paragraphs usually contain up to four types of sentences: transition, topic, support, and point sentences.

TRANSITION SENTENCE The purpose of a transition sentence is to make a smooth bridge from the previous paragraph to the present paragraph. For example, a transition sentence might state the following:

> With these facts about West Nile virus in mind, let us consider the actions we should take.
>
> What should we do to help these learning disabled children in our classrooms?

By referring back to the previous paragraph or paragraphs, transition sentences provide a smoother bridge into the new paragraph.

Transition sentences are typically used when the new paragraph handles a significantly different topic than the previous paragraph. They help close the gap between the two paragraphs or redirect the discussion.

TOPIC SENTENCE The topic sentence is the claim or statement that the rest of the paragraph is going to prove or support:

> To combat the spread of the West Nile virus in Pennsylvania, we recommend three immediate steps: vaccinate all horses, create a public relations campaign to raise public awareness, and spray strategically.
>
> Children with learning disabilities struggle to cope in the normal classroom, so teachers need to be trained to recognize their special needs.

In technical documents, topic sentences typically appear in the first or second sentence of each paragraph. They are placed up front in each paragraph for two reasons.

- The topic sentence sets a goal for the paragraph to reach. It tells readers the claim you are trying to prove. Then, the remainder of the paragraph proves that claim.
- The topic sentence is the most important sentence in any given paragraph. Since readers, especially scanning readers, tend to pay the most attention to

More examples of these kinds of sentences are available at **www.ablongman.com/johnsonweb/7.10**

the beginning of a paragraph, placing the topic sentence up front ensures they will read it. If the topic sentence comes later in the paragraph, they are less likely to read it.

SUPPORT SENTENCES The bulk of any paragraph is typically made up of support sentences. Usually these sentences contain examples and reasoning. Other support sentences will present facts, data, definitions, and descriptions.

> First, we recommend that all horses be vaccinated against West Nile virus, because they are often the most significant victims of the virus. Last year, nearly 2000 horses died from the virus. Second, we believe a public relations campaign . . .

> Learning disabled children often do not easily understand their teacher, especially if the teacher is lecturing.

Support sentences are used to prove the claim made in the paragraph's topic sentence.

Four Kinds of Sentences in a Paragraph

- Transition Sentence (optional)
- Topic Sentence
- Support Sentence
- Point Sentence (optional)

POINT SENTENCES Point sentences restate the paragraph's main point toward the end of the paragraph. They are used to reinforce the topic sentence by restating the paragraph's original claim in new words. Point sentences are especially useful in longer paragraphs where readers may not fully remember the claim stated at the beginning of the paragraph. They often start with transitional devices like "Therefore," "Consequently," or "In sum" to signal that the point of the paragraph is being restated.

> These three recommendations represent only the first steps we need to take. Combating West Nile virus in the long term will require a more comprehensive plan.

> Again, learning disabled children often struggle undetected in the classroom, so teachers need to be trained to recognize the symptoms.

Point sentences are optional in paragraphs, and they should be used only occasionally when a particular claim needs to be reinforced. Too many point sentences will cause your text to sound too repetitious and even condescending.

Using the Four Types of Sentences in a Paragraph

Of these four kinds of sentences, only the topic and support sentences are needed to construct a good paragraph. Transition sentences and point sentences are useful in situations where bridges need to be made between paragraphs or specific points need to be reinforced.

Here are the four kinds of sentences used in a paragraph:

> 8a. How can we accomplish these five goals (transition sentence)? Universities need to study their core mission to determine whether distance education is a viable alternative to the traditional classroom (topic sentence). If universities can maintain their current standards when moving their courses on-line, then distance education may

Want some more advice about organizing paragraphs? Go to **www.ablongman.com/johnsonweb/7.11**

provide a new medium through which nontraditional students can take classes and perhaps earn a degree (support sentence). Utah State, for example, is reporting that students enrolled in its on-line courses have met or exceeded the expectations of their professors (support sentence). If, however, standards cannot be maintained, we may find ourselves returning to the traditional on-campus model of education (support sentence). In sum, the ability to meet a university's core mission is the litmus test to measure whether distance education will work (point sentence).

Here is the same paragraph with the transition sentence and point sentence removed:

8b. Universities need to study their core mission to determine whether distance education is a viable alternative to the traditional classroom (topic sentence). If universities can maintain their current standards when moving their courses on-line, then distance education may provide a new medium through which nontraditional students can take classes and perhaps earn a degree (support sentence). Utah State, for example, is reporting that students enrolled in its on-line courses have met or exceeded the expectations of their professors (support sentence). If, however, standards cannot be maintained, we may find ourselves returning to the traditional on-campus model of education (support sentence).

As you can see in paragraph 8b, a paragraph works fine without transition and point sentences. Nevertheless, transition and point sentences can make texts easier to read while amplifying important points.

Aligning Sentence Subjects in a Paragraph

Now let's discuss how you can make paragraphs plainer by weaving sentences together effectively.

Have you ever read a paragraph in which each sentence seems to go off in a new direction? Have you ever run into a paragraph that actually feels "bumpy" as you read it? More than likely, the problem was a lack of alignment of the paragraph's sentence subjects. To illustrate, consider this paragraph:

Notice how the subjects of these sentences (underlined) are not in alignment, making the paragraph seem rough to readers.

9. The lack of technical knowledge about the electronic components in automobiles often leads car owners to be suspicious about the honesty of car mechanics. Although they might be fairly knowledgeable about the mechanical workings of their automobiles, car owners rarely understand the nature and scope of the electronic repairs needed in modern automobiles. For instance, the function and importance of a transmission in a car are generally well known to all car owners; but, the wire harnesses and printed circuit boards that regulate the fuel consumption and performance of their car are rarely familiar. Repairs for these electronic components can often run over $400—a large amount for a customer who cannot even visualize what a wire harness or printed circuit board looks like. In contrast, a $400 charge for the transmission on the family car, though distressing, is more readily understood and accepted.

There is nothing really wrong with this paragraph—it's just hard to read. Why? The paragraph is difficult to read because the subjects of the sentences change with each new sentence. Look at the underlined subjects of the sentences in this paragraph. They are all different, causing each sentence to strike off in a new direction. As a result, each sentence forces readers to shift focus to concentrate on a new subject.

With your word processor, you can easily revise paragraphs to avoid this bumpy, unfocused feeling. The secret is lining up the subjects in the paragraph.

To line up subjects, first ask yourself what the paragraph is about. Then, cut and paste words to restructure the sentences to line up on that subject. Here is a revision of paragraph 9 that focuses on the "car owners" as subjects:

Here, the paragraph has been revised to focus on people.

9a. Due to their lack of knowledge about electronics, some car owners are skeptical about the honesty of car mechanics when repairs involve electronic components. Most of our customers are fairly knowledgeable about the mechanical features of their automobiles, but they rarely understand the nature and scope of the electronic repairs needed in modern automobiles. For example, most people recognize the function and importance of a transmission in an automobile; but, the average person knows very little about the wire harnesses and printed circuit boards that regulate the fuel consumption and performance of his or her car. So, for most of our customers, a $400 repair for these electronic components seems like a large amount, especially when these folks cannot even visualize what a wire harness or printed circuit board looks like. In contrast, most car owners think a $400 charge to fix the transmission on the family car, though distressing, is more acceptable.

In this revised paragraph, you should notice two things:

- First, the words "car owners" are not always the exact two words used in the subject slot. Synonyms and pronouns should be used to add variety to the sentences.
- Second, not all the subjects need to be car owners. In the middle of the paragraph, for example, "$400 repair" is the subject of a sentence. This deviation from "car owners" is fine as long as the majority of the subjects in the paragraph are similar to each other. In other words, the paragraph will still sound focused, even though an occasional subject is not in alignment with the others.

Of course, the subjects of the paragraph could be aligned differently to stress something else in the paragraph. Here is another revision of paragraph 9 in which the focus of the paragraph is "repairs."

This paragraph focuses on repairs.

9b. Repairs to electronic components often lead car owners, who lack knowledge about electronics, to doubt the honesty of car mechanics. The nature and scope of these repairs are usually beyond the understanding of most nonmechanics, unlike the typical mechanical repairs with which customers are more familiar. For instance, the importance of fixing the transmission in a car is readily apparent to most car owners, but adjustments to electronic components like wire harnesses and printed circuit boards are foreign to most customers—even though these electronic parts are crucial in regulating their car's fuel consumption and performance. So, a repair to these electronic components that costs $400 seems excessive, especially when the repair can't even be visualized by the customer. In contrast, a $400 replacement of the family car's transmission, though distressing, is more readily accepted.

In this paragraph, the subjects are aligned around words associated with "repairs."

One important item you should notice is that paragraph 9a is easier for most people to read than paragraph 9b. Paragraph 9a is more readable because sentences that have "doers" in the subject slots are easier for people to read. In paragraph 9a, the car owners are active subjects, while in paragraph 9b the car repairs are inactive subjects.

Want to practice revising a few paragraphs? Go to **www.ablongman.com/johnsonweb/7.12**

The Given/New Method

Another way to write plain paragraphs is to use the "given/new" method to weave sentences together. Developed by Susan Haviland and Herbert Clark in 1974, the given/new method is based on the assumption that readers try to process new information by comparing it to information they already know. Therefore, every sentence in a paragraph should contain something the readers already know (the given) and something that the readers do not know (the new).

Consider these two paragraphs:

10a. Santa Fe has many beautiful places. Artists choose to strike off into the mountains to work, while others enjoy working in local studios. The landscapes are brilliant.

10b. Santa Fe offers many beautiful places for artists to work. Some artists choose to strike off into the mountains to work, while others enjoy working in local studios. Everyone agrees, though, that the landscapes are brilliant.

Both of these paragraphs are readable, but paragraph 10b is easier to read because the word "artists" appears in the first two sentences. Example 10a is a little harder to read, because there is nothing "given" that carries over from each sentence to its following sentence.

TAKE NOTE To use given/new techniques to smooth out your writing, you can look back to the previous sentence to find something you can use to begin the next sentence.

Using the given/new method is not difficult. In most cases, the given information should appear early in the sentence and the new information should appear later in the sentence. The given information will provide a familiar anchor or context, while the new information will build on that familiar ground. Consider this larger paragraph:

Notice how the chaining of words in this paragraph makes it smoother to read.

11. Recently, an art gallery exhibited the mysterious paintings of Irwin Fleminger, a modernist artist whose vast Mars-like landscapes contain cryptic human artifacts. One of Fleminger's paintings attracted the attention of some young school children who happened to be walking by. At first, the children laughed, pointing out some of the strange artifacts in the painting. Soon, though, the strange artifacts in the painting drew the students into a critical awareness of the painting, and they began to ask their bewildered teacher what the artifacts meant. Mysterious and beautiful, Fleminger's paintings have this effect on many people, not just schoolchildren.

In this paragraph, the beginning of each sentence provides something given—typically an idea, word, or phrase drawn from the previous sentence. Then, the remainder of each sentence adds something new to that given information. By chaining together given and new information, the paragraph builds readers' understanding gradually, adding a little more information with each sentence.

In some cases, however, the previous sentence in a paragraph does not offer a suitable subject for the sentence that follows it. In these cases, transitional phrases can be used to provide readers given information in the beginning of the sentence. To illustrate,

For more information on the given/new method, go to **www.ablongman.com/johnsonweb/7.13**

12. This public relations effort will strengthen Gentec's relationship with leaders of the community. With this new relationship in place, the details of the project can be negotiated in terms that are fair to both parties.

In this sentence, the given information in the second sentence appears in the transitional phrase, not the subject. Transitional phrases are a good place to include given information when the subject cannot be drawn from the previous sentence.

When Is It Appropriate to Use Passive Voice?

Before discussing the elements of persuasive style, we should expose a writing boogeyman as a fraud. Since childhood, you have probably been warned against using *passive voice*. You have been told to write in *active voice*.

One problem with this prohibition on passive voice is that passive voice is very common in technical communication. In some scientific fields, passive voice is the standard way of writing. So, when is it appropriate to use passive voice? Let's start with some definitions.

Passive Voice: The subject of the sentence is being acted upon, so the verb is in the passive voice.

Active Voice: The subject of the sentence is acting, so the verb is an action verb.

Passive voice occurs when the subject of the sentence is being acted upon, so the verb is in passive voice. Active voice occurs when the subject of the sentence is acting, so the verb is an action verb. Here is an example of a sentence written in passive voice and active voice:

13a. The alloy was heated to a temperature of 300°C. (passive)

13b. Andy James heated the alloy to a temperature of 300°C. (active)

Written in passive voice, sentence 13a lacks a doer. The subject of the sentence, the alloy, is being acted upon, but it's not really doing anything. As a result, sentence 13b might be a bit easier to understand. But, as you can see, sentence 13a is not difficult to read and understand.

Passive voice does have a place in technical documents. A passive sentence is fine if

- the readers do not need to know who or what is doing something in the sentence;
- the subject of the sentence is what the sentence is about.

For example, in Sentence 13a, the person who heated the alloy might be unknown or irrelevant to the readers. Is it really important that we know that Andy James heated the alloy? Or, do we simply need to know that the alloy was heated? If the alloy is what the sentence is about and who heated the alloy is unimportant, then the passive voice is fine.

If you are wondering whether the passive voice should be used, consider your readers. Do they expect you to use the passive voice? Do they need to know who did something in a sentence? Go ahead and use the passive voice if your readers expect you to use passive voice or they don't need to know who did what.

Do you want more information on using passive voice? See **www.ablongman.com/johnsonweb/7.14**

Consider these other examples of passive and active sentences:

14a. The shuttle bus will be driven to local care facilities to provide seniors with shopping opportunities. (passive)

14b. Jane Chavez will drive the shuttle bus to local care facilities to provide seniors with shopping opportunities. (active)

15a. The telescope was moved to the Orion system to observe the newly discovered nebula. (passive)

15b. Our graduate assistant, Mary Stewart, moved the telescope to the Orion system to observe the newly discovered nebula. (active)

In both these sets of sentences, the passive voice may be more appropriate, unless there is a reason Jane Chavez or Mary Stewart needs to be singled out for special consideration.

When developing a focused paragraph, passive sentences can often help you align the subjects and use given/new strategies (i.e., make the paragraph more cohesive). For example, compare the following paragraphs, one written in passive voice and the other in active voice:

Paragraph written in passive voice

16a. We were shocked by the tornado damage. The downtown of Hampton had been completely flattened. Brick houses had been torn apart. A school bus was tossed into a farm field, as though some fairytale giant had crumpled it like a soda can. The town's fallen water tower had been dragged a hundred yards to the east. Amazingly enough, no one had been killed. Clearly, lives were saved by the early warning that the weather bureau had sent out.

Same paragraph written in active voice

16b. The tornado damage shocked us. The tornado completely flattened the downtown of Hampton. It had torn apart brick houses. The tornado had tossed a school bus into a farm field, as though some fairytale giant had crumpled it like a soda can. The tornado had knocked over the town's water tower and dragged it a hundred yards to the east. Amazingly enough, the tornado did not kill anyone. Clearly, the early warning that the weather bureau sent out saved lives.

Most people would find paragraph 16a more interesting and readable because it uses passive voice to put the emphasis on the *damage* caused by the tornado. Paragraph 16b is not as interesting, because the emphasis is repeatedly placed on the *tornado.*

Used properly, passive voice can be a helpful tool in your efforts to write plain sentences and paragraphs. Passive voice is misused only when the readers are left wondering who or what is doing the action in the sentence. In these cases, the sentences should be restructured to put the doers in the subject slots.

Persuasive Style

There are times when you will need to do more than simply present information clearly. You will need to influence your readers to take action or make a decision. In these situations, you should shift to *persuasive style.* When used properly, the persuasive style can add emphasis, energy, color, and emotion to your writing.

There are, of course, many ways to be persuasive or amplify your ideas. But in technical documents the following four persuasion techniques will help give your writing more impact:

- Elevate the tone.
- Use similes and analogies.
- Use metaphors.
- Change the pace.

A combination of these persuasive techniques, properly used, will make your writing more influential and vivid.

As you consider whether to use persuasive style techniques, think about the needs of your readers. Novice readers will appreciate the well-placed simile, analogy, or metaphor, while experts might find them trite. Likewise, altering the tone or pace of a document won't work if that tone or pace does not suit the subject.

LINK For more information on reader analysis, go to Chapter 3, page 55.

Elevate the Tone

Tone is the resonance or pitch that the readers will "hear" as they look over your document. Of course, most people read silently to themselves, but all readers have an inner voice that sounds out the words and sentences. By paying attention to tone, you can influence the readers' inner voice in ways that persuade them to read the document with a specific emotion or attitude.

One easy way to elevate tone in written texts is to first decide what emotion or attitude you want the readers to adopt. Then, use logical mapping to find words and phrases that evoke that emotion or attitude.

For example, let us say you want your readers to feel excited as they are looking over your document. You would first put the word "excitement" in the middle of your screen or sheet of paper. Then, as shown in Figure 7.3, you can map out the feelings associated with that emotion.

You can use these words at strategic moments in your document. Subconsciously, the readers will detect this tone in your work. Their inner voice will begin reinforcing the sense of excitement you are trying to convey.

Similarly, if you want to create a specific attitude in your text, map out the words associated with it. For instance, let us say you want your document to convey a feeling of "security." A map around "security" might look like the diagram in Figure 7.4.

If you use these words associated with "security" in your text at strategic points, the readers will perceive the sense of security that you are trying to convey.

TAKE NOTE The best approach is to choose one emotion tone and/or one attitude tone you want to seed into your document. Don't choose more than two tones, though, because the conflicting emotions and attitudes will create confusion.

One warning: Use these words sparingly throughout the text. If you overuse them, the emotion or attitude you are setting will be too obvious. You need to use only a few words at strategic moments to set the tone you are seeking.

Another thing to keep in mind is that your tone needs to reflect the subject of your document. If you try to set a tone that is inappropriate for your subject, that tone will just sound hollow and artificial.

For more information on improving tone, go to **www.ablongman.com/johnsonweb/7.15**

Mapping an Emotional Tone

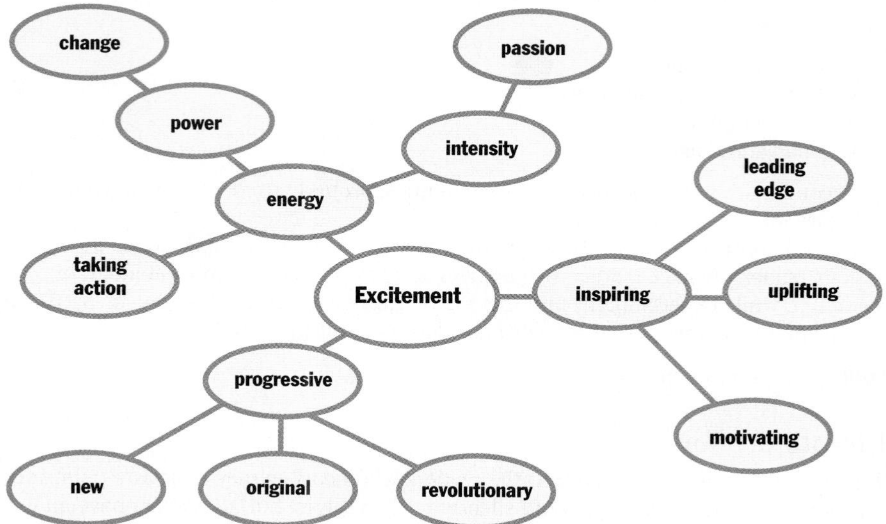

Figure 7.3: You can set an emotional tone by mapping it out and "seeding" the text with words associated with that tone.

Mapping an Authoritative Tone

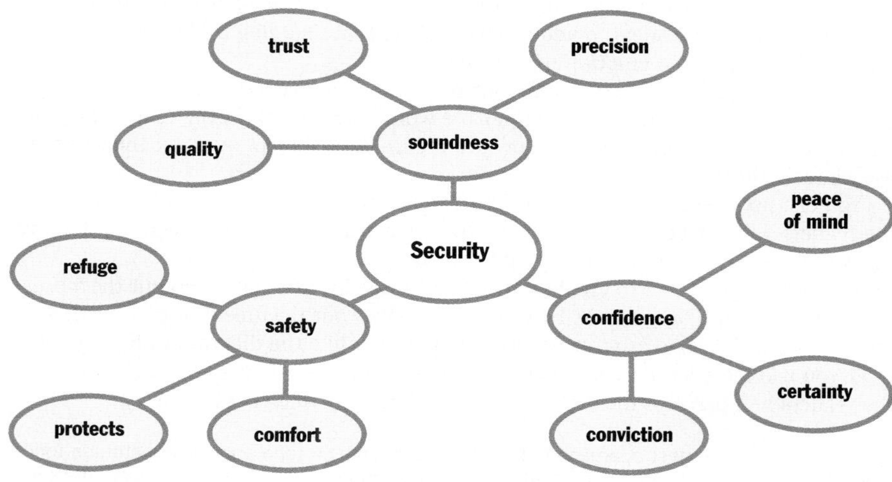

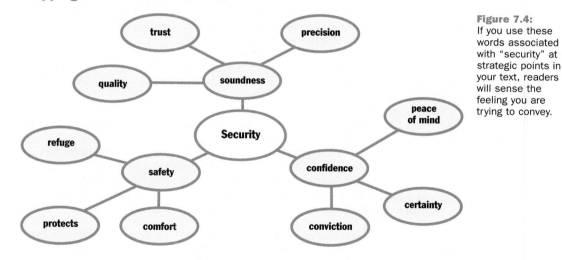

Figure 7.4: If you use these words associated with "security" at strategic points in your text, readers will sense the feeling you are trying to convey.

Use Similes and Analogies

Similes and analogies are rhetorical devices that help writers define difficult concepts by comparing them to familiar things. For example, let us say you are writing a brochure that describes an integrated circuit to people who know almost nothing about these devices. Using a simile, you might describe it this way:

AT WORK

Patrick Ward

PARTNER, CLARITY CORPORATION

The Clarity Corporation is a consulting firm that specializes in high-technology industries.

What is the secret to writing persuasively?

Much of persuasion is knowing your audience. For example, when I write proposals, I usually have to persuade several audiences at the same time. Most often, particularly with larger organizations, many people are involved in the decision to hire my company, and it is impossible for me to meet with every one of those people personally. My proposal may be my company's only chance to persuade these people.

My best persuasion strategy is to focus on the customers' needs more than my company's strengths. It's a good idea to replace every "I" or "we" with "you" or the magical phrase "you get." By using these kinds of words and phrases, you see things through the eyes of your customer. For example, here is how you can make a sentence persuasive:

> **Original:** We provide the finest customer service and quickest turnaround in the industry.
>
> **Revised:** You get the finest customer service along with the industry's quickest turnaround times.

By putting the focus on the readers' needs, you can make a lasting impression on them. The remainder of your document should show how your company can meet the customer's needs. This approach is especially important for documents like proposals, which must be persuasive.

In the end, remember that persuasive documents like proposals are sales tools. A persuasively written proposal will interact with many more prospects than your company's salesperson ever will.

> An integrated circuit is like a miniature Manhattan Island crammed into a space less than one centimeter long.

In this case, the simile ("X is like Y") creates an image that helps readers visualize the complexity of the integrated circuit. Here are a couple of other examples of similes:

> A cell is like a tiny cluttered room with a nucleus inside and walls at the edges. The nucleus is like the cell's brain. It's an enclosed compartment that contains all the information that cells need to form an organism. This information comes in the form of DNA. It's the differences in our DNA that make each of us unique. (Source: Genetic Science Learning Center, http://gslc.genetics.utah.edu/units/cloning/whatiscloning)

> Everything that you see is made of things called atoms. An atom is like a very, very small solar system. As you know, the sun is at the center of the solar system. Just like the solar

system, there is something at the center of an atom. It is called the nucleus. So the nucleus of an atom is like the sun in our solar system. Around the sun there are planets. They orbit the sun. In an atom, there are things called electrons that orbit the nucleus. (Source: Chickscope, http://chickscope.beckman.uiuc.edu/about/overview/mrims.html)

Analogies are also a good way to help readers visualize difficult concepts. An analogy follows the structure "A is to B as X is to Y." For example, a medical analogy might be

Like police keeping order in a city, white blood cells seek to control viruses and bacteria in the body.

In this case, both parts of the analogy are working in parallel. "Police" is equivalent to "white blood cells," and "keeping order in the city" is equivalent to "control viruses in the body."

TAKE NOTE A good rule of thumb is to use similes and analogies when readers are new to the subject. Expert readers don't need as many similes and analogies, because they can already visualize what you are talking about.

Use Metaphors

Though comparable to similes and analogies, metaphors work at a deeper level in a document. Specifically, metaphors are used to create or reinforce a particular perspective that you want readers to adopt toward your subject or ideas. For example, a popular metaphor in Western medicine is the "war on cancer."

While the three decades since the start of the American "war on cancer" have witnessed many innovative offensive strategies to treat the disease, a new key battle that may well be the turning point has emerged—the battle of prevention. As part of this attempt to keep the enemy from even entering the field, the Cancer Prevention Program at New York–Presbyterian Hospital has launched a national newsletter and website to keep both consumers and health professionals abreast of the latest developments in this new field of cancer prevention. (Source: Weill Cornell Medical Center, http://www.med.cornell.edu/news/press/2003/03_17_03.html)

As shown in this example, by employing this metaphor throughout the document, the writer can reinforce a particular perspective about cancer research. A metaphor like "war on cancer" would add a sense of urgency, because it suggests that cancer is an enemy that must be defeated, at almost any cost. Metaphors are very powerful tools in technical writing because they tend to work at a subconscious level.

But what if a commonly used metaphor, like "war on cancer," is not appropriate for your document? Perhaps you are writing about cancer in a hospice situation, where the patient is no longer trying to "defeat" cancer. In these cases, you might develop a new metaphor and use it to invent a new perspective. For example, perhaps we want our readers to view cancer as something to be managed, not fought.

To manage their cancer, patients become supervisors who set performance goals and lead planning meetings for teams of doctors and nurses. Our doctors become consultants who help patients become better managers of their own care.

For more examples of similes and analogies, go to **www.ablongman.com/johnsonweb/7.16**

In a hospice situation, a "management" metaphor would be much more appropriate than the usual "war" metaphor. It would invite the readers to see their cancer from a less confrontational perspective.

If you look for them, you will find that metaphors are commonly used in technical documents. Once you are aware of these metaphors, you can use the ones that already exist in your field, or you might create new metaphors to shift your readers' perspective about your subject. In Figure 7.5, for example, the writers use metaphors to recast the debate over genetically engineered crops.

Change the Pace

You can also control the readers' pace as they read through your document. Longer sentences tend to slow down the reading pace, while shorter sentences tend to speed it up. By paying attention to the length of sentences, you can increase or decrease the intensity of your text.

These long sentences slow the reading down.

Long Sentences (low intensity)

The effect of food insecurity on behavior problem indices suggests that children who experience food insecurity suffer more psychological and emotional difficulties than other children. Children who have experienced food insecurity exhibit more aggressive and destructive behaviors, and more withdrawn and distressed behaviors than other children. It appears that children are experiencing a great deal of psychological and emotional distress in response to issues of food insecurity within the family, and that children respond to this distress with a range of behavioral responses. The results clearly indicate that food insecurity negatively impacts children's psychological and emotional well-being, which is another important indicator of the well-being of children that may also have implications for other child development outcomes. Children who experience higher levels of psychological and emotional distress may also experience more difficulties in other areas such as school achievement, and they may interfere with a number of other activities in which children may be involved.

Shorter sentences raise the "heartbeat" of the text.

Short Sentences (high intensity)

The effect of children's food insecurity can be crucial. Food-insecure children have more psychological and emotional difficulties. They exhibit more aggressive and destructive behaviors. They are more withdrawn and distressed. As a result, they respond to this distress with a range of behavioral responses. The results show that food insecurity negatively affects children's well-being. This insecurity may have implications for other child development outcomes. These children may have more trouble at school. They may also struggle at a number of other activities.

Persuasive Style Techniques

- Elevate the tone.
- Use similes and analogies.
- Use metaphors.
- Change the pace.

If a situation is urgent, using short sentences is the best way to show that something needs to be done right away. Your readers will naturally feel compelled to take action. On the other hand, if you want your readers to be cautious and deliberate, longer sentences will decrease the intensity of the text, giving readers the sense that there is no need to rush.

Want to learn more about metaphors? Go to **www.ablongman.com/johnsonweb/7.17**

A Persuasive Argument

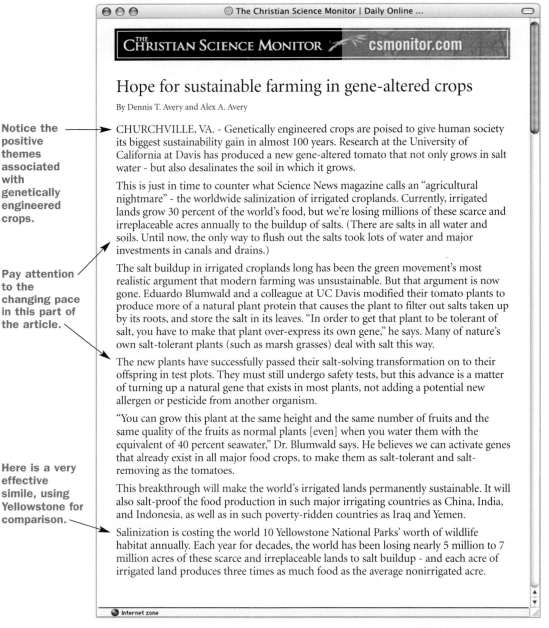

Figure 7.5: In this article, the authors use numerous persuasive style techniques to add energy to their writing. Are the authors persuasive?

Hope for sustainable farming in gene-altered crops

By Dennis T. Avery and Alex A. Avery

Notice the positive themes associated with genetically engineered crops.

CHURCHVILLE, VA. - Genetically engineered crops are poised to give human society its biggest sustainability gain in almost 100 years. Research at the University of California at Davis has produced a new gene-altered tomato that not only grows in salt water - but also desalinates the soil in which it grows.

This is just in time to counter what Science News magazine calls an "agricultural nightmare" - the worldwide salinization of irrigated croplands. Currently, irrigated lands grow 30 percent of the world's food, but we're losing millions of these scarce and irreplaceable acres annually to the buildup of salts. (There are salts in all water and soils. Until now, the only way to flush out the salts took lots of water and major investments in canals and drains.)

Pay attention to the changing pace in this part of the article.

The salt buildup in irrigated croplands long has been the green movement's most realistic argument that modern farming was unsustainable. But that argument is now gone. Eduardo Blumwald and a colleague at UC Davis modified their tomato plants to produce more of a natural plant protein that causes the plant to filter out salts taken up by its roots, and store the salt in its leaves. "In order to get that plant to be tolerant of salt, you have to make that plant over-express its own gene," he says. Many of nature's own salt-tolerant plants (such as marsh grasses) deal with salt this way.

The new plants have successfully passed their salt-solving transformation on to their offspring in test plots. They must still undergo safety tests, but this advance is a matter of turning up a natural gene that exists in most plants, not adding a potential new allergen or pesticide from another organism.

"You can grow this plant at the same height and the same number of fruits and the same quality of the fruits as normal plants [even] when you water them with the equivalent of 40 percent seawater," Dr. Blumwald says. He believes we can activate genes that already exist in all major food crops, to make them as salt-tolerant and salt-removing as the tomatoes.

Here is a very effective simile, using Yellowstone for comparison.

This breakthrough will make the world's irrigated lands permanently sustainable. It will also salt-proof the food production in such major irrigating countries as China, India, and Indonesia, as well as in such poverty-ridden countries as Iraq and Yemen.

Salinization is costing the world 10 Yellowstone National Parks' worth of wildlife habitat annually. Each year for decades, the world has been losing nearly 5 million to 7 million acres of these scarce and irreplaceable lands to salt buildup - and each acre of irrigated land produces three times as much food as the average nonirrigated acre.

Internet zone

Source: The Christian Science Monitor, http://www.csmonitor.com/2001/0830/p9s1-coop.html.

The Christian Science Monitor | Daily Online ...

For at least 100 years, plant breeders have tried to cross conventional crop plants with salt-tolerant ones, with virtually no success. But Dr. Blumwald says he's already got both tomatoes and canola plants flourishing in 40 percent seawater. Since the canola plants are tall, with lots of leaves, they remove 12 grams of salt per plant.

The desalinating-tomato breakthrough should be important proof for the world's consumers and governments that biotech is agriculture's most important 21st-century tool. It will help farmers grow the additional food crops the world wants and needs - through higher yields on the land they're already farming. Biotech opponents claim that bioengineered crops help only big corporations.

The UC Davis research on desalinating crops proves these critics massively wrong. It also proves that if we aren't comfortable with big corporations doing all the biotech crop research, we can productively increase funding for public research.

Instead, the green movement has been demanding no agricultural research at all. The activists prefer an organic retreat to the famines and worn-out farms of the 19th century. Back then, of course, we could cut down virgin forests and get enough additional cropland to cover traditional farming's shortcomings.

The last comparable step in human sustainability was the Haber-Bosch process, developed in 1909, which allows us to get nitrogen fertilizer from the air. (Both Fritz Haber and Carl Bosch worked for a private company, and won Nobel Prizes.) Without their fertilizer, we'd have long since cleared all of the world's existing forests, to pasture 10 million cattle instead of 1.3 million, so we'd have enough manure to fertilize current food production.

Biotechnology is already demonstrating such power in agriculture that by 2050 it may ensure fully sustainable grain and meat production for the world's projected peak population of 9 billion people - without taking any more land from nature. How many Nobel Prizes should that be worth?

Dennis T. Avery is director of Global Food Issues for the Hudson Institute in Indianapolis. Alex A. Avery, a biologist, is research director for the center.

From the August 30, 2001 edition, http://www.csmonitor.com/2001/0830/p9s1-coop.html.

Internet zone

The "research is an argument" metaphor is used here and elsewhere in this article.

Figure 7.5: (*continued.*)

Balancing Plain and Persuasive Style

When you are drafting and revising a document for style, look for appropriate places to use plain and persuasive style. Minimally, a document should exhibit plain style. Sentences should be clear and easy to read. Your readers should not have to struggle to figure out what a sentence or paragraph is about.

Persuasive style should be used to add energy and color. It should also be used in places in the document where readers are expected to make a decision or take action. The use of tone, similes, analogies, and metaphors in strategic places should encourage them to do what you want. You can use short or long sentences to adjust the intensity of your prose.

In the end, developing good style takes practice. At first, revising a document to make it plain and persuasive might seem difficult. Before long, though, you will start writing better sentences in the first place. You will have internalized the style guidelines presented in this chapter.

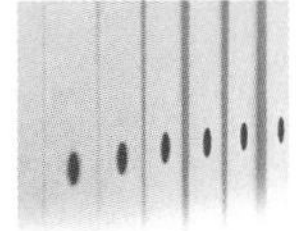

CHAPTER REVIEW

- The style of your document can convey your message as strongly as the content; *how you say something is often what you say.*
- Style can be classified as plain, persuasive, or grand. Technical communication is most often concerned with plain or persuasive style.
- To write plain sentences, you can follow eight guidelines, which include (1) identifying your subject, (2) making the "doer" the subject of the sentence, (3) putting the doer's action in the verb, (4) moving the subject of the sentence close to the beginning of the sentence, (5) reducing nominalizations, (6) removing excess prepositional phrases, (7) eliminating redundancy, and (8) making sure that sentences are "breathing length."
- Writing plain paragraphs involves the use of four types of sentences: transitional and point sentences, which are optional, and topic and support sentences, which are necessary.
- Check sentences in paragraphs for subject alignment, and use the given/new method to weave sentences together.
- Techniques for writing persuasively include elevating the tone; using similes, analogies, and metaphors; and changing the pace.

Individual or Team Projects

1. In your workplace or on campus, find three sentences that seem particularly difficult to read. Use the eight "plain style" guidelines discussed in this chapter to revise them to improve their readability. Make a presentation to your class in which you show how using the guidelines helped make the sentences easier to read.

2. In your workplace or on campus, find two paragraphs that seem difficult to read. First, align the subject in each paragraph to see if this technique improves the readability of the paragraph. Then, use given/new strategies to revise the paragraphs. Which method works better? Would a combination of subject alignment and given/new strategies be more appropriate? Present your revisions to your class, showing how you used the subject alignment and given/new techniques to improve the readability of the paragraphs.

3. Find a document and analyze its style. Underline key words and phrases in the document. Based on the words you underlined, what is the tone used in the document? Does it use any similes or analogies to explain difficult concepts? Can you locate any metaphors woven into the text? How might you use techniques of persuasive style to improve this proposal? Write a memo to your instructor in which you analyze the style of the document. Do you believe the style is effective? If so, why? If not, what stylistic strategies might improve the readability and persuasiveness of the document?

4. While revising some of your own writing, try to create a specific tone by mapping out an emotion or character that you would like the text to reflect. Weave a few concepts from your map into your text. At what point is the tone too strong? At what point is the tone just right?

5. Look over the case study at the end of this chapter. Imagine you are Henry and write a memo to the managers of NewGenSport expressing your concerns about the safety of the product and its advertising campaign. Make sure you do some research on ephedra, so you can add some technical support to your arguments.

Collaborative Project

Metaphors that use "war" are widely used in American society. We have had wars on poverty, cancer, drugs, and even inflation. More recently, "war on terrorism" has been a commonly used metaphor. With a group of others, choose one of these war metaphors and find examples of how it is used. Identify places where the metaphor seems to fit the situation. Locate examples where the metaphor seems to be misused.

For these exercises, sample sentences, paragraphs, and documents are available at **www.ablongman.com/johnsonweb/7.18**

With your group, discuss the ramifications of the war metaphor. For example, if we accept the metaphor "war on drugs," who is the enemy? What weapons can we use? What level of force is necessary to win this war? Where might we violate civil rights if we follow this metaphor to its logical end?

Then, try to create a new metaphor that invites people to see the situation from a different perspective. For example, what happens when we use "managing drug abuse" or "healing drug abuse" as new metaphors? How do these new metaphors, for better or worse, change how we think about illegal drugs and react to them?

In a presentation to your class, use examples to show how the war metaphor is used. Then, show how the metaphor could be changed to see the situation from a different perspective.

For sites that use or discuss war metaphors, go to **www.ablongman.com/johnsonweb/7.19**

CASE STUDY

Going Over the Top

Henry Wilkins was a nutritionist who worked for NewGenSport, a company that made sport drinks. His company's best-seller, Overthetop Sport Drink, was basically a fruit-flavored drink with added carbohydrates and salts. As a nutritionist, Henry knew that sport drinks like Overthetop did not really do much for people, but they didn't harm them either. Overwhelmingly, the primary benefit of Overthetop was the water, which people need when they exercise. The added carbohydrates and salts were somewhat beneficial, because people depleted them as they exercised.

The advertisements for Overthetop painted a different picture. They showed overly muscular athletes drinking Overthetop and then dominating opposing players. Henry thought the advertisements were a bit misleading, but harmless. The product would not lead to the high performance promised by the advertising. But, Henry reasoned, that's advertising.

One day, Henry was asked to evaluate a new version of Overthetop, called Overthetop Extreme. The new sport drink would include a small amount of an ephedra-like herbal stimulant that was supposed to enhance performance and cause slight weight loss. Henry also knew that ephedra had been linked to a few deaths, especially when people took too much. Ephedra was banned by the FDA in 2003, but this new "ephedra-like" herbal stimulant was different enough to be legal. Henry suspected this stimulant would have many of the same problems as its sister drug, ephedra.

Henry's preliminary research indicated that 36 ounces (three bottles) a day of Overthetop Extreme would have no negative effect. But, more than a few bottles a day might lead to complications, perhaps even death in rare cases.

But before Henry could report his findings, the company's marketing team pitched its new advertising campaign in preparation for the debut of the new product. In the campaign, as usual, overly muscular athletes were shown guzzling large amounts of Overthetop Extreme and then going on to victory. It was clear that the advertisements were overpromising—as usual. But, they were also suggesting the product be used in a way that would put people at risk. The advertisements were technically accurate, but their style gave an inflated sense of the product's capabilities. In some cases, Henry realized, the style of the advertisements might lead to overconsumption, which could lead to death.

The company's top executives were very excited about the new product. The advertisements only made them more enthusiastic. They began pressuring Henry to finish his evaluation of Overthetop right away, so they could start advertising and putting the product on store shelves.

Henry, unfortunately, was already concerned about the product's safety. But he was especially concerned that the advertisements were clearly misleading consumers in a dangerous way.

If you were Henry, how would you handle this situation? In your report to the company's management, how would you express your concerns? What else might you do?

Want to know more about ephedra? Go to
www.ablongman.com/johnsonweb/7.20

CHAPTER

8

Designing Documents and Interfaces

CHAPTER CONTENTS

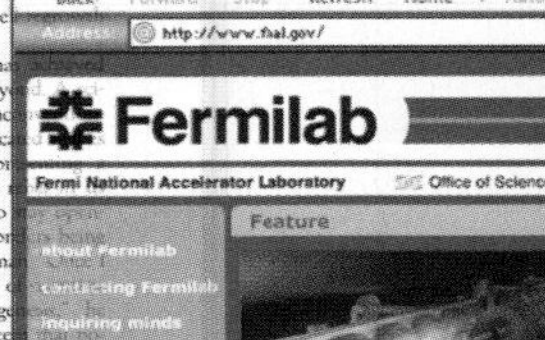

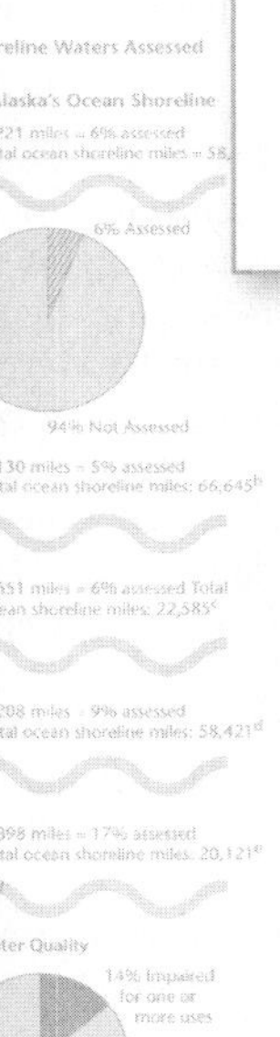

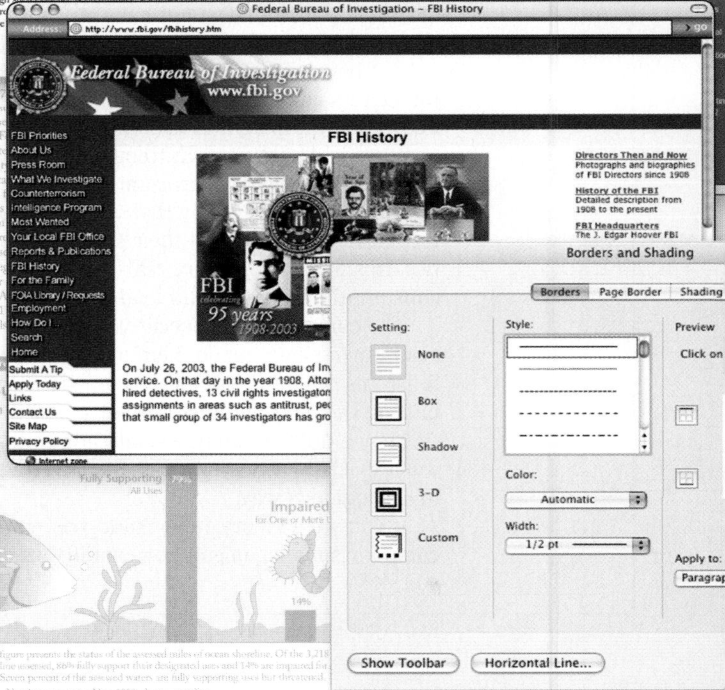

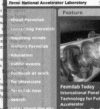

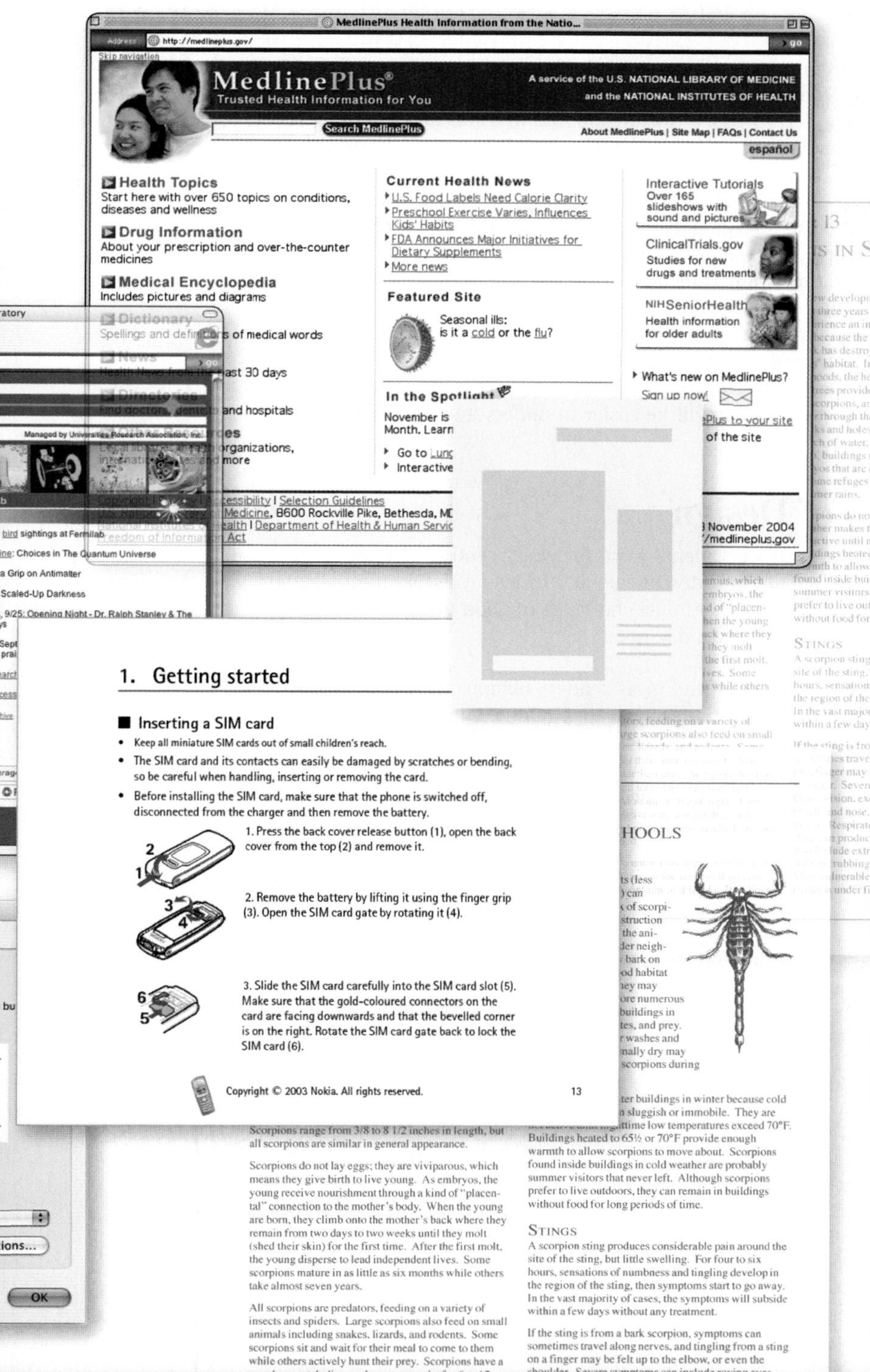

CHAPTER OBJECTIVES

In this chapter, you will learn:

- How to design technical documents for electronic and paper-based media.
- To use the five principles of design: balance, alignment, grouping, consistency, and contrast.
- To use balancing techniques that enhance readability in a document.
- To use alignment and grouping strategies to add visual structure to text.
- To use consistency and contrast to balance uniformity and difference in the design of a document or interface.
- How to use a process to design documents.
- About options for binding documents, as well as the sizes, colors, and weights of paper available.

Document design has become increasingly important with the use and availability of computers. Today, readers expect paper-based documents to be attractive and easy to read. Meanwhile, screen-based documents like websites and multimedia texts are highly visual, so readers expect the interface—how the page looks on the screen—to be well-designed. Readers don't just *prefer* well-designed documents—they *expect* the design to highlight important ideas and concepts.

Imagine your own reaction to a large document with no headings, graphics, or lists. If you're like most people, you wouldn't even want to start reading. Your readers feel the same way. Like you, they want your document's design to make information easy to find and understand.

In this chapter, you will learn how to design technical documents for paper-based and electronic media. The principles of design discussed in this chapter are rather easy to learn, but they are very powerful tools that will help you capture and keep your readers' attention. Your documents will be easier to understand and scan, while also being more attractive.

Five Principles of Design

Few people actually read technical documents word for word, sentence by sentence. Instead, they tend to look over technical documents at various levels, skimming some parts and paying closer attention to others. The challenge of good design is to let readers choose how *they* want to read the document.

Good design creates a sense of order and gives readers obvious "access points" to begin reading and locating the information they need.

Good design creates a sense of order and gives readers obvious "access points" to begin reading and locating the information they need. Good design is not something to be learned in a day. However, you can master some basic principles that will help you make better decisions about how your document should look. Here are five principles that you can follow as you design documents:

Balance—The document looks balanced from left to right and top to bottom.

Alignment—Images and words on the page are aligned to show the document's structure, or hierarchy.

Grouping—Related images and words are placed near each other on the page.

Consistency—Design features in the document are used consistently, so the document looks uniform.

Contrast—Items in the document that are different look significantly different.

These five principles offer a simple list of concepts that you can use while designing a document. They are based on theories of Gestalt psychology, which studied how the mind recognizes patterns and wholes as more than the sum of their parts (Arnheim, 1969; Koffka, 1935). Designers of all kinds, including architects,

To learn more about Gestalt psychology and design, go to **www.ablongman.com/johnsonweb/8.1**

fashion designers, and artists, have used gestalt principles in a variety of ways (Bernhardt, 1986). You will find these five principles helpful as you learn about designing documents.

Design Principle 1: Balance

Balance is perhaps the most prominent feature of design in technical documents. On a balanced page or screen, the design features should offset each other to create a feeling of stability.

To balance a text, pretend your page or screen is balanced on a point. Each time you add something to the left side of the page, you need to add something to the right side to maintain balance. Similarly, when you add something to the top of the page, you need to add something to the bottom. Figure 8.1, for example, shows an example of a balanced page and an unbalanced page.

Balanced and Unbalanced Page Layouts

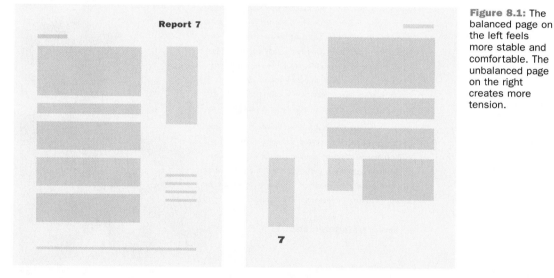

Figure 8.1: The balanced page on the left feels more stable and comfortable. The unbalanced page on the right creates more tension.

In Figure 8.1, the page on the left is balanced because the design features offset each other. The page on the right is unbalanced because the items on the right side of the page are not offset by items on the left. Meanwhile, the right page is top-heavy because the design features are bunched at the top of the page.

Balanced page layouts can take on many different forms. Figures 8.2 and 8.3 show examples of balanced layouts in magazines. The idea is not to create symmetrical pages (i.e., where left and right, top and bottom mirror each other). Instead, you want to balance the pages by putting text and images on all sides.

GO TO THE NET

To see samples of unbalanced pages, go to **www.ablongman.com/johnsonweb/8.2**

A Balanced Design with Energy

The elements on the page offset each other to create balance.

What does an outdoor clothing company know about genetic engineering?

Not enough, and neither do you.

Even the scientists working on genetic engineering admit they don't know the full story. But despite the fact that we know so little about the impacts, a salmon has already been engineered that grows at twice the rate of normal salmon, a strain of corn has been created with pesticide in every cell, and trees have been engineered with less lignin to break down more easily in the pulping process. What will be the impact on our health, and the health of the ecosystem, once these new species make it out into the wild or into our food supply? No one knows.

Let's not repeat the mistakes we've made in the past with such inadequately tested technologies as DDT and nuclear energy. We don't know enough about the dangers of genetic engineering. Let's find out all the risks before we turn genetically modified organisms loose on the world, or continue to eat them in our food.

Find out more at
www.patagonia.com/enviroaction

patagonia

Photo: Topher Donahue © 2001 Patagonia, Inc.

Figure 8.2: The graphic designers who design magazine pages and advertisements are especially careful about balance. Notice this advertisement places elements on the page to create energy, or tension.

Notice how your eyes are drawn to the person. People in a picture usually attract the most attention from readers.

By design, you end up here with the product name.

Source: Patagonia, 2003.

A Simpler Design That Is Elegant

Figure 8.3: This page from a magazine is simpler in design than the advertisement shown in Figure 8.2. Notice how the elements on the page offset each other to create a balanced, stable look.

This image draws your eyes to it.

The two-column format balances the text.

Your eyes naturally flow down the page.

Birdbrain Breakthrough

STARTLING EVIDENCE THAT THE HUMAN BRAIN CAN GROW NEW NERVES BEGAN WITH UNLIKELY STUDIES OF BIRDSONG

BY EDWIN KIESTER, JR., & WILLIAM KIESTER

THE BARN WHERE HE WORKS IS IN the horse country of Millbrook, New York, but it echoes with trills, tweets and obbligatos—the raucous music of more than a thousand caged canaries and finches. "Hear that one singing his heart out?" Fernando Nottebohm asks. "He has more than a dozen songs. He's telling the males, 'This is my territory.' He's telling the females, 'Hey, look at me.'"

Deconstructing birdsong may seem an unlikely way to shake up biology. But Nottebohm's research has shattered the belief that a brain gets its quota of nerve cells shortly after birth and stands by helplessly as one by one they die—a "fact" drummed into every schoolkid's skull. On the contrary, the often-rumpled Argentina-born biologist demonstrated two decades ago that the brain of a male songbird grows fresh nerve cells in the fall to replace those that die off in summer.

The findings were shocking, and scientists voiced skepticism that the adult human brain had the same knack for regeneration. "Read my lips: no new neurons," quipped Pasko Rakic, a Yale University neuroscientist doubtful that a person, like a bird, could grow new neurons just to learn a song.

Yet, inspired by Nottebohm's work, researchers went on to find that other adult animals—including human beings—are indeed capable of producing new brain cells. And in February, scientists reported for the first time that brand-new nerves in adult mouse brains appeared to conduct impulses—a finding that addressed lingering concerns that newly formed adult neurons might not function. Though such evidence is preliminary, scientists believe that this growing body of research will yield insights into how people learn and remember. Also, studying neurogenesis, or nerve growth, may lead them to better understand, and perhaps treat, devastating diseases such as Parkinson's and Alzheimer's, caused by wasted nerves in the brain.

Few would have predicted that canary courtship would lead to such a breakthrough. Nottebohm's bird studies "opened our eyes that the adult brain does change and develops new cells throughout life," says neurobiologist Fred Gage of the Salk Institute in La Jolla, California, whose lab recently found evidence of nerve cell regrowth in the human brain.

Nottebohm's research has achieved renown in biology and beyond. A scientist who advances an unconventional view and is later vindicated makes for compelling drama, presenting a hero who appeals to the rebel in us and a cautionary lesson to stay open-minded. Yet Nottebohm prefers being a revolutionary to a statesman. "Once I was in the 5 or 10 percent of scientists who believed in neurogenesis," he says. "Now 95 percent accept that position. I rather liked it better being in the minority."

He has been a bird lover since his boyhood, in Buenos Aires. "Listening to birds was sort of my hobby," he says. "Other boys had cars, I had birds. I liked to try to identify them by their songs." He obtained a doctorate at the University of California at Berkeley—yes, studying birds—before moving to Rockefeller University.

A key moment came in 1981 when he showed that the volume of the part of a male canary's brain that controls song-making changes seasonally. It peaks in the spring, when the need to mate demands the most of a suitor's musical ability, and shrinks in the summer. It then starts expanding again in the fall—a time to learn and rehearse new

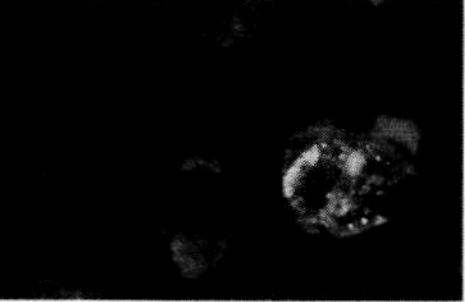

In a study of adult male canaries by Fernando Nottebohm (top), new brain cells appear green, while older ones are red.

MARTIN SCHOELLER/CORBIS OUTLINE; FERNANDO NOTTEBOHM

SMITHSONIAN 36 JUNE 2002

Source: Smithsonian, *2002, p. 36.*

Balance is also important in screen-based documents. In Figure 8.4, the screen interface is balanced because the items on the left offset the items on the right.

A Balanced Interface

The banner at the top of the screen offsets text and images lower on the page.

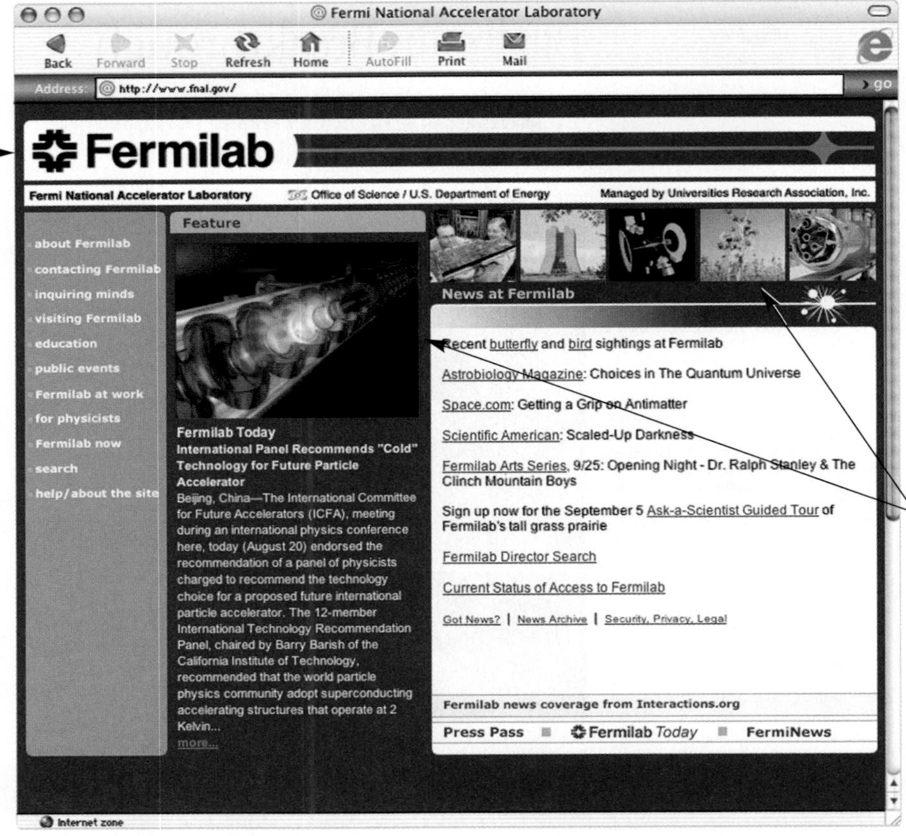

Figure 8.4: This screen is balanced, even though it is not symmetrical. The items on the left offset the items on the right.

Left and right, images and text are used to balance the screen.

Source: Fermilab National Accelerator Laboratory, http://www.fermilab.gov.

Weighting a Page or Screen

When balancing a page or screen, graphic designers will talk about the "weight" of the items on the page. What they mean is that some items on a page or screen attract readers' eyes more than others—these features have more weight. A picture, for example, has more weight than printed words because readers' eyes tend to be drawn toward pictures. Similarly, an animated figure moving on the screen will capture much more attention than static items.

GO TO THE NET

To see samples of unbalanced webpages, go to **www.ablongman.com/johnsonweb/8.3**

Here are some basic weighting guidelines for a page or screen:

- Items on the right side of the page weigh more than items on the left.
- Items on the top of the page weigh more than items on the bottom.
- Big items weigh more than small items.
- Pictures weigh more than written text.
- Graphics weigh more than written text.
- Colored items weigh more than black and white items.
- Items with borders around them weigh more than items without borders.
- Irregular shapes weigh more than regular shapes.
- Items in motion weigh more than static items.

As you are balancing a page, use these weight guidelines to help you offset items. For example, if an image appears on the right side of the page, make sure that there is something on the left to create balance.

TAKE NOTE In some cases, designers purposefully create an unbalanced page layout to add tension to the document. In technical writing, these situations are rare. Unless you are a graphic designer, you should use balanced pages.

Using Grids to Balance a Page Layout

When designing a page or screen, your challenge is to create a layout that is balanced but not boring. A time-tested way to devise a balanced page design is to use a *page grid* to evenly place the written text and graphics on the page. Grids divide the page vertically into two or more columns. Figure 8.5 shows some standard grids and how they might be used.

Grids for Designing Page Layouts

One-column grid

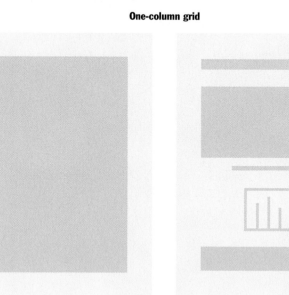

Figure 8.5: Grids can help you place items on a page in a way that makes it look balanced.

One-column grids offer simplicity, but not much flexibility.

Continued on following page

Figure 8.5: (*continued.*)

Two-column grid

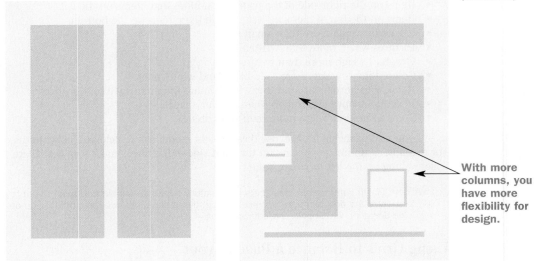

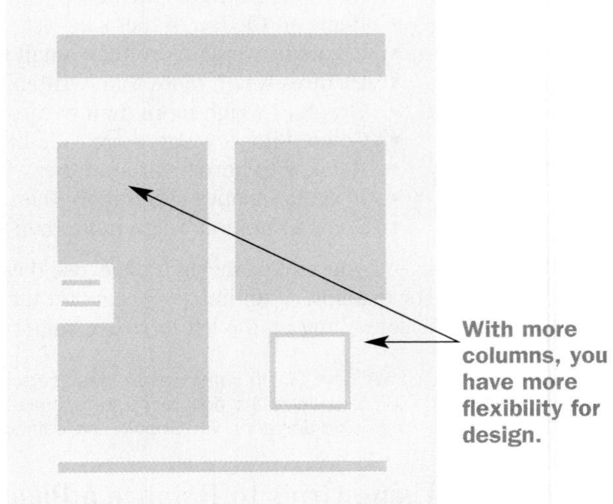

With more columns, you have more flexibility for design.

Three-column grid

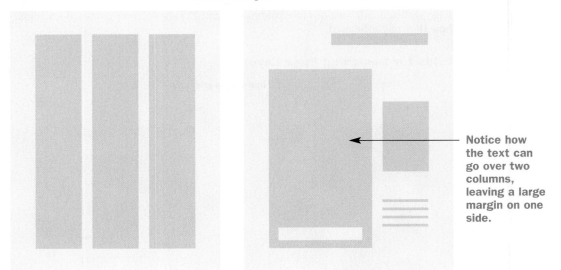

Notice how the text can go over two columns, leaving a large margin on one side.

Four-column grid

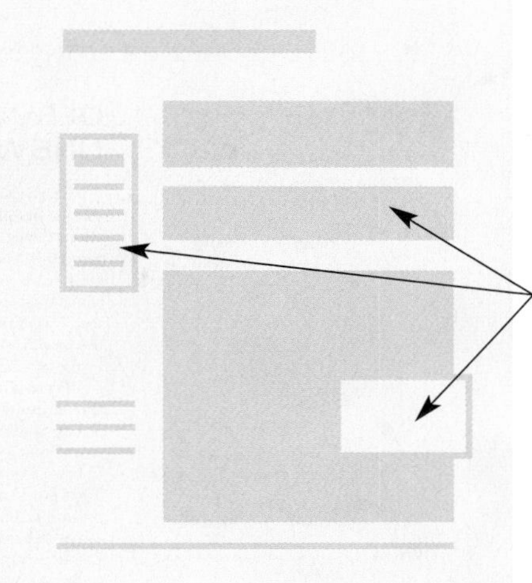

Figure 8.5: (*continued.*)

A four-column grid offers plenty of opportunities for creativity.

Figure 8.6 shows the use of a three-column grid in a report. Notice how the graphics and text offset each other in the page layout.

Grids are also used to lay out screen-based texts. Even though screen-based texts tend to be wider than they are long, readers still expect the material to be balanced on the interface. Figure 8.7 (on page 203) shows possible designs using three- and four-grid templates. The webpages shown in Figures 8.4 and 8.9 demonstrate the use of a six-column grid. Notice how the features of the interface are spaced consistently to create an orderly feel to the text.

In most cases, as shown in these paper-based and screen-based grids, the columns on a grid do not translate directly into columns of written text. Instead, columns of text and pictures often overlap one or more columns in the grid.

LINK To learn more about designing screen interfaces, go to Chapter 14, page 398.

Using Other Balance Techniques

Word-processing programs and desktop publishing software also give you the ability to use professional design features like *margin comments, sidebars,* and *pullouts* (Figure 8.8). These features provide "access points" where people can begin reading the text.

For downloadable templates, go to **www.ablongman.com/johnsonweb/8.4**

Figure 8.6: A three-column grid was used to lay out this text in a balanced manner.

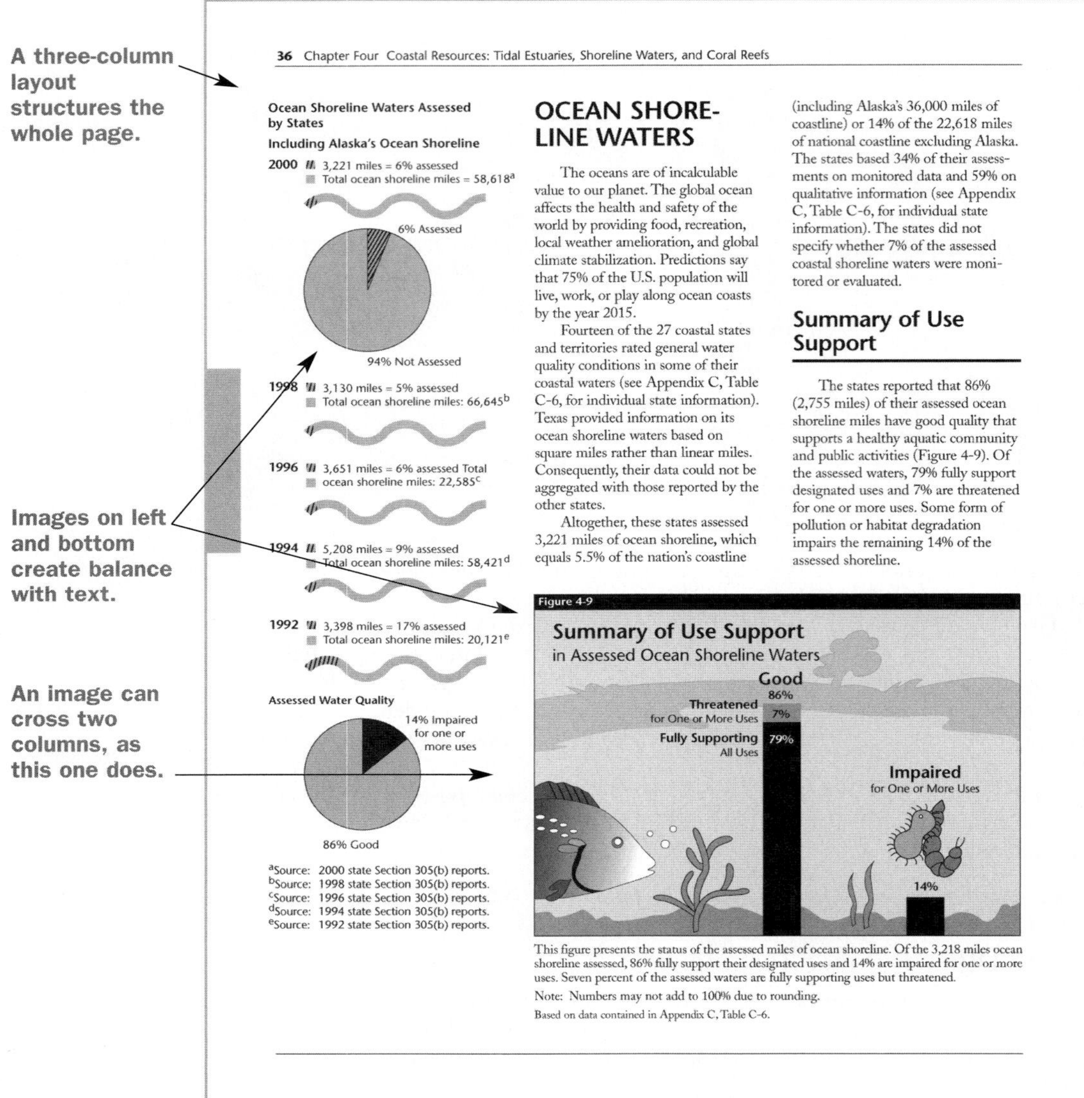

A three-column layout structures the whole page.
Images on left and bottom create balance with text.
An image can cross two columns, as this one does.

36 Chapter Four Coastal Resources: Tidal Estuaries, Shoreline Waters, and Coral Reefs

Ocean Shoreline Waters Assessed by States

Including Alaska's Ocean Shoreline

2000 3,221 miles = 6% assessed
Total ocean shoreline miles = 58,618[a]

6% Assessed

94% Not Assessed

1998 3,130 miles = 5% assessed
Total ocean shoreline miles: 66,645[b]

1996 3,651 miles = 6% assessed Total ocean shoreline miles: 22,585[c]

1994 5,208 miles = 9% assessed
Total ocean shoreline miles: 58,421[d]

1992 3,398 miles = 17% assessed
Total ocean shoreline miles: 20,121[e]

Assessed Water Quality

14% Impaired for one or more uses

86% Good

[a]Source: 2000 state Section 305(b) reports.
[b]Source: 1998 state Section 305(b) reports.
[c]Source: 1996 state Section 305(b) reports.
[d]Source: 1994 state Section 305(b) reports.
[e]Source: 1992 state Section 305(b) reports.

OCEAN SHORE-LINE WATERS

The oceans are of incalculable value to our planet. The global ocean affects the health and safety of the world by providing food, recreation, local weather amelioration, and global climate stabilization. Predictions say that 75% of the U.S. population will live, work, or play along ocean coasts by the year 2015.

Fourteen of the 27 coastal states and territories rated general water quality conditions in some of their coastal waters (see Appendix C, Table C-6, for individual state information). Texas provided information on its ocean shoreline waters based on square miles rather than linear miles. Consequently, their data could not be aggregated with those reported by the other states.

Altogether, these states assessed 3,221 miles of ocean shoreline, which equals 5.5% of the nation's coastline (including Alaska's 36,000 miles of coastline) or 14% of the 22,618 miles of national coastline excluding Alaska. The states based 34% of their assessments on monitored data and 59% on qualitative information (see Appendix C, Table C-6, for individual state information). The states did not specify whether 7% of the assessed coastal shoreline waters were monitored or evaluated.

Summary of Use Support

The states reported that 86% (2,755 miles) of their assessed ocean shoreline miles have good quality that supports a healthy aquatic community and public activities (Figure 4-9). Of the assessed waters, 79% fully support designated uses and 7% are threatened for one or more uses. Some form of pollution or habitat degradation impairs the remaining 14% of the assessed shoreline.

Figure 4-9

Summary of Use Support
in Assessed Ocean Shoreline Waters

Good 86%
Threatened for One or More Uses 7%
Fully Supporting All Uses 79%
Impaired for One or More Uses 14%

This figure presents the status of the assessed miles of ocean shoreline. Of the 3,218 miles ocean shoreline assessed, 86% fully support their designated uses and 14% are impaired for one or more uses. Seven percent of the assessed waters are fully supporting uses but threatened.

Note: Numbers may not add to 100% due to rounding.

Based on data contained in Appendix C, Table C-6.

Source: U.S. Environmental Protection Agency, National Water Quality Inventory, 2000 Report, *2000, p. 36.*

Grids for Interfaces

Figure 8.7: Items on screen-based pages should also be evenly placed. This approach creates a sense of stability.

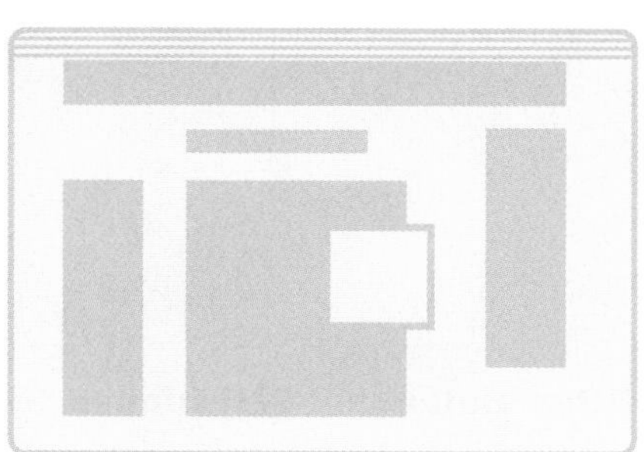

MARGIN COMMENTS Margin comments summarize key points or highlight quotations in the margin of the proposal. When a grid is used to design the page, one of the margins often leaves enough room to include an additional list, offer a special quote, or provide a simple illustration (Figure 8.8). Also, the page shown in Figure 8.6 shows how a whole column can be used to offer supplementary information to the main text.

SIDEBARS Sidebars are used to provide examples or information that supports the main text. Sidebars should never contain essential information that readers must have in order to understand the subject or make a decision. Rather, they should offer supplemental information that enhances the readers' understanding.

PULLOUTS Pullouts are quotes or paraphrases drawn from the body text and placed in a special text box to catch the readers' attention. A pullout should draw its text from the page on which it appears. Often the pullout is framed with rules (lines) or a box, and the text wraps around it (Figure 8.8).

Screen-based documents, like websites, offer some additional design features that you can use to balance the text.

NAVIGATION BARS In screen interfaces navigation bars are used to highlight links, usually as buttons or text (Figure 8.9). Navigation bars are typically placed along the left side of the screen or the top, but they can appear anywhere on the page.

Using Other Balancing Techniques for Print Documents

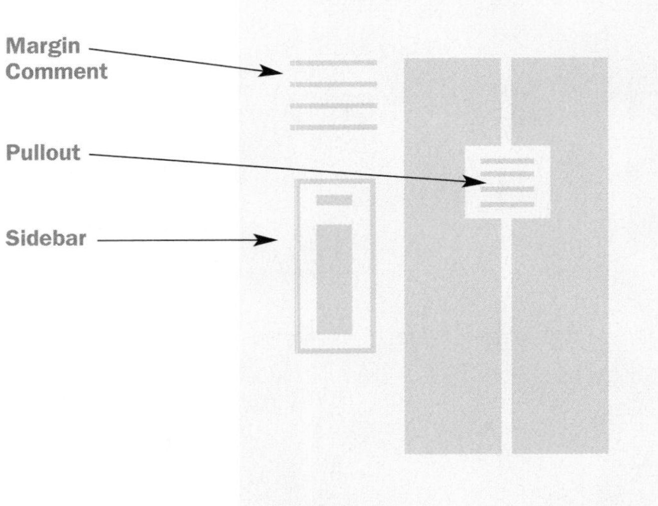

Figure 8.8: Margin comments, pullouts, and sidebars can make a text more accessible.

Other Balancing Techniques for Interfaces

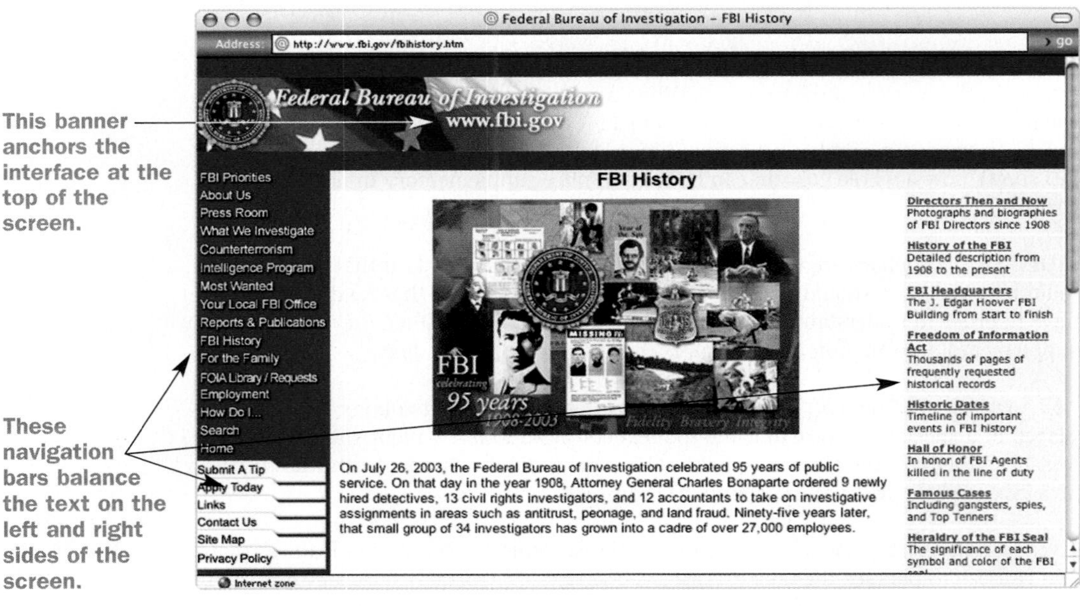

Figure 8.9: Usually, what works on paper can work on-screen. You can use pullouts, margin comments, and sidebars on a screen interface. Navigation bars and banners can also be added to screen-based documents to make them look balanced.

Source: Federal Bureau of Investigation, http://www.fbi.gov/fbihistory.htm.

GO TO THE NET

For more advice about interface design, go to **www.ablongman.com/johnsonweb/8.5**

AT WORK

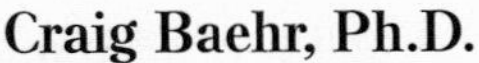

Craig Baehr, Ph.D.

ASSISTANT PROFESSOR OF TECHNICAL COMMUNICATION,
TEXAS TECH UNIVERSITY

Dr. Baehr previously worked for 10 years as a web developer and technical writer for the U.S. Army Corps of Engineers.

How do you know whether the interface of your website is working correctly?

Not all interface layouts are created equal.

Once you've designed the ideal interface for your customer or company, you should test it for usability. Try it out with a few different browsers, platforms, Internet connections, screen resolutions, and color depth settings. This is especially important if you lack the funding, manpower, or time for formal user testing.

I usually test my interface designs with two browsers (such as Internet Explorer and Netscape), PC and Mac platforms, two resolution settings (usually 1024 x 768 and 800 x 600), two color depth settings (256K and 16- or 32-bit), and both a 56K dial-up and a network connection (such as a T1 connection). These basic tests will help you ensure that your interface design is consistent in its content and layout to users.

These quick, but important, tests will also help you identify design problems in your interface layout. Based on the results, you may need to make changes to frames, tables, or CSS scripts. In more extreme cases, you may have to design part of your site in an alternate layout (one compatible with Explorer and Netscape). In this case you can use a script to detect the user's settings and redirect him or her to the compatible version.

The main point is to ensure your interface design provides the same quality and consistency to all your users. These quick usability checks will also save you time in responding to e-mails addressed to you, the web designer.

TAKE NOTE When placed along the right side or the bottom of the page, navigation bars can cause the screen to look unbalanced.

BANNERS Banners run along the top of the screen interface. In websites, they may include the name of the organization or an advertiser. In multimedia documents, they may identify the name of the document. Banners often include links, but they don't need to.

LINK To learn more advanced strategies for designing interfaces, turn to Chapter 14, page 398.

Design Principle 2: Alignment

Items on a page or screen can be aligned vertically and horizontally. By aligning items *vertically* on the page, you can help readers identify different levels of information in a document. By aligning items *horizontally,* you can connect them visually, so readers view them as a unit.

Want to learn about other balancing techniques? Go to **www.ablongman.com/johnsonweb/8.6**

In Figure 8.10, for example, the page on the left gives no hint about the hierarchy of information, making it difficult for a reader to scan the text. The page on the right, meanwhile, uses alignment to clearly signal the hierarchy of the text.

Using Vertical Alignment

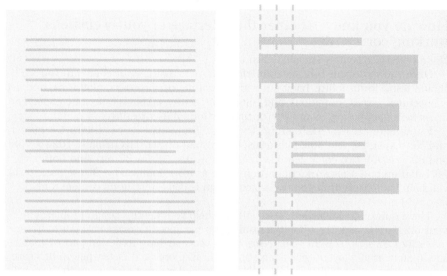

Figure 8.10: Alignment allows readers to see the hierarchy of information in a text.

Alignment takes advantage of readers' natural tendency to search out visual relationships among items on a page. If a picture, for example, is aligned with a block of text on a page, readers will naturally assume that they go together.

TAKE NOTE If items on a page are not related, they should not be aligned on the page. Readers will make a natural connection between two visually aligned items, even if they are not connected. These unexpected connections can cause some confusion.

In paper-based documents, look for ways you can use margins, indentation, lists, headings, and graphics to create two or three levels in the text. If you use a consistent alignment strategy throughout the text, you will design a highly readable and accessible document.

In technical documents, items are usually aligned on the left side. In rare cases, you might try aligning titles and headings on the right side. But, you should use centering only for titles, because it causes alignment problems in the text. Figure 8.11 shows how centering can create unpredictable vertical lines in the text.

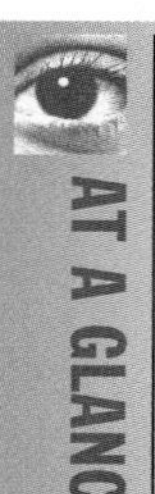

AT A GLANCE

Balancing the Design of a Page or Screen

- Weight items on the page or screen.
- Use grids to balance the layouts.
- Use design features, such as margin comments, sidebars, pullouts, navigation bars, and banners.

Alignment Problems with Centered Text

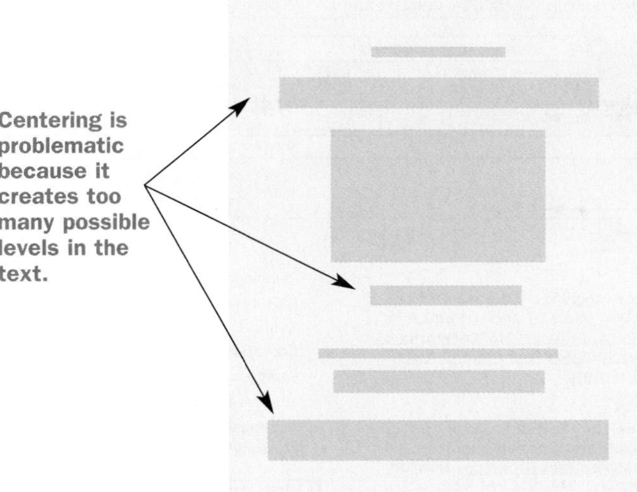
Centering is problematic because it creates too many possible levels in the text.

Figure 8.11: Centering is fine for headings, but too much centered material can make the text look chaotic.

Alignment is also very important in on-screen documents. To create a sense of stability, pay attention to the horizontal and vertical alignment of features on the interface. For example, the screen in Figure 8.12 shows how you can align text and graphics to make an interface look stable.

Design Principle 3: Grouping

The principle of grouping means that items on a page that are near each other will be seen as one unit. Grouping allows you to break up the information on a page by dividing the text into scannable blocks.

Humans naturally see items that are placed near each other as a whole unit. So, if two items are placed near each other, like a picture and a caption, readers will assume that they belong together. In Figure 8.12, notice how pictures are put near paragraphs, so they are seen as units. The banner at the top of the page is supposed to be seen as a block unto itself.

Grouping is also referred to as "using white space" to frame items on the page. White spaces are places where no text or images appear on the page, including

- the margins of the document.
- the space around a list.
- the area between an image and the body text.
- the space between two paragraphs.

These spaces create frames around the items on the page, so readers can view them as groups. For example, the white space around a vertical list (like the one in Figure 8.12) helps readers see that list as one unit.

GO TO
THE NET

To see sample texts that use alignment well and badly, go to **www.ablongman.com/johnsonweb/8.7**

An Interface That Uses Alignment and Grouping Well

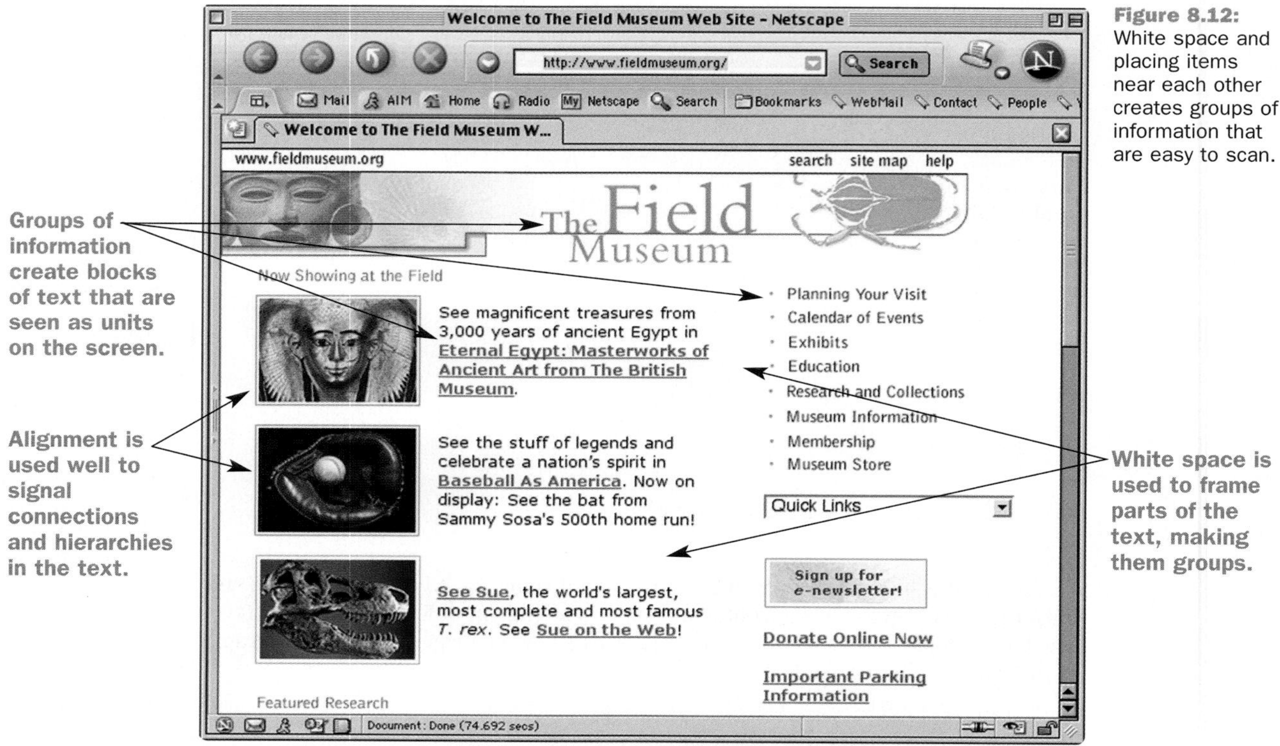

Figure 8.12: White space and placing items near each other creates groups of information that are easy to scan.

Source: The Field Museum, Chicago, Illinois, http://www.fieldmuseum.org.

Using Headings

One way to group information is to use headings. When used properly, headings will help your readers quickly understand the structure of your document and how to use it.

Your computer's word-processing program makes it easy for you to use headings by changing fonts and font sizes. Or, you can use your word processor's "Style" feature to create standard headings for your documents (see the "Help" feature in this chapter).

LINK For more advice on using headings to organize a document, see Chapter 6, page 132.

Different types of headings should signal the various levels of information in the text.

- **First-level headings** should be sized significantly larger than second-level headings. In some cases, first-level headings might use all capital letters (all caps) or small capital letters (small caps) to distinguish them from the font used in the body text.

- **Second-level headings** should be significantly smaller and different from the first-level headings. Whereas the first-level headings might be in all caps, the second-level headings might use bold lettering.
- **Third-level headings** might be italicized and a little larger than the body text.
- **Fourth-level headings** are about as small as you should go. They are usually bolded or italicized and placed on the same line as the body text.

Figure 8.13 shows various levels of headings and how they might be used.

Levels of Headings

Figure 8.13: The headings you choose for a document should be clearly distinguishable from the body text and from each other, so the readers can see the levels in the text.

DOCUMENT TITLE

FIRST-LEVEL HEADING

This first-level heading is 18-point Avant Garde, bolded with small caps. Notice that it is significantly different from the second-level heading, even though both levels are in the same typeface.

Second-Level Heading

This second-level heading is 14-point Avant Garde with bolding.

Third-Level Heading

This third-level heading is 12-point Avant Garde italics. Often no space appears between a third-level heading and the body text, as shown here.

Fourth-Level Heading. This heading appears on the same line as body text. It is designed to signal a new layer of information without standing out too much.

In most technical documents the headings should use the same typeface throughout (e.g., Avante Garde, Times, Helvetica). In other words, the first-level heading should use the same typeface as the second-level and third-level headings. Only the font size and style (bold, italics, small caps) should be changed.

TAKE NOTE Often people confuse the words "typeface" and "font." The typeface is the design of type you are choosing (e.g., Times, Helvetica, Arial). The font specifies not only the typeface, but other style attributes, like type size, bolding, italics, or small caps.

Headings should also follow consistent wording patterns. A consistent wording pattern might use gerunds (*-ing* words) to lead off headings. Or, to be consistent, questions might be used as headings.

Inconsistent Headings

Global Warming

What Can We Do about Global Warming?

Alternative Energy Sources

Consistent Headings Using Gerunds

Defining Global Warming

Doing Something about Global Warming

Finding Alternative Energy Sources

Consistent Headings Using Questions

What Is Global Warming?

How Can We Do Something about Global Warming?

Are Alternative Energy Sources Available?

Headings should also be specific, clearly signaling the content of the sections that follow them:

Unspecific Headings

Kinkajous

Habitat

Food

Specific Headings

Kinkajous: Not Rare, But Hard to Find

The Habitat of Kinkajous

The Food and Eating Habits of Kinkajous

Headings serve as *access points* for readers, breaking a large text into smaller groups. If headings are used consistently, readers will be able to quickly access your document to find the information they need. If they are used inconsistently, readers may have difficulty understanding the structure of the document.

Using Borders and Rules

In document design, borders and straight lines called *rules* can be used to carve a page into smaller groups of information. They can also help break the text into more manageable sections for the readers.

Borders completely frame parts of the document (Figure 8.14). Whatever appears in a border should be able to stand alone. For example, a bordered warn-

Want more information on using headings? Go to **www.ablongman.com/johnsonweb/8.8**

ing statement should include all the information readers need to avoid a dangerous situation. Similarly, a border around several paragraphs (like a sidebar) suggests that they should be read separately from the main text.

Rules are often used to highlight a banner or carve a document into sections. They are helpful for signaling places to pause in the document. But, when they are overused, they can make the text look and feel too fragmented.

TAKE NOTE Too many rules can make a document look chopped up and impede the flow of the text. Rules should be used sparingly to highlight titles and headings or segregate text into topics.

Using Rules and Borders to Group Information

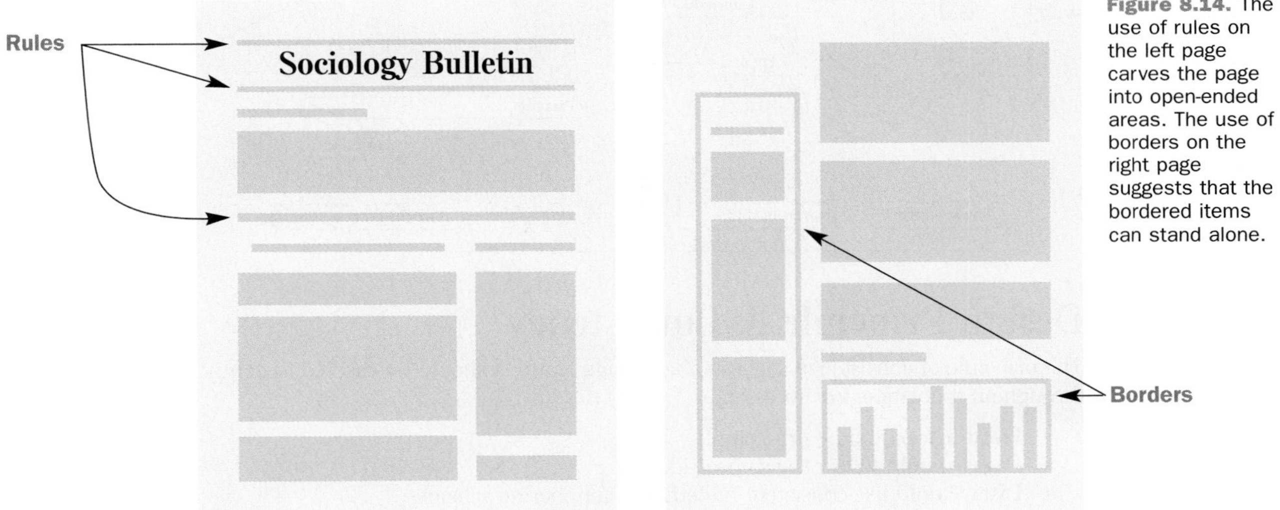

Figure 8.14. The use of rules on the left page carves the page into open-ended areas. The use of borders on the right page suggests that the bordered items can stand alone.

Borders and rules are usually easy to create with any word processor. To put a border around something, highlight that item and find the "Borders" command in your word processor (Figure 8.15). In the window that appears, you can specify what kind of border you want to add to the text.

Rules are usually a bit more difficult to use, requiring a desktop layout program, like Adobe FrameMaker or QuarkXpress. However, if your document is not *too* large or complex, you can use the "Draw" function of your word processor to draw horizontal or vertical rules in your text.

One warning: Borders and rules can be helpful in documents, but they can also be overused. If used occasionally, they will draw readers' attention to important information. If overused, though, they will slow readers down. In some cases, readers grow immune to these grouping techniques, preferring to skip the information that has been framed off.

More advice about using rules and borders is available at **www.ablongman.com/johnsonweb/8.9**

Making Borders

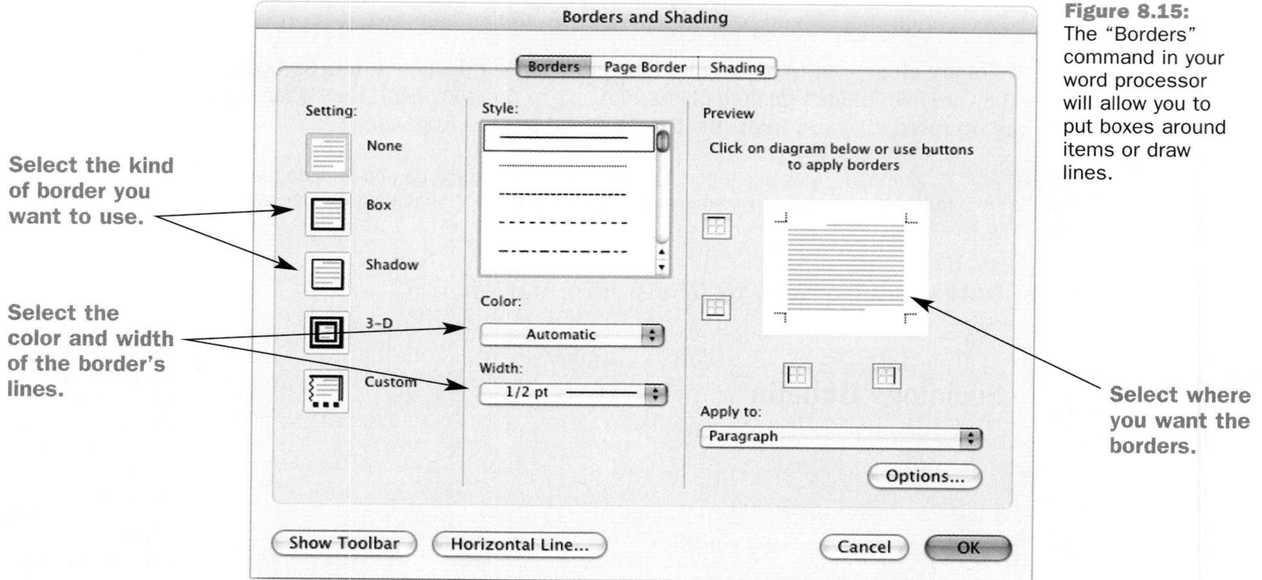

Figure 8.15: The "Borders" command in your word processor will allow you to put boxes around items or draw lines.

Design Principle 4: Consistency

The principle of consistency suggests that design features should be used consistently throughout a document or website:

- Headings should be predictable.
- Pages should follow the same grid.
- Lists should use consistent bulleting or numbering schemes.
- Page numbers should appear in the same place on each page.

Consistency is important because it creates a sense of order in a document while limiting the amount of clutter. A consistent page design helps readers access information quickly, because each page is similar and predictable. When design features are used inconsistently, readers will find the document erratic and hard to interpret.

Figure 8.16 shows a consistent layout for a wireless phone user manual. In this sample text, notice how the consistent use of design features creates a predictable, useful document.

Choosing Typefaces

Consistency should be an important consideration when you choose typefaces for your document. As a rule of thumb, a document should not use more than two different typefaces. Most page designers will choose two typefaces that are very different from each other, usually a *serif* typeface and a *sans serif* typeface.

Consistent Layout

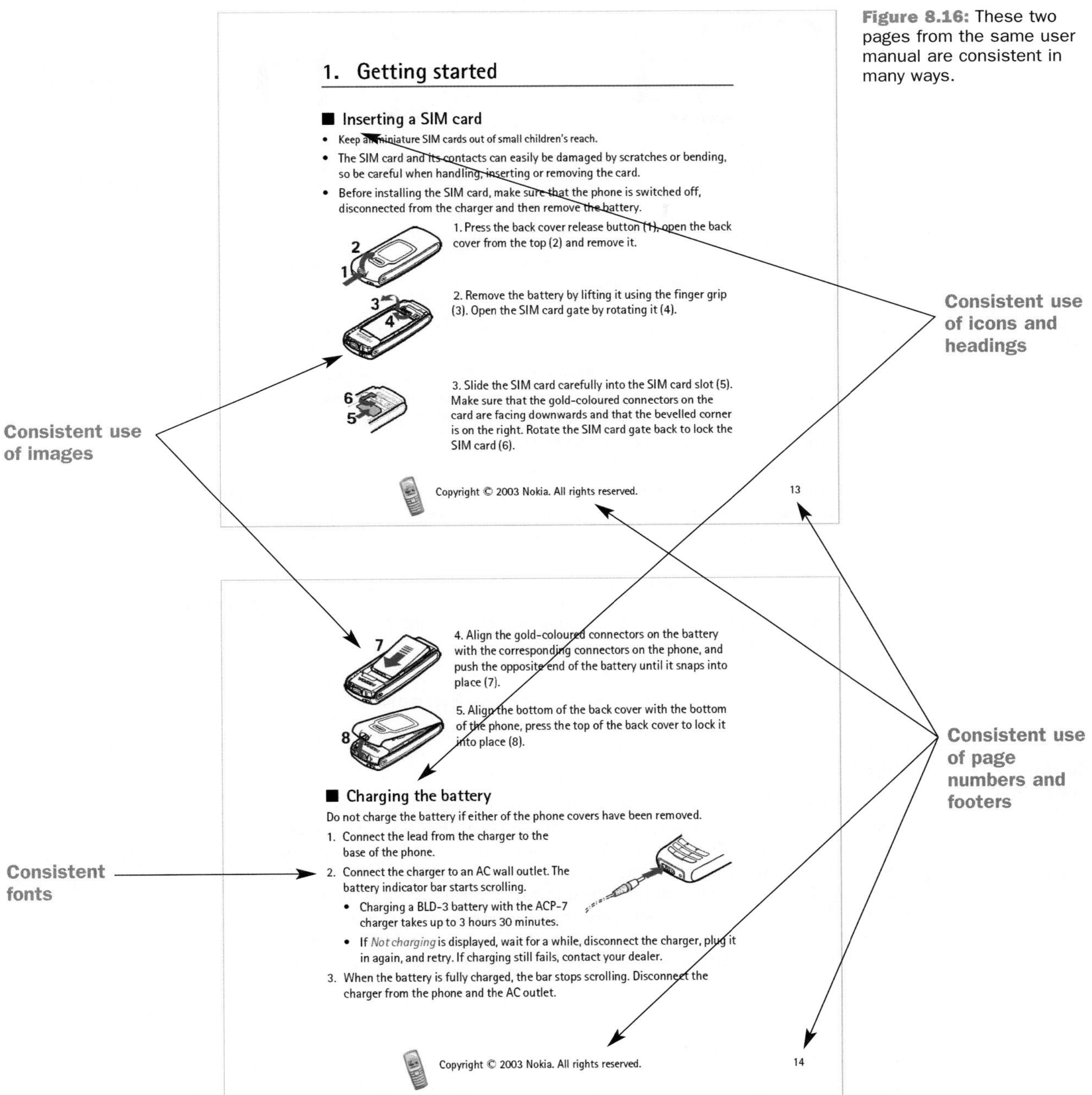

Figure 8.16: These two pages from the same user manual are consistent in many ways.

Source: Nokia, Nokia 2100 User Guide, *2003, pp. 13–14.*

Serif and Sans Serif Typefaces

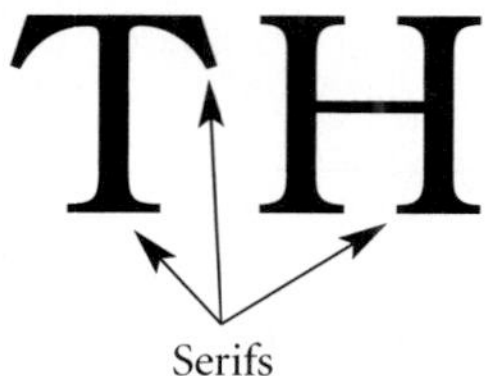

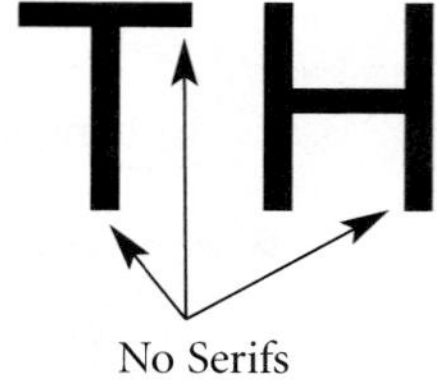

Figure 8.17: A serif typeface like Bookman (left) includes the small tips at the ends of letters. A sans serif typeface like Helvetica (right) does not include these tips.

A serif typeface like Times Roman, New York, or Bookman has small tips (serifs) at the ends of main strokes in each letter (Figure 8.17). Sans serif typefaces like Arial or Helvetica do not have these small tips.

TAKE NOTE Sans means "without" in French, so sans serif means "without serifs."

Serif fonts, like Times, New York, or Bookman, are usually perceived to be more formal and traditional. Sans serif typefaces like Helvetica seem more informal and progressive. So, designers often use serif fonts for the body of their text and sans serif in the headings, footers, captions, and titles (Figure 8.18). Using both kinds of typefaces gives a design a progressive and traditional feel at the same time.

Your choice of typefaces, of course, depends on the needs and expectations of your readers. Use typefaces that fit their needs and values, striking the appropriate balance between progressive and traditional design.

TAKE NOTE On computer screens, typefaces often look different than they do on paper. Serif typefaces, for example, can often look grainy and inconsistent, making them hard to read at length. Sans serif typefaces, because they are simpler, often seem cleaner and easier to read on-screen.

Using Different Typefaces for Different Purposes

Helvetica (sans serif font) →

Bookman (serif font) →

Which Is Better? Serif or Sans Serif?

Serif typefaces are often used in traditional-looking texts, especially for the body text. Studies have suggested inconclusively that serif typefaces like Bookman are more legible, but there is some debate as to why. Some researchers claim that the serifs create horizontal lines in the text that make serif typefaces easier to follow. These studies, however, were mostly conducted in the United States, where serif typefaces are common. In other countries, like Britain, where sans serif typefaces are often used for body text, the results of these studies might be quite different.

Figure 8.18: Often designers will choose a sans serif typeface for headings, while a serif typeface is used for the main text.

Labeling Graphics

Graphics such as tables, charts, pictures, and graphs, should be labeled consistently in your document. In most cases, the label for a graphic will include a number.

For websites that discuss the history of typefaces, go to **www.ablongman.com/johnsonweb/8.10**

Graph 5: Growth of Sales in the Third Quarter of 2004

Table C: Data Set Gathered from Beta Radiation Tests

Figure 10: Diagram of Wastewater Flow in Hinsdale's Treatment Plant

The label can be placed above the graphic or below. Use the same style for labeling every graphic in the document. Then locate the labels consistently on each graphic.

LINK For more information on labeling graphics, see Chapter 9, page 241.

Creating Sequential and Nonsequential Lists

Early in the design process, you should decide how lists will be used and how they will look. Lists are very useful for showing a sequence of tasks or setting off a group of items. But, if they are not used consistently, they can create confusion for the readers.

When deciding how lists will look, first make decisions about the design of *sequential* and *nonsequential* lists (Figure 8.19).

Using Sequential and Nonsequential Lists

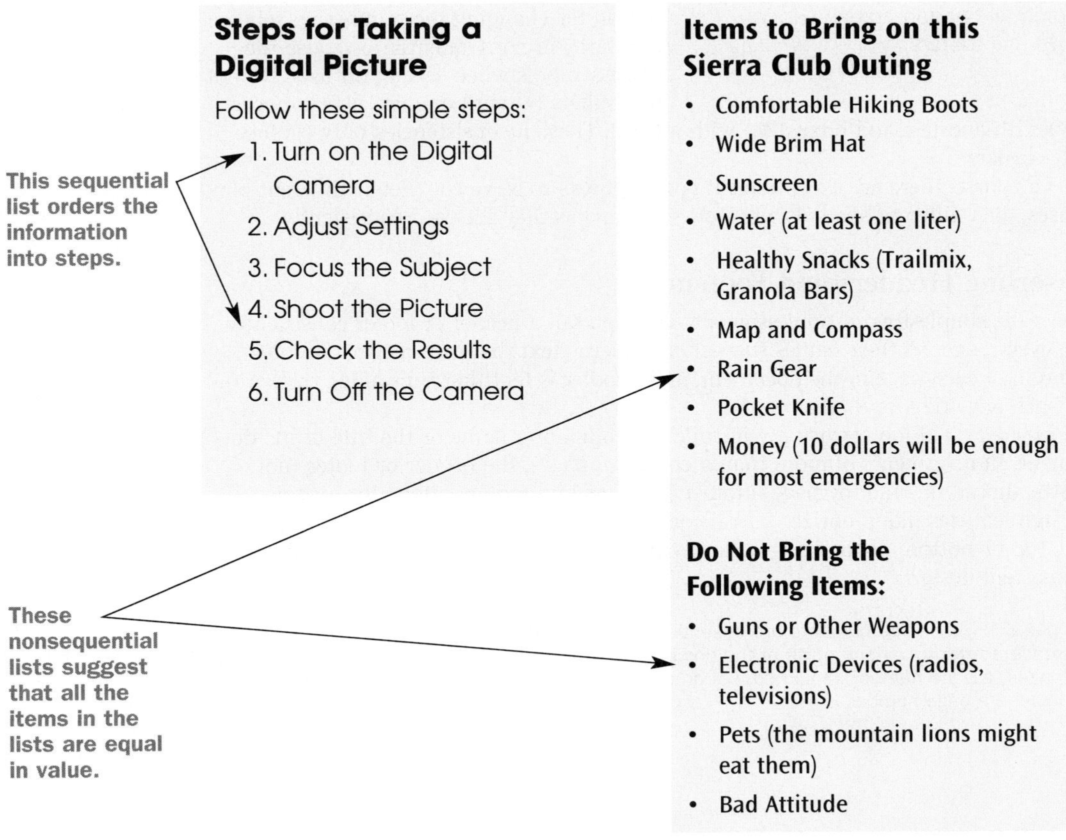

Steps for Taking a Digital Picture
Follow these simple steps:
1. Turn on the Digital Camera
2. Adjust Settings
3. Focus the Subject
4. Shoot the Picture
5. Check the Results
6. Turn Off the Camera
This sequential list orders the information into steps.
Items to Bring on this Sierra Club Outing
• Comfortable Hiking Boots
• Wide Brim Hat
• Sunscreen
• Water (at least one liter)
• Healthy Snacks (Trailmix, Granola Bars)
• Map and Compass
• Rain Gear
• Pocket Knife
• Money (10 dollars will be enough for most emergencies)
Do Not Bring the Following Items:
• Guns or Other Weapons
• Electronic Devices (radios, televisions)
• Pets (the mountain lions might eat them)
• Bad Attitude
These nonsequential lists suggest that all the items in the lists are equal in value.

Figure 8.19: The sequential list on the left shows an ordering of the information, so it requires numbers. The nonsequential list on the right uses bullets because there is no particular ordering of these items.

Sequential (numbered) lists are used to present items in a specific order. In these lists, you can use numbers or letters to show a sequence, chronology, or ranking of items.

Nonsequential (bulleted) lists include items that are essentially equal in value or have no sequence. You can use bullets, dashes, or check boxes to identify each item in the list.

Lists make information more readable and accessible. So, you should look for opportunities in your documents to use them. If, for example, you are describing how to do something by listing steps, you have an opportunity to use a sequential list. Or, if you are creating a list of four items or more, ask yourself whether a nonsequential list would present the information in a more accessible way.

Just make sure you use lists consistently. Sequential lists should follow the same numbering scheme throughout the document. For example, you might choose a numbering scheme like (1), (2), (3). If so, do not number the next list 1., 2., 3. and others A., B., C., unless you have a good reason for changing the numbering scheme. Similarly, in nonsequential lists, use the same symbols when setting off lists. Do not use bullets (•) with one list, check marks (✓) with another, and boxes (■) with a third. These inconsistencies only confuse the readers.

Of course, there are exceptions. If you need lists to serve completely different purposes, then different symbols will work—as long as they are used consistently.

AT A GLANCE

Checking for Consistency

The following items should be used consistently in your document:

- Typefaces (serif and sans serif)
- Labeling of graphics
- Lists (sequential and nonsequential)
- Headers and footers

Inserting Headers and Footers

Even the simplest word-processing software can put a header or footer consistently on every page. As their names suggest, a header is text that runs across the top margin of each page in the document, and a footer is text that runs along the bottom of each page (Figure 8.20).

Headers and footers usually include the company's name or the title of the document. In documents of more than a couple of pages, the header or footer (not both) should include the page number. Headers and footers often also include design features like a horizontal rule or a company logo. If these items appear at the top or bottom of each page, the document will tend to look like it is following a consistent design.

TAKE NOTE Page numbers are essential in technical documents, because they help readers refer to various parts of the document. If your document includes more than three pages, a page number should appear on each page. Sometimes the first page does not include a page number.

For advice about using lists, go to
www.ablongman.com/johnsonweb/8.11

Headers and Footers

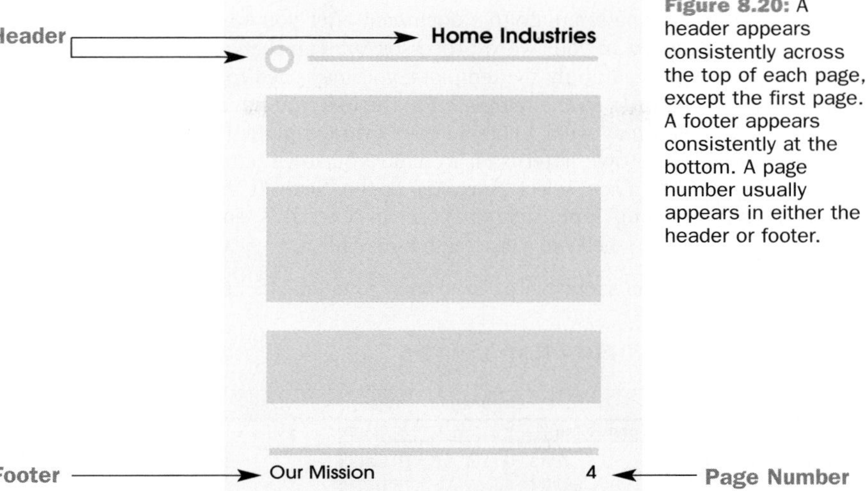

Figure 8.20: A header appears consistently across the top of each page, except the first page. A footer appears consistently at the bottom. A page number usually appears in either the header or footer.

HELP

Using Styles and Templates

Word-processing programs like Word or WordPerfect include "Styles" functions that will help you manage the formatting of your documents. When you apply a style to, say, a heading, paragraph, caption, or list, the computer will automatically change the text into a predetermined format.

For example, let us say you want all your second-level headings to set in 12-point Helvetica Italics. You can define that style with the "Styles" function. From then on, whenever you use a second-level heading, you can simply select that style from the style menu (Figure A). The computer will do the formatting for you.

That's handy, but here is the real advantage of using styles: If you ever need to change the styles in your document (say you decide to use the Arial typeface instead of Helvetica for headings), you simply need to redefine that style. The computer will go through your document and change all the headings to that new style.

Styles are useful when you are designing a larger document, especially when you are working with a team. If you and your group decide on the styles of the document up front, each member of the team can follow the same formatting. Then, when you put the document together, you will need to do only some minimal editing to make the format consistent.

Word processors also include style *templates* that you can follow. These templates will help you format letters, memos, resumes, reports, and presentations.

The easiest way to use a template is to start with one. When you create a document with the "New" function, your word processor will usually offer you a selection

For websites that will help you use the "Styles" menu, go to **www.ablongman.com/johnsonweb/8.12**

of templates you can use. Choose the template you need. Then, you can modify the template to handle the individual needs of the document you are writing.

If you want to apply a template to a document after you have started writing, find the "Templates" function in your word processor. You can then apply the template to your document. After you apply the template, you may need to make some adjustments to words or images so the template fits the text you have written.

In some cases, you may want to create your own template. Perhaps, for example, you want to make your own letterhead. To make your own template, design a document on a blank page. Then, select "Save As." In the "Save As" box, you can "Save Document as a Template." From then on, you can select this template from the collection of other templates whenever you create a new file.

LINK For more information about making templates, go to Chapter 16, page 475.

Using the "Styles" Function

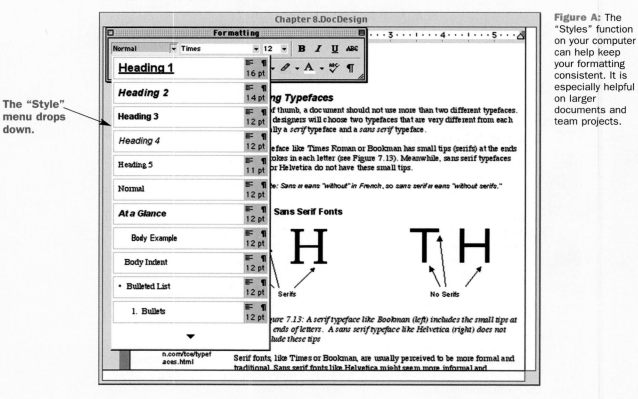

The "Style" menu drops down.

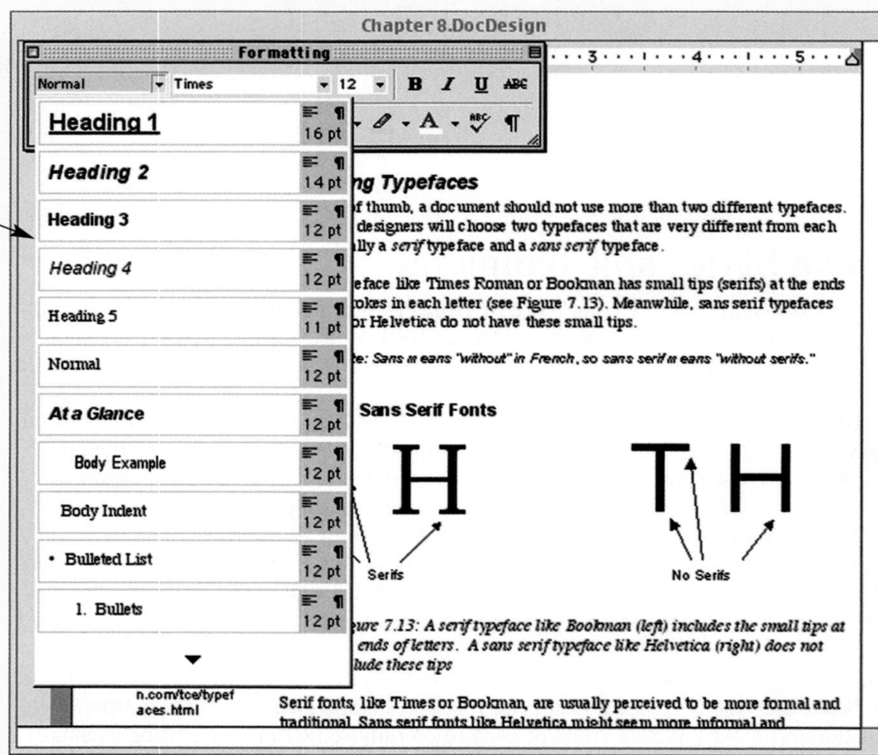

Figure A: The "Styles" function on your computer can help keep your formatting consistent. It is especially helpful on larger documents and team projects.

Need a template? Go to **www.ablongman.com/johnsonweb/8.13**

Design Principle 5: Contrast

Contrast makes the items on a page more distinct and thus more readable. Contrast makes things look different, adding energy and sharpening boundaries among items on the page screen.

For example, on a page or screen, you might contrast green text with black and white text. You might use a bolder 18-point sans serif font for headings, so they look different from the smaller 12-point serif font used in the main text. Contrast means making things noticeably different, so readers see these items as separate entities. Contrast, as shown in Figure 8.21, makes design elements lively.

Contrast in a Webpage

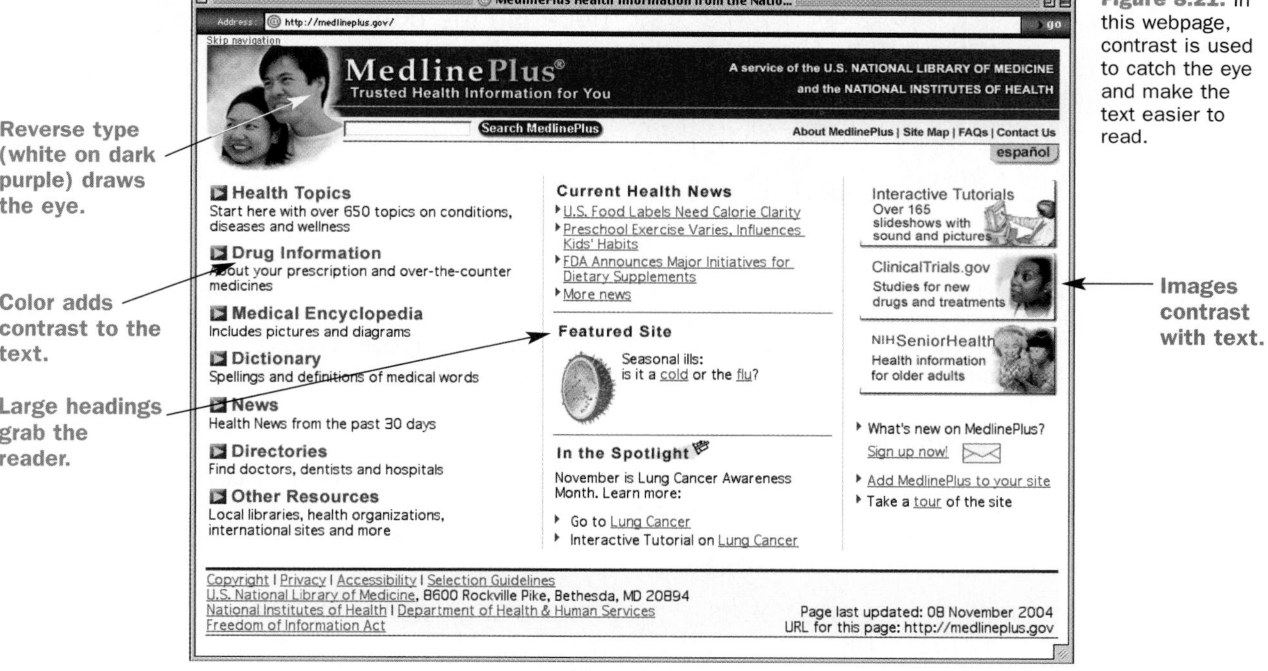

Figure 8.21: In this webpage, contrast is used to catch the eye and make the text easier to read.

Source: Medline Plus, http://medlineplus.gov.

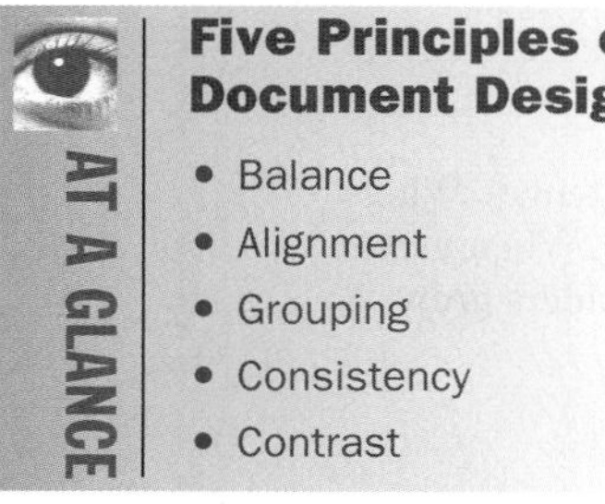

AT A GLANCE
Five Principles of Document Design
• Balance
• Alignment
• Grouping
• Consistency
• Contrast

Your uses of contrast when designing a page should be considered carefully. Word processors offer many tools for adding contrast in ways that capture readers' attention. Sometimes, though, you can accidentally create contrast problems with different colors or shading in the background—or just too much clutter on the page. In this section there are some helpful tips for using contrast in documents and interfaces.

Adding Shading and Background Color

When used properly, shading and background color can help highlight important text in a document. However, these design features can also make texts hard to read. For example, Figure 8.22 shows how a lack of contrast can make a text hard to read.

Shading and Contrast

This text is set against a gray background. It is difficult to read because there is not enough contrast between the words and the background against which they are set.	This text is set against a white background. It is easier to read, because the contrast between the words and the white background is sharp and definite.

Figure 8.22: Shaded versus unshaded text: The shaded text on the left is more difficult to read because the words do not contrast significantly with the background shading.

Background color or images can also cause text on computer screens to be difficult to read. So, use them carefully and be sure to check how backgrounds work (or don't work) with the text on the screen.

It is fine to use shading, background color, and background images. However, make sure the words on the page contrast significantly with their background.

Highlighting Text

There are several different ways to use highlighting, a form of contrast, to make words stand out in a text. With a computer, you can highlight text with ALL CAPS, SMALL CAPS, **bolding**, *italics*, underlining, and color. You can even use outline and shadow to make information stand out.

> ALL CAPS SHOULD BE USED IN MODERATION, USUALLY ONLY FOR HEADINGS OR SHORT WARNING STATEMENTS. AS YOU CAN SEE, THESE SENTENCES ARE HARD TO READ. ALSO, ALL CAPITALS SUGGESTS THAT THE DOCUMENT IS SHOUTING.
>
> SMALL CAPS ARE A LITTLE EASIER TO READ THAN ALL CAPS BUT STILL SHOULD BE USED SPARINGLY. THEY ARE ESPECIALLY USEFUL FOR HEADINGS.
>
> **Bolding is useful for highlighting individual words or whole sentences. When used in moderation, bolding helps words or sentences stand out. When used too often, however, the highlighting effect is reduced because readers grow used to the bolded text.**

To view webpages with contrast problems, go to
www.ablongman.com/johnsonweb/8.14

Italics, similarly, can be used to add emphasis to words or sentences. A whole paragraph in italics, however, is hard to read, so use italics sparingly—only when you need it.

Color is perhaps the most prominent way to highlight words or sentences. It is especially helpful for making headings more prominent. However, color can be distracting when it is overused. So, use it selectively and consistently.

Underlining is rarely used now that we have other ways to highlight text. Underlining was useful when typewriters were widely used, because it was the only way to highlight information. Computers give us better options for highlighting, like italics, bolding, and color.

Outline and shadow should be rarely used, if ever. Used at length, they are both difficult to read. You may choose to use them for headline text. But use outline and shadow only in very special cases.

Highlighting is an easy way to create contrast in a document. When used selectively, it can make important words or sentences pop off the page. If it is overused, though, readers become immune to it.

Using Font Size and Line Length

Your decisions about font size, line length, and line spacing should depend on the readers of your document.

APPROPRIATE FONT SIZE The font size you choose for your text depends on how your readers will be using the text.

For most readers, text in 11-point or 12-point font sizes is easy to read.

Fonts smaller than 10 points can be difficult to read, especially when used for a longer stretch of text. When a page length is required for a document, it is often tempting to use a smaller font, so more words can be put on a page. But most people quickly grow tired of straining their eyes to read the text. You are better off using a normal-sized font with fewer words. That way, at least the readers will actually read the text.

Large font sizes (above 14 points) should be used only for special purposes, such as texts written for older readers or situations where the text needs to be read at a distance. When used at length, these larger sizes can make a text look childish. Extra-large font sizes usually suggest that the writer is simply filling a page to hide a lack of content.

PROPER LINE LENGTH Line length is also an important choice in a document. Scanning readers usually prefer a shorter line length, making columns especially useful for documents that will be read quickly.

However, when lines are
short, they force
readers to quickly look
back and forth.
Eventually, these short
lines will frustrate
readers because they
require such rapid
movement of the eyes.
They also seem to suggest
fragmented thinking on
the part of the writer,
because the sentences
look fragmented.

Longer line lengths can have the opposite effect. Readers quickly grow tired of following the same line for several inches across the page. The lines seem endless, giving readers the impression that the information is hard to process. More important, though, readers find it difficult to locate the next line on the left side. A line should never be wider than six inches across.

As you consider font size and line length, you should anticipate how your readers will use your document. If they are scanning, shorter line lengths will help them read quickly. If they are not scanning, longer lines are fine. Then, choose the font size or line length that suits their needs.

Using the Principles of Design

Balance, alignment, grouping, consistency, contrast: These five basic design principles should help you create easy-to-read page layouts and screen interfaces that highlight important information and attract your readers.

In many ways, designing documents is like drafting the written text in your documents. You should go through a process—in this case, a *design process.*

Analyze readers and the document's context of use.

Use thumbnails to rough out the design.

Design the document.

Revise and edit the design.

As with drafting, each stage in the process takes you closer to a finished document.

Analyze Readers and the Document's Context of Use

You should begin designing a document by first looking back at your analyses of your readers and the contexts in which they will use your document. Different kinds of readers will have different expectations and preferences for the design of your document. A more traditional reader, for example, might prefer a conservative layout with simple headings and a one-column or two-column layout. A less orthodox, more progressive reader might prefer a bolder layout with more color and contrast.

As you consider your readers, try to match your design to the tone you are setting in the written part of the document. Positive, excited themes in the written text should be reflected in the color and boldness of the design. Serious or somber themes should be reflected with subdued color schemes and a plainer design.

The context in which your document will be used should also play an important role in helping you decide the appropriate design. Pay special attention to the *physical context* in which the readers will be looking over your document. Will they be in a meeting? At a worksite? In their living room?

Each of these contexts might shape how you will design the document. For example, if you know your readers will be discussing your document in a meeting, you might use margin notes and lists to highlight information that will be discussed at the meeting. That way, your readers will be able to quickly find "talking points" in your document that they can share with the others in the meeting.

Use Thumbnails to Rough Out the Design

Next, start roughing out how you think the design will look. Professional graphic designers will often use *thumbnails* to sketch out possible layouts for their documents or screen interfaces. Thumbnails can be sketched by hand, usually with pencil, or with a computer drawing program (see Figure 8.23). They are miniature versions of the pages.

Thumbnailing is especially helpful when you are working with a team. With little effort, you can quickly sketch out sample pages or screen interfaces. When the group agrees, you can then use these sketches to help you lay out the page.

TAKE NOTE The simplest way to thumbnail pages is to fold a regular sheet of paper into four parts (i.e., fold it once lengthwise and once horizontally). The folds will then give you four separate "pages" to practice on. For computer interfaces, simply turn the sheet sideways. You will have four rectangles that are each about the shape of a screen.

One nice thing about designing with thumbnails is how easy they are to use. It takes only seconds to sketch out a few layouts. If you or people on your team don't like the sketches, you can toss them in the recycling bin. Then, start again. Thumbnails require much less effort than an attempt to develop a full layout of a page or document. As a result, you will feel more freedom to be creative and try out a few different designs for the document.

Design the Document

Usually, the design is developed as the document is drafted. You don't need to wait until the entire document is written out before you start laying out some of the pages. In fact, it is better to begin designing *before* the whole document is drafted.

For more advice on thumbnailing, go to
www.ablongman.com/johnsonweb/8.15

Sample Page and Interface Thumbnails

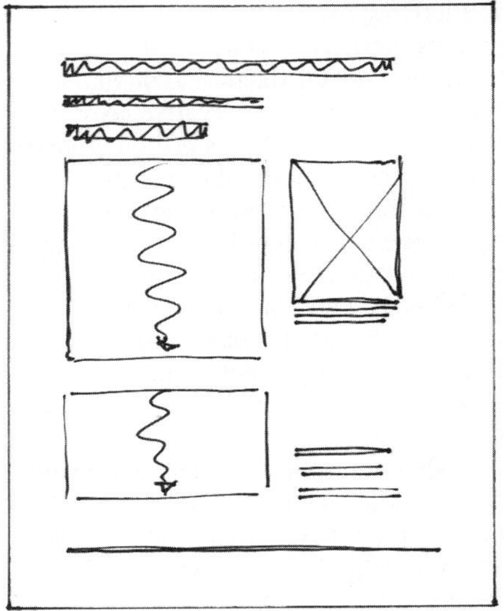

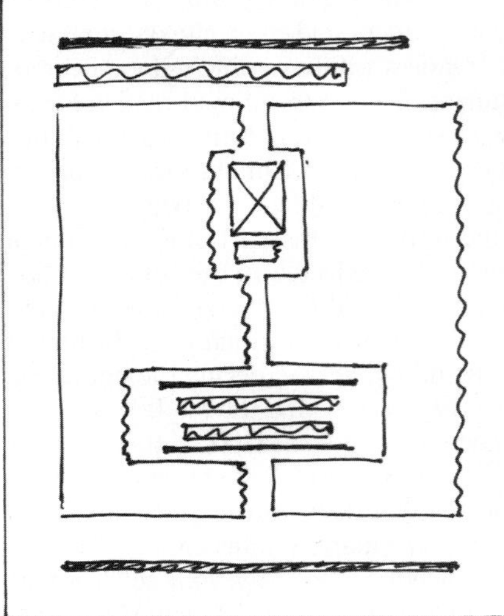

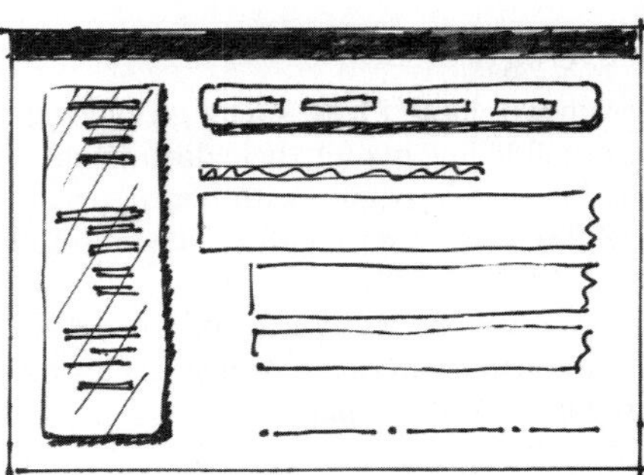

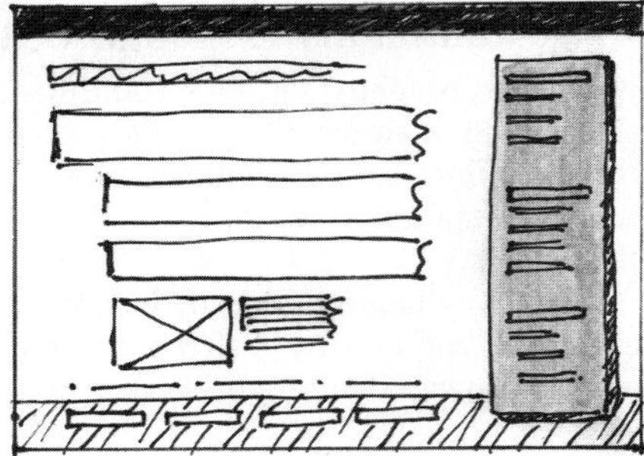

Figure 8.23: Thumbnails are miniature sketches of possible page or screen layouts. They help you find a good design with a minimum amount of effort.

That way, the design might help you (1) identify gaps in the content, (2) find places where visuals, like graphs and charts, are needed, and (3) locate places in the document where supplemental text, like sidebars or margin notes, might be added to enhance the reading.

When you or your team have drafted a few pages of written text, you should try turning your thumbnails into page designs. Of course, the text will change as the drafting continues. But, the actual written words will help you see how the document is going to look on paper or on the screen. Then, you can start using templates and/or grids to lay out information on the page. Make sure you create a flexible enough design for your document, leaving plenty of room for graphics and other visual elements.

As the drafting of the document nears completion, use these sample pages to guide the design of the rest of the pages.

Revise and Edit the Design

Revising and editing a document's design is similar to revising and editing the written text. Start looking over the document globally, paying attention to issues of balance, alignment, grouping, consistency, and contrast. Then, revise the document page by page to see if the visuals are consistent and appropriate.

Finally, proofread the design. While proofreading, you are going to find places where smaller alterations and corrections are needed. You should take time to iron out these little wrinkles. After all, small inconsistencies and errors in the design will draw readers' attention away from the written text.

A Primer on Binding and Paper

In the technical workplace, most documents are printed on standard-sized paper—the kind you use in your copier or printer. There will be times, though, when you need to make decisions about how your document will be bound, as well as decisions about the size, weight, and color of the paper. As copiers grow more sophisticated, you may find yourself making these decisions with more frequency.

Here is a quick primer to help you choose binding and paper.

Binding

In most cases, technical documents are bound with a paper clip, staple, or perhaps a three-ring binder. But if you are interested in a more permanent type of binding, here are a few of your options.

PERFECT BINDING In this method, glue is applied to the back edge of the pages, and then the pages are inserted into a cover. Perfect binding is a relatively inexpensive option for binding larger documents. The downside is that the binding will eventually come loose with time. When the binding begins to fail, pages will fall out of the document (Figure 8.24).

COMB BINDING For comb binding, slots are punched along the back edge of the pages. Then, a machine is used to insert a plastic comb binder. This kind of binding is usually the least expensive, especially for smaller documents. However, comb binding often comes loose with heavy usage. Also, over time, the plastic combs become brittle and break.

SPIRAL BINDING This method is similar to a comb binding, except holes are drilled or punched and a wire spiral is threaded through them. Compared to comb binding, spiral binding is much more secure, though a little more expensive. This binding is great for documents that need to lay flat when open.

SEWN OR STITCHED BINDING Here, sheets of paper are sewn or stapled together down the middle fold. Most hardback books use one of these two methods. Expensive hardbacks that are meant to last will be sewn with thread. Less expensive hardbacks will have two or three staples down the middle fold. In large books, sections of the book (called *signatures*) are sewn or stitched together. Then, these signatures are placed side by side and bound into the cover.

To learn more about binding documents, go to **www.ablongman.com/johnsonweb/8.16**

Different Kinds of Bindings

Perfect Binding

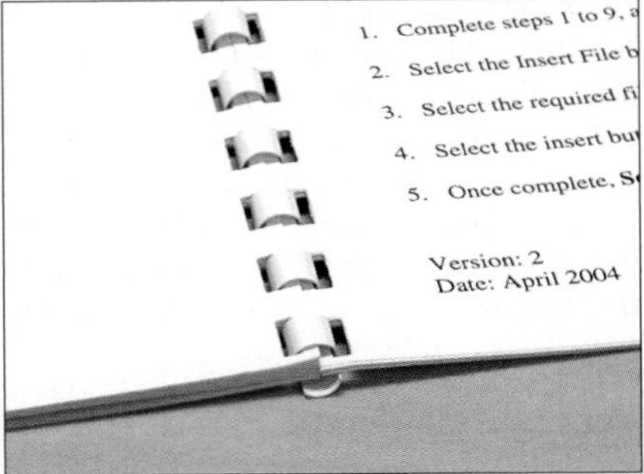

Comb Binding

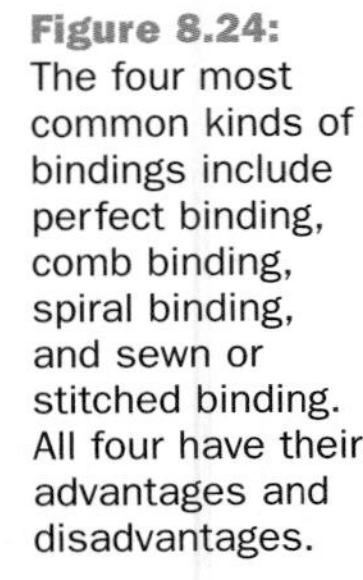
Figure 8.24: The four most common kinds of bindings include perfect binding, comb binding, spiral binding, and sewn or stitched binding. All four have their advantages and disadvantages.

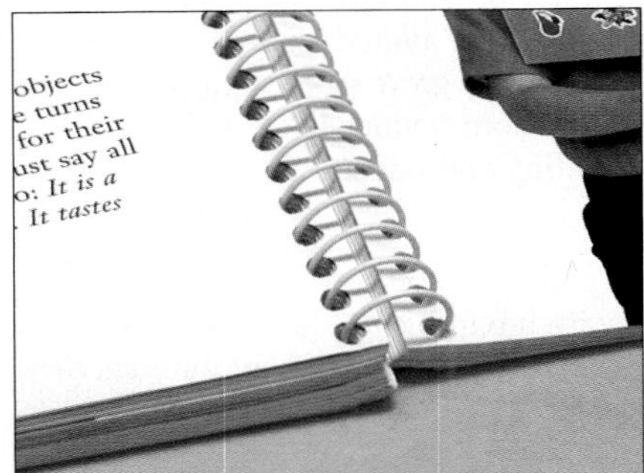
Spiral Binding

Stitched Binding

Selecting the Paper

Your choice of paper depends on the needs of your readers and the contexts in which the document will be used.

SIZE A standard sheet of paper in North America is 8.5 × 11 inches, while most other countries use a metric sheet of paper called A4, which is slightly larger at 21.0 centimeters × 29.7 centimeters. You can fold these sheets in half lengthwise or widthwise to make a nice booklet. You can fold a sheet into three panels to make a brochure. Meanwhile, a double sheet (11 × 17 inches or 42 centimeters × 29.7 centimeters) is available in many copiers. You can fold one of these sheets in half to make two standard sheets.

Other sizes can be used. You should talk to a printer or copy shop about what kinds of sizes are available. If you want an irregular size of paper, the cost of printing will go up significantly.

WEIGHT Paper comes in a variety of weights. The paper that you use in your printer or a copier is probably *20-pound bond*, which is a relatively lightweight paper. You can choose heavier paper if you want to add firmness to the document. For a cover, you might even choose a 60-pound or more card stock to make the document stiffer.

GO TO THE NET

Interested in learning more about paper? Go to **www.ablongman.com/johnsonweb/8.17**

COLOR Paper comes in a variety of different colors. When choosing a color, you should keep two things in mind. First, you want the words and images to *contrast* significantly with the page. Black words printed on a gray-colored sheet of paper, for instance, will often be difficult to read. Second, you do not want the color to interfere with any duplicating the readers might need to do. A mauve-colored document, for example, is not going to duplicate well on a copy machine. To avoid problems with contrast and duplication, in most cases your best choices for paper colors are white or off-white.

GLOSSY OR MATTE Increasingly, people are able to choose whether the paper is glossy or matte. Glossy paper is smooth and shiny and is often used in magazines or annual reports. Matte paper has a rougher texture and is not shiny, like the paper in your printer or copier. Most professional printers can offer you glossy paper for more expensive documents.

Computers are giving us more options for binding and paper choices. As copiers grow more sophisticated, you might find yourself making these kinds of choices about the binding and paper used in your documents.

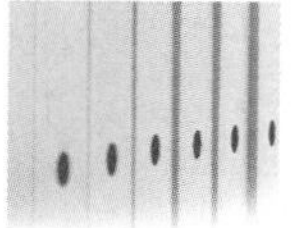

CHAPTER REVIEW

- Good design allows readers to (a) easily scan a document, (b) access the document to find the information they need, (c) understand the content more readily, and (d) appreciate the appearance of the document.
- The five principles of document design are balance, alignment, grouping, consistency, and contrast.
- On a balanced page, elements offset each other to create a stable feeling in the text. Unbalanced pages add tension to the reading process.
- Alignment creates relationships among items in a text and helps readers determine the hierarchical levels in the text.
- Grouping divides text into "scannable" blocks by using headings, rules, and borders to make words and images easier to comprehend.
- Consistency makes documents more accessible by making features predictable; inconsistent documents are harder to read and interpret.
- Contrast can cause text features or elements to stand out, but contrast should be used with restraint so that text elements work together rather than compete with each other.
- Use a design process that includes (a) analyzing your readers and your document's context of use, (b) using thumbnails to rough out the design, (c) designing the document, and (d) revising and editing the design.
- You can choose from a variety of bindings and paper sizes, weights, and colors.

EXERCISES AND PROJECTS

Individual or Team Projects

1. On campus or at your workplace, find a poorly or minimally designed document. If you look on any posting board, you will find several good ones you can use. In a memo to your instructor, critique this document using the five design principles discussed in this chapter. Explain why the document is ineffective because it does not follow the principles.

2. Read the case study at the end of this chapter. Using the design principles discussed in this chapter, sketch out some thumbnails of a better design for this document. Then, using your wordprocessor, develop an improved design. In a memo to your instructor, compare and contrast the old design with the new, showing why the new is superior.

3. On the Internet, find examples of badly designed websites and well-designed websites. In a memo to your instructor, compare two websites, using the design principles in this chapter to point out their strengths and weaknesses. In your memo, make some suggestions for improvements.

4. Find a document that illustrates good design. Then, change some aspect of its rhetorical situation (purpose, readers, context). Redesign the document to fit that altered rhetorical situation. For example, you might redesign a user manual to accommodate the needs of eight-year old children. You might turn a normal-sized document into a poster that can be read from a distance. In a memo to your instructor, discuss the changes you made to the document. Show how the changes in the rhetorical situation led to alterations in the design of the document. Explain why you think the changes you made were effective.

Collaborative Project

With other members of your class, choose a provider of a common product or service (for example, a car manufacturer; mobile phone, computer, clothing, or music store; museum, or theater). Then, find the websites of three or four competitors for this product or service.

Using the design principles you learned in this chapter, critique these websites by comparing and contrasting their design. Considering its target audience, which website design seems the most effective? Which is the least effective? Explain your positive and negative criticisms in some depth.

Then, using thumbnails, redesign the weakest site so that it better appeals to its target audience. How can you use balance, alignment, grouping, consistency, and contrast to improve the design of the site?

In a presentation to your class, discuss why the design of one site is stronger and others are weaker from a design perspective. Then, using an overhead projector, use your thumbnails to show how you might improve the design of the weakest site you found.

Sample documents and websites for critiquing and redesigning are available at **www.ablongman.com/johnsonweb/8.18**

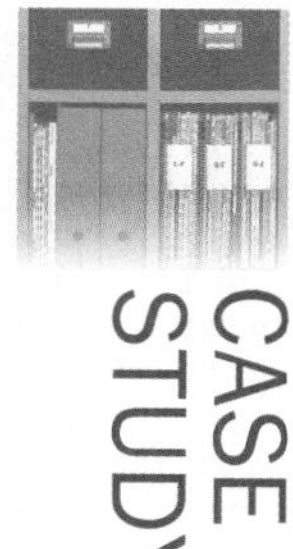

CASE STUDY

Scorpions Invade

Lois Alder is a U.S. Environmental Protection Agency biologist in western Texas who specializes in Integrated Pest Management (IPM). The aim of IPM is to use the least toxic methods possible for controlling insects, rodents, and other pests. Lois' job is to help companies, schools, and government organizations develop IPM strategies to minimize their use of pesticides, especially where people live, work, and play.

Recently, a rural school district outside El Paso, Texas, was experiencing an infestation of scorpions. Scorpions were finding their way into classrooms, and they were regularly found in the schoolyards around the buildings. Teachers and children were scared, though no one had been stung yet. Parents were demanding the school district do something immediately.

In a conference call, Lois talked to the principals and custodians of the schools. The custodians said they just wanted to "nuke" the scorpions by spraying large amounts of pesticides in and around the buildings. But this solution would expose the children, teachers, and staff at the schools to high levels of toxins. Lois convinced them that they should first try an IPM approach to the problem. She told them she would quickly set up a training program to help them. Lois knew, though, that she needed to send these school officials some information right away.

From the EPA website, Lois downloaded a well-written chapter, "IPM for Scorpions in Schools," (Figure A) (http://www.epa.gov/pesticides/ipm/schoolipm/). The design of the document, however, was a bit bland, and Lois was concerned that school officials, especially the custodians, would not read it carefully. Moreover, she knew that some of the custodians had poor reading skills.

So, she decided she would need to redesign the document to make it more accessible. She wanted to highlight the important need-to-know information while stripping out any information that the school officials would not need. If you were Lois, how would you redesign this document, using the principles of design discussed in this chapter? How could you make this document more accessible to the people who need to use it? How can you make it more attractive and interesting to its likely readers?

A Document Design That Could Be Improved

Chapter 13
IPM for Scorpions in Schools

Introduction

Scorpions live in a wide variety of habitats from tropical to temperate climates and from deserts to rain forests. In the United States they are most common in the southern states from the Atlantic to the Pacific. All scorpions are beneficial because they are predators of insects.

The sting of most scorpions is less painful than a bee sting. There is only one scorpion of medical importance in the United States: the sculptured or bark scorpion, Centruroides exilicauda (=sculpturatus). The danger from its sting has been exaggerated, and its venom is probably not life-threatening. This species occurs in Texas, western New Mexico, Arizona, northern Mexico, and sometimes along the west bank of the Colorado River in California.

Identification and Biology

Scorpions range from 3/8 to 8 1/2 inches in length, but all scorpions are similar in general appearance.

Scorpions do not lay eggs; they are viviparous, which means they give birth to live young. As embryos, the young receive nourishment through a kind of "placental" connection to the mother's body. When the young are born, they climb onto the mother's back where they remain from two days to two weeks until they molt (shed their skin) for the first time. After the first molt, the young disperse to lead independent lives. Some scorpions mature in as little as six months while others take almost seven years.

All scorpions are predators, feeding on a variety of insects and spiders. Large scorpions also feed on small animals including snakes, lizards, and rodents. Some scorpions sit and wait for their meal to come to them while others actively hunt their prey. Scorpions have a very low metabolism and some can exist for 6 to 12 months without food. Most are active at night. They are shy creatures, aggressive only toward their prey. Scorpions will not sting humans unless handled, stepped on, or otherwise disturbed.

It is rare for scorpions to enter a building since there is little food and temperatures are too cool for their comfort. There are some exceptions to this rule. Buildings in new developments (less than three years old) can experience an influx of scorpions because the construction work has destroyed the animals' habitat. In older neighborhoods, the heavy bark on old trees provide good habitat for scorpions, and they may enter through the more numerous cracks and holes in buildings in search of water, mates, and prey. Also, buildings near washes and arroyos that are normally dry may become refuges for scorpions during summer rains.

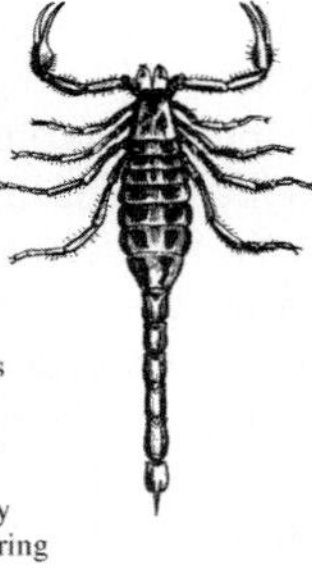

Scorpions do not enter buildings in winter because cold weather makes them sluggish or immobile. They are not active until nighttime low temperatures exceed 70°F. Buildings heated to 65½ or 70°F provide enough warmth to allow scorpions to move about. Scorpions found inside buildings in cold weather are probably summer visitors that never left. Although scorpions prefer to live outdoors, they can remain in buildings without food for long periods of time.

Stings

A scorpion sting produces considerable pain around the site of the sting, but little swelling. For four to six hours, sensations of numbness and tingling develop in the region of the sting, then symptoms start to go away. In the vast majority of cases, the symptoms will subside within a few days without any treatment.

If the sting is from a bark scorpion, symptoms can sometimes travel along nerves, and tingling from a sting on a finger may be felt up to the elbow, or even the shoulder. Severe symptoms can include roving eyes, blurry vision, excessive salivation, tingling around the mouth and nose, and the feeling of having a lump in the throat. Respiratory distress may occur. Tapping the sting can produce extreme pain. Symptoms in children also include extreme restlessness, excessive muscle activity, rubbing at the face, and sometimes vomiting. Most vulnerable to the sting of the bark scorpion are children under five years and elderly persons who have

IPM for Schools 103 **Chapter 13 • Scorpions**

Figure A: This document's design is rather minimal, and perhaps less than effective for the readers who might use it. How could the design be improved?

Source: U.S. Environmental Protection Agency, http://www.epa.gov/pesticides/ipm/schoolipm.

Figure A:
(*continued.*)

Avoiding Scorpion Stings

Schools in areas where encounters with scorpions are likely should teach children and adults how to recognize scorpions and to understand their habits. Focus on scorpion biology, behavior, likely places to find them, and how to avoid disturbing them.

At home, children and parents should be taught to take the following precautions to reduce the likelihood of being stung:

- Wear shoes when walking outside at night. If scorpions are suspected indoors, wear shoes inside at night as well.
- Wear leather gloves when moving rocks, boards, and other debris.
- Shake out shoes or slippers before they are worn, and check beds before they are used.
- Shake out damp towels, washcloths, and dishrags before use.
- When camping, shake out sleeping bags, clothes, and anything else that has been in contact with the ground before use.
- Protect infants and children from scorpions at night by placing the legs of their cribs or beds into clean wide-mouthed glass jars and moving the crib or bed away from the wall. Scorpions cannot climb clean glass.

an underlying heart condition or respiratory illness. The greatest danger to a child is the possibility of choking on saliva and vomit.

Antivenin for the bark scorpion is produced at Arizona State University in Tempe and is available in Arizona but not in other states. The therapeutic use of antivenin is still experimental. People have been treated without antivenin for many years, and in areas where antivenin is unavailable, people are monitored closely by medical professionals until the symptoms subside.

Detection and Monitoring

To determine where scorpions are entering, inspect both the inside and outside of the building at night (when scorpions are active) using a battery-operated camp light fitted with a UV (black) fluorescent bulb. Scorpions glow brightly in black light and can be spotted several yards away.

Always wear leather gloves when hunting scorpions. Places to check inside the building include under towels, washcloths, and sponges in bathrooms and kitchen; under all tables and desks, since the bark scorpion may climb and take refuge on a table leg or under the lip of a table; and inside storage areas. Outside, check piles of rocks and wood, under loose boards, and in piles of debris. After the following treatment strategies have been implemented, monitoring with the black light should continue to verify population reduction.

Management Options

Physical Controls

In most cases, physical controls will be adequate to manage a scorpion problem.

Removal of Scorpions

Any scorpions found during monitoring can be picked up using gloves or a pair of kitchen tongs, and transferred to a clean, wide-mouthed glass jar. Scorpions cannot climb clean glass. You can also invert a jar over a live scorpion and then slide a thin piece of cardboard under the mouth of the jar to trap the scorpion inside. Once a scorpion is captured, drop it into a jar of alcohol or soapy water (water without soap will not work) to kill it.

Habitat Modification

If you discover areas near school buildings that harbor a number of scorpions, you can try to alter the habitat to discourage them. Wood piles, rocks, loose boards, and other debris should be removed from the immediate vicinity of the building.

If there are slopes on the school grounds that are faced with rip rap (large rocks placed on a slope or stream bank to help stabilize it), or other similar areas highly attractive to scorpions, place a barrier of aluminum flashing between the rip rap and the school to prevent scorpion access. The flashing must be bent in an "L" shape away from the building. The other edge of the flashing should be buried a short distance from the rocks, deep enough in the soil so that the L shape will not fall over and lean on the rip rap. Make the height of the barrier before the bend greater than the height of the rip rap to prevent scorpions from standing on the rocks and jumping over the barrier.

Continued on following page

Figure A:
(*continued.*)

Carry a caulking gun during nighttime inspections inside and outside the building to seal any cracks and crevices found. If scorpions are entering through weep holes in windows or sliding doors, cover the holes with fine-mesh aluminum screening, available from hardware or lumber stores. The ends of pipes that are designed as gray water drains should be fitted with loose filter bags, or makeshift end-pieces made from window screen. The screened end will prevent scorpion access to the drainpipe, sink, and other parts of the building.

It is important to continue nighttime patrols and caulking until all entryways have been located and sealed, and all the scorpions in the building have been captured and killed. Once the access routes are sealed, and all indoor scorpions have been removed, only doorways provide access, unless the scorpions ride in on logs and other materials. Glazed tiles can be placed around the perimeter of the buildings, and under or around doors and windows as part of the decor and as practical scorpion barriers. Scorpions have difficulty crossing smooth tiles unless the grout line is wide. Wood storage should be elevated above the ground since scorpions like contact with moist soil. Before bringing materials such as logs inside, bang them on a stone to dislodge any scorpions.

Traps

A simple trap made of damp gunny sacks laid down near the building in the evenings may be useful for monitoring and trapping. Scorpions may seek out the moist environment under the sacks where they can be collected in the morning. This trap is most effective when used before summer rains.

Chemical Controls

In general, preventing scorpion problems is better than trying to kill these creatures with pesticide. Spraying the perimeters of buildings is not only unnecessary but also ineffective. Scorpions can tolerate a great deal of pesticide in their environment. The pesticide will only be harmful to humans, especially children, and to other wildlife. Using physical controls along with education to reduce the fear of scorpions will help prevent unnecessary treatments.

First Aid for Scorpion Stings

Most scorpion stings are similar to a bee or wasp sting, and like bee or wasp stings, the majority of scorpion stings can be treated at school or the victim's home.

First aid for a scorpion sting includes the following:

- Calm the victim.
- Do not use a tourniquet.
- Wash the area with soap and water.
- Apply a cool compress (an ice cube wrapped in a wet washcloth), but do not apply ice directly to the skin or submerge the affected limb in ice water.
- To reduce pain, over-the-counter pain relievers such as aspirin or acetaminophen can help.
- Elevate or immobilize the affected limb if that feels more comfortable.
- Do not administer sedatives such as alcohol.

Bibliography

Bio-Integral Resource Center (BIRC). 1996. 1997 directory of least-toxic pest control products. IPM Practitioner 18(11/12):1-39.

Cloudsley-Thompson, J.L. 1968. Spiders, Scorpions, Centipedes and Mites. Pergamon Press, New York. 278 pp.

Keegan, H.L. 1980. Scorpions of Medical Importance. Jackson University Press, Jackson, MS. 140 pp.

Mallis, A. 1982. Handbook of Pest Control. Franzak and Foster, Cleveland, OH. 1101 pp.

Olkowski, W., S. Daar, and H. Olkowski. 1991. Common-Sense Pest Control: Least-toxic solutions for your home, garden, pets and community. Taunton Press, Newtown, CT. 715 pp.

Polis, G.A., ed. 1990. The Biology of Scorpions. Stanford University Press, Stanford, CA. 587 pp.

Smith, R.L. 1982. Venomous animals of Arizona. Cooperative Extension Service, College of Agriculture, University of Arizona, Tucson, AZ, Bulletin 8245. 134 pp.

Smith, R.L. 1992. Personal communication. Associate Professor, Entomology Dept., University of Arizona at Tucson.

This document was produced for USEPA (Document #909-B-97-001) by the Bio-Integral Resource Center, P.O. Box 7414, Berkeley, CA 94707, March 1997.

CHAPTER

9

Creating and Using Graphics

CHAPTER CONTENTS

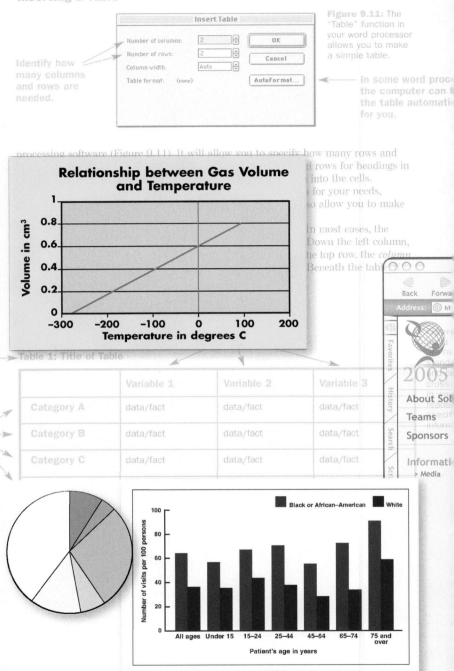

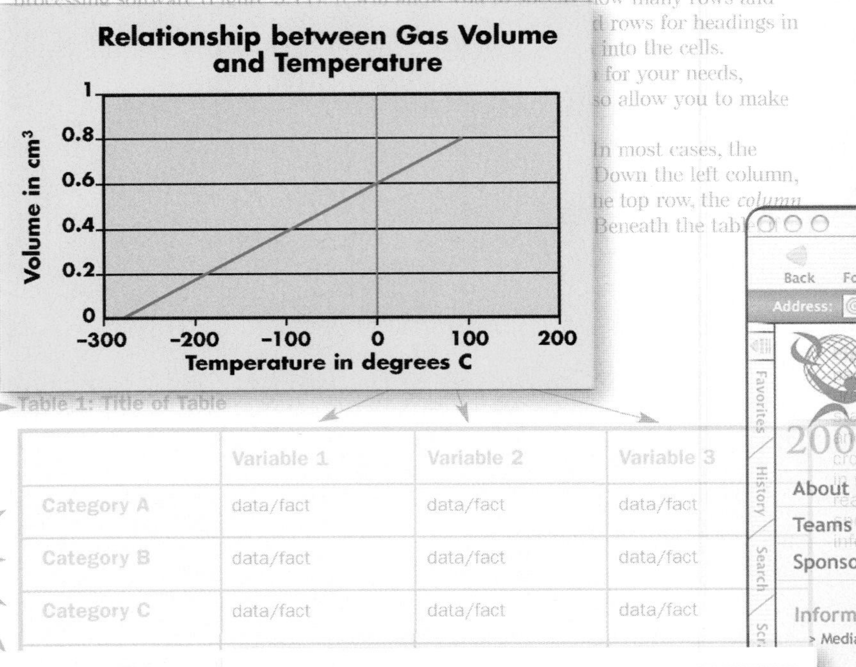

Inserting a Table
Insert Table
Number of columns:
Number of rows:
Column width:
Table format:
OK
Cancel
AutoFormat...
Identify how many columns and rows are needed.
Figure 9.11: The "Table" function in your word processor allows you to make a simple table.
Relationship between Gas Volume and Temperature
Volume in cm³
Temperature in degrees C
Table 1: Title of Table
Variable 1
Variable 2
Variable 3
Category A
Category B
Category C
data/fact

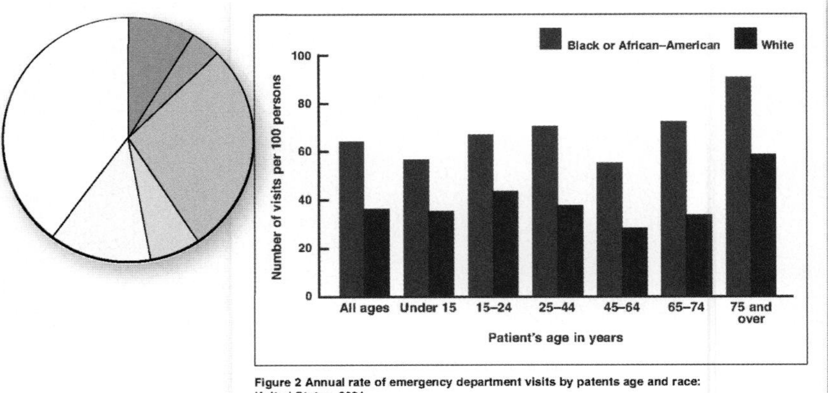

Black or African-American
White
Number of visits per 100 persons
All ages
Under 15
15–24
25–44
45–64
65–74
75 and over
Patient's age in years
Figure 2 Annual rate of emergency department visits by patents age and race: United States. 2001

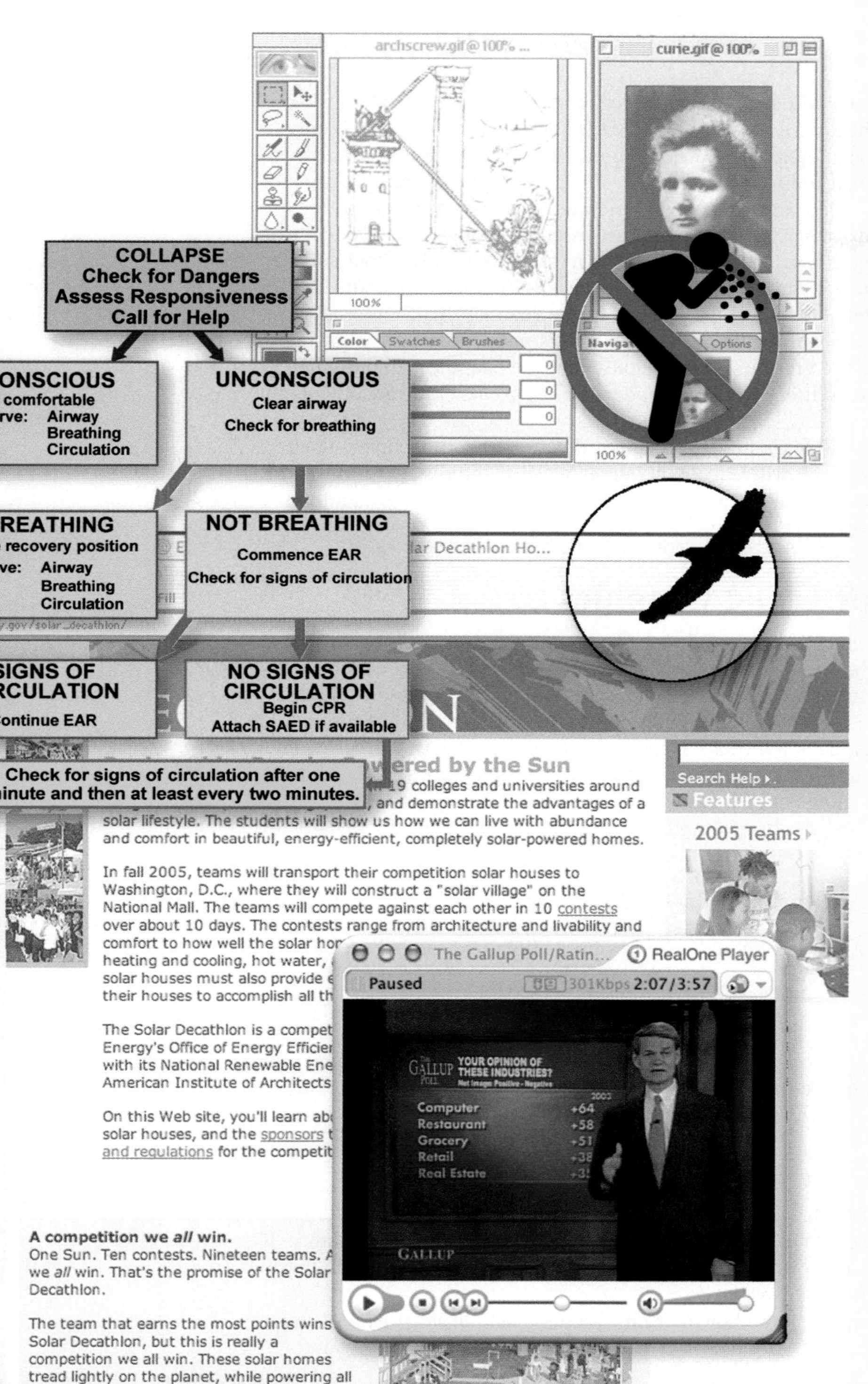

CHAPTER OBJECTIVES

In this chapter, you will learn:

- The importance of visuals in documents and presentations.
- Four guidelines for using visuals effectively.
- How to use tables, charts, and graphs.
- Strategies for taking photographs and using them in documents and presentations.
- How drawings, icons, and clip art can be used effectively to enhance understanding.
- How to use video and audio in multimedia and Internet-based documents.

Graphics are an essential part of any technical document or presentation. Before computers, graphics were considered a luxury. They were typically used only in the most important documents. Today, with computers, you can create charts, graphs, diagrams, and even illustrations with much less effort. Also, with little time, you can use digital cameras or video cameras to insert pictures and movies into your documents. Graphics are no longer a luxury—they are essential in any kind of document.

Why are graphics so important? Today, we live in an increasingly visual culture. Your readers will often pay more attention to the visuals in your document than the written text. For example, think about how you began reading this chapter. More than likely, you did not begin reading at the top of this page. Instead, you probably took a quick glance at the graphics to figure out what the chapter is about. Then, you started reading the written text. Your readers will approach your documents the same way.

In this chapter, you will learn how to use graphics effectively in technical documents and presentations. You will learn how to create charts, graphs, diagrams, and illustrations. Finally, you will learn how to use digital photography, video, and music to enhance and reinforce your writing.

Guidelines for Using Graphics

As you draft your document, you should look for places where graphics could be used to support the text. Graphics are especially helpful in places where you want to reinforce important ideas or help your readers understand complex concepts or trends.

Graphics should be used to enhance and clarify your message, to slice through the details and numbers to make information easier to process. In Figure 9.1, for example, the written text and the table provide essentially the same information. And yet, the information in the table is much easier to access.

To help you create and use graphics effectively, here are four guidelines that you should keep in mind:

> **Guideline One:** A graphic should tell a simple story.
>
> **Guideline Two:** A graphic should reinforce the written text, not replace it.
>
> **Guideline Three:** A graphic should be ethical.
>
> **Guideline Four:** A graphic should be labeled and placed properly.

These four guidelines are worth committing to memory to help you create and use graphics properly.

Guideline One: A Graphic Should Tell a Simple Story

Graphics should tell the "story" about your data in a concise way. In other words, your readers should be able to figure out what the graphic says in a quick glance. If readers need to pause longer than a moment, there is a good chance they will not understand what the graphic means.

Figure 9.2, for example, shows how a graph can tell a simple story. Almost immediately, a reader recognizes that reported AIDS cases in Ghana show a dramatic

To see more examples of text-to-table conversions, go to **www.ablongman.com/johnsonweb/9.1**

Written Text

In its August 27, 2003, count, the Utah Raptors Alliance reported slight increases in most species of raptors (p. 21). In the Uinta Mountains, its members counted 87 Red-Tailed Hawks (avg. 82.5), 48 Golden Eagles (avg. 32.6), 8 Bald Eagles (avg. 9.1), 21 American Kestrels (avg. 22.4), and 8 Cooper's Hawks (avg. 7.2). In the Bear River Refuge, they counted 45 Red-Tailed Hawks (avg. 30.2), 91 Golden Eagles, (avg. 87.4), 10 Bald Eagles (avg. 7.4), 34 American Kestrels (36.5), and 15 Cooper's Hawks (avg. 13.5). At Mt. Marvine, they counted 129 Red-Tailed Hawks (avg. 110.8), 62 Golden Eagles (avg. 58.2), 12 Bald Eagles (avg. 13.2), 36 American Kestrels (22.0), and 63 Cooper's Hawks (avg. 63.4).

Table 5.1: Raptor 12-Hour Counts in Utah (2003)

Raptor	Uinta Mountains		Bear River Refuge		Mt. Marvine	
	2003	Average	2003	Average	2003	Average
Red-Tailed Hawk	87	82.5	45	30.2	129	110.8
Golden Eagle	48	32.6	91	87.4	62	58.2
Bald Eagle	8	9.1	10	7.4	12	13.2
American Kestrel	21	22.4	34	36.5	36	22.0
Cooper's Hawk	8	7.2	15	13.5	63	63.4

Source: Utah Raptors Alliance 2003 Annual Report, p. 21.

Figure 9.1: Putting figures into a table makes them much easier to access. The table cannot completely replace the written text, but it can reinforce it by organizing the information more effectively.

spike among people in their twenties and thirties. Another, perhaps surprising, story is that women have significantly higher rates of reported AIDS than do men.

This first guideline—tell a story—also applies to photographs in a document (Plotnik, 1982). At a glance, your readers should be able to figure out what story a photograph is telling. The photograph in Figure 9.3, for example, is not complex, but it tells a clear story about the markings on a Desert Checkerspot butterfly.

When you begin developing a graphic to illustrate a concept or trend, first ask yourself what story you want the chart, table, graph, picture, or video to tell. For instance, you might write:

> This graph will show how reported cases of AIDS have spiked among women in Ghana who are in their twenties and thirties.

Once you have decided on the story you want to tell, you can then identify the best way to illustrate that story. Later in this chapter, we will discuss a variety of graphics. Each one allows you to tell a different story with the data available.

For websites that offer other guidelines for using graphics, go to **www.ablongman.com/johnsonweb/9.2**

A Graph That Tells a Simple Story

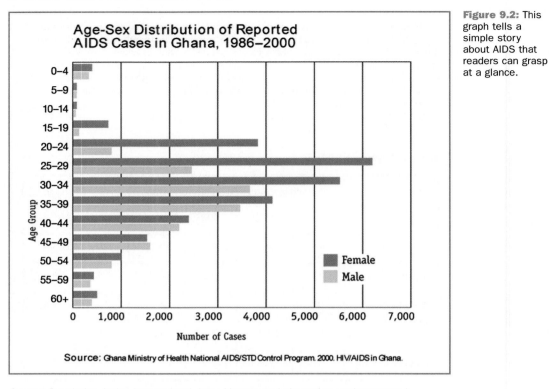

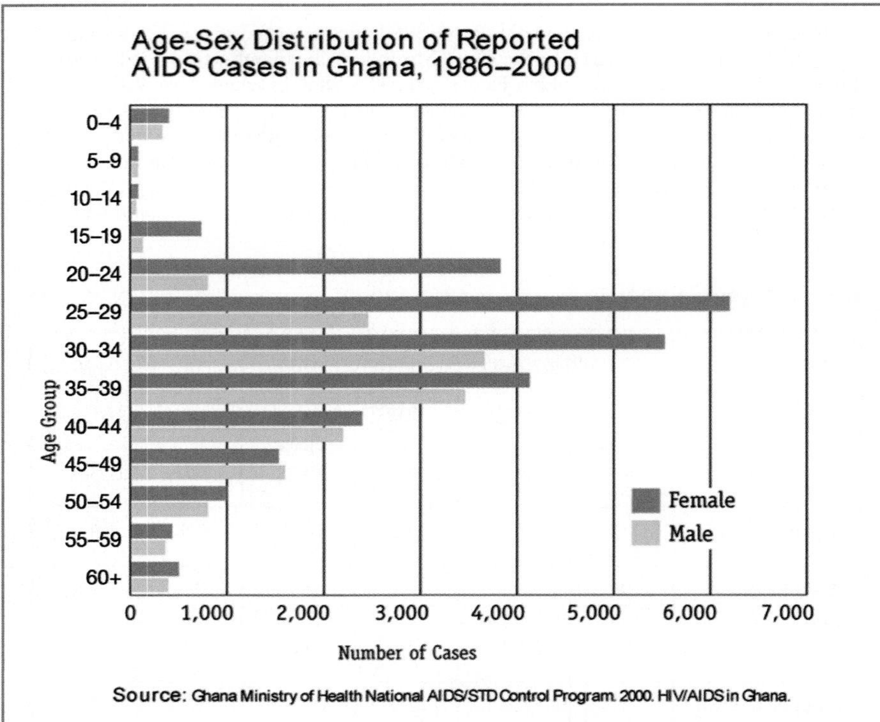

Figure 9.2: This graph tells a simple story about AIDS that readers can grasp at a glance.

Source: Population Action International, http://www.populationaction.org/resources/publications/condomscount/images/graphs.htm.

Guideline Two: A Graphic Should Reinforce the Written Text, Not Replace It

Graphics should be used to support the written text, but they cannot replace it altogether. Since technical documents often handle complex ideas or relationships, it is tempting to simply refer the readers to a graphic (e.g., "See Chart 9 for an explanation of the data"). Chances are, though, if you cannot explain something in writing, you won't be able to explain it in a graphic either.

Instead, your written text and visual text should work side by side. The written text should refer readers to the graphics, and the graphics should support the written information. For example, the written text might say, "As shown in Graph 2.3, despite talk about cutting the federal budget, some of the largest federal spending increases began when 'smaller government' politicians became president." The graph would then support this written statement by illustrating this trend (Figure 9.4).

To see other graphs that tell simple stories, go to **www.ablongman.com/johnsonweb/9.3**

A Photograph Should Tell a Simple Story Too

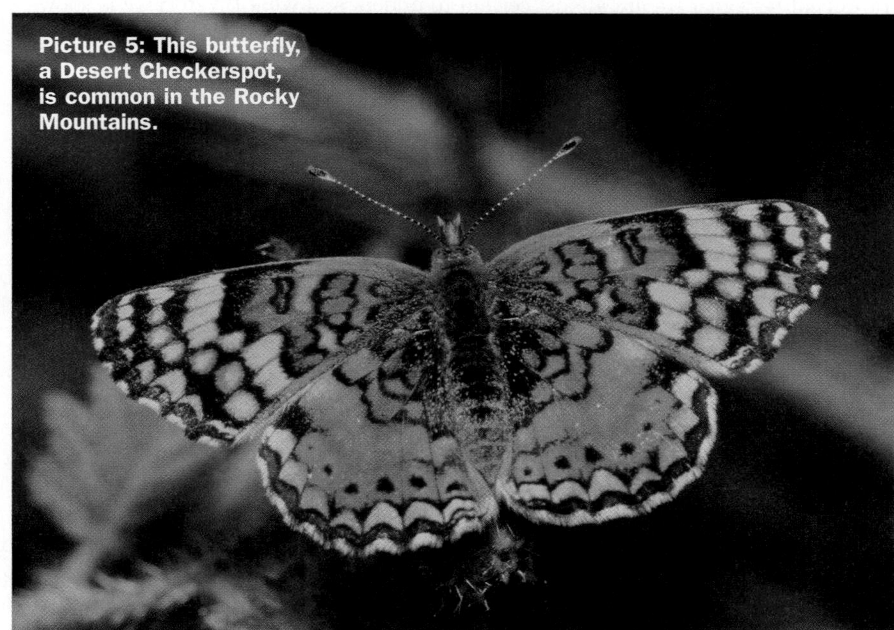

Picture 5: This butterfly, a Desert Checkerspot, is common in the Rocky Mountains.

Source: Corel.

Figure 9.3: This photograph tells a simple story that reinforces the written text.

A Graph That Reinforces the Written Text

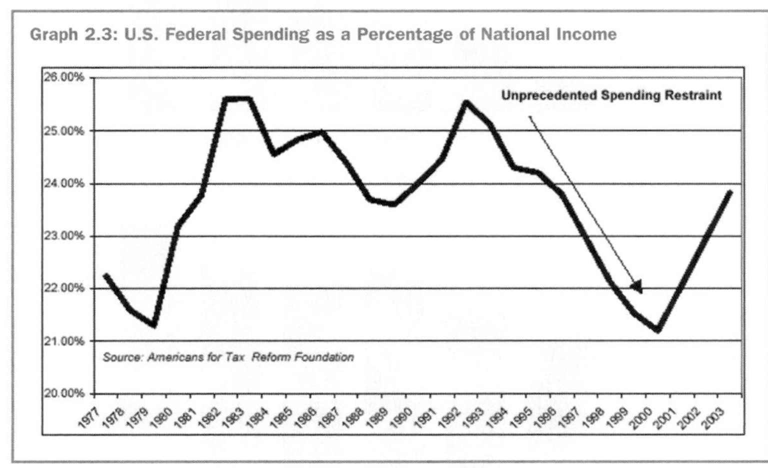

Source: Americans for Tax Reform Foundation, http://www.atr.org/pdffiles/cogdcharts2003.pdf.

Figure 9.4: A line graph typically shows a trend over time.

The written text should tell readers the story that the graphic is trying to illustrate. That way, readers are almost certain to understand what the graphic is showing them.

Keep in mind that people learn both visually and verbally. They process words and images differently. So, by providing both visual and written forms of information, you will greatly enhance their ability to process the information you are offering.

Guideline Three: A Graphic Should Be Ethical

Graphs, charts, tables, illustrations, and photographs can be used to hide information, distort facts, or exaggerate trends. In a bar chart, for example, the scales can be altered to suggest more growth than is actually the case (Figure 9.5). In a line graph, it is tempting to leave out data points that resist attempts to draw a smooth line. Meanwhile, with computers, photographs can be distorted or doctored.

A good rule of thumb with graphics—and a safe principle to follow in technical communication altogether—is to always be absolutely honest with the readers. Your readers are not fools, so attempts to use graphics to distort or stretch the truth will eventually be detected.

Unethical and Ethical Bar Charts

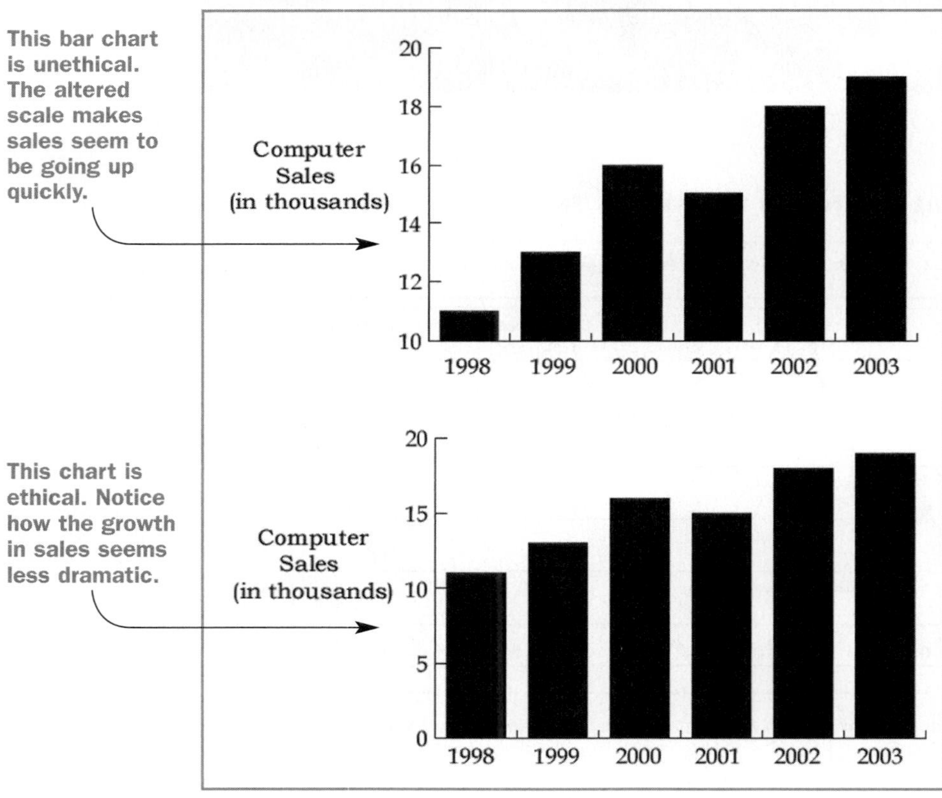

Figure 9.5: The top bar chart is unethical because the y-axis has been altered to exaggerate growth in sales of computers. The second bar chart presents the data ethically.

Once detected, unethical graphics can erode the credibility of an entire document or presentation (Kostelnick & Roberts, 1998). Even if your readers only suspect deception in your graphics, they will begin to doubt the honesty of the whole text.

LINK For more information on the ethical use of data, see Chapter 4, page 70.

Guideline Four: A Graphic Should Be Labeled and Placed Properly

Proper labeling and placement of graphics help readers move back and forth between the written and visual text. Each graphic should be labeled with an informative title (Figure 9.6). Other parts of the graphic should also be carefully labeled:

- The x- and y-axes of graphs and charts should display standard units of measurement.
- Columns and rows in tables should be labeled so the readers can easily locate specific data points.
- Important features of drawings or illustrations should be identified with arrows or lines and some explanatory text.
- The source of the data used to make the graphic should be clearly identified underneath.

Assuming you include a title with the graph, an explanatory caption is not needed. Nevertheless, a sentence or two of explanation in a caption can often help reinforce or clarify the story the graphic is trying to tell.

Labeling of a Graphic

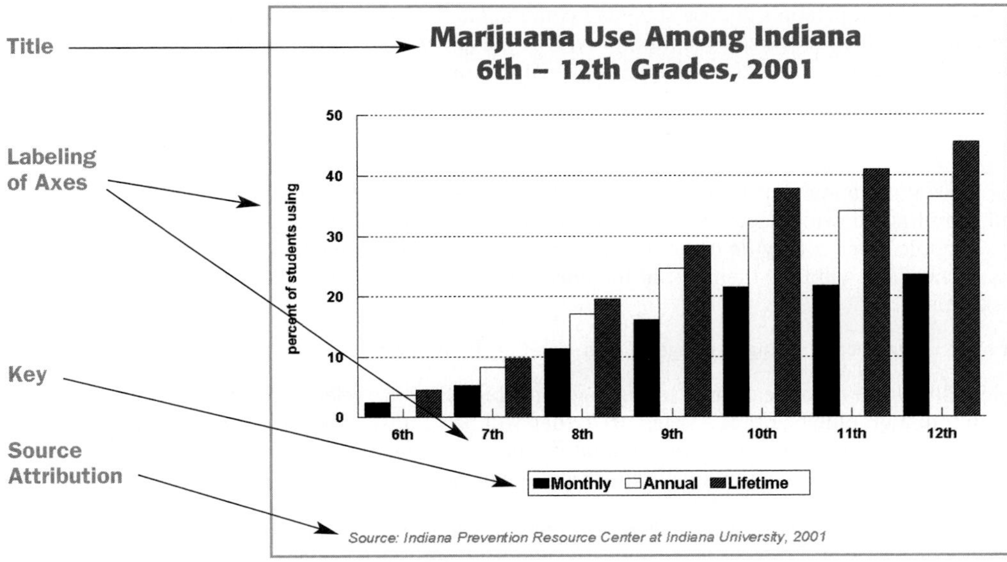

Figure 9.6: Good labeling of a graphic is important so that readers can understand it.

Source: Indiana Prevention Resource Center, http://www.iprc.indiana.edu/drug_stats/youth2001/s2001-26.gif.

Want to see other unethical graphics? Go to **www.ablongman.com/johnsonweb/9.4**

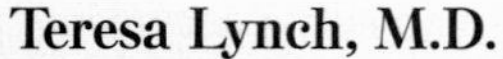

AT WORK

Teresa Lynch, M.D.
RESIDENT PHYSICIAN OF INTERNAL MEDICINE AND PEDIATRICS,
RUSH UNIVERSITY MEDICAL CENTER, CHICAGO, ILLINOIS

Rush University Medical Center is an important research hospital that specializes in urban needs.

How do you use graphics in your everyday work?

Over the course of my training as a resident physician and even in undergraduate and medical school, I have used graphics to supplement and amplify my written communication.

Probably the most important area in which I use graphics is in the presentation of research data. Most data collected in the lab or in clinical studies get written up as a document and submitted for publication in a scientific or medical journal. Because these journals have very strict criteria regarding document length and content, a well-placed graphic and short explanatory paragraph can substitute for monotonous lists of research data that, in addition to occupying precious space, can be both boring to read and difficult to interpret.

Furthermore, graphics often determine who will actually read your article. Most people will scan the graphics first to determine whether or not they have any interest in the subject matter. If there are no graphics, there is a good chance they won't read the text. Well-placed graphics, on the other hand, will gather an audience for your research.

I also use graphics in my day-to-day work as a physician. On one occasion, I may create an algorithm, or diagram, to explain the proper treatment of a disease process to the medical students I work with. Later that same day, I may create a line graph to show the trend in a person's disease state as I contemplate the next step in his/her treatment. In all of these instances, graphics provide a useful visual summary of the information I need to convey to the various people with whom I work.

When placing a graphic, put it on the page where it is referenced or, at the farthest, put it on the following page. Your readers will rarely flip more than one page to look for a graphic. Even if they *do* make the effort to hunt down a graphic that is pages away, the effort will take them out of the flow of the document, inviting them to start skimming.

LINK For more information on designing page layouts, see Chapter 8, page 192.

Graphics should never appear prior to the page on which they are referenced. If readers run into a graphic before it is referenced, they will be confused because they will lack the information necessary to understand it. The written text should provide that information first; then the graph can reinforce the message.

TAKE NOTE Before computers, people often put all their graphics in an appendix at the back of a large document. Today's readers will rarely turn that many pages to see a graph, chart, or photograph. So, if the graphic is important, include it inside the main document.

For more information on labeling and placing graphs, go to **www.ablongman.com/johnsonweb/9.5**

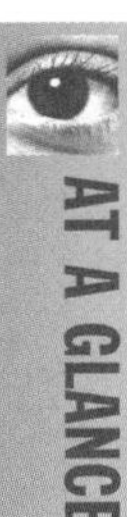

AT A GLANCE

Guidelines for Using Graphics

- A graphic should tell a simple story.
- A graphic should reinforce the written text, not replace it.
- A graphic should be ethical.
- A graphic should be labeled and placed properly.

Readers should be able to locate a graphic with a quick glance. Then, they should be able to quickly return to the written text to continue reading. When labeled and placed properly, graphics work seamlessly into the reading of the whole text.

Displaying Data with Graphs, Tables, and Charts

Various kinds of graphics are available for displaying information and data. Each type of graphic allows you to tell a different story with your data. The challenge is to find the type of graph, table, chart, illustration, or photograph that best presents your interpretation of the data.

To decide which graphic is best for the data you want to display, first ask what story you want to tell. Then, choose the type of graphic that best fits that story. The chart in Figure 9.7 will help you make your decision.

Choosing the Appropriate Graphic

The Story to Be Told	Best Graphic	How Data Are Displayed
"I want to show a trend."	Line graph	Shows how a quantity rises and falls, usually over time
"I want to compare two or more quantities."	Bar chart	Shows comparisons among different items or the same items over time
"I need to present data or facts for analysis and comparison."	Table	Displays data in an organized, easy-access way
"I need to show how a whole is divided into parts."	Pie chart	Shows data as a pie carved into slices
"I need to show how things, people, or steps are linked together."	Flowchart	Illustrates the connections among people, parts, or steps
"I need to show how a project will meet its goals over time."	Gantt chart	Displays a project schedule, highlighting the phases of the work

Figure 9.7: Different kinds of graphics tell different stories. Think about what story you want to tell. Then, locate the appropriate graph, table, or chart for that story.

Line Graphs

Line graphs are perhaps the most familiar way to display data. They are best used to show measurements over time. Some of their more common applications include the following:

> **Showing trends**—Line graphs are especially good at showing how quantities rise and fall over time (Figure 9.8). Whether you are illustrating trends in the stock market or the temperature of a chemical reaction, a line graph can show how the quantity gradually increases or decreases. When two or more lines are charted on a line graph, you can show how quantities rise and fall in tandem (or don't).

A Line Graph Showing a Trend

This graph shows the number of wolves over time.

Number of Wolves
50
40
30
20
10
1960 1970 1980 1990 2000

Number of Moose
2500
1875
1250
625
0
1960 1970 1980 1990 2000

Figure 4.5: Population of Wolves and Moose on Isle Royale

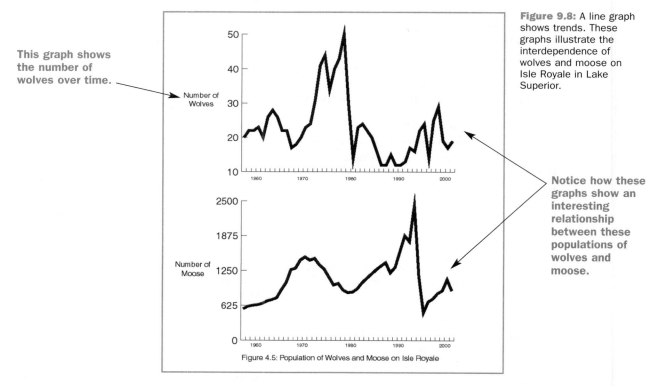

Figure 9.8: A line graph shows trends. These graphs illustrate the interdependence of wolves and moose on Isle Royale in Lake Superior.

Notice how these graphs show an interesting relationship between these populations of wolves and moose.

> **Showing relationships between variables**—Line graphs are also helpful when charting the interaction of two different variables. Figure 9.9, for example, shows a line graph that illustrates how a rise in the temperature of a gas is accompanied by a rise in the volume.

Examples of these documents are available at **www.ablongman.com/johnsonweb/9.6**

In a line graph, the vertical axis (y-axis) displays a measured quantity such as sales, temperature, production, growth, etc. The horizontal axis (x-axis) is usually divided into time increments such as years, months, days, or hours.

The x-axis in a line graph usually represents the "independent variable," which has a consistently measurable value. For example, in most cases, time marches forward steadily, independent of other variables. So, time is often measured on the x-axis. The y-axis often represents the "dependent variable." The value of this variable fluctuates over time.

As shown in Figure 9.8, in a line graph, the x- and y-axes do not need to start at zero. Often, by starting one or both axes at a nonzero number, you can better illustrate the trends you are trying to show. The drawback of line graphs is their inability to present data in exact numbers. For example, in Figure 9.8, can you tell exactly how many wolves were counted in 1997? Line graphs are most effective when the trend you are showing is more significant than the exact figures.

A Line Graph Showing a Relationship Between Variables

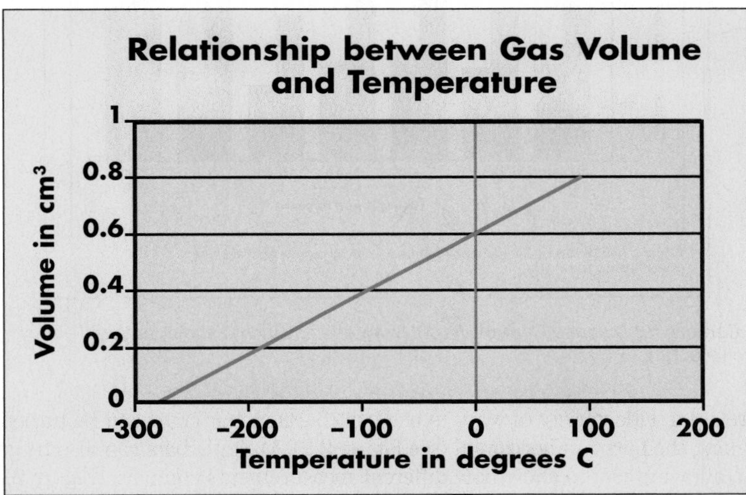

Figure 9.9: Here, the volume of a gas is plotted against the temperature. In this case, an extrapolation of the line allows us to estimate "absolute zero," the temperature at which all molecular activity stops.

Source: The Safetyline Institute, http://www.safetyline.wa.gov.au/institute/level2/course16/lecture47/l47_02.asp.

You can use more than one line to illustrate trends in a line graph. Depending on your printer, computers also give you the ability to use colors to distinguish the lines. Or, you can use dashes, dots, and solid lines to help your readers distinguish one line from the others.

Keep in mind that computers give us the ability to connect all the data points with a smooth line. However, the line needs to reflect the actual progression of the data. Just because there are two points on a graph doesn't mean they should have a line drawn between them.

For more examples of line graphs, go to **www.ablongman.com/johnsonweb/9.7**

Bar Charts

Bar charts are used to show quantities, allowing readers to make visual comparisons among measurements. The width of the bars is kept the same, while the length of the bars usually varies to represent the quantity measured. Like line graphs, bar charts can be used to show trends over time. They are also useful for showing increases or decreases in volume. For example, Figure 9.10 demonstrates how a bar chart can plot a volume over time.

A Bar Chart

A bar chart allows readers to make direct comparisons between quantities.

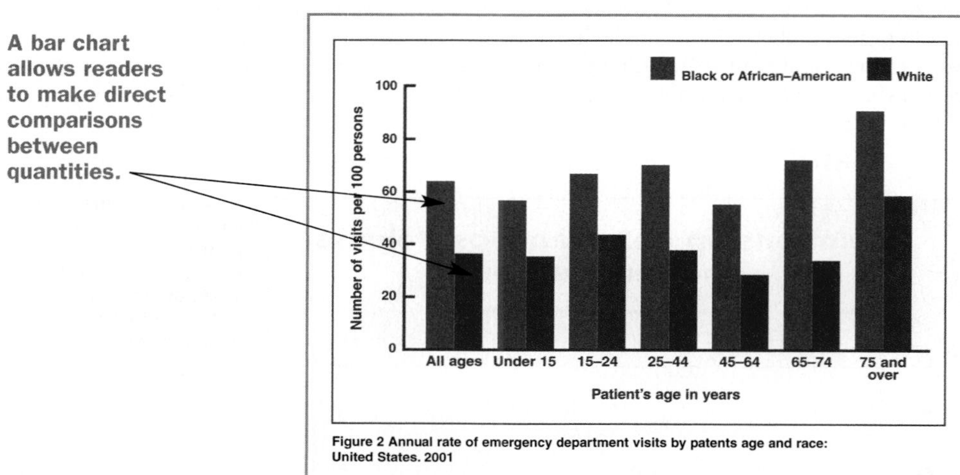

Figure 2 Annual rate of emergency department visits by patents age and race: United States. 2001

Figure 9.10: A bar chart is especially effective for showing volumes.

Source: Centers for Disease Control, http://www.cdc.gov/nchs/about/major/ahcd/ercharts.htm.

There are a wide variety of ways to use bar charts. A bar chart can be turned on its side, making the bars run horizontal (see Figure 9.2). Multiple bars can also be grouped together, allowing you to show how different measurements compare (Figure 9.10).

Computers can be used to enhance bar charts even further. Coloring and shading the bars will enhance the readers' ability to interpret the data and identify trends.

TAKE NOTE Some graphing programs allow you to create bar charts that look three-dimensional. In almost all cases, this additional trait only makes the graph harder to interpret. You are better off using only two-dimensional bar charts.

Tables

Tables provide the most efficient way to display data or facts in a small amount of space. In a table, information is placed in horizontal rows and vertical columns, allowing readers to quickly find specific numbers or words that address their needs.

Creating a table takes careful planning, but computers can do much of the hard work for you. For simpler tables, you can use the "Table" function on your word-

To see more sample bar charts, go to **www.ablongman.com/johnsonweb/9.8**

Inserting a Table

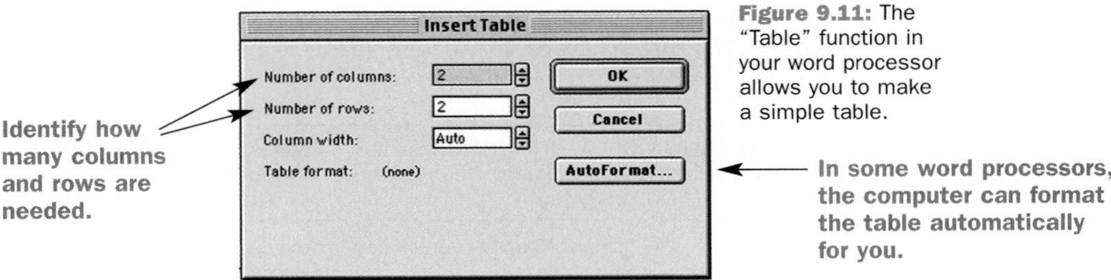

Figure 9.11: The "Table" function in your word processor allows you to make a simple table.

processing software (Figure 9.11). It will allow you to specify how many rows and columns you need (make sure you include enough columns and rows for headings in the table). Then, you can start typing your data or information into the cells.

If the "Table" function in your word processor is not enough for your needs, spreadsheet programs like MS Excel and Corel Quattro Pro also allow you to make quick tables (see the "Help" feature in this chapter).

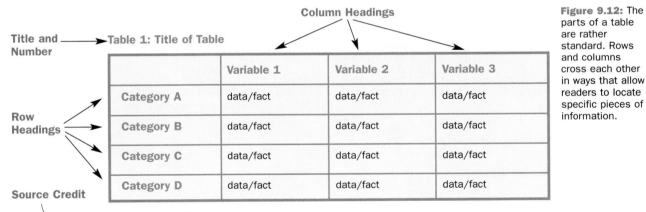

After creating the basic table, you should properly label it. In most cases, the table's number and title should appear above it (Figure 9.12). Down the left column, the *row headings* should list the items being measured. Along the top row, the *column headings* should list the qualities of the items being measured. Beneath the table, if needed, a citation should identify the source of information.

Parts of a Table

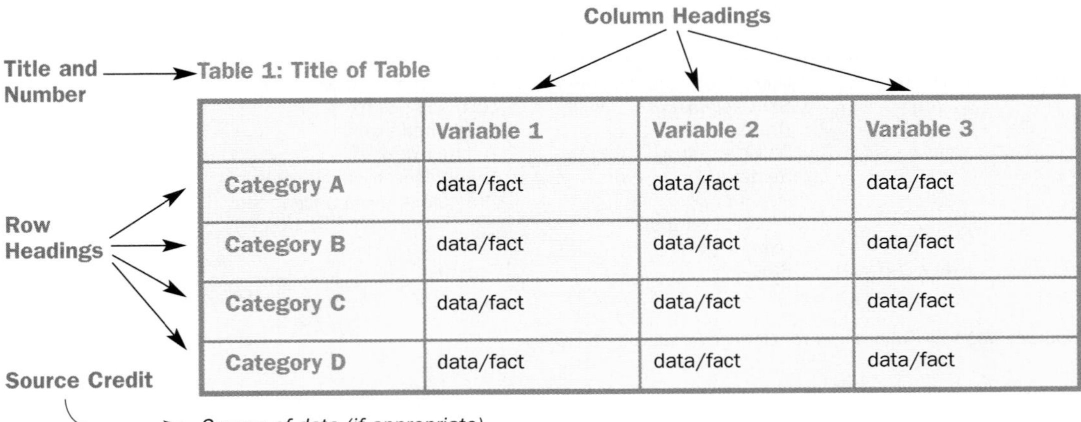

	Variable 1	Variable 2	Variable 3
Category A	data/fact	data/fact	data/fact
Category B	data/fact	data/fact	data/fact
Category C	data/fact	data/fact	data/fact
Category D	data/fact	data/fact	data/fact

Figure 9.12: The parts of a table are rather standard. Rows and columns cross each other in ways that allow readers to locate specific pieces of information.

In some cases, tables can be used to present verbal information rather than numerical data. In Figure 9.13, for example, a table is being used to verbally provide health information. With this table, readers can quickly locate their age and find the cancer test they need.

Need more help making a table? Go to **www.ablongman.com/johnsonweb/9.9**

A Table that Presents Verbal Information

CANCER DETECTION

		TEST OR PROCEDURE	
Age	**Frequency**	**Females**	**Males**
18–20	One time Monthly Yearly	Complete health examination Skin self-exam Pap smear	Complete health examination Skin self-exam Testis self-exam
20–40	Every 5 years Monthly Yearly	Complete health examination Skin self-exam Breast self-exam Pelvic exam Pap smear	Complete health examination Skin self-exam Testis self-exam
40–50	Every 3 years Monthly Yearly Every 1–2 years	Complete health examination Skin self-exam Breast self-exam Pelvic exam Pap smear Stool blood test Mammogram	Complete health examination Skin self-exam Testis self-exam Rectal exam Stool blood test
50–65	Every 2 years Monthly Yearly Every 3–5 years	Complete health examination Skin self-exam Breast self-exam Pelvic exam Rectal exam Pap smear Stool blood test Mammogram Procto	Complete health examination Skin self-exam Testis self-exam Rectal exam Stool blood test Prostate specific antigen test Procto
65+	Every Year Monthly Yearly Every 3–5 years	Complete health examination Skin self-exam Breast self-exam Pelvic exam Rectal exam Pap smear Stool blood test Mammogram Procto	Complete health examination Skin self-exam Testis self-exam Rectal exam Stool blood test Prostate specific antigen test Procto

Source: National Foundation for Cancer Research, 1998.

Figure 9.13: Tables can also present verbal information concisely. In this table a great amount of information is offered in an easy-to-access format.

When adding a table to your document, think about what your readers need to know. It is often tempting to include tables that hold all your data. However, these large tables might clog up your document, and readers will find it hard to locate specific information in them. You are better off creating small tables that focus on the specific information you want to present. Move larger tables to an appendix, especially if they present data not directly referenced in the document.

Pie Charts

Pie charts are useful for showing how a whole divides into parts (Figure 9.14). Pie charts are popular, but they should be used sparingly. They take up a great amount of space in a document, but they usually present only a small amount of data. The pie chart in Figure 9.14, for instance, uses a third of a page to plot a mere six data points.

A Pie Chart

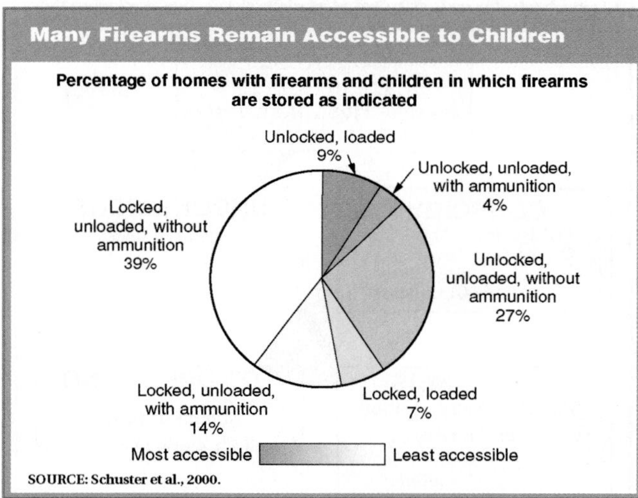

Figure 9.14: A pie chart is best for showing how a whole can be divided into parts.

Source: Rand Corporation, http://www.rand.org/publications/randreview/issues/rr.12.00/tranquil.html.

Pie charts are difficult to construct by hand, but your computer's spreadsheet program (Excel or Quattro Pro) can usually help you create a basic pie chart from your data. When labeling a pie chart, you should try to put titles and specific numbers in the graphic itself. For instance, in Figure 9.14, each slice of the pie chart is labeled and includes a measurement to show how the pie was divided. These labels and measurements help readers compare the data points plotted in the chart.

The key to a good pie chart is a clear story. For example, what story is the pie chart in Figure 9.14 trying to tell? At first glance the pie chart might appear to be

Do you want to see other pie charts? Go to **www.ablongman.com/johnsonweb/9.10**

saying that a large portion of gun owners (39%) responsibly lock away their guns with no ammunition. But then, you might also notice a worrisome statistic that almost one out of ten gun owners (9%) leaves his or her guns unlocked and loaded. This pie chart is effective, but many pie charts, including this one, risk telling an unclear story.

Flowcharts

Flowcharts are used to visually guide readers through a series of decisions, actions, or steps. They typically illustrate a description of a process described in the written text. Arrows are used to connect parts of the flowchart, showing the direction of the process (Figure 9.15).

A Flowchart

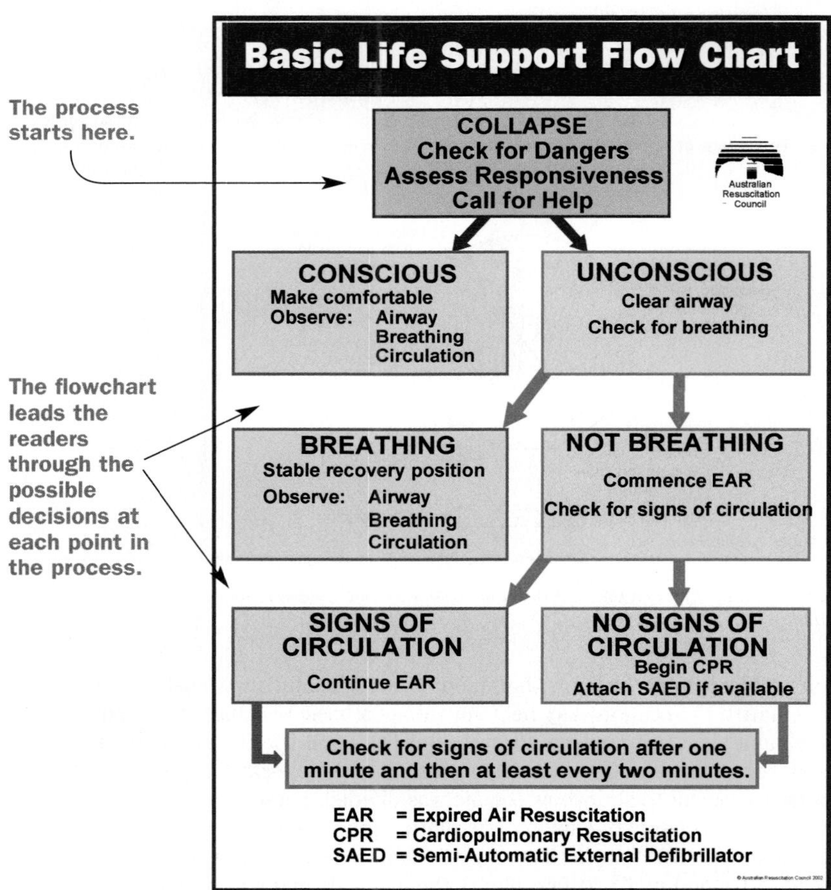

Figure 9.15: A flowchart is often useful for illustrating a process.

Source: Australian Resuscitation Council, http://www.resus.org.au/public/bls_flow_chart.pdf.

Other flowcharts can be viewed at **www.ablongman.com/johnsonweb/9.11**

As shown in Figure 9.15, flowcharts are helpful for illustrating instructions, especially when decisions need to be made by the user of the instructions. A flowchart typically cannot replace written instructions, especially if the steps are complex. But, it can illustrate the steps in the process to help readers understand the written text.

LINK For more information on writing instructions, go to Chapter 19, page 557.

Flowcharts can be used for a variety of other purposes, such as organization charts or circuit diagrams. An organization chart can illustrate the hierarchy of decision making in an organization. In a circuit diagram, meanwhile, a flowchart might be used to chart the path of electricity.

Gantt Charts

Gantt charts have become quite popular in technical documents, especially proposals and progress reports. Gantt charts, like the one in Figure 9.16, are used to illustrate a project schedule, showing when the phases of the project will begin and end.

LINK For more help with planning a project timeline, go to Chapter 13, page 359.

Visually, these charts illustrate the interrelations among different aspects of a large project. That way, people who are working on one part of the project will know what other teams are doing. Another benefit to a Gantt chart is that it gives readers an overall sense of how the project proceeds from beginning to end.

A Gantt Chart

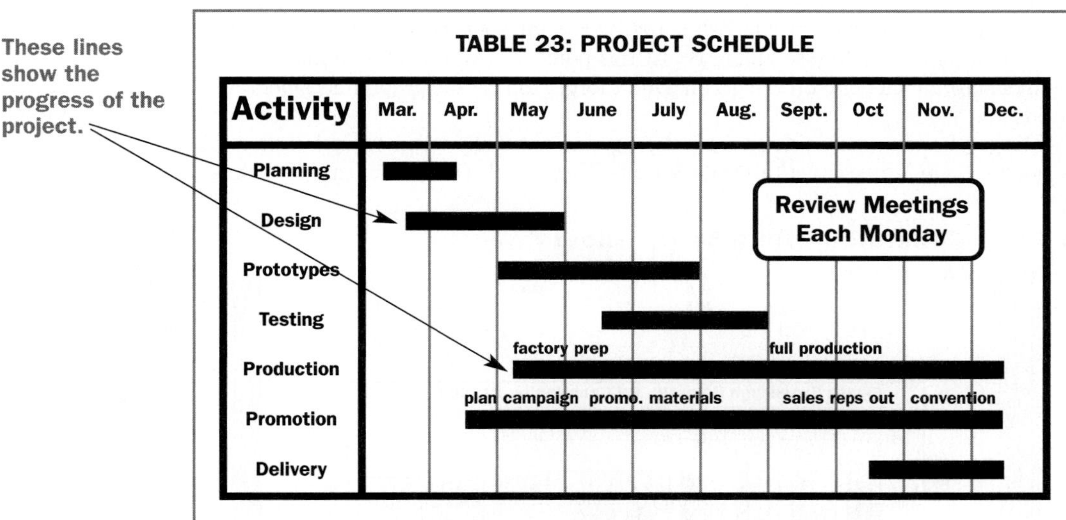

Figure 9.16: A Gantt chart is often used to show how the stages in a project interrelate. In this chart, you can easily see how the promotion of the product is started, even before the product is fully tested.

Gantt charts are becoming increasingly simple to create, because project planning software like ArrantSoft, Artemis Project Management, and MS Project can easily generate them for use in technical documents.

To find out more about project management software, go to **www.ablongman.com/johnsonweb/9.12**

HELP

Making Visuals with a Spreadsheet Program

Most word-processing programs, like Word or WordPerfect, come bundled with a companion spreadsheet program, like Excel or Quattro Pro. These spreadsheet programs allow you to make quick graphs and charts from a data set.

You can then insert these charts and graphs directly into your document. Overall, making and inserting graphics is not too difficult, because your word-processing program and spreadsheet programs are designed to be compatible. With a few clicks of the mouse, you can copy the graph you made in the spreadsheet program directly into your document.

To make a chart or graph, open a spreadsheet and enter the data you would like graphed (Figure A).

Data in a Spreadsheet

Data points are entered here.

Labels are added in the spreadsheet.

Average Rainfall

	A	B	C	D	E	F	G	H
1		1940	1950	1960	1970	1980	1990	2000
2	La Plata County	17.9	20.5	22.8	24.5	15.2	13.3	12.8
3	Montezuma County	18.8	19.2	21.6	23.4	16.1	13.6	11.7
4	Archuleta County	19.9	20.3	23.6	22.1	14	12.7	9.2
5								
6								

Figure A: To make a table, chart, or graph in a spreadsheet, begin by entering the data into the "worksheet."

Then, in the toolbar, usually at the top of the page, the spreadsheet program will have a button that allows you to look at the variety of graphs and charts available. You might even be able to preview different graphs and charts to see which one best illustrates the story you want the data to tell.

Once you find a graph or chart style that works for your data, let the spreadsheet program graph the data (see Figure B). At this point, you will be able to make adjustments to the labeling on the graph. Don't forget to title the graph and label the x-axis and y-axis.

The graphs you make in a spreadsheet program are usually rather plain, but they will be good enough for most workplace documents. They are also quick and easy to make.

A Graph Generated with a Spreadsheet Program

Here, the graphic is inserted into the spreadsheet.

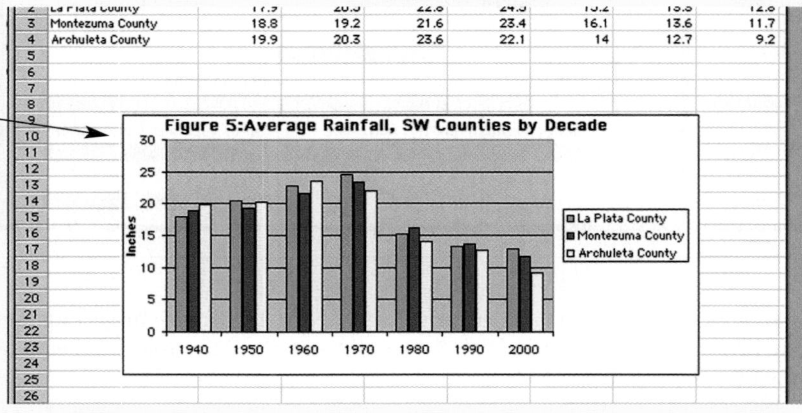

Figure B: The spreadsheet will generate the visual for you. This visual can then be pasted into a document.

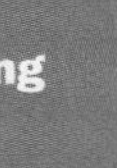

Need help using your spreadsheet program to make graphs? Go to **www.ablongman.com/johnsonweb/9.13**

Using Pictures, Drawings, and Video

Increasingly, computers give you the ability to include pictures, drawings, and video in documents. Even if you are not artistic, you can quickly use a digital camera, a drawing program, a scanner, or a video camera to add life to your documents.

The purpose of a picture, drawing, or video is to show what something looks like. These kinds of visuals are especially helpful when your readers may not be familiar with something, like an animal or a piece of equipment. They are also helpful for showing the condition of something, like damage to a car or a building under construction.

Photographs

Digital cameras and scanners are making the placement of photographs in technical documents easier than ever. Unfortunately, like those slides of your last vacation, photographs rarely capture the essence of what you are trying to show. A good first step is to ask what *story* you want the photograph to tell. Then, set up a shot that tells that story.

PHOTOGRAPHING PEOPLE When photographing people, a good rule of thumb is to only show people doing what they actually do in the workplace (Figure 9.17). Hint: people rarely huddle in a cubicle, pointing at a computer screen. If you need to include a picture of a person or a group of people standing still, take them outside and photograph them against a simple but scenic background. Photographs taken in the office tend to look dark, depressing, and dreary. Photographs taken outdoors, on the other hand, imply a sense of openness and free thinking.

A Photograph of People

Figure 9.17: Try to capture people in action, close up.

Source: Corel.

If you need to photograph people inside, put as much light as possible on the subjects. If your subjects will allow it, use facial powder to reduce the glare off their cheeks, noses, and foreheads. Then, take their picture against a simple backdrop to reduce background clutter.

If you are photographing an individual, take a picture of his or her head and shoulders. People tend to look uncomfortable in full-body pictures.

PHOTOGRAPHING THINGS When taking pictures of objects, try to capture a close-up shot while minimizing any clutter in the background (Figure 9.18). It is often a good idea to put a white drop cloth behind the object to block out the other items and people in the background. Make sure you put as much lighting as possible on the object, so it will show up clearly in your document.

A Photograph of an Object

Note the plain background behind the subject of the photograph.

Figure 9.18: When photographing things, try to reduce the amount of clutter around your subject.

Source: The Internet Public Library, http://www.ipl.si.umich.edu/div/pottery/image15.htm.

When photographing machines or equipment, try to capture them in action. After all, a picture of equipment sitting idle on the factory floor is rather boring. But if you show the machine being used or you focus on the moving parts, you will have a much more dynamic picture.

PHOTOGRAPHING PLACES Places are especially difficult to photograph. When you are at the place itself, snapping a picture seems simple enough. But, the pictures often come out flat and uninteresting. Moreover, unless people are in the picture, it is often difficult to tell the scale of the place being photographed.

When photographing places, focus on people doing something in that place. For example, if you need to photograph a factory floor, include people doing their jobs. If you are photographing an archaeological site, include someone working on the site. The addition of people will add a sense of action and scale to your photograph.

To learn more tips about photography, go to **www.ablongman.com/johnsonweb/9.14**

Inserting Photographs and Other Images

A digital camera will usually allow you to save your photographs in a variety of memory sizes. High-resolution photographs (lots of pixels) require a lot of memory in the camera and in your computer. They are usually saved in a format called a .tiff file. Lower-resolution photographs (fewer pixels) are saved as .gif or .jpg files. Usually, .gif and .jpg files are fine for print and on-line documents. However, if the photograph needs to be high quality, a .tiff file might be the best choice.

Once you have downloaded an image to your computer, you can work with it using software programs like Adobe Photoshop or MS Paint (Figure 9.19). These programs will allow you to touch up the photographs or, if you want, completely alter them.

Working with Images

Here are tools for altering the images.

Colors can be added or altered with this tool.

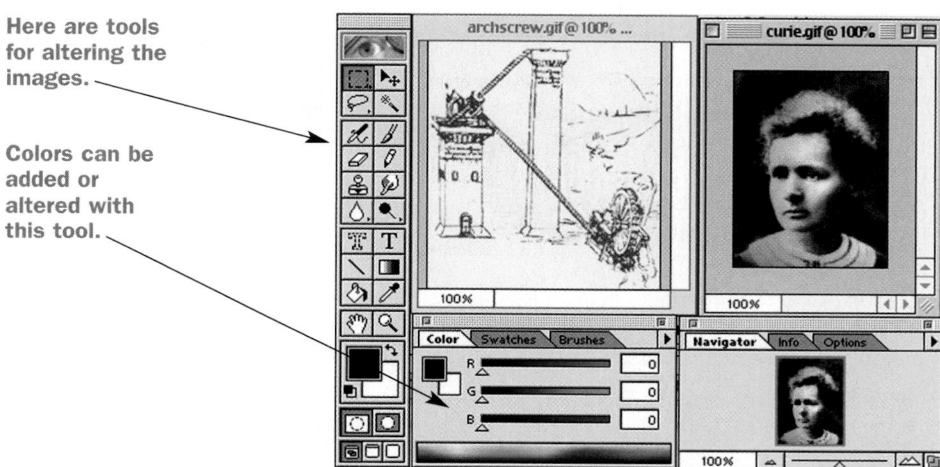

Figure 9.19: Software such as Adobe Photoshop allows you to work with photographs and other kinds of images.

Once you have finished touching up or altering the image, you can then insert it into your document or presentation. Most word-processing programs have an "Insert Picture" command. To insert the picture, put your cursor where you want the image to appear in the document. Then, select "Insert Picture." A box will open that allows you to locate the image on your computer's hard drive. Find and select the image you want to insert.

At this point, your computer will insert the image into your document. Usually, you can then do a few simple alterations to the file, like cropping, with the Picture toolbar in your word processor.

Illustrations

Illustrations are often better than photographs for depicting buildings, equipment, maps, and schematic designs. Whereas photographs usually include more detail than needed, a good illustration highlights only the most important features of the subject.

To learn more about using electronic images, go to **www.ablongman.com/johnsonweb/9.15**

LINE DRAWINGS AND DIAGRAMS A line drawing or diagram is a semirealistic illustration of the subject being described. You can usually create simple drawings and diagrams with the "Draw" function of most word-processing programs. As the drawings grow more complex, however, most writers will hire professional artists to transform rough sketches into finished artwork.

Line drawings offer several advantages. They can provide a close-up view of important features or parts. They can also be easily labeled, allowing you to point out important features to the readers.

In some ways, drawings and diagrams are less than realistic. For example, the diagram of the rabies virus in Figure 9.20 does not look exactly like the actual virus. Instead, it shows only how the larger parts of the virus are interconnected and work together.

A Diagram

Labels are added to identify features.

Explanatory text can be added to clarify the meaning of the diagram.

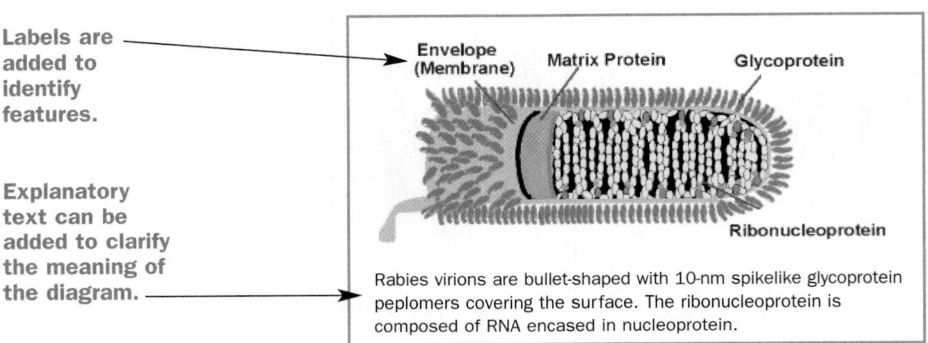

Figure 9.20: A drawing is only partially realistic. It concentrates on relationships instead of showing exactly what the subject looks like.

Source: Centers for Disease Control, http://www.cdc.gov/ncidod/dvrd/rabies/the_virus/virus.htm.

MAPS Maps offer a view of the subject from above. You can use them to show geographic features like streets, buildings, or rivers. They can be used to portray the rooms in a building or illustrate a research site (Figure 9.21). In some cases, they can be used to show the readers where a particular event occurred, allowing them to see where a particular place fits into an overall geographic area. When including a map, zoom in on the area that is being discussed in the text.

Maps may be easier to generate than you think. Today, Internet sites like Mapquest (www.mapquest.com) and Topozone (www.topozone.com) allow you to create very detailed maps. Mapquest is especially helpful for making maps with driving directions. Before trying to draw your own map, you might let the computer do it for you.

Keep in mind, though, that the maps you find, especially on the Internet, might be copyrighted. If you are using the map for personal or educational uses, you should be able to use it without asking permission. But, if you are using the map for professional reasons, especially for publication, you will likely need to ask permission from the owner of the map.

LINK For more information on copyright law, go to Chapter 4, page 80.

To see examples of line drawings, go to **www.ablongman.com/johnsonweb/9.16**

A Map

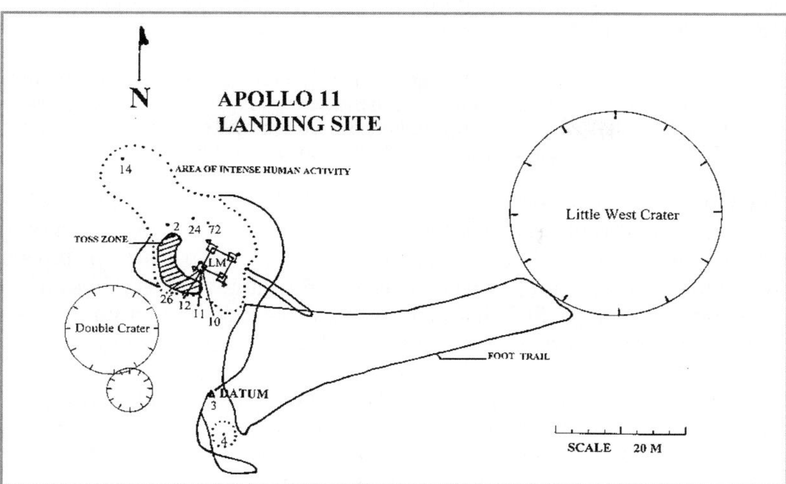

Figure 9.21: A map is useful for showing a place from above.

Source: New Mexico State University, http://www.spacegrant.nmsu.edu/lunarlegacies/images/scan2.gif.

ICONS AND CLIP ART Icons play an important role in technical documentation. In some documents, they are used as warning symbols. They can also serve as signposts in a text to help readers quickly locate important information (Figure 9.22). If you need an icon, standard sets of symbols are available on the Internet for purchase or for free.

TAKE NOTE Many icons are not copyrighted, so you can use them for free in your documents. If an icon is copyrighted (or you aren't sure whether it is), you need to ask permission to use it.

Common Icons

Figure 9.22: Icons are widely available on the Internet. The person in the middle is supposed to be sneezing, but icons can often have unexpected meanings.

Sources: Centers for Disease Control, http://www.cdc.gov/diabetes/pubs/images/balance.gif, and International Association for Food Protection, http://www.foodprotection.org.

GO TO THE NET

For links to mapping Internet sites, go to **www.ablongman.com/johnsonweb/9.17**

Clip art drawings are commercially produced illustrations that can be purchased or used for free. Usually, when you purchase a collection of clip art, you are also purchasing the rights to use that clip art in your own documents.

TAKE NOTE Before using a clip art image, you should be careful about who "owns" it. Clip art websites and clip art collections will usually tell you how the images can be used legally. Some sites will limit you to 10 free images. Others may require you to ask permission. And others will allow you to use the clip art as long as you do not alter it.

It is tempting to advise you not to use clip art at all. When desktop publishing first came into the workplace, clip art was an original way to enhance the message and tone of a document. But now, most readers are tired of those little pictures of people shaking hands, pointing at whiteboards, and climbing ladders. In some cases, clip art becomes decorative fluff that takes readers' attention away from the document's message. Use it sparingly and only when it truly contributes to your message.

SCREEN SHOTS With a combination of keystrokes, most computers will allow you to make a *screen shot* of your computer screen. Essentially, a screen shot is a picture of whatever is on the screen (Figure 9.23). You can use screen shots to insert a variety of different images into your document or presentation.

You can make screen shots quickly with a PC or Mac. With a PC, press the "Print Screen" button on your keyboard. The computer will put an image of the screen on your "clipboard." Look for it there. With a Mac, press the three keys Apple-Shift-3 at the same time. You should hear a camera clicking sound. The picture will then appear on your hard drive in the Finder menu.

A Screen Shot

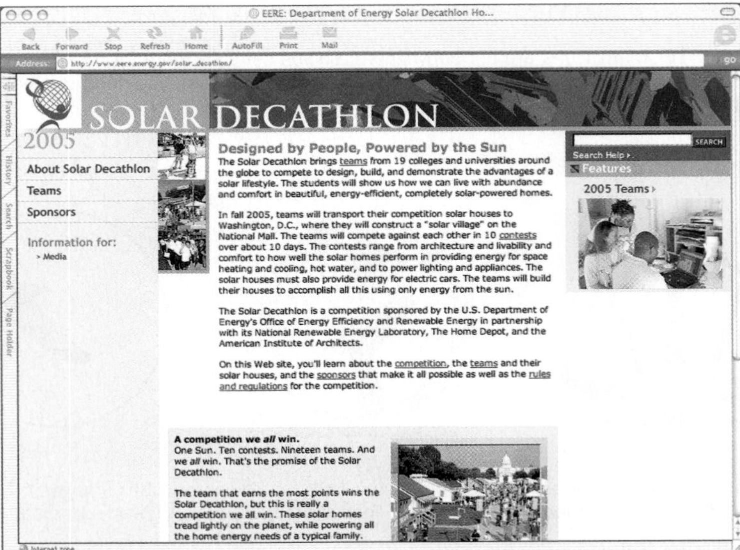

Figure 9.23: Screen shots are a simple way to add images to your document or presentation.

Source: Office of Energy Efficiency and Renewable Energy, http://www.eere.energy.gov/solar_decathlon.

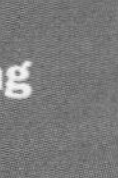

For websites that offer free icons and clip art, go to **www.ablongman.com/johnsonweb/9.18**

You can then insert the screen shot into your document with the "Insert Picture" command in your word processor. Using your Drawing tools, you can then "crop" the image (trim it) to remove any extra details.

Screen shots are useful in a variety of situations. For example, you can use a screen shot to put a picture of a webpage in a printed document. Or, after drawing an illustration, you can make a screen shot and add it to a webpage.

TAKE NOTE Just a warning: Images of webpages or other interfaces are copyrighted. If you are using a screen shot for other than personal or educational purposes, you should ask permission from the owner of the site or software.

Using Video and Audio

One of the options available with computers is the use of video and audio in multimedia or Internet-based documents. With a digital video or audio recorder, you can add movies, music, and sound to any on-screen document.

VIDEO The ability to add video to a text is a mixed blessing. On one hand, nothing will grab readers' attention like a video presentation. A video, for example, might show an important experiment or a current event. It might show an expert talking about a product or service (Figure 9.24). On the other hand, poorly made videos seem amateurish and quickly become tedious. Another problem is that videos can take a long time to download.

Using a Video

Keep the background in a video simple.

Videos should be used to make short, simple points.

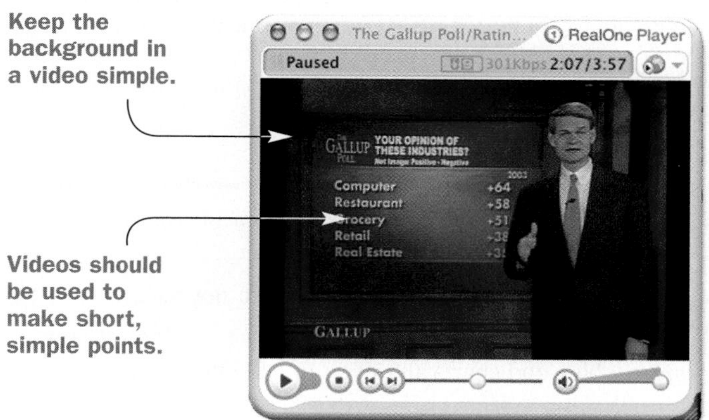

Figure 9.24: Video editing software can help you make videos to be inserted into multimedia documents, websites, or presentations.

Source: The Gallup Organization, http://www.gallup.com.

If you want to add a video to your document, first decide exactly what you want the video to do. Then, write a script that achieves that goal in the least amount of time, because long videos take up lots of memory (and they can get a little boring).

As you create a video, think of it as a moving photograph in your document. Many of the same guidelines for shooting photographs will work with videos. For example, try to reduce background clutter, so the viewers can focus on the subject.

To learn more about video editing software, go to **www.ablongman.com/johnsonweb/9.19**

Also, put as much light on the subject as possible. And, if you are filming people, you might want to put some powder on their foreheads, noses, and cheeks to reduce glare.

After you shoot the video, you can use video editing software like Apple iMovie, Adobe Premiere, and Microsoft Movie Maker to edit the final piece into a concise format.

When you are finished filming and editing the video, you can place it into your document, usually with the "Insert Movie" command. Even simple word processors like MS Word and Corel WordPerfect will allow you to put a video into the text. Videos can also be inserted into presentations made with MS PowerPoint, Corel Presentation, and other presentation software packages.

MUSIC AND AUDIO You can also add audio clips or background music to your electronic documents, including websites. Most word-processing programs, presentation software, and website design software will allow you to "Insert" music or audio.

The best way to insert music or audio is from a compact disc (CD) or by downloading a song in MP3 format off the Internet. Several software applications, like Apple iTunes or Windows Media Player, will allow you to download music legally off the Internet, usually at minimal cost.

TAKE NOTE There are many websites that will allow you to download music "free." These sites may be illegally offering music, thus violating copyrights. Make sure you check the source of the music to ensure that you are not breaking copyright laws.

When you select "Insert," the computer will usually ask you to select a file. Choose the audio track or music file to be inserted. Then, if you want the music to play continuously, tell the computer to "Loop" the music when you are in the "Audio" text box.

Music or audio can add a nice touch to a multimedia document. It can also annoy the readers or audience. So, you should choose music or audio that best complements the material you are providing. Avoid using music or audio to simply ornament the document or presentation.

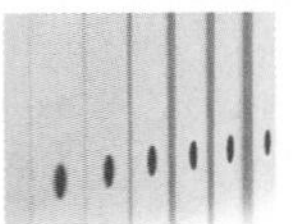

- Computers have made the use of graphics in technical documents easier. So, readers have come to expect them.
- Graphics should: (a) tell a simple story, (b) reinforce the text, not replace it, (c) be ethical, and (d) be properly labeled and placed on the page.
- Various kinds of graphs, tables, and charts allow you to tell different stories with data or facts.
- Digital cameras and scanners are making the placement of photographs in documents easier than ever.
- Use icons and clip art only when they enhance the readability and comprehension of the document. Clip art, especially, can simply clutter a document.
- Increasingly, computers allow us to insert video and audio into documents. Treat these items as you would treat photographs.

GO TO THE NET

For links to legal music download sites, go to **www.ablongman.com/johnsonweb/9.20**

EXERCISES AND PROJECTS

Individual or Team Projects

1. On the Internet, find a chart or graph that you can analyze. Using the four guidelines for using graphics discussed in this chapter, critique the chart or graph by discussing its strengths and places where it might be improved. Present your findings to your class.

2. Find a set of data. Then, use different kinds of charts and graphs to illustrate trends in the data. For example, you might use a bar chart, line graph, and pie chart to illustrate the same data set. How does each type of graphic allow you to tell a different "story" with the data? What are the strengths and limitations of each kind of graphic? Which kind of chart or graph would probably be most effective for illustrating your data set?

3. On the Internet, find some written text that could be turned into a graph, table, or chart. Then, create a visual that would complement this text. In a memo to your instructor, show how the text and visual could complement each other by referring readers back and forth between text and visual.

4. Using a digital camera, practice taking pictures and inserting those pictures into documents. Take pictures of people, things, and places. When taking pictures of people, compare pictures taken inside and outside. Take full-body pictures and head shots. When taking pictures of things, first leave the background behind the object cluttered. Then, use a backdrop to unclutter the picture. When photographing places, try to make images that tell a story about the place.

   When you are finished, compare and contrast your photographs. Which types of photographs seem to work best in a document? What kinds of photographs tend not to work?

Collaborative Project

With a group of others, locate a large document that has few or no visuals. Then, do a "design makeover" in which you find ways to use visuals to support and clarify the written text. Try to include at least one visual for every two pages in the document. Use graphs, photographs, and drawings to illustrate important points in the document. Then, add icons and clip art to reinforce important points or themes in the document.

When you are finished, write a brief report to your instructor in which your group discusses how you made over the document. Critique the original draft of the document, showing how the lack of adequate visuals made the information in the document hard to access. Then, discuss the ways in which your revised version improved on the original. Finally, discuss some of the following issues about the amount and types of visuals used in this kind of document:

- At what point are there too many graphics?
- Do some graphics work better than others?

Sample charts, graphs, and data sets for these exercises are available at **www.ablongman.com/johnsonweb/9.21**

- How can you balance the written text with visuals to avoid making the document too text-heavy or visual-heavy?
- How do the needs and characteristics of the expected readers of the document shape the kinds of visuals that are used?

Your report might offer some additional guidelines, beyond the ones discussed in this chapter, for using visuals more effectively.

You can find sample documents for this assignment at **www.ablongman.com/johnsonweb/9.22**

CASE STUDY

The Illusion

Things weren't going well at Kate Bunson's company. Sales were down, and expenses were up. Everyone knew that the downturn in sales was due to the current recession, but the company's investors didn't seem to have much sympathy for those sorts of explanations. They wanted to see rising profits. If they didn't see rising profits, they might pull their money out of the company.

It was time for the company to put together the annual report before the stockholders' meeting in Chicago. Kate's responsibility was to write the part of the report that discussed product sales and expenses. She needed to find a way to show that the current downturn in sales was simply a result of the overall downturn in the economy. She needed to illustrate that sales would return as soon as the current recession was over.

But her supervisor had some other ideas. "Listen, Kate," she said, "it's simple. Investors rarely look closely at these annual reports. They scan the graphics, looking for rising lines. So, all you need to do is design your graphs so that they describe losses, not gains." On a sheet of paper, Kate's boss sketched a quick idea of the kind of graphs she had in mind.

The Boss's Drawings

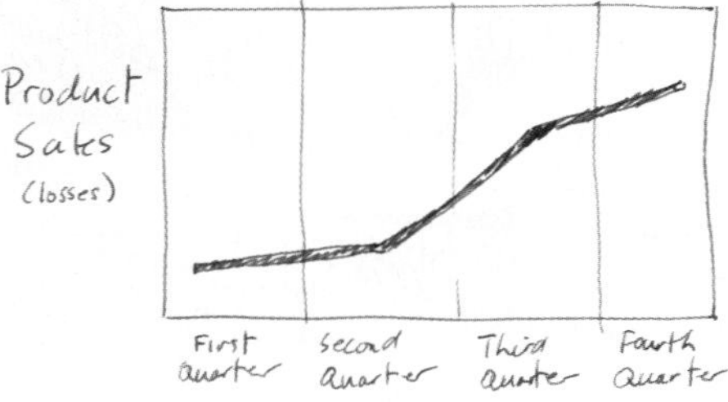

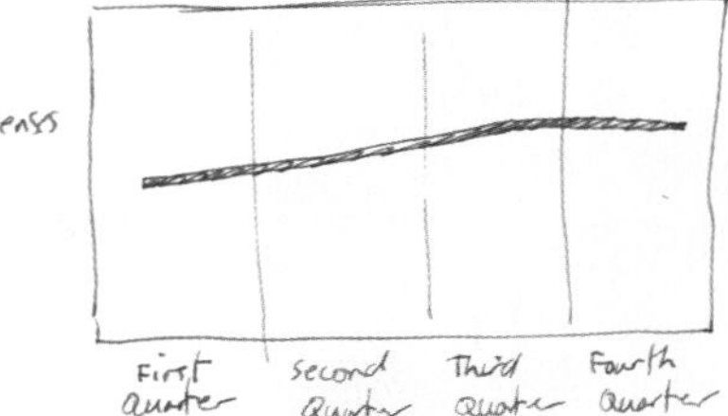

In other words, Kate's boss was suggesting that her graphs use rising lines to reflect increased losses and increased expenses. It was, at best, a visual trick to give the investors the impression that sales were going up. At worst, it was dishonest, because the "product sales" graph would imply the opposite of the truth, even though it technically was telling the truth.

Kate knew there must be a better way to express the sales data ethically without also giving the impression that the company was dramatically losing money. What are some ways she could use graphs ethically without also spooking the company's investors?

GO TO THE NET

For examples of annual reports, go to **www.ablongman.com/johnsonweb/9.23**

CHAPTER

10

Revising and Editing for Usability

CHAPTER CONTENTS

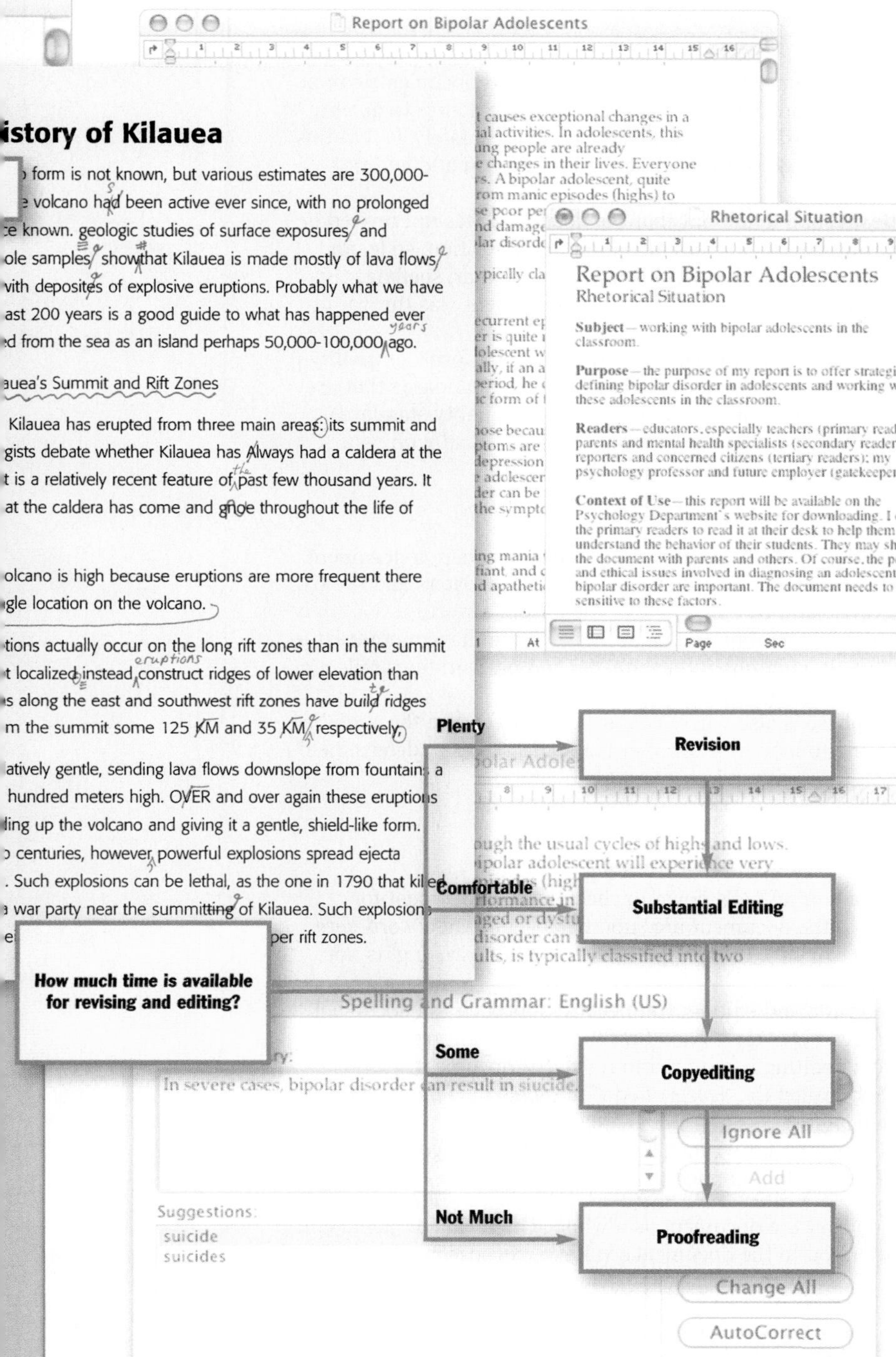

CHAPTER OBJECTIVES

In this chapter, you will learn:

- How to use revising and editing as a form of quality control.
- How to apply the four levels of editing to your text.
- How to use copyediting symbols.
- Strategies for revising your draft by reviewing its rhetorical situation (subject, purpose, readers, and context of use).
- Strategies for reviewing the content, organization, and design of your document.
- How to revise your sentences, paragraphs, headings, and graphics.
- To proofread for grammar, punctuation, and word usage.
- Strategies for document cycling and usability testing.

The computer's ability to make rough drafts look like finished documents can be both an advantage and a disadvantage. An advantage is the ability to quickly produce a document that looks finished. The disadvantage, unfortunately, is that this finished-looking document might fool you into skipping the revision and editing phase of the writing process.

Of course, your readers will soon figure out that the document was not revised or edited. They will begin spotting the weaknesses in content, organization, style, and design. Meanwhile, your document will include mistakes in grammar, spelling, and usage. It doesn't take long for readers to figure out that a document was thrown together and hurried out the door.

Put simply, revising and editing are important because they are forms of ***quality control*** in a document. Your documents should reflect the quality standards that are held by your company (or they should maintain an even higher quality standard). After all, if your documents demonstrate poor quality, they reflect badly on your company's products or services. They will also reflect badly on you.

LINK For more information on quality control, turn to Chapter 13, page 375.

In technical communication, the revising and editing phase is where a document or presentation goes from "adequate" to "excellent." Accomplished writers and speakers leave plenty of time to revise and edit their work. The payoff is worth it. Not only will your readers appreciate your attention to detail, they will also have more trust in you and your company. After all, high-quality documents and presentations reflect a personal or corporate commitment to excellence.

The revising and editing phase is also where you are going to ensure the ***usability*** of the document. You are going to look at it through the readers' eyes to determine if it fits their needs.

Levels of Edit

For some people, revising and editing boils down to checking the document for errors. Mostly, their changes to the document are superficial—altering a word here and there, rewriting a sentence or two. In most cases, this kind of "revising" is not sufficient for technical documents.

Instead, when you are revising and editing, you need to approach the document from a variety of levels and perspectives. A good first step toward improving a document is to decide what level of editing is required to finish the document. Professional editors use a tool called the "levels of edit" to assess how much editing a document needs before the deadline.

The levels of edit are also helpful for writers, because they encourage them to go deeper into the text to make important changes:

Level 1: Revision—looks at the document as a whole. Often called "global editing," revision pays attention to the document's subject, purpose, readers, and context of use.

Level 2: Substantive editing—pays special attention to the content, organization, and design of the document.

To see other variations of the levels of edit, go to **www.ablongman.com/johnsonweb/10.1**

Level 3: Copyediting—concentrates on revising the style for clarity, persuasion, and consistency, especially at the sentence and paragraph levels.

Level 4: Proofreading—catches only the grammar mistakes, misspellings, and usage problems.

TAKE NOTE Depending on the editor, the levels of edit can be numbered different ways. Sometimes proofreading is called a "level 1 edit." In this book, we have decided to number the levels of edit in the order that would likely be followed by a writer.

Given enough time, a writer will ideally go through all four levels, beginning with revision and ending with proofreading (Figure 10.1). But, sometimes the deadline is looming. In those cases, you may need to start at substantive editing, copyediting, or even proofreading.

Which level of edit is appropriate for your document? The answer to this question depends on how much time you have and the quality needed in the document. If you have plenty of time and the quality needs to be high, you should do a full revision (level 1) of the document. However, if you don't have much time or high quality is not essential, you should at least proofread (level 4) the document.

So, as you begin the revising and editing phase, start out by determining what level of editing is possible and/or needed to produce the desired quality of document. Then, begin reworking the document at that level.

The Levels of Editing

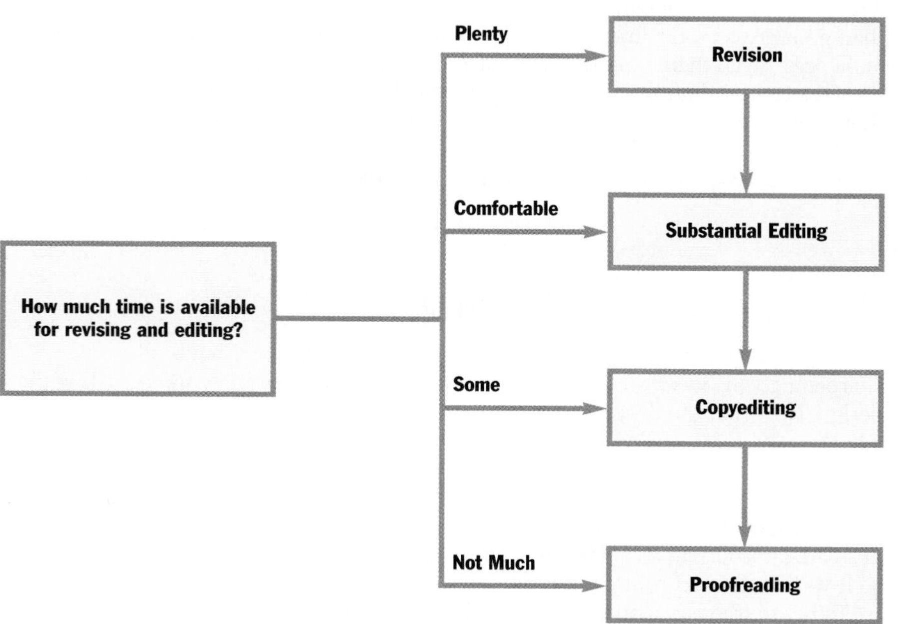

Figure 10.1: Ideally, writers would always go through all four levels of edit, from revision to proofreading. Sometimes, though, limited time determines what level is appropriate to finish a document.

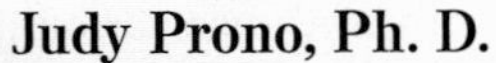

AT WORK

Judy Prono, Ph. D.

TECHNICAL EDITOR AND TEAM LEADER, LOS ALAMOS NATIONAL LABORATORY, LOS ALAMOS, NEW MEXICO

Los Alamos National Laboratory is a U.S. Government Research Facility.

How are the levels of edit used at your workplace?

Since the mid-1980s, we have used levels of edit at Los Alamos National Laboratory to define editing expectations for both authors and editors. In 1986, a panel of authors helped us condense our four levels into three: proofreading, grammar, and full edit.

- A proofreading edit corrects glaring errors that could prove embarrassing to author and laboratory alike.
- The grammar edit ensures document clarity for audiences of technical peers.
- The full edit polishes technical documents and is the recommended edit for documents intended for broader audiences than those of technical peers.

Beyond simply improving the writing, our levels also encompass issues such as the sequencing and clarity of tables, figures, and equations; acronym and number usage; and lab publication policies.

Our levels of edit help authors control the cost of editing: The more comprehensive an edit, the more time consuming and hence more costly it is. However, we encourage authors to choose their edits in terms of a document's purpose and audience. An archival report intended simply to document an experiment requires less editing than a progress report intended to update key research sponsors. Similarly, technical audiences, given their expertise, will infer the significance and relationships of data that less technical audiences need stated explicitly. Documents intended for the latter thus need more extensive editing to ensure clarity.

Although, occasionally, editing comes down to the money or time available, in general, our levels of edit have proved helpful in defining editing needs and priorities.

Revising: Level 1 Editing

While you were drafting the document, you were probably revising it all along, sharpening your ideas on the subject, reconsidering your purpose, and adjusting the design. Now that the document is completely drafted, you can start revising it as a whole.

Revision is a process of "revisioning" the document.

Revision is a process of "revisioning" the document. In other words, you are trying to gain a new perspective on your text, so you can ensure that its subject and purpose are appropriate for your intended readers. You also want to revise the document so it will work in its context of use.

To revise (re-vision) your document, look back at your initial decisions about the rhetorical situation that defined your document at the beginning of your writing process:

SUBJECT Check whether your subject area needs to be narrowed or broadened.

- Has your subject changed or evolved?
- Did you limit or expand your subject area?
- Has your document strayed from the subject anywhere?

PURPOSE Make sure the document is achieving its purpose.

- What do you want the document to achieve?
- Is your document's purpose still the same?
- Has your purpose become more specific or has it broadened?

READERS Think about the characteristics of the primary readers and other possible readers.

- Do you now know more about your primary readers' needs?
- How will your readers' values and attitudes influence how they will interpret your text?
- Do you need to more thoroughly address the secondary, tertiary, and gatekeeper readers in your document?

CONTEXT OF USE Consider the contexts in which your document might be read or used.

- Do you better understand the physical places where your readers will read or use the document?
- Do you better understand the economic, political, and ethical issues that will influence how your readers will interpret your document?
- Have you anticipated the personal, corporate, and industry-related issues that will also shape your readers' interpretation?

LINK For more information on defining the rhetorical situation, go to Chapter 2, page 22.

Guidelines for Revising (Level 1)

- Subject—Is the subject too narrow or too broad?
- Purpose—Does the document achieve its stated purpose?
- Readers—Is the document appropriate for the readers?
- Context of Use—Is the document appropriate for its context of use?

Certainly, these are difficult questions to ask as you finish your document. But, they are worth asking. You need to make adjustments to the document if its purpose has evolved or you have gained a better understanding of your readers.

Once you have reconsidered the document's rhetorical situation, you can work through the text to see if it stays focused on the subject and achieves its purpose (Figure 10.2). Look for places where you can revise the text to better suit your readers and the contexts in which they will use the information you are providing them.

TAKE NOTE Your computer is the perfect tool for revising. Don't let the finished-looking document on your screen fool you into thinking the document is almost done and will be hard to revise. With the computer, you can easily move text around and make adjustments to the content, organization, style, and design to suit the rhetorical situation.

Need more help with revision? Go to **www.ablongman.com/johnsonweb/10.2**

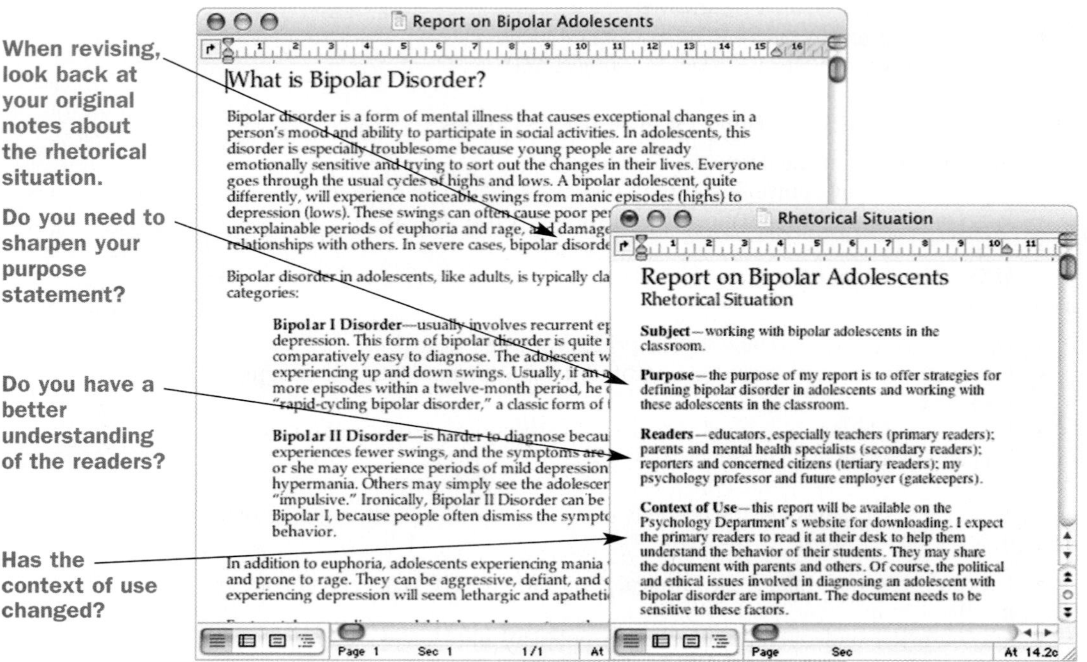
When revising, look back at your original notes about the rhetorical situation.
Do you need to sharpen your purpose statement?
Do you have a better understanding of the readers?
Has the context of use changed?
Report on Bipolar Adolescents
What is Bipolar Disorder?
Bipolar disorder is a form of mental illness that causes exceptional changes in a person's mood and ability to participate in social activities. In adolescents, this disorder is especially troublesome because young people are already emotionally sensitive and trying to sort out the changes in their lives. Everyone goes through the usual cycles of highs and lows. A bipolar adolescent, quite differently, will experience noticeable swings from manic episodes (highs) to
Rhetorical Situation
Report on Bipolar Adolescents
Rhetorical Situation
Subject—working with bipolar adolescents in the classroom.
Purpose—the purpose of my report is to offer strategies for defining bipolar disorder in adolescents and working with these adolescents in the classroom.
Readers—educators, especially teachers (primary readers); parents and mental health specialists (secondary readers); reporters and concerned citizens (tertiary readers); my psychology professor and future employer (gatekeepers).
Context of Use—this report will be available on the Psychology Department's website for downloading. I expect the primary readers to read it at their desk to help them understand the behavior of their students. They may share the document with parents and others. Of course, the political and ethical issues involved in diagnosing an adolescent with bipolar disorder are important. The document needs to be sensitive to these factors.

Figure 10.2: Look back at your original notes about the rhetorical situation. Have any of these elements changed or evolved? Does your document reflect these original decisions? If not, does the document need to change or does your understanding of the rhetorical situation need to change?

Revision at this global level requires some courage. You may discover that parts of your document need to be completely rewritten. In some cases, the whole document may need to be reconceived. But, it is better to be honest with yourself at this point and make those changes. After all, a document that fails to achieve its purpose was a waste of your time.

Substantive Editing: Level 2 Editing

Substantive editing assumes that your document is essentially meeting its purpose and is appropriate for your readers. At this level of edit, you should concentrate on the content, organization, and design of the document (Figure 10.3).

A good approach to substantive editing is to consider the document from three different perspectives:

CONTENT Look for any gaps or digressions in the content.

- Are there any places (gaps) where you are lacking proof or support for your claims?
- Do you need to conduct more research to support your points?
- Are there any places where you have included additional information that the readers do not need to know to make a decision or take action?

Substantive Editing

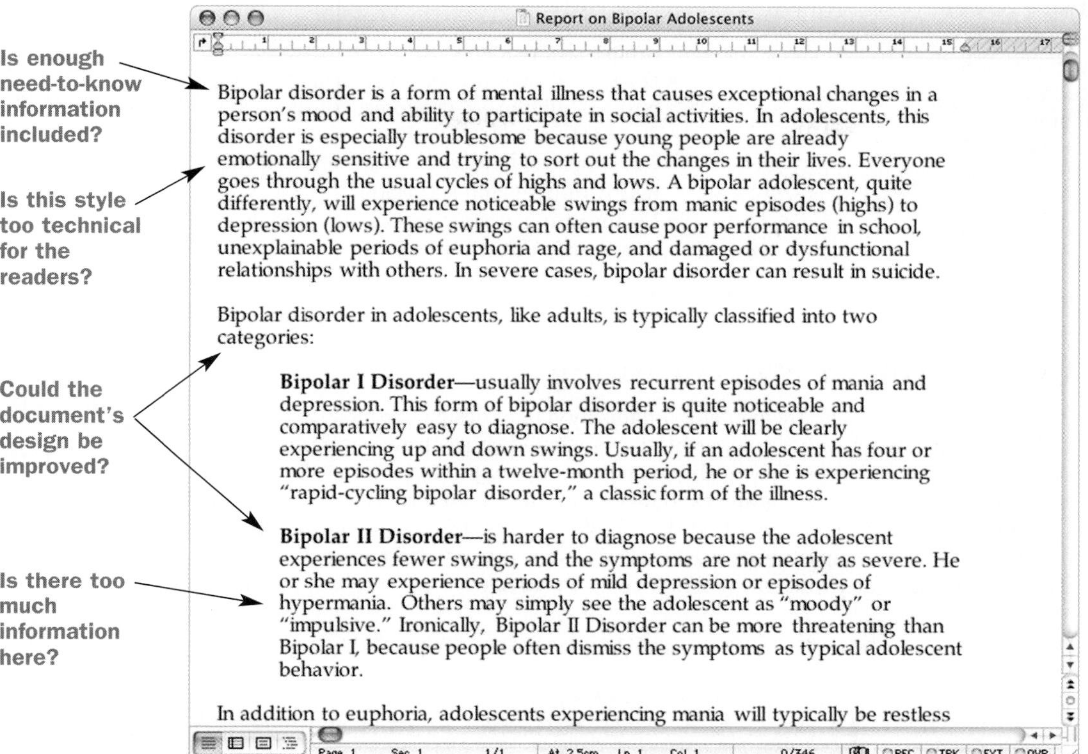

Is enough need-to-know information included?
Is this style too technical for the readers?
Could the document's design be improved?
Is there too much information here?
Report on Bipolar Adolescents
Bipolar disorder is a form of mental illness that causes exceptional changes in a person's mood and ability to participate in social activities. In adolescents, this disorder is especially troublesome because young people are already emotionally sensitive and trying to sort out the changes in their lives. Everyone goes through the usual cycles of highs and lows. A bipolar adolescent, quite differently, will experience noticeable swings from manic episodes (highs) to depression (lows). These swings can often cause poor performance in school, unexplainable periods of euphoria and rage, and damaged or dysfunctional relationships with others. In severe cases, bipolar disorder can result in suicide.
Bipolar disorder in adolescents, like adults, is typically classified into two categories:
Bipolar I Disorder—usually involves recurrent episodes of mania and depression. This form of bipolar disorder is quite noticeable and comparatively easy to diagnose. The adolescent will be clearly experiencing up and down swings. Usually, if an adolescent has four or more episodes within a twelve-month period, he or she is experiencing "rapid-cycling bipolar disorder," a classic form of the illness.
Bipolar II Disorder—is harder to diagnose because the adolescent experiences fewer swings, and the symptoms are not nearly as severe. He or she may experience periods of mild depression or episodes of hypermania. Others may simply see the adolescent as "moody" or "impulsive." Ironically, Bipolar II Disorder can be more threatening than Bipolar I, because people often dismiss the symptoms as typical adolescent behavior.
In addition to euphoria, adolescents experiencing mania will typically be restless
Page 1 Sec 1 1/1 At 2.5cm Ln 1 Col 1 0/346 REC TRK EXT OVR

Figure 10.3: Substantive editing urges you to ask questions about the appropriateness of the content, style, and design of your document. When answering these questions, think about the needs of your readers.

ORGANIZATION A document should conform to a recognizable genre, and it should have an identifiable introduction, body, and conclusion.

- Are there any places where you have deviated from the organizational pattern of the genre you are following? Are these deviations helpful toward achieving your purpose? Or, should you reorganize the document to suit the genre?
- Does the introduction clearly identify the subject, while stating your purpose and your main point? Should the introduction include more background information or stress the importance of the subject?
- Does the conclusion restate your main point, restress the importance of the subject, and look to the future?

LINK For more information about organization, see Chapter 6, page 132.

DESIGN The document should be designed for the readers and the contexts in which it will be used. Think about your readers and where they will use the document.

- Is the text readable (scannable) in the situations and places where people will use it?

To learn more about editing design, go to **www.ablongman.com/johnsonweb/10.3**

AT A GLANCE

Guidelines for Substantive Editing (Level 2)

- Content—Are there any digressions or gaps in content?
- Organization—Does the document conform to a recognizable genre or pattern?
- Design—Do the page layout and graphics enhance the readability of the document?

- Does the design reflect your readers' values and attitudes? Is it straightforward for conservative readers, or is it more innovative for progressive readers?
- Does the design properly use principles of balance, alignment, grouping, consistency, and contrast?
- Does the design clarify the structure of the text with titles and subheads?
- Do the graphics support the text, and do they clarify difficult points?

LINK For more document design strategies, see Chapter 8, page 192.

Copyediting: Level 3 Editing

Copyediting assumes that the content, organization, and design are mostly set and will not change. When you are copyediting, you should concentrate on improving style and consistency, especially at the sentence and paragraph level. You should also check over the headings and graphics to make sure they are appropriate and accurate.

While copyediting, you might find the "Track Changes" feature of your word processor especially helpful (Figure 10.4). It will show places where changes were made to the text. Later, you can decide if you want to keep those changes. This fea-

Tracking Changes While Copyediting and Proofreading

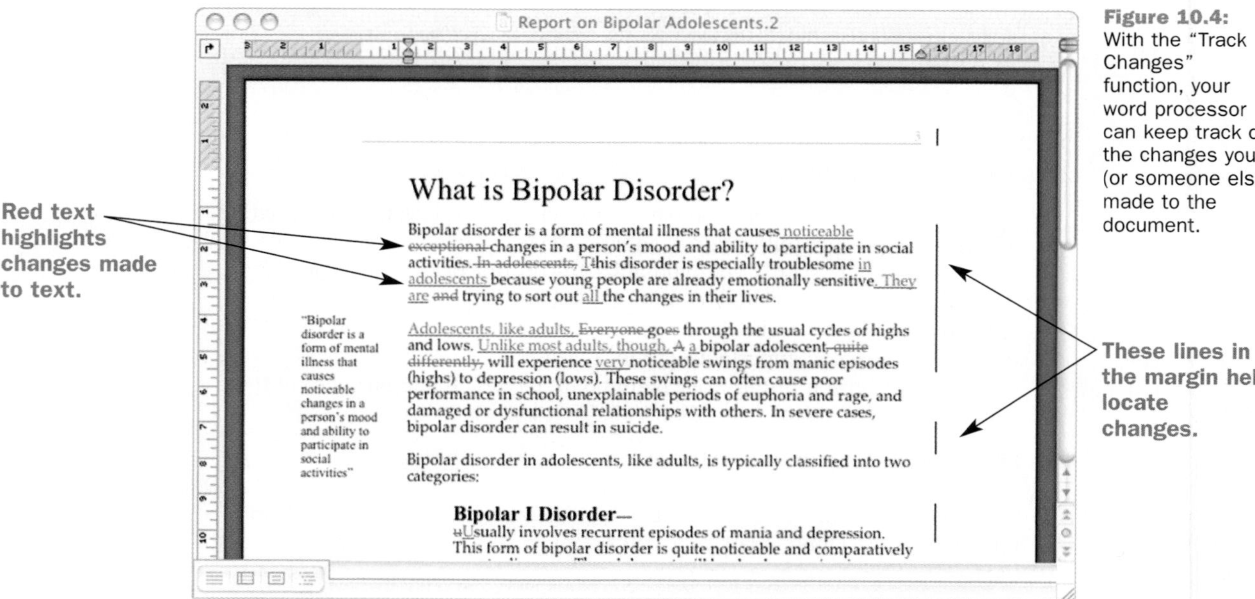

Figure 10.4: With the "Track Changes" function, your word processor can keep track of the changes you (or someone else) made to the document.

ture is especially helpful if you are working with a team on a document. That way, any changes are recorded for the others in the group to approve.

When copyediting, pay close attention to the sentences, paragraphs, headings, and graphics.

SENTENCES Look over the sentences to make sure they are clear and concise.

- Are the subjects of the sentences easy to locate?
- Do the verbs of the sentences express the actions of the sentences?
- Can you eliminate any unnecessary prepositional phrases?
- Are the sentences breathing length?

LINK For help on improving style at the sentence level, go to Chapter 7, page 165.

PARAGRAPHS Make sure the paragraphs support specific claims. Rework the sentences so that each paragraph flows smoothly.

- Does each paragraph have a clear topic sentence (a claim) and enough support to back it up?
- Would any paragraphs be stronger if you included a transition sentence at the beginning or a point sentence at the end?
- Are the subjects in the paragraph aligned, or could you use given/new strategies to smooth out the text?
- Would transitions or transitional phrases help bridge any gaps between sentences?

LINK For help on improving style at the paragraph level, go to Chapter 7, page 173.

AT A GLANCE

Guidelines for Copyediting (Level 3)

- Sentences—Are the sentences clear and concise?
- Paragraphs—Do the paragraphs have a clear topic sentence and support?
- Headings—Do the headings help the readers scan for important information?
- Graphics—Do the graphics support the written text?

HEADINGS The headings should be easy to understand and consistently used.

- Do the headings in the document properly reflect the information that follows them?
- Do the headings make the document scannable, highlighting places where important information can be found?
- Are there clear levels of headings that help readers identify the structure of the document and the importance of each part of the document?

LINK For more information on using headings, see Chapter 8, page 208.

GRAPHICS Look over the graphics in the document to make sure they support the written text. Check the graphics for accuracy.

- Does each graphic tell a simple story?
- Does each graphic support the written text without replacing it?
- Are the graphics clearly titled and referred to by number in the written text?

LINK For more help using graphics, turn to Chapter 9, page 236.

To learn more about copyediting, go to **www.ablongman.com/johnsonweb/10.4**

Proofreading: Level 4 Editing

Proofreading assumes that the document is complete in almost every way. Now, you need to focus only on the mechanical details of the document, like the grammar, spelling, punctuation, and word usage. While proofreading you should focus on marking "errors" and making only very minor stylistic changes to the text.

Grammar

In technical documents, correct grammar and punctuation are expected. You should always remember that your readers will notice the grammatical mistakes, even if they do not always know the difference between a well-written document and a weaker one. So, these mistakes can often sabotage an otherwise solid document.

Most word-processing programs have a grammar checker, but these checkers are notoriously unreliable. So, use suggestions from the grammar checker cautiously, because your computer will often flag perfectly grammatical sentences. There is no substitute for knowing grammar rules completely.

TAKE NOTE Many grammar checkers will flag passive sentences as though these sentences are grammatically incorrect. They are not. It is fine to suggest that a passive sentence be turned into an active sentence, but the passive form is not grammatically wrong.

As you are proofreading, pay attention to your own reactions to the text. When you stumble or pause, chances are good that you have found a grammatical mistake. Mark these places, and then identify what grammatical mistake has been made. Figure 10.5 describes some of the more common grammatical errors.

Common Grammatical Errors

Error	Explanation
comma splice	Two or more distinct sentences are joined only by a comma.
run-on sentence	The sentence is composed of two or more distinct sentences.
fragment	The sentence is incomplete, usually missing a subject or verb.
dangling modifier	A modifier (usually an introductory phrase) implies a different subject than the one in the sentence's subject slot.
subject-verb disagreement	A singular or plural subject does not agree with the verb form.
misused apostrophe	An apostrophe is used where it doesn't belong (usually confusing it's and its).
misused comma	A comma signals an unnecessary pause in a sentence.
pronoun-antecedent disagreement	A pronoun does not agree with a noun earlier in the sentence.
faulty parallelism	A list of items in a sentence is not parallel in structure.
pronoun case error	The case of a pronoun is incorrect (usually due to confusion about when to use "I" or "me").
shifted tense	Sentences inconsistently use past, present, and future tenses.
vague pronoun	It is unclear what the pronoun refers to.

Figure 10.5: Here are the usual grammatical culprits. If you avoid these simple errors, you will have almost no grammatical problems.

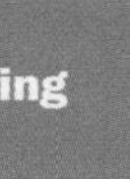

Want to improve your grammar? Go to **www.ablongman.com/johnsonweb/10.5**
For quick answers to grammar issues, go to **www.ablongman.com/johnsonweb/10.6**

In the Grammar and Punctuation Guide (Appendix A in the back of this book), you will find examples and remedies for these common errors. If you are not familiar with one or more of these errors, you should spend a little time learning how to avoid them.

LINK For more on grammar rules, go to Appendix A, page A-2.

In the end, most readers can usually figure out the meaning of a document, even when it has errors in grammar. But these errors make the author of the document look sloppy or unintelligent. After stumbling over a few errors, readers will begin to doubt the quality or soundness of the document and the information it contains. Even worse, they may doubt your abilities or commitment to quality. Readers expect correct grammar in technical documents.

Punctuation

Punctuation is designed to mirror the way we speak. For example, a *period* is supposed to reflect the amount of time (a period) that it takes to say one sentence. If you were to read a document out loud, the periods would signal places to breathe. Similarly, commas are used to signal a pause. When you come across a comma in a sentence, you pause slightly.

By understanding these physical characteristics of punctuation, you can learn how to use the marks properly (Figure 10.6).

Punctuation is intended to help readers understand your text. But, when the marks are misused they can create a great amount of confusion. The Grammar and Punctuation Guide (Appendix A) includes a more detailed explanation of punctuation usage.

LINK For a more detailed discussion of punctuation, go to Appendix A, page A-10.

Physical Characteristics of Punctuation

Punctuation Mark	Physical Characteristic
capitalization	signals a raised voice to indicate the beginning of a sentence or a proper name
period [.]	signals a complete stop after a statement
question mark [?]	signals a complete stop after a question
exclamation mark [!]	signals a complete stop after an outcry or objection
comma [,]	signals a pause in a sentence
semicolon [;]	signals a longer pause in a sentence and connects two related, complete statements
colon [:]	signals a complete stop but joins two equal statements; or, it indicates the beginning of a list
hyphen [-]	connects two or more words into a compound word
dash [—]	sets off a comment by the author that is an aside or interjection
apostrophe [']	signals possession, or the contraction of two words
quotation marks [" "]	signal a quotation, or when a word or phrase is being used in a "unique way"
parentheses [( )]	enclose supplemental information like an example or definition

Figure 10.6: Punctuation mirrors the physical characteristics of speech.

For short essays on grammar issues, go to **www.ablongman.com/johnsonweb/10.7**
For websites that discuss the history of punctuation marks, go to **www.ablongman.com/johnsonweb/10.8**

Spelling and Typos

Spelling errors and typos can be jarring. One or two spelling errors or typos in a document may be forgiven, but several errors will cause your readers to seriously question your commitment to quality. Here are some ways to avoid those errors.

Use the spell check feature on your computer—Most word-processing programs come with a spell checker that is rather reliable. Even if you are a good speller, the spell checker will often catch those annoying typos that inevitably find their way into texts (Figure 10.7).

Running the Spell Checker

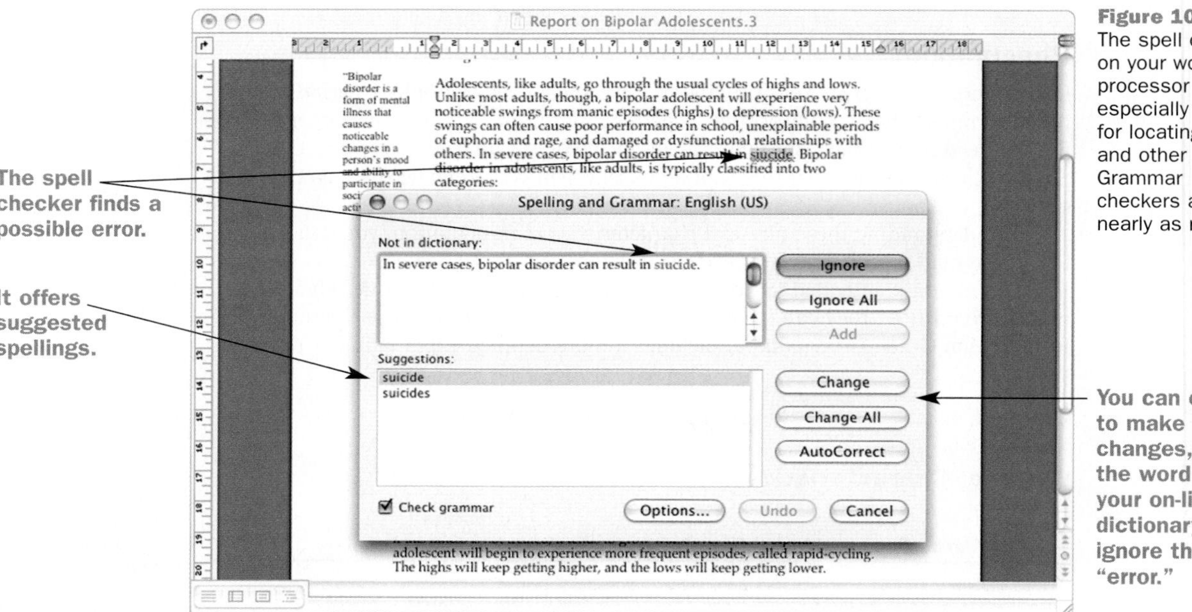

Figure 10.7: The spell checker on your word processor is especially helpful for locating typos and other errors. Grammar checkers are not nearly as reliable.

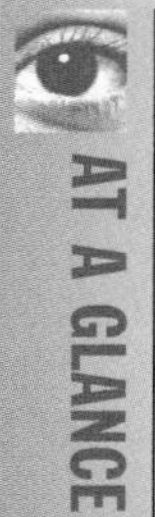

AT A GLANCE

Guidelines for Proofreading (Level 4)

- Grammar—Are all the sentences grammatically correct?
- Punctuation—Are the sentences properly punctuated?
- Spelling and Typos—Are there any spelling errors or typos?
- Word Usage—are all the words used properly?

However, a spell checker is not perfect, so you will need to still pay careful attention to spelling in your documents. After all, words may be spelled correctly, but they may not be the words you intended. Here is a sentence, for example, that has no errors according to a spell checker:

> Eye sad, they're our many places four us to sea friends and by good she's stakes in Philadelphia.

To avoid these embarrassing errors, don't rely exclusively on the spell checker.

GO TO THE NET

For help using your spell checker, go to **www.ablongman.com/johnsonweb/10.9**

Keep a dictionary close by—Technical documents often use terms and jargon that are not in your computer's spell checker. A dictionary is helpful in these cases for checking spelling and usage.

On the Internet, you may be able to find dictionaries that define specialized words in your field. There are, for example, numerous on-line dictionaries for engineering or medicine.

Overall, though, you should be quick to grab the dictionary or look up a word on-line when there are any doubts about the spelling of a word. The best editors are people who do not hesitate to look up a word, even when they are almost certain that they know how the word is spelled.

Word Usage

In English, many words seem the same, but they have subtle differences in usage that you should know (Figure 10.8). There are several good *usage guides* available in print that explain the subtle differences among words. More usage guides are becoming available over the Internet.

Common Usage Problems in Technical Documents

Confused Words	Explanation of Usage
accept, except	*accept* means to receive or agree to; *except* means to leave out
affect, effect	*affect* is usually used as a verb; *effect* is usually used as a noun
anyone, any one	*anyone* is a pronoun that refers to a person; *any one* means any one of a set
between, among	*between* is used for two entities; *among* is used for more than two entities
capitol, capital	*capitol* is the seat of a government; *capital* is money or goods
criterion, criteria	*criterion* is singular; *criteria* is plural
ensure, insure	*ensure* means to make certain; *insure* means to protect with insurance
complement, compliment	to *complement* is to complete something else or make it whole; a *compliment* is a kind word or encouragement
discreet, discrete	*discreet* means showing good judgment; *discrete* means separate
farther, further	*farther* refers to physical distance; *further* refers to time or degree
imply, infer	*imply* means to suggest indirectly; *infer* means to interpret or draw a conclusion
its, it's	*its* is always possessive; *it's* is always a contraction that means "it is." Its' is not a word
less, fewer	*less* refers to quantity; *fewer* refers to number
personal, personnel	*personal* refers to an individual characteristic; *personnel* refers to employees
phenomenon, phenomena	*phenomenon* is singular; *phenomena* is plural
precede, proceed	*precede* means to come before; *proceed* means to move forward
principle, principal	a *principle* is a firmly held belief or law; a *principal* is someone who runs a school
their, there, they're	*their* is a possessive pronoun; *there* is a place; *they're* is a contraction that means "they are"
whose, who's	*whose* is a possessive pronoun; *who's* is the contraction of "who is"
your, you're	*your* is a possessive pronoun; *you're* is a contraction that means "you are"
who, whom	*who* is a subject of a sentence; *whom* is used as the object of the sentence

Figure 10.8: You can avoid some of the most common usage problems in technical documents by consulting this list.

Want to try out some great on-line dictionaries? Go to **www.ablongman.com/johnsonweb/10.10**
Need to check the usage of a word? Go to **www.ablongman.com/johnsonweb/10.11**

If you are unsure about how a word should be used, look it up. Dictionaries are usually sufficient for answering usage questions. When a dictionary cannot answer your question, though, look up the word in an on-line or print usage guide.

Using Copyediting Symbols

While editing, you might find it helpful to use the same editing symbols as professional editors. To mark any stylistic changes and inconsistencies, editors have developed a somewhat universal set of copyediting marks (Figure 10.9). They are easy to use and widely understood. See Figure 10.10 for examples of how they are used.

Editing Symbols and Their Uses

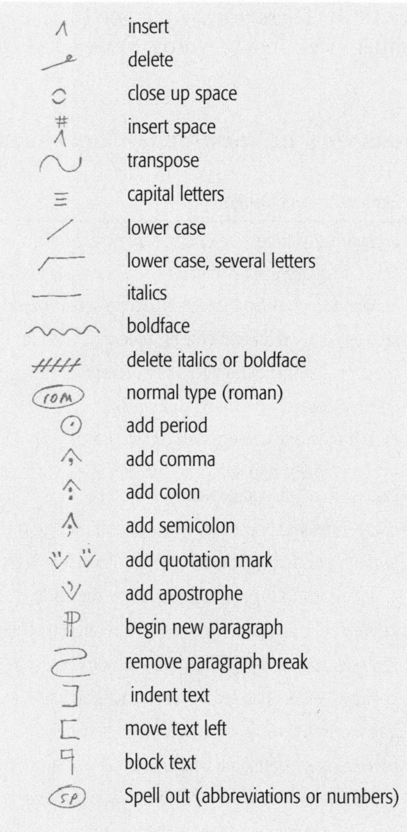

Figure 10.9: Copyediting symbols offer a standardized method for editors and writers to work together on a document.

GO TO THE NET

Want to learn more about using copyediting symbols? Go to **www.ablongman.com/johnsonweb/10.12**

Figure 10.10: Copyediting marks can be used to identify changes in the text.

Eruption History of Kilauea

Can you be more specific here? Are more accurate estimates available?

When Kilauea began to form is not known, but various estimates are 300,000-600000 years ago. The volcano had been active ever since, with no prolonged periods of quiescence known. geologic studies of surface exposures and examination of drillhole samples showthat Kilauea is made mostly of lava flows locally interbedded with deposites of explosive eruptions. Probably what we have seen hapen in the past 200 years is a good guide to what has happened ever since Kilauea emerged from the sea as an island perhaps 50,000-100,000 ago.

Lava Erupts from Kilauea's Summit and Rift Zones

Define the jargon in this paragraph.

Thrughout its history Kilauea has erupted from three main areas: its summit and two rift zones. Geologists debate whether Kilauea has Always had a caldera at the summit or whether it is a relatively recent feature of past few thousand years. It seems most likely that the caldera has come and gnoe throughout the life of Kilauea.

The summit of the volcano is high because eruptions are more frequent there than at any other single location on the volcano.

However, more eruptions actually occur on the long rift zones than in the summit area, but they are not localized instead construct ridges of lower elevation than the summit. Eruptions along the east and southwest rift zones have build ridges reaching outward from the summit some 125 KM and 35 KM respectively.

It's hard to figure out what you're describing here. A diagram would help.

Most eruption are relatively gentle, sending lava flows downslope from fountains a few meters to a few hundred meters high. OVER and over again these eruptions occur, gradually building up the volcano and giving it a gentle, shield-like form. Every few decades to centuries, however powerful explosions spread ejecta across the landscape. Such explosions can be lethal, as the one in 1790 that killed scores of people in a war party near the summitting of Kilauea. Such explosions can take place from either the bridging summit or the upper rift zones.

Source: U.S. Geological Survey, http://hvo.wr.usgs.gov/kilauea/history/main.html. Errors in text were added and did not appear in original.

HELP

On-line Copyediting

Professional editors still usually prefer to edit text on paper. On paper, it is easy to suggest changes and keep track of the version being edited. Increasingly, though, editing is being handled on-line, because it is easier to collaborate with other writers and cycle documents in electronic form. Also, people are becoming more comfortable with electronic documents, so they prefer to work on-line.

Here are some on-line editing techniques you can use.

Creating and using styles—As you edit the document, use the "Styles" feature of your word-processing program to define the formatting of headings, indented text, numbered and bulleted lists, examples, etc. At first, setting styles might slow your editing down, but as you move through the document, you will find that the preset styles will save you time in the long run.

Version control—Most word-processing programs will allow you to keep track of multiple versions of your document. Each version is automatically saved under a slightly different name. Then, you can use the "Versions" feature of your program to see past versions and the dates they were created. You can also use the "Save As" feature to save different versions. For example, with each major revision, you might change the name slightly (e.g., ProReport.1, ProReport.2, and so on).

Track changes—You can also track any changes automatically with the "Track Changes" feature. Your word processor will highlight various changes in a different color, allowing the original author or your coworkers to see the changes you made.

Copyediting symbols—With the "Font" feature of your computer, you can simulate some of the copyediting symbols used on paper (Figure A).

On-line Copyediting Symbols

Copyedit Action	On-line Mark	How to Do It
delete	~~delete this text~~	strike through
italicize	italicize this text	underline
bold	bold this text	underline wave
capitalize	capitalize this text	double underline
insert text	insert word	hidden text
insert text	insert more than one word	"Comments" feature

Figure A: Computers can simulate many of the copyediting symbols.

Hidden text—This feature allows you to make small insertions into text (Figure B). All word-processors will allow you to "hide" text. Hidden text appears on the screen, but it does not appear in the printed version of the document. Usually the "Hidden Text" feature is located in the same place as the "Bold" or "Italics" option. You simply highlight text you want hidden and select "Hidden."

Copyediting on the Screen

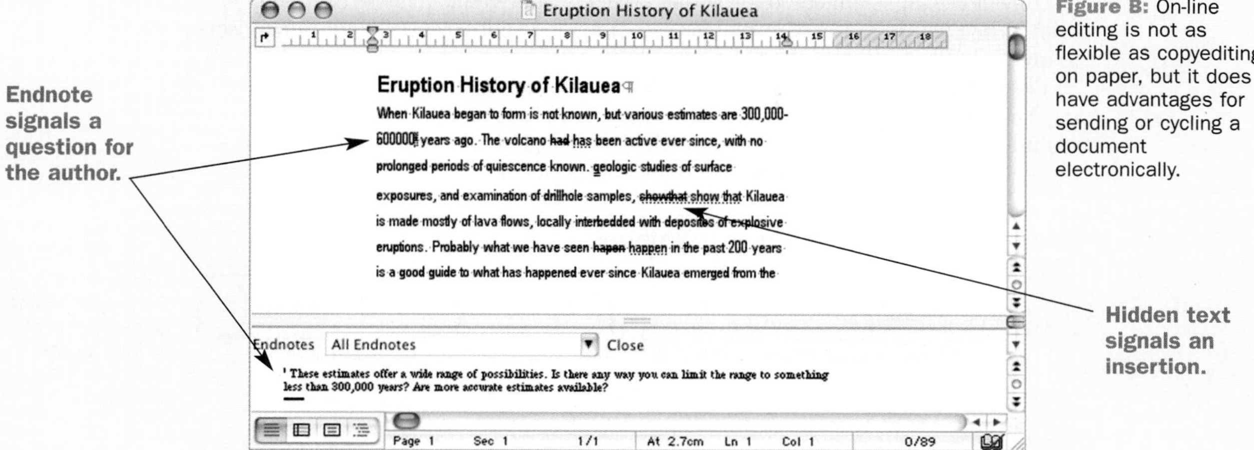

Figure B: On-line editing is not as flexible as copyediting on paper, but it does have advantages for sending or cycling a document electronically.

Comments—To signal that you want to insert more than a word or phrase, use the "Comments" feature in your word-processing program. This feature lets you add a hidden note, which does not appear in the printed text. When readers put their cursors over that highlighted text, the note pops up on the screen.

Endnotes and footnotes—Some on-line editors like to signal queries by using endnotes or footnotes. Use the "Insert Footnote" feature of your word processor. You can then type a note in the footnote window at the bottom of your screen.

Outlining view—The "Outline View" feature is helpful for gaining an overall sense of the structure of the text. As discussed in Chapter 6, you can use electronic outlining to help you organize the document into a logical structure.

LINK For more help on using "Outline View," see Chapter 6, page 139.

TAKE NOTE Depending on the amount of editing you need to do, you might use the "Customize" feature of your word-processor to put these editing features on your word processor's toolbar (the buttons on your screen). That way, as you are editing a document, each of these features will be available with the click of the mouse.

For more on-line editing tips, go to **www.ablongman.com/johnsonweb/10.13**

Document Cycling and Usability Testing

When you are completing your document, it is important that you gain an outside perspective. Often, while drafting, we become too close to our documents. Consequently, we can no longer edit or assess our own work objectively. Two ways to gain that outside perspective are *document cycling* and *usability testing.*

Document Cycling

Document cycling is a method for letting others at your company look over your draft. When you *cycle* a document, you pass it among your coworkers and supervisors to obtain feedback (Figure 10.11).

Computers give you the ability to quickly send your document around to others for their suggestions for improvement. You can send it as an attachment or make multiple paper copies. Then, your supervisors, colleagues, and even your primary readers can look it over and offer ideas for improvement.

Document Cycling in the Workplace

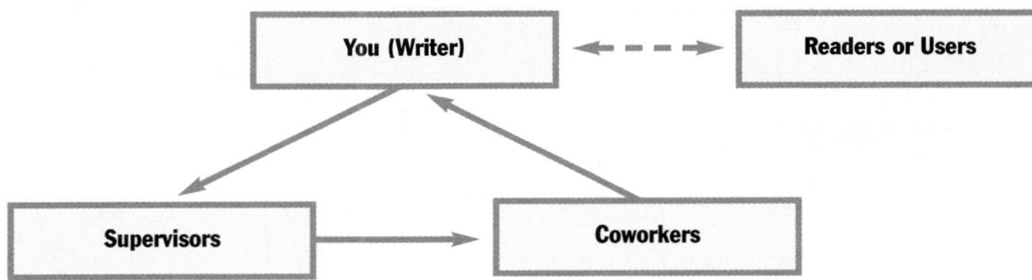

Figure 10.11: Document cycling gathers feedback by letting others look over the document.

When revising and editing your documents, it is important to let others look over your work. Document cycling is an important part of a *quality feedback loop,* a central principle of quality management. If you rely on yourself alone to edit your work, the quality of your document might be low.

TAKE NOTE The more people you bring into your quality feedback loop, the more chances you will have to improve your document.

Usability Testing

Usability testing means trying out your document on real readers. This kind of testing on real readers can be informal or formal, depending on the importance of your document and the time you have to test it.

Informal usability testing—Minimally, you should ask other people to look over the document, marking places where they stumble or find it difficult to understand its meaning.

Formal usability testing—More formally, you can run experiments on sample readers to measure how well they can understand and use your document. Instructions and user manuals are commonly user tested in this way. But, increasingly, focus groups of readers are being used to test persuasive documents like proposals. In some cases, companies will create *usability laboratories* to observe the reactions of readers as they use documents.

The more your test subjects are like your target readers, the better the results of your testing will be. In other words, the folks around the office are fine for an informal test. But, if you want a more accurate assessment of how your document will be used, you should try to locate a group of people who are most like your readers. Figure 10.12 lists some of the more common methods for testing usability, ranking them from informal methods to formal methods.

Types of Usability Testing

	Usability Test	How it is Conducted
informal testing	document markup	Readers are asked to read through a document, marking places where they stumble or fail to understand.
↑	read and locate test	Readers are asked to locate specific kinds of information in a document. They are timed and videotaped.
	summary test	Readers are asked to summarize the important information in a document.
	protocols	Readers are asked to talk out loud as they are using the text. Their comments are taped and transcribed.
	journal or tape recording	Readers are asked to keep a written or taped journal at their workplace to record their experiences with the document.
	surveying	Readers are given a questionnaire after they use the document, asking them about their experience.
	interviewing	Readers are interviewed about their experiences using a document.
↓	focus groups	Groups of readers look over a document and discuss their reactions to the work.
formal testing	laboratory testing	Through cameras and a one-way mirror, readers are carefully observed using a text.

Figure 10.12: There is a variety of different ways to user-test a document.

Most usability testing is designed to answer four questions (Figure 10.13):

Can they find it?—*Read-and-locate tests* are used to determine whether users can locate important parts of the document and how quickly they can do so. Often, the users are videotaped and timed while using the document.

Can they understand it?—*Understandability tests* are used to determine if the users retain important concepts and remember key terms. Users are often asked to summarize parts of the document or define concepts.

GO TO THE NET

To learn more about usability testing, go to **www.ablongman.com/johnsonweb/10.14**

Usability Questions

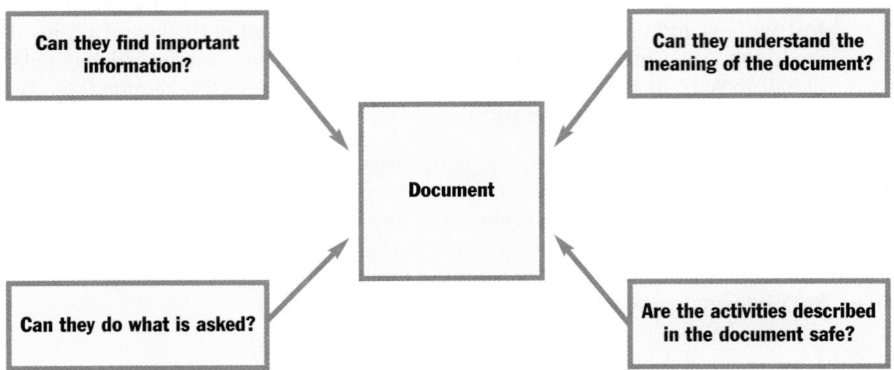

Figure 10.13: Usability testing is designed to answer four simple questions.

Can they do it?—*Performance tests* are used to determine whether users can perform the actions the document describes. These tests are often used with instructions and other kinds of procedures.

Is it safe?—*Safety tests* are used to study whether the activities described in the document, especially in instructions or user manuals, are safe. These tests carefully watch for possible safety problems by having sample readers use the product documentation.

As you devise a usability test or series of usability tests, you should set quantifiable objectives that will allow you to measure *normal* and *minimal* user performance with the document. In other words, first define what you would expect *typical* users to be able to accomplish with your document. Then, define what you would *minimally* expect them to be able to do.

READ-AND-LOCATE TESTS: CAN THEY FIND IT? To run a read-and-locate test, list five to seven important pieces of information that you want readers to locate in the document. Then, while timing and/or videotaping, see how long it takes them to find that information.

Videotaping your subjects is especially helpful, because you can observe how readers will go about accessing the information in your document (Figure 10.14). Do they go right to the beginning or the middle? Do they flip through the text looking at the headings or graphics? Do they look at the table of contents or index (if these features exist)?

After your subjects locate the major pieces of information you asked for, have them tell you about these major points orally or in writing. Then, check their answers against your original list. If they successfully found four or five items easily, your document is likely well-written and well-designed. If, however, they struggled to find even a few of your major points, you probably need to revise your document to ensure that the important information is easy to locate.

Usability-Testing a Document

Figure 10.14: To test the usability of a document, you can run experiments on people who represent real readers. Videotaping is an especially good way to see how people actually use a document.

UNDERSTANDABILITY TESTS: CAN THEY UNDERSTAND IT? When running an understandability test, you want to determine how well the users of your document grasped its meaning. Before running the test, write down your document's purpose and main point. Then, write down three important concepts or points that anyone should retain after reading the document.

Give your readers a limited amount of time to read through the document or use it to perform a task. Then have them put the document away, so they cannot use it. Orally or in writing, ask them—

- What is the purpose of this document?
- What is the document's main point?
- Can you tell me three major points that are made in the document?

If their answers to these questions are similar to the ones you wrote down, your document is likely understandable. If, however, your readers struggle to answer these questions or even get them wrong, you should think seriously about revising the document to highlight the information you intended them to retain.

PERFORMANCE TESTS: CAN THEY DO IT? Almost all technical documents are written to help readers take some kind of action. A set of instructions, obviously, asks readers to follow a procedure. A report might make some recommendations for change.

To do a performance test, have the users perform the procedure the document describes. Or, ask them to react to your recommendations. Here again, videotaping the users is a good way to keep a record of what happened. Did they seem to find the document easy to use? Where did they stumble or show frustration? When did they react positively or negatively to the tasks or ideas described in the document?

Ultimately, performance tests are designed to find out whether the users can do what the document asks of them. But it is also important to determine their attitude toward performing these tasks. You want to ensure not only that they *can* do it, but also that they *will* do it.

SAFETY TESTS: IS IT SAFE? Above all, you want your documentation and products to be safe. It is impossible to reduce all risk of injury, but you should try to reduce the risk as much as possible. Today, it is common for companies to be sued when their documentation or products are shown to be inadequate. Often, in product liability lawsuits, documents like instructions and user manuals are used to prove or deny negligence for an injury.

Without putting test subjects at risk themselves, safety tests are usually designed to locate places where users may make potentially injurious mistakes. They also ask readers about the warnings and cautions in the document to determine whether these features were noticed and understood.

Setting Objectives and Measuring Results

The challenge to effective usability testing is to first identify some objectives for the document. These objectives could refer to (a) how well the users can find information, (b) how well they understand important ideas, and (c) how well they perform tasks described in the document. Then, measure the results of your usability testing against these objectives.

It's often quite sobering as you watch people fumble around with your document, misunderstand its meaning, and not follow the directions. But, the results of your tests should help you revise the document to improve its usability.

No form of usability testing will ensure that your document is a success. However, feedback from users is usually the best way to gain new insights into your document and solicit suggestions for improvement.

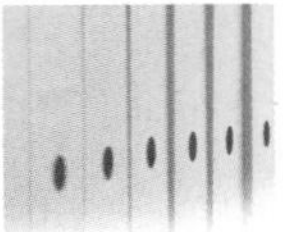

CHAPTER REVIEW

- Revising and editing are forms of quality control that should be a regular part of your writing process.
- Documents and presentations can be edited at four different levels: revision, substantive editing, copyediting, and proofreading.
- Editorial tools like copyediting marks are helpful even for noneditors.
- Document cycling is a process of circulating your text among colleagues and your supervisor. You can use e-mail attachments to send your work out for review by others.
- Usability testing can offer informal or formal methods to test the effectiveness of your document. You might use sample readers to test documents that will be used by a broad readership.

EXERCISES AND PROJECTS

Individual or Team Projects

1. Find a document on campus or at your workplace that needs editing. Edit the document by working backward from a level 4 edit (proofreading) to a level 1 edit (revision). As you apply each level of edit to the document, pay attention to the different kinds of actions and decisions you make at each level. How is the document evolving as you edit it?

2. Exchange a text with a member of your class. Then, do a level 2 edit (substantive editing) of the draft, using copyediting symbols to reflect the changes you think should be made to the document. Write a cover letter to the author in which you explain the changes you want made. Hand in a copy of this documentation to your instructor.

3. Find a text on the Internet that needs to be edited. Do a level 3 edit (copyediting) of the text, using the copyediting marks shown in this chapter. Write a memo to your instructor in which you discuss how you edited the text. Discuss some of the places where you struggled to mark the changes you wanted to make.

4. Find a text that needs editing and use the on-line editing strategies discussed in this chapter to edit it. In an e-mail to your instructor (with edited version attached), discuss some of the differences and difficulties between paper-based and on-line editing. Discuss which form of editing you prefer and why.

Collaborative Project

With your group, locate a longer document on the Internet that needs editing. The document might be a report or proposal, or perhaps even a website. Then, do a level 2 edit (substantive edit) on the document, including the two levels of edit below it (copyediting and proofreading).

As you are editing, set up a document cycling routine that allows you to send your document among your group members. You can cycle the document in a paper version or use on-line editorial strategies to keep track of versions and changes in the document.

When your group has finished editing the document, revise it into a final form. Then, conduct usability tests on the document, using other members of your class as subjects.

In a memo to your instructor, discuss the evolution of the document at each stage of the editorial process. Tell your instructor (1) how you used the substantive edit to improve the document, (2) how you cycled the document among your group, and (3) how you used usability testing to identify places where the document might be improved.

Longer documents you can use for these exercises are available at **www.ablongman.com/johnsonweb/10.15**

Just My (Bad) Style

Gerald Williams was a manager for an electrical engineering company. The company was growing rather quickly. Only a couple years out of college, he found himself a project manager in charge of a new microelectronics design team.

The first thing he did was hire some new members for his team. He was fortunate to find some of the best and brightest electrical engineers available in his area.

He had one problem, though. One of the employees he hired, Darren Yander, was an intense guy who did not handle criticism well. He was one of those people who believed he did everything right.

Unfortunately, he also was a mediocre writer. His documents were hard to understand, and they were loaded with grammatical errors and typos. Gerald spoke with Darren about his writing skills on a couple occasions. He told Darren that he needed to spend more time revising and editing his drafts.

Darren would usually dismiss Gerald's concerns by saying he had a unique "style" that worked well for him. Darren said, "All you want is for me to write like you. Well, I have a style of my own. I don't think my creativity should be smothered with some boring bureaucratic style." Gerald liked Darren, so he didn't take offense.

Gerald decided that arguing with Darren about revising and editing his documents wasn't worth it. Almost all of Darren's documents were used exclusively within their team, so it didn't seem to matter.

Unfortunately, one of Darren's documents did go beyond the design team. One of his reports found its way into the hands of Margie Washington, one of the vice presidents of the company. Margie came to Gerald's office and asked him about an "odd" report that was written by someone in Gerald's group. Gerald immediately recognized that the report was written by Darren in his "style." The report was difficult to read and was very convoluted. It was also littered with grammatical errors and typos.

Gerald explained to Margie that he and Darren were working on improving his writing skills. Margie said, "You better do something about this problem—quick. I don't care if the guy is bright and hardworking. If another report like this one is sent to upper management, they're going to start looking for someone to fire. You're the boss. Do something."

After Margie left, Gerald called Darren to his office and explained the problem with the report. Darren grew very defensive. "It's not my fault upper management is too stupid to know good writing when they see it." He refused to make revisions to the report. He also refused to work on improving his writing skills.

If you were Gerald, what would you do in this situation? How would you handle it in a way that allows all sides to be satisfied?

CHAPTER

11

Preparing and Giving Presentations

CHAPTER CONTENTS

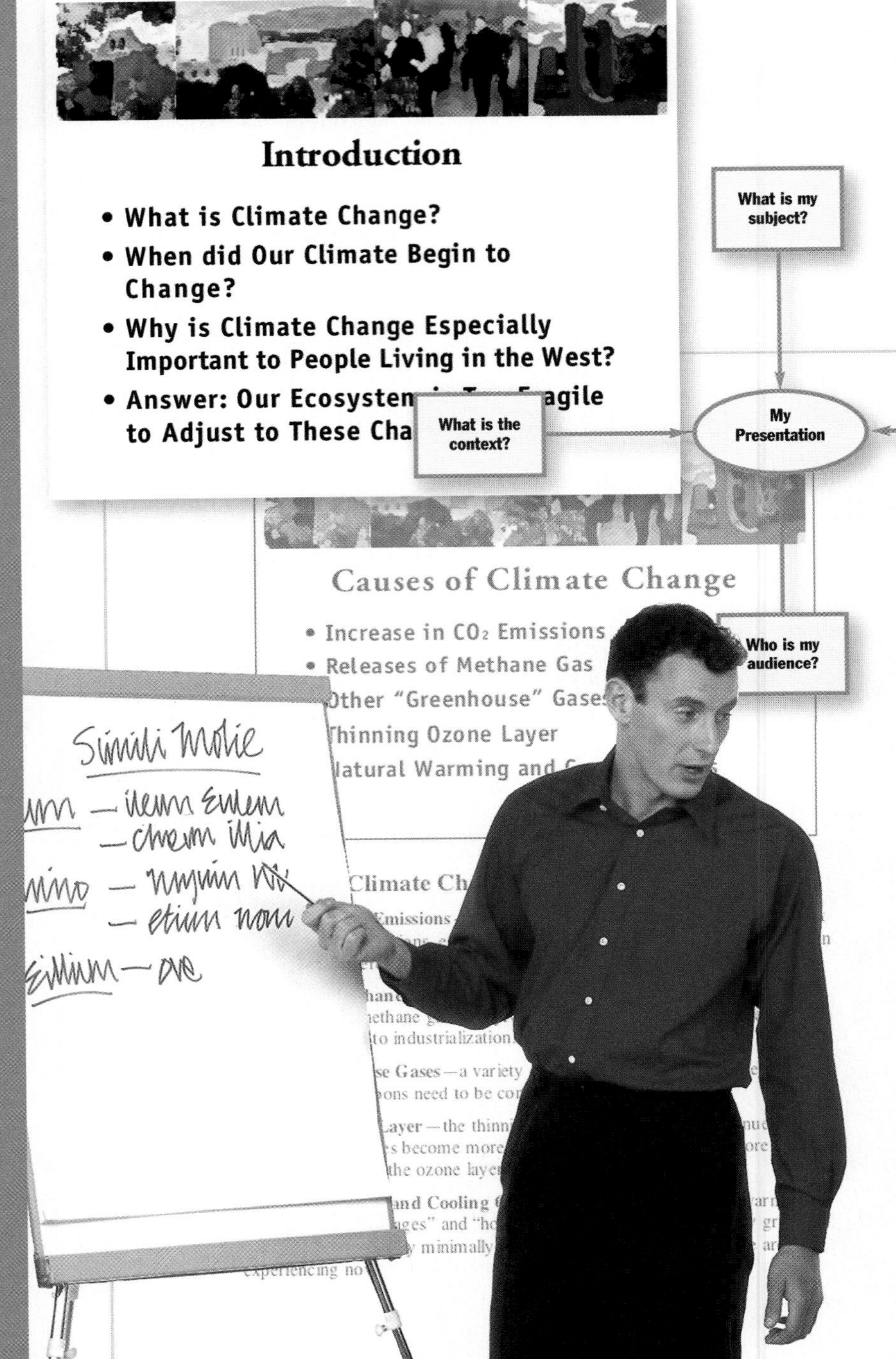

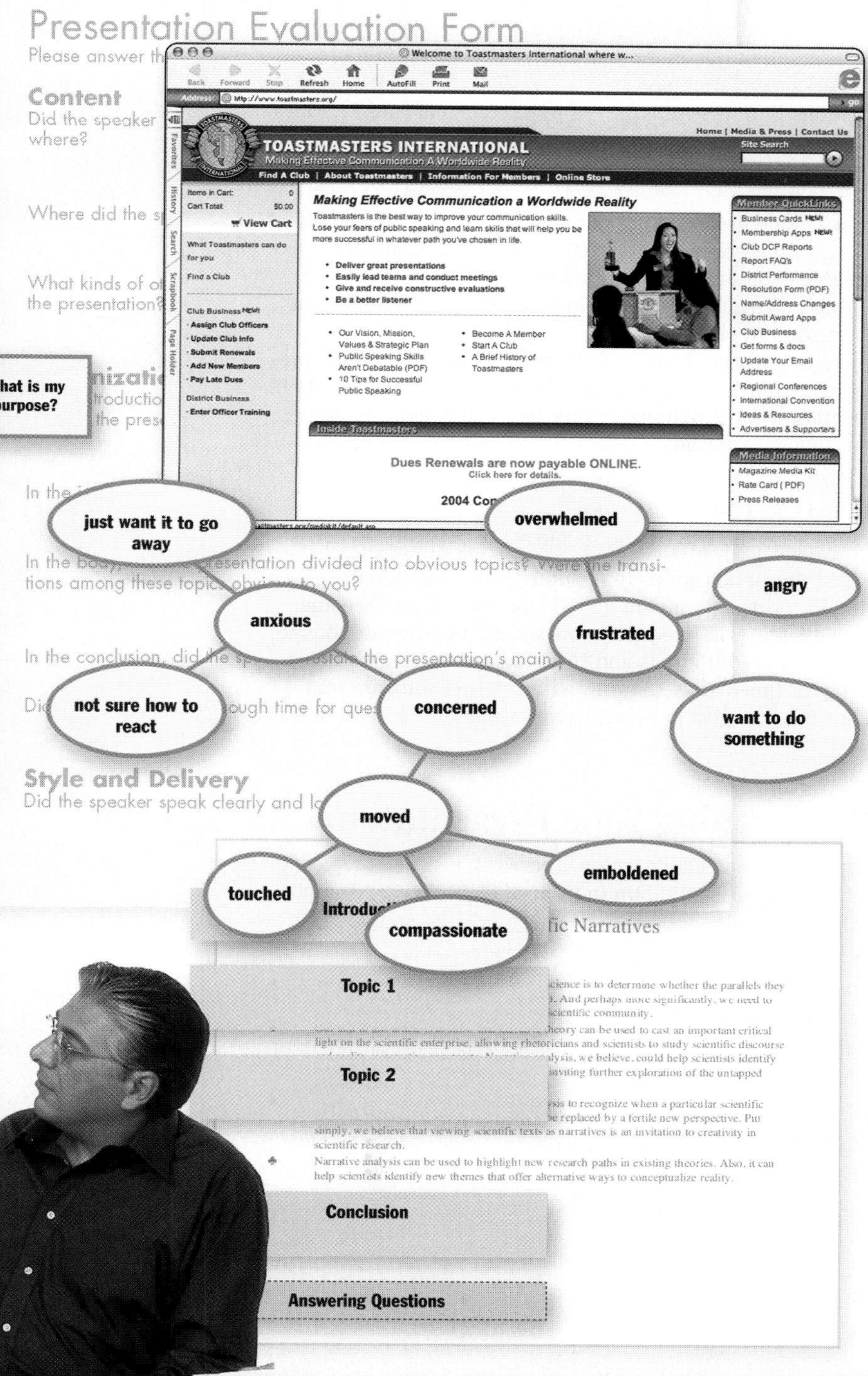

CHAPTER OBJECTIVES

In this chapter, you will learn:

- The importance of being able to prepare and deliver public presentations.
- How to define the rhetorical situation (subject, purpose, audience, and context of use) for your presentation.
- Strategies for organizing the content of presentations.
- How to create an effective presenting style.
- How to create and use visuals in presentations.
- The importance of practice and rehearsal.

If you don't like giving public presentations, you are not alone. Each year, surveys show that people fear speaking in public more than anything else—even more than death.

Yet giving public presentations is an essential part of most technical careers. More than likely, you will find yourself regularly giving presentations to clients, supervisors, and colleagues. Presenting information in public is a crucial skill in today's technical workplace.

Fortunately, computers have made presenting in public a bit easier. With the aid of computers, you can develop a polished presentation with professional visual aids that help reinforce your message. Presentation software like MS PowerPoint, Corel Presentations, and Serious Magic's Visual Communicator can help you organize and design your information for maximum effect.

Keep in mind that computers have only increased the need for making public presentations. In today's computer-centered workplace, people are more visual, and they often prefer interactive presentations over written documents. As a result, public presentations are more common than ever because

- they are more visual than print documents.
- they require less time and effort for the audience.
- they allow people to interact directly with the speakers.

In fact, the need to present information publicly will likely only increase as computer-based technologies, like video streaming and videoconferencing, become regular features of the workplace. You will likely find yourself giving presentations to a digital camera for audiences that are hundreds and thousands of miles away. You might also find yourself giving "distance" presentations with Internet software that uses a wired whiteboard. The ability to give presentations is an important skill in the technical workplace.

Planning and Researching Your Presentation

The preparation of a public presentation should follow a process, just like preparing written documents (Figure 11.1). Before getting up in front of an audience, you will need to

- plan and research your subject.
- organize your ideas.
- choose an appropriate presentation style.
- create graphics and slides.
- practice and rehearse your presentation.

Even if you are already an accomplished public speaker, it is important that you spend significant time on each stage of the process. After all, a presentation cannot succeed on content alone. You need to pay attention to issues of organization, style, design, and delivery.

The Preparation Process for a Presentation

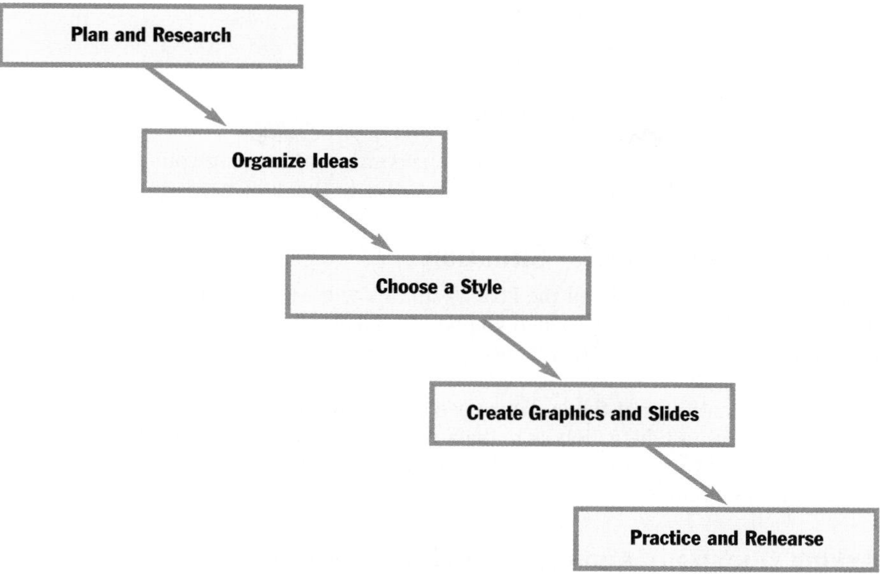

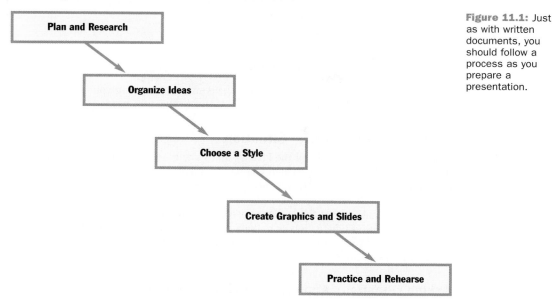

Figure 11.1: Just as with written documents, you should follow a process as you prepare a presentation.

As you begin planning your talk, remember that presentations can be either formal or informal:

> **Formal presentations**—These presentations often include the use of a podium, speaking notes, and slides made with presentation software. Formal presentations include speeches, workshops, trainings, briefings, demonstrations, and panel discussions. They are made to clients, management, and colleagues.
>
> **Informal presentations**—Most presentations at work are informal. At monthly meetings, you will be asked to report on your team's progress. If you have a new idea, you will need to pitch it to your boss. And, if your supervisor stops by and says, "Hey, in 10 minutes, could you come by my office to tell the vice president how the project is going?" you are about to make an informal public presentation.

In both formal and informal presentations, good planning is the key to success. A good way to start the planning phase is to analyze the rhetorical situation of your presentation. Begin by asking some strategic questions about your subject, purpose, audience, and the context in which you will be giving your talk (Figure 11.2).

The Five-W and How Questions give you a good place to start.

> *Who will be in my audience?*
>
> *What kind of information do they need or want?*

For websites that discuss giving presentations, go to **www.ablongman.com/johnsonweb/11.1**

Why am I presenting this information to this audience?

Where will I be presenting?

When will I need to give my talk?

How should I present this information?

Your answers to these questions should help you start crafting your materials into something that will be interesting and informative to the audience.

Defining the Rhetorical Situation

Now that you have considered the Five-W and How questions, you can think a little deeper about the situation in which you will be speaking (Figure 11.2). Consider closely your subject, purpose, audience, and the context of use.

SUBJECT Identify what the audience needs to know, putting emphasis on information they require to take action or make a decision. You might also ask yourself what the audience does *not* need to know. That way, you can keep your presentation concise and to the point.

Asking Questions About the Rhetorical Situation

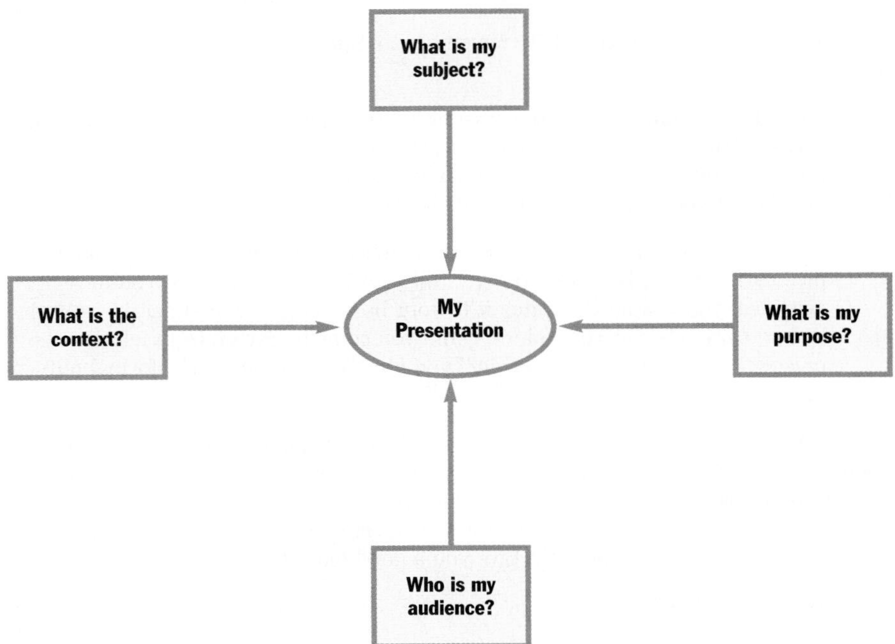

Figure 11.2: Asking questions about the subject, purpose, audience, and context will help you anticipate the information you need and how that information should be presented.

GO TO THE NET To learn more about analyzing audiences, go to **www.ablongman.com/johnsonweb/11.2**

PURPOSE You need to know exactly what you want to achieve in your presentation. Try to state your purpose in one sentence.

> My goal is to persuade elected officials that global warming is a looming problem for our state.
>
> We need to demonstrate the G290 Robot Workstation to the CEO of Geocom Industries, showing her how it can be used to clean up toxic spills.
>
> I need to motivate our technical staff to improve quality, so we can meet the standards demanded by our new client.

If you need more than one sentence to express your purpose, you are probably trying to do too much in your presentation.

LINK For more help defining your purpose, go to Chapter 2, page 23.

AUDIENCE Members of your audience will come to your presentation with various needs, values, and attitudes. You should anticipate these characteristics and shape your presentation to their specific requirements and interests.

Primary audience (action takers)—For most presentations, the primary audience is the most important, because they are the people who will be making a decision or taking action.

Secondary audience (advisors)—These members of the audience might advise the primary audience on what actions to take. They might be experts in your field or people who have information or opinions on your subject.

Tertiary audience (evaluators)—Others in the audience may have an interest in what you are saying. They might be journalists, lawyers, activists, or concerned citizens.

Gatekeepers (supervisors)—Your supervisors and others at your company will often need to see your presentation before you give it to an audience. They will be looking for accuracy and checking whether you are achieving your purpose and fulfilling the mission of the company.

LINK For more audience analysis techniques, go to Chapter 3, page 44.

CONTEXT OF USE Context is always important in technical communication, but it is especially important in public presentations. You need to be fully aware of the physical, economic, ethical, and political factors that will shape your presentation and how your audience will react to it.

Physical context—Take time to familiarize yourself with the room in which you are speaking and the equipment you will be using. You will need to adjust your presentation, including your visuals, to the size, shape, and arrangement of the room. Also, find out what kind of furniture and equipment you will have available.

Will you be using a podium or lectern?

Will you be sitting behind a table or standing out in the open?

For worksheets to help you analyze your audience and the context of the document's use, go to **www.ablongman.com/johnsonweb/11.3**

Is there a microphone?

Is a projector available, and are you able to use it?

Do you need to bring your own projector and computer?

Are other visual aids like whiteboards, flip charts, and large notepads available?

Will the audience be eating, and are drinks available?

When will the audience need breaks?

It is astonishing how many speakers fail because they were not prepared for the physical characteristics of the room. They show up with slides that can't be read because the room is too large, or they try to talk to a large room with no public address system.

These sorts of problems might be someone else's fault, but it's *your* presentation. You will be the person who looks unprepared. Proper preparation for the physical context will help you avoid these problems.

Economic context—As always, money will be a central concern for your audience. So, consider the microeconomic and macroeconomic factors that will influence how they receive your presentation. Microeconomic issues might include budgetary concerns or constraints. Macroeconomic issues might include your audience's economic status, economic trends in the industry, or the state of the local or national economy.

Ethical context—Presentations almost always touch on ethical issues in one way or another. As you prepare, identify and consider any rights, laws, or issues of common concern that might shape your presentation.

LINK For more information on ethics, go to Chapter 4, page 70.

Political context—Politics also play a role in presentations. In some cases, politics might simply involve the usual office politics that shape how people react to you and your subject. In other situations, larger political issues may come into play. National issues (energy policies, conservation, gun rights, privacy, security, etc.) will evoke different political responses among members of your audience.

LINK To learn more about analyzing context of use, go to Chapter 3, page 53.

Allotting Your Time

As the speaker, you have an unstated contract with the audience. According to this contract, your audience is allowing you a specific number of minutes. It's your responsibility to fill that time productively—*and not go over*. Nothing annoys an audience more than a speaker who runs past the time allotted.

So, as you are planning your presentation, first determine how much total time you have to speak. Then, *scale* your presentation to fit the time you have allotted. Figure 11.3 shows how a few common time periods might be properly budgeted.

For more time-management strategies, see **www.ablongman.com/johnsonweb/11.4**

Allotting Your Presentation Time

	15-Minute Presentation	30-Minute Presentation	45-Minute Presentation	One-Hour Presentation
Introduction	1 minute	1–2 minutes	2 minutes	2–3 minutes
Topic 1	2 minutes	5 minutes	8 minutes	10 minutes
Topic 2	2 minutes	5 minutes	8 minutes	10 minutes
Topic 3	2 minutes	5 minutes	8 minutes	10 minutes
Topic 4	2 minutes	5 minutes	8 minutes	10 minutes
Conclusion	1 minute	1–2 minutes	2 minutes	2 minutes
Questions	2 minutes	5 minutes	5 minutes	10 minutes

Figure 11.3: When planning your presentation, carve up your time carefully to avoid going over.

Of course, if you have fewer or more than four topics, you should make adjustments to the times allowed for each. Also, you might need to spend more time on one topic than another. If so, adjust your times accordingly.

There are two things you should notice about the times listed in Figure 11.3.

- Longer presentations do not necessarily allow you to include substantially longer introductions and conclusions. No matter how long your presentation is scheduled to run, keep your introductions and conclusions concise.
- You should not budget all the time available. Always leave yourself some extra time in case something happens during your talk or you are interrupted.

In the end, the contract with your audience is that you will finish within the time scheduled. If you finish a few minutes early, you won't hear any complaints. However, if you run late, your audience will not be pleased.

Choosing the Right Presentation Technology

As you plan your presentation, it is a good idea to think about what presentation technology you will use for your talk. Will you use presentation software with a digital projector? Are you going to use a whiteboard? Are you going to make transparencies for an overhead projector? The kind of presentation technology you need depends on the type of presentation you are making (Figure 11.4).

Fortunately, presentation software like PowerPoint, Presentations, and Visual Communicator make it easy to create slides and use graphics in your presentations. These programs will help you create visually interesting presentations for a variety of situations.

Each kind of visual aid offers specific advantages and disadvantages. Here are some pros and cons of the more common types of visuals:

To find a selection of presentation software packages, go to **www.ablongman.com/johnsonweb/11.5**

Presentation Technologies

Type of presentation	Visuals
Presentation to a group of more than 10 people	Digital projector with computer
	Overhead projector with transparencies
	35mm-slide projector
	Whiteboard or chalkboard
Presentation to a group of fewer than 10 people	Digital projector with computer
	Overhead projector with transparencies
	35-mm slide projector
	Flip charts
	Large notepads
	Digital video on TV monitor (DVD or CD-ROM)
	Posters
	Handouts
	Computer screen
	Whiteboard or chalkboard

Figure 11.4: There are many different ways to present materials. You should choose the one that best fits your subject and audience.

Digital projector with computer—Most companies have a digital projector available for your use (Figure 11.5). The projector can display the slides from your computer screen on a large screen. The advantage of digital projectors is their ease of use and the ability to create highly attractive, colorful presentations. The disadvantage is that the projected slides often dominate the room, because the lights need to be turned down. As a result, the audience can become fixated on the slides and stop listening to what you need to say.

Overhead projector with transparencies—The overhead projector is the tried-and-true method for giving presentations. You can use an overhead projector to project slides on a large screen. Transparencies for the projector can be made on a paper copier. The advantage of overhead projectors is that they are commonly available in most workplaces and are more reliable than digital projectors. The disadvantage is that presentations using overheads often seem more static and lifeless than ones made with digital projectors. The colors are not as sharp, and the pictures can be blurry.

Whiteboard, chalkboard, or large pad of paper—People often forget about the possibility of using the whiteboard, chalkboard, or large pad of paper in a room. But if you are planning a presentation that requires interaction with the audience, you can use these items to make visuals on the fly. The advantage is that you can create your visuals in front of the audience. Your listeners won't feel like they are receiving a canned presentation in which they have little input. The disadvantage is that you need to think on your feet. You need

Using a Digital Projector

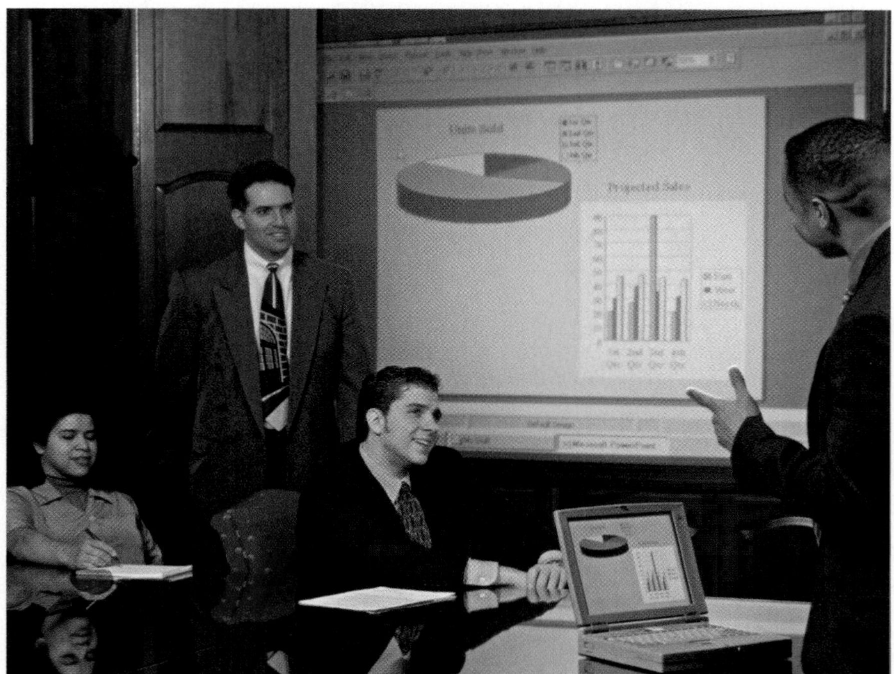

Figure 11.5: Digital projectors are increasingly common. They project your computer screen onto a large screen.

to find ways to translate your and the audience's comments into visuals on the board.

Flip charts—For small, more personal presentations, a flip chart is a helpful tool (Figure 11.6). As the speaker talks, he or she flips a new page forward or behind with each new topic. The advantage of flip charts is their closeness with the audience. You can give a flip chart presentation to a small group. The disadvantage of flip charts is that they are too small to be seen at a distance. If you have more than a handful of people in the audience, the flip chart won't work.

Posters—In some cases, you might be asked to present a poster. A poster usually includes about five to seven slides that describe a product or show a procedure. They are often used to summarize an experiment. The advantage of a poster is that everything covered in the talk is visually available to the audience. In some cases, the posters might be left alone on a display or a wall, so readers can inspect them on their own time. Posters, however, have the same disadvantages as flip charts. They can be used only in presentations to a handful of people. They are too hard to see from a distance.

GO TO THE NET

Want to know more about presentation technologies? Go to **www.ablongman.com/johnsonweb/11.6**

Using a Flip Chart

Figure 11.6: Flip charts are low-tech, but they are especially effective in presentations to a few people.

Handouts—Handouts can be helpful in some cases, but they are not always appropriate. When used properly, they can reinforce points made in your presentation or provide data that won't be visible with a projector. Also, handouts made with presentation software can be formatted to leave room for note taking. Handouts, though, can also be very distracting. In a large room, handouts can take a few minutes to be passed around, causing the speaker to lose momentum. The speaker might also need to spend a few minutes making sure everyone has a copy. Meanwhile, people in the audience are often distracted by the handouts, looking at them rather than at the presentation.

These kinds of technology decisions are best made up front—while you are planning—because your choice of technology will often shape your decisions about the content, organization, style, and design/delivery of your information. For some presentations, you might have the freedom to make choices about the technology you will use. In other cases, you may be restricted to using one kind of format (e.g., when only an overhead projector is available). You will need to adjust to these constraints.

GO TO THE NET

To see sample poster presentations, go to **www.ablongman.com/johnsonweb/11.7**

AT WORK

Karen Paone

OWNER AND LEAD CONSULTANT, PAONE & ASSOCIATES, ALBUQUERQUE, NEW MEXICO

Paone and Associates is a communication consulting firm that trains employees to make effective presentations.

How can I overcome my fear of speaking in public?

The fear of public speaking routinely ranks ahead of death among the top three fears of most adults. Indeed, there's nothing quite as horrifying as the rush of adrenaline, the rising sense of panic in the pit of your stomach, increased heart rate, dry mouth, and complete loss of memory that assault you the very moment you realize your turn at the podium is imminent.

So what's the difference between these terrifying sensations and the energy, enthusiasm, and engaging style of a polished public speaker? In fact, there is very little difference between the physical reaction of an inexperienced individual presenting at a conference for the first time and an engaged, seasoned presenter. Both experience a heightened sense of awareness and both feel a strong sensation in the solar plexus. Why is one likely to succeed and the other to fail at presenting the information at hand?

- **Rehearsal**—The seasoned speaker creates a realistic rehearsal situation in which she either enlists a practice audience or visualizes her anticipated audience. She knows that simply thinking about the presentation is insufficient to prepare for a real audience. She stands in front of a practice audience or a mirror and runs the entire presentation at least three times. She reworks the rough spots until the entire presentation flows smoothly, anticipates questions, and practices responding with the answers.
- **Heightened awareness**—The seasoned presenter recognizes that the rush of adrenaline is very normal, and that a heightened sense of awareness is essential for an interesting presentation. She uses deep breathing, stretching, and smiling to focus that energy into a confident and enthusiastic stage presence.
- **Communication and connection**—Finally, the seasoned presenter knows that the secret to managing stage fright is focusing on communicating the content at hand in a sincere and personal way to each and every member of the audience. By focusing on connecting with each listener, the presenter's concern is with the audience rather than the way she or he looks, sounds, or feels.

The prescription for stage fright is to rehearse until you can deliver your presentation smoothly from start to finish. Recognize that the physical response to fear and excitement is essentially the same. What differs is how you use those sensations. And finally, when all of your energy is focused on connecting with your audience rather than observing yourself—you, too, will be perceived as a polished and successful public speaker.

Organizing the Content of Your Presentation

One problem with organizing public presentations is that you usually end up collecting more information than you can talk about in the time allowed. Of course, you cannot tell the audience everything, so you need to make some hard decisions about what they need to know and how you should organize that information.

LINK To help you collect content for your presentation, see Chapter 5, page 96.

Keep your purpose and audience foremost in your mind as you make decisions about what kind of content you will put in the presentation (Figure 11.7). You want to include only need-to-know information, cutting out any want-to-tell information that is not relevant to your purpose or audience. As you make decisions about what to include or cut, you should keep the following saying in mind: *The more you say, the more they will forget.*

Audiences rarely complain about short presentations. As with wedding ceremonies, the longer your presentation, the less your listeners are going to pay attention. If you go on too long, they will simply stop listening altogether.

Remember Your Audience

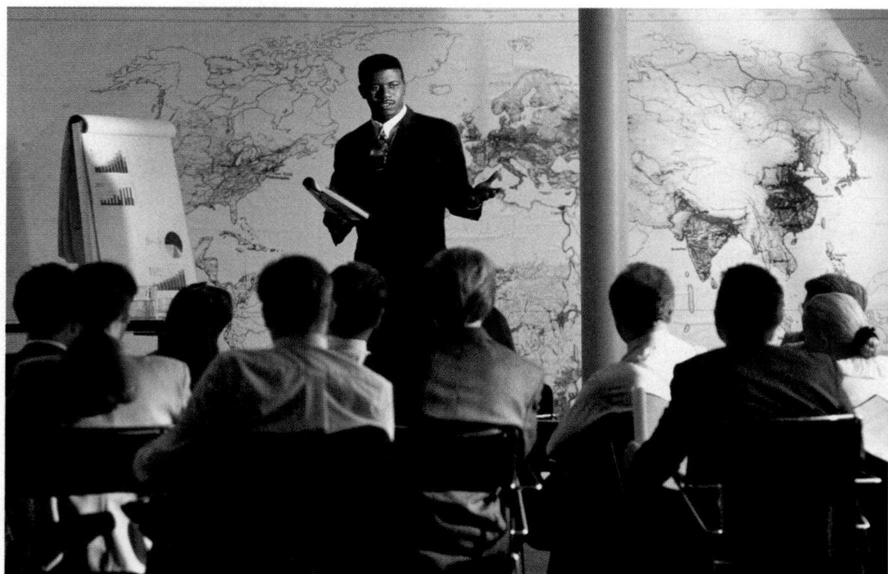

Figure 11.7: Always keep your audience in mind. Most audiences prefer a concise, to-the-point presentation.

Building the Presentation

There is an old adage about public presentations that has been used successfully for years: *Tell them what you're going to tell them. Tell them. Tell them what you told them.*

In other words, like any document, your presentation should have a beginning (introduction), a middle (body), and an end (conclusion) (Figure 11.8).

Basic Pattern for a Presentation

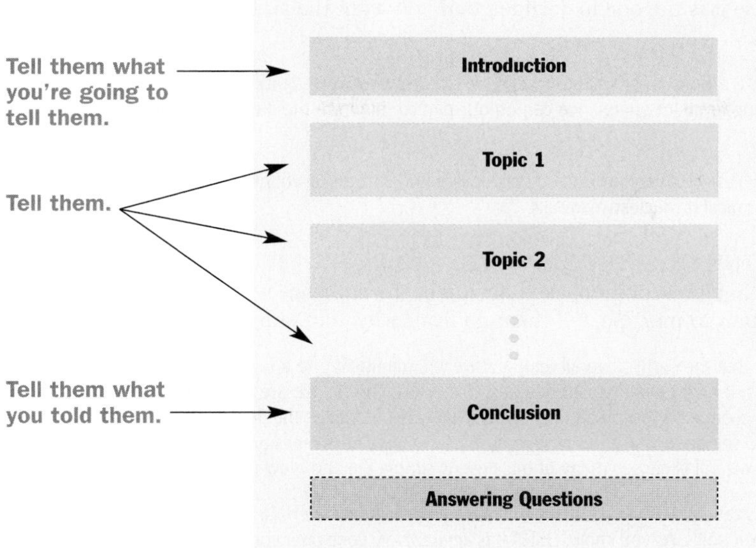

Figure 11.8: A presentation has a beginning, middle, and end. Usually time is left for questions at the end.

The Introduction: Tell Them What You're Going to Tell Them

The beginning of your presentation is absolutely crucial. If you don't catch the audience's attention in the first few minutes, there is a good chance they are going to tune out and not listen to the rest of your presentation. Your introduction slide(s) for the presentation should present at least the subject, purpose, and main point of your presentation (one slide). As shown in Figure 11.9, you might also use a second slide to forecast the structure of the talk.

Like the introduction to a document, your presentation should begin by making up to six moves:

MOVE 1: DEFINE THE SUBJECT Make sure the audience clearly understands the subject of your presentation. You might want to use a *grabber* to introduce your subject (see the "Help" feature in this chapter). A grabber can quickly help you identify your subject while capturing your audience's attention.

MOVE 2: STATE THE PURPOSE OF YOUR PRESENTATION In public presentations, you can be as blunt as you like about what you are trying to achieve. Simply tell the audience your purpose up front.

> The purpose of this presentation is to prove to you that global warming is real and it is having a serious impact on Nevada's ecology.
>
> In this demonstration, our aim is to show you that the G290 Robot is ideal for cleaning up toxic spills.

Want more ideas on how to start a presentation, go to **www.ablongman.com/johnsonweb/11.8**

MOVE 3: STATE YOUR MAIN POINT Before moving into the body of your presentation, you should also state your main point. Your main point holds the presentation together, because it is the one major idea that you want the audience to take away from your talk.

> Global warming is a serious problem for our state, and it is growing worse quickly. By switching to nonpolluting forms of energy, we can do our part to minimize the damage to our ecosystem.

> The G290 Robot gives you a way to clean up toxic spills without exposing your hazmat team to dangerous chemical or nuclear materials.

MOVE 4: STRESS THE IMPORTANCE OF THE SUBJECT TO THE AUDIENCE At the beginning of any presentation, each person in the audience wants to know "Why is this important to me?" So, tell them up front why your subject is important.

> Scientists predict that the earth's overall temperature will minimally rise a few degrees in the next 30 years. It might even rise 10 degrees. If they are correct, we are likely to see major ecological change on this planet. Oceans will likely rise a foot as the polar ice caps melt. We will also see an increase in the severity of storms like hurricanes and tornados. Here in Nevada, we will likely see many of our deserts simply die and blow away.

> OSHA regulations require minimal human contact with hazardous materials. This is especially true with toxic spills. As you know, OSHA has aggressively gone after companies that expose their employees to toxic spills. To avoid these lawsuits and penalties, many companies are letting robot workstations do the dirty work.

TAKE NOTE Don't bother telling the audience why the subject is important to you. Since you are the speaker, they will assume it is important to you.

AT A GLANCE

Opening Moves in a Presentation

- Define the subject.
- State the purpose of your presentation.
- State your main point.
- Stress the importance of the subject to the audience.
- Offer background information on the subject.
- Forecast the structure of the presentation.

MOVE 5: OFFER BACKGROUND INFORMATION ON THE SUBJECT Providing background information on your subject is a good way to build a framework for understanding what you are going to say. You can give the audience a little history on the subject or perhaps tell them about your relationship with it.

MOVE 6: FORECAST THE STRUCTURE OF THE PRESENTATION In your introduction, tell the audience how you have organized the rest of the presentation. If you are going to cover four topics, say something like,

> In this presentation, I will be going over four issues. First, I will discuss. . . Second, I will take a look at. . . . Third, I will identify some key objectives that . . . And finally, I will offer some recommendations for . . .

Forecasting gives your audience a mental framework they can use to follow your presentation. A major advantage to forecasting is that it helps the audience pay

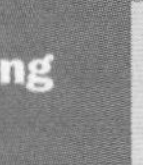

How do you set the right tone at the start of your presentation? For answers, go to **www.ablongman.com/johnsonweb/11.9**

Introduction Slides for a Presentation

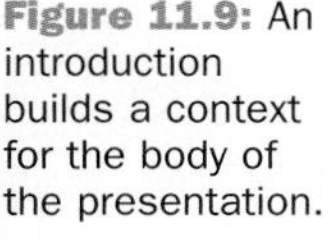

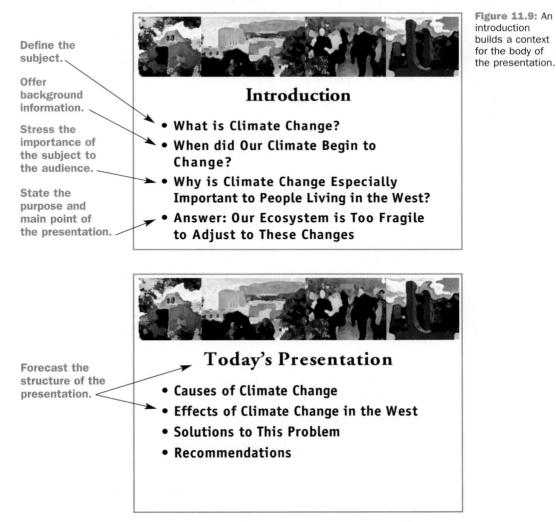

Introduction

- What is Climate Change?
- When did Our Climate Begin to Change?
- Why is Climate Change Especially Important to People Living in the West?
- Answer: Our Ecosystem is Too Fragile to Adjust to These Changes

Define the subject.

Offer background information.

Stress the importance of the subject to the audience.

State the purpose and main point of the presentation.

Figure 11.9: An introduction builds a context for the body of the presentation.

Today's Presentation

- Causes of Climate Change
- Effects of Climate Change in the West
- Solutions to This Problem
- Recommendations

Forecast the structure of the presentation.

attention. If you tell them up front that you will be discussing four topics, they will always know where you are in the presentation. They will stay more alert.

LINK For more help creating your introduction, go to Chapter 6, page 141.

The Body: Tell Them

The body of your presentation is where you are going to do the heavy lifting. Start out by dividing your subject into two to five major topics that you want to discuss.

HELP

Surfing the Net for Grabbers

You already know it's usually a bad idea to start your presentation with a joke—unless you're a professional comedian. But you still want to capture the audience's attention and get them to stop thinking about other things. So, you need a grabber.

A grabber states something interesting or challenging to capture the audience's attention. Some effective grabbers include

A rhetorical question—"Have you ever thought about changing your career and becoming a professional chef?"

A startling statistic—"A recent survey shows that 73 percent of children aged 15 to 18 in the Braynard area have tried marijuana. Almost a third of Braynard teens are regular users of drugs."

A compelling statement—"Unless we begin to do something about global warming soon, we will see dramatic changes in our ecology within a couple decades."

An anecdote—"A few years ago, I walked into a computer store to find the place empty. I looked around and didn't find a salesperson anywhere. Now, I've always been an honest person, but it did occur to me that I could pocket thousands of dollars of merchandise without being caught."

A quotation—"William James, the famous American philosopher, once said, 'Many people believe they are thinking, but they are merely rearranging their prejudices.' "

A show of hands—"How many of you think children watch too much violent television? Raise your hands." Follow this question with an interesting startling statistic or compelling statement.

But where do you find grabbers? The Internet is a source of endless material for creating grabbers. There are plenty of reference websites like Bartleby.com or Infoplease.com that will help you find quotations, statistics, and anecdotes.

Otherwise, you can use search engines like Google.com, Altavista.com, or Yahoo.com to find interesting information for grabbers. Type in your subject and a keyword or phrase that sets a specific tone that you want your grabber to establish. The search engine will locate stories, quotes, statistics, and other information that you can use to create an interesting grabber.

The best grabber is one that (1) identifies your subject, (2) states something that intrigues the audience in some way, and (3) makes the point of your presentation in a concise way.

Need a good grabber? Go to
www.ablongman.com/johnsonweb/11.10

Why only two to five topics? Why not eight or twelve topics? Experience and research show that people can usually only remember five to seven items comfortably. So, if a presentation goes beyond five topics, the audience will start feeling overwhelmed or restless. If you have more than five topics, try to consolidate smaller topics into larger ones (Figure 11.10).

Presenting the Content

Figure 11.10: Keep your presentation to two to five major points. You risk losing the audience if you try to cover more than five points.

If you have already written a document, you might follow its organizational structure. If you are starting from scratch, some of the basic organizational patterns listed below can be followed.

PROBLEM, NEED, SOLUTION This pattern is most effective for proposing new ideas. After your introduction, offer a clear definition of the problem or opportunity you are discussing. Then, specify what is needed to solve the problem or take advantage of the opportunity. Finally, offer a solution/plan that achieves the objective.

CHRONOLOGICAL When organizing material chronologically, divide the subject into two to five major time periods. Then, lead the audience through these time periods, discussing the relevant issues involved in each. In some cases, a three-part *past, present, future* pattern is a good way to organize a presentation.

SPATIAL You might be asked to explain or demonstrate visual spaces, like building plans, organizational structures, or diagrams. In these cases, divide the subject into two to five zones. Then, walk the audience through these zones, showing each zone individually and discussing how it relates to the zones around it.

NARRATIVE Audiences always like stories, so you might organize your presentation around the narrative pattern. Narratives typically (1) set a scene, (2) introduce a complication, (3) evaluate the complication, (4) resolve the complication, and (5) explain what was learned through the experience.

METHODS, RESULTS, DISCUSSION This pattern is commonly used to present the results of research. This pattern (1) describes the research plan or methodology, (2) presents the results of the study, (3) discusses and interprets the results, and (4) makes recommendations.

LINK For more information on presenting research, go to Chapter 22, page 664.

CAUSES AND EFFECTS This pattern is common for problem solving. Begin the body of the presentation by discussing the causes for the current situation. Then, later in the body, discuss the effects of these causes and their likely outcome. You can also alternate between causes and effects. In other words, discuss a cause and its effect together. Then discuss another cause and its effect, and so on.

DESCRIPTION BY FEATURES OR FUNCTIONS If you are demonstrating a product or process, divide your subject into its two to five major features or functions. Then, as you discuss each of these major features/functions, you can discuss the minor features/functions that are related to them.

LINK To learn more about describing products or processes, turn to Chapter 18, page 533.

COMPARISON AND CONTRAST Usually this pattern is followed when the speaker is comparing something new or unfamiliar with something that the audience knows well. Choose two to five major points on which these two things can be compared and contrasted. Then, compare and contrast them point by point.

There are countless patterns available for organizing the body of your presentation. The ones shown in Figure 11.11 are some of the most common in technical communication. These patterns are not formulas to be followed in lockstep. Rather, they can be manipulated to fit a variety of different speaking situations.

LINK For more ideas about organizing body information, go to Chapter 6, page 145.

The Conclusion: Tell Them What You Told Them

The conclusion is often the most important part of any presentation, and yet speakers consistently make mistakes at the ends of their talks. They end the presentation by shrugging their shoulders and saying, "Well, that's all I have to say. Any questions?"

Common Patterns for Public Presentations

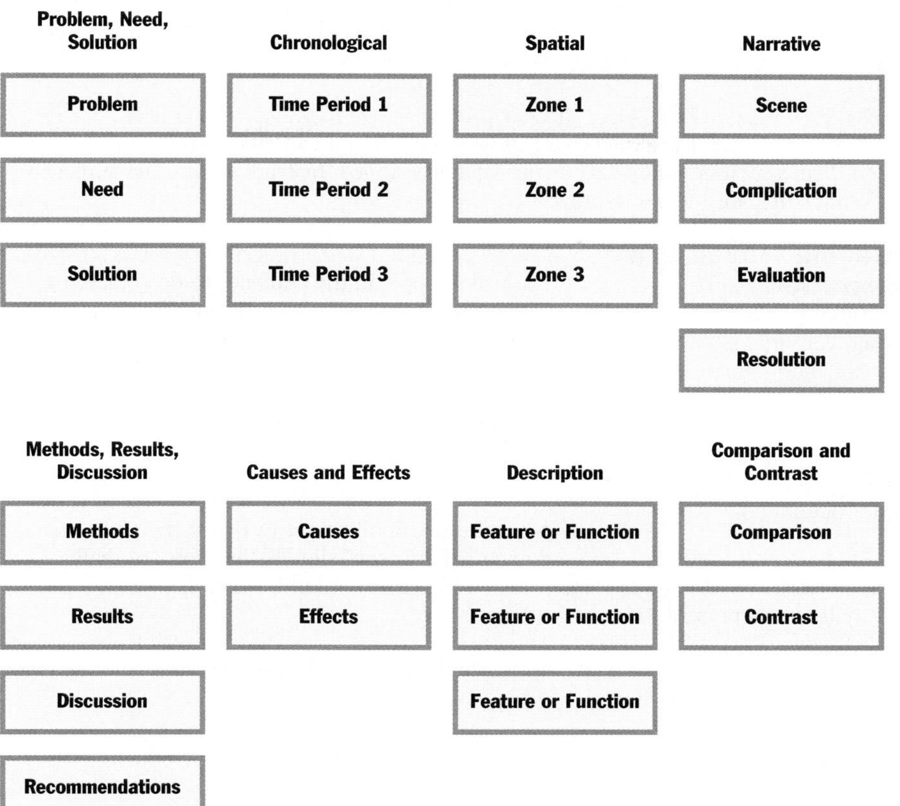

Figure 11.11: These common patterns will help you organize the body of your presentation.

Your conclusion needs to do much more. Specifically, you want to summarize your key points, while leaving the people in your audience in a position to say "yes" to your ideas. But you don't have much time to do all these things. Once you signal that you are concluding, you probably have about one to three minutes to make your final points. If you go beyond a few minutes, your audience will become agitated and frustrated.

Like the introduction, a conclusion should make some standard moves.

MOVE 1: SIGNAL CLEARLY THAT YOU ARE CONCLUDING When you begin your conclusion, use an obvious transition like "In conclusion," "Finally," "To summarize my main points," or "Let me wrap up now." When you signal your conclusion, your audience will sit up and pay attention because they know you are going to tell them your main points.

TAKE NOTE Do not use conclusion-oriented transitions anywhere else in your presentation. You never want to give the audience the impression that you are concluding if you aren't.

MOVE 2: RESTATE YOUR KEY POINTS Summarize your key points for the audience, including your overall main point. Minimally, you can simply list them and go over them one last time. That way, if your audience remembers anything about your presentation, they will remember these most important items.

MOVE 3: RESTRESS THE IMPORTANCE OF YOUR SUBJECT TO THE AUDIENCE Tell the audience again why they should care about this subject. Don't tell them why it is important to you—that's assumed. Instead, answer the audience's "What's in it for me?" questions.

AT A GLANCE

Concluding Moves in a Presentation

- Signal that you are concluding.
- Restate your main points.
- Restress the importance of the subject.
- Call the audience to action.
- Look to the future.
- Say thank you.
- Ask for questions.

MOVE 4: CALL THE AUDIENCE TO ACTION If you want people in the audience to do something, here is the time to tell them. Be specific about what action they should take.

MOVE 5: LOOK TO THE FUTURE Briefly, offer a vision of the future, usually a positive one, that will result if they agree with your ideas.

MOVE 6: SAY THANK YOU At the end of your presentation, don't forget to thank the audience. Saying "Thank you" signals that you are really finished. Often, it will also signal the audience to applaud, which is always a nice way to end a presentation.

MOVE 7: ASK FOR QUESTIONS Once the audience has stopped applauding, you can ask for questions.

Figure 11.12 shows how the concluding moves might look on a slide.

Conclusion Slide for a Presentation

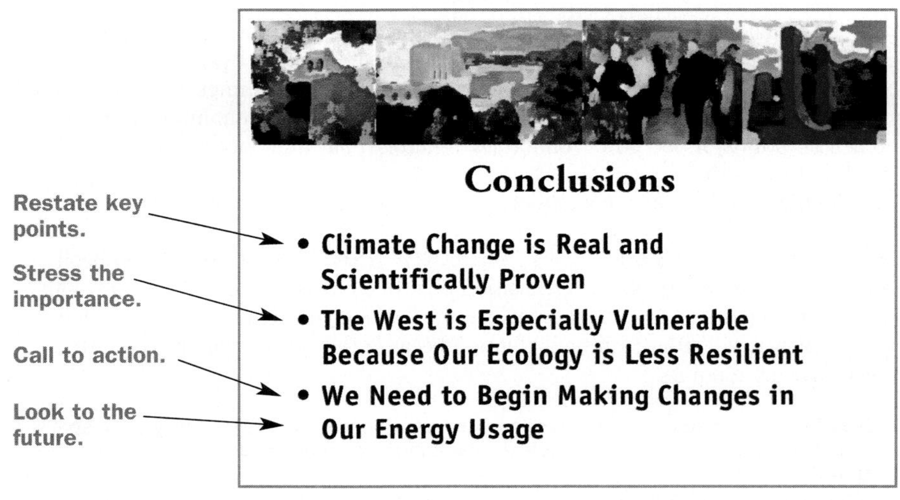

Figure 11.12: The conclusion slide should drive home your main points and look to the future.

Not sure how to end your presentation? For ideas, go to **www.ablongman.com/johnsonweb/11.11**

Preparing to Answer Questions

While preparing and researching your presentation, you should spend some time anticipating the kinds of questions you might be asked. Questions are an opportunity to interact with the audience and clarify your ideas (Figure 11.13). You will generally be asked three types of questions: elaboration, hostile, and heckling questions.

Answering Questions

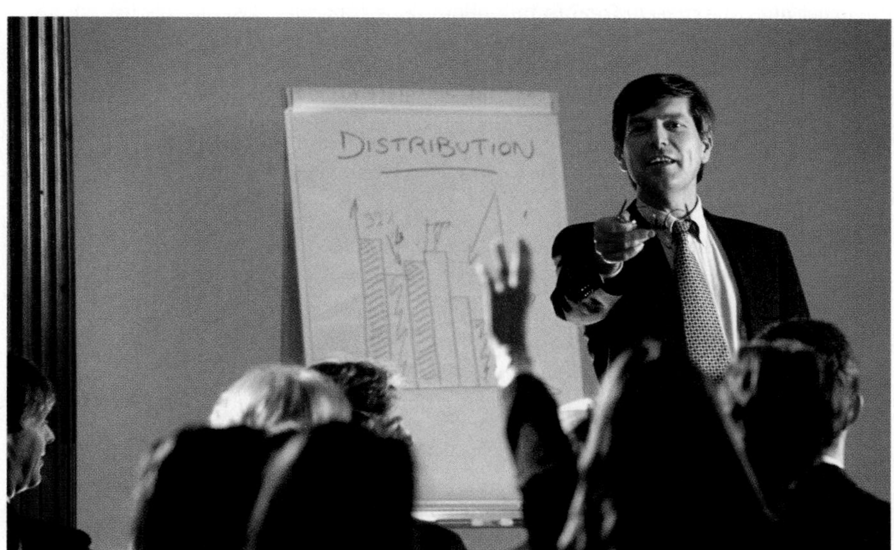

Figure 11.13: Prepare in advance for the kinds of questions you might be asked after your presentation.

THE ELABORATION OR CLARIFICATION QUESTION Members of the audience might ask you to expand on your ideas or explain some of your concepts. These questions offer you an opportunity to reinforce some of your key points.

THE HOSTILE QUESTION Occasionally, an audience member will ask you a question that calls your ideas into doubt. For example, "I really don't trust your results. Do you really expect us to believe that you achieved precision of .0012 millimeters?"

Here is a good three-step method for deflecting these kinds of questions:

1. *Rephrase the question*—"The questioner is asking whether it is possible that we achieved a precision of .0012 millimeters."
2. *Validate the question*—"That's a good question, and I must admit that we were pleasantly surprised that our experiment gave us this level of precision."
3. *Elaborate and move forward*—"We achieved this level of precision because . . ."

You should allow a hostile questioner only one follow-up remark or question. After giving the hostile questioner this second opportunity, do not look at that

For helpful question-answer strategies, go to **www.ablongman.com/johnsonweb/11.12**

person. If you look elsewhere in the room, someone else in the audience will usually raise his or her hand and bail you out.

HECKLING In rare cases, a member of the audience will be there to heckle you. He or she will ask rude questions or make blunt statements like "I think this is the stupidest idea we've heard in 20 years."

In these situations, you need to recognize that the heckler is *trying* to sabotage your presentation and cause you to lose your cool. Don't let the heckler do that to you. You simply need to say something like "I'm sorry you feel that way. Perhaps we can meet after the presentation to talk about your concerns." Usually, at this point, others in the audience will step forward to ask more constructive questions.

It is rare that a heckler will actually come to talk to you later. That's not why he or she was there in the first place. If a heckler does manage to dominate the question-and-answer period, simply end your presentation. Say something like

> Well, we are out of time. Thank you for your time and attention. I will stick around in the room for more questions.

Then, step away from the podium or microphone and walk off the stage. This kind of ending may seem abrupt, but it shows the audience that you are not going to tolerate rude behavior.

Choosing Your Presentation Style

Your speaking style is very important. In a presentation, you can use style to add flavor to your materials while gaining the trust of the audience. Poor style, on the other hand, can bore the audience, annoy them, and even turn them against you.

There are many ways to create an appropriate style for your presentation, but four techniques seem to work best for technical presentations. Each of these techniques will help you project a special tone in your speaking.

DEVELOP A PERSONA In ancient Greek, the word *persona* meant "mask." So, as you consider your presentation style, think about the mask you want to wear in front of the audience.

Then, step into this character. Put on the mask. You will find that choosing a persona will make you feel more comfortable, because you are playing a role, much like an actor plays a role. The audience isn't seeing/judging *you*. They are seeing the mask you have chosen to wear for the presentation.

TAKE NOTE Accomplished presenters, like actors, can actually choose a persona that fits the presentation. They might adopt one persona for one kind of presentation and a completely different persona for another.

SET A THEME A theme is a consistent tone you want to establish in your presentation. The best way to set a theme is to decide which one word best characterizes how you want the audience to *feel* about your subject. Then, use logical mapping to find words and phrases associated with that feeling (Figure 11.14).

Need to improve your style? Go to
www.ablongman.com/johnsonweb/11.13

A Theme in a Presentation

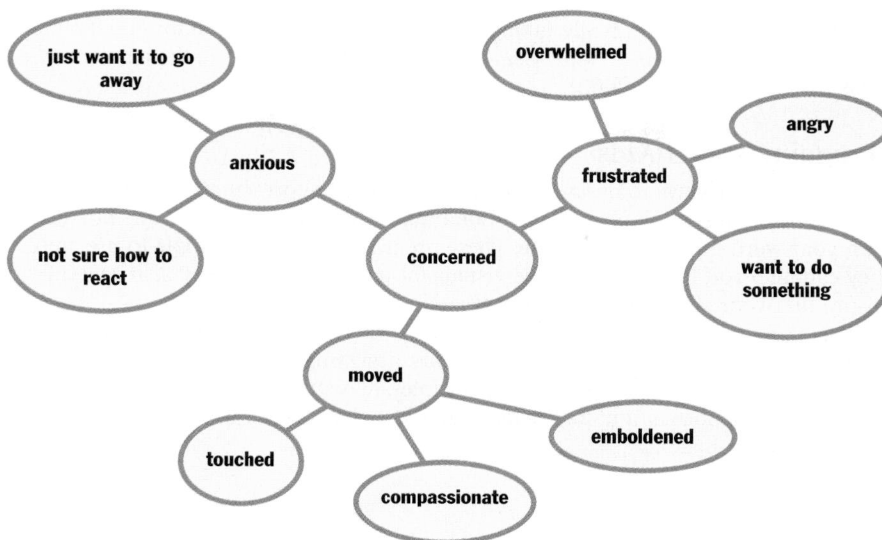

Figure 11.14: By mapping around a key term and seeding your text with related words, you can set a theme that creates a specific feeling in your audience.

As you prepare your talk, use these words regularly. When you give your presentation, the audience will naturally pick up on your theme.

For example, let's say you want your audience to be "concerned." Map out words and phrases related to this feeling. Then, as you prepare your presentation, weave these words and phrases into your text. If you do it properly, your audience will feel your concern.

SHOW ENTHUSIASM If you're not enthusiastic about your subject, your audience won't be either. So, get excited about what you have to say. Be intense. Get pumped up. Show the audience that you are enthusiastic about this subject (even if you aren't) and they will be too.

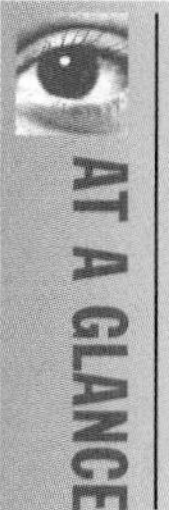

AT A GLANCE

Creating Your Presentation Style

- Develop a persona.
- Set a theme.
- Show enthusiasm.
- KISS: Keep It Simple (Stupid).

KISS: KEEP IT SIMPLE (STUPID) The KISS principle is always something to keep in mind when you are presenting, especially when it comes to style. Speak in plain, simple terms. Don't get bogged down in complex details and concepts.

In the end, good style is a choice. You may have come to believe that some people just have good style and others don't. That's not true. In reality, people with good style are just more conscious of the style they want to project.

GO TO THE NET

To see websites that offer advice for presenting well, go to **www.ablongman.com/johnsonweb/11.14**

Creating Visuals

In this visual age, you are really taking a risk if you try to present information without visuals. People not only want visuals, they *need* them to fully understand your ideas (Munter & Russell, 2002).

Designing Visual Aids

One of the better ways to design visual aids is to use the presentation software (PowerPoint, Presentations, or Visual Communicator) that probably came bundled with your word-processing software. These programs are rather simple to use, and they can help you quickly create the visuals for a presentation. They also generally ensure that your slides are designed well and are readable from a distance.

The design principles discussed in Chapter 8 (balance, alignment, grouping, consistency, and contrast) work well when you are designing visual aids for public presentations. In addition to these design principles, here are some special considerations concerning format and font choices that you should keep in mind as you are creating your visuals:

FORMAT CHOICES

- Title each slide with an action-oriented heading.
- Put five or fewer items on each slide. If you have more than five points to state about a topic, divide the topic into two slides.
- Use left-justified text in most cases. Centered text should be used infrequently and right-justified text almost never.
- Use lists instead of paragraphs or sentences.
- Use icons and graphics to keep your slides fresh for the audience.

FONT CHOICES

- Use a simple typeface that is readable from a distance. Sans serif fonts are often more readable from a distance than serif fonts.
- Use a minimum of a 36-point font for headings and a minimum of a 24-point font for body text.
- Use color to keep slides interesting and improve retention.
- Do not use ALL UPPERCASE letters because they are hard to read at a distance.

Overall, it is best to keep your slides as simple as possible (Figure 11.15). After all, if your audience needs to puzzle through your complex slides, they probably won't be listening to you.

Using Graphics

Graphics are also helpful, especially when you are trying to describe something to the audience. An appropriate graph, chart, diagram, picture, or even a movie will help

For advice about creating visuals, go to
www.ablongman.com/johnsonweb/11.15

Sample Slides from a Presentation

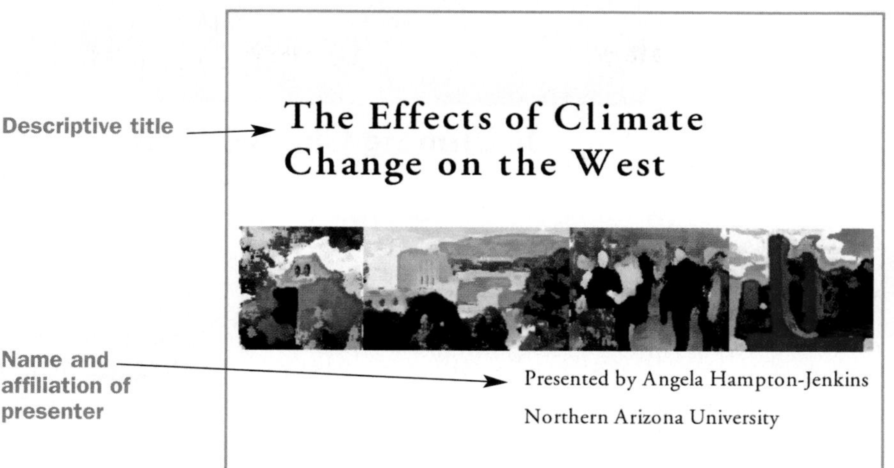

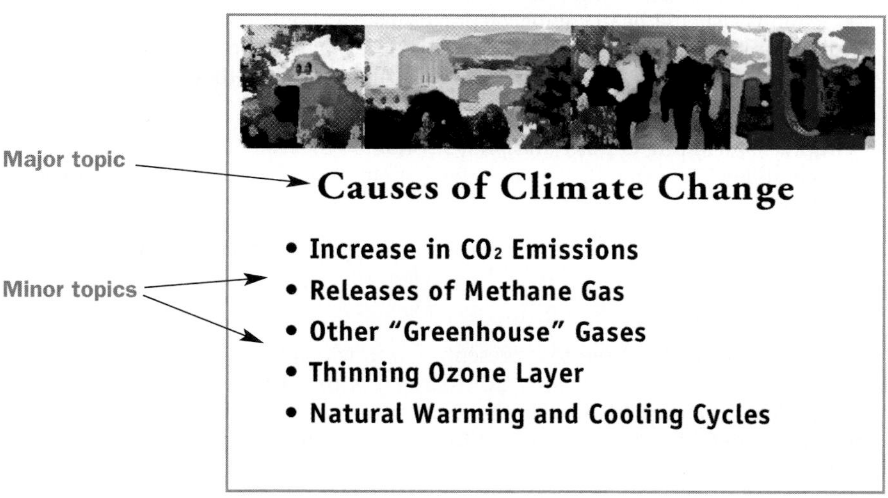

Figure 11.15: These two slides show good balance, simplicity, and consistency. Keep the slides simple so the audience doesn't need to work too hard to read them.

support your argument (Figure 11.16). Chapter 9 in this book discusses the use of graphics in documents. Most of those same guidelines apply in presentations.

Here are some guidelines that pertain specifically to using graphics in a presentation:

Using a Graphic on a Presentation Slide

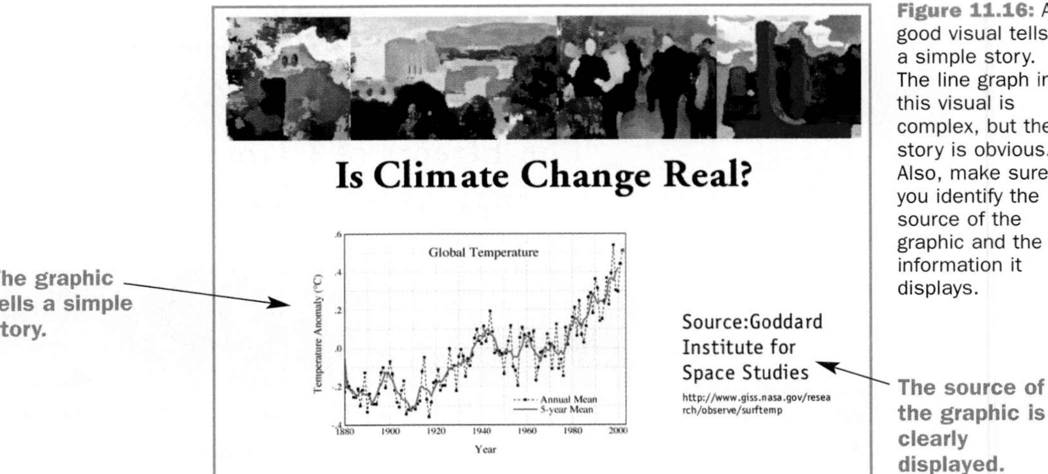

Figure 11.16: A good visual tells a simple story. The line graph in this visual is complex, but the story is obvious. Also, make sure you identify the source of the graphic and the information it displays.

- Words or figures in the graphic should be large enough to be read from a distance.
- Label each graphic with a title.
- Graphics should be uncomplicated and make a simple point.
- Keep tables small and simple. Large tables full of data do not work well as visuals, because the audience cannot read them—nor do they want to.
- Use clip art or photos to add life to your slides.

Graphics, including clip art and photos, should never be used to merely decorate your slides. They should reinforce the content, organization, and style of your presentation.

LINK For more information on creating and using graphics, go to Chapter 9, page 234.

Slides to Avoid

All of us have probably been to a presentation in which the speaker created ineffective slides. He or she put up a transparency with a 12-point type font and minimal design (Figure 11.17). Or, the speaker put up a table or graph that was completely indecipherable because the font was too small or the graphic was too complex.

These kinds of slides are nothing short of painful. The only thing the audience wants from such a slide is for the speaker to remove it—as soon as possible. Always remember that we live in a visual culture. People are sensitive to bad design. So, take the time to properly create slides that enhance your presentation—not send the audience running for the door.

An Ineffective Slide

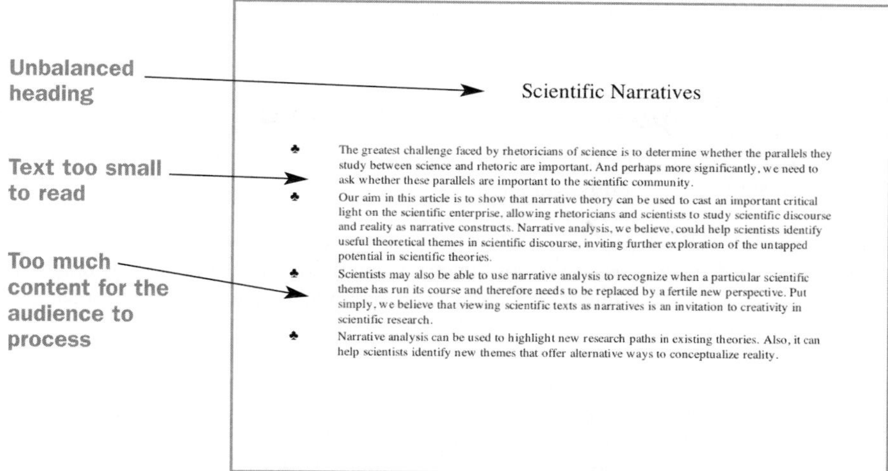

Unbalanced heading
Text too small to read
Too much content for the audience to process
Scientific Narratives
♣ The greatest challenge faced by rhetoricians of science is to determine whether the parallels they study between science and rhetoric are important. And perhaps more significantly, we need to ask whether these parallels are important to the scientific community.
♣ Our aim in this article is to show that narrative theory can be used to cast an important critical light on the scientific enterprise, allowing rhetoricians and scientists to study scientific discourse and reality as narrative constructs. Narrative analysis, we believe, could help scientists identify useful theoretical themes in scientific discourse, inviting further exploration of the untapped potential in scientific theories.
♣ Scientists may also be able to use narrative analysis to recognize when a particular scientific theme has run its course and therefore needs to be replaced by a fertile new perspective. Put simply, we believe that viewing scientific texts as narratives is an invitation to creativity in scientific research.
♣ Narrative analysis can be used to highlight new research paths in existing theories. Also, it can help scientists identify new themes that offer alternative ways to conceptualize reality.

Figure 11.17: An ineffective slide often says a great amount—if the audience could read it.

Delivering the Presentation

Why do people come to presentations, especially if a paper version of your work is available?

People attend presentations because they want to see you perform the material. They want to see how you act and interact with them. They want you to put a human face on the material. For this reason, you should pay close attention to your delivery, so the audience receives a satisfying performance.

The usual advice is to "be yourself" when you are presenting. Of course, that's good advice if you are comfortable talking to an audience. Better advice is to "be the person the audience expects." In other words, like an actor, play the role that seems to fit your material and your audience.

Body Language

The audience will pay close attention to your body language. So use your body to reflect and highlight the content of your talk (Figure 11.18).

DRESS APPROPRIATELY How you dress should reflect the content and importance of your presentation. A good rule of thumb is to dress a level better than you expect your audience to dress. For example, if the audience is in casual attire, dress a little more formally than them. A female speaker might wear a blouse and dress pants or nice skirt. Men might wear a shirt, tie, and dress pants. If the audience is wearing suits and "power" dresses, you will need to wear an even *nicer* suit or even *better* power dress.

GO TO THE NET

Want more advice on delivering presentations? Go to **www.ablongman.com/johnsonweb/11.16**

STAND UP STRAIGHT When people are nervous, they have a tendency to slouch, lean, or rock back and forth. To avoid these problems when you speak, keep your feet squarely under your shoulders with the knees slightly bent. Keep your shoulders back and your head up.

DROP YOUR SHOULDERS Under stress, people also have a tendency to raise their shoulders. Raised shoulders restrict your airflow and make the pitch of your voice go up. By dropping your shoulders, you will improve airflow and lower your voice. A lower voice sounds more authoritative.

USE OPEN HAND AND ARM GESTURES For most audiences, open hand and arm gestures will convey trust and confidence. If you fold your arms, keep your arms at your sides, or put both hands in your pockets, you will convey a defensive posture that audiences do not trust.

MAKE EYE CONTACT Everyone in the audience should believe you made eye contact with him or her at least once during your presentation. As you are presenting, make it a point to look at all parts of the room at least once. If you are nervous about making eye contact, look at the audience members' foreheads instead. They will think you are looking them directly in the eye.

TAKE NOTE There are exceptions to generally accepted guidelines about gestures and eye contact. In some cultures, like some Native American cultures, open gestures and eye contact might be considered rude and even threatening. If you are speaking to an unfamiliar audience, find out which gestures and forms of eye contact are appropriate for that audience.

Delivering the Presentation

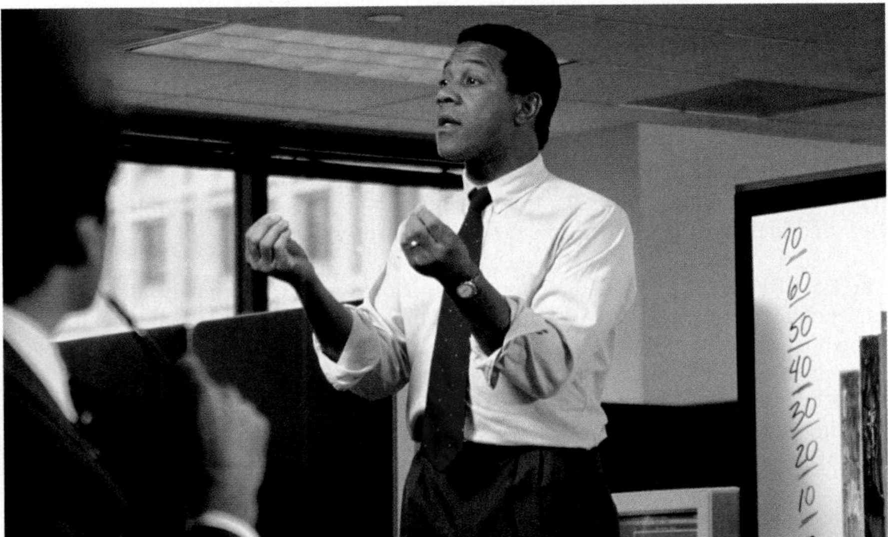

Figure 11.18: You can use your body and hands to highlight important parts of your presentation.

MOVE AROUND THE STAGE If possible, when you make important points, step forward toward the audience. When you make transitions in your presentation from one topic to the next, move to the left or right on the stage. Your movement across the stage will highlight the transition.

POSITION YOUR HANDS APPROPRIATELY Nervous speakers often strike a defensive pose by using their hands to cover specific parts of their body (perhaps you can guess which parts). Keep your hands away from these areas.

Voice, Rhythm, and Tone

A good rule of thumb about voice, rhythm, and tone is to *speak lower and slower than you think you should.*

Why lower and slower? When you are presenting, you need to speak louder than normal. As your volume goes up, the pitch of your voice will go up. So, your voice will seem unnaturally high (even shrill) to the audience. By consciously lowering your voice, you should sound just about right to the audience.

Meanwhile, nervousness usually encourages you to speak faster than you normally would. By consciously slowing down, you will sound more comfortable and more like yourself.

USE PAUSES TO HIGHLIGHT MAIN POINTS When you make an important point, pause for a moment. Your pause will signal to the audience that you just made an important point that you want them to consider and retain.

USE PAUSES TO ELIMINATE VERBAL TICS Verbal tics like "um," "ah," "like," "you know," "OK?" and "See what I mean?" are simply nervous habits that are intended to fill gaps between thoughts. If you have problems with a verbal tic (who doesn't?) train yourself to pause when you feel like using one of these sounds or phrases. Before long, you will find them disappearing from your speech altogether.

Using Your Notes

The best presentations are the ones that don't require notes. Notes on paper or index cards are fine, but you need to be careful not to keep looking at them. Nervousness will often lead you to keep glancing at them instead of looking at the audience. Some speakers even get stuck looking at their notes, glancing up only rarely at the audience. Looking down at your notes makes it difficult for the audience to hear you. You want to keep your head up at all times.

The following are some guidelines for making and using notes.

USE YOUR SLIDES AS MEMORY TOOLS You should know your subject inside out. So you likely don't need notes at all. Practice rehearsing your presentation with your slides alone. Eventually, you will be able to completely dispense with your written notes and work solely off your visual aids while you are speaking.

Need help improving your voice, rhythm, and tone? Go to **www.ablongman.com/johnsonweb/11.17**

TALK TO THE AUDIENCE, NOT YOUR NOTES OR THE SCREEN Make sure you are always talking to the audience. It is sometimes tempting to begin looking at your notes while talking. Or, in some cases presenters end up talking to the screen. You should only steal quick glances at your notes or the screen. Look at the audience instead.

PUT WRITTEN NOTES IN A LARGE FONT If you need to use notes, print them in a large font on the top half of a piece of paper. That way, you can quickly find needed information without effort. Putting the notes on the top half of the paper means you won't need to glance at the bottom of the piece of paper, restricting airflow and taking your eyes away from the audience.

USE THE "NOTES VIEW" IN YOUR PRESENTATION SOFTWARE Presentation software usually includes a "Notes View" that allows you to type notes under or to the side of slides (Figure 11.19). These notes can be helpful, but be wary of using them too much. First, they force you to look at the bottom of a sheet of paper, restricting your airflow. Second, if you are nervous, you may be distracted by these notes or start reading them to the audience.

IF SOMETHING BREAKS, KEEP GOING If your overhead projector dies, your computer freezes, or you drop your notes all over the front row, don't stop your presentation to fumble around trying to fix the situation. Keep going. You only need to acknowledge the problem, "Well, it looks like some gremlins have decided to sabotage my projector. So, I will move forward without it." Then, do so.

Practice and Rehearsal

The common advice for improving your ability to present in public is "practice, practice, practice." This is good advice, but it doesn't tell the whole story. You also need to "rehearse, rehearse, rehearse."

Practice, Practice, Practice

Practicing your presentation is a good way to work out the kinks and revise your materials. While you are practicing, you should

- look for problems with the content, organization, and style of your presentation.
- edit and proofread your visuals and handouts.
- decide how you are going to deliver the information by paying attention to your body language and voice.

Practice is something you usually do at your desk or in your office. It involves talking through your presentation and fixing problems as you find them.

Rehearse, Rehearse, Rehearse

Once you have ironed out the kinks in your presentation, though, it is time to rehearse, rehearse, rehearse. Unlike practicing, rehearsal means giving the presentation from beginning to end *without stopping*. While practicing, you can stop to make

The "Notes View"

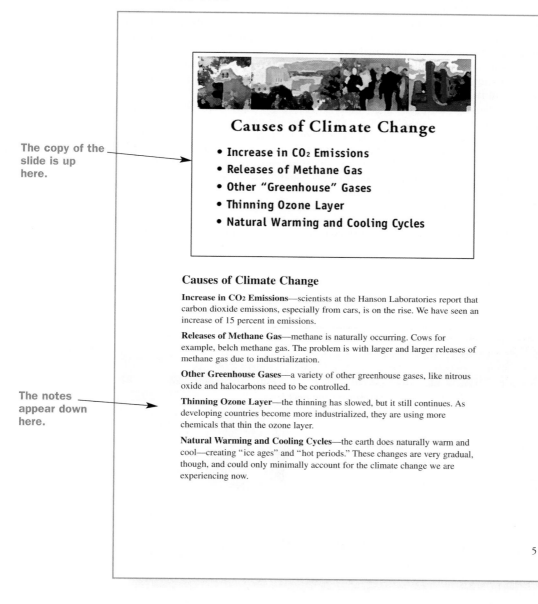

The copy of the slide is up here.

Causes of Climate Change

- Increase in CO_2 Emissions
- Releases of Methane Gas
- Other "Greenhouse" Gases
- Thinning Ozone Layer
- Natural Warming and Cooling Cycles

Causes of Climate Change

Increase in CO_2 Emissions—scientists at the Hanson Laboratories report that carbon dioxide emissions, especially from cars, is on the rise. We have seen an increase of 15 percent in emissions.

Releases of Methane Gas—methane is naturally occurring. Cows for example, belch methane gas. The problem is with larger and larger releases of methane gas due to industrialization.

Other Greenhouse Gases—a variety of other greenhouse gases, like nitrous oxide and halocarbons need to be controlled.

Thinning Ozone Layer—the thinning has slowed, but it still continues. As developing countries become more industrialized, they are using more chemicals that thin the ozone layer.

Natural Warming and Cooling Cycles—the earth does naturally warm and cool—creating "ice ages" and "hot periods." These changes are very gradual, though, and could only minimally account for the climate change we are experiencing now.

5

The notes appear down here.

Figure 11.19: The "Notes View" in your presentation software allows you put notes below or to the side of a slide. Your audience won't be able to see them, but you can print out each slide with your notes so you don't need to keep looking up at your slides.

To locate groups like Toastmasters, go to **www.ablongman.com/johnsonweb/11.18**

corrections or changes to your presentation. When rehearsing, just keep talking, even if you make a mistake. You won't have time to make changes or correct mistakes when you are in front of an audience, so use rehearsal to practice getting past those weak parts of your presentation. The closer you can replicate the actual speaking situation, the better your rehearsal will be.

As you rehearse, you will notice smaller problems with your presentation. Don't stop to fix them. You can fix them when you have finished the rehearsal. Pretend you are in front of the audience and you can't stop to fix small problems.

For some presentations, you might want to rehearse your presentation in front of a test audience. At your workplace, you can draft people into listening to your talk. These practice audiences might be able to give you some advice about improving your presentation. If anything, though, they will help you anticipate the reactions of the audience to your materials. Groups like Toastmasters also can provide a test audience (Figure 11.20). At Toastmasters meetings, people give presentations and work on improving their public speaking skills.

The more you rehearse, the more comfortable you are going to feel with your materials and the audience. Practicing will help you find the errors in your presentation, but rehearsal will help you put the whole package together into an effective presentation.

Places to Improve Your Presentation Skills

Source: Toastmasters International, http://www.toastmasters.org.

Figure 11.20: Toastmasters is a group that helps people practice and improve their public speaking skills. Local groups can be found in almost any larger town.

GO TO THE NET

For more practice and rehearsing tips, go to **www.ablongman.com/johnsonweb/11.19**

Evaluating Your Performance

Evaluation is an important way to receive feedback on your presentation. In some speaking situations, the audience will expect to evaluate your presentation. If an evaluation form is not provided, they will expect you to hand one out, so they can give you and perhaps supervisors feedback on your performance. Figure 11.21 shows a typical evaluation form that could be used for a variety of presentations.

An Evaluation Form for Presentations

Presentation Evaluation Form

Please answer these questions with one or more sentences

Content

Did the speaker include more information than you needed to know? If so, where?

Where did the speaker not include enough need-to-know information?

What kinds of other facts, figures, or examples would you like to see included in the presentation?

Organization

In the introduction, was the speaker clear about the subject, purpose, and main point of the presentation?

In the introduction, did the speaker grab your attention effectively?

In the body, was the presentation divided into obvious topics? Were the transitions among these topics obvious to you?

In the conclusion, did the speaker restate the presentation's main point clearly?

Did the speaker leave enough time for questions?

Style and Delivery

Did the speaker speak clearly and loudly enough?

Figure 11.21: An evaluation form is a good way to receive feedback on your presentation. Have your audience comment on your presentation's content, organization, style, and delivery.

Continued on following page

To download this evaluation form, go to **www.ablongman.com/johnsonweb/11.20**

Figure 11.21: (continued.)

Did the speaker move effectively, using hands and body language to highlight important points?

Did the speaker have any verbal tics (uh, um, ah, you know)?

Did the speaker look around the room and make eye contact?

Was the speaker relaxed and positive during the presentation?

Visuals

Did the visuals effectively highlight the speaker's points in the presentation?

Did the speaker use the visuals effectively, or did they seem to be a distraction?

Did the speaker use notes effectively? Was the speaker distracted by his or her notes?

Concluding Remarks

List five things you learned during this presentation.

What did you like about this presentation?

What did you not like about this presentation?

What suggestions can you offer to make this presentation better?

CHAPTER REVIEW

- Public presentations are an essential part of communicating effectively in technical workplaces.
- Planning is the key to successful public presentations. You need a firm understanding of the subject, purpose, audience, and context for your presentation.
- A well-organized presentation "tells them what you're going to tell them, tells them, and tells them what you told them."
- Visual aids are essential in presentations. Software programs are available to help you make these aids.
- People come to presentations because they want to see you *perform* the material.
- Good delivery means paying attention to your body language, appearance, voice, rhythm, and tone, and effective use of notes.
- Practice is important, but so is rehearsal. Use practice to iron out the rough spots in the presentation, then rehearse, rehearse, rehearse.

EXERCISES AND PROJECTS

Individual or Team Projects

1. Using the Internet, locate a professional in your major or field of study and ask that person about the kinds of oral presentations he or she makes. Avoid simply asking "Do you make oral presentations?" Instead, ask about informal presentations at meetings. Ask about product demonstrations. And, of course, ask about more formal presentations in which your interviewee needs to inform or persuade an audience. Write a one- to two- page memo to your instructor in which you report on your findings about oral presentations in your field.

2. Attend a public presentation at your campus or workplace. Instead of listening to the content of the presentation, pay close attention to the speaker's use of organization, style, and delivery. In a memo to your instructor, discuss some of the speaker's strengths and suggest places where he or she might have improved the presentation.

3. Using presentation software, turn a document you have written for this class into a short presentation with slides. Your document might be a set of instructions, a report, a proposal, or any other document. Your task is to make the presentation interesting and informative to the audience.

4. Pick a campus or workplace problem that you think should be solved. Then, write a three-minute oral presentation in which you (1) identify the problem, (2) discuss what is needed to solve the problem, and (3) offer a solution. As you develop your presentation, think about the strategies you might use to persuade the audience to accept your point of view and solution.

5. Interview one of your classmates for 10 minutes about his or her hometown. Then, from your notes alone make an impromptu presentation to your class in which you introduce your classmate's hometown to the rest of the class. In your presentation, you should try to persuade the audience that they should visit this town.

Collaborative Projects

Group presentations are an important part of the technical workplace because projects are often team efforts. Therefore, a whole team of people often needs to make the presentation, with each person speaking about their part of the project.

Turn one of the collaborative projects you have completed in this class into a group presentation. Some of the issues you should keep in mind are the following:

- Who will introduce the project and who will conclude it?
- How will each speaker "hand off" to the next speaker?
- Where will each presenter stand while presenting?
- Where will the other team members stand or sit while someone else is presenting?
- How can you create a consistent set of visuals that will hold the presentation together?
- How will questions and answers be handled during and after the presentation?
- How can the group create a coherent, seamless presentation that doesn't seem to change style when a new speaker steps forward?

Something to keep in mind with team presentations is that members of the audience are often less interested in the content of the presentation than they are in meeting your team. So, think about some ways in which you can convey that personal touch in your team presentation while still talking effectively about your subject.

CASE STUDY

The Coward

Jennifer Sandman is an architect who works for a small firm in Kansas City. The firm is completing its largest project ever, designing a building for a GrandMart superstore. Jennifer and her boss, a lead architect, Bill Voss, are in charge of the project.

At an upcoming city council meeting, they will need to give a public presentation on their plans for the new GrandMart. At the meeting, the mayor and city council will look over the plans and listen to comments from people in the audience. Failure to persuade the audience might mean the loss of the project as well as future business with GrandMart, an important client.

Everyone knows that the meeting will be contentious, with many angry residents arguing against the plans. And yet, GrandMart's surveys have shown that a silent majority would welcome the new store in their neighborhood. Meanwhile, GrandMart's marketing specialists are certain that the store would be profitable, because there are no competitors within 10 miles.

Jennifer has one problem—her boss is a coward. He scheduled a business trip to coincide with the meeting, and he promptly delegated the presentation to her. As her boss walked out the door, he said, "It will be a good growth opportunity for you. Just don't make the firm look bad."

For Jennifer, making presentations isn't a problem. She is a whiz with PowerPoint, and she has given enough presentations to feel comfortable in front of people. Her concern, though, is the hostile people in the audience. There might even be some hecklers. These hostile people will never accept the idea that a GrandMart should be built in their neighborhood. Others in the audience, though, will be undecided. So, she wants to devise ways in her presentation to soften and deflect some of the hostile criticisms, while gaining the support of people in the community who might want the store built.

If you were Jennifer, how would you craft your presentation to fit these difficult conditions? How would you prepare for the hostile and heckling questions? How would you design and deliver your presentation to avoid being seen as defensive?

PART 3

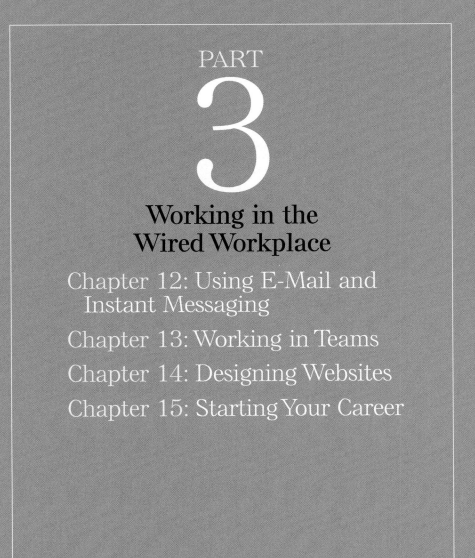

Working in the Wired Workplace

CHAPTER

12

Using E-Mail and Instant Messaging

CHAPTER CONTENTS

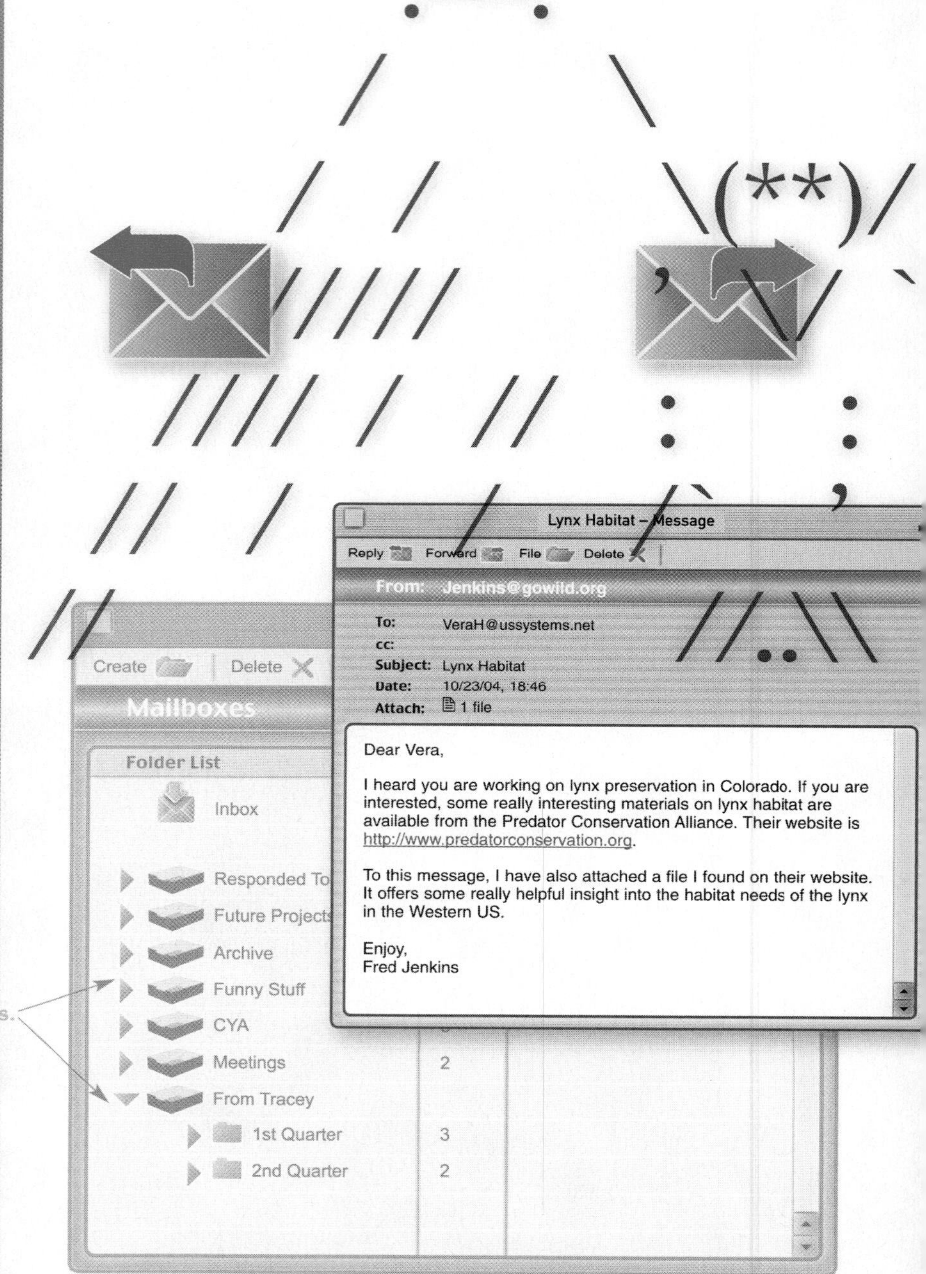

\(**)/
Lynx Habitat – Message
Reply
Forward
File
Delete
From: Jenkins@gowild.org
To: VeraH@ussystems.net
cc:
Subject: Lynx Habitat
Date: 10/23/04, 18:46
Attach: 1 file
Dear Vera,
I heard you are working on lynx preservation in Colorado. If you are interested, some really interesting materials on lynx habitat are available from the Predator Conservation Alliance. Their website is http://www.predatorconservation.org.
To this message, I have also attached a file I found on their website. It offers some really helpful insight into the habitat needs of the lynx in the Western US.
Enjoy,
Fred Jenkins
Create
Delete
Mailboxes
Folder List
Inbox
Responded To
Future Projects
Archive
Funny Stuff
CYA
Meetings
2
From Tracey
1st Quarter
3
2nd Quarter
2

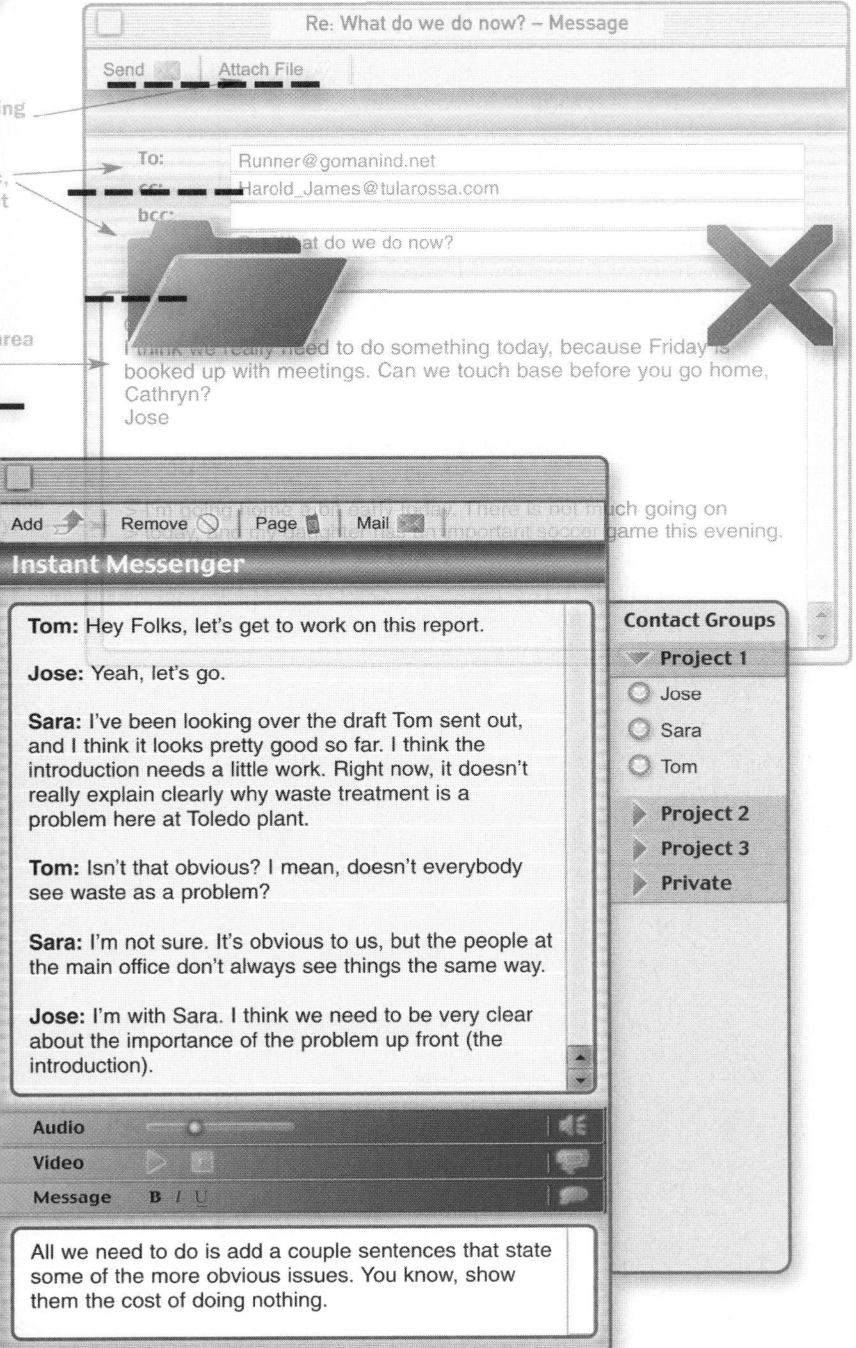

Re: What do we do now? – Message
Send
Attach File
To:
Runner@gomanind.net
Harold_James@tularossa.com
bcc:
booked up with meetings. Can we touch base before you go home,
Cathryn?
Jose
Add
Remove
Page
Mail
Instant Messenger
Tom: Hey Folks, let's get to work on this report.
Jose: Yeah, let's go.
Sara: I've been looking over the draft Tom sent out, and I think it looks pretty good so far. I think the introduction needs a little work. Right now, it doesn't really explain clearly why waste treatment is a problem here at Toledo plant.
Tom: Isn't that obvious? I mean, doesn't everybody see waste as a problem?
Sara: I'm not sure. It's obvious to us, but the people at the main office don't always see things the same way.
Jose: I'm with Sara. I think we need to be very clear about the importance of the problem up front (the introduction).
Contact Groups
Project 1
Jose
Sara
Tom
Project 2
Project 3
Private
Audio
Video
Message
All we need to do is add a couple sentences that state some of the more obvious issues. You know, show them the cost of doing nothing.

CHAPTER OBJECTIVES

In this chapter, you will learn:

- The importance of e-mail and instant messaging in the workplace.
- To define e-mail as a communication tool and understand how it is used in work settings.
- The basic features and functions of e-mail.
- Strategies for managing large amounts of e-mail in the workplace.
- The use of netiquette to avoid problems with e-mail.
- To use instant messaging and understand its basic functions.

E-mail has become such a common part of our lives that we no longer see it as something new. But not too long ago, e-mail was something quite new and quite different. Before e-mail, if you wanted to communicate with other people, you picked up the phone, sent them a memo, or walked over to their offices. Today, e-mail is used to handle many of these kinds of communications.

Instant messaging (IM) is one of the latest communication tools to arrive in the workplace. Today, instant messaging provides a way to chat with others in "real time" through computers, wireless phones, and personal digital assistants (PDAs). The simplest forms of instant messaging allow you to have written conversations with others. More advanced forms allow you to use webcams (computer-mounted video cameras) and microphones to talk to people face to face through your computer. Increasingly, instant messaging is finding applications in the workplace.

E-mail is already an essential communication tool in the workplace, and instant messaging is quickly becoming yet another helpful tool. They allow you to communicate, collaborate, and connect with others. In almost all workplaces, it is assumed that you are ready to use e-mail effectively (Figure 12.1). Soon, the ability to use instant messaging will also be important.

What Is E-Mail?

Today, we take e-mail for granted. Checking e-mail has become part of our daily routine, as normal as getting dressed in the morning. If you are like many people, you don't think twice about using e-mail to communicate with your friends, classmates, and teachers. In the workplace, e-mail is a common way to interact with supervisors, coworkers, and clients.

E-mail is an electronic system for sending and receiving messages and files over a computer network.

E-mail is an electronic system for sending and receiving messages and files over a computer network. Compared to the phone or paper-based documents, though, e-mail is still relatively new in the workplace. As a result, its purpose and uses are not clearly defined. Should it completely replace memos and letters in the workplace? Does it replace the phone? Can e-mail be used to convey crucial information or confidential details? Is e-mail secure? When is e-mail better than face-to-face communication? When is it worse?

Test Your E-Mail Readiness

	Yes	No	Score
• Can you receive and send messages?	❑	❑	Yes Marks
• Can you forward documents to others?	❑	❑	Expert 8–10
• Can you receive and send attached documents?	❑	❑	Achiever 7–8
• Do you sort your e-mail messages into folders?	❑	❑	Beginner 5–6
• Can you send one e-mail message to multiple people?	❑	❑	Get to Work 0–5
• Have you created and used a signature file?	❑	❑	
• Are you on any mailing lists (listservs)?	❑	❑	
• Do you know what flaming is?	❑	❑	
• Do you know what spam is?	❑	❑	
• Do you know what netiquette is?	❑	❑	

Figure 12.1: Are you e-mail ready? Here is a quick test to see if you're up to speed.

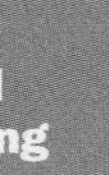

To learn about the history of e-mail, go to **www.ablongman.com/johnsonweb/12.1**

E-mail is a hybrid technology—it combines and utilizes some of the functions of a phone, a memo, and a letter. It can be used like a phone (or voice mail) to leave a short message. Or, like a memo, it can be used to send information, like drafts of a document, new policies, or progress reports. Like a letter, it can be used to request information, notify clients about a decision, or promote a new idea or service.

Different expectations for e-mail usage can lead to misunderstandings. Each company, it seems, has its own e-mail culture. In some companies, e-mail is the preferred way to communicate. Sending an e-mail to people sitting only a few cubicles away is considered normal. In other companies, the phone and face-to-face interactions are still preferred over e-mail. E-mail is used only for minor, usually secondary, communications ("Do you want to go to lunch today?").

These different expectations for e-mail can lead to problems. For example, someone at another company might consider e-mail an informal way to send a note. So, your formal request for information might be overlooked, or it might receive an informal response (Figure 12.2). Or, perhaps your reader does not use e-mail for informal purposes, so your casual or flippant response to his or her message would seem highly unprofessional.

Different Expectations for E-Mail Usage

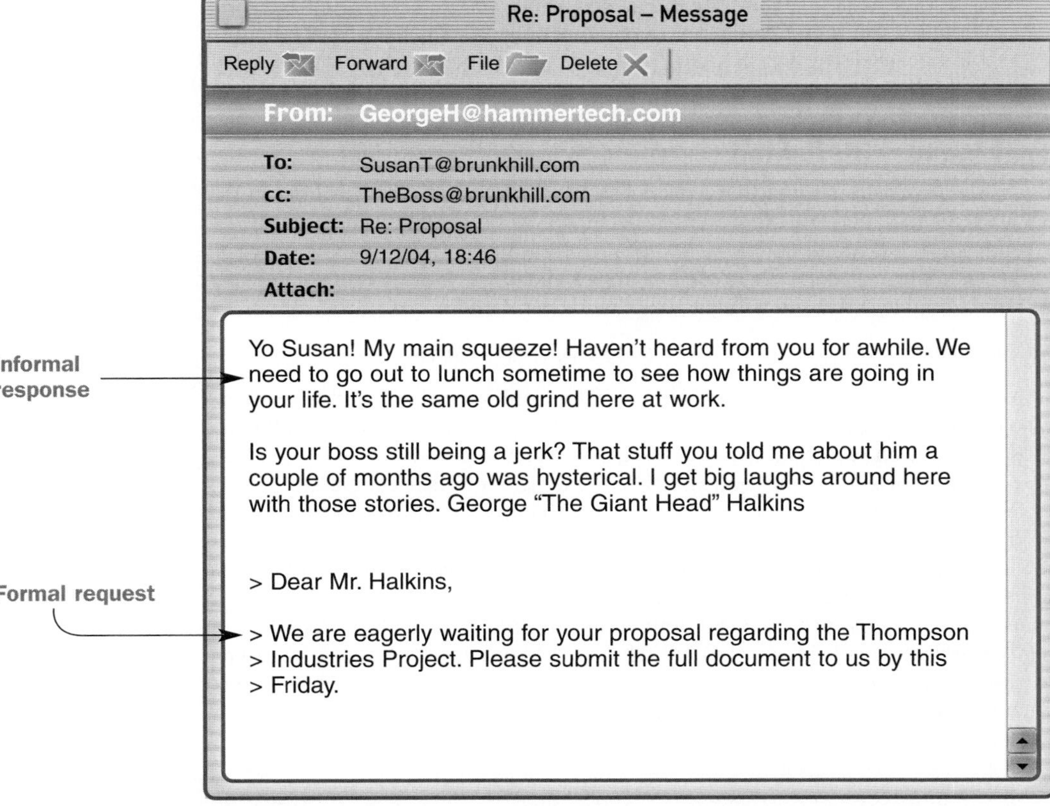

Figure 12.2: People often view the importance of e-mail differently. Here, a formal request is misunderstood. Also, Susan might have a bigger problem (notice the cc line).

GO TO THE NET

For websites that offer free e-mail service, go to **www.ablongman.com/johnsonweb/12.2**

When you are using e-mail, here are a few guidelines to keep in mind:

E-mail is increasingly used for professional purposes—Not long ago, e-mail was considered a secondary form of communication. It was spontaneous and chatty, used mostly for quick comments or nonessential information. Today, e-mail is a principal form of communication in most workplaces, so people expect e-mail messages to be professional and nonfrivolous.

E-mail is a form of public communication—Your readers can purposely or mistakenly send your e-mail messages to countless others. So, you should not say things with e-mail that you would not say openly to your supervisors, coworkers, or clients.

E-mail is increasingly formal—In the past, readers forgave typos, misspellings, and bloopers in e-mail messages, especially when e-mail was new and difficult to use. Today, readers expect e-mails to be more formal, reflecting the quality expected in other forms of communication.

E-mail standards and conventions are still being formed—How e-mail should be used in the workplace is still being worked out. People hold widely different views about the appropriate (and inappropriate) use of e-mail. So, you need to pay close attention to how e-mail is used in your company and your readers' companies. Many companies are developing policies explaining how e-mail should be used. If your company has a policy on e-mail usage, you should read it and follow it.

Legal Issues with E-Mail

You should also keep in mind that legal constraints shape how e-mail is used in the workplace. E-mail, like any other written document, is protected by copyright law. So, you need to be careful not to use e-mails in any way that might violate copyright law. For example, if you receive an e-mail from a client, you cannot immediately post it to your company's website without that client's permission.

LINK For more information on copyright law, go to Chapter 4, page 80.

Also, lawyers and courts treat e-mail as written communication, equivalent to a memo or letter. For example, much of the antitrust case against Microsoft in the late 1990s was built on recovered e-mail messages in which Bill Gates and other executives chatted informally about aggressively competing with other companies.

Legally, any e-mail you send via the employer's computer network belongs to the employer. So, your employers are within their rights to read your e-mail without your knowledge or permission. Also, deleted e-mails can be retrieved from the company's servers, and they can be used in a legal case.

Increasingly, harassment and discrimination cases hinge on evidence found in e-mails. Careless e-mails about personal relationships or appearances can be saved and used against the sender in a court case. Indiscreet comments about gender, race, or sexual orientation can also have unexpected consequences. Your "harmless" dirty

For examples of corporate e-mail policies, go to **www.ablongman.com/johnsonweb/12.3**
To read some classic e-mail blunders, go to **www.ablongman.com/johnsonweb/12.4**

jokes sent to your coworkers might end up being used by a lawyer to prove that you are creating a "hostile workplace environment."

LINK For more information on ethics, see Chapter 4, page 70.

Basic Features of E-Mail

An e-mail is formatted similarly to a memo (Figure 12.3). Typical e-mail messages will have a *header* and *body*. They also have additional features like *attachments* and *signatures*.

Basic Features of an E-Mail

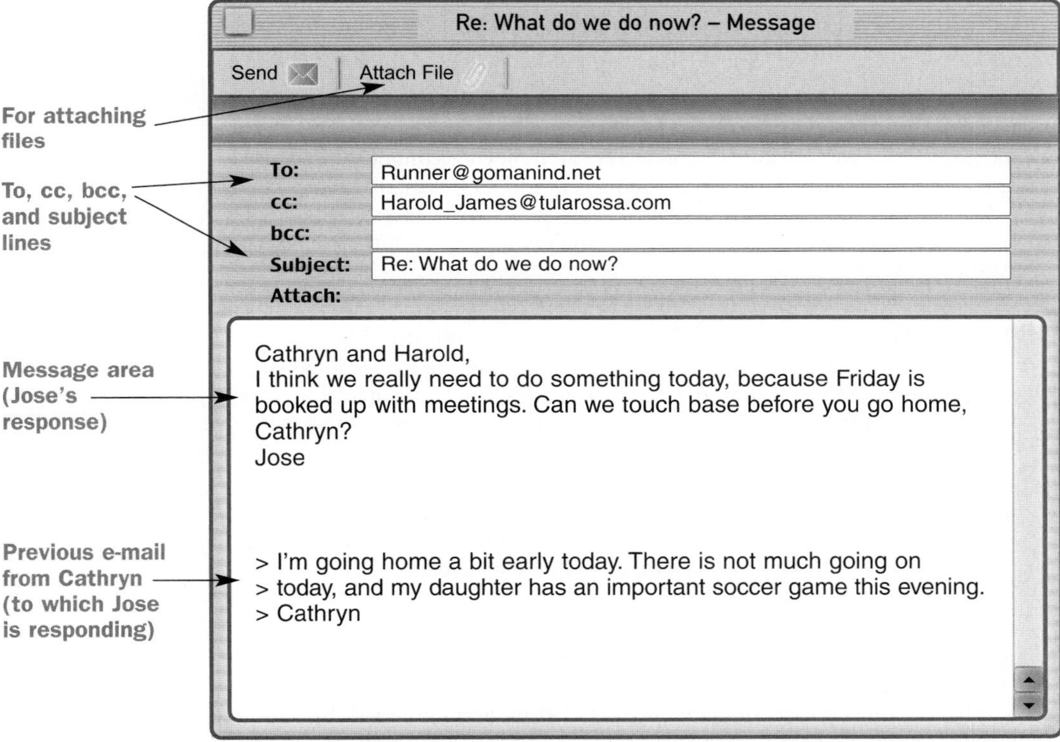

Re: What do we do now? – Message
Send
Attach File
For attaching files
To, cc, bcc, and subject lines
To: Runner@gomanind.net
cc: Harold_James@tularossa.com
bcc:
Subject: Re: What do we do now?
Attach:
Message area (Jose's response)
Cathryn and Harold,
I think we really need to do something today, because Friday is booked up with meetings. Can we touch base before you go home, Cathryn?
Jose
Previous e-mail from Cathryn (to which Jose is responding)
> I'm going home a bit early today. There is not much going on
> today, and my daughter has an important soccer game this evening.
> Cathryn

Figure 12.3: A basic message has several different parts. In this message, the sender, Jose, is replying to Cathryn and copying (cc'ing) the message to Harold.

Header

The header has lines for *To* and *Subject*. Usually, there are also lines like *cc*, *bcc*, and *Attachments*, which allow you to expand the capabilities of the message.

> **To line**—Here is where you type the e-mail address of the person to whom you are sending the e-mail. You can put multiple addresses on this line, allowing you to send your message to many people.

To learn about the history of e-mail, go to **www.ablongman.com/johnsonweb/12.5**

cc and bcc lines—The cc and bcc lines are used to copy the message to people who are not the primary readers, like your supervisors or others who might be interested in your conversation. The cc line shows your message's recipient that others are receiving copies of the message too. The bcc line ("blind cc") allows you to copy your messages to others without anyone else knowing.

Subject line—The subject line signals the topic of the e-mail. Usually a small phrase is used. If the message is a response to a prior message, e-mail programs usually automatically insert a "Re:" into the subject line. If the message is being forwarded, a "Fwd:" is inserted in the subject line.

Attachments line—This line signals whether there are any additional files, pictures, or programs attached to the e-mail message. You can attach whole documents created on your word processor, spreadsheet program, or presentation software. An attached document retains its original formatting and can be downloaded right to the reader's computer.

Message Area

After the header, the *message area* is where you can type your comments to your readers. Messages can be short or long. However, a good rule of thumb is to keep the length of your message under one and a half screens. A shorter message minimizes the amount of scrolling required. Most people don't want to read long e-mail messages. Many just won't.

The body should have a clear introduction, body, and conclusion.

- The *introduction* should minimally (1) define the subject, (2) state your purpose, and (3) state your main point. Also, if you want the reader to do something, you should mention it up front, not at the end of the e-mail.
- The *body* should provide the information needed to prove or support your e-mail's purpose.
- The *conclusion* should restate the main point and look to the future. Again, here is not the place to first tell someone to do something. Most readers of e-mail never reach the conclusion, so you should tell them any action items early in the message and then *restate* them in the conclusion.

The message area might also include these other kinds of text:

Reply text—When you reply to a message, most e-mail programs allow you to copy parts of the original message into your message. These parts are often identified with > arrows running down the left margin (Figure 12.3).

Links—You can also add in direct links to websites (Figure 12.4). Most programs will automatically recognize a webpage address like "http://www.predatorconservation.org" and make it a live link in the e-mail's message area.

Attachments—If you attach a file to your e-mail message, you should tell the readers in the message area that a file is attached. Otherwise, they may not notice it.

For helpful advice about using e-mail, go to **www.ablongman.com/johnsonweb/12.6**

Using the Message Area

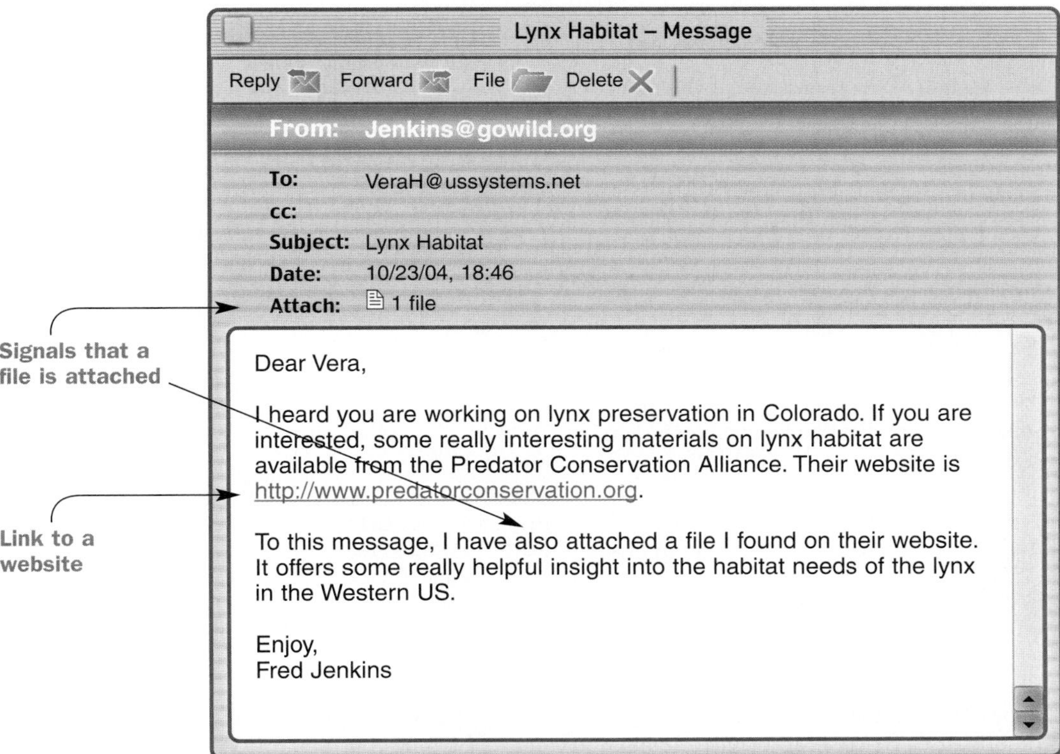

Lynx Habitat – Message
Reply
Forward
File
Delete
From: Jenkins@gowild.org
To: VeraH@ussystems.net
cc:
Subject: Lynx Habitat
Date: 10/23/04, 18:46
Attach: 1 file
Dear Vera,
I heard you are working on lynx preservation in Colorado. If you are interested, some really interesting materials on lynx habitat are available from the Predator Conservation Alliance. Their website is http://www.predatorconservation.org.
To this message, I have also attached a file I found on their website. It offers some really helpful insight into the habitat needs of the lynx in the Western US.
Enjoy,
Fred Jenkins
Signals that a file is attached
Link to a website

Figure 12.4: This e-mail includes a live link that will take the readers directly to the website mentioned. Also, an attached file is mentioned in the second paragraph.

Emoticons—Another common feature in the message area is the use of *emoticons.* Used sparingly, they can help you signal emotions that are hard to convey in written text (Figure 12.5). When overused, emoticons can become annoying to some readers. In most workplace situations, emoticons should not be used. They are playful and informal; therefore, they are really only appropriate in e-mails between close colleagues or friends. In any other e-mail, their meaning or the text around them could be misunderstood, leading to confusion.

Signature

E-mail programs usually let you create a *signature file* that automatically puts a *signature* at the end of your messages. Some signatures are simple:

Frank Randall, Marketing Assistant

Genflex Microsystems

612-555-9876

frandall@genflexmicro.net

For more emoticons, go to **www.ablongman.com/johnsonweb/12.7**

Common Emoticons

Symbols	Meaning
:-)	Happy face
;-)	Winking happy face
:-o	Surprised face
:-|	Grim face
:-\	Smirking face
:-(	Unhappy face
;-(	Winking unhappy face
:-x	Silent face
:'-(	Crying

Figure 12.5: Emoticons are best used in e-mails between close colleagues and friends. They can show emotions that are hard to convey in writing.

Some signature files can be complex:

```
Elizabeth Smith
Physics Goddess
Los Alamos NL
P.O. Box 4501345E
Los Alamos, NM 87545
eliza455@lanl.gov
```

Signature files allow you to personalize your message and add in additional contact information. By creating a signature file, you can avoid typing your name, title, phone number, etc. at the end of each message you write.

Attachments

One of the advantages of e-mail over other forms of communication is the ability to send and receive attachments. Attachments are files, pictures, or programs that readers can download to their own computer. So, you can send your readers a 20-page report that you promised them. Or, you can send out pictures of the office holiday party.

Sending attachments—If you would like to add an attachment to your e-mail message, click on the button that says "Attach Document" or "Attachment" in your e-mail software program. Most programs will then open a textbox that allows you to find and select the file you want to attach.

Receiving attachments—If someone sends you an attachment, your e-mail program will use an icon to signal that a file is attached to the e-mail message. Click on that icon. Most e-mail programs will then allow you to save the document to your hard disk. From there, you can open the file.

Want to find a cool signature file? Go to **www.ablongman.com/johnsonweb/12.8**

AT WORK

A. J. Hyland

PRESIDENT AND CEO OF HYLAND SOFTWARE, WESTLAKE, OHIO

Hyland Software develops software for document management at the department and corporate levels.

How do you ensure your readers don't misread your e-mail messages?

E-mail is a fantastic business tool, but misreading someone's voice and tone in an e-mail is very easy. So, you have to play close attention to the words you choose. Otherwise, your readers might glean the wrong message and take actions that you didn't expect.

When I write e-mail, I do a couple of different things to make sure I get my message across. If I'm communicating company policy, particularly if the issue is a sensitive one, then I will pick a group of people to review the e-mail before I send it out. That way, the group can edit any mistakes and give me substantive feedback about the policy. Most important, the extra time lets me reread the message before it goes out. Letting others review the e-mail makes the process a bit longer, but it is essential for making sure we don't send the wrong message.

We tell our employees a few simple things about e-mail:

- Be concise and get to the point quickly.
- Don't use slang. With slang, you can't be sure that your readers will understand. Also, you can't be sure they won't be offended by it.
- If you have the slightest doubt about what the e-mail is communicating, have a mentor or a coworker look it over.

With e-mail, it is always better to be safe than sorry.

TAKE NOTE Before opening attachments, make sure your virus protection software is working properly. Many viruses enter computers through attachments. To avoid a virus, it is also a good idea to set your e-mail program so that attachments are not downloaded automatically when they are received.

Managing E-Mail

Before long, e-mail messages begin to pile up in your Inbox, and you start to feel like you will never be able to read them all. This is called an *e-mail glut.* To avoid an e-mail glut, you need to develop a strategy for managing your e-mail messages.

A good e-mail management strategy follows four steps:

Step 1: Prioritize—Which messages are the most important?

Step 2: Read and Reply—Which messages need action right now?

Step 3: Sort and Archive—Which messages should be saved?

Step 4: Purge—Which messages can be deleted?

Tired of unwanted e-mail? For help, go to **www.ablongman.com/johnsonweb/12.9**

If you follow these four steps regularly, you will be able to work more efficiently with e-mail.

Step 1: Prioritize

When you open your e-mail Inbox, begin by mentally prioritizing your messages. Some messages, like those from your supervisor or clients, probably need immediate attention. Other messages, like personal messages and those looking for information, can be handled later.

TAKE NOTE Some people like to answer their e-mail from oldest to most recent. This approach only ensures that the most recent, and perhaps most urgent, do not receive much attention. By prioritizing your e-mails, you can handle first things first. Messages that are most significant and urgent receive the most attention. The least important messages appropriately receive less attention.

Step 2: Read and Reply

E-mail messages tend to fall into three basic categories: action-item messages, informational messages (FYI), and junk. As you begin to read and respond to your messages, ask yourself three key questions:

Who is the message from?

Why did they send me this message?

What do they want me to do?

The *who* question will help you determine whether an e-mail is junk. If you don't have a professional or personal relationship with the sender, the message is probably junk. So, you can begin by deleting frivolous messages, spam, and messages from listserv mailing lists that you don't have time to read.

The *why* question will help you mentally prioritize your messages. If the sender is asking you to take action ("Please send me that report I asked for"), you should give the message high priority. If the message is informational ("Hey, folks, don't forget I'm going on vacation next week"), you can simply read and delete. If you are being copied (cc:) from someone else's conversation, you can also read and delete.

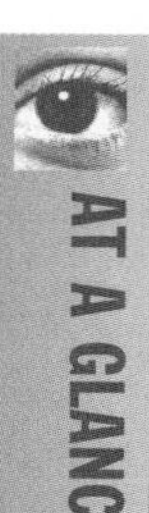

Reading and Replying to E-Mail

- Delegate—Should someone else do this work?
- Do Now—Can I handle this request right now?
- Do Later—Should I handle this request later?

The *what* question helps you determine if you need to take action. Specifically, determine what the sender wants you to do and decide whether you can do it.

Now that you have prioritized your e-mails, start working with the e-mails that require you to take action. You should take one of the following actions:

Delegate—Determine whether someone else can or should do this work for you. If someone can, hit the Forward button and send him or her the message.

Do now—Most requests from e-mail senders can be handled right now. If you can do what is being requested in two minutes or less, do it now. That way, these small requests won't start cluttering up your day.

Do later—Some actions require more time. Jot down these tasks on your "To Do" list. Also, write down a day and time when you will do these tasks.

As you answer your e-mail, be careful when you are replying. We have probably all received embarrassing messages from people who intended to reply individually to a sender but instead replied to everyone who received the original message. To avoid these mishaps, keep in mind the difference between the "Reply" and the "Reply All" functions of your e-mail program.

Reply—Sends a message back to the person who sent you the message. The other receivers of the original message will not receive your reply.

Reply All—Sends a message back to that person and all the other people who received the original message.

So, to avoid sending unnecessary or embarrassing e-mail to all those other people, you might set your e-mail program to reply *only to the sender* when you select the "Reply" button. If you want to reply to everyone, you would then need to make the additional effort to select "Reply All" from your e-mail program's menu.

Step 3: Sort and Archive

When you are finished reading and responding, you should then sort the remaining e-mails into folders. Even the simplest e-mail program will allow you to create folders for sorting your messages (Figure 12.6).

In most e-mail programs, you can also create *filters* for your e-mail that will automatically sort your messages into mailboxes. In your e-mail program, look for a function called "Create Filter" or something similar. When "Create Filter" is selected, a box will usually appear that allows you to create filters that sort e-mails by *sender, subject line*, or *keywords*. Filters are a handy way to sort e-mail, especially if you receive many messages each day.

You may decide to keep a few e-mails in your Inbox because you need to respond to them soon. But most of the time, your Inbox should be as empty as possible. If a message is not worth sorting, it is probably not worth keeping.

TAKE NOTE Your e-mail folders should reflect your activities at work. Each project should have its own folder. Each of your coworkers, especially your supervisor, should have a separate folder too.

You might also decide to *archive* more important e-mail messages as a way to keep a record of your decisions and actions. E-mails can be archived in electronic or print form. Less important e-mails can be stored in one of your e-mail account's folders. You might even create a special folder called "Archive" for these messages. Or, you might save them on your computer's hard drive. The most important e-mails should be printed out and filed in paper form.

To learn more about creating filters, go to **www.ablongman.com/johnsonweb/12.10**

Mailboxes and Folders

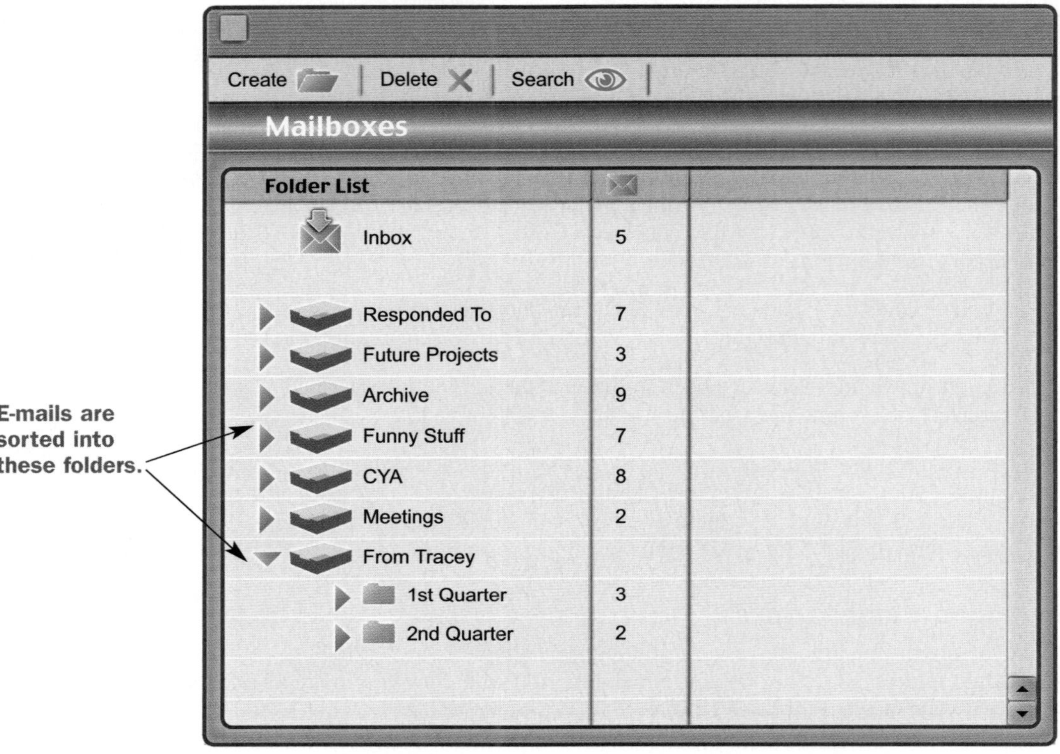

Figure 12.6: Your e-mail program will allow you to sort your e-mails into folders. Some programs allow you to sort files automatically by keywords.

Step 4: Purge

No matter how hard you try, e-mail messages will still pile up. So, once a week, you should set aside time to *purge* your e-mail account. Start the purging process by doing something about all the messages piling up in your Inbox.

Delete—Scrap any messages that have been superceded by the week's events. If you didn't respond to them in time, you might as well delete them and stop feeling guilty about not replying.

Take action—Don't procrastinate any more. Do it now. Take action on e-mails that require you to do something.

Write it down—If you cannot take action now, write down the senders' requests on your "To Do" list, put them on your calendar, or enter them into your personal digital assistant (PDA). Then, delete the e-mails or sort them.

When you are finished with the Inbox, go through your e-mail account's other folders one by one, purging any old messages that are no longer important or relevant. Be realistic about the messages you save. Will you ever need these messages

To find more tips on managing e-mail, go to **www.ablongman.com/johnsonweb/12.11**

AT A GLANCE

Managing E-Mail

- Step 1: Prioritize—Which messages are the most important?
- Step 2: Read and reply—Which messages need action right now?
- Step 3: Sort and archive—Which messages should be saved?
- Step 4: Purge—Which messages can be deleted?

again? If these old messages are important, ask yourself whether they should then be archived on your hard drive or printed out and filed. If they don't need to be kept, delete them.

TAKE NOTE E-mail messages may be purged from your system, but they are often retained on the server. An expert or investigator can recover these e-mails long after they were deleted. Here is another reason why e-mail should not be used to do anything that might be unethical. You never know when those e-mails will resurface.

E-Mail Netiquette

Because e-mail is relatively new, we are still figuring out how it should be used. When using the phone, you have a rather firm idea about the etiquette to be followed. You answer the phone with "Hello." Or, perhaps at work you might say something like, "Hello, this is Brenda Chavez." At work, if someone answered the phone with "Yo" or "Speak to me," you might be a bit annoyed.

Like the phone, e-mail also has its own etiquette, or "netiquette." Here are a dozen netiquette guidelines that you can follow in the workplace:

Guideline 1: Be concise—Keep the length of your messages under a screen and a half. If you have more than a screen and a half of content, pick up the phone, write a memo, or send the information as an attachment.

Guideline 2: Provide only need-to-know information—Decide who needs to know what you have to say, and send them only information they need—nothing more.

Guideline 3: Treat the security of the message about the same as a message on a postcard—Recognize that anyone can pass your message along or use it against you. If the message is confidential or proprietary, e-mail is not an appropriate way to send it.

Guideline 4: Don't say anything over e-mail that you would not say in a meeting or to legal authorities—Hey, accidents happen. You might send something embarrassing to the entire organization. Or, someone might do it for you.

Guideline 5: Never immediately respond to a message that made you angry or upset—Give yourself time to cool off. You should never write e-mail when you are angry, because these little critters have a way of returning to bite you.

Guideline 6: Avoid using too much humor, especially irony or sarcasm—In an e-mail, your witty quips rarely come off exactly as you intended. Remember that your e-mails can be easily misinterpreted or taken out of context. The American sense of humor, meanwhile, is often disturbing to international audiences; so if your message is going overseas, keep the humor to a minimum.

GO TO THE NET

Want to learn more about e-mail netiquette? Go to **www.ablongman.com/johnsonweb/12.12**

Guideline 7: Be extra careful about excerpting or forwarding the e-mail of others—If there is the remotest chance that someone else's message could be misunderstood, paraphrase it in your own message instead of forwarding his or her message or copying parts of it.

Guideline 8: Don't be unpleasant over e-mail if you would not be so face to face—E-mail provides a false sense of security, much like driving in a car. But in reality, those are real people on that information superhighway, and they have feelings. Plus they can save your abusive or harassing messages and use them as evidence against you.

Guideline 9: Never send anything that could be proprietary—There have been some classic cases where classified information or trade secrets have been sent to competitors or the media over e-mail. If the information is proprietary, use traditional routes like the regular mail or a courier.

Guideline 10: Don't flame, spam, or chain-letter people at work—If you must, do these things from your personal computer at home. At work, assume these activities will ultimately be used against you. You might be fired.

Guideline 11: Think twice before you send "urgent" messages that are making their way across the net—There are lots of hoaxes about computer viruses "erasing your hard drive." These hoaxes gain new life when some gullible person starts sending them to all their friends. Some viruses are real, but most are not. Wait a day or two to see if the threat is real before you warn all your friends and colleagues.

Guideline 12: Be forgiving of the grammatical mistakes of others—but don't make them yourself—Some people treat e-mail informally, so they rarely revise (though they should). Grammar mistakes and misspellings happen—so you can forgive the sender. On the other hand, these misteaks and typos make the sender look stoopid. Don't send out sloppy e-mail messages.

HELP

Using Mailing Lists (Listservs)

Mailing lists, commonly called listservs or list servers, are a special feature of e-mail. When an e-mail is sent to a mailing list, the message is broadcasted to everyone who subscribes to the list.

Mailing lists are available on just about any topic, allowing specialists to communicate widely with others in their field. In the technical workplace, mailing lists are essential for keeping up to date with new developments and conversations in your field. If you are a medical doctor, for example, a mailing list might alert you to a new drug. If you are a scientist, a mailing list might help you find others who are working in your research area.

TAKE NOTE Listserv is actually the brand name of a software product. Like Kleenex, Twinkies, and Saran Wrap, however, the brand name has become synonymous with the product. Other mailing list software products include ListSTAR, Lyris, Majordomo, and UnityMail.

Want to know if a virus is a hoax? Go to **www.ablongman.com/johnsonweb/12.13**

Some people like to participate actively in mailing lists, conversing with specialists with similar backgrounds or interests. Others like to "lurk" on mailing lists, reading the messages but almost never participating in the discussions.

Finding a Mailing List

The best place to search for a specific mailing list is at the CataList site (http://www.lsoft.com/catalist.html). This site catalogs all the mailing lists that use Listserv software. Here, you can use keywords to search for lists on a specific topic that interests you.

You can also ask other students, professors, colleagues, and supervisors about the mailing lists they subscribe to. They can probably send you information about how to subscribe.

Subscribing and Unsubscribing to a List

To subscribe to a list, you will need to send a message to the owner of the mailing list. In the "To" line of your e-mail, you should type the address of the mailing list owner. Then, in the message area of the e-mail, type

```
subscribe <listname>
```

Some mailing lists require you to state your name also.

```
subscribe <listname> <yourfirstname> <yourlastname>
```

To unsubscribe from a list, you can use the "unsubscribe" or "signoff" command.

```
unsubscribe <listname>

signoff <listname>
```

Most mailing lists will immediately send you a file that describes how the list works, including a list of Frequently Asked Questions. A mailing list may also automatically ask you to confirm your subscription by e-mail.

Participating in the Discussion

Once you have subscribed to a mailing list, you will begin receiving mail. Usually, the best way to participate is to reply to someone else's e-mail. Hitting the "Reply" button will automatically put the address of the mailing list in the "To" line of your e-mail message. You can then type your message.

TAKE NOTE Because your message to a mailing list may be sent to thousands of other people, you need to be especially aware of netiquette. It's one thing to look careless or foolish in e-mails sent to your friends, but thoughtless e-mails to a mailing list can do quick damage to your reputation.

Flaming is a special problem on mailing lists. Sometimes debates on mailing lists turn ugly when two or more subscribers start quarreling with each other. For the most part, flaming is a waste of time for the participants, and flames are annoying to the other subscribers on the list. You should avoid participating in these kinds of e-mail exchanges.

To find a listserv on a subject that interests you, go to **www.ablongman.com/johnsonweb/12.14**

What Is Instant Messaging?

Variations of instant messaging (IM) have been around for some time. Recently, however, instant messaging has started to become an important tool in the technical workplace. People now use instant messaging to communicate through their computers, wireless phones, and personal digital assistants (PDAs).

Unlike e-mail, instant messaging happens in real time. When you hit "Send," your message will immediately appear on the receivers' screens (Figure 12.7). They can then respond, sending you an instant message.

An Instant Messaging Interface

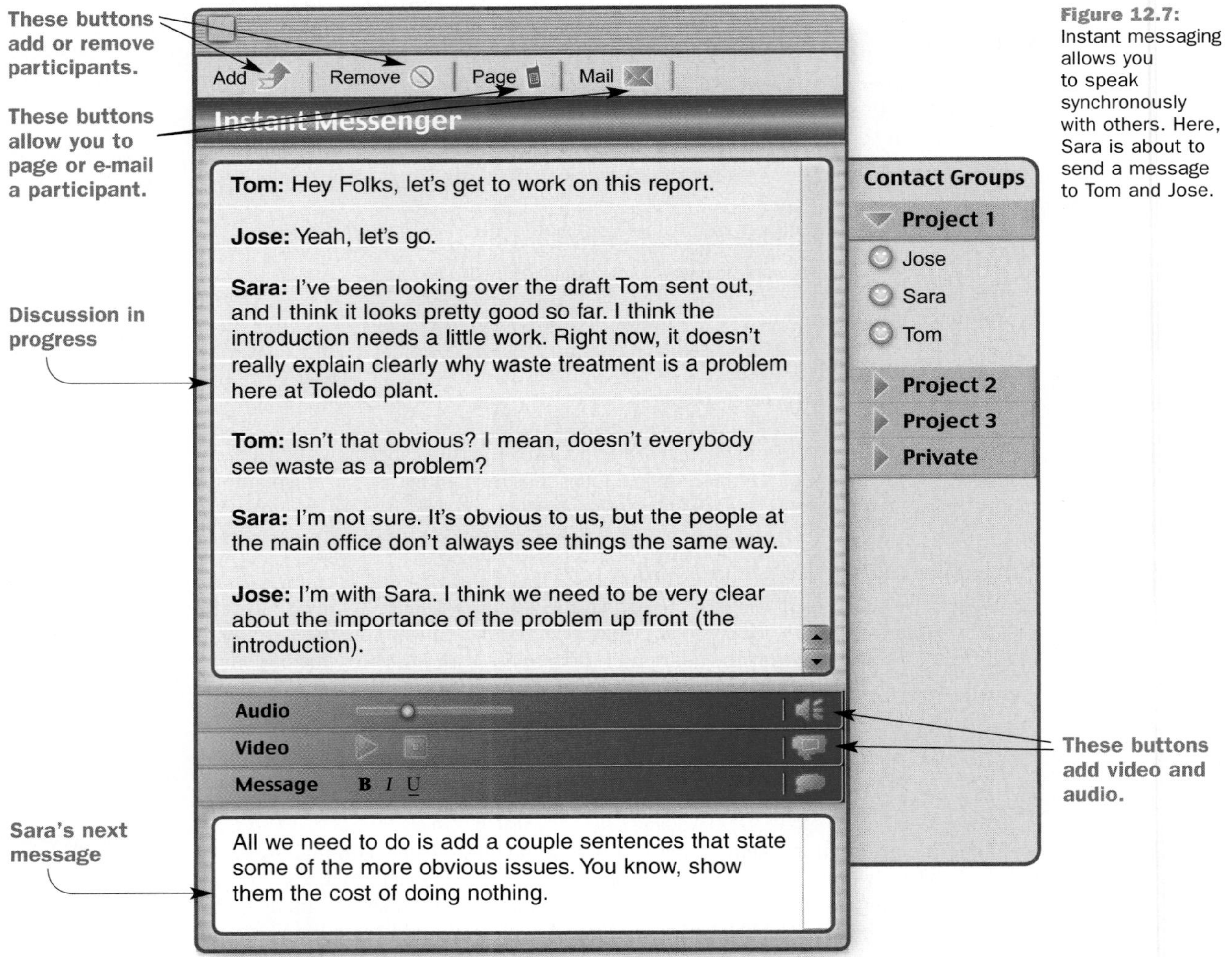

These buttons add or remove participants.
These buttons allow you to page or e-mail a participant.
Discussion in progress
Sara's next message
These buttons add video and audio.
Add
Remove
Page
Mail
Instant Messenger
Tom: Hey Folks, let's get to work on this report.
Jose: Yeah, let's go.
Sara: I've been looking over the draft Tom sent out, and I think it looks pretty good so far. I think the introduction needs a little work. Right now, it doesn't really explain clearly why waste treatment is a problem here at Toledo plant.
Tom: Isn't that obvious? I mean, doesn't everybody see waste as a problem?
Sara: I'm not sure. It's obvious to us, but the people at the main office don't always see things the same way.
Jose: I'm with Sara. I think we need to be very clear about the importance of the problem up front (the introduction).
Contact Groups
Project 1
Jose
Sara
Tom
Project 2
Project 3
Private
Audio
Video
Message
B I U
All we need to do is add a couple sentences that state some of the more obvious issues. You know, show them the cost of doing nothing.

Figure 12.7: Instant messaging allows you to speak synchronously with others. Here, Sara is about to send a message to Tom and Jose.

GO TO THE NET

Want to go to websites that offer free instant messaging? Go to **www.ablongman.com/johnsonweb/12.15**

TAKE NOTE Instant messaging is called "synchronous" because the conversation happens in real time. E-mail is "nonsynchronous" because messages are saved in the receiver's Inbox until he or she is ready to read and respond to them.

Some instant messaging programs include the ability to send live audio or video. With a microphone and/or digital camera (webcam) mounted on your computer, you can let other people see and hear you talking.

Using Instant Messaging

Several companies, such as Yahoo, AOL, and Microsoft, will give you free access to instant messaging. To use their systems, you will need to download their software to your computer from their websites. If you have a wireless phone or PDA, you might also have access to instant messaging.

Most instant messaging programs ask you to sign in with a password. Then, you can create a list of other people with whom you want to converse. To begin a conversation, send a message to the others to see if they are at their computers. Their computers will open a window or make a sound to tell them you want to talk to them. If they respond, you can start writing back and forth.

TAKE NOTE Instant messaging has always been a good way to stay in touch with friends. Increasingly, though, coworkers are starting to use instant messaging to hold meetings and collaborate on projects. With video and audio capabilities, people across the country can have a live meeting through their computers.

Here are a few guidelines for using instant messaging at work:

Schedule instant messaging meetings—If you are going to use instant messaging to meet with people, you should schedule a time for the virtual meeting. Without a scheduled meeting, you may not be able to pull people away from their other work.

Write concise messages—Instant messaging conversations usually move along quickly, especially if you are talking with more than one person. So, keep your messages short and easy to read. You might learn to use some of the common instant messaging abbreviations to keep the messages short (Figure 12.8). The common emoticons for e-mail also work well.

Cut the chitchat—Instant messaging can be fun, but time is valuable. When collaborating with others, keep the personal comments to a minimum.

Use instant messaging at work only when necessary—The problem with instant messaging is that others need to respond right away. If they are doing other important work, they may not be able to write back. So, if your message does not need immediate attention, you might send an e-mail instead.

Instant messaging is a valuable new tool that is becoming as common as e-mail in the workplace. As companies look to cut costs and improve efficiency, instant messaging is a good way to replace in-person meetings, while allowing people to collaborate through their computers.

For websites that discuss using IM, go to **www.ablongman.com/johnsonweb/12.16**

Instant Messaging Abbreviations

Abbreviation	Meaning
AFAIK	as far as I know
B4N	bye for now
BBL	be back later
BTDT	been there done that
CUL	see you later
F2F	face to face
G	grin
GDR	grinning, ducking, and running
HTH	hope this helps
IMO	in my opinion
JTLYK	just to let you know
LOL	laughing out loud
TTFN	ta ta for now
WFM	works for me

Figure 12.8: Common instant messaging abbreviations. These abbreviations can help you communicate more efficiently and add emotion to your messages.

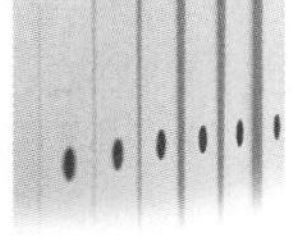

CHAPTER REVIEW

- E-mail is an essential tool in the workplace, and instant messaging is quickly growing in popularity. These tools allow you to communicate, collaborate, and connect to others.
- E-mail is a hybrid technology that combines the functions of the phone, memos, and letters.
- In the workplace, e-mail should be used for professional purposes only, because it is a public form of communication and should always convey professionalism.
- An e-mail has the following features: header, message area, signature, and attachments.
- Managing e-mail requires some discipline and diligence. To avoid e-mail glut, you need to (1) prioritize, (2) read and reply, (3) sort and archive, and (4) purge your messages.
- As you are reading and responding to your e-mails, remember to delegate, do now, or do later.
- Netiquette involves guidelines for politeness and ethical behavior with e-mail.
- Mailing lists, often called listservs, are good ways to stay in touch with your colleagues and professional groups.
- Instant messaging is a synchronous way to communicate and hold "virtual meetings" with other people.

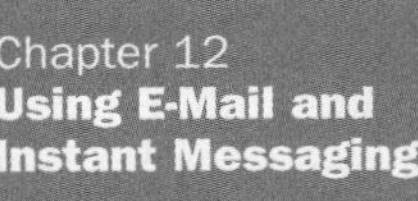

More IM abbreviations can be found at
www.ablongman.com/johnsonweb/12.17

Individual and Team Projects

1. Use a search engine to find a special interest group on the Internet that you might like to join. Send the webmaster a polite e-mail that requests information about the group. Pay special attention to the content and organization of your e-mail. Also, pay attention to netiquette, trying to avoid any statements that might annoy the recipient of your e-mail.

2. One of your coworkers likes to send out jokes to your whole work team. The jokes are tasteful, but they clog up your Inbox. Write an e-mail to your coworker in which you tactfully ask him to reduce the amount of jokes he sends out.

3. Look at your own e-mail account. Use the e-mail management scheme described in this chapter (prioritizing, reading and replying, sorting and archiving, and purging) to clean up your Inbox and purge old e-mails. Try to manage your e-mail for three weeks with this system, spending some time once a week cleaning up your e-mail account.

4. Find an e-mail that could be misread to sound angry or nasty. Then, rewrite the e-mail in a way that avoids the potential nastiness. If you do not have a sample e-mail, ask one of your classmates or coworkers to send you an e-mail that uses an angry tone.

5. Read the case study at the end of this chapter. Pretend you are Jane. Write an e-mail to your coworkers in which you explain your complaint to your former supervisor. Trade your e-mail with one of the other people in your class to see whether you would handle Jane's problem similarly.

Collaborative Projects

1. With a group of others, develop your own set of netiquette guidelines for your class or university. What are some guidelines for proper use of e-mail in the classroom? How can these guidelines be expanded to suit the entire university?

2. Using instant messaging, collaborate with a group of others to compose a memo to your college or company. In the memo, discuss a specific problem on campus (e.g., the need for better food, safety, faster computers) and ask what can be done about the problem. Pay attention to how instant messaging helps or impedes your ability to compose the memo. When you are finished with the memo, discuss how the writing process changes with instant messaging.

3. With a group, use e-mail or instant messaging to have a practice "flame" about a specific topic (e.g., television program, politician, musician, college team). Without being offensive, argue emphatically about the topic, trying to score points on the other members of the flame. As you flame with your group, notice your reactions to their e-mails. Does each e-mail add to the stress and tension, rarely resolving the conflict? What are some ways you might stop the flaming? How can you and your group restore good relations using e-mail?

CASE STUDY

Exposed

Jane Fredrickson was asked to join an existing work team at her office. The group had been together for a long time, and they all got along really well. Jane was excited about working with some new people.

Unfortunately, the closeness of her new work team soon became a problem. The others on the team seemed to have their own language and code of conduct. They joked around with each other with instant messaging. They sent each other pictures of their children in the bathtub. They sent each other dirty jokes. Frankly, Jane was fed up with their behavior.

So, one day she decided to vent her annoyance in an e-mail to Harold Grimes, her former supervisor. After listing her complaints using words like "annoying," "idiotic," and "unprofessional," she wrote "I wish someone would do something about these people and their juvenile antics."

A week later, Jane's supervisor, Alisha Jones, sent an e-mail to the work team, asking them to put an end to the fun and games (Figure A). To her e-mail, she attached a couple of files that spelled out the company's rules for using e-mail and instant messaging. She also mentioned that there had been a complaint.

To Jane's horror, part of her original message to Harold was copied below Alisha's message. Harold had apparently forwarded part of Jane's original message to Alisha. And, Alisha had simply forwarded it to the group with her own message added on.

Needless to say, the other members of Jane's work team were going to be upset. Even though her name was not on the message, Jane knew they would figure out who complained about them. What should she do now?

Problems with Using E-Mail

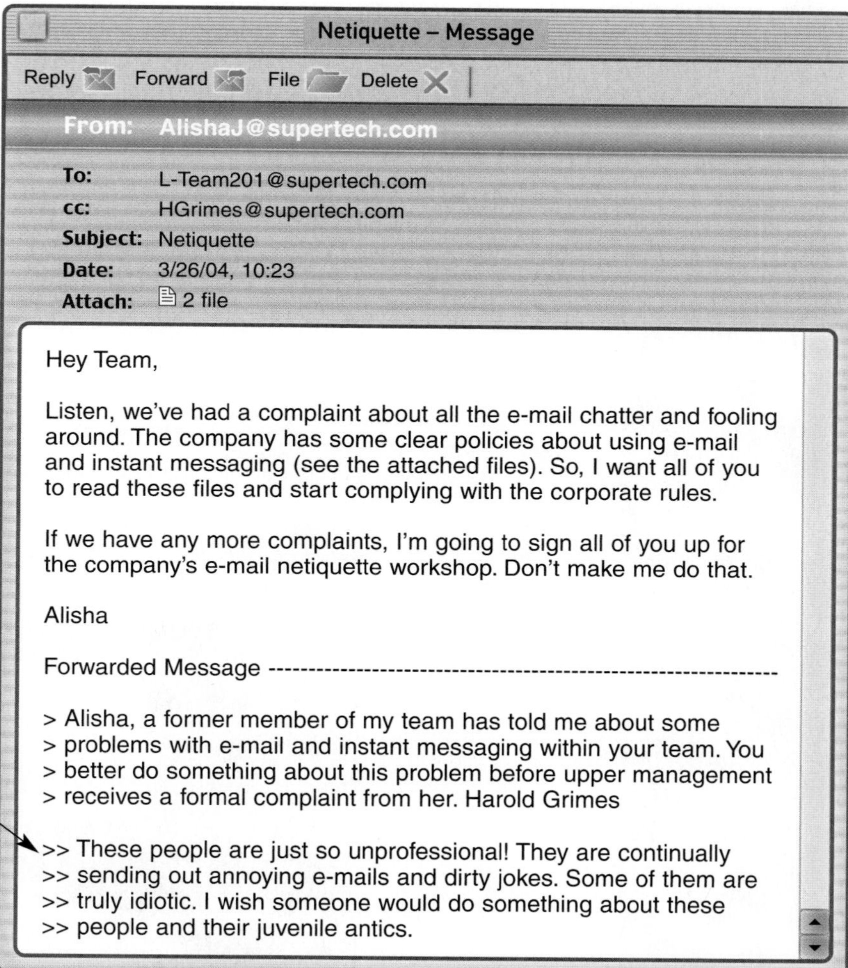
Netiquette – Message

Reply Forward File Delete

From: AlishaJ@supertech.com

To: L-Team201@supertech.com
cc: HGrimes@supertech.com
Subject: Netiquette
Date: 3/26/04, 10:23
Attach: 2 file

Hey Team,

Listen, we've had a complaint about all the e-mail chatter and fooling around. The company has some clear policies about using e-mail and instant messaging (see the attached files). So, I want all of you to read these files and start complying with the corporate rules.

If we have any more complaints, I'm going to sign all of you up for the company's e-mail netiquette workshop. Don't make me do that.

Alisha

Forwarded Message ---

> Alisha, a former member of my team has told me about some
> problems with e-mail and instant messaging within your team. You
> better do something about this problem before upper management
> receives a formal complaint from her. Harold Grimes

>> These people are just so unprofessional! They are continually
>> sending out annoying e-mails and dirty jokes. Some of them are
>> truly idiotic. I wish someone would do something about these
>> people and their juvenile antics.

Oops. Here is part of Jane's original message.

Figure A: Part of Jane's message is copied below Harold's.

CHAPTER

13

Working in Teams

CHAPTER CONTENTS

St. Thomas Medical Center
Date: April 28, 2004
To: Staph Infection Task Force (M. Franks, C. Little, J. Archuleta, L. Drew, V. Yi, J. Matthews, J. McManus)
From: Alice Falsworthy, Infections Specialist
Re: Work Plan for
Last week, we met to disc the increase in st infections here at the St. Thomas Medical Center. We defined our mission, defined major tasks, and developed a project calendar. The purpose of this memo is to summarize those decisions and lay out a work plan that we will follow.
I cannot overstate the importance of this project. Staph infections are becoming an increasing problem at hospitals around the country. Of particular concern are antibiotic-resistant bacteria called methicillin-resistant Staphylococcus an kill patients w otherwise routine injuri do everything in power to control MRSA and other forms of staph.
Please post this work plan in your office to keep you on task and on schedule.
Project Mission
Our Mission: Th
determine the le
Center and to d
staph, especially
Secondary Objec
• Gain a better
treatments.
• Raise awaren
patients.
• Assess the lev
• Develop meth
Forming
Storming
Norming
Performing

Tasks
A1: Collect Curr
A2: Create Pamphlets 7/13
A3: Conduct Experiments 6/2
A4: Develop Contingency Plans 7/13
A5: Develop Training Modules 8/29

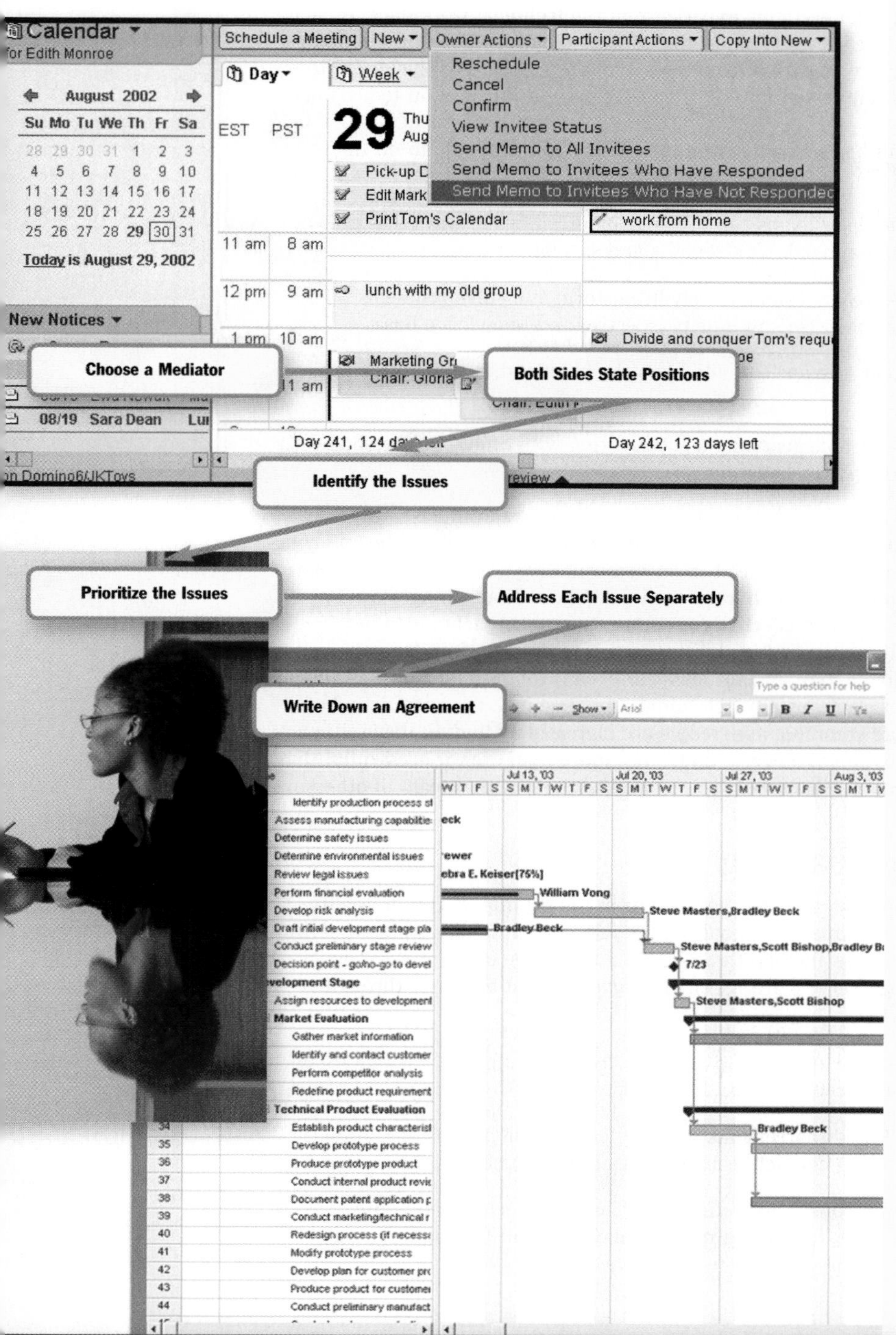

CHAPTER OBJECTIVES

In this chapter, you will learn:

- Why working in teams is essential in technical workplaces.
- The four stages of teaming in the workplace: forming, storming, norming, performing.
- How to use strategic planning to form a team and begin a project.
- Strategies for managing team conflict in the storming stage.
- To define team roles in the norming stage to improve productivity.
- How to improve performance with Total Quality Management (TQM) strategies.
- How to work as part of a "virtual" team.

Working in teams is an everyday experience in the technical workplace. In fact, when managers are surveyed about the abilities they look for in new employees, they often put "works well with a team" near the top of their list. The ability to collaborate with others is an essential skill if you are going to succeed in today's networked workplace.

Working with a team has several advantages. Teams allow people to do the following:

Concentrate strengths—Teams can divide responsibilities in ways that take advantage of each team member's strengths and special abilities.

Foster creativity—Teams can creatively brainstorm and solve problems by drawing from each member's unique perspective and knowledge base.

Share the workload—Teams can divide up the workload, making large projects possible.

Improve morale—People generally enjoy working with others, so the shared responsibilities and rewards boost morale.

Of course, there are also disadvantages to working in teams. Conflicts with team members can be frustrating. Poor planning and time mismanagement can be annoying and demoralizing. Collaboration can lead to uneven-sounding documents. Sometimes other team members don't do their share of the work. These disadvantages can be avoided with good planning and effective communication among team members.

Computers have only increased the ability and necessity to work in teams. Communication tools like e-mail, instant messaging, chat rooms, and websites allow people to work together electronically. Telecommuting, teleworking, and "virtual offices" are becoming more common, and people are increasingly finding themselves working outside the traditional office setting. Now, more than ever, it is essential that you learn how to use computers to help you work effectively with a team of others.

The Stages of Teaming

It would be nice if people worked together well from the beginning. But, in reality, team members often need time to set goals and adjust to each other's working styles and abilities. In 1965, Bruce Tuckman introduced a four-stage description of how teams learn to work together (Figure 13.1). He suggested that teams go through four stages:

Forming—defining the mission, setting objectives, defining responsibilities, establishing a project schedule with deadlines.

Storming—handling disagreements, sensing tension and anxiety, doubting the leadership, addressing conflict, feeling uncertain and frustrated.

Norming—forming consensus, refining the team's objectives and outcomes, solidifying team roles, focusing group members around the mission and objectives.

To learn about Tuckman's teaming theories, go to **www.ablongman.com/johnsonweb/13.1**

Performing—sharing a vision, delegating tasks, feeling a high degree of autonomy, resolving conflicts and differences constructively.

When *forming* and *storming*, a team will be reliant on the team leader or manager for guidance and negotiation. As the team reaches the *norming* and *performing* stages, members will become more autonomous, because they will better understand the project's goals and how each person fits into the overall project mission.

Tuckman's Four Stages of Teaming

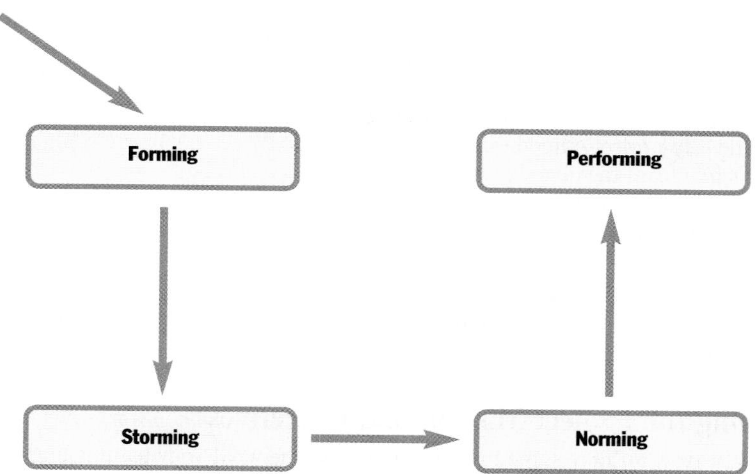

Figure 13.1: A team will typically go through four stages: forming, storming, norming, and performing. Give your team time to properly evolve as a unit.

These stages are not rigid. Instead, teams will tend to move back and forth among the stages as the project evolves and moves forward. For example, a performing group might need to return to the norming stage if the objectives change or someone joins or leaves the group. Occasionally, norming and performing groups return to the storming stage when personalities or goals come into conflict.

You should keep the four stages in mind as you work with your group. Here are a few hints about working well in a group:

- Give your group time to properly form at the beginning of a project by setting objectives, defining responsibilities, and establishing a work plan.
- When storming, don't become too upset or frustrated, because a period of conflict is normal soon after a group is formed.
- When norming, take time to revise team expectations and recognize each member's role in the team, as well as strengths and weaknesses.
- When performing, look for ways to improve quality and refine your group's ability to work at a high level.

TAKE NOTE Not all teams reach the performing stage. Some teams fall apart at the storming phase, becoming dysfunctional and ineffective. Other teams reach only the norming stage, successfully achieving the goals of the project but perhaps not at the optimum level of efficiency or quality.

By keeping these four stages in mind, you can gain a broader understanding of the team's progress and your role within the team's mission.

Forming: Strategic Planning

Forming is an important part of the team-building process. When a team is first created, the members are usually excited and optimistic about the new project. They are often also a little anxious, because each person is uncertain about the others' expectations. So, in the forming stage, members should spend time getting to know each other and assessing each other's strengths and abilities.

When forming a new team, strategic planning is the key to effective teamwork. You should use the following steps to develop a strategic plan for the project:

1. Define the project mission and objectives.
2. Identify project outcomes.
3. Define team members' responsibilities.
4. Create a project calendar.
5. Write out a work plan.
6. Agree on how conflicts will be resolved.

By working through these steps at the beginning of a project, you will give your team time to form properly, often saving yourselves time and frustration as the project moves forward.

Defining the Project Mission and Objectives

Don't just rush in. Take some time to first determine what your team is being asked to accomplish. A good way to start forming is to define the rhetorical situation in which your team is working:

Subject—What are we being asked to develop? What are the boundaries of our project? What are we not being asked to do?

Purpose (mission statement)—What is the mission of the project? Why are we being asked to do this? What are the end results (deliverables) that we are being asked to produce?

Readers—Who are our clients? What are their needs, values, and attitudes? Who will be evaluating our work?

Context—What are the physical, economic, political, and ethical factors that will influence this project? How should we adjust to them?

LINK For more help defining the rhetorical situation, turn to Chapter 2, page 22.

For websites that describe other planning strategies, go to **www.ablongman.com/johnsonweb/13.2**

Your statement of the purpose is your *mission statement* for the project. You and your team members should first agree on this mission statement before you do anything else.

TAKE NOTE Like any statement of purpose, your team should be able to state its mission in one sentence. If you need more than one sentence, your mission is probably not focused enough.

After you have defined the rhetorical situation, set some specific objectives that your team will strive to reach. Your mission statement (purpose) is your primary objective. Now, list two to five secondary objectives that you intend to reach along the way.

Our Mission

The purpose of this Staph Infection Task Force is to determine the level of staph infection vulnerability at St. Thomas Medical Center and to develop strategies for limiting our patients' exposure to staph, especially MRSA.

Secondary Objectives:

- Gain a better understanding of current research on staph and its treatments.
- Raise awareness of staph infections among our medical staff and patients.
- Assess the level of staph risk, especially MRSA, here at the hospital.
- Develop methods for controlling staph on hospital surfaces.

By first agreeing on your team's mission and objectives, you can clarify the goals of the project. You can also avoid misunderstandings about what the team was formed to accomplish.

Identifying Project Outcomes

Outcomes are the tangible results of your project. They describe the measurable results of the team's efforts. To identify the outcomes of your project, simply convert your objectives into measurable results (Figure 13.2). Then specify the *deliverables* that the project will produce. Deliverables are the products or services that you will deliver to the client during the project and after it is completed.

As shown in Figure 13.2, your goal is to transform the project's rather abstract objectives into measurable outcomes. Then, you should go one step further and specify how the outcomes become products (deliverables).

This process of turning objectives into outcomes and deliverables can take some time. But, in the end, the process will save you time, because everyone in your team will have a clear understanding of the expected results and the products that will be created.

To see other sample mission statements, go to **www.ablongman.com/johnsonweb/13.3**

Defining the Project Objectives and Outcomes

Objectives	Outcomes
Gain a better understanding of current research on staph and its treatments	Outcome: Collection of current data and research literature on staph infections. Interview other hospitals about successful control methods. Deliverable: A report on the findings, due May 25
Raise awareness of staph infections among our medical staff and patients	Outcome: More awareness of the problem here at the hospital. People taking precautions against staph. Deliverable: Pamphlets and posters raising awareness and describing staph control procedures
Assess the level of staph risk, especially MRSA, here at the hospital	Outcome: Experiments that quantify the staph risk at the hospital Deliverable: Report to the administration about the extent of the problem, due June 2
Develop methods for controlling staph on hospital surfaces	Outcome: Develop strategies for preventing and containing staph infections Deliverable: Create a contingency plan that offers concrete steps for controlling staph. Make training modules to educate staff about the problem.

Figure 13.2: By transforming your objectives into outcomes and deliverables, you can show how abstract goals become measurable results.

Defining Team Member Responsibilities

Everyone isn't good at everything. So, once you have defined your objectives, ask everyone to talk about their previous experiences and abilities. Ask them to identify ways they could contribute to the project. Discuss any time limitations or potential conflicts with other projects.

TAKE NOTE A good way to define responsibilities in the team is to look at the outcomes and deliverables. Put team member names next to each of these items, making people responsible for completing that part of the project.

When writing a document as a team, you might identify responsibilities while dividing up the writing task. For example, here are four jobs that you might consider:

Coordinator—The coordinator is responsible for maintaining the project schedule and running the meetings. The coordinator is not the "boss." Rather, he or she is a facilitator who helps keep the project on track.

Researchers—One or two people in the group should be assigned to collect information. They are responsible for doing Internet searches, digging up materials in the library, and coordinating the team's empirical research.

For more on sharing team responsibilities, go to **www.ablongman.com/johnsonweb/13.4**

Editor—The editor is responsible for the organization and style of the document. He or she identifies places where the document is missing content or where information needs to be reorganized to achieve the project's purpose.

Designer—The designer is responsible for laying out the document, collecting images, and making tables, graphs, and charts.

Notice that there is not a "writer" among these roles. Instead, everyone in the group is responsible for writing some part of the document. These specific responsibilities are the additional duties of each group member.

Creating a Project Calendar

Project calendars are essential for meeting deadlines. Numerous project management software packages like Microsoft Project, ArrantSoft, and Artemis Project Management help teams lay out calendars for completing projects. These programs are helpful for setting deadlines and specifying when interrelated parts of the project need to be completed (Figure 13.3).

But you don't need project management software for smaller projects. A good time management technique is to use *backwards planning* to determine when you need to accomplish specific tasks and meet smaller and final deadlines.

Project-Planning Software

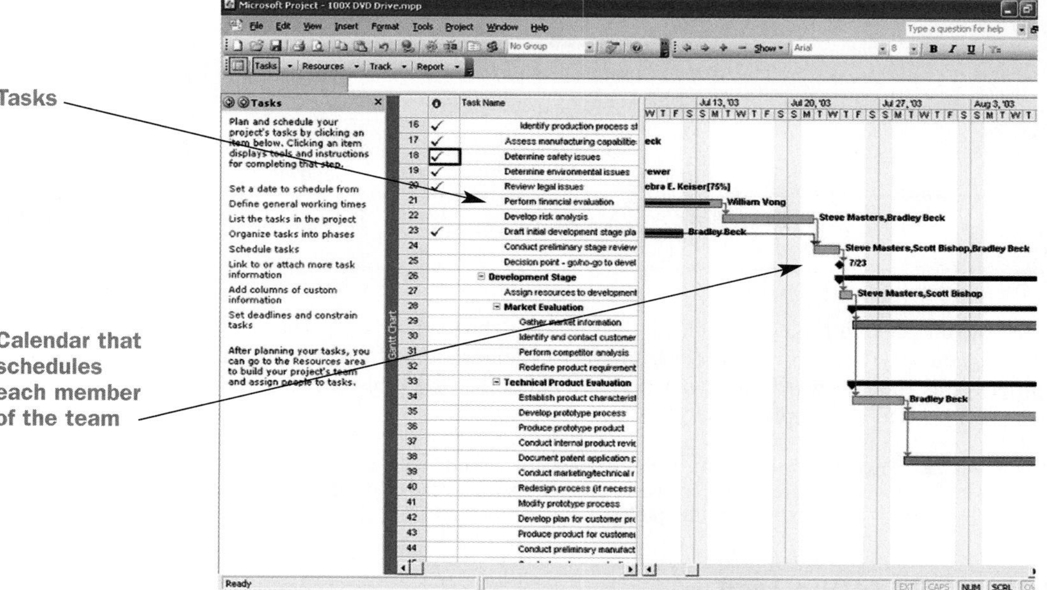

Figure 13.3: Project-planning software can be helpful when setting a calendar for the team. In this screen, the calendar is represented visually to show how tasks relate to each other over time.

Source: Microsoft Inc., http://www.microsoft.com/office/project/prodinfo/standard/overview.mspx.

To do backwards planning, start out by putting the project's deadline on a calendar. Then, work backward from that deadline, writing down the dates when specific project tasks need to be completed (Figure 13.4).

Backwards-Planning Calendar

Project Calendar: Staph Infection Training Modules				
Monday	**Tuesday**	**Wednesday**	**Thursday**	**Friday**
4 Staff Meeting	**5**	**6**	**7**	**8** Complete Collection of Data
11	**12**	**13** Report on Findings Due	**14**	**15**
18	**19** Brochures, Pamphlets Printed	**20**	**21**	**22** Proofread Training Materials
25	**26**	**27** Training Modules Completed	**28**	**29** Deadline: Training Day

Figure 13.4: Backwards planning is a process of working backwards from the deadline. Starting with the deadline, chart out when each task needs to be completed.

Here is the deadline. Work backwards from this date.

The advantage of a project calendar is that it keeps the team on task. The calendar shows the milestones for the project, so everyone knows how the rest of the team is progressing. In other words, everyone on the team knows when his or her parts of the project need to be completed.

TAKE NOTE Project calendars help avoid procrastination and that mad rush to finish a project right before deadline. With a calendar, everyone knows when a deadline has been missed, and he or she can help put the project back on track.

Writing Out a Work Plan

A work plan is a description of how the project will be completed (Figure 13.5). Some work plans are rather simple, using an outline to describe how the project will go from start to finish. Other work plans, like the one in Figure 13.5, are very detailed and thorough.

Figure 13.5: A work plan specifies the who, what, where, when, why, and how of the project.

St. Thomas Medical Center

Date: April 28, 2004
To: Staph Infection Task Force (M. Franks, C. Little, J. Archuleta, L. Drew, V. Yi, J. Matthews, J. McManus)
From: Alice Falsworthy, Infections Specialist
Re: Work Plan for Combating Staph Infection

Last week, we met to discuss how we should handle the increase in staph infections here at the St. Thomas Medical Center. We defined our mission, defined major tasks, and developed a project calendar. The purpose of this memo is to summarize those decisions and lay out a work plan that we will follow.

I cannot overstate the importance of this project. Staph infections are becoming an increasing problem at hospitals around the country. Of particular concern are antibiotic-resistant bacteria called methicillin-resistant *Staphylococcus aureus* (MRSA), which can kill patients with otherwise routine injuries. It is essential that we do everything in our power to control MRSA and other forms of staph.

Please post this work plan in your office to keep you on task and on schedule.

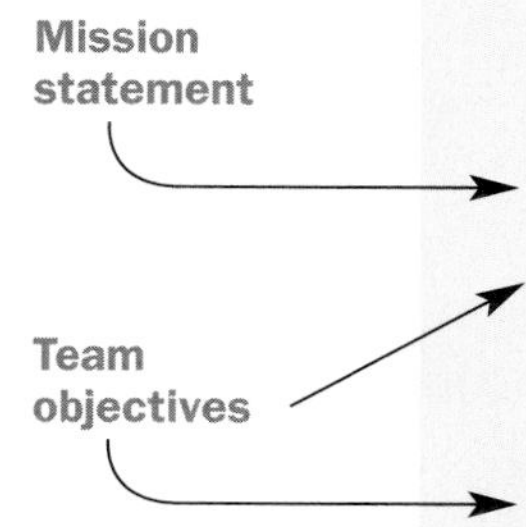

Project Mission
Our Mission: The purpose of this Staph Infection Task Force is to determine the level of staph infection vulnerability at St. Thomas Medical Center and to develop strategies for limiting our patients' exposure to staph, especially MRSA.

Secondary Objectives:
- Gain a better understanding of current research on staph and its treatments.
- Raise awareness of staph infections among our medical staff and patients.
- Assess the level of staph risk, especially MRSA, here at the hospital.
- Develop methods for controlling staph on hospital surfaces.

Continued on following page

Figure 13.5: (*continued.*)

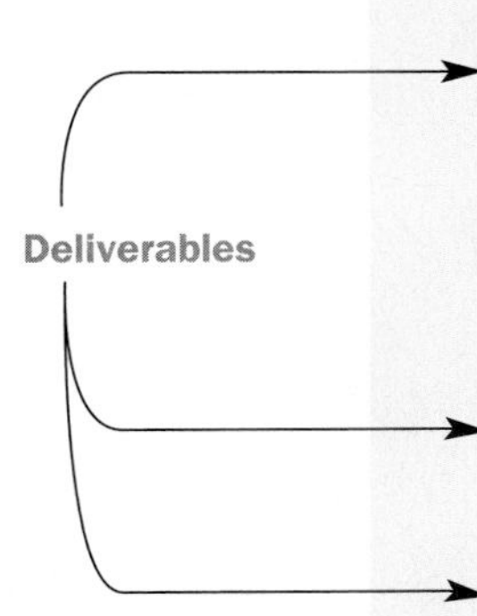

Project Plan

To achieve our mission and meet the above objectives, we developed the following five-action plan:

Action One: Collect Current Research on Staph Infections and Their Treatments

Mary Franks and Charles Little will collect and synthesize current research on staph infections, the available treatments, and control methods. They will run Internet searches, study the journals, attend workshops, and interview experts. They will also contact other hospitals to collect information on successful control methods. They will write a report on their findings and present it to the group on May 25 at our monthly meeting.

Action Two: Create Pamphlets and Other Literature That Help Medical Personnel Limit Staph Infections

Juliet Archuleta will begin developing a series of pamphlets and white papers that stress the importance of staph infections and offer strategies for combating them. These documents will be aimed at doctors and nurses to help them understand the importance of using antibiotics responsibly. Juliet will begin working on these documents immediately, but she will need the information collected by Mary and Charles to complete her part of the project. The documents will be completed by July 13, so we can have them printed to hand out at training modules.

Action Three: Conduct Experiments to Determine Level of Staph Risk at St. Thomas Medical Center

Lisa Drew and Valerie Yi will collect samples around the medical center to determine the risk of staph infection here at St. Thomas. The Center's Administration has asked us to develop a measurable knowledge of the problem at the Center. Lisa and Valerie will write a report to the Administration in which they discuss their findings. The report will be completed and delivered to the task force by June 2.

Action Four: Develop Contingency Plans for Handling Staph Instances

John Matthews and Joe McManus will develop contingency plans for handling instances of staph infections at the Center. These plans will offer

Figure 13.5: (*continued.*)

concrete steps that the staph task force can take to limit exposure to staph bacteria. These plans will be based on the research collected by Mary and Charles. They will be completed by July 13.

Deliverables

Action Five: Develop Training Modules
When our research is complete, all of the members of this task force will develop training modules to raise awareness of staph infections, offer prevention strategies, and provide information on proper use of antibiotics. Different modules will be developed for doctors, nurses, and custodial staff. Alice Falsworthy and Charles Little will coordinate the development of these training modules. The training module will be ready by August 29.

Project Calendar
Here is a chart that illustrates the project calendar and its deadlines:

Timeline for achieving goals

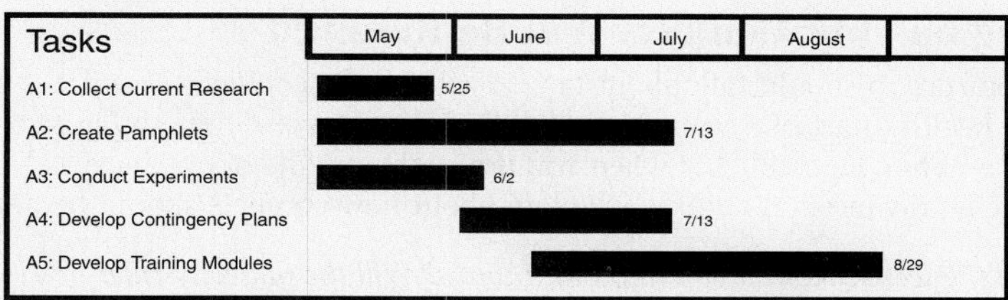

Conclusion
If anyone on the task force would like to change the plan, we will call a meeting to discuss the proposed changes. If you wish to call a meeting, please contact me at ext. 8712, or e-mail me at Alice_Falsworthy@stthomasmc.com.

A work plan will do the following:

- Identify the mission and objectives of the project.
- Lay out a step-by-step plan for achieving the mission and objectives.
- Establish a project calendar.
- Estimate a project budget if needed.
- Summarize the results/deliverables of the project.

A work plan is necessary for small and large projects, because people need to see the project in writing. Otherwise, team members will walk away from meetings with very different ideas about what needs to be accomplished.

By writing up a work plan, your team specifies how the project will be completed and who is responsible for which parts of the project. That way, team members can review the work plan if they are uncertain about (1) what tasks are being completed, (2) when the tasks will be finished, and (3) who is responsible for completing them.

LINK For help writing work plans as proposals, turn to Chapter 20, Page 580.

Agreeing on How Conflicts Will Be Resolved

Finally, your group should talk about how it will handle conflicts. Conflict is a natural, even healthy, part of a team project. However, the worst time to figure out how the group will handle conflict is when you are in the middle of one.

Instead, in advance, talk with your team about how conflicts should be handled.

Will the team take votes on important topics? Will the majority rule?

Will the team rely on the judgment of the team coordinator or the supervisor?

Does the team need to reach full consensus on decisions?

Can any team member call a team meeting to discuss conflicts?

Should agreements be written down for future reference?

You should not shy away from conflict in your team, because conflict is a natural part of the teaming process. Constructive conflict often leads to more creativity and closer bonds among team members. But, destructive conflict can often lead to dysfunctional working relationships, frustration, and lower morale. To foster constructive conflict and minimize destructive conflict, you should spend some time during the forming phase talking about the ways conflicts will be resolved.

AT A GLANCE

Six Steps for Strategic Planning

- Define the project mission and objectives.
- Identify project outcomes.
- Define team member responsibilities.
- Create a project calendar.
- Write out a work plan.
- Agree on how conflicts will be resolved.

Want to learn more about resolving conflict in the workplace? Go to **www.ablongman.com/johnsonweb/13.6**

Storming: Managing Conflict

Not long after the forming stage, a team will typically go through a storming phase. Once the project is planned out and the actual work begins, some tension will usually surface among team members. People may even become upset. In most cases, though, team members eventually find themselves negotiating, adapting, and compromising to achieve the team's mission.

During the storming stage, team members may:

- resist suggestions for improvement by other members.
- have doubts about the work plan's ability to succeed.
- compete for resources or recognition.
- resent that others are not listening to their ideas.
- want to change the team's objectives.
- raise issues of ethics or politics that need to be addressed.
- believe they are doing more than their share of the work.

Storming is rarely pleasant, but it is a natural part of the teaming process. When storming, the team starts to realize that work plans are never perfect and that people don't always work the same way or have the same expectations. The important thing is to not let small conflicts or disagreements sidetrack the project.

Running Effective Meetings

One way to constructively work through the storming phase is to conduct effective meetings. Nothing is more frustrating to team members than wasting their time and effort sitting around in an unproductive meeting. By running organized meetings, your team can maintain the structure needed to keep people on track.

CHOOSE A MEETING FACILITATOR In the workplace, usually a manager or supervisor runs the meeting, so he or she is responsible for setting the time and agenda. An interesting workplace trend, though, is to rotate the facilitator role among team members. That way, everyone has a chance to run the meeting, allowing others in the team to take on leadership roles. In classroom situations, your team should rotate the facilitator role to maintain a more democratic approach.

SET AN AGENDA An agenda is a list of topics that will be discussed at the meeting (Figure 13.6). The meeting coordinator should send out the meeting agenda at least a couple days before the meeting. That way, everyone knows what issues will be discussed and decided upon. Begin each meeting by first making sure everyone agrees to the agenda. Then, during the meeting, use the agenda to avoid going off track into nonagenda topics.

START AND END MEETINGS PROMPTLY If team members are not present, start the meeting anyway. Waiting around for late people can be frustrating, so you should insist that people arrive on time. If people know the meeting will start on time, they will be there on time. Likewise, end meetings on time. Meetings that drag on endlessly can be equally frustrating.

Want to know more about running effective meetings? Go to **www.ablongman.com/johnsonweb/13.7**

An Agenda

Meeting Agenda

Staph Infection Task Force

May 15, 2004, 3:00 p.m., Memorial Room

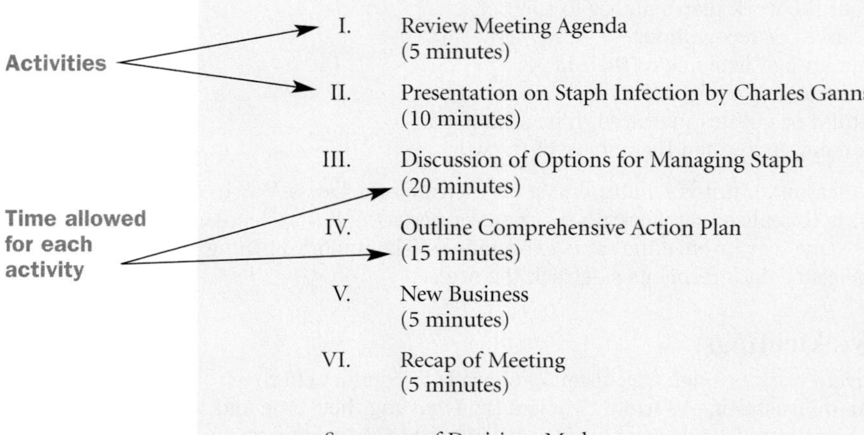
I. Review Meeting Agenda (5 minutes)

II. Presentation on Staph Infection by Charles Ganns (10 minutes)

III. Discussion of Options for Managing Staph (20 minutes)

IV. Outline Comprehensive Action Plan (15 minutes)

V. New Business (5 minutes)

VI. Recap of Meeting (5 minutes)

Summary of Decisions Made:

If possible, leave space for recording discussions.

Figure 13.6: A simple agenda is a helpful tool for keeping the meeting on track.

ADDRESS EACH AGENDA ITEM SEPARATELY Discuss each agenda item before moving on to the next one. Bouncing around among items on the agenda only ensures that the meeting will be inefficient. If someone wants to move ahead to a future agenda item, first make sure the current item of discussion has been addressed.

ENCOURAGE PARTICIPATION Everyone on the team should say something about each item. If one of the team members has not spoken, the facilitator should give that person an opportunity to speak.

ALLOW DISSENT At meetings, it is fine to disagree. Active debate about issues will help everyone consider the issues involved. In fact, if the team reaches consensus too quickly on an issue, someone might raise possible objections, allowing a consideration of alternatives.

REACH CONSENSUS AND MOVE ON People can talk endlessly about each agenda item, even after the team has reached consensus. So, allow any group member to "call the question" when they feel consensus has been reached. At that point, you can take an informal or formal vote to determine the course of action.

RECORD DECISIONS During meetings someone is usually responsible for keeping the minutes. The minutes record the team's decisions. Minimally, all decisions should be written down. After the meeting, the facilitator should send these notes or the full minutes to the team members, usually via e-mail.

RECAP EACH AGENDA ITEM At the end of the meeting, leave a few minutes to recap the decisions made by the team and clarify who is doing what. It is not uncommon for teams to have a "great meeting" that still leaves people unsure of what was decided and who is doing what. So, go through each agenda item, summarizing (1) what action will be taken and (2) who is responsible for taking that action.

LOOK AHEAD Discuss when the team will meet again and the expectations for that meeting. If necessary, clarify what should be accomplished before the next meeting. Also, decide who will be responsible for facilitating the next meeting.

Typically, storming becomes most evident during meetings. People grow frustrated and even angry as the team struggles to accomplish its objectives. That's why running effective meetings is so important. By creating a predictable structure for the meeting, your team will lower the level of frustration, allowing you to get work done.

Mediating Conflicts

Smaller conflicts should be handled through the conflict resolution methods you discussed when the team was forming. Make sure everyone is encouraged to express his or her views openly. Make sure everyone is heard. Then, use the conflict resolution methods (vote, decision of the team leader, appeal to supervisor, reach full consensus) to decide which way to go.

TAKE NOTE No decision will please everyone. If you find yourself outvoted, accept the decision of the group and do your best to support the group's decision. Don't let issues of winning and losing or personality differences sidetrack the team's overall pursuit of its mission.

To learn more about mediation, go to **www.ablongman.com/johnsonweb/13.8**

The Steps of Mediation

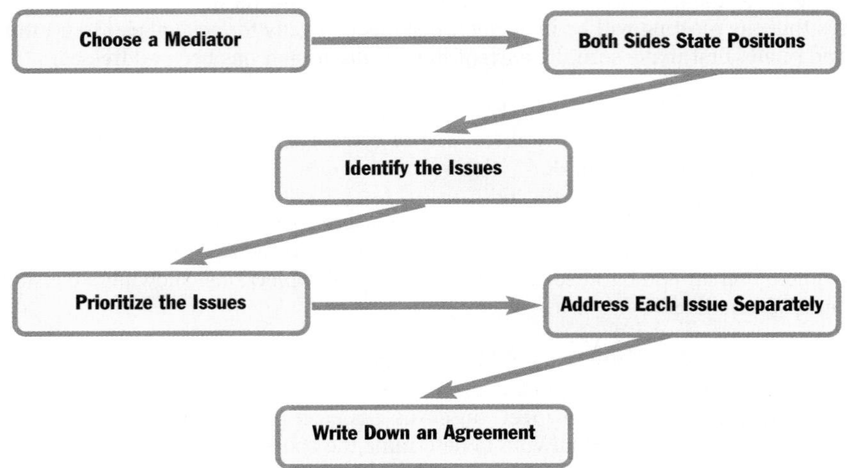

Figure 13.7: When more formal mediation is needed, you can follow these steps toward a resolution of the problem.

There will be times, however, when personalities or ideas clash in ways that cannot be easily resolved. At these times, you may want to use mediation techniques to help your team move forward (Figure 13.7).

1. **Choose a mediator**—A mediator is like a referee in a game. He or she does not take sides. Instead, it is the mediator's job to keep both sides of an argument talking about the issue. When the dialogue becomes unfriendly or goes off track, the mediator brings both sides back to the issues. In formal mediation, the mediator is someone who is not part of the group and has no connections to either side. In informal mediation, one of the group members who hasn't chosen sides can often serve as the referee.
2. **Ask both sides to state their positions**—Often, conflicts arise simply because each side has not clearly stated its position. Once both sides have a chance to make their case and be heard by the other side, the conflict should seem more resolvable.
3. **Identify the issues**—Both sides should discuss and identify the "issues" they disagree about. Often, disputes hinge on a small number of issues. Once these issues are identified, it becomes easier to talk about them.

TAKE NOTE Identifying the issues also has an added benefit. Once both sides begin agreeing on the issues, they have already begun to work together toward a solution.

4. **Prioritize the issues from most to least important**—Some issues are more important than others. By prioritizing the issues, both sides can usually find places where they already agree. Or, in some cases, a top priority to one side is not a foremost concern to the other side. By prioritizing the issues, both sides often begin to see room for negotiation.
5. **Address each issue separately, trying to find a middle ground that is acceptable to both sides of the dispute**—Focus on each issue separately and keep looking

for middle ground between both sides. When both sides realize that they have many common interests, they will usually come up with solutions to the conflict.

6. **Write down an agreement that both sides can accept**—As the mediation continues, both sides usually find themselves agreeing on ways to resolve some or all of the issues. At this point, it helps to write down any compromises that both sides can accept.

LINK Ethical issues are often a source of conflict. To learn more about ethics, go to Chapter 4, page 70.

The secret to successful mediation is a focus on issues, not personalities or perceived wrongs. The mediator's job is to keep the two sides discussing the issues, steering the discussion away from other distractions.

Firing a Team Member

Sometimes a team member does not do his or her share of the work. When this happens, the team might consider removing that person from the project. The best way to handle these situations is to first mediate the problem. The members of the team should meet to talk with this person about their expectations, give the person a chance to explain the situation.

After hearing this person's side of the story, the team might decide to give him or her a second chance. At that point, a work contract should be written that specifies exactly what this person needs to do for the project.

HELP

Virtual Teaming

Increasingly, technical workplaces are turning to virtual teaming to put the right people on any given project. Virtual teaming allows people to work on projects collaboratively through electronic networks, using e-mail, instant messaging, and phones to stay in contact. Electronic networks, called *intranets,* are often used to share information and documents among team members.

Virtual teams, just like teams working together in an office, schedule regular meetings, share ideas and documents, and work toward achieving specific goals. The main differences between on-site teams and virtual teams are how people communicate and where they are located.

Trends in the technical workplace support virtual teaming. Today, more people are "telecommuting" or "teleworking" from home or remote sites. Also, the global economy means people on the same project are sometimes working thousands of miles away from each other. Meanwhile, wireless technologies allow people to work just about anywhere. Chances are you will find yourself working with a virtual team in the near future—if you aren't already.

Interestingly, virtual teaming does not change traditional teaming strategies—it only makes them more necessary. A virtual team will go through the forming, storming, norming, and performing stages, just like an on-site team. If good planning, communication, and conflict resolution are important with on-site teaming, they are even

Need to fire a team member? Go to
www.ablongman/johnsonweb/13.9

more important in virtual teams. After all, communicating with your virtual team is a little more difficult, because you cannot physically visit each other.

Here are some strategies for managing a successful virtual team:

Develop a work plan and stick to it—Members of virtual teams do not bump into each other in the hallway or the break room. So, they need a clear work plan to keep everyone moving together toward the final goal. Your team's work plan should (1) define the mission, (2) state objectives and measurable outcomes, (3) spell out each stage and task in the project, (4) specify who is responsible for each task, and (5) lay out a project calendar.

Communicate regularly—In virtual teams, the old saying "out of sight, out of mind" now becomes "out of communication, out of mind." Each member of the virtual team should agree to communicate with the others regularly (e.g., two times a day, two times a week). Your team can use e-mail, phones, instant messaging, or chat rooms to contact each other. You and your team members should constantly keep the others up to date on your progress. And, if someone falls out of communication for a day or two, the team leader should track him or her down and urge the team member to begin communicating again.

Hold teleconferences and videoconferences—There are many ways to hold real-time virtual meetings with team members. Your team members can teleconference over the phone, or you can use a chat room or instant messaging to exchange ideas with each other. Increasingly, broadband technology is allowing people to hold videoconferences in which people meet and see each other with Internet cameras (webcams) through computer screens (Figure A). Like on-site meetings, virtual meetings should be preplanned and follow an agenda. The only significant difference between on-site meetings and virtual meetings is that people are not in the same room.

Build trust and respect—One of the shortcomings of virtual teaming is the lack of nonverbal cues (smiles, shrugs, scowls) that help avoid misunderstandings. As a result, people in virtual teams can feel insulted or disrespected much more easily than people in on-site teams. So, it is doubly important that team members learn how to build trust with others and show respect. Trust is built by communicating effectively, meeting deadlines, and doing high-quality work. Respect is fostered by giving compliments and using "please" and "thank you" in messages. When conflicts do arise (and they will), focus on issues and problem solving, not personalities or perceived slights.

Keep regular hours—Time management is always important, even if you are working in a virtual team. During regular office hours, team members should be confident that they can contact each other. You and your team members should be ready to answer the phone, use instant messaging, or answer e-mail as though you were all working together in a typical office.

More than likely, virtual offices, teleworking, and virtual teams will be a common part of the workplace in the near future. Like on-site teaming, virtual teaming requires you to learn how to effectively work with others.

To find websites that discuss virtual teaming, go to **www.ablongman.com/johnsonweb/13.9**

Teleconferencing

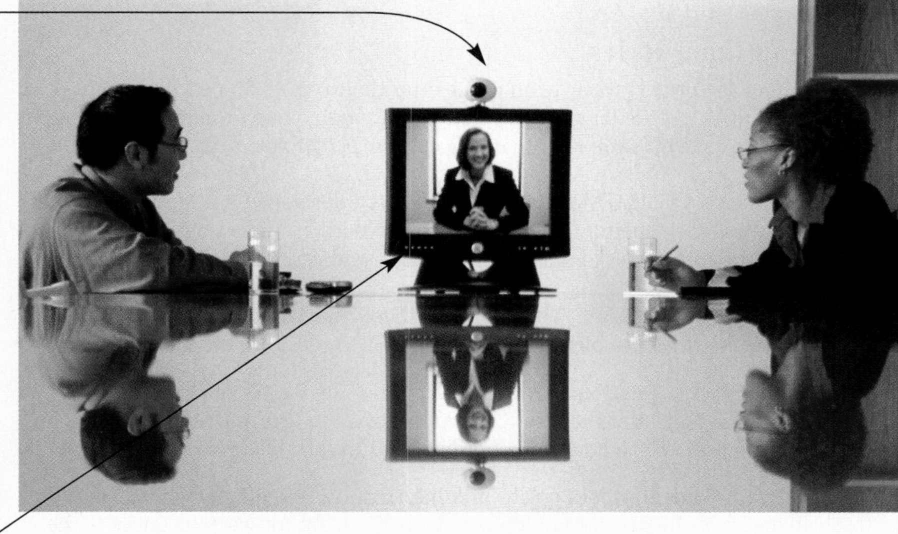

A webcam projects images to the other participants in the virtual meeting.
The monitor can be used to show people, documents, or presentations.

Figure A: Teleconferencing allows team members to hold meetings virtually.

If the team still wants to let the person go, the supervisor (perhaps your instructor) should be asked about removing the person from the team. The supervisor should be present when the team tells the problematic team member that he or she is being removed.

Norming: Determining Team Roles

The storming period can be frustrating, but soon afterward your team should enter the *norming* stage. In this stage, members of your team will begin to accept their responsibilities and their role in the project. A sense of team unity will develop as people begin to trust each other. Criticism will become increasingly constructive as team members strive to achieve the project's mission and objectives.

Revising Objectives and Outcomes

The storming stage often reveals the flaws in the work plan. So, when norming, you might find it helpful to revisit and refine the team's original decisions about objectives and outcomes. The team may also want to revise the project schedule and redivide the workload.

You don't need to completely rewrite the work plan from scratch. After the storming stage, it might be tempting to just start over. But, you should stay with your original work plan in most cases. The plan probably just needs to be revised and refined, not completely redone.

Identifying Team Roles

When planning the project, your team divided up the work, giving each person specific responsibilities. As the team begins norming, though, you will notice that team members tend to take on *team roles* that reflect their personalities, capabilities, and interests.

A management specialist, Meredith Belbin (1981), developed a description of nine team roles that people generally follow. He also carved these nine roles into three categories: people-oriented roles, action-oriented roles, and cerebral roles.

PEOPLE-ORIENTED ROLES The people in these roles are responsible for managing the activities of the team members:

> **Coordinator** sets the agenda and keeps track of the team's objectives; asks broader questions and occasionally summarizes the team's decisions; keeps an eye on the project calendar and coordinates the work of various team members
>
> **Resource investigator** goes out to find information, bringing new ideas and strategies into the discussion; looks outside the team for ways to improve the project
>
> **Team worker** focuses on getting the work done; may not be fully invested in planning the project but will do his or her part of it.

ACTION-ORIENTED ROLES The people in these roles are responsible for getting things done:

> **Shaper** focuses on team tasks, while looking for patterns in team discussions; emphasizes completing the project
>
> **Implementor** stresses the "how to" nature of the project and is eager to develop methods for turning abstract objectives and plans into real actions
>
> **Completer finisher** stresses attention to details and the overall quality of the project; concerned about meeting deadlines and maintaining a schedule.

CEREBRAL ROLES The people in cerebral roles are responsible for planning, creating, and providing expertise in a project:

> **Monitor evaluator** keeps the team on task by critiquing poor decisions or pointing out any flaws in reasoning; tends to focus on achieving outcomes
>
> **Plant** thinks creatively, often providing original suggestions and innovative solutions to problems; tends to stress the big picture over smaller details
>
> **Specialist** contributes special skills and knowledge to the team; masters a specific topic or area of research, adding depth to the team's discussions.

To learn more about team role theory, go to **www.ablongman.com/johnsonweb/13.10**

As the team begins norming, you and the other members of the team might take some time to identify the roles each of you are playing. By identifying team roles, you can take advantage of each member's natural strengths and interests.

Not all the roles will be filled, especially in a smaller team. Instead, each team member might take on two or three roles, depending on the project. Indeed, roles may change and evolve as the project moves forward.

In other words, let the "completer finisher" in the team worry about the deadlines and quality issues. Let the "team workers" concentrate on achieving specific tasks. Encourage the "coordinators" and "shapers" to keep an eye on the overall objectives and mission of the team.

Using Groupware to Facilitate Work

When working in a team, you might need to use groupware, a class of software products that helps move information and documents around. Groupware allows team members to communicate and work collaboratively through a local area network (LAN) or an intranet.

Perhaps the most common use of groupware is sharing documents and sending messages. The two most popular groupware packages include IBM's Lotus Notes and Microsoft's Outlook. These software packages and ones like them support the following kinds of activities:

> **Scheduling and calendaring**—One of the more powerful features of groupware is the ability to schedule meetings and keep a common calendar for the team (Figure 13.8). With this feature, team members can regularly check the calendar to stay on task.

Using Groupware

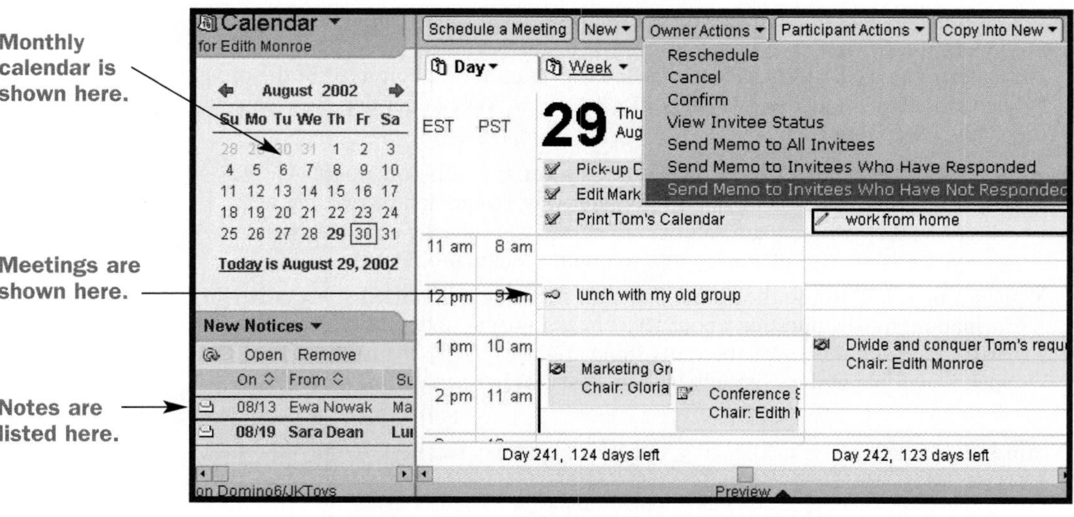

Figure 13.8: With groupware software like Lotus Notes, shown here, members of a group can keep a common calendar, post notes, and send e-mails to each other. In this screen, the user can do a variety of functions using the menu at the top.

Source: IBM, http://www.lotus.com/products/product4nsf/wdocs/notes_name_page.

GO TO THE NET

To learn more about groupware, go to **www.ablongman.com/johnsonweb/13.11**

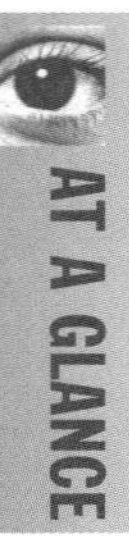

AT A GLANCE

Norming

- Revise objectives and outcomes.
- Identify team roles.
- Use groupware to facilitate work.

Discussion lists and instant messaging—Groupware offers easy access to discussion lists (usually using e-mail) and instant messaging. With these tools, team members can post notes to each other on a discussion list or have real-time discussions through instant messaging.

Document posting and commenting—Files can be posted to a common site, allowing team members to view documents, comment on them electronically, or download them.

Using groupware effectively takes some practice. Eventually, the team will use the groupware as a meeting place and a posting board. It will become an integral part of the project.

AT WORK

Grant Moser

TECHNICAL WRITER, VIDIOM SYSTEMS

Vidiom Systems specializes in interactive television (iTV) application design, consulting, technical documentation, and training.

How can you keep that "team feeling" when you're working virtually?

I've been a telecommuting technical writer for about four years now. It's 750 miles to the home office and 2,000 miles to the engineering office, so I don't get a lot of face time with other writers or engineers on projects. So, when I am writing documents, I need to work a little harder at being part of the team.

Actually, it has been pretty easy to develop the "team feeling" on most of the projects I've worked on. Here are two suggestions on how you can develop that team feeling while telecommuting:

- Always try to make a trip for face time as soon as you become part of a project. Even if a project is already well under way, seeing and interacting with other members of the team can help integrate you into the group, as well as give you a good idea of who to talk to about specific problems.
- Don't be afraid to use the phone. Just call someone if you need substantive information, have a question about style, or are unsure about a deadline. It's better to spend a few dollars to get things right from the beginning than to spend a couple of days fixing text. And it's a lot better than missing a crucial deadline.

Writing from a distance is challenging, but it can be done. You need to stay in touch.

Performing: Improving Quality

Some teams never reach the *performing* stage. Your team is performing when members are comfortable with the project and their roles in it. When performing, team members will recognize the other members' talents and weaknesses. They also begin to anticipate each other's needs and capabilities.

When your team is performing, you can start looking for ways to improve the quality of your work. One of the gurus of quality was W. Edwards Deming, who developed many of the principles behind Total Quality Management (TQM) and Continuous Quality Improvement (CQI) that are used in technical workplaces. Deming argued that teams should put an emphasis on improving the process rather than simply exhorting people to improve the product (Deming, 2000).

An important part of performing is an emphasis on improving quality. To enhance quality, team members proactively seek out ways they can improve their teamwork and the quality of the final product. Quality has perhaps become an overused buzzword in technical workplaces, but the emphasis on continuous improvement of processes and products is still essential.

How can you improve quality in your team? While performing, a helpful technique is to develop *quality feedback loops* in which your team regularly compares outcomes to the project objectives (Figure 13.9). The team should systematically take the time to objectively consider whether it is meeting or exceeding the outcomes.

Quality Feedback Loop

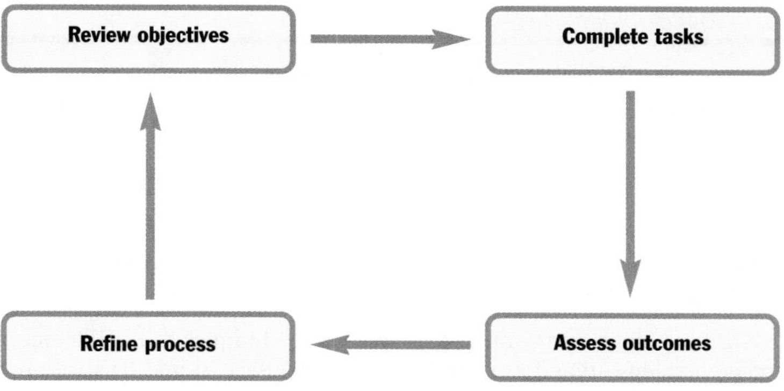

Figure 13.9: A quality feedback loop continually searches out ways to refine the work process and thus improve quality.

An effective quality feedback loop should incorporate methods to collect feedback on the outcomes of the project. To collect feedback, perhaps "focus groups" of customers might be consulted. Perhaps the product might be thoroughly user tested. Perhaps supervisors or outside experts might be called in to offer suggestions for improvement. The type of feedback you need depends on the kind of product your team is being asked to produce.

LINK For more ideas about improving quality through document cycling, see Chapter 10, page 282.

Are you interested in TQM or CQI? Go to **www.ablongman.com/johnsonweb/13.12**

A performing team will have its ups and downs. There may even be times when the team regresses into the norming or even the storming stages. Eventually, though, the performing team usually regroups and puts the focus back on quality.

The Keys to Teaming

The keys to good teaming are good planning and effective communication. The planning strategies discussed in this chapter might seem like extra work, especially when time is limited and your team is eager to start working on the project. But all that planning will save your team time in the long run. Each person needs a clear understanding of the mission and the steps in the project. Then, you need to keep the communication lines open. Plan to communicate regularly over the phone, e-mail, and instant messaging. Not long from now, it will be common to hold meetings through computers. Internet cameras (webcams) will allow you to see your fellow team members right on the computer screen. The secret, no matter what technology you are using, is to communicate regularly. You will find that keeping the communication lines open will solve many teaming problems.

Finally, telecommuting, or teleworking, is becoming much more common in today's technical workplace. Sometimes your team members will be working a few days per week at home. Or, they may be working while they are on the road. Some telecommuters work almost exclusively from home, going to the office only when absolutely necessary. Again, good planning and effective communication are the keys to success in these virtual workplace environments.

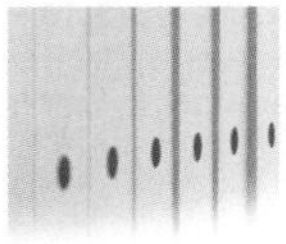

CHAPTER REVIEW

- Teams have several advantages. They can concentrate strengths, foster creativity, share the workload, and improve morale. Disadvantages include conflict with other members and disproportionate workloads.
- Tuckman's four stages of teaming are forming, storming, norming, and performing.
- In the forming stage, use strategic planning to define the mission, set objectives, define roles, and establish a project schedule.
- In the storming stage, use conflict management techniques to handle emerging disagreements, tension and anxiety, leadership challenges, and frustration.
- In the norming stage, revise the work plan to form consensus, refine the team's objectives and outcomes, and solidify team roles.
- In the performing stage, pay attention to improving the process in ways that improve the quality of the work.
- W. Edwards Deming developed many of the principles behind Total Quality Management (TQM) and Continuous Quality Improvement (CQI), which are common in technical workplaces today.
- Virtual teaming, or working together from a distance, requires good planning and effective communication.

Want to learn more about telecommuting? Go to **www.ablongman.com/johnsonweb/13.13**

EXERCISES AND PROJECTS

Individual or Team Projects

1. When working with a team, keep a journal that describes your experiences working with others. In a memo to your instructor, discuss whether your team went through the four stages of teaming (forming, storming, norming, performing). Describe how each stage brought up new issues to be resolved. How did you and your team members handle them?

   In your memo to your instructor, discuss how you might use awareness of Tuckman's four stages to improve your ability to work with a team.

2. If you have a job now, write a report to your instructor in which you talk about how your coworkers fluctuate among Tuckman's four stages. What are some of the indications that the team is forming, storming, norming, and performing? How does the team tend to react during each of these stages? Does your team aid in forming strategic planning? How do you mediate conflicts during storming? How is norming achieved at your workplace? What does performing look like?

3. Imagine your class is a workplace. What are the objectives and outcomes of the course? How are conflicts resolved in the classroom? Do members of your class take on various team roles in the classroom? How could you and your instructor create quality feedback loops to improve your learning experience? In a class, discuss how teaming strategies might be helpful in improving how the class is managed.

4. On the Internet, research the theories and writings of Tuckman, Belbin, or Deming. Write a report to your instructor in which you summarize the principles of one of their theories. Then, discuss the ways in which you might apply their theories to your own life and work.

Collaborative Project

With a team, try to write a report in one hour on a topic of interest to all of you. For example, you might write about a problem on campus or at your workplace. While you are writing the report, pay attention to how your group forms, norms, storms, and performs. Pay attention to how the group plans and divides up the responsibilities. Then, as the project moves forward, pay attention to the ways conflict is resolved. Finally, identify the different roles that group members tend to play as the group develops norms for the project.

When your one-hour report is finished, talk among yourselves about how the group project went. Did the group form and plan properly? Did you handle conflict well? Did the group members take on identifiable roles? Do you think you ever reached the performing stage? If you were going to do the project over, how might your group do things differently?

CASE STUDY

Not a Sunny Day

Veronica Norton liked working on teams. She liked interacting with others, and she liked how working with others allowed her to accomplish projects that were too big for one person. Veronica also had a strong interest in building solar motors, like the ones that went into solar-powered cars. She hoped to find a job in the automotive industry building solar vehicles.

So, she decided to join her university's team in the American Solar Challenge race competition. She and a group of 10 other students were going to build a solar car to race across the country against cars from other universities.

The first meeting of the team went great. Everyone was excited, and they found out there were funds to support the project. Several engineering departments were contributing money, and a local aerospace engineering firm was matching university funds dollar for dollar. The firm was also letting the team use its wind tunnel to help streamline the car. There were other local sponsors who wanted to participate.

The team rushed right into the project. They sketched out a design for the car, ordered supplies, and began welding together a frame for the car. Veronica began designing a solar motor that would be long lasting and have plenty of power. She noticed that the overall plan for the car was not well thought out, but she went along with it anyway. At least they were making progress.

One of the team members, George Franks, began emerging as the leader. He was finishing up his engineering degree, so he had only a couple of classes left to take. He had plenty of time to work on the solar car.

Unfortunately, he didn't like to follow plans. Instead, he just liked to tinker, putting things together as he saw fit. It wasn't long before the team started running into problems. George's tinkering approach was creating a car that would be rather heavy. Also, each time Veronica visited the shop, the dimensions of the car had changed. So, the motor she was building needed to be completely redesigned.

Perhaps the worst problem was that they were running out of money. George's tinkering meant lots of wasted materials. As a result, they had almost used up their entire budget. Sally, who was in charge of the finances for the project, told everyone they were going to be out of money in a month. Everyone was getting anxious.

Finally, at one of the monthly meetings, things fell apart. People were yelling at each other. After being blamed for messing up the project, George stormed out of the meeting. Things looked pretty hopeless. Everyone left the meeting unsure what to do.

Veronica still wanted to complete the project, though. If you were her, how might you handle this situation? What should she do to get the project going again?

For more information on the American Solar Challenge, go to **www.ablongman.com/johnsonweb/13.14**

CHAPTER

14 Designing Websites

CHAPTER CONTENTS

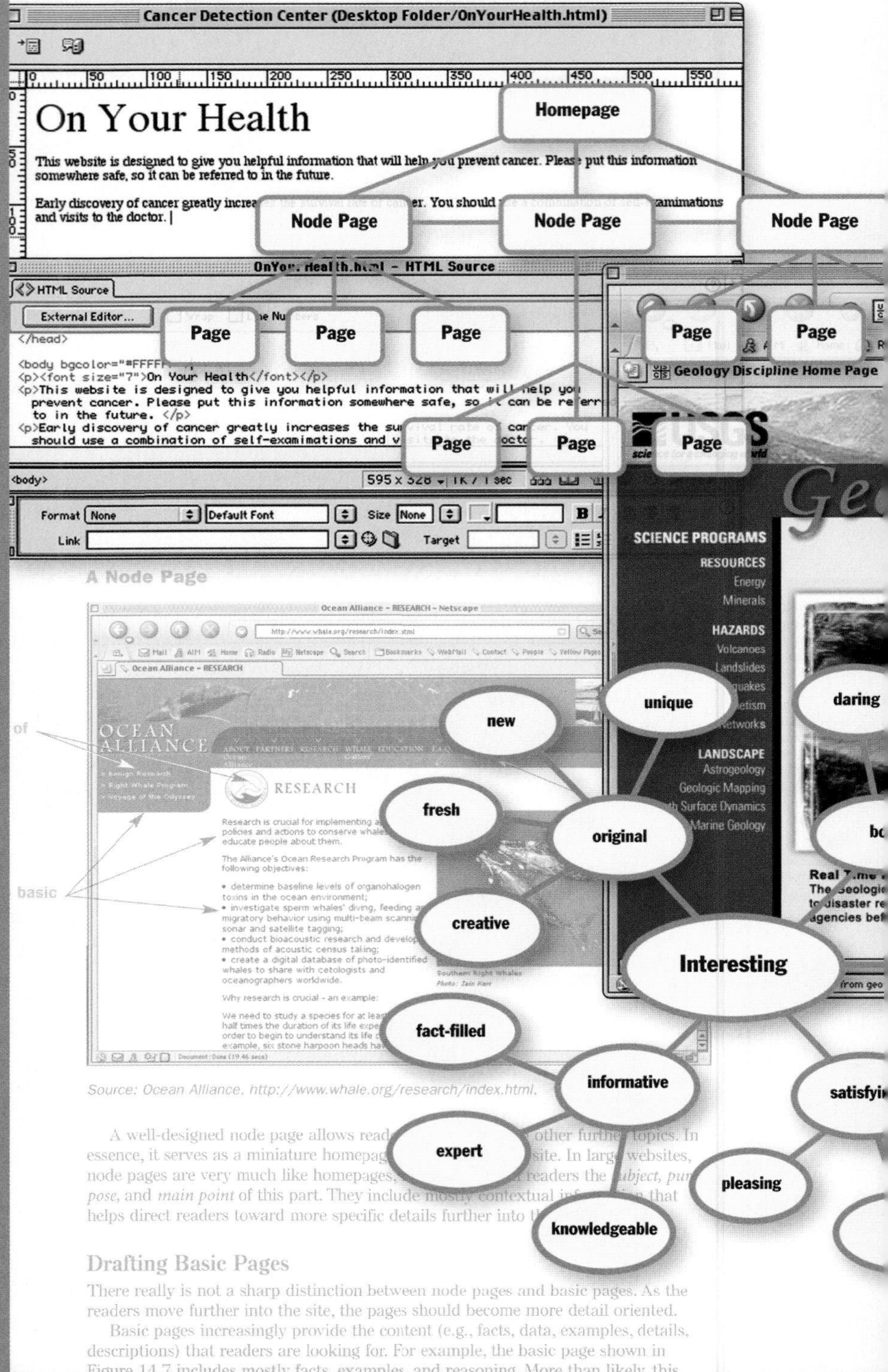

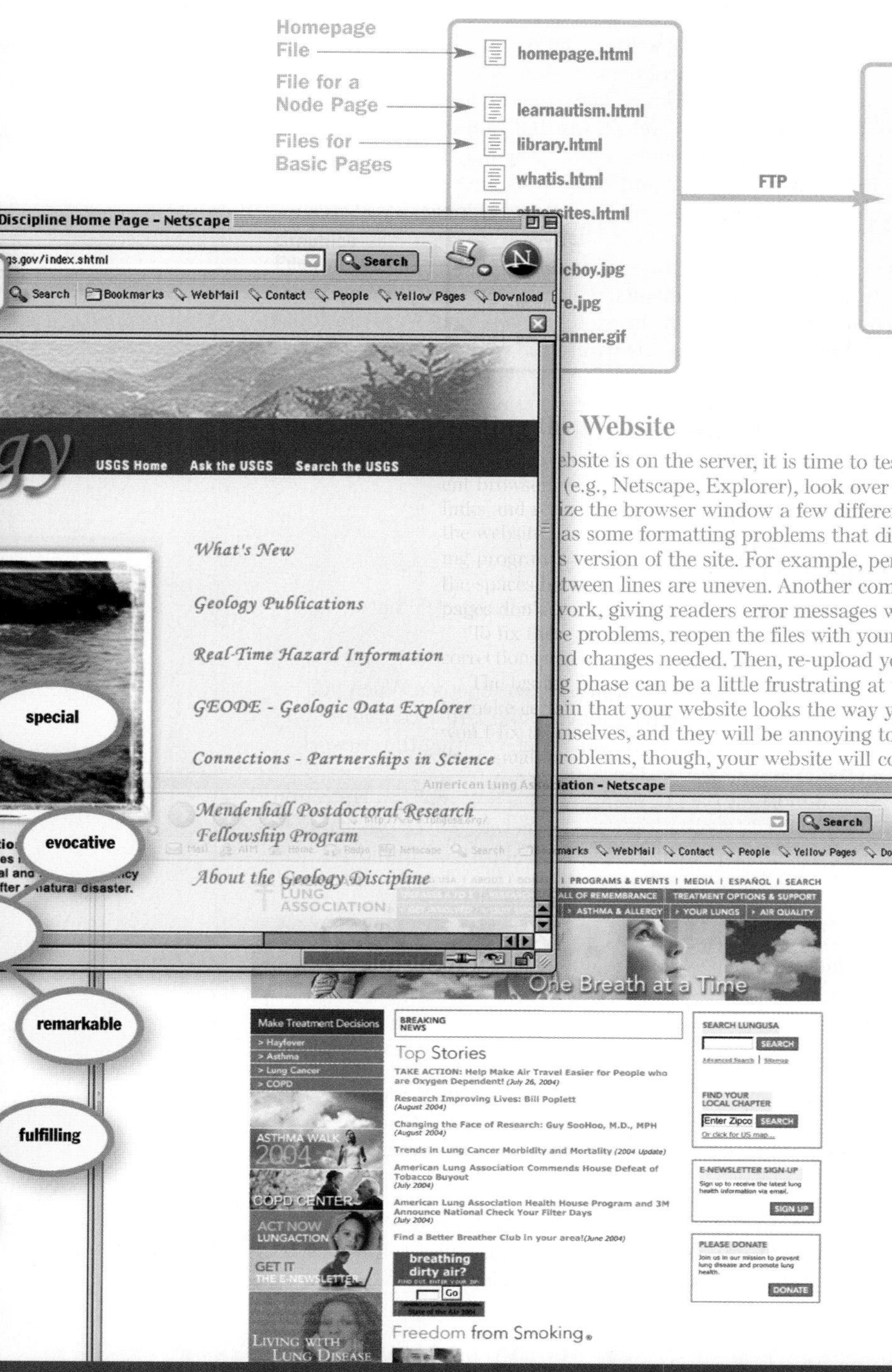

CHAPTER OBJECTIVES

In this chapter, you will learn:

- How websites are both similar to and different from print documents.
- The basic features of websites.
- How to plan and research a website.
- How to organize and draft a website.
- How to use web-authoring software.
- How to use style to make a website more readable.
- How design principles are used to create an effective website interface.
- The importance of good revision, editing, and proofreading in creating websites.

Increasingly, websites are becoming an important form of communication in technical workplaces. Today, websites are regularly written and designed by engineers, scientists, medical personnel, and other technical personnel. In some high-tech fields it is already expected that new employees will know how to write for the web.

The first thing you should remember about websites is that *they are documents.* They may look different from paper-based documents and they may be used differently, but they are still *written texts* with words and images. As a result, you can apply many of the communication strategies in this book to developing these screen-based documents.

An important difference though, is that websites are *visual-spatial* documents, making them different from paper-based documents, which are generally *linear* documents. In other words, people tend to read websites spatially, scanning from block of information to block of information. They *navigate* within the website, among these blocks of information. Quite differently, paper-based documents tend to be read linearly, with readers moving from concept to concept along a linear path (Ong, 1982).

Because they are visual-spatial texts, websites are composed and designed differently than paper-based documents. This chapter will show you how to take advantage of the visual-spatial qualities of on-screen documents.

Basic Features of a Website

Let's begin with a few definitions. First, websites exist on the Internet. What is the Internet?

The Internet is a vast network of computers that are linked together.

The Internet is a vast network of computers that are linked together. When you use the Internet, your desktop or laptop computer connects to a larger computer called a *server.* In turn, servers are all connected to each other, creating the network of computers called the Internet.

The World Wide Web is an important part of the Internet. The "web" is a collection of billions of webpages that are linked together on the Internet. Software programs called *browsers,* like Netscape, Mozilla, Opera, and Explorer, allow you to move around in the web, accessing the webpages available.

So, what is a website?

A website is a group of related webpages that are linked together into a whole document.

A website is a group of related webpages that are linked together into a whole document. Like any document, the pages are held together by a common topic and organized in a pattern that is familiar and logical to the readers. Websites have a few different kinds of pages.

Homepage—The homepage is usually the first page of the site. It identifies the subject and purpose of the website, while forecasting its overall structure. In many ways, the homepage is like an introduction and table of contents for the website (Figure 14.1).

Node pages—These pages tend to be links that follow the homepage. They subdivide the website's content into smaller topics. For example, a university's homepage will have links that go to node pages like *Academics, Colleges and*

Want to learn more about how the Internet works? Go to **www.ablongman.com/johnsonweb/14.1**

Departments, Libraries, Students, and *Faculty and Staff.* These are all nodes in the website.

Pages—Individual pages in the website contain the content of the site. These pages contain the facts, details, images, and other information that readers are seeking.

Navigation pages—The navigation pages help readers find the information they are seeking. These pages include site maps that show the contents of the site, like an index in a book. They also include search engines that allow readers to search the site with keywords.

Splash page—This optional page comes before the homepage and acts like a cover to a book. Splash pages are mostly decorative, using animation or images to set a specific tone before readers enter the website through the homepage.

In the end, though, don't be too concerned about all the website jargon. If you think about it, all these features of websites are very similar to features in paper-based documents.

A Homepage

Buttons allow users to navigate the site spatially.

Navigation bar offers links to other parts (nodes) of the website.

Text presents information for the website readers.

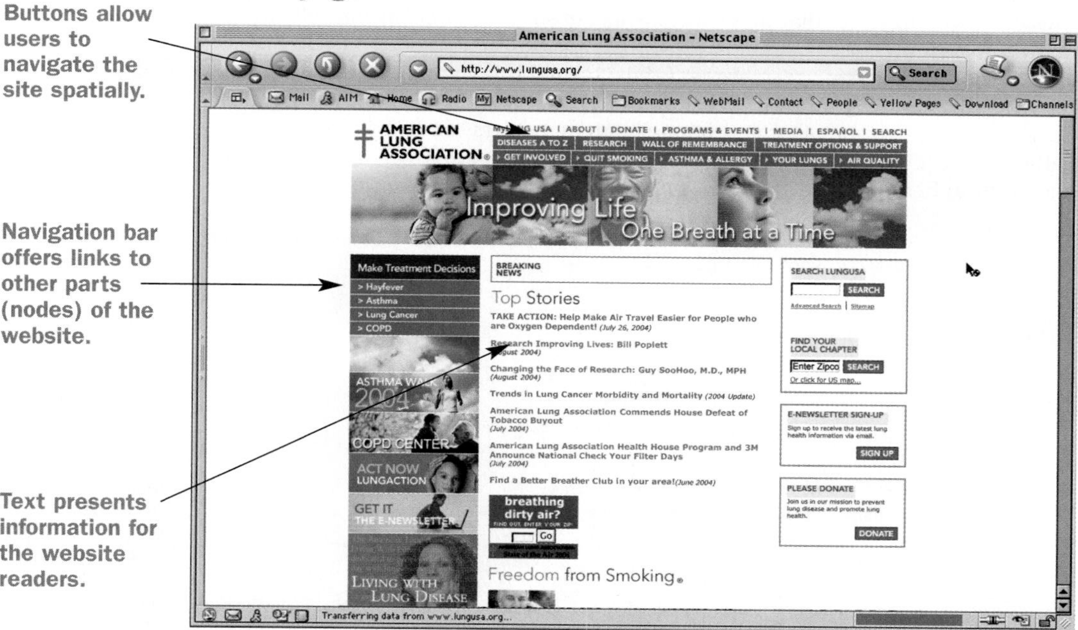

Figure 14.1: A website uses visual-spatial strategies to organize information. In this website from the American Lung Association, notice how the text and images are presented visually for easy access to the information on the page and site.

Source: American Lung Association, http://www.lungusa.org.

GO TO THE NET

To see some interesting splash pages, go to **www.ablongman.com/johnsonweb/14.2**

Similarities Between Websites and Other Documents

Website	Paper-based document
Homepage	Introduction, or table of contents
Node page	Chapter or section
Page	Page
Linking	Turning a page
gifs, jpegs, tiffs	Graphics (pictures, drawings)
Navigating	Reading and scanning
Surfing the web	Reading and scanning
Site map	Index
Search engine	Index
Navigation bar	Table of contents or tabs

Pointing out these similarities does not diminish the unique qualities of websites as texts. They *are* different from paper-based texts. Just don't let the jargon of websites cause you to forget that these documents are similar to documents you already know well.

One significant difference, however, is how the information in websites is organized. As shown in Figure 14.2, websites usually use spatial (or nonlinear) organizational patterns. In a spatial organizational pattern, readers can follow a variety of paths after reading the homepage. The website lets readers move around freely in the document to find the information they need.

TAKE NOTE Websites come in many shapes and sizes, and they are usually not as symmetrical as the diagram shown in Figure 14.2.

Basic Pattern for a Website

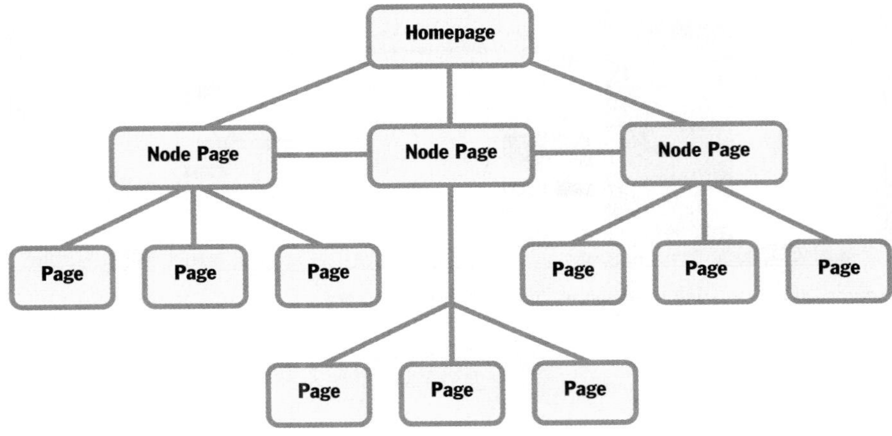

Figure 14.2: Websites are spatial rather than linear. They allow readers to move in a variety of different directions. In this example, readers can go directly to any of three pages from the node page without going through other pages.

GO TO THE NET To see other organizational structures for websites, go to **www.ablongman.com/johnsonweb/14.3**

AT WORK

Mitch Curren

COPY WRITER, BEN AND JERRY'S ICE CREAM

Mitch Curren writes copy for both printed materials and the web for Ben and Jerry's Ice Cream in South Burlington, Vermont.

How is writing for a website different from other kinds of writing?

In many ways, writing for the web is similar to other kinds of writing. As Ben & Jerry's Chief S'creamwriter & LexiConeHead, I'm responsible for writing and editing copy that communicates the company's *brand* voice, the voice that conveys Ben & Jerry's unique brand personality in ways our customers find recognizable and engaging.

Whether I'm creating copy for the back of an ice cream pint, a printed brochure, or for our website and e-newsletter audience, ensuring brand voice and style are constant objectives. I also work closely with various project managers to meet project-specific goals and objectives.

Quite often that means I put on my "Composition 101" instructor's hat to make sure folks' copy requests contain basic copy-directional essentials:

- What's the message?
- Who's your audience?
- What do you want the message to achieve?

While the same writing basics and project dynamics apply to copy written for our website, the dynamics of the web itself are different. Most people don't "read" web content the same way they'd read a printed document; hypertext and hyperlinks enable all kinds of browsing, surfing, and scanning opportunities no linear print materials can match.

Those and other opportunities are what I keep in mind when I write stuff for our website. Whatever topic I'm covering or message I'm conveying needs to be organized into informational "chunks"—often with links to additional chunks—that offer a quick gist for folks who like to scan as well as links for folks who want further information.

Planning and Researching a Website

One of the advantages (and disadvantages) of websites is that they are widely available for anyone to use. If you create a website, for example, anyone around the world can potentially use it. So, it is difficult to anticipate who might visit the site and why they might be interested in it.

Nevertheless, you can make some strategic guesses about the kinds of people who might be interested in using your website. You should start by first answering the Five-W and How questions about your website:

Who will be visiting this website?

Why are they using this website?

What information or kind of information are they looking for?

Where will the information in the site be useful to them?

When will they use this information?

How might readers use this website?

With these questions answered, you can think more deeply about how the website should be developed. Specifically, define its rhetorical situation by considering the website's subject, purpose, readers, and context of use.

Subject

Clearly define the boundaries of your website by determining what is inside the subject area and what is outside. Websites offer nearly limitless amounts of space for text. So, the content of the site needs to be scaled to fit the needs of your readers. If you define your subject too broadly, they won't be able to easily locate the information they need.

LINK For more information on defining the boundaries of a subject, go to Chapter 2, page 22.

Purpose

Like any other document, you should define the purpose of your website. Otherwise, the site will seem unfocused.

For the most part, websites are meant to be informative, meaning they are developed to provide information to readers, rather than persuade them. Consequently, some keywords in a statement of purpose might include the following:

to inform	to demonstrate
to explain	to illustrate
to train	to report
to acquaint	to summarize
to teach	to collect
to instruct	to familiarize
to present	to show
to display	to exhibit

A purpose statement for a website might read something like:

This site is designed to present the life and works of Richard Feynman.

Our intent is to familiarize you with Dinosaur National Park, so you can help us protect this treasure for future generations.

This training CD-ROM will teach you how to be a stronger, more confident public speaker.

Make sure you clearly define your purpose *before* you start creating the website. You should be able to state your website's purpose in one sentence. If you cannot boil it down into one sentence, your website will probably not be focused enough.

LINK For more tips on defining a document's purpose, see Chapter 2, page 23.

For worksheets that will help you analyze your website's rhetorical situation, go to **www.ablongman.com/johnsonweb/14.4**

Readers

Even though just about anyone can use your website, you should target specific readers by paying attention to their needs.

PRIMARY READERS (ACTION TAKERS) Are the people who will use the information on your website to take some kind of action. What are their needs, values, and attitudes? Are they researching your company's services or products? Are they collecting information for a report? Are they looking for help to solve a problem?

SECONDARY READERS (ADVISORS) Will have a wide variety of reasons for visiting your site. Are they reviewing your information for accuracy, or are they confirming facts? Are they advising the primary readers about a decision? What level of expertise do they have concerning your website's subject?

TERTIARY READERS (EVALUATORS) May also have various reasons for visiting your site. Have you included any inappropriate proprietary or personal information? How might your company's competitors use the information on your website? How might a journalist or lawyer use the information?

GATEKEEPER READERS (SUPERVISORS) Are mostly interested in the accuracy of the site and whether it reflects the image and values the company wants to promote. Specifically, your supervisors will want to check your site to make sure it reflects the mission of the company.

LINK For more information on analyzing readers, see Chapter 3, page 41.

Context of Use

Websites require access to a computer, which limits the number of contexts in which they can be used. But as laptops are more common and wireless networks are more available, computers will be usable in more contexts than ever.

PHYSICAL CONTEXT Involves the physical places where people might use your website. Will readers be accessing the site from home, in their office, in a meeting room, or perhaps at an Internet cafe? Are they at a desk, on the factory floor, or on an airplane? Are they accessing through a wireless network? How fast is their connection to the Internet?

TAKE NOTE A common mistake is to design a website so that it runs well on the company's computers but does not run well on the computers of clients or customers. Most companies have relatively fast connections to the Internet. But clients and customers may use a slower connection. The time needed to navigate a slow website can be frustrating to these readers.

ECONOMIC CONTEXT Involves the financial issues that shape how readers will interpret your website. What is their likely economic situation or status? What are the trends in the industry or national economy? What kinds of economic decisions are you asking readers to make?

For worksheets that will help you analyze your website's audience, go to **www.ablongman.com/johnsonweb/14.5**

ETHICAL CONTEXT Involves the personal, social, and environmental issues that your website touches. Where does your website involve issues of rights, laws, utility, or caring? How should these ethical issues be handled?

LINK For more information on ethics, see Chapter 4, page 72.

POLITICAL CONTEXT Involves the micropolitical and macropolitical issues that might influence how your readers interpret the website. For example, does your website step on some toes within the company? How does it fit in with your company's overall political stance?

LINK For more information on defining a document's context of use, go to Chapter 3, page 53.

As you think about the rhetorical situation for your website, put yourself in your readers' place. Try to anticipate the kinds of information they are looking for and why they need that information. Putting yourself in your readers' place will help you enormously as you are writing and designing the website.

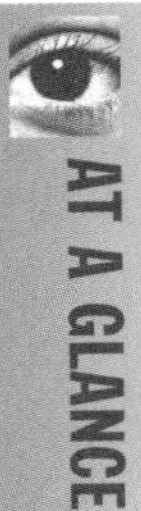

AT A GLANCE

Defining a Website's Rhetorical Situation

- Subject: What information is inside the scope of the website, and what isn't?
- Purpose: In one sentence, what is the purpose of the website?
- Readers: Who will be reading the site, and what kind of information are they looking for?
- Context: What physical, economic, ethical, and political factors will shape how the website is written and read?

Websites for International and Cross-Cultural Readers

It is becoming increasingly important to design websites for international and cross-cultural readers. In a sense, English has been the unofficial language of the web, but increasingly, non-English-speaking users around the world are accessing the Internet.

Of course, it would be impossible to anticipate all the needs of potential readers around the world, but you can make your website more usable in a few important ways.

Translate the website—The global marketplace means attracting potential clients and customers around the world. So, if your company regularly does business with people from another culture, it might be a good idea to make your website, or at least parts of your website, available in their language.

Use common words—Try to use words that are commonly defined in English. The meanings of slang and jargon words change quickly, sometimes leaving international readers confused.

Avoid cliches and colloquialisms—Informal American English includes phrases like "piece of cake" or "miss the boat" that might be meaningless to people from other cultures. Also, sports metaphors like "kickoff meeting" or "hit a home

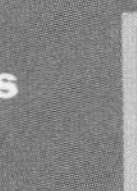

Want to know more about ethical issues involving websites? Go to **www.ablongman.com/johnsonweb/14.6**

run" sound very odd to people who are not familiar with American football and baseball.

Avoid cultural icons—Symbols, especially religious symbols, should be avoided where possible and carefully used where necessary.

Minimize humor—American humor does not translate well into other cultures. So, attempts to be funny on a website might be offensive or just confusing.

Chapter 3 discusses writing for international and cross-cultural audiences in more depth. In most cases, these guidelines for writing to cross-cultural readers are applicable to websites also.

LINK For more information on writing for international and cross-cultural readers, go to Chapter 3, page 60.

Organizing and Drafting a Website

Organizing and drafting a website is not all that different from organizing and drafting a print document.

Organizing the Website

Professional website developers often like to start creating a new site by mapping out its contents on a whiteboard, piece of paper, or screen. Logical mapping is a good way to sort out the content of the website and develop an efficient organizational scheme for it.

To map out the contents of the site, start by writing your website's purpose in the middle of your screen or a sheet of paper. Then, use lines and circles to begin identifying the contents of the site (Figure 14.3).

You might also try a low-tech method that is popular with professional web designers—using sticky notes. On a blank wall, use sticky notes to map out your website. This low-tech method allows you and your team to move the notes around to look at different ways to organize the site. Another advantage to sticky notes is that they can be added or eliminated with ease.

Creating Levels in the Website

Your logical map and research will likely give you a sense of how many levels are needed in the website. The map shown in Figure 14.3, for example, might translate into a structure like the one shown in Figure 14.4.

How many levels should you create in a website? Professional website designers use the following guidelines to determine the number of levels in a website:

- a maximum of three links for the most important information.
- a maximum of five links for 80 percent of all information.
- a maximum of seven links for all information.

These guidelines are helpful, because if you force your readers, especially customers, to wade through too many pages, you risk losing them. If you make them work too hard, they will grow frustrated and give up.

For more advice about writing websites for international and cross-cultural readers, go to **www.ablongman.com/johnsonweb/14.7**

Mapping Out a Website

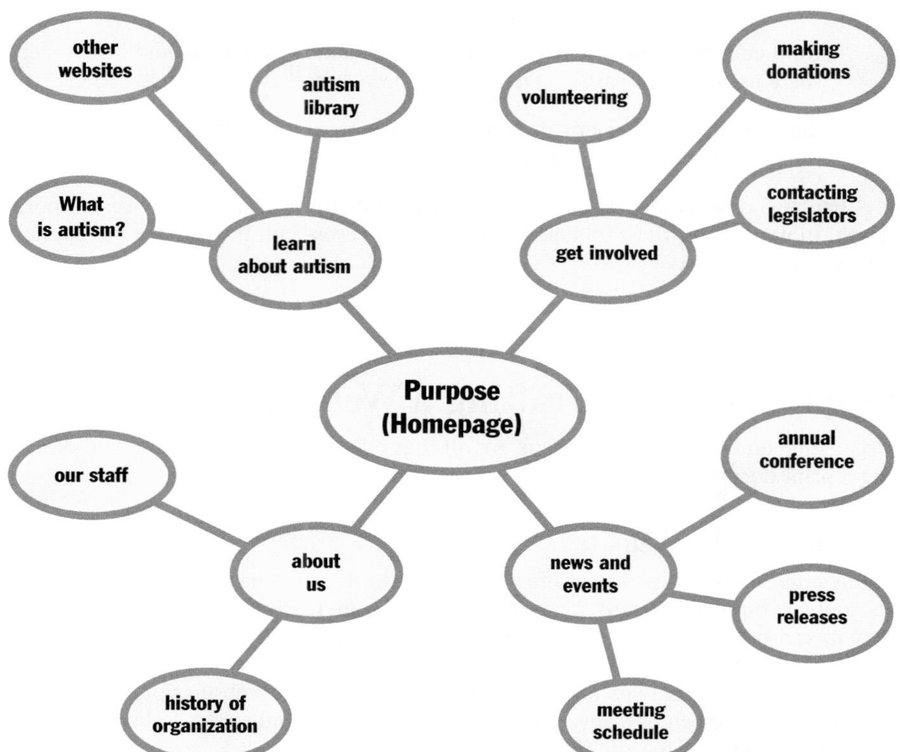

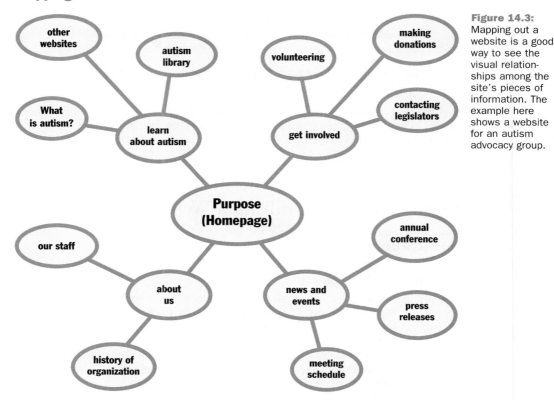

Figure 14.3: Mapping out a website is a good way to see the visual relationships among the site's pieces of information. The example here shows a website for an autism advocacy group.

Like any guidelines, though, these are not rules to be followed absolutely, because some situations resist your efforts to adhere to them. Nevertheless, if you notice that your website's structure goes against these guidelines, you might look for ways to reduce the number of levels to reach the information your readers need.

Drafting the Homepage

Your homepage should be similar to the introduction in a paper-based document. The homepage should set a context by clearly signaling the subject, purpose, and main point of the site. It might also stress the importance of the site's information to your readers.

SUBJECT Your site's subject should be clear as soon as readers access the homepage. You don't need to say something like "The subject of this site is . . .," but you can use a title or a graphic to quickly indicate what the site is about (Figure 14.5).

PURPOSE Likewise, the purpose of your site should also be clear on the homepage.

Levels in a Website

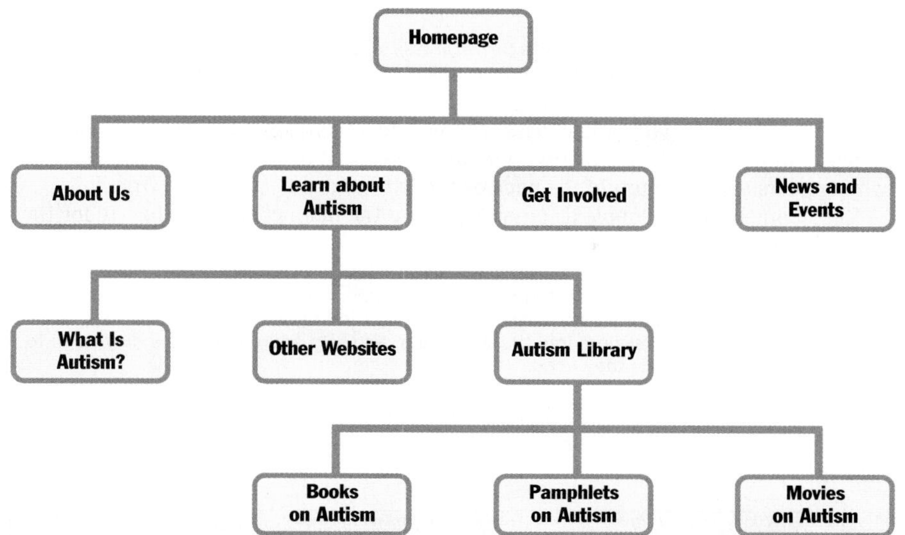

Figure 14.4: A logical map can be turned into levels of information. Each level offers information that is increasingly specific.

> In this website, you will find information about autism and the many services and activities supported by the Southeastern Autism Network.

Often, websites do not directly state their purpose on the homepage. In these situations, the purpose of the website should be obvious without the purpose statement.

MAIN POINT Since websites tend to be more informative than persuasive, they often do not state a main point on the homepage. Nevertheless, you should think about the overall point you want to stress.

> An autistic family member can be a challenge, but that challenge can be rewarding too. Autistic children and adults have unique qualities and personalities that make them fascinating individuals.

STATEMENT OF IMPORTANCE The importance of your subject may be immediately obvious to you, but it might not be obvious to your website's readers. You should stress the importance of the subject on the homepage, to encourage readers to use the site.

> Being informed about autism is one of the best ways to handle this unique mental condition. The information on this site allows you to draw on the experiences and knowledge of the autism community.

In some cases, a picture or graphic can also signal the importance of the site.

BACKGROUND INFORMATION On most homepages, background information is limited, because there is only a limited amount of space on the screen. You might,

To see good and bad examples of homepages, go to **www.ablongman.com/johnsonweb/14.8**

however, give the readers some background information on your subject to help them gain a quick overview.

> The Southeastern Autism Network was created in 1986 to foster communication among families and professionals about autism. Since then, it has grown to include over 200 members.

FORECASTING Usually homepages will include a navigation bar that forecasts the structure of the site. The navigation bar can appear just about anywhere, but it typically appears along the top of the page or the left-hand side (Figure 14.5).

Keep homepages as simple as possible, saving the main content of the site for the pages that follow.

> **TAKE NOTE** You might be tempted to tell your readers everything on the homepage. But they are not ready for all that information at this point. Instead, think of the homepage as a way to welcome your readers into the site. Then, let them figure out where they need to go to find the information they need.

LINK For more information on writing introductions, see Chapter 6, page 141.

A Homepage

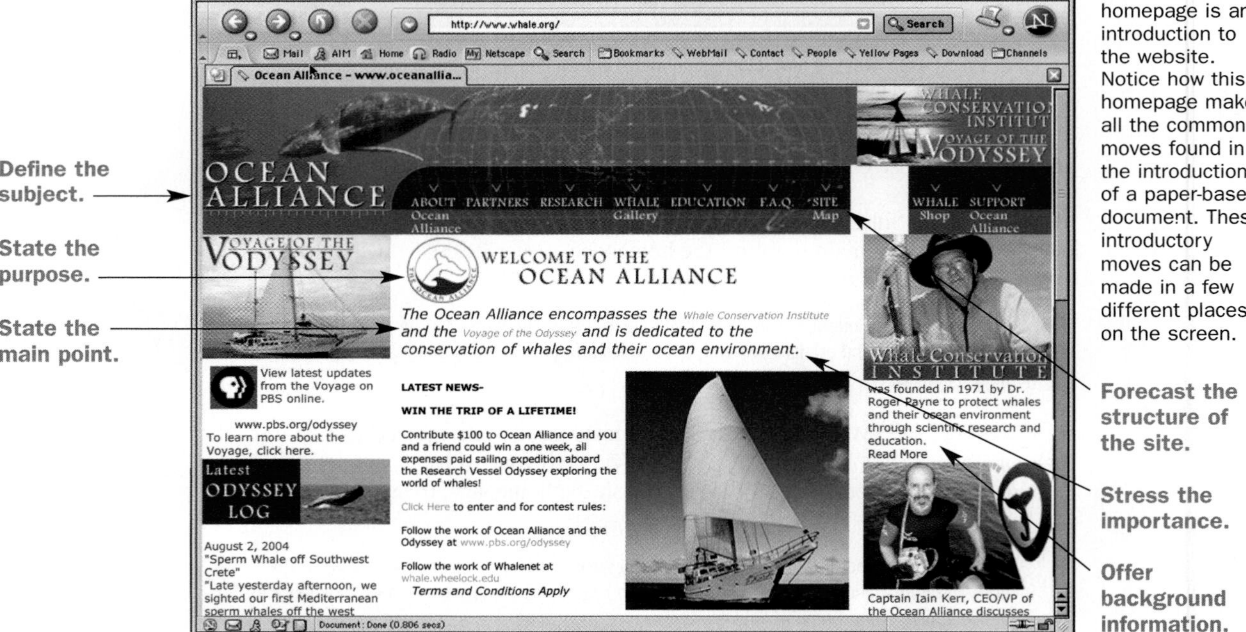

Figure 14.5: The homepage is an introduction to the website. Notice how this homepage makes all the common moves found in the introduction of a paper-based document. These introductory moves can be made in a few different places on the screen.

Source: Ocean Alliance, http://www.whale.org.

Drafting Node Pages

Node pages direct traffic in your website. These pages typically follow links on the homepage, further dividing the subject of the site. Each node page introduces readers to one of the site's major topics. For example, the node page shown in Figure 14.6 was a link displayed on the homepage in Figure 14.5.

A Node Page

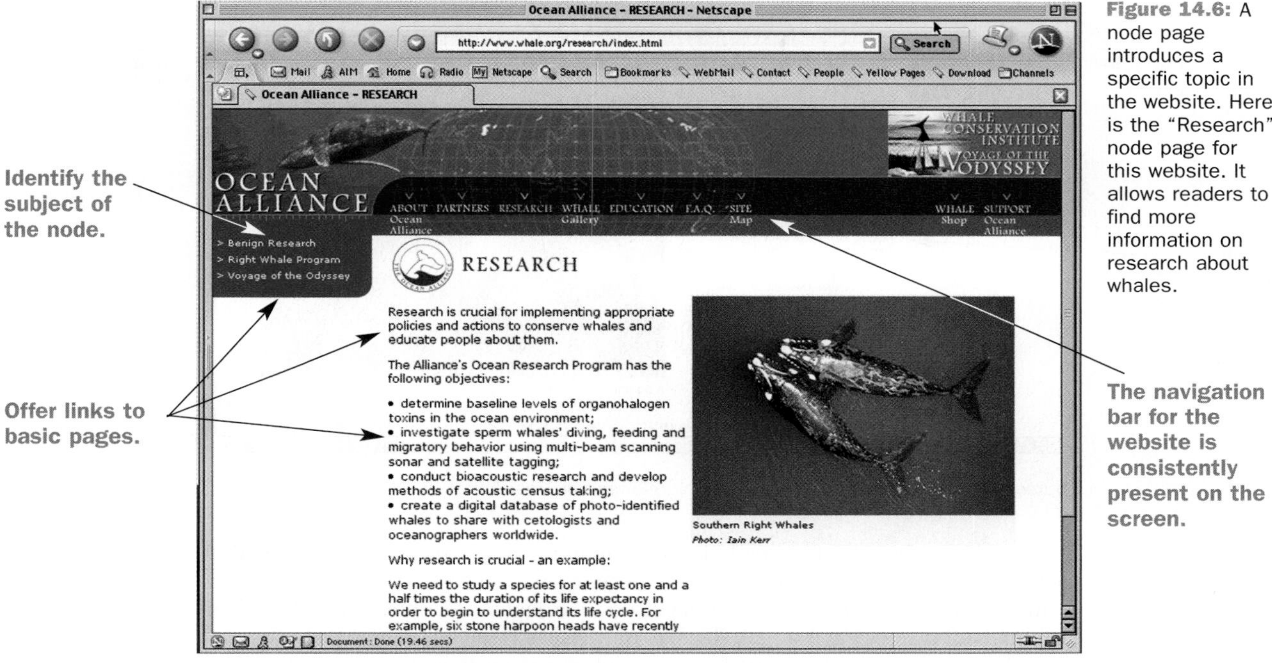

Figure 14.6: A node page introduces a specific topic in the website. Here is the "Research" node page for this website. It allows readers to find more information on research about whales.

Source: Ocean Alliance, http://www.whale.org/research/index.html.

A well-designed node page allows readers to select among other further topics. In essence, it serves as a miniature homepage to this part of the site. In large websites, node pages are very much like homepages, because they tell readers the *subject, purpose,* and *main point* of this part. They include mostly contextual information that helps direct readers toward more specific details further into the website.

Drafting Basic Pages

There really is not a sharp distinction between node pages and basic pages. As readers move further into the site, the pages should become more detail oriented.

Basic pages increasingly provide the content (e.g., facts, data, examples, details, descriptions) that readers are looking for. For example, the basic page shown in Figure 14.7 includes mostly facts, examples, and reasoning. More than likely, this

To see other sample node pages, go to **www.ablongman.com/johnsonweb/14.9**

A Basic Page

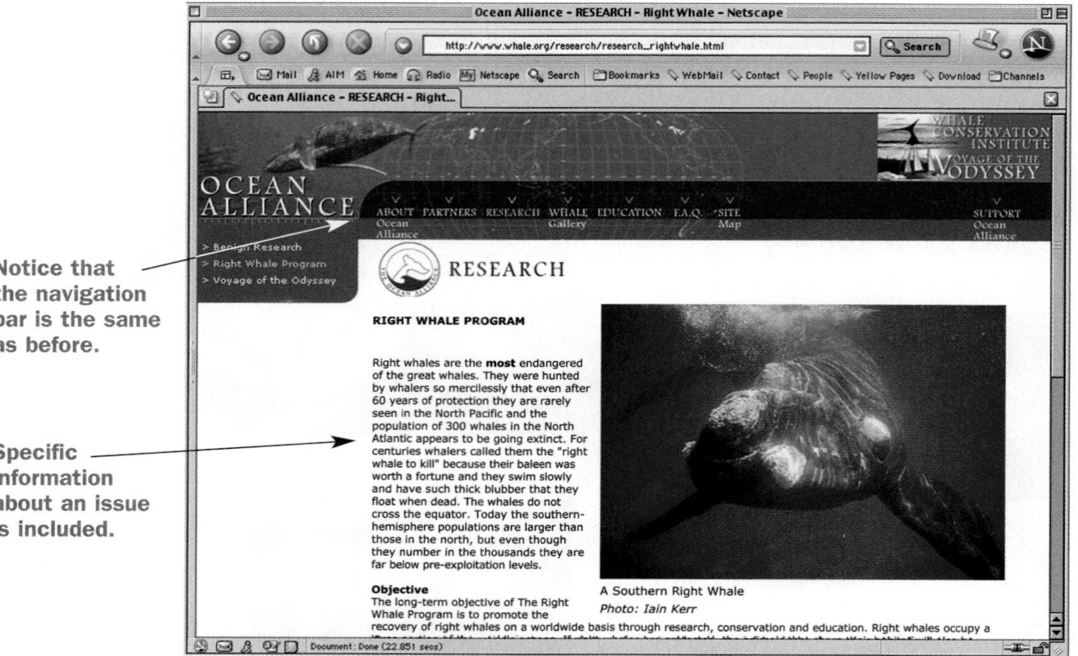

Figure 14.7: A basic page contains the specific information that readers are looking for. This basic page (linked from the "Research" node page) discusses a specific issue, the right whale.

Source: Ocean Alliance, http://www.whale.org/research/research_rightwhole.html.

information is why readers came to the website in the first place. But, they needed to navigate through the homepage and node pages to arrive here.

In a basic page, you should keep the discussion limited to a topic that can be handled in about a screen and a half. Readers of websites rarely have enough patience to scroll down farther. So, if you need more than two screens to cover a particular issue, you might need two or more basic pages.

Drafting Navigational Pages

Websites also include other kinds of pages that can help readers navigate the site.

SITE MAPS A site map is like an index in a book. The site map lists all the topics covered in a website, large or small. That way, if readers are having trouble finding what they want, they can go straight to the site map to find a link to the information they need.

HELP Software programs like RoboHelp, DocToHelp, and Help & Manual will allow you to create Help areas for your website. Usually, only the largest websites include a Help area. These Help areas offer additional information and usually provide definitions or tutorials that readers may need.

GO TO THE NET

For more information about Help software packages, go to **www.ablongman.com/johnsonweb/14.10**

SEARCH ENGINE Larger websites may also include an internal search engine that helps readers locate information by typing in keywords. Search engines are available, usually free, from the larger search engines like Excite (excite.com), Lycos (lycos.com), and Google (google.com). However, putting a search engine into your website takes a little more programming skill.

A well-designed website gives readers a few different ways to access information. Preferably, they can start out at the homepage and work their way down to the information they need. If they are having trouble locating information, perhaps they can use a site map to find it. And, as a last resort, the search engine might help them find pages that discuss the topic they are looking for.

AT A GLANCE

Types of Website Pages

- Homepage—introduction page for the website
- Node pages—opening pages for sections of the website
- Basic pages—pages containing the content readers are looking for
- Navigational pages—site maps, Help area, search engine

A Warning About Copyright and Plagiarism

Inventing and collecting the content for websites is not much different from inventing the content for print documents. You still need to thoroughly research the subject through electronic, print, and empirical sources. The information you invent or collect will be the substance of the site.

To avoid any problems with copyright or plagiarism, remember that the same rules apply to websites as to print documents. If you want to use an image off someone else's website, you need to ask permission. Also, you cannot take passages of text from other sites and use them in your own, unless they are properly cited.

LINK To learn more about copyright and plagiarism, go to Chapter 4, page 82.

HELP

Using Web-Authoring Software

Not long ago, website developers needed to know how to use HTML (hypertext mark-up language) to write a website. HTML is a coding language that tells the computer how something should look on the screen.

Fortunately, there is now a variety of "web-authoring" software packages available that make designing webpages much easier. Web-authoring software is similar to word-processing software, except it's used for writing webpages. The web-authoring software writes the HTML code while you type on the screen (see Figure A). You can also lay out pages, insert images, choose colors, and include graphics.

Countless different web-authoring software packages are available. The most popular commercial packages include Adobe GoLive, Macromedia Dreamweaver, and MS FrontPage. Figure A shows Dreamweaver in use.

For more information on copyright issues and the Internet, go to **www.ablongman.com/johnsonweb/14.11**

Using Web-Authoring Software

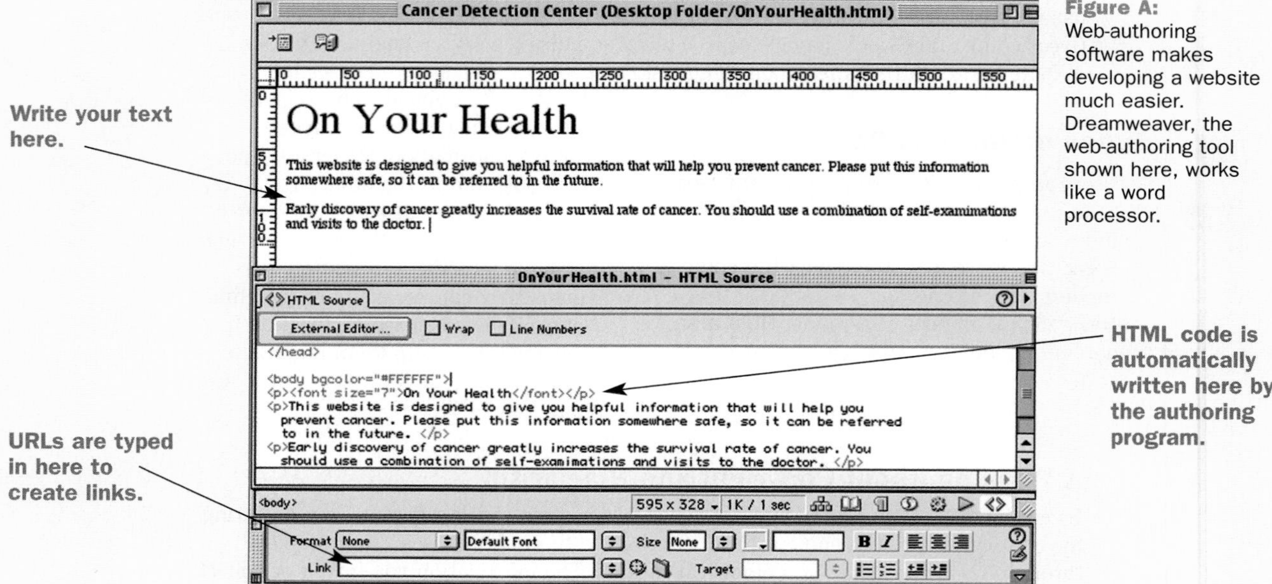

Figure A: Web-authoring software makes developing a website much easier. Dreamweaver, the web-authoring tool shown here, works like a word processor.

Creating a webpage is not difficult with web-authoring software.

1. Open a file by choosing "New Page."
2. Type in a title for the page.
3. Type in some text.
4. Highlight text for links and type in the address of the link.
5. Use the buttons on the design pallette for color, lines, and shapes to add some pizzazz.
6. Use the "Insert" command from the pull-down menu to add images.
7. Save the page.

Each software package uses different commands, but if you know how to use a word processor, you can figure out these commands.

When you have finished creating your webpage, you will need to "FTP" (file transfer protocol) the file to a server (a large computer that connects your computer to the Internet). Here's where things become a bit more complicated. Your university, company, or Internet service provider (ISP) should have instructions available for putting your webpage on the Internet.

The first time you create a website might be a little difficult, because there is much to learn. You will find, though, that creating websites is not difficult once you learn how to do it. Web-authoring software eliminates many of the complications that once existed with HTML coding.

For links to web-authoring software sites, go to **www.ablongman.com/johnsonweb/14.12**

Using Style in a Website

As in any other document, the style of a website is important. The difference between websites and paper-based documents, though, is that readers are even more likely to scan the website. More than ever, your readers will be "raiders" for the information they need. They will rarely read every word on a page.

Consequently, few readers have the patience to actually read long paragraphs or sentences on a screen. So, here are some strategies for improving the readability of your website.

Links should reflect titles—When readers click on a link, the title of the page that follows should be the same as the link they selected. So, when readers click on a link saying "Research on Whales," "Research on Whales" should be the title of the next page. If, instead, the phrase, "Information on Whales" appears in the title, readers are going to be confused for a moment.

Keep sentences short—On average, sentences in websites should be shorter than sentences in paper-based documents. Website readers usually scan faster than readers of paper documents. By shortening sentences, you will help them maintain that faster pace.

Keep paragraphs short—Paragraphs should also be shorter to aid scanning. Paragraphs should be kept to a few sentences or less. That way, readers can scan the paragraph in a glance. If you find yourself writing a long paragraph, you might be offering too much support for your point. If so, you may need to make a separate webpage for this supporting information.

Use mapping to develop themes—You can use logical mapping to set themes or create a tone for a website. Since websites are more modular than paper-based documents, it is a good idea to use themes to set an overall tone that will keep the website sounding consistent.

To create a theme, put the word that best represents the theme or tone you desire in the center of your screen or a sheet of paper. Then, use mapping to find words that are associated with that theme or tone (Figure 14.8). As with paper-based documents, use these words strategically throughout the site. When your readers experience these words, they will subconsciously sense the theme or tone you are trying to create.

AT A GLANCE

Using Style in a Website

- Links should reflect titles.
- Keep sentences short.
- Keep paragraphs short.
- Use mapping to develop themes.

LINK For more information about using logical mapping to improve style, turn to Chapter 7, page 181.

GO TO THE NET

For websites that discuss Internet style, go to **www.ablongman.com/johnsonweb/14.13**

Mapping a Theme for a Website

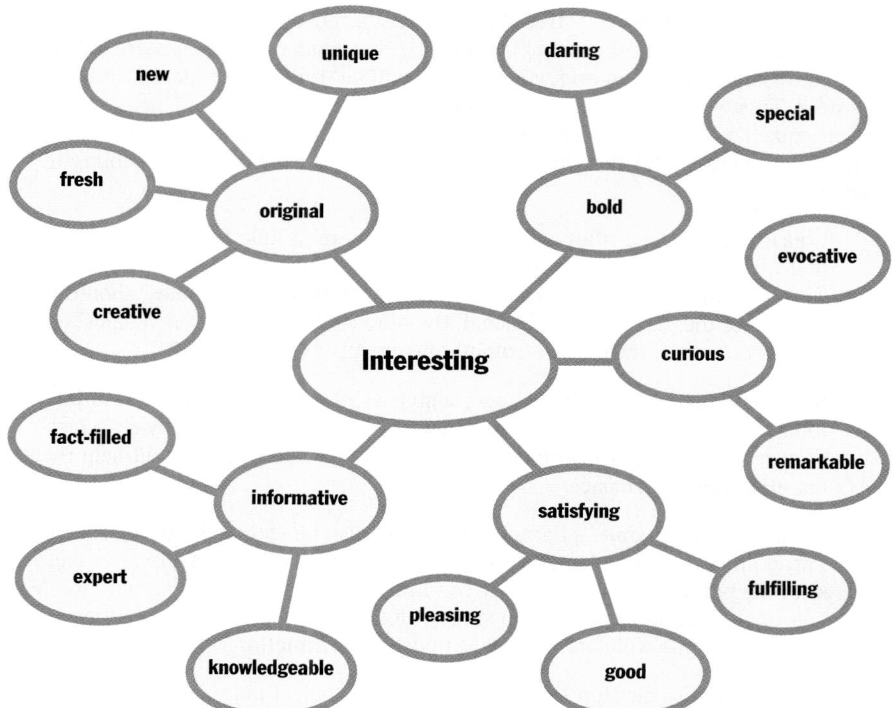

Figure 14.8: Logical mapping can help you create themes or tones that hold the website together. A website that uses this map would set the theme "interesting."

Designing the Website

The design of your website is important for a few reasons. First, readers prefer attractive webpages. They are more likely to spend time looking over webpages that please their eyes. Second, well-designed websites are more navigable and more effective. If readers can quickly locate paths to the information or products they are seeking, they will be much more willing to do business with your company. And finally, a well-designed website gives readers confidence in the information you are providing. A professional-looking website will make your company and its products seem more trustworthy.

Designing the Interface

The look of your website on the screen is called the *interface.* The interface involves how the pages look on the screen and how readers interact with them. The *interface design* determines in great part whether your website will be attractive to readers and whether it will be easy to use.

As discussed in Chapter 8, when designing an interface for a website, you should use the *Five Principles of Design:*

Want to know more about designing interfaces? Go to **www.ablongman.com/johnsonweb/14.14**

Balance—Items placed on the screen offset each other to create a feeling of stability on the page or screen.

Alignment—Aligned items on the page help readers identify different levels of information in the interface.

Grouping—Items that are near each other on the screen will be seen as being "grouped."

Consistency—Each page in the website looks similar to the others, minimizing chaos in the interface.

Contrast—Contrast sharpens the relationships between images and text, headings and text, and text and the background.

The interface shown in Figure 14.9, for example, demonstrates all of these principles well. Notice how the text is balanced from side to side. Information is also aligned in clear vertical lines, and you can see where specific information has been grouped into larger blocks. Meanwhile, consistency and contrast are used to make the interface interesting but consistent.

A Well-Designed Interface

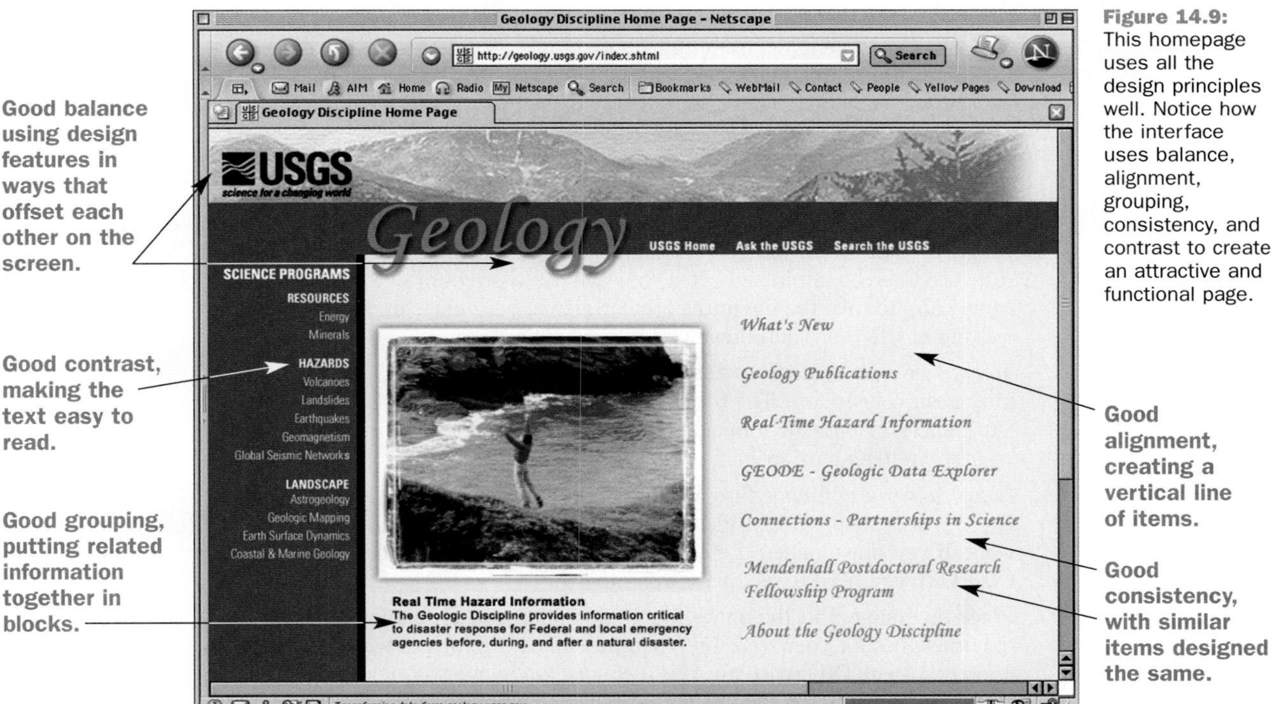

Figure 14.9: This homepage uses all the design principles well. Notice how the interface uses balance, alignment, grouping, consistency, and contrast to create an attractive and functional page.

Source: U.S. Geological Survey, http://geology.usgs.gov/index.shtml.

These five principles should be used to develop a consistent interface, called a template, for your webpages. The template will give your website a standardized, predictable design, allowing your readers to quickly locate the information they are seeking.

LINK For more information on designing interfaces, see Chapter 8, page 192.

Adding Images

Images in websites are saved as separate files on your computer and on the server. These files include any pictures, graphs, or drawings you might want to include on your webpage. They also include elements of the interface, such as banners on the screens, corporate logos, and icons. Images can be saved in a variety of *file formats.* The two formats most commonly used for websites are called *jpeg* and *gif* formats.

> **jpeg (joint photographic experts group)**—The jpeg file format is widely used for photographs and illustrations with many colors. Images in jpeg format can use millions of different colors, allowing them to better capture the subtleties of photographs. The main limitation of jpeg images is their higher memory requirements, causing them to download more slowly, especially on computers with slow connections to the Internet.
>
> **gif (graphic interchange format)**—The gif format is primarily used for illustrations, logos, and simple graphics. Gif images can use a maximum of only 256 colors, making them less useful for photographs. Their advantage is that they use less memory than jpeg files, making them quicker to download.

You can tell the difference between jpeg and gif files by the extensions on their file names. A jpeg file will have the extension .jpg added to it (e.g., cougarpic.jpg), and a gif file will have the extension .gif added (e.g., cougarpic.gif).

To put images in your webpages, you need to do three things.

1. **Create or capture the image.** You can use a digital camera, scanner, or drawing software to create an image. Then, you can use a program like Adobe Photoshop to turn the file into a jpeg or gif. Or, you might find an image on a website or CD-ROM and download it to your hard drive.

   To capture an image off a website, put your cursor on it and hold down the button on your mouse. The browser will bring up a window that gives you a few choices. Choose **Save Image As.** The browser will let you save the image to your computer's hard drive.

   Again, a warning: Image files are the property of the people who created them. If you made the picture or illustration yourself, you can use it however you want. If you downloaded the image off someone else's website or off a CD-ROM, you need to ask permission to use it.

2. **Place the image.** Put the image file in the folder on your computer that holds the other files for your website. When possible, keep images with the pages that reference them. Otherwise, there is a chance the image will not be transferred with the page, causing the image to not be shown on the page (see "Uploading, Testing, and Maintaining Your Website" later in this chapter).

Want to learn more about using images? Go to **www.ablongman.com/johnsonweb/14.15**

3. **Insert the image.** Using your web-authoring software, open the page in which the image will appear. Then, use the **Insert Image** command (or equivalent) to add the image to the page.

When you are finished inserting the image, it should appear on the webpage.

Revising, Editing, and Proofreading

Before you put your website on the Internet, you should revise the document just as you would a paper-based document. As discussed in Chapter 4, you should use the four levels of edit (revising, substantive editing, copyediting, and proofreading) to improve the quality of the document.

Revising and editing are not simply ways to polish the website. You need to look closely at the content, organization, style, and design to see if they are appropriate for your readers and the context in which the site will be used. When you are revising the website, ask yourself these questions:

Does the website highlight information my readers want?

Are the subject and purpose of the website obvious on the homepage?

Is the site organized in a way that is logical to my site's readers?

Is the style of the website appropriate? Does it set the right tone?

Is the interface easy to navigate?

Is the interface attractive and interesting?

Proofreading a website is especially difficult, because seeing errors on the screen is often harder than seeing them on paper. So, you might print out the screens of your website and proofread them on paper.

When you have finished revising and editing the website, do some usability testing to ensure quality. Minimally, you should ask your coworkers and friends to try using the site, most likely on your computer.

Depending on the importance of the site, you might also set up formal usability tests with sample readers. You can do the same kinds of read-and-locate tests, understandability tests, performance tests, and even safety tests that you might use to test the usability of a print document.

LINK For more information on usability testing, go to Chapter 10, page 282.

Uploading, Testing, and Maintaining Your Website

Once you are finished revising, editing, and testing the usability of your website, it is time to upload it to the Internet. Once it is uploaded, you will need to test the site yet again to see if it works the way you expected. After correcting any problems, you will need to maintain the website to keep it up to date.

For more editing strategies for websites, go to **www.ablongman.com/johnsonweb/14.16**

Uploading the Website

To make your website available to the public, you need to upload it to a server—a large computer that connects your computer to the Internet. Before uploading to the server, most writers of websites prefer to create a finished website on their own computer. Then, they use an FTP software program to copy the whole website to the server at once (Figure 14.10).

When you copy your webpage files, don't forget to also copy your image files with them. If the image files are not included with the webpage files, the pictures, illustrations, graphs, and other images you placed on your website will not appear.

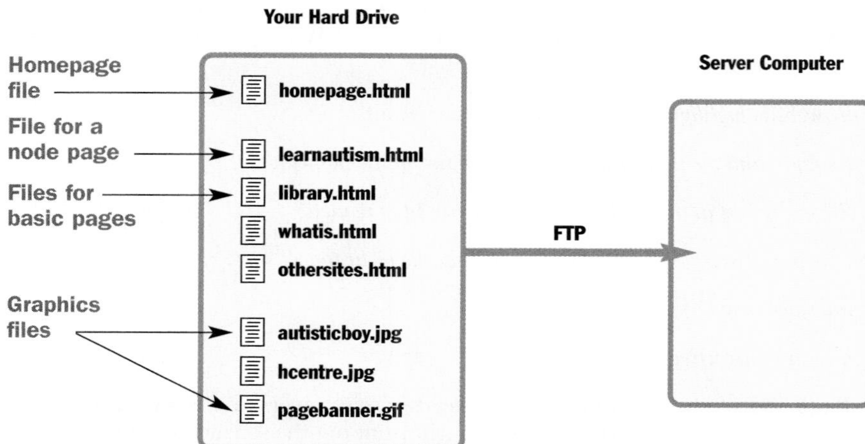

Figure 14.10: An FTP program copies files from your computer's hard drive to a server. The server is connected to the Internet.

Testing the Website

Once your website is on the server, it is time to test it. Using at least a couple different browsers (e.g., Netscape, Explorer), look over the pages on the site. Try out the links and resize the browser window a few different ways. Usually, you will find that the website has some formatting problems that did not show up in your web-authoring program's version of the site. For example, perhaps an image failed to appear or the spaces between lines are uneven. Another common problem is that links to other pages don't work, giving readers error messages when links to these pages are chosen.

To fix these problems, reopen the files with your web-authoring program. Make the corrections and changes needed. Then, re-upload your revised pages to the website.

The testing phase can be a little frustrating at times, but it is important. You want to make certain that your website looks the way you intended. Those little problems won't fix themselves, and they will be annoying to your readers. Once you have fixed these small problems, though, your website will convey the sense of quality you are looking for.

Want to learn more about FTP? Go to **www.ablongman.com/johnsonweb/14.17**

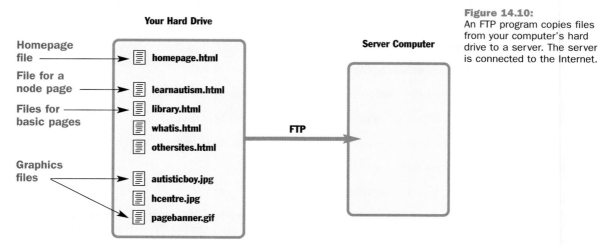

Maintaining the Website

Few websites can be left alone for long. You will find yourself wanting to add or change information on your website regularly. You will find new links to other websites, or you will discover that information on your site is out of date. You should plan to regularly update your website, usually weekly or monthly.

To update a page, go back to the file saved on your hard drive and open it in your web-authoring program. Make the changes you want and then use FTP to copy the updated file onto the server. The server will replace the old file with your new file.

Most companies hire or contract a webmaster to maintain their site. The webmaster regularly updates the site, incorporating any changes and eliminating any dated materials. When you are your own site's webmaster, though, you should revisit your site regularly, looking for items to update.

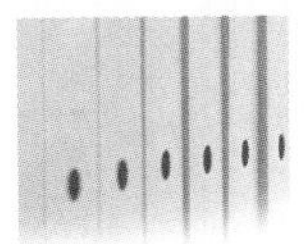

CHAPTER REVIEW

- Websites are important for both corporate and personal communication.
- The basic features of a website are a homepage, node pages, basic pages, and navigational pages.
- Prepare to develop a website by first defining its subject, purpose, readers, and context of use. Consider international and multicultural audiences as well.
- Mapping is a good way to plan the content of a website; an initial map can be followed by a more detailed plan of the levels that will be used in the website.
- The homepage of a website is like the introduction to a document. Tell readers the subject, purpose, and main point of the site, while offering background information, stressing the importance of the subject, and forecasting the site's structure.
- Choose a consistent style to make your website easier to scan and interpret.
- The design of the website should follow the five principles of design: balance, alignment, grouping, consistency, and contrast.
- Revising and editing are especially important in websites, because the text is regularly changed.
- When your website is finished, you need to upload, test, and maintain it.

EXERCISES AND PROJECTS

Individual or Team Projects

1. Write a critique of a website you found on the Internet. Look at its content, organization, style, and design. Does it achieve its purpose? Is it appropriate for its intended readers? Can you find information easily? Write your critique in memo or e-mail form for your instructor.

2. On paper or whiteboard, map out the contents of a website. Identify its homepage, node pages, basic pages, and navigational pages. Then, in a presentation to your class, discuss the organization of the website. Show its strengths and identify any places where the website's organization might be improved.

3. Imagine a website you would like to build. Identify its subject, purpose, readers, and context of use. Then, diagram the site on paper or a whiteboard, showing how it would be organized. Thumbnail some sample pages on paper, sketching out how a few pages would look on the site (homepage, node pages, basic pages). Attach these drawings to a memo of transmittal to your instructor in which you discuss the content and organization of your website.

4. Find a website at your college or workplace that is ineffective. Write a report to a specific reader in which you discuss the shortcomings of the site and make recommendations for improvement. Consider the content, organization, style, and design in your report. In your recommendations, show your sample reader pages and organizational schemes that would improve the site.

Collaborative Project

You and your group have been asked to develop a "virtual tour" of your university or workplace. The tour will allow visitors to your website to look around and familiarize themselves with important places on your campus or in your office. The tour should allow them to quickly navigate the site, finding information they need to locate important places and people.

Start out by first discussing and describing in depth the subject, purpose, readers, and context of use for the site. Then, using paper or a whiteboard, describe how the site would be organized. Make decisions about the style of the site and how the various pages should be designed.

Write a proposal to your university administrators or managers, showing how this virtual tour would be a nice addition to the university's or company's website. Your proposal should offer a work plan for making the website a reality and describe its costs and benefits, especially the benefits to visitors.

Then, if the software and hardware are available, create part of the website. Of course, you probably cannot create the whole virtual tour, but you can make the homepage, a couple of node pages, and a few sample basic pages that show important places on campus or your workplace.

Sample websites for you to critique are available at **www.ablongman.com/johnsonweb/14.18**

Invasion of Privacy

Kate Harold was thrilled that her company hired a new intern to create a website for her work group. The intern's name was Vance Lawrence, and he seemed like a really bright person. He was also the beloved son of the company's CEO, looking for some experience in the "real world."

Vance was a little quirky, but he knew how to create great websites. When he was interviewing for the internship, he showed Kate and her supervisor some very slick websites he had created for his fraternity and for a couple of clubs to which he belonged. They were very professional. After the interview, Vance's mother, the company's CEO, called Kate to see how Vance did at the interview and when he would be able to start. Kate said Vance would work out great and that he could start right away.

So, they let him take the lead on developing the new website for the group. Vance made the rounds several times to interview Kate and the other members of the group. He asked some insightful questions about the group's current project and each of their roles in it.

After a month, Vance announced he was finished and he gave Kate the web address so she could check out the new website. Kate was busy that week, so she didn't look it over right away. A couple of days later, the CEO called her again and said, "What do you think of Vance's website? I think it looks great." Kate lied a bit and said, "Yeah, I looked it over, and it's just what we needed."

After hanging up the phone, Kate went immediately to her computer to look at the website. At first, the site looked great. The homepage was nice, and most of the website was easy to navigate.

There was one problem, though. Under a link called "About Us," Vance had gone a little overboard. In this link he had developed separate pages for each of the employees in the group. These pages included home addresses, pictures from the company picnic, names of children, and home phone numbers. Vance also included a section called "Vance's Impressions" in which he offered some flattering commentary showing how much he liked each of the group members.

One other problem was apparent. Vance also made a link to his own website, which featured pictures of himself at fraternity parties, Goth dances, and speedmetal concerts.

Kate wasn't sure how to handle the situation. She didn't want to upset Vance or his mother, but she felt this personal information was not appropriate on the website. If you were Kate, how would you handle this situation?

CHAPTER

15

Starting Your Career

CHAPTER CONTENTS

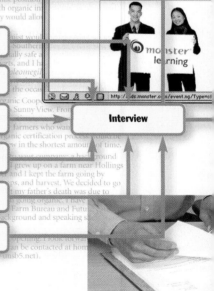

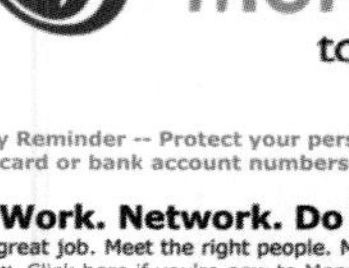

Privacy Reminder -- Protect your persona credit card or bank account numbers to

Get Work. Network. Do it

Find a great job. Meet the right people. Make you how. Click here if you're new to Monster

- Search hundreds of thousands of jobs
- Apply online

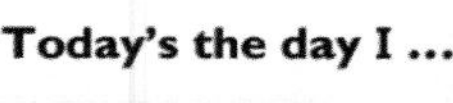

April 2, 2004

834 County Line Rd.
Hollings Point, Illinois 62905

Valerie Sims, Human Resources Manager
Sunny View Organic Products
1523 Cesar Chavez Lane
Sunny View, California 95982

Dear Ms. Sims:

Opening paragraph states the subject, purpose, and main point.

Education is discussed up front with examples.

Work experience is used to show potential contributions to employer.

Other skills are highlighted to show unique abilities.

Conclusion ends on a positive note and offers contact information.

Sincerely,

Anne Franklin

Job Search Engines

Personal Networking

Professional Networking

College Placement

Targeting

Classified Ads

Interview

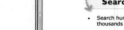

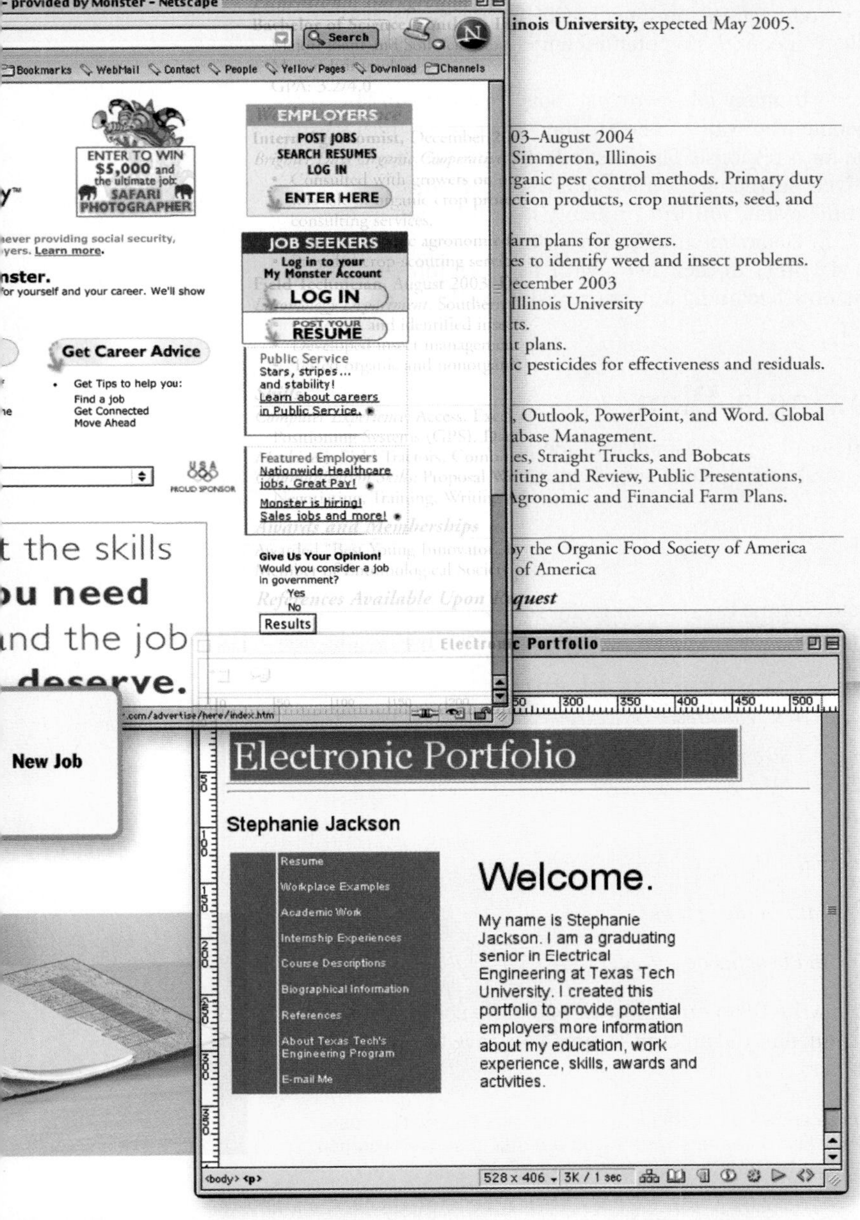

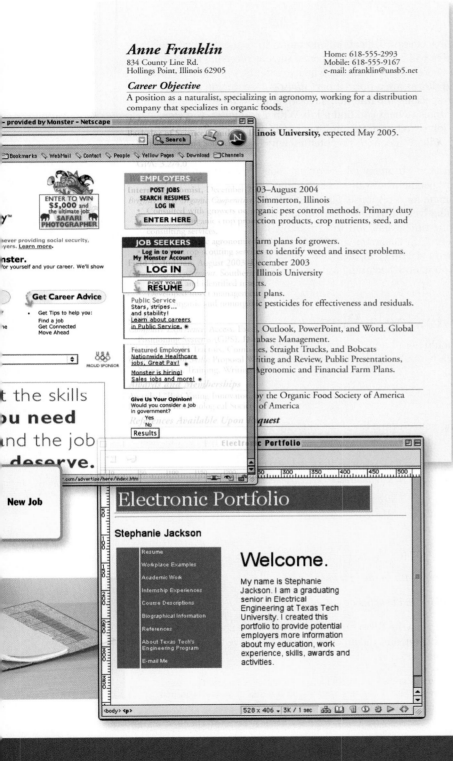

CHAPTER OBJECTIVES

In this chapter, you will learn:

- How to set goals and make a plan for finding a career position.
- Methods for using both web-based and traditional networking.
- How to design and prepare a resume in both electronic (scannable) and traditional formats.
- Techniques for writing a persuasive application letter.
- How to create a targeted, professional portfolio that highlights your background and experience.
- Strategies for effective interviewing.

Finding a good job requires energy, dedication, and optimism—and readiness to hear the word "no." There will be exciting moments, such as receiving a call for an interview or a job offer. There will be moments when you doubt yourself and your abilities. Emotional ups and downs are a normal part of the job-searching process.

Fortunately, computers, especially the Internet, have streamlined the searching process, and they offer new tools to help you find a job. With a computer, you can contact employers through job search engines, on-line classifieds, professional group websites, and e-mail. Meanwhile, you can locate potential employers worldwide through the Internet.

Your computer gives you access to many job-searching tools, so use them. The least effective job seekers are people who wait for the Sunday paper, send out a few resumes, and wait by the phone for a response. The most effective job seekers are people who spend each day actively searching for a job along several paths.

Job searching is not a one-time event. You will probably look for a job at least a few times during your career. U.S. Department of Labor statistics show that people tend to change careers four to six times in their lives, with 12 to 15 job changes. In other words, finding a job is now becoming a necessary skill in the technical workplace.

Setting Goals, Making a Plan

From the beginning, you should adopt a professional attitude about looking for a job. You should start your search by first setting some specific goals and developing a plan for reaching those goals.

Setting Goals

So, before you begin sending your resume to every company you can think of, spend a couple of days setting goals for your search. Begin by answering the Five-W and How questions:

What are my needs and wants in a job/career?

Who would I like to work for?

Where would I like to live?

When do I need to be employed?

Why did I choose this career path in the first place?

How much salary, vacation, and benefits do I need?

Type your answers in a file, or write them on a piece of paper. Be specific about things like location, salary, vacation, and the amount of time you have to find a position. Be realistic about your answers.

TAKE NOTE If you don't know the answers to specific questions, like salary, then use Internet search engines to help you find these answers. If you are still in school, you can ask your professors to give you a good estimate of salaries in your field.

To see statistics and trends about employment, go to **www.ablongman.com/johnsonweb/15.1**

Your first reaction to unemployment might be to say, "I'll take any job available." That's a common response to an uncomfortable situation like job searching. Instead, spend a little time thinking about what quality of life you desire and how your career fits into that overall life picture. Once you know your goals, you can start looking for a job that suits you.

Using Job-Seeking Tools

With your goals in mind, you can begin developing a plan for finding a good job. Fortunately, your computer offers a door into an amazing number of job-seeking tools.

JOB SEARCH ENGINES Some good places to start looking for jobs are the numerous job search engines available on the Internet. By entering a few keywords in these search engines, you can usually locate a variety of job opportunities. Most search engines will allow you to limit your search by region, job title, or industry. Some search engines will even allow you to post your resume so employers can find you. For example, Monster.com, shown in Figure 15.1, has a button on its homepage that takes you to its resume-posting area.

An Internet Job Search Engine

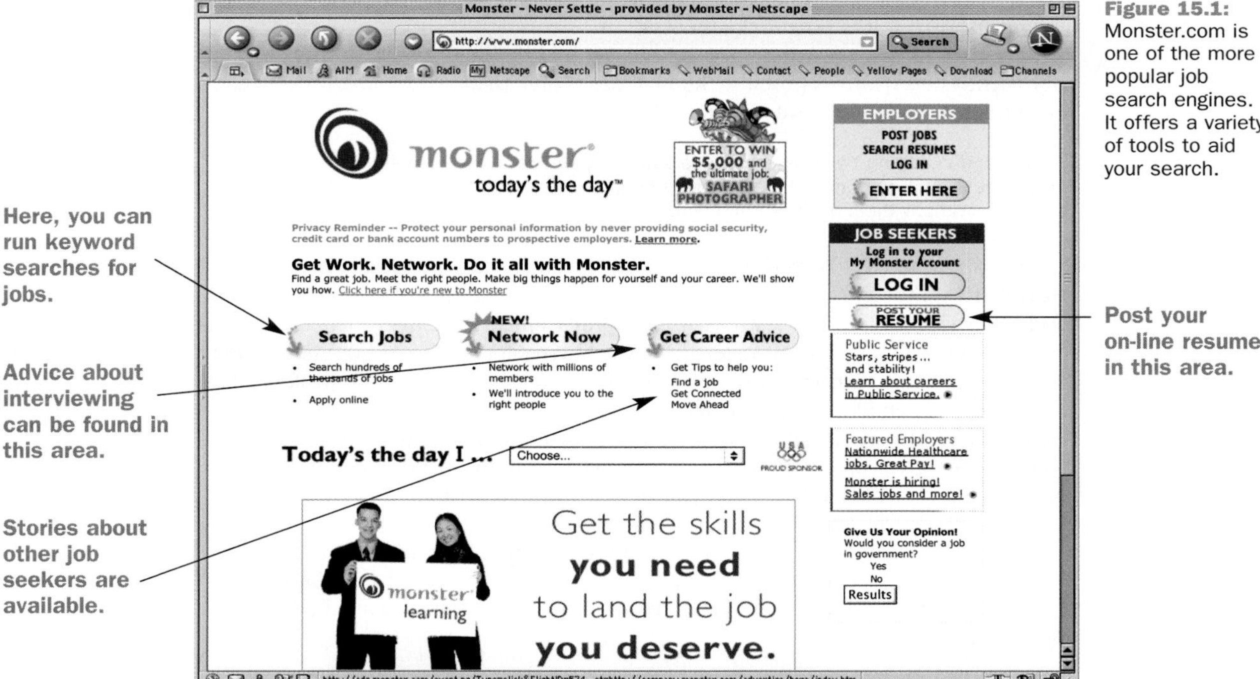

Figure 15.1: Monster.com is one of the more popular job search engines. It offers a variety of tools to aid your search.

Source: Monster, http://www.monster.com.

For websites that will help you answer these goal-setting questions, go to **www.ablongman.com/johnsonweb/15.2**

Here are some of the more popular job search engines:

4Jobs.com	fedworld.gov/jobs/jobsearch.html
ajb.dni.us	hotjobs.com
brassring.com	Jobbankusa.com
careerbuilder.com	job-hunt.org
careercity.com	monster.com
collegegrad.com	worktree.com

Also, popular Internet search engines like Yahoo.com, Altavista.com, and Google.com have job search areas that you can use.

Job search engines are fast becoming the best way to find a job, especially in technical fields. They offer easy and inexpensive contacts between job seekers and employers. Employers are beginning to prefer search engines because these tools can be used to run a nationwide search for the best people. You should check these sites regularly and post your resume on at least a few.

PERSONAL NETWORKING Someone you know probably is aware of a job available in your area. Or, they know someone who knows about a job. Make a list of your friends, relatives, and professors who might be able to help you find a job. Then, send each of these people an e-mail that tells them you are "on the market," looking for a job. You might even attach a resume to your e-mail, so they can look it over and perhaps forward it to a potential employer.

More than likely, you will be pleasantly surprised by the response to these e-mails. Your friends and relatives probably know more people than you realize. Meanwhile, your professors are often aware of opportunities available in your area.

TAKE NOTE Don't be reluctant to use your personal network. People are usually very willing to help. And, perhaps, someday these people will be asking you to help them find a job. You can pay them back at that point.

PROFESSIONAL NETWORKING Most career tracks have professional groups associated with them. Engineers, for example, have the Institute of Electrical and Electronics Engineers (IEEE), while medical practitioners have the American Medical Association (AMA). Technical writers have the Society for Technical Communication (STC). These kinds of groups are especially helpful for networking with professionals who are already employed in your field.

Most large professional groups have a local chapter you can join. These groups' meetings offer great opportunities to contact people who have jobs similar to the one you want. Also, local chapters usually have websites that post job advertisements in your area.

You should become involved with these groups as soon as possible, even if you have not graduated from college yet. It takes awhile to become a regular at meetings, but once people get to know you, they can be very helpful toward finding job opportunities.

COLLEGE PLACEMENT OFFICE Most colleges have a placement office that is available to students. The placement office may have jobs posted on its website, or

For links to job search sites, go to **www.ablongman.com/johnsonweb/15.3**
Want to find a professional group in your major? Go to **www.ablongman.com/johnsonweb/15.4**

you can visit the office itself. There, you can sign up for interviews and speak with a counselor about improving your job-searching skills.

TARGETING Make up a list of 10 to 20 "target" companies for which you might want to work. Then, look at their websites, paying special attention to each company's Human Resources office.

TAKE NOTE You can probably already list some of the major companies in your industry. Can you name any medium and small companies? Using keywords in a search engine, identify large, medium, and small companies that you could work for.

If one of your targeted companies does not have a job available, send its human resources department a copy of your resume and an application letter. In your letter, tell the company you are sending materials for its files in case a position becomes available.

CLASSIFIED ADVERTISEMENTS In the classifieds section of a newspaper, especially the Sunday edition, you will find job advertisements. Keep in mind, though, that newspapers carry advertisements for only a few jobs in any given area. *Most jobs are not advertised in the paper.* But these ads are worth checking once a week.

Most major newspapers now have companion websites that list on-line classifieds. If you are looking for a position in a specific city or state, these on-line classifieds may be helpful. They include all the jobs that are printed in the classified advertisements in the newspaper.

The secret to effective job searching is to set clear goals and have a strategy for reaching those goals. With the variety of electronic tools available, you have many different paths to follow to find a position (Figure 15.2). The most effective job seekers use them all.

The Job-Searching Cycle

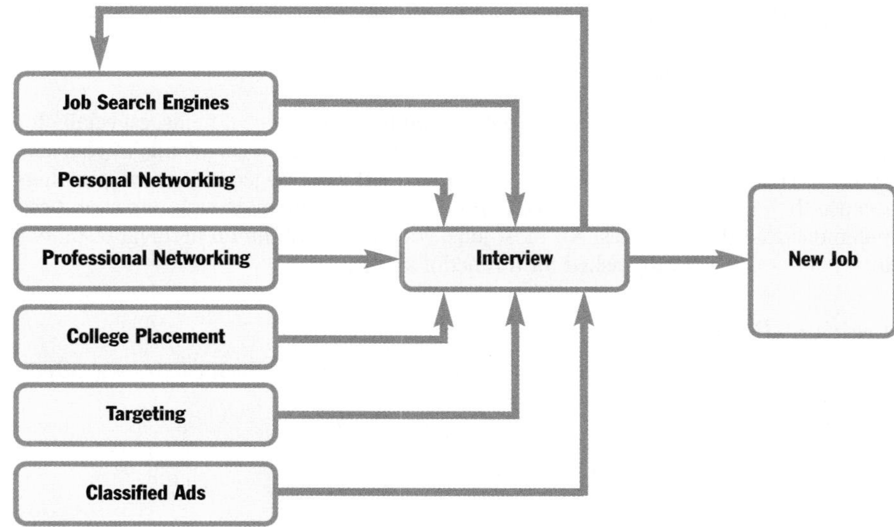

Figure 15.2: There are many tools available for finding a job and building a career. You should take advantage of all of them.

For links to several major newspapers' classified ads, go to **www.ablongman.com/johnsonweb/15.5**

Preparing a Resume

A resume is a summary of your background, experience, and qualifications. Usually, resumes for entry-level jobs fit on one page. Resumes for advanced positions might extend to two pages. You need to spend some serious time and effort on your resume, because it may be one of the most important documents you will ever write.

Your resume needs to make a good first impression, because it is usually the first item that employers will look over. If your resume reflects the qualifications and experience they are looking for, they may read your other materials and give you a call. If your resume is poorly written or poorly designed, chances are slim that you will ever get your foot in the door.

TAKE NOTE On the first cut, employers are usually looking for any reason to put resumes in the "No" pile. So resumes that are poorly written, poorly designed, or mechanically flawed are almost always eliminated immediately.

An important benefit of using your computer to write and design your resume is the ability to change the resume to suit the needs of each job. You can make strategic decisions about how to organize the materials on your resume to highlight the strengths an employer is looking for.

Types of Resumes

Resumes tend to follow one of two organizational approaches: the *archival* approach and the *functional* approach.

Archival approach—Organizes the resume according to education and work experiences, highlighting your responsibilities in a few different areas. An archival resume might be organized into sections such as Education, Work Experience, Skills, and Awards/Activities.

Functional approach—Organizes the resume according to your talents, abilities, skills, and accomplishments. A functional resume would be organized into sections such as Leadership, Design Experience, Communication Skills, and Training Abilities.

By far, the archival approach is the most common in technical fields, especially for entry-level jobs. The archival approach allows you to highlight the details of your experience. The functional approach, in contrast, is not advantageous for new graduates because their experiences are limited and have not given them enough opportunities to demonstrate their abilities. For most jobs, you should submit an archival resume unless you are specifically asked for a functional resume.

Archival Resume

Archival resumes can be divided into the following sections, which can be organized a variety of different ways:

- Name and Contact Information
- Career Objective/Career Summary
- Educational Background

- Related Work Experience
- Other Work Experience
- Skills
- Awards and Activities
- References

The headings in this list are the ones commonly used in archival resumes, but other headings are available. Use headings that best highlight your qualifications and set the specific tone you desire (Figures 15.3 and 15.4).

NAME AND CONTACT INFORMATION At the top of the page, your resume should include a heading that states your name, address, phone number, e-mail address, and a website address. To catch the reader's eye, your name should use a larger font size. The heading can be left justified, centered, or right justified, depending on the design of your resume. Some people like to add a *rule* (a line) to set off the heading from the rest of the information in the resume.

LINK For more information on using fonts effectively, go to Chapter 8, page 212.

CAREER OBJECTIVE OR CAREER SUMMARY A *career objective* is a phrase or sentence that describes the career you are seeking. Your career objective should specify the type of position you are applying for and the industry in which you want to work.

> Seeking a position as a Physicians Assistant in a large research hospital.
>
> A computer programming career working with mainframes in the defense industry.
>
> Looking for a career as a school psychologist working with children who have behavioral problems.
>
> Mechanical Engineer developing nanotechnological solutions for biotech applications.

Avoid writing a career objective that is too narrow or too broad. A career objective that is too narrow might eliminate you from some potential jobs. For example, if you specify that you are looking for a job at a "large engineering firm," a small engineering firm might assume you are not interested in its job. On the other hand, an objective that is too broad, like "a job as an electrical engineer," might give the impression that you are not sure what kind of career will suit your talents.

A *career summary* is a brief sentence or paragraph that describes your career to this point. Career summaries are typically used by people with years of experience.

> I have been employed in the hospitality industry for 10 years, working my way up from desk clerk to assistant manager at a major downtown hotel. My specialty is coordinating catering services for large conferences.
>
> My experience as a webmaster includes designing and managing a variety of interactive sites that have been used by large companies to promote their products and solicit new business.

Your career objective or career summary should try to convey a sense of what makes you unique or interesting as an applicant.

To find other headings for your resume, go to **www.ablongman.com/johnsonweb/15.6**

Archival Resume

Name and contact information are prominently displayed.

Career objective describes position sought.

Education is highlighted by placement high in the resume.

Work experience lists duties with bullets.

Skills are listed separately for emphasis.

Awards and memberships are placed later in the resume.

Anne Franklin

834 County Line Rd.
Hollings Point, Illinois 62905

Home: 618-555-2993
Mobile: 618-555-9167
e-mail: afranklin@unsb5.net

Career Objective

A position as a naturalist, specializing in agronomy, working for a distribution company that specializes in organic foods.

Educational Background

Bachelor of Science, Southern Illinois University, expected May 2005.
Major: Plant and Soil Science
Minor: Entomology
GPA: 3.2/4.0

Work Experience

Intern Agronomist, December 2003–August 2004
Brighter Days Organic Cooperative, Simmerton, Illinois

- Consulted with growers on organic pest control methods. Primary duty was sale of organic crop protection products, crop nutrients, seed, and consulting services.
- Prepared organic agronomic farm plans for growers.
- Provided crop-scouting services to identify weed and insect problems.

Field Technician, August 2003–December 2003
Entomology Department, Southern Illinois University

- Collected and identified insects.
- Developed insect management plans.
- Tested organic and nonorganic pesticides for effectiveness and residuals.

Skills

Computer Experience: Access, Excel, Outlook, PowerPoint, and Word. Global Positioning Systems (GPS). Database Management.
Machinery: Field Tractors, Combines, Straight Trucks, and Bobcats
Communication Skills: Proposal Writing and Review, Public Presentations, Negotiating, Training, Writing Agronomic and Financial Farm Plans.

Awards and Memberships

Awarded "Best Young Innovator" by the Organic Food Society of America
Member of Entomological Society of America

References Available Upon Request

Figure 15.3: Anne Franklin's resume uses a list style. Hers achieves a classic look by using a serif font (Garamond) and left justification.

Archival Resume

Name and contact information are prominently displayed.

A career summary is used to summarize experiences.

Work experience is featured because it is placed high in the resume.

Other work experience is listed concisely.

A paragraph lists classes taken.

Resume ends by mentioning where a dossier can be obtained.

James L. Mondragon

576 First Avenue, Rolla, Missouri 65408
Phone: 573-555-4391, e-mail: bigmondy12@umr.edu

Career Summary

My interests in chemical engineering started with a childhood fascination with plastic products. I enrolled at the University of Missouri–Rolla because of their strong chemical engineering program, especially in the area of applied rheology and polymeric materials. I have also completed a co-op with Vertigo Plastics in St. Louis. My background includes strong computer modeling skills, especially involving polymeric materials.

Work Experience

Vertigo Plastics, Inc., St. Louis, Missouri, 5/02–1/03, 5/03–1/04
Co-op Chemical Engineer

Performed inspections of chemical equipment and plant equipment affected by chemical systems (such as boilers and condensers). Monitored the performance of chemical systems at various sites throughout the Vertigo Plastics system. Performed calculations and wrote reports summarizing the performance of those systems. Helped troubleshoot problems.

Other Work Experience

To Go Pizza, *Server,* Springfield, Missouri, 2/97–7/99

Educational Background

University of Missouri–Rolla, BSE, Expected May 2005
Major: Chemical Engineering

Advanced Coursework included Chemical Engineering Fluid Flow, Chemical Engineering Heat Transfer, Chemical Engineering Thermodynamics I & II, Process Dynamics and Control, Chemical Engineering Reactor Design, Chemical Engineering Economics, Chemical Process Safety, Chemical Process Design, Chemical Process Materials.

Activities and Awards

Awarded Stevenson Scholarship in Engineering, 2004
Treasurer, UMR Student Chapter, American Institute of Chemical Engineers (AIChE), 2003–Present
Member, Tau Beta Pi, 2003

Dossier with References Available at UMR Placement Services (573-555-2941)

Figure 15.4: James Mondragon's resume uses a paragraph style. It uses a sans serif font for headings (Myriad) and serif font (Times) for the body text. The headings are centered.

TAKE NOTE A career objective or career summary is not required in a resume. In fact, some resume experts suggest that these statements unnecessarily take up space on a resume, because they rarely do more than describe the position being applied for. If you are looking for room to put more information on your resume, you might leave out the career objective or career summary.

EDUCATIONAL BACKGROUND Your educational background should list your most recent college degree and other degrees in reverse chronological order—most recent to least recent (Figure 15.3). Name each college and list your major, minor, and any distinctions you earned (e.g., scholarships or summa cum laude or distinguished-scholar honors). You might also choose to mention any coursework you have completed that is related to your career (Figure 15.4).

TAKE NOTE If you are looking for your first job out of college, you may want to include your grade point average (GPA) if it is a B average or better. If your GPA in your major is higher than your overall GPA, you can list it instead. After their first job, few people list their GPA on their resume.

Any specialized training, such as welding certification, experience with machinery, or military training, might also be listed here with dates of completion.

If you are new to the technical workplace, you should put your educational background high on your resume. Your degree is probably your most prominent achievement, so you want to highlight it. If you have years of professional work experience, you may choose to put your educational background lower on your resume, allowing you to highlight your years of experience.

RELATED WORK EXPERIENCE Any career-related jobs, internships, or co-ops that you have held should be listed, starting with the most recent and working backward chronologically. For each job, include the title of the position, the company, and dates of employment (month and year).

Below each position, list your workplace responsibilities. You can describe these responsibilities in a bulleted list or in a brief paragraph (see Figures 15.3 and 15.4). Use action verbs and brief phrases to add a sense of energy to your work experience. Also, where possible, add any numbers or details that reflect the importance of your responsibilities.

> Coordinated a team of 15 student archaeologists in the field.
>
> Participated in the development of UNIX software for a Cray supercomputer.
>
> Worked with over 100 clients each year on defining, updating, and restoring their water rights in the Wilkins Valley.

Verb-first phrases are preferable over full sentences, because they are more scannable and require less space. Some action verbs you might consider for your resume include the following:

adapted	devised	organized
analyzed	directed	oversaw
assisted	equipped	planned

Want more advice on writing resumes? Go to **www.ablongman.com/johnsonweb/15.7**

collaborated	examined	performed
collected	exhibited	presented
compiled	implemented	proposed
completed	increased	recorded
conducted	improved	researched
constructed	instructed	studied
coordinated	introduced	supervised
corresponded	investigated	taught
designed	managed	trained
developed	observed	wrote

Keep in mind that it might be tempting to exaggerate your responsibilities at a job. For example, a cashier at a fast food restaurant might say he "conducted financial transactions." In reality, nobody is fooled by these kinds of puffed-up statements. They simply draw attention to a lack of experience and, frankly, a mild lack of honesty. There is nothing wrong with simply and honestly describing your experiences.

OTHER WORK EXPERIENCE Almost everyone has worked at jobs that were not related to his or her desired career. If you have worked at a pizza place, waited tables, or painted houses in the summer, those jobs can be listed in your resume. But, they should not be placed more prominently than your related work experience, nor should they receive a large amount of space on your resume. Instead, simply list these jobs in reverse chronological order, with names, places, and dates.

Pizza Chef. Giovanni's. Lincoln, Nebraska. September 2000–August 2001

Painter. Campus Painters. Omaha, Nebraska. May 2000–September 2000

Server. Crane River Brewpub and Cafe. Omaha, Nebraska. March 1998–May 2000

Do not offer any additional description. After all, most people are well aware of the responsibilities of a pizza chef, painter, and server. Any more description will only take up valuable space on your resume.

TAKE NOTE Don't underestimate the importance of non-career-related jobs on your resume. If you do not have much professional work experience, any kind of job shows an ability to work and hold a job. As your career progresses, these items should be removed from your resume.

SKILLS Resumes will often include a section that lists career-related skills. In this area, you may choose to mention your abilities with computers, including any software or programming languages you know how to use. If you have been trained on any specialized machines or know how to do bookkeeping, these skills might be

For sample resumes, go to **www.ablongman.com/johnsonweb/15.8**

worth mentioning. If you have proven leadership abilities or communication skills (technical writing or public speaking) you might also list them.

> *Computer Skills:* word processing (Word, WordPerfect), desktop publishing (Pagemaker, Quark), web design (Dreamweaver, Frontpage), and data processing (Excel, Access).
>
> *Leadership Abilities:* President of Wilkins Honor Society, 2003–2004. Treasurer of Lambda Kappa Kappa Sorority, 2001–2004. Volunteer Coordinator at the Storehouse Food Shelter, 2002 to present.

The skills section is a good place to list any training you have completed that does not fit under the "Educational Background" part of your resume.

AWARDS AND ACTIVITIES List any awards you have won and any organized activities in which you have participated. For example, if you won a scholarship for your academic performance, list it here. If you are an active member of a club or fraternity, show those activities in this part of your resume. Meanwhile, volunteer work is certainly worth mentioning, because it shows your commitment to the community and your willingness to take the initiative.

REFERENCES Your references are the three to five people who potential employers can call to gather more information about you such as current or former supervisors, professors, colleagues, and professionals who know you and your work. Your references should be people you trust to offer a positive account of your abilities. Each reference should include a name, title, address, phone number, and e-mail address.

AT A GLANCE

Sections in an Archival Resume

- Name and Contact Information
- Career Objective/Career Summary
- Educational Background
- Related Work Experience
- Other Work Experience
- Skills
- Awards and Activities
- References

References can take up a large amount of space on a resume, so they are typically not listed on the resume itself. Instead, a line at the bottom of the resume states "References available upon request." Then, the references—listed on a separate sheet of paper under the heading "References"—can be sent to any employer who requests them.

If the employer asks for references to appear on the resume, you can add them at the bottom of your resume, usually on a second page (Figure 15.5). To avoid causing your resume to go over two pages, you may need to list them in two or three columns on the second page.

Do not list someone as a reference until you have asked whether he or she will serve as a positive reference. Your professors, former supervisors, and others will usually be willing to serve as references. But if you did not ask them first, they will be a little shocked when an employer calls to ask some questions.

In some cases, someone may not want to be a reference for you. If so, it is better that he or she tell you up front. That way, this person will not give a negative review to a potential employer.

Helpful services for managing your letters can be found at **www.ablongman.com/johnsonweb/15.9**

Listing References

References

George Roberts, Ph. D.
Professor of Agronomy
Southern Illinois University
Department of Plant, Soil, and Agriculture
Carbondale IL 62901
groberts394@siu.edu
618-555-2314

Jane Falters, Ph. D.
Assistant Professor of Agronomy
Southern Illinois University
Department of Plant, Soil, and Agriculture
Carbondale IL 62901
faltersj7@siu.edu
618-555-2312

Shelly Winters
Insect Management Supervisor
Brighter Days Organic Cooperative
229 115th Street
Simmerton IL 63912
swinters@bdorganic.net
618-555-9638

Timothy Hanson
Owner, CEO
Brighter Days Organic Cooperative
229 115th Street
Simmerton IL 63912
thanson@bdorganic.net
618-555-9630

Figure 15.5: Preferably, references should be listed on a separate sheet of paper. They can also appear at the end of a resume.

Your resume needs to be a fair representation of your qualifications, experience, and skills. It might be tempting to exaggerate, stretch the truth, or even lie on a resume. These deceptions usually come back to haunt applicants, because human resource officers are experts at detecting misrepresentations.

Your best strategy is to be completely honest on your resume. That way, you won't need to keep track of the places where you were not fully honest. You can just tell the truth.

Functional Resume

The functional resume is less common than the archival resume, especially for new college graduates. This type of resume is designed to highlight the job applicant's abilities and skills by placing them up front in the resume. The advantage of this type of resume is its ability to boil years of experience into a few strengths that the job applicant would like to highlight.

Figure 15.6 shows an example of a functional resume. In this example, note how the resume places the applicant's strengths, including awards, high in the document. Then, the remainder of the resume concisely lists details about the applicant's employment background, education background, and professional memberships.

The functional resume is best for someone who has held several jobs or has some notable life experiences. This type of resume allows the applicant to highlight the qualities an employer is seeking, rather than burying them among the many other points throughout the resume.

A Functional Resume

Figure 15.6: A functional resume puts the applicant's abilities and skills up front where an employer will see them. Other features, such as employment history and education, are minimized.

Name and contact information are placed up front.

The objective describes the position sought.

Qualities are summarized here, including awards.

Employment and education history is very concise, listing only the details.

Walter David Trimbal
818 Franklin Drive
Atlanta, Georgia 30361
404-555-2915

Objective: Senior architect position in a firm that specializes in urban revitalization projects.

Leadership Experience

- Managed design team for additions/alterations for small-scale commercial buildings in downtown Atlanta.
- Led planning charette for Bell Hill Neighborhood renovation.
- Awarded a 2002 "Archi" for design of Delarma Commerce Center.

Technical Expertise

- Experienced with the latest developments in computer-aided drafting hardware and software.
- Able to resolve conflicting building and zoning codes, while preparing site plans.

Community Involvement

- Founding Member of the Better Atlanta Commission in 1995 and served as board member until 2003.
- Served on the Architecture Public Involvement Committee for Atlanta Metropolitan Council of Governments from 1998–2002.

Employment Background

Vance & Lipton—Architects, Senior Architect, Atlanta, GA, 1999 to present.
Fulton County Planning Department, Planning Architect, Atlanta, GA, 1995–1999.
Ronald Alterman—Architect, Intern, Boston, MA, 1994–1995.

Education Background

B.A. in Architecture, Boston College, Boston, MA, 1995.
A.A. in Computer-Aided Drafting, Augusta Technical College, Augusta, GA, 1992.

Professional Memberships

National Council of Architectural Registration Boards
Greater Atlanta Architects Guild

References Available Upon Request

AT WORK

Kim Isaacs

EXECUTIVE DIRECTOR, ADVANCED CAREER SYSTEMS, INC.

Advanced Careers systems is a resume writing and career development company.

How has the Internet changed the job search process?

The Internet has changed the job search dramatically, and while the web makes the search easier in many ways, it also requires the seekers to be savvier than ever. Because job seekers can search the Internet for positions, they can now apply for jobs that they might not have even heard about in the past. They can also apply for international jobs as easily as national ones. Seekers can go to major on-line career sites and set up "job agents" that enable them to enter information about their goals and receive notifications when matching jobs become available.

Because the candidate pool is so large, many employers report that they're constantly inundated with new resumes, and employers have the luxury of seeking a "perfect match." That's why it's so important, for example, to make sure that electronic resumes use keywords that will be found on a resume-tracking system. Even if your resume is excellent, it can be overlooked without the right wording and phrasing.

Despite the increase in using the Internet to search for positions, on-line recruiting still accounts for a small fraction of actual hires (the figure keeps changing, but it's around 5%). Of course job seekers should use the Internet to research companies, scour for opportunities, and take advantage of the career tools available, but they still need to focus most of their effort on old-fashioned job search methods such as networking.

Designing the Resume

The design of your resume should reflect your personality and the industry in which you want to work. A resume for an engineering firm, for example, will probably be somewhat plain and straightforward. A resume for a graphic artist position at a magazine, on the other hand, might demonstrate some of your skills as a designer. There are, of course, exceptions. Some progressive engineering firms, for example, might prefer a layout that reflects your innovative qualities.

Most word-processing programs include resume templates that you can use to lay out your information. If you choose to use one of these templates, alter the design of the template in some way. Unfortunately, many thousands of people have access to the same templates, so employers often receive several resumes that look identical. You want yours to stand out.

If you decide to design your own resume, Chapter 8 in this book offers design principles that are helpful toward creating a design for your resume. These principles are balance, alignment, grouping, consistency, and contrast. All of these principles should be used to design your resume.

Balance—Pay attention to the vertical and horizontal balance of the page. Your resume should not be weighted too heavily toward the left or right, top or bottom.

Alignment—Different levels of information should be consistently indented to make the resume easy to scan. Don't just align everything at the left margin; instead, use vertical alignment to create two or three levels in the text.

TAKE NOTE Center justification and right justification can create alignment problems in a resume, making it look odd. It is recommended that you center only headings and never right-justify anything on your resume.

Grouping—Use white space to frame groups of information. For example, a job listed on your resume with its responsibilities should be identifiable as a chunk of text. Sometimes using rules, especially horizontal lines, is a good way to carve a resume into quickly identifiable sections (Figure 15.3).

Consistency—The design of your resume should be internally consistent. Use bolding, italics, and font sizes consistently. Bullets or other symbols should also be used consistently throughout the resume.

Contrast—Titles and headings should be noticeably different from the body text. To contrast with the body text, you might choose a different serif or sans serif font for your titles and headings. You can increase the font sizes and/or use bolding to make the resume more scannable.

LINK For more information on page design, turn to Chapter 8, pages 192–233.

A helpful strategy for designing your resume is to collect resumes from other people. There are also numerous books and websites available that offer ideas about designing resumes. You can use these sample resumes as models for designing your own.

Writing Effective Application Letters

Your resume will provide the employer with facts and details about your education, work experience, and skills. It cannot, however, convey a full sense of your individuality or your interest in the company that has a job available. Your application letter, which will accompany your resume, should offer a fuller picture of you as a person. It should also discuss how you can make a contribution to the company.

An effective application letter strives to prove that you are uniquely qualified for the available position. Job applicants regularly make two mistakes in application letters. First, they simply restate the information available on their resume, failing to demonstrate why they are the right person for the job. Second, they discuss why the position would be good *for them* (e.g.,"A job at Gurson Industries would help me reach my goal to become an electrical engineer working with sensors.").

Instead, your letter should prove to potential employers that your education, experience, and skills will allow you to make a valuable contribution to *their* company. Put the emphasis on *their* needs, not yours.

You can download resume templates at **www.ablongman.com/johnsonweb/15.10**

Designing a Scannable/Searchable Resume

Companies are increasingly asking for *scannable* resumes (Figure A). These kinds of resumes are scannable by computers, which sort and rank the best candidates. Also, resumes posted on job search engines need to be searchable through keywords.

How are scannable/searchable resumes used by employers? Usually, after all the resumes are scanned by the computer, a human resources officer or recruiter will enter 10 keywords that describe the position. Then, the computer returns a ranked list of the applicants who matched the most keywords. To survive the cut, you need to find a way to anticipate the keywords that will be entered. Here's a hint: The job advertisement probably contains many of the keywords the employer will be looking for.

Here are some ideas for developing a scannable/searchable resume:

- Use well-known keywords to describe your skills and experience.
- Use terms in predictable ways. For example, you should write "Managed a team of technicians" rather than "Responsible for guiding a contingent of technical specialists."
- Include acronyms specific to your field (e.g., CAD, TQM, APA, IEEE).
- Use common headings found in resumes: Career Objective, Work Experience, Skills, Qualifications, Education, Honors, Publications, Certifications.
- At the end of your resume, make a list of any additional traits or skills: time management skills, dependability, efficiency, leadership, responsibility. These may be used as keywords.

If you are making a paper-based scannable resume:

- Use white 8½-by-11-inch paper, printed on one side only.
- Do not fold or staple the paper.
- Place your name on its own line at the top of the page.
- Use a standard address format below your name.
- List each phone number on its own line.
- Use standard typefaces like Arial, Helvetica, Times, New York, or Garamond. The computer may have trouble reading other fonts.
- Don't use a font size below 10 point for any of the text.
- Don't use italics, underlining, shadows, or reverse type (white type on a black background) because scanners have trouble reading them.
- Don't use vertical and horizontal lines, graphics, boxes, and shading.
- Don't use two- or three-column formats. One column is easier for the computer to scan.

Fortunately, scanning machines and computers do not care about the length or design of your resume. Your scannable resume should be plain in design, and it can be longer than your regular resume, if needed.

If you suspect your resume will be scanned, your best strategy is to make two resumes. One should be your regular resume and the other should be a scannable resume. Clearly identify the scannable resume for the employer by placing a cover note on it. That way, the employer won't be confused by the submission of two resumes.

Need more help making a scannable resume? Go to **www.ablongman.com/johnsonweb/15.11**

A Scannable Resume

Figure A: A scannable resume removes much of the formatting and design, allowing the computer to more easily locate keywords. Compare this resume with the regular resume in Figure 15.3.

Name and contact information is plainly presented.

Anne Franklin
834 County Line Rd.
Hollings Point, IL 62905
Home: 618-555-2993
Mobile: 618-555-9167
e-mail: afranklin@unsb5.net

Career Objective: A position as a naturalist, specializing in agronomy, working for a distribution company that specializes in organic foods.

Headings are predictable and easy to locate.

Educational Background
Bachelor of Science, Southern Illinois University, expected May 2005.
Major: Plant and Soil Science
Minor: Entomology
GPA: 3.2/4.0

Work Experience

Italics have been removed for easier scanning.

Intern Agronomist, December 2003–August 2004
Brighter Days Organic Cooperative, Simmerton, IL
Consulted with growers on organic pest control methods. Primary duty was sale of organic crop protection products, crop nutrients, seed, and consulting services.
Prepared organic agronomic farm plans for growers.
Provided crop-scouting services to identify weed and insect problems.

Bullets have been removed to simplify text.

Field Technician, August 2003–December 2003
Entomology Department, Southern Illinois University
Collected and identified insects.
Developed insect management plans.
Tested organic and nonorganic pesticides for effectiveness and residuals.

Skills
Computer Experience: Access, Excel, Outlook, PowerPoint, and Word. Global Positioning Systems (GPS). Database Management.
Machinery: Field Tractors, Combines, Straight Trucks, and Bobcats
Communication Skills: Proposal Writing and Review, Public Presentations, Negotiating, Training, Writing Agronomic and Financial Farm Plans.

Awards and Memberships
Awarded "Best Young Innovator" by the Organic Food Society of America (OFSA)
Member of Entomological Society of America (ESA)

Content and Organization

Like any letter, an application letter will have an introduction, body, and conclusion (Figure 15.7). It will also include common features of a letter such as the header (your address and the employer's address), a greeting (Dear), and a salutation (Sincerely) with your signature.

LINK For more information on writing letters, go to Chapter 16, page 446.

The Basic Pattern of an Application Letter

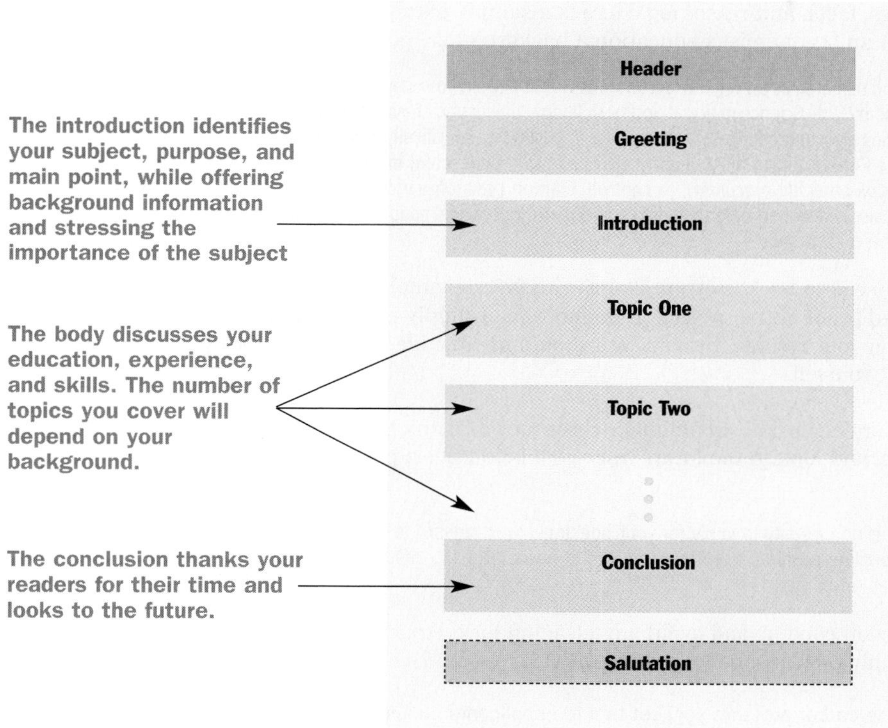
Header
Greeting
Introduction
Topic One
Topic Two
Conclusion
Salutation
The introduction identifies your subject, purpose, and main point, while offering background information and stressing the importance of the subject
The body discusses your education, experience, and skills. The number of topics you cover will depend on your background.
The conclusion thanks your readers for their time and looks to the future.

Figure 15.7: An application letter includes the common features of a letter, like a head, introduction, body, and salutation.

INTRODUCTION You should begin your letter by making up to five moves commonly found in introductions of any document: identify your subject, state your purpose, state your main point, stress the importance, and offer background information.

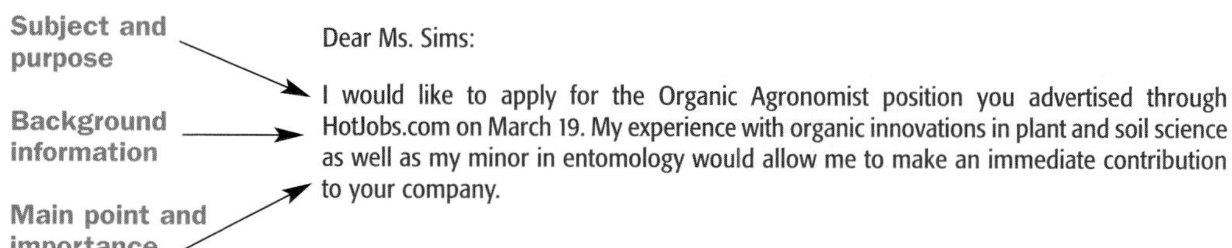
Subject and purpose
Background information
Main point and importance

Dear Ms. Sims:

I would like to apply for the Organic Agronomist position you advertised through HotJobs.com on March 19. My experience with organic innovations in plant and soil science as well as my minor in entomology would allow me to make an immediate contribution to your company.

This introduction makes all five introductory moves, but all five moves are not needed for a successful introduction. Minimally, an introduction should identify the subject (the position being applied for), state your purpose (I would like to apply for your job), and state your main point (I am uniquely qualified for your position).

BODY In the body of the letter, you should include two to three paragraphs that show how your educational background, work experience, and skills fit the employer's needs. Remember that you are making an argument, so each paragraph should start out with a claim, and the rest of the paragraph should support that claim with examples, facts, and reasoning. Here is a sample paragraph that supports a claim about an agronomist's educational background.

A clear claim. →
Support for that claim. →

> My education and research as an organic agronomist would benefit your company significantly. As a Plant and Soil Science major at Southern Illinois University, I have been studying and researching environmentally safe alternatives to pesticides. Specifically, my mentor, Professor George Roberts, and I have been working on using benevolent insects, like ladybird beetles (*Coleomegilla maculata*), to control common pests on various vegetable plants. We have also developed several varieties of organic insecticidal soaps that handle the occasional insect infestation.

In the body, you need to back up your claims with facts, examples, details, and reasoning—you need proof. You also need to do more than simply restate information that can be found on your resume. Instead, you should breathe life into your letter by telling stories about yourself.

CONCLUSION Your conclusion should make three moves: thank the reader, offer contact information, and look to the future. Your goal is to leave a positive impression at this point.

Thank-you statement →
Look to the future →
Contact information →

> Thank you for this opportunity to apply for your opening. I look forward to hearing from you about this exciting position. I can be contacted at home (618-555-2993) or through e-mail (afranklin@unsb5.net).

Keep the conclusion concise, and avoid any pleading for the position. Employers will not look favorably on someone who is begging for the job.

TAKE NOTE Some career advisors suggest that job applicants should state that they will be "calling to schedule an interview" in their conclusion. Some employers may like this assertive approach; others won't. You need to decide if you want to be that assertive in your letter.

You should organize your letter to highlight your strengths. If your educational background is your best asset, put that paragraph right after the letter's introduction (Figure 15.8). If your work experience is stronger than your education, put that information up front (Figure 15.9). Your letter should fit on one page for an entry-level position. Two pages would be a maximum for almost any job.

To see sample application letters, go to **www.ablongman.com./johnsonweb/15.12**

Letter of Application Emphasizing Education

Figure 15.8: A letter of application should make an argument, not simply restate items that can be found on the resume.

April 2, 2004

834 County Line Rd.
Hollings Point, Illinois 62905

Valerie Sims, Human Resources Manager
Sunny View Organic Products
1523 Cesar Chavez Lane
Sunny View, California 95982

Dear Ms. Sims:

Opening paragraph states the subject, purpose, and main point. →

I would like to apply for the Organic Agronomist position you advertised through HotJobs.com on March 19. My experience with organic innovations in plant and soil science as well as my minor in entomology would allow me to make an immediate contribution to your company.

Education is discussed up front with examples. →

My education and research as an organic agronomist would benefit your company significantly. As a Plant and Soil Science major at Southern Illinois University, I have been studying and researching environmentally safe alternatives to pesticides. Specifically, my mentor, Professor George Roberts, and I have been working on using benevolent insects, like ladybird beetles (*Coleomegilla maculata*), to control common pests on various vegetable plants. We have also developed several varieties of organic insecticidal soaps that handle the occasional insect infestation.

Work experience is used to show potential contributions to employer. →

I also worked as an intern for Brighter Days Organic Cooperative, a group of organic farmers, who have an operation similar to Sunny View. From your website, I see that you are currently working toward certification as an organic farm. At Brighter Days, I wrote eleven agronomic plans for farmers who wanted to change to organic methods. My work experience in the organic certification process would be helpful toward earning certification for Sunny View in the shortest amount of time.

Other skills are highlighted to show unique abilities. →

Finally, I would bring two other important skills to your company: a background in farming and experience with public speaking. I grew up on a farm near Hollings Point, Illinois. When my father died, my mother and I kept the farm going by learning how to operate machinery, plant the crops, and harvest. We decided to go organic in 1997, because we always suspected that my father's death was due to chemical exposure. Based on our experiences with going organic, I have given numerous public speeches and workshops to the Farm Bureau and Future Farmers of America on organic farming. My farming background and speaking skills would be an asset to your operation.

Conclusion ends on a positive note and offers contact information. →

Thank you for this opportunity to apply for your opening. I look forward to hearing from you about this exciting position. I can be contacted at home (618-555-2993) or through e-mail (afranklin@unsb5.net).

Sincerely,

Anne Franklin

Anne Franklin

Letter of Application Emphasizing Work Experience

Figure 15.9: The organization of the application letter should highlight strengths. Here, work experience is being highlighted, because it appears immediately after the introduction.

March 12, 2004

576 First Avenue
Rolla, Missouri 65408

Mr. Harold Brown, Human Resources Director
Farnot Plastic Solutions
4819 Renaissance Lane
Rochester, New York 14608

Dear Mr. Brown:

Opening paragraph uses background information to make personal connection. →

Last week, you and I met at the University of Missouri–Rolla engineering job fair. You mentioned that Farnot Plastics might be interviewing entry-level chemical engineers this spring to work on applications of polymeric materials. If a position becomes available, I would like to apply for it. With my experience in applied rheology and polymeric materials, I would be a valuable addition to your company.

Work experience with co-op is highlighted. →

My work experience includes two summers as a co-op at Vertigo Plastics, a company similar in size, products, and services to Farnot. My responsibilities included inspecting and troubleshooting the plant's machinery. I analyzed the production process and reported on the performance of the plant's operations. While at Vertigo, I learned to work with other chemical engineers in a team-focused environment.

Paragraph on education makes connections to employer's needs. →

My education in chemical engineering at University of Missouri–Rolla would allow me to contribute a thorough understanding of plastics engineering to Farnot. In one of the best programs in the country, I have excelled at courses in thermodynamics, chemical process design, and chemical process materials. In addition, my work in the university's state-of-the-art chemical laboratories has prepared me to do the prototype building and vacuum forming that is a specialty of your company. I also have experience working with the CAD/CAM systems that your company uses.

Conclusion indirectly requests the interview. →

The enclosed resume highlights my other qualities. I enjoyed speaking with you at the job fair, and I would appreciate an opportunity to talk to you again about opportunities at Farnot. If you would like more information or you would like to schedule an interview, please call me at 573-555-4391 or e-mail me at bigmondy12@umr.edu.

Sincerely,

JL Mondragon

James L. Mondragon

Enclosure: Resume

Style

Another way your application letter differs from your resume is in its style. You want to adopt a style that conveys a sense of your own personality and your interest in the position available.

"YOU" ATTITUDE Put the emphasis of the letter on your readers by using the *"you" attitude.* By strategically using the words "you" and "your" in the letter, you can discuss your qualifications from your readers' point of view.

> Your company would benefit from my mechanical engineering training and hands-on experience as a certified welder.
>
> Your group is one of the fastest growing medical practices in Denver. As a member of your medical staff, I would be reliable and hardworking.

Another way to help put the stress on the readers rather than yourself is to change "I" sentences to "my" sentences. The first sentence above, for example, uses the phrase "My mechanical engineering training" rather than, "I have been trained in mechanical engineering." This subtle change lessens the overuse of "I" in the letter, allowing you to put more emphasis on your readers.

LINK For more information on using the "you" sytle in letters, go to Chapter 16, page 469.

ACTIVE VOICE In stressful situations, like writing application letters, you might be tempted to switch to passive voice. Passive voice in application letters will make you sound detached and even apathetic.

> *Passive:* Attending the University of Arizona was a good way to pursue my interests in anthropology.
>
> *Active:* I attended the University of Arizona to pursue my interests in anthropology.
>
> *Passive:* The proposal describing the need for a new bridge over the Raccoon River was completed by my team, and it was presented to the Franklin City Council.
>
> *Active:* My team completed the proposal for a new bridge over the Raccoon River, and we presented it to the Franklin City Council.

Why active voice? The active voice shows that you took action. You were in charge, not just a passive observer of the events around you.

LINK For more help on using active voice, see Chapter 7, page 179.

NONBUREAUCRATIC TONE Avoid using business cliches to adopt a highly formal tone.

> *Bureaucratic:* Pursuant to your advertisement in the *Chicago Sun Times,* I am tendering my application toward your available position in marketing.
>
> *Nonbureaucratic:* I am applying for the marketing position you advertised in the *Chicago Sun Times.*

For more advice about letter-writing style, go to **www.ablongman.com/johnsonweb/15.13**

Business cliches such as "per your advertisement" or "in accordance with your needs" only make you sound stuffy and pretentious. Employers are not interested in hiring people who have these qualities—unless they are looking for a butler.

THEMES Think of one quality that sets you apart from others. Do you work well in teams? Are you a self-starter? Are you able to handle pressure? Are you quality minded? These are some themes that you might mention throughout your letter.

> While working on the Hampton County otter restoration project, I was able to communicate clearly and effectively, helping us win over a skeptical public that was concerned about the effects of introducing these animals to local streams.

Beyond simply telling readers that you have a special quality, use examples to show them how you have used this quality to solve problems in the past. Some qualities you might mention include the ability to

- make public presentations.
- be a leader.
- manage time effectively.
- motivate yourself and others.
- follow instructions.
- meet deadlines.
- work well in a team.
- write clearly and persuasively.
- handle stressful situations.

Weave one of these qualities into your letter or address it directly. One theme should be enough, because more than one theme will make you sound like you are boasting.

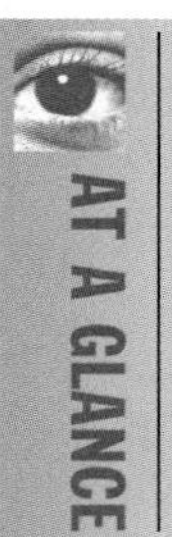

Style in an Application Letter

- "You" attitude—Put the emphasis on the employer.
- Active voice—Put yourself in charge, not events.
- Nonbureaucratic tone—No one wants to hire bureaucrats, so don't sound like one.
- Themes—What makes you different, makes you attractive.

Revising and Proofreading the Resume and Letter

When you are finished writing your resume and application letter, you should spend a significant amount of time revising and proofreading your materials. These documents need to be revised just like any other documents.

> **Content**—Did you provide enough information? Did you provide too much? Do the resume and letter answer the interviewers' basic questions about

your qualities and abilities? Do your materials match up with the job advertisement?

Organization—Are your materials organized in ways that highlight your strengths? Can readers easily locate information about your education, work experience, and skills?

Style—Do your materials set a consistent theme? Is the tone projected in the letter of application appropriate?

Design—Is the resume highly scannable? Does it follow the design principles of balance, alignment, grouping, consistency, and contrast?

When you are finished revising, proofread everything carefully. Have as many people as possible look over your materials, especially your resume. For most jobs, even the smallest typos can become good excuses to pitch an application into the recycle bin. Your materials need to be nearly flawless in grammar and spelling.

Remember that the quality of your application materials says something about your own attention to quality. If your materials are well crafted and error free, employers are going to feel much more confident about your ability to do high-quality work for them.

Creating a Professional Portfolio

Increasingly, it is becoming common for employers to ask applicants to bring a *professional portfolio* to their interview. Sometimes, interviewers will ask that a portfolio be sent ahead of the interview, so they can familiarize themselves with the applicant and his or her abilities.

A portfolio is a collection of materials that you can use to demonstrate your qualifications and abilities. Portfolios include some or all of the following items:

- Resume
- Samples of Written Work
- Examples of Presentations
- Descriptions and Evidence of Projects
- Diplomas and Certificates
- Awards
- Letters of Reference

These materials can be placed in a three-ring binder with dividers (Figure 15.10), and/or they can be put on-line on a website or CD-ROM.

Your personal portfolio should not be left with the interviewer, because it may be lost or not returned. Instead, you should create a "giveaway" portfolio with copies of your work, which you can send to or leave with the employer. If the interviewer asks to keep the portfolio, you can hand over your giveaway portfolio, explaining that you have original documents in your main portfolio.

Even if interviewers do not ask you to bring a portfolio, putting one together is still well worth your time. After all, if you walk into an interview with just your resume and a smile, you will have little evidence to prove that you are the right

A Print Portfolio

Figure 15.10: A portfolio is a useful tool for interviewing. It holds all your work in an accessible package, so you can support your claims during your interview.

person for the job. If you bring a portfolio, you will have a collection of materials to show your interviewers, allowing you to make the strongest case possible that you have the qualifications and experience they need.

Collecting Materials

If you have not made a portfolio before, you are probably wondering if you have enough material to put one together. Here are some steps you can take toward finding materials for your portfolio:

Search your hard drive—Your personal computer probably holds documents and presentations that you can print out. The materials may be old, but a little revision and polishing will make these materials suitable for your portfolio.

Look in your current files—You will probably be surprised by the amount of materials you have stuffed away in file cabinets, boxes, and closets. Start pulling out old projects, presentations, reports, certificates, diplomas, and awards that you have saved over the years.

Start saving materials—From now on, any projects you do in your classes or at your workplace should be saved for your portfolio. Save copies of these items on your hard drive in a special folder called "Portfolio." For print documents, you should locate a desk drawer or file cabinet where you can store items for your portfolio.

Find opportunities to create materials—If you do not have much material for your portfolio, then you will need to create some. Internships and co-ops are especially good ways to gain experience and fill your portfolio. If these oppor-

To see samples of on-line portfolios, go to **www.ablongman.com/johnsonweb/15.14**

tunities are not available, look for ways to volunteer with worthy organizations or causes. Volunteering is a good way to gain experience, while adding materials to your portfolio.

Ask for letters of reference—When you have good experiences with professors or other people, ask them for letters of reference for your portfolio. Even if you are not actively looking for a job, ask for these letters when the relationship with a reference is still strong. Good letters are hard to obtain after your class or other experience fades into the past.

Something to keep in mind is that much of your written work in college is suitable for your portfolio. These materials do not need to be directly applicable to the job you are applying for. Instead, potential employers are interested in seeing evidence of your success and your everyday abilities. Your documents written for class will be that evidence of success.

Even nontechnical materials—like that critical analysis of Beethoven's Ninth Symphony that you wrote for a course in classical music—still show your ability to do research, adopt a critical perspective, and write at length. These materials are appropriate if you are looking for things to fill a portfolio.

Of course, not everything you do is suitable for your portfolio. As you select items for the portfolio, you should ask yourself whether each item demonstrates a quality you might want to discuss in an interview.

Once you have collected together all the materials you can find, you will likely have too many documents for a three-ring binder. So, depending on your interview, you will need to make some strategic choices about what will go with you to the interview and what will stay home. The rest of the items can be saved in a file or scanned for placement in your electronic portfolio.

Organizing Your Portfolio

Now that you have your materials and portfolio ready, you can organize the materials according to various schemes. For example, you might organize materials year by year or job by job. A particularly good way to organize a portfolio is to follow the categories in a typical resume:

Cover sheet—A sheet that includes your name, address, phone, and e-mail address.

Educational background—Diplomas you have received, workshops you have attended, a list of relevant courses, and college transcripts.

Related work experience—Printed materials from your previous jobs, internships, co-ops, and volunteer work. You might include performance reviews, news articles that mention you, or brochures about specific projects. You might even include photographs of places you have worked, projects you have worked on, and people you have worked with.

Specialized skills—Certificates of completion for training, including any coursework you have completed outside of your normal college curriculum.

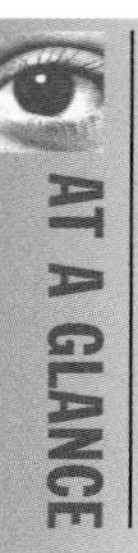

Organizing Your Portfolio

- Cover sheet
- Educational background
- Related work experience
- Specialized skills
- Awards
- Other interests
- References

Writing samples and publications—Examples of your written work, presentations, or websites.

Awards—Any awards certificates or letters of congratulation. You might include letters that mention any scholarships you have received.

Other interests—Materials that reflect your other activities, such as volunteer work, sports, or hobbies. Preferably, these materials would be relevant to the kinds of jobs you are seeking.

References—Letters of reference from teachers, colleagues, coworkers, or employers. Also, it is helpful to keep a list of references with phone numbers and addresses where they can be contacted.

Assembling the Portfolio in a Binder

You want to organize your portfolio so that it is easy to use at an interview. If your portfolio is nothing more than a hodgepodge of paper stuffed in a three-ring binder, you will find it difficult to locate important information.

So, go to an office supply store and purchase the following items:

Three-ring binder—The binder should be at least two inches wide to hold your materials, but it should not be too large to be comfortably carried into an interview. Find one that is suitable for a formal situation like an interview.

Dividers with tabs—Tabbed dividers are helpful for dividing your materials into larger categories. The tabs are helpful for finding materials quickly, especially in a stressful situation like an interview.

Pocketed holders or clear plastic sleeves—Put your materials in pockets or clear plastic sleeves. That way, you can easily insert and remove your materials when you want.

As you assemble the portfolio, keep professionalism and ease of use in mind:

- Copies of your resume can usually be placed in a pocket on the inside cover of the three-ring binder. You may need them if interviewers do not have a copy of your resume in front of them.
- Labels can be used to provide background information on each item in the portfolio. These labels are helpful in jogging your memory during a stressful interview.

Creating an Electronic Portfolio

If you know how to make a basic website, you can create an electronic portfolio for yourself (Figure 15.11). An electronic portfolio has several advantages. It can

Where can these materials be found? Go to **www.ablongman.com/johnsonweb/15.15**

Electronic Portfolio

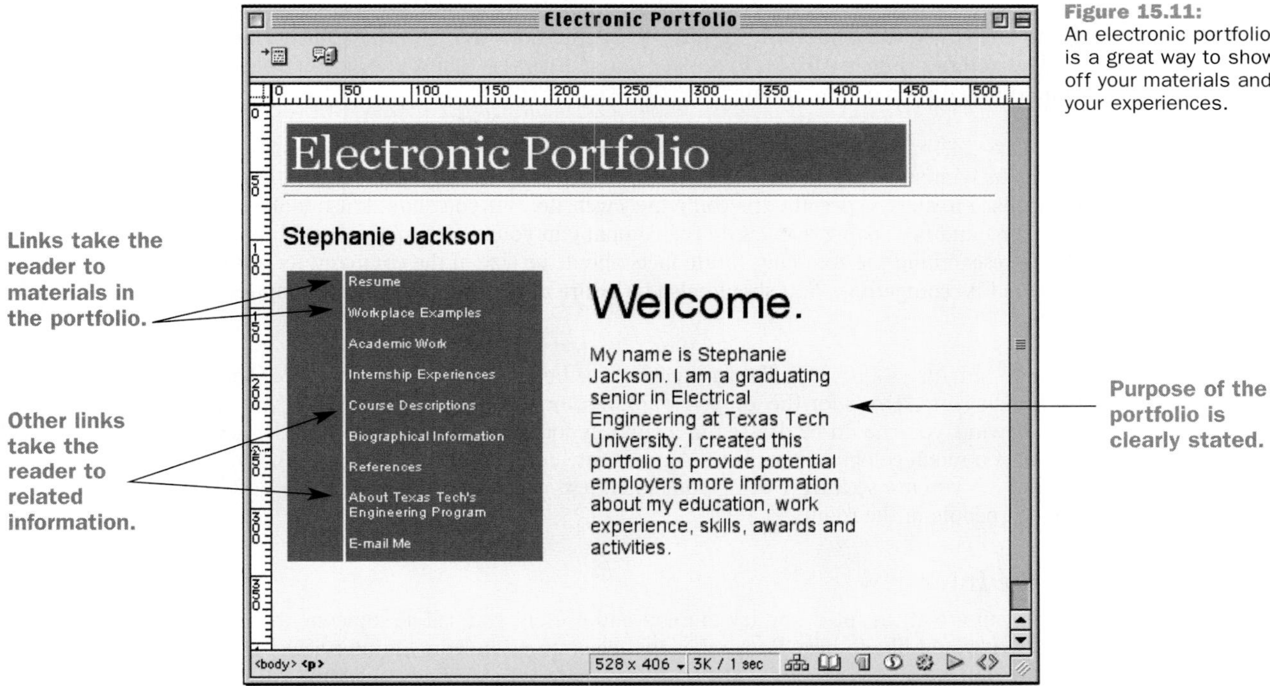

Figure 15.11: An electronic portfolio is a great way to show off your materials and your experiences.

- be accessed from anywhere there is a networked computer, including an interviewer's office.
- include multimedia texts, like movies, presentations, and links to websites you have created.
- include materials and links to information that would not typically be found in a nonelectronic portfolio. For example, you might put links to your university and academic department to help interviewers learn about your educational background.

You should keep your electronic portfolio separate from any personal websites you might create. The materials you include in your electronic portfolio should be appropriate for discussion in a professional interview. (Your vacation photos from that Aspen skiing trip with your friends probably aren't appropriate.)

Interviewing Strategies

When you are called for an interview, an employer is already telling you that you are qualified for the position. Now, you just need to compete with the other qualified candidates who are also interviewing for the position.

Preparing for the Interview

In many ways, interviewing is a game. It's not an interrogation. The interviewers are going to ask you some questions or put you in situations that will test your problem-solving abilities (Figure 15.12). To answer the interviewers' questions appropriately, you need to do some preparation:

RESEARCH THE COMPANY Before the interview, find out as much as possible about the company and the people who will be interviewing you. The Internet is a great place to start, especially the company's website. But you should also look for magazine and newspaper articles on the company in your university or local library. While researching the company, locate facts about the size of the company, its products, and its competitors. You should also be aware of major trends in the company's market.

DRESS APPROPRIATELY The interview game begins with your appearance. There is an old saying, "Dress for the job you want, not for the job you have." When you are interviewing, you should be dressed in a suitably formal manner avoiding flashy jewelry or too much cologne or perfume. Appropriate clothing usually depends on the kind of job you are seeking. During your interview, you want to be among the best-dressed people in the room.

At the Interview

When you are at the interview, try to relax and present yourself as someone the interviewer would like to hire. Remember that each question from an interviewer is a move in the game, and there are always appropriate countermoves available.

GREET PEOPLE WITH CONFIDENCE When you meet people at the interview, you should greet them with confidence. Most North Americans will expect you to shake their hand firmly and make eye contact. Let the interviewer indicate where you are going to sit. Then, set your briefcase and/or portfolio at your side on the floor. Don't put things on the interviewer's desk, unless he or she asks you to.

LINK For more information on making a professional presentation, see Chapter 11, page 290.

When time permits, write down the name and title of each interviewer you meet. You will need these names and titles when you are writing thank-you e-mails and letters after the interview.

ANSWER QUESTIONS Interviewers will usually work from a script of questions. However, most interviews go off the script as interesting topics come up. You should be prepared with answers to some of the following questions:

Tell me about yourself—Spend about two minutes talking about your work experience, education, and skills, relating them to the position.

What about this position attracted you?—Talk about the strengths of the company and what qualities of the position are interesting to you.

To visit sites that discuss interviewing strategies, go to **www.ablongman.com/johnsonweb/15.16**

The Interview

Figure 15.12: Interviewing is more like a game than an interrogation. Once you know how the game is played, you can win it.

Why should we hire you?—Stress qualifications and skills appropriate to the job that set you apart from the other candidates.

Where do you want to be in three to five years?—Without being too ambitious, talk about doing your job well and moving up in the company.

What is your greatest strength?—Discuss a strong qualification, skill, or knowledge area relevant to the job.

What is your greatest weakness?—Discuss something you would like to learn that would enhance your ability to do your job (e.g., a new language, more advanced computer skills, greater communication skills).

TAKE NOTE The "weakness" question is not the time to admit your shortcomings. If you are a procrastinator or you have a bad temper, now is not the time to confess your sins. Also, answers like "I just work too hard" or "I am too committed to doing an excellent job" don't really fool anyone. Instead, mention something you want to learn to improve yourself.

What are your salary requirements?—This question is uncommon, but you should have a salary figure in your head in case it is asked. You don't want to fumble this question or ask for too little.

LINK For more strategies on answering questions, see Chapter 11, page 311.

Want some strategies for answering tough questions? Go to **www.ablongman.com/johnsonweb/15.17**

USE YOUR PORTFOLIO You have no doubt been told numerous times "show, don't just tell." In your interview, use your professional portfolio to back up your answers to interview questions. The materials in your portfolio should be used to reinforce your claims about your background, qualifications, and skills.

ASK QUESTIONS As the interview comes to an end, the interviewer will usually ask if you have any questions. You should be ready to ask two or three insightful questions. Here are a few example questions that will demonstrate your interest in the company and the job:

Where do you see the company going in the next five years?

What can you tell me about your customers/clients?

What kinds of additional learning opportunities are available?

What happens in a normal day at this position?

TAKE NOTE Avoid asking questions about salary, vacation, and benefits at this point. Usually these items are discussed after an offer is made.

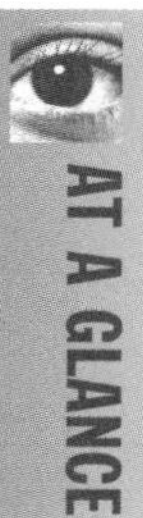

Interviewing Strategies

- Research the company.
- Dress appropriately.
- Greet people with confidence.
- Answer questions.
- Use your portfolio.
- Ask questions.
- Leave with confidence.

After the interviewer answers each question, you might offer a follow-up statement that reinforces one of your strengths.

LEAVE WITH CONFIDENCE When the interview is finished, thank the interviewers for their time and say you are looking forward to hearing from them. Also, ask if they would like you to send them any other information. Then, shake each interviewer's hand firmly and go.

As soon as possible after the interview, find a place where you can write down everything you can remember about what you and the interviewers talked about. These notes may be helpful later, especially if a week or two lapses before you hear about the job. Your notes should mention any important discussion points that developed during the interview.

Writing Thank-You Letters and/or E-Mails

After an interview, it is polite to write a thank-you letter to the people who interviewed you. A basic thank-you letter shows your appreciation for the interviewers' time while expressing continued interest in the job. A more sophisticated letter could reinforce one or more of your strengths, in addition to saying thank you and expressing continued interest in the job (Figure 15.13).

TAKE NOTE When writing thank-you letters, make sure you have the correct spelling of the interviewers' names and their titles. It might be a good idea to check names and titles at the company's website before sending an e-mail or letter. Send the letter the day of the interview.

A Thank-You Letter

Figure 15.13: A thank-you letter can be used to reinforce an important point that came out during the interview.

May 15, 2004

834 County Line Rd.
Hollings Point, Illinois 62905

Inside address

Valerie Sims, Human Resources Manager
Sunny View Organic Products
1523 Cesar Chavez Lane
Sunny View, California 95982

Dear Ms. Sims:

A thank-you statement leads off the letter.

Thank you for interviewing me for the Organic Agronomist position at Sunny View Organic Products. I enjoyed meeting you and the others at the company. Now, after speaking with you, I am more interested than ever in the position.

The main point is clearly stated in introduction.

I noticed during the interviews that my experience with benevolent insects was a recurring topic of discussion. This area of pest control is indeed very exciting, and my work with Professor George Roberts is certainly cutting edge. By paying attention to release times and hatching patterns, we have been able to maximize the effectiveness of predator insects. I would bring the latest research in this area to Sunny View's crops.

An important point is reinforced with details from the interview.

Again, thank you for the opportunity to interview with you for this position. I can be contacted at home (618-555-2993) or through e-mail (afranklin@unsb5.net). Please call or e-mail me if you need more information or would like to speak with me further about this position.

The conclusion thanks the reader and offers contact information.

Sincerely,

Anne Franklin

Anne Franklin

The signature includes the full name.

If you send a thank you through e-mail, follow it with a letter through the mail. An e-mail is nice for giving the interviewers immediate feedback, but the letter shows more professionalism.

If all goes well, you will be offered the position. At that point, you can decide if the responsibilities, salary, and benefits fit your needs.

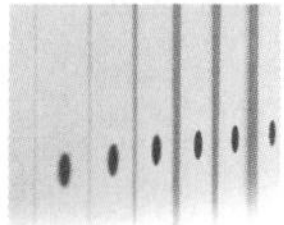

CHAPTER REVIEW

- From the beginning, you should adopt a professional approach to job seeking: set goals and make a plan.
- Your job-searching plan should include job search engines, personal networking, professional networking, targeting, and classified advertisements.
- Two types of resumes are commonly used, the archival resume and the functional resume. Either should summarize your background, experience, and qualifications.
- A well-designed resume should use the basic principles of design; a scannable resume incorporates keywords so it can be sorted electronically.
- An effective application letter can be more individual and specifically targeted than the resume. It shares the common features of a letter, including the appropriate "moves" for the introduction, body, and conclusion.
- An appropriate style for an application letter uses the "you" attitude, active voice, and nonbureaucratic language; you can also work in a theme that will make your resume stand out.
- A professional portfolio is a helpful tool for an interview to demonstrate your qualifications and abilities; material for your "master" portfolio can be assembled from classroom work, volunteer projects, related work experience, awards, certificates, or letters of reference.
- When you go to your interview, be prepared with information about the company and appropriate questions to ask; dress appropriately for the position; greet people with confidence; answer questions thoughtfully; and obtain the name and title of each interviewer (and its correct spelling) for a follow-up note to be sent that day.

EXERCISES AND PROJECTS

Individual or Team Projects

1. Find a sample resume on the Internet. Write a one-page critique of the resume to your instructor. Discuss its strengths and make suggestions for improvement. Pay special attention to the content, organization, style, and design of the resume.

2. Imagine you are looking for an internship in your field. Write a one-page resume that summarizes your education, work experience, skills, awards, and activities. Then, pay attention to issues of design, making sure the text uses principles of balance, alignment, grouping, consistency, and contrast.

3. Using an Internet job search engine, use keywords to find a job for which you might apply after college. Underline the qualifications required and the responsibilities of the position. Then, write a resume and application letter suitable for the job. In a cover memo addressed to your instructor, discuss some of the reasons why you would be a strong candidate for the position. Then, discuss some areas where you might need to take more courses or gain more experience before applying for the position.

4. Contact a human resources manager at a local company. In person or through e-mail, request an "informational interview" with this manager. If the interview is granted, ask him or her the best way to approach the company about a job. Ask what kinds of qualifications the company is usually looking for in a college graduate. Ask what you can do now to enhance your chances of obtaining a position at the company. After your interview, present your findings to your class.

5. Create a professional portfolio that you could use at an interview. Pull together examples of your work as well as any other relevant materials that you can find. What are the gaps in your portfolio? What kinds of materials should you try to obtain before you leave college? How can you go about filling in these gaps?

6. Make an electronic portfolio from your materials. Besides the materials available in your regular portfolio, what are some other links and documents you might include in your electronic portfolio?

Collaborative Project

With a group of people pursuing similar careers, develop some career goals and a job search plan for reaching those goals. Then, send each member of the group out to collect information. One group member should try out the job search engines, while another should explore personal and professional networking opportunities. One member should create a target list of potential employers in your area. Another should explore on-line classified advertisements in newspapers.

For good and bad samples of resumes, go to **www.ablongman.com/johnsonweb/15.18**

After your group has collected the information, write a small report to your class and instructor in which you discuss the results of your research. What did your group discover about the job market in your field? Where are the hottest places to find jobs? How can professional groups and personal networking help you make contacts with potential employers?

Individuals in your group should then choose one job that seems interesting. Each member of the group should write an application letter for that job and create a resume that highlights qualifications and strengths.

Then, the group should come up with five questions that might be asked at an interview. Write down four questions that are usually asked at an interview. Write down one question that is meant to trip up an interviewee.

Finally, take turns interviewing each other. Ask your questions and note down good answers and bad answers to the questions. Discuss how each member of the group might improve his or her interviewing skills based on your experiences in this project.

The Lie

Henry Romero wanted to be an architect his whole life. Even as a child, he was fascinated by the buildings in nearby Chicago. So, he spent years preparing himself to be an architect, winning top honors at his university in design.

So, he was thrilled when one of the top architectural firms, Goming and Cooper, announced it would be visiting his campus, looking to interview promising new talent. This firm was certainly one of the top ones in the country. It had designed some of the most innovative buildings in recent years. Henry was especially interested in the firm's international projects, which would allow him to work and live abroad. Recently, in *Architecture Times,* he had read about the firm's successful relationships with companies in France.

So, Henry gave his resume and a letter of application to the university's placement office personnel and told them to send it to Goming and Cooper. A month later, his placement counselor called him and said he had an interview with the firm. Henry was very pleased.

There was only one problem. On his resume, Henry had put down that he "read and spoke French fluently." But, in all honesty, he had taken French for only a few years in high school and one year in college. He hadn't really mastered the language. If necessary, he could muddle his way through a conversation in French, but he was hardly fluent.

To make things worse, last week Henry's roommate, Paul, had a nightmare interview. He too had reported on his resume that he spoke a language fluently—in this case, German. When he arrived at the interview, the interviewer decided to break the ice by talking in German. Paul was stunned. He stammered in German for a few minutes. Then the interviewer, clearly angry, showed him the door.

Henry was worried. What if the same thing happened to him? He knew the ability to speak French was probably one of the reasons he had received the interview in the first place. If the interviewer found out he was not fluent in French, there was a good chance the interview would go badly.

If you were Henry, how would you handle this touchy situation?

PART 4

Genres of Technical Communication

CHAPTER

16

Letters and Memos

CHAPTER CONTENTS

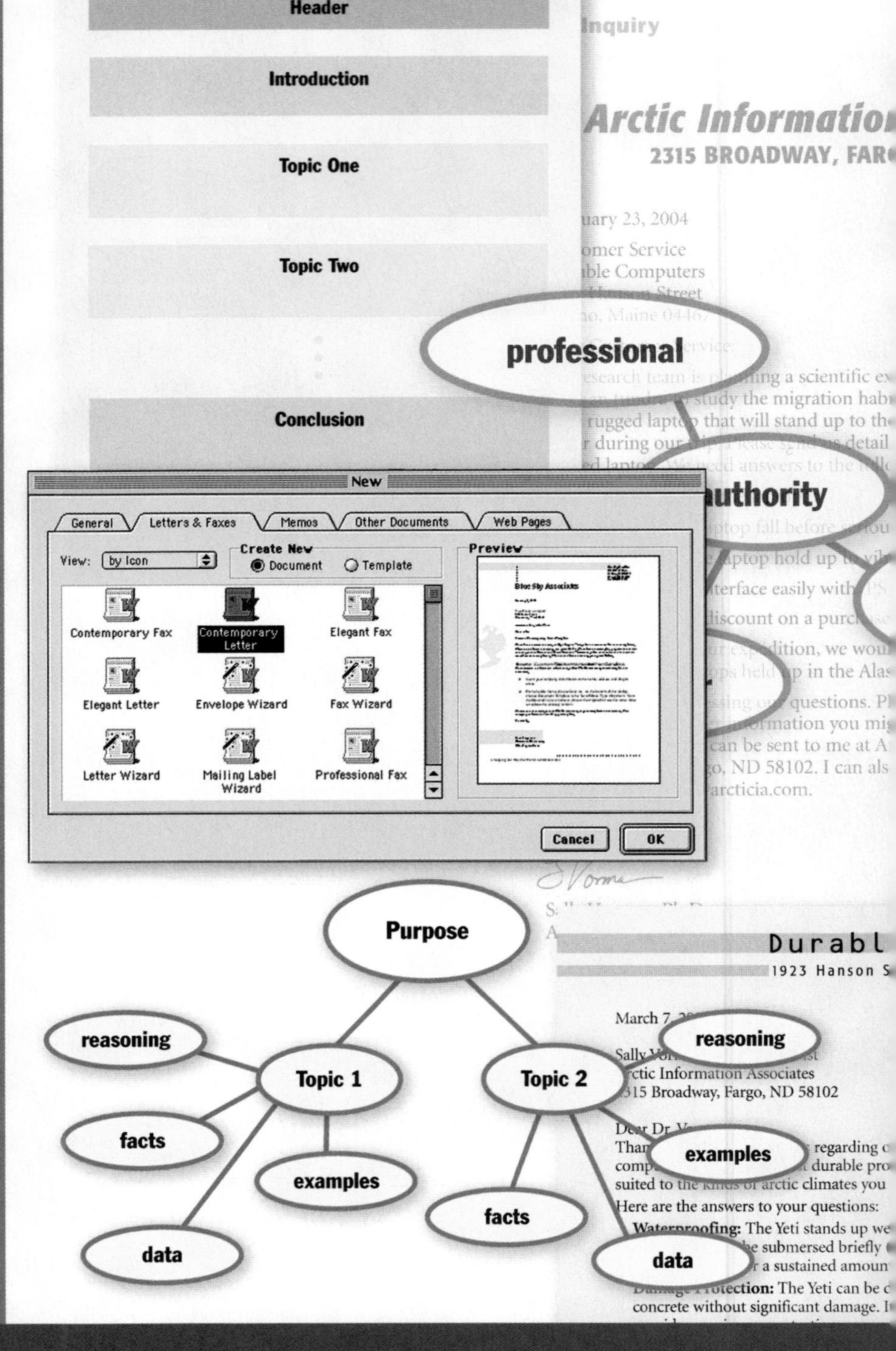

CHAPTER OBJECTIVES

In this chapter, you will learn:

- The role of correspondence in the technical workplace.
- The basic features of letters and memos.
- How to plan, organize, and draft letters and memos.
- Common patterns for letters and memos.
- How to choose an appropriate style for correspondence.
- How to design and format letters and memos.
- How to revise, review, and proofread letters and memos.

With the invention of e-mail, the role of letters and memos in the workplace has evolved. Not long ago, letters and memos were the predominant way to communicate with others in writing. They were used to convey both informal and formal messages. Today, e-mail is rapidly replacing memos and letters as the preferred way to communicate *informal* messages. Letters and memos are now used mostly to communicate *formal* messages, such as

- results of important decisions.
- formal inquiries.
- refusals.
- new policies.
- transmittals of important documents or materials.

The increasingly formal nature of letters and memos ensures that your readers will take them seriously. The paper-based format of a letter or memo signals that the message is too important or too proprietary to send as e-mail.

TAKE NOTE Even form letters or sales letters are taken more seriously than they once were because websites and e-mail have given companies less expensive ways to broadcast their messages to clients.

In many ways, letters and memos will be a regular part of your working life, but they can also be a drain on your workday. The key is to learn how to write these documents quickly and efficiently, within the natural flow of your workday. Once you have mastered workplace correspondence, writing letters and memos will become a part of your daily routine.

Basic Features of Letters and Memos

For the most part, letters and memos are very similar. They look different because they follow different *formats*. But, in the end, they tend to use the same kinds of content, organization, style, and design to put across their message. So, how are they different?

Letters are written to people *outside* the company or organization. Primarily, you will need to use letters in formal situations, in which you are acting as a representative of your company. Letters can be used to make requests, make inquiries, accept or refuse claims, communicate important information, record agreements, and apply for jobs.

Memos are written to people *inside* your company or organization. They contain meeting agendas, set policies, present internal reports, and offer short proposals. When a message is too important or proprietary for e-mail, most people will send a memo instead. Memos are still more reliable than e-mails for information that should not be broadly released.

Letters and memos are also regularly used as *transmittal documents*. In the workplace, a letter or memo of transmittal is placed on top of another document to explain the document's purpose and clearly state who should receive it. A letter or memo of transmittal helps make sure your document reaches its intended readers.

When is e-mail not appropriate? Go to **www.ablongman.com/johnsonweb/16.1**

Letters and memos share many of the same basic features, though these features look different on the page. Both kinds of documents include the following elements:

- Header
- Introduction
- Body
- Conclusion

However, letters and memos differ in formatting:

- The format for a letter usually includes these five features: a letterhead, the date, an inside address, a greeting, and a closing with signature.
- The format for a memo usually includes these five features: a header, the date, and lines for the addressee (To:), the sender (From:), and the subject (Subject:).

TAKE NOTE A memo does not include a signature, though many people sign their initials next to their name on the From line.

Figure 16.1 illustrates the way most letters and memos are organized.

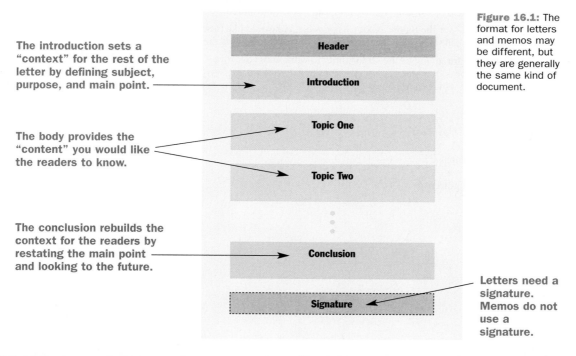

As you will notice in Figure 16.1, the basic pattern for letters and memos is similar to other documents. They have an introduction, body, and conclusion. The introduction sets a context up front by telling the readers the subject, purpose, and main point of the letter. The body provides the details the readers need to make a decision or take action. The conclusion often restates the main point and looks to the future.

Basic Pattern for Letters and Memos

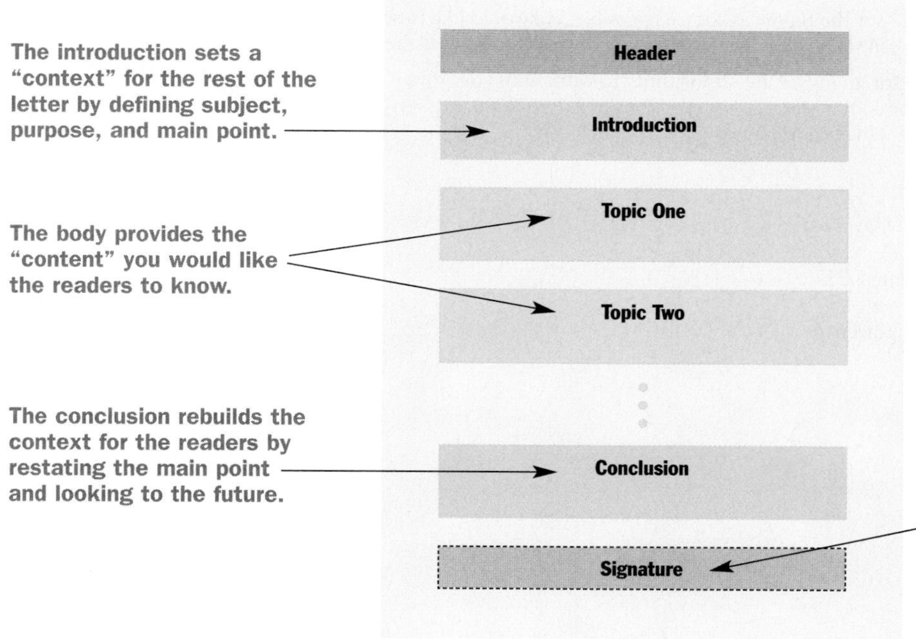

The introduction sets a "context" for the rest of the letter by defining subject, purpose, and main point.
The body provides the "content" you would like the readers to know.
The conclusion rebuilds the context for the readers by restating the main point and looking to the future.
Header
Introduction
Topic One
Topic Two
Conclusion
Signature
Letters need a signature. Memos do not use a signature.

Figure 16.1: The format for letters and memos may be different, but they are generally the same kind of document.

Figure 16.2 shows a letter and memo with basically the same content, so you can see the differences in formatting.

Letter and Memo

(a)

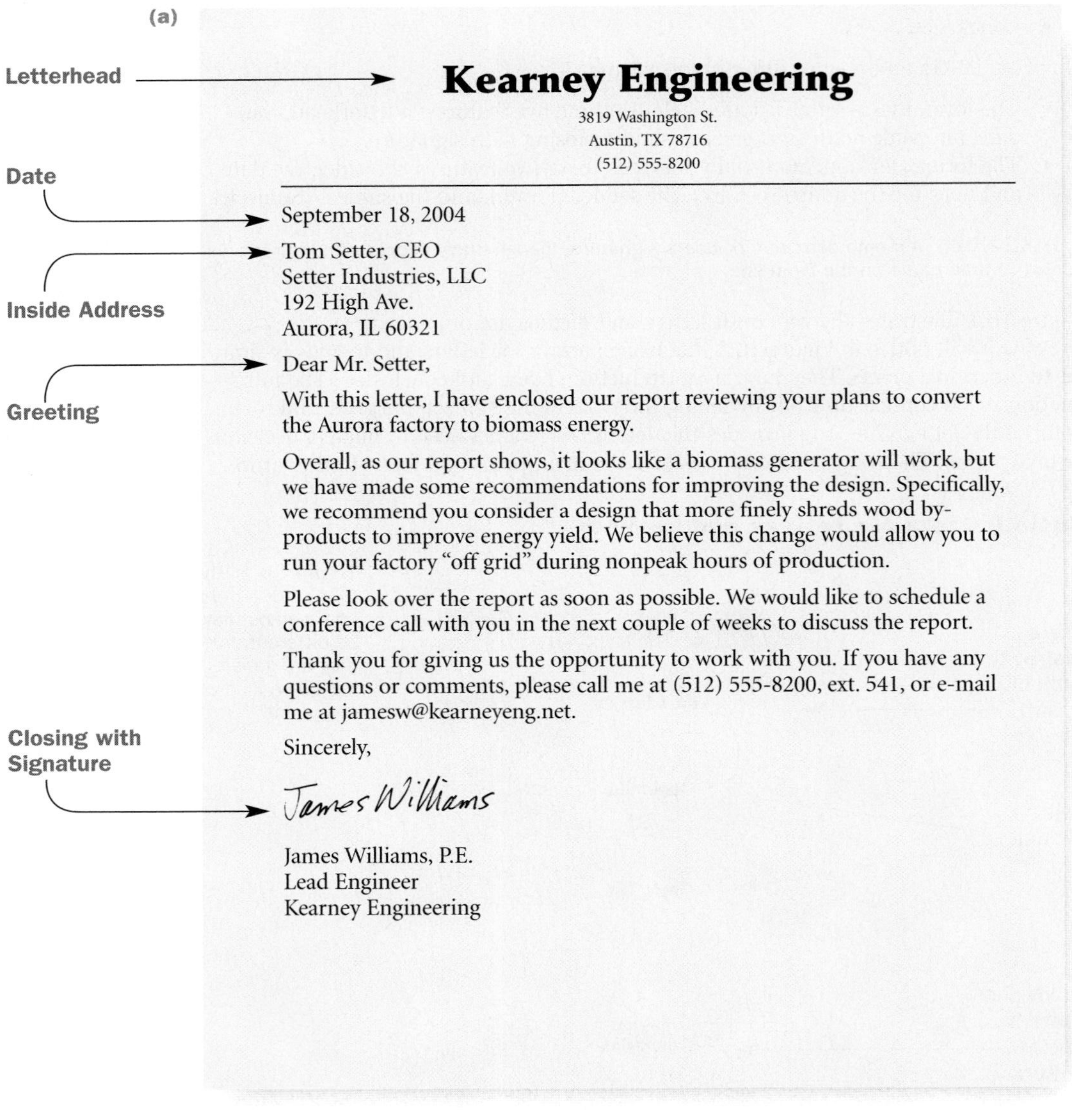

Letterhead →

Kearney Engineering

3819 Washington St.
Austin, TX 78716
(512) 555-8200

Date → September 18, 2004

Inside Address → Tom Setter, CEO
Setter Industries, LLC
192 High Ave.
Aurora, IL 60321

Greeting → Dear Mr. Setter,

With this letter, I have enclosed our report reviewing your plans to convert the Aurora factory to biomass energy.

Overall, as our report shows, it looks like a biomass generator will work, but we have made some recommendations for improving the design. Specifically, we recommend you consider a design that more finely shreds wood by-products to improve energy yield. We believe this change would allow you to run your factory "off grid" during nonpeak hours of production.

Please look over the report as soon as possible. We would like to schedule a conference call with you in the next couple of weeks to discuss the report.

Thank you for giving us the opportunity to work with you. If you have any questions or comments, please call me at (512) 555-8200, ext. 541, or e-mail me at jamesw@kearneyeng.net.

Closing with Signature → Sincerely,

James Williams

James Williams, P.E.
Lead Engineer
Kearney Engineering

Figure 16.2: Letters (a) and memos (b) are basically the same, except in formatting. The main differences are that letters are written to readers outside the company. Memos are written to readers inside the company.

To see other sample letters and memos, go to **www.ablongman.com/johnsonweb/16.2**

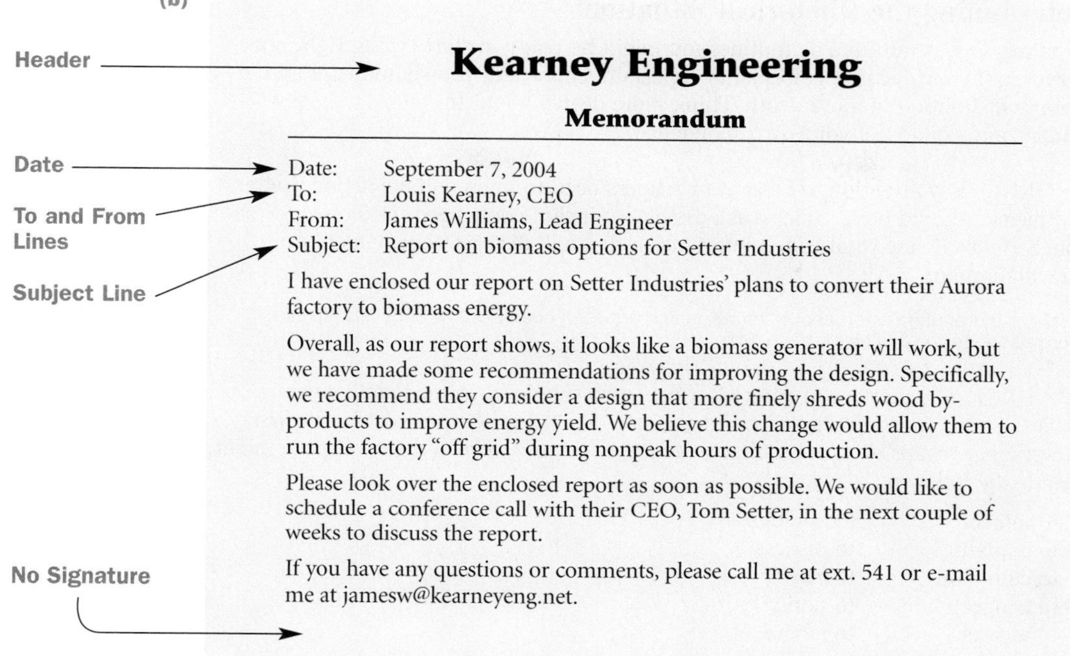

Kearney Engineering

Memorandum

Date: September 7, 2004
To: Louis Kearney, CEO
From: James Williams, Lead Engineer
Subject: Report on biomass options for Setter Industries

I have enclosed our report on Setter Industries' plans to convert their Aurora factory to biomass energy.

Overall, as our report shows, it looks like a biomass generator will work, but we have made some recommendations for improving the design. Specifically, we recommend they consider a design that more finely shreds wood by-products to improve energy yield. We believe this change would allow them to run the factory "off grid" during nonpeak hours of production.

Please look over the enclosed report as soon as possible. We would like to schedule a conference call with their CEO, Tom Setter, in the next couple of weeks to discuss the report.

If you have any questions or comments, please call me at ext. 541 or e-mail me at jamesw@kearneyeng.net.

Figure 16.2: (*continued.*)

Planning and Researching Letters and Memos

Most people don't prepare to write a letter or memo. They just do it. This approach works fine for routine letters and memos.

But, as the message becomes more important, you should prepare more thoroughly to write the correspondence. Otherwise, your message may lack a clear point or tend to ramble. In some cases, you may mistakenly provide too much information or use an inappropriate tone.

When you begin writing a letter or memo on your computer, first consider how your readers will use the information you are providing. You might start by answering the Five-W and How questions that describe the writing situation.

Who is the reader of my letter or memo?

Why am I writing to this person?

What is my point? What do I want my reader to do?

Where will the letter or memo be read?

When will the letter or memo be used?

How will readers use this document now and in the future?

To learn about the history of letter writing, go to **www.ablongman.com/johnsonweb/16.3**

Determining the Rhetorical Situation

If the message is informal or routine, you might be ready to start typing right now. However, if your message is formal or especially important, you should explore the rhetorical situation in more depth. Think more deeply about the subject, purpose, readers, and context of your correspondence.

SUBJECT Pay attention to what your readers need to know to take action. Letters and memos should be as concise as possible, so include only need-to-know information. Strip away any want-to-tell information that will distract your readers from your main point.

LINK For more information about distinguishing between need-to-know and want-to-tell information, go to Chapter 5, page 115.

PURPOSE The purpose of your letter or memo should be immediately obvious. To be clear about your intentions, you might include a purpose statement in the first paragraph, perhaps even in the first sentence. Some key words for the purpose might include the following.

to inform	to apologize
to explain	to discuss
to complain	to clarify
to congratulate	to notify
to answer	to advise
to confirm	to announce
to respond	to invite

Your purpose statement might sound like one of the following:

> We are writing to inform you that we have accepted your proposal to build the Washington Street overpass.
>
> I would like to congratulate the Materials Team for successfully patenting the fusion polymer blending process.
>
> This memo explains and clarifies the revised manufacturing schedule for the remainder of this year.

Your purpose should be immediately obvious to your readers as soon as they read the first paragraph. They should not need to guess why you are writing to them.

LINK For more guidance on defining a document's purpose, see Chapter 2, page 23.

READERS Letters and memos can be written to individuals or whole groups of people. Since these documents are often shared or filed, you need to anticipate all possible readers who might want a copy of your document.

Primary readers (action takers) are the people who will take action after they read your message. Your letter or memo needs to be absolutely clear about what you want these readers to do. It should also be tailored to their individual motives, values, and attitudes about the subject.

To read about some letter and memo blunders, go to **www.ablongman.com/johnsonweb/16.4**

Secondary readers (advisors) are the people to whom your primary readers will turn if they need advice. They may be experts in the area, support staff, supervisors, or colleagues. You should anticipate these readers' concerns, but your focus should still be on the primary readers' needs.

Tertiary readers (evaluators) are any other people who may have an interest in your letter or memo. These readers may be more important with correspondence than other kinds of documents. Letters and memos have a strange way of turning up in unexpected places. For example,

- that "confidential" memo you wrote to your research team might end up in your competitor's hands.
- the local newspaper might get hold of a letter you sent to your company's clients explaining a problem with an important new product.

Before sending any correspondence, you should think carefully about how the document would look if it were made public. Anticipate how it might be used against you or your company.

Gatekeeper readers (supervisors), such as your supervisor or legal counsel may want to look over an especially important correspondence before it is sent out. You should always keep in mind that you are representing your company in your letters and memos. Your supervisor or the corporate lawyer may want to ensure that you are communicating appropriately with clients.

LINK For more information on analyzing readers, turn to Chapter 3, page 55.

CONTEXT OF USE Imagine all the different places your letter or memo may be used. Where will readers use this document now and in the future? Where will the document be kept (if at all) after it is read?

Be sure to consider the physical, economic, political, and ethical factors that will influence how your readers will interpret and respond to your message. Put yourself in their place, imagining their concerns as they are reading your document.

LINK For strategies to help identify contextual issues, go to Chapter 3, page 53.

The personal nature of letters and memos means that every word, sentence, and paragraph should be carefully measured to reflect the needs of your readers and the documents' context of use. Well-written letters and memos can persuade readers to take action in positive ways. They can build important personal relationships that go beyond the corporate connections.

Organizing and Drafting Letters and Memos

Some messages require more time and care than others, but in most cases you should be able to generate them within the natural flow of your workday.

The introductions and conclusions of these texts tend to make some predictable moves. If you memorize these moves, you can spend more time concentrating on what you need to say in the body.

Introduction That Sets a Context

In the introduction, a letter or memo should make at least three moves. The introduction should be immediately clear about its *subject, purpose*, and *main point.* By telling the readers at least these three items up front, you will set a context for the information that follows in the body of the correspondence.

Depending on your message, you might also make two additional moves. You might offer some *background information* and stress the *importance of the subject.*

SUBJECT Your subject should be stated or signaled in the first or second sentence of the introduction. Simply tell your readers what you are writing about, and *do not assume* they know what you are writing about.

> Recently, the Watson Project has been a source of much concern for our company.

> This memo discusses the equipment thefts that have occurred in our office over the last few months.

PURPOSE Your reason for writing should also be stated almost immediately in the first paragraph, preferably in the first or second sentence.

> Now that we have reached Stage Two of the Oakbrook Project, I would like to refine the responsibilities of each team member.

> The purpose of this letter is to inform you about our new transportation policies for low-level nuclear waste sent to the WIPP Storage Facility in New Mexico.

MAIN POINT All letters and memos should have a point that you want your reader to grasp or remember. In many cases, "the point" is something you want your readers to do when they are finished reading. In other words, state the *action* you want readers to take.

> We request the hiring of three new physician's assistants to help us with the recent increases in emergency room patients.

> Put bluntly, our subcontractors must meet ISO-9001 quality standards. It is our job to make sure that they comply.

It may seem odd to state your main point up front. Wouldn't it be better to lead up to the point, perhaps putting it in the conclusion? No. Most of your readers will only scan your message. By putting your main point (the action item) up front, you will ensure that they do not miss it.

BACKGROUND INFORMATION Writers often like to start their letters and memos with a statement that gives some background or makes a personal connection with readers.

> Our staff meeting on June 28 was very productive, and I hope we all came away with a better understanding of the project. In this memo . . .

> When you and I met at the NEPSCORE Convention last October, our company was not ready to provide specifics about our new ceramic circuit boards. Now, we are ready . . .

Need help writing an introduction? Go to **www.ablongman.com/johnsonweb/16.5**

IMPORTANCE OF THE SUBJECT In some cases, you might also want your introduction to stress the importance of the subject.

> This seems like a great opportunity to expand our network into the Indianapolis market. We may not see this opportunity again.

> If we don't start looking for a new facility now, we may find ourselves struggling to keep up with the demand for our products.

Introductions in letters and memos should be as concise as possible. At a minimum, the introduction should tell readers your subject, purpose, and main point. Background information and statements about the importance of the subject should be used where needed.

LINK For more information on writing introductions, see Chapter 6, page 141.

Body That Provides Information

The body is where you will provide your readers with the information they need to know to make a decision or take action. The body is the largest part of the letter, and it will take up one or more paragraphs.

As you begin drafting the body of your letter, think about the *topics* you need to discuss with your readers. For example, an application letter for a job might be divided into three discussion topics: education, work experience, and skills. Each of these topics would likely receive a paragraph or two of discussion.

If you are struggling to develop the content, use mapping to put your ideas on the screen or on a piece of paper (Figure 16.3). Start out by putting the purpose statement in the center of the screen. Then, branch out into topics. You can use mapping to identify any supporting information that will be needed for those topics.

Using Mapping to Generate Content

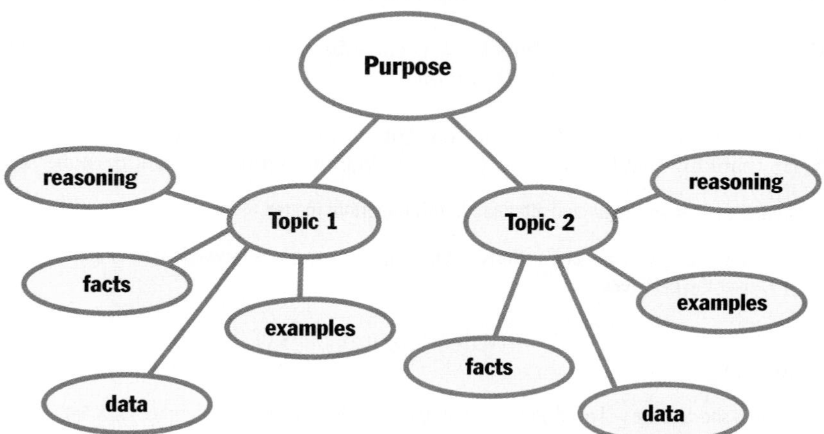

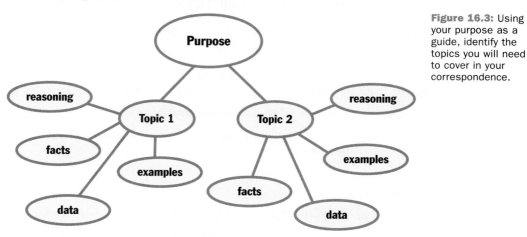

Figure 16.3: Using your purpose as a guide, identify the topics you will need to cover in your correspondence.

As you are drafting the body, keep looking back at your purpose statement in the introduction. Ask yourself, "What information do I need to achieve that purpose?" Then, include any facts, examples, data, and reasoning that will help you support your argument. In other words, make sure you provide all the need-to-know information that your readers require to make a decision or take action.

TAKE NOTE Put your claims up front in each paragraph of the body. Then, use facts, examples, data, and reasoning to support those claims.

AT A GLANCE

Elements of a Letter or Memo

- Header
- Introduction—subject, purpose, main point, background information, importance of the subject
- Body—discussion topics, with usually one paragraph per topic
- Conclusion—thank you, main point (restated), and a look to the future

Also, keeping your purpose statement in mind, you should slice away any want-to-tell information that does not help you achieve that purpose. Because letters and memos are usually short, they should provide only need-to-know information. All that extra want-to-tell information will only cloud your message.

LINK For more information on using logical mapping, go to Chapter 5, page 103.

Conclusion

The conclusion of your letter or memo should be short and to the point. Nothing essential should appear in the conclusion that has not already been stated in the introduction or body.

Conclusions in these documents tend to make three moves: *thank the readers, restate your main point*, and *look to the future.*

THANK THE READERS Tell them that you appreciate their attention to your message. By thanking them at the end, you leave them with a positive impression.

> Thank you for your time and attention to this important matter.
>
> We appreciate your company's efforts on this project, and we look forward to working with you over the next year.

RESTATE YOUR MAIN POINT Remind your readers of the action you would like them to take.

> Time is short, so we will need your final report in our office by Friday, September 15, at 5:00.
>
> Please discuss this proposal right away with your team, so we can make any final adjustments before the submission deadline.

LOOK TO THE FUTURE Try to end your letters by looking forward in some way.

> When this project is completed, we will have taken the first step toward renovating our whole approach to manufacturing.
>
> If you have questions or comments, please call me at 555-1291 or e-mail me at sue.franklin@justintimecorp.com.

Your conclusion should run a maximum of one to three sentences. If you find yourself writing a conclusion that is more than a small paragraph, you probably need to trim the added information, or you need to move some of it into the body of the letter or memo.

Remember that any action items can be restated in the conclusion, but they should not appear there for the first time. Put them early in the letter or memo. In general, conclusions should not include any new information. They should only summarize points that were made earlier in the letter.

LINK To learn more about writing conclusions, go to Chapter 6, page 155.

Types of Letters and Memos

Technical professionals send out letters and memos regularly. This section discusses some of the more common patterns for letters and memos in the technical workplace:

- Inquiry
- Response
- Transmittal
- Claim
- Adjustment
- Refusal.

Inquiry Letters and Memos

The purpose of an *inquiry* is to gather information, especially answers to questions about important or sensitive subjects. In these situations, you could use e-mail, but a printed document is sometimes preferable because the recipients will view it as a formal request.

Here are some guidelines to follow when writing an inquiry:

- Clearly identify your subject and purpose.
- State your questions clearly and concisely.
- Limit your questions to five or less.
- If possible, offer something in return.
- Thank readers in advance for their response.
- Provide contact information (address, e-mail address, or phone number).

Figure 16.4 shows a typical letter of inquiry. Notice how the author of the letter is specific about the kinds of information she wants.

For more examples of inquiry and response letters, go to **www.ablongman.com/johnsonweb/16.6**

Letter of Inquiry

Figure 16.4: A letter of inquiry needs to be clear about the information it is seeking. In this letter, notice how the writer has listed out her questions in an unmistakable way.

Arctic Information Associates
2315 BROADWAY, FARGO, ND 58102

February 23, 2004

Customer Service
Durable Computers
1923 Hanson Street
Orono, Maine 04467

Dear Customer Service:

State the subject and purpose of the letter. →

My research team is planning a scientific expedition to the northern Alaskan tundra to study the migration habits of caribou. We are looking for a rugged laptop that will stand up to the unavoidable abuse that will occur during our trip. Please send us detailed information on your Yeti rugged laptop. We need answers to the following questions:

State questions clearly and concisely. →

- How waterproof is the laptop?
- How far can the laptop fall before serious damage will occur?
- How well does the laptop hold up to vibration?
- Does the laptop interface easily with GPS systems?
- Can we receive a discount on a purchase of 20 computers?

Offer something in return. →

Upon return from our expedition, we would be willing to share stories about how your laptops held up in the Alaskan tundra.

Thank the readers. →

Provide contact information. →

Thank you for addressing our questions. Please respond to these inquiries and send us any other information you might have on the Yeti rugged laptop. Information can be sent to me at Arctic Information Associates, 2315 Broadway, Fargo, ND 58102. I can also be contacted at 701-555-2312 or salvorman@arcticia.com.

Sincerely,

S Vorman

Sally Vorman, Ph.D.
Arctic Specialist

Response Letters and Memos

A *response* is written to answer an inquiry. A response letter should answer each of the inquirer's questions in specific detail. The amount of detail you provide will depend on the kind of questions asked. In some situations, you may need to offer a lengthy explanation. In other situations, a simple answer or referral to the corporate website or enclosed product literature is sufficient.

Here are some guidelines to follow when writing a response:

- Thank the writer for the inquiry.
- Clearly state the subject and purpose of the letter or memo.
- Answer any questions point by point.
- Offer more information, if available.
- Provide contact information (address, e-mail, or phone number).

Figure 16.5 shows an example of a response letter. Pay attention to the author's point-by-point response to the questions in the original letter of inquiry.

Transmittal Letters and Memos

When sending documents or materials through the mail, you should include a letter or memo of transmittal. Also called "cover letters" or "cover memos," the purpose of these documents is to explain the reason the enclosed materials are being sent. For example, if you were sending a proposal to the vice president of your company, you would likely add a memo of transmittal like the one shown in Figure 16.6. Earlier in this chapter, the documents in Figure 16.2 also showed a letter and memo of transmittal.

A transmittal letter or memo should do the following:

- Identify the materials enclosed.
- State the reason the materials are being sent.
- Summarize the information being sent.
- Clearly state any action requested or required of readers.
- Provide contact information.

You should limit your comments in transmittal letters or memos and assume your readers will not read them closely. After all, readers are mostly interested in the enclosed materials, not your transmittal letter or memo.

Why should you include a letter or memo of transmittal in the first place? There are a few good reasons:

- If a document, like a report, shows up in your readers' mail without a transmittal letter or memo, they may not understand why it is being sent to them and what they should do with it.
- Transmittal letters and memos give you an opportunity to make a personal connection with the readers.
- They also give you an opportunity to set a specific tone for readers, motivating them to respond positively to the document or materials you have enclosed.

An effective transmittal letter or memo welcomes your readers to the materials you have sent.

Response Letter

Figure 16.5: A response letter should answer the inquirer's questions point by point and offer additional information, if available.

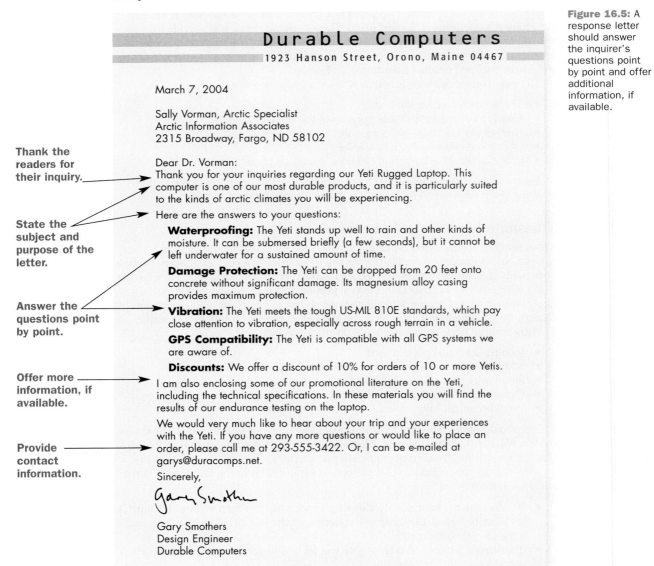

Thank the readers for their inquiry.

State the subject and purpose of the letter.

Answer the questions point by point.

Offer more information, if available.

Provide contact information.

Durable Computers

1923 Hanson Street, Orono, Maine 04467

March 7, 2004

Sally Vorman, Arctic Specialist
Arctic Information Associates
2315 Broadway, Fargo, ND 58102

Dear Dr. Vorman:

Thank you for your inquiries regarding our Yeti Rugged Laptop. This computer is one of our most durable products, and it is particularly suited to the kinds of arctic climates you will be experiencing.

Here are the answers to your questions:

Waterproofing: The Yeti stands up well to rain and other kinds of moisture. It can be submersed briefly (a few seconds), but it cannot be left underwater for a sustained amount of time.

Damage Protection: The Yeti can be dropped from 20 feet onto concrete without significant damage. Its magnesium alloy casing provides maximum protection.

Vibration: The Yeti meets the tough US-MIL 810E standards, which pay close attention to vibration, especially across rough terrain in a vehicle.

GPS Compatibility: The Yeti is compatible with all GPS systems we are aware of.

Discounts: We offer a discount of 10% for orders of 10 or more Yetis.

I am also enclosing some of our promotional literature on the Yeti, including the technical specifications. In these materials you will find the results of our endurance testing on the laptop.

We would very much like to hear about your trip and your experiences with the Yeti. If you have any more questions or would like to place an order, please call me at 293-555-3422. Or, I can be e-mailed at garys@duracomps.net.

Sincerely,

Gary Smothers

Gary Smothers
Design Engineer
Durable Computers

Memo of Transmittal

Figure 16.6: A transmittal memo should be concise. Make sure any action items are clearly stated.

Identify the enclosed materials.

State the reason materials are being sent.

Summarize the enclosed materials.

State the action item clearly.

Provide contact information.

Rockford Services

MEMORANDUM

Date: May 8, 2004
To: Brenda Young, VP of Services
From: Valerie Ansel, Outreach Coordinator
cc: Hank Billups, Pat Roberts
Re: Outreach to Homeless Youth

Enclosed is the Proposal for the Rockford Homeless Youth Initiative, which you requested at the Board Meeting on January 16, 2004. We need you to look it over before we write the final version.

The proposal describes a broad-based program in which Rockford Services will proactively reach out to the homeless youth in our city. In the past, we have generally waited for these youths to find their way to our shelter on the west side of town. We always knew, though, that many youths are reluctant or unable to come to the shelter, especially the ones who are mentally ill or addicted to drugs. The program described in this proposal offers a way to reach out to these youths in a non-threatening way, providing them a gateway to services or treatment.

Please look over this proposal. We welcome any suggestions for improvement you might have. We plan to submit the final version of this proposal to the Board on May 25th at the monthly meeting.

Thank you for your help. You can contact me by phone at 555-1242, or you can e-mail me at valansel@rockfordservices.org.

Enclosed: Proposal for the Rockford Homeless Youth Initiative

For more samples of claim and adjustment letters, go to **www.ablongman.com/johnsonweb/16.7**

Claim Letters and Memos

In the technical workplace, products break and errors happen. For these situations, you may need to write a claim, also called a complaint. The purpose of a claim letter or memo is to explain a problem and ask for amends. Here are some guidelines to follow when writing a claim:

- State the subject and purpose of the letter clearly and concisely.
- Explain the problem in detail.
- Describe how the problem inconvenienced you.
- State what you would like the receiver of the letter to do to address the problem.
- Thank your reader for his or her response to your request.
- Provide contact information.

Figure 16.7 shows a claim letter with these features.

A claim should always be professional in tone. You might be tempted to send an angry message when errors are made. Angry letters and memos might give you a temporary sense of satisfaction, but they are less likely to achieve your purpose—to have the problem fixed. If possible, you want to avoid putting readers on the defensive, because they may choose to ignore you or halfheartedly try to remedy the situation.

Adjustment Letters and Memos

If you receive a claim letter or memo, you may need to respond with an *adjustment* letter or memo. The purpose of an adjustment letter or memo is to resolve the problem described by the client, customer, or coworker. These documents, though, need to do more than simply respond to the problem. They should also try to rebuild a potentially damaged relationship with the client, customer, or colleague.

Here are some guidelines to follow when writing an adjustment letter:

- Express regret for the problem without directly taking blame.
- State clearly what you are going to do about the problem.
- Tell your reader when he or she should expect results.
- Show appreciation for his or her continued business with your company.
- Provide contact information.

Figure 16.8 shows an adjustment letter with these features.

Why shouldn't you take direct blame? Several factors are usually involved when something goes wrong. So, it is fine to acknowledge that something unfortunate happened. For example, you can say, "We are sorry to hear about your injury when using the Zip-2000 soldering tool." But, it is something quite different to say, "We accept full responsibility for the injuries caused by our Zip-2000 solder tool." This kind of statement could make your company unnecessarily liable for damages.

Ethically, your company may need to accept direct blame for an accident. In these situations, legal counsel should be involved with the writing of the letter.

Claim Letter

Figure 16.7: A claim letter should explain the problem in a professional tone and describe the remedy being sought.

State the subject and purpose of the letter.

Explain the problem in detail.

Describe how the problem inconvenienced you.

State what the reader should do.

Thank the reader for the anticipated response.

Provide contact information.

Outwest Engineering

2931 Mission Drive, Provo, UT 84601 (801) 555-6650

June 15, 2004

Customer Service
Optima Camera Manufacturers, Inc.
Chicago, IL 60018

Dear Customer Service:

We are requesting the repair or replacement of a damaged ClearCam Digital Camcorder (#289PTDi), which we bought directly from Optima Camera Manufacturers in May 2004.

Here is what happened. On June 12, we were making a promotional film about one of our new products for our website. As we were making adjustments to the lighting on the set, the camcorder was bumped and it fell ten feet to the floor. Afterward, it would not work, forcing us to cancel the filming, causing us a few days' delay.

We paid a significant amount of money for this camcorder because your advertising claims it is "highly durable." So, we were surprised and disappointed when the camcorder could not survive a routine fall.

Please repair or replace the enclosed camcorder as soon as possible. I have provided a copy of the receipt for your records.

Thank you for your prompt response to this situation. If you have any questions, please call me at 801-555-6650, ext. 139.

Sincerely,

Paul Williams

Paul Williams
Senior Product Engineer

Adjustment Letter

Figure 16.8: An adjustment letter should express regret for the problem and offer a remedy.

O C M

Optima Camera Manufacturers, Inc.
Chicago, IL 60018 312-555-9120

July 1, 2004

Paul Williams, Senior Product Engineer
Outwest Engineering Services
2931 Mission Drive
Provo, UT 84601

Dear Mr. Williams,

We are sorry that the ClearCam Digital Camcorder did not meet your expectations for durability. At Optima, we take great pride in offering high-quality, durable cameras that our customers can rely on. We will make the repairs you requested.

After inspecting your camera, our service department estimates the repair will take two weeks. When the camera is repaired, we will return it to you by overnight freight. The repair will be made at no cost to you.

We appreciate your purchase of a ClearCam Digital Camcorder, and we are eager to restore your trust in our products.

Thank you for your letter. If you have any questions, please contact me at 312-555-9128.

Sincerely,

Ginger Faust

Ginger Faust
Customer Service Technician

Express regret for the problem.

State what will be done.

Tell when results should be expected.

Show appreciation to the customer.

Provide contact information.

Refusal Letters and Memos

Refusals, also called "bad news" letters or memos, always need to be carefully written. In these documents, you are telling someone something they don't want to hear. Yet, if possible, you want to maintain a professional or business relationship with the customer or client.

When writing a refusal, guide your readers logically to your decision. In most cases, you do not want to start out immediately with the bad news (e.g., "We have finished interviewing candidates and have decided not to hire you"). You also do not want to make readers wait too long for the bad news.

Here are some guidelines for writing a refusal:

- State your subject.
- Summarize your understanding of the facts.
- Deliver the bad news, explaining your reasoning.
- Offer any alternatives, if they are available.
- Express a desire to retain the relationship.
- Provide contact information.

In a refusal letter or memo, keep the apologizing to a minimum. Some readers will see your apologies as an opening to negotiate or complain further. An effective refusal logically explains the reasons for the turndown, leaving your reader satisfied with your response—if a bit disappointed. Figure 16.9 shows a sample refusal letter with these features.

Using Style in Letters and Memos

The style of a letter or memo can make a big difference. One thing to keep in mind as you consider your style in a letter or memo is—*All letters and memos are personal.* They make a one-to-one connection with readers.

Even if you are writing a memo to the whole company or sending out a form letter to your company's customers, you are still making a personal, one-to-one connection with each of those readers. People will read your correspondence as a message written to them individually.

The personal nature of these documents has its advantages and disadvantages:

- The advantage is that your readers will view your letter or memo as something sent directly to them. They will assume that they are interacting with a person (you), not just a faceless company.
- A potential disadvantage is that readers tend to react to these documents in an emotional way. If your correspondence sounds indifferent, rude, or uncaring, your readers are likely to become irritated or angry. Readers respond very negatively to memos or letters that they perceive to be angry, insulting, or condescending.

To see samples of refusal letters, go to
www.ablongman.com/johnsonweb/16.8

Refusal Letter

Figure 16.9: A refusal letter should deliver the bad news politely and offer alternatives if available. You should strive to maintain the relationship with the person whose request is being refused.

O C M

Optima Camera Manufacturers, Inc.
Chicago, IL 60018 312-555-9120

July 1, 2004

Paul Williams, Senior Product Engineer
Outwest Engineering Services
2931 Mission Drive
Provo, UT 84601

Dear Mr. Williams,

State the subject. →

We are sorry that the ClearCam Digital Camcorder did not meet your expectations for durability. At Optima, we take great pride in offering high-quality, durable cameras that our customers can rely on.

Summarize what happened. →

According to the letter you sent us, the camcorder experienced a fall and stopped working. After inspecting your camcorder, we have determined that we

Deliver the bad news, explaining your reasoning. →

will need to charge for the repair. According to the warranty, repairs can only be made at no cost when problems are due to manufacturer error. A camcorder that experienced a fall like the one you described is not covered under the warranty.

Offer alternatives. →

We sent your camcorder to the service department for a repair estimate. After inspecting your camera, they estimate the repair will take two weeks at a cost of $156.00. When it is repaired, we will return it to you by overnight freight.

Provide contact information. →

If you would like us to repair the camcorder, please send a check or money order for $156.00. If you do not want us to repair the camcorder, please call me at 312-555-9128. Upon hearing from you, we will send the camcorder back to you immediately.

Express a desire to retain the relationship. →

Again, we are sorry for the damage to your camcorder. We appreciate your purchase of a ClearCam Digital Camcorder, and we are eager to retain your business.

Sincerely,

Ginger Faust

Ginger Faust
Customer Service Technician

Enclosed: Warranty Information

Strategies for Developing a Style

Letters and memos are personal documents, so their style needs to be suited to their readers and context of use. Here are some strategies for projecting the appropriate style in your letter or memo:

- Use the "you" style.
- Create a tone.
- Avoid bureaucratic phrasing.

USE THE "YOU" STYLE When you are conveying neutral or positive information, use the word "you" to address your readers. The "you" style puts the emphasis on readers rather than on you, the author.

> Well done. Your part of the project went very smoothly, saving us time and money.
>
> We would like to update your team on the status of the Howards Pharmaceutical case.
>
> You are to be congratulated for winning the Baldrige Award for high-quality manufacturing.

In most cases, negative information should not use the "you" style, because the readers will tend to react with more hostility than you expect.

> **Offensive:** Your lack of oversight and supervision on the assembly line led to the recent work stoppage.
>
> **Improved:** Increased oversight and supervision will help us avoid work stoppages in the future.
>
> **Offensive:** At our last meeting, your ideas for new products were not thought through completely. In the future, you should come more prepared.
>
> **Improved:** Any ideas for new products should be thoroughly considered before they are presented. In the future, we would like to see presenters more prepared.

Don't worry about whether your readers will notice you are criticizing them. Even without the "you" style, they will figure out that you are conveying negative information or criticisms. By avoiding "you" in these negative situations, though, you will create a constructive tone and avoid an overly defensive reaction from your readers.

TAKE NOTE In most cases, you should go out of your way to project a positive tone in letters and memos. Your readers will tend to view a neutral or objective tone as negative.

LINK For more advice about choosing an appropriate style, see Chapter 7, page 187.

CREATE A TONE As you write your letter or memo, think about the image you want to project. Put yourself into character as you compose your message. Are you satisfied, hopeful, professional, pleased, enthusiastic, or annoyed? Write your message with that tone in mind.

Mapping is an especially good way to project a specific tone in your letter or memo (Figure 16.10). For example, perhaps you want to establish yourself as the "expert" on a particular subject. Put the word *expert* in the middle of the screen or a piece of paper. Then, start writing down words associated with this word.

Mapping a Tone

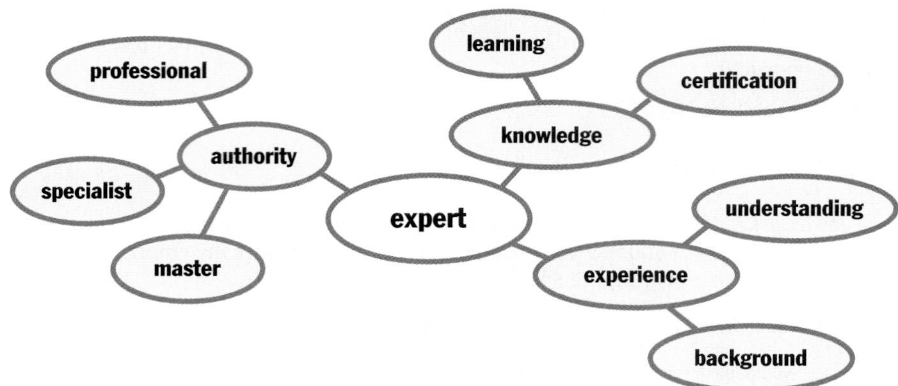

Figure 16.10: To set a tone, you can use mapping to develop words that are associated with the tone you want to create.

Once you have mapped out the tone you want to create, you can then weave these words into your letter or memo. If the words are used strategically, your readers will subconsciously sense the tone you want to create.

AVOID BUREAUCRATIC PHRASING When writing correspondence, especially a formal letter, some people feel a strange urge to use phrasing that sounds bureaucratic:

> **Bureaucratic:** Pursuant to your request, please find the enclosed materials.
>
> **Nonbureaucratic:** We have included the materials you requested.

Do you notice how the bureaucratic phrasing in the first sentence only makes the message harder to understand? It doesn't add any information. Moreover, this phrasing depersonalizes the letter, undermining the one-to-one relationship between writer and reader.

Here are a few other bureaucratic phrases and ways they can be avoided:

Bureaucratic phrase	Nonbureaucratic phrase
Per your request	As you requested
In lieu of	Instead of
Attached, please find	I have attached
Enclosed, please find	I have enclosed
Contingent upon receipt	When we receive
In accordance with your wishes	As you requested
In observance with	According to
Please be aware that	We believe
It has come to our attention	We know
Pursuant to	In response to
Prior to receipt of	Before receiving

A simple guideline is not to use words in letters and memos that you would not use in everyday speech. If you would not use words like *lieu, contingent*, or *pursuant* in a conversation, you should not use them in a letter or memo.

Designing and Formatting Letters and Memos

Letters and memos are usually rather plain in design. In the workplace, they typically follow standardized formats and templates that prescribe how they will look.

Most companies have premade word-processing templates for letters and memos, which you can use on your computer. These templates allow you to type your letter or memo directly into a word processor file. When you print out the document, the letterhead or memo header appears at the top.

In some cases, especially with formal letters, you may need to leave room in your document for the letterhead at the top of the paper. That way, when you print the letter on the company's stationery, the text will appear below the letterhead.

Formatting Letters

As stated earlier in this chapter, letter formats typically include some predictable features: a header (letterhead), an inside address, a greeting, the message, and a closing with signature (Figure 16.11).

LETTERHEAD Companies typically have letterhead available as a premade word processor template or as stationery. Letterhead includes the company name and address. If letterhead is not available, you should enter your return address under the date. Do not include your name in the return address.

> December 19, 2004
>
> 1054 Kellogg Avenue, Apt. 12
> Hinsdale, Illinois 60521

The return address is best set along the left margin of the letter.

INSIDE ADDRESS The address of the person to whom you are sending the letter (called the *inside address*) should appear two lines below the date or return address.

> George Falls, District Manager
> Optechnical Instruments
> 875 Industrial Avenue, Suite 5
> Starkville, New York 10034

The inside address should be the same as the address that will appear on the letter's envelope.

TAKE NOTE Your word processor can print the inside address on an envelope. Look for the Envelopes and Labels function. In some cases, your word processor will pick up the inside address automatically. You just need to print the envelope.

Want more advise about avoiding bureaucratic language? Go to **www.ablongman.com/johnsonweb/16.9**

AT WORK

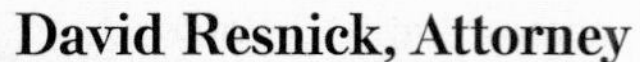

David Resnick, Attorney

ATTORNEY, FOLEY & LARDNER, LLP, CHICAGO OFFICE

Foley & Lardner LLP is a national law firm providing interdisciplinary services that result in high-value legal counsel.

How important is the tone in letters and memos?

As an attorney, I spend a substantial amount of time every day drafting correspondence to clients and other attorneys, usually in the form of letters and memoranda. Because I know that words on a page can often be construed in a manner not intended by their author, I am always conscious of my tone and word choice whenever I write.

To establish an unambiguous tone in a piece of correspondence, I try to "step into the shoes" of my reader and choose words and phrases that will most effectively evoke the reaction I am seeking. For instance, if I am writing a letter to my client reminding her of an impending deadline, I want to convey a sense of urgency without conveying a sense of panic. She is my client, and my professional duties include both counseling her on the law and protecting her from it. However, when I correspond with opposing counsel regarding a dispute between our clients, I will use a considerably more direct and insistent tone to achieve my result, for I do not owe him the same kind of duty I owe my client.

I will often write to lawyers in a more aggressive manner when defending my client's position, for my objective in that circumstance is to instill doubt, rather than a sense of security. In both cases, I have to be aware of the words I am choosing, for a single word can affect tone.

GREETING Include a greeting two lines below the inside address. It is common to use the word *Dear*, followed by the name of the person to whom you are sending the letter. A comma or colon can follow the name, although in business correspondence a colon is preferred.

If you do not know the name of the person to whom you are sending the letter, choose a gender-neutral title like Human Resources Director, Production Manager, or Head Engineer. A generic greeting like *To Whom It May Concern* is inappropriate because it is too impersonal. With a little thought, you can usually come up with a gender-neutral title that better targets the reader of your letter.

Also, remember that it is no longer appropriate to use gender-biased terms like *Dear Sirs* or *Dear Gentlemen*. You will offend at least half the receivers of your letters with these kinds of gendered titles.

MESSAGE The message of your letter will make up its bulk. As discussed earlier in this chapter, it includes an introduction, body, and closing.

The message should begin two lines below the greeting. Today, most letters are set in *block format*, meaning the message is set against the left margin with no indentation. In block format, a space appears between each paragraph.

Formatting a Letter

GeneriTech
1032 Hunter Ave., Philadelphia, PA, 19130

March 29, 2004

George Falls, District Manager
Optechnical Instruments
875 Industrial Avenue, Suite 5
Starkville, New York 10034

Dear Mr. Falls:

I enjoyed meeting with you when we visited Starkville last week. I am writing to follow up on our conversation about building a wireless network for your business. A wireless network would certainly avoid many of the Internet access problems your company is experiencing in its current facility. It would also save you the wiring costs associated with a more traditional system.

With this letter, I have enclosed a variety of materials on GeneriTech's wireless systems. You will find that there is a range of options to suit your company's needs. I have tried to include only brochures and white papers that you would find useful. I have also included testimonials written by a few of our customers about their wireless systems and their experiences with our company.

I will call you in a couple of weeks to see if you have any questions about our wireless products. If you want answers sooner, please call me at 547-555-5282 or e-mail me at lhampton3712@generitechsystems.com. I can answer your questions.

Again, thank you for meeting with us last week. I look forward to speaking with you again.

Sincerely,

Lisa Hampton
Senior Engineer, Wireless Division

Figure 16.11: The format of a letter has predictable features, like the letterhead, inside address, greeting, message, and a closing with signature.

To download letter and memo templates, go to **www.ablongman.com/johnsonweb/16.10**

TAKE NOTE Other formats than block, such as like semiblock (in which the first line of each paragraph is indented), are available. But, block format is considered the standard in most technical workplaces.

CLOSING WITH SIGNATURE Two lines below the message, you should include a closing with a signature underneath. In most cases, the word *Sincerely,* followed by a comma, is preferred. Sometimes writers will be more creative and say *Best Wishes, Respectfully,* or *Cordially*. Avoid being cute with the closing. Phrases like *Your Next Employee* or *Respectfully Your Servant* are unnecessarily risky. Just use the word *Sincerely* in almost all cases.

Your signature should appear next, with your name and title typed beneath it. To save room for your signature, you should leave about three lines between the closing and your typed name.

Sincerely,

Lisa Hampton
Senior Engineer, Wireless Division

If you are sending the letter electronically, you can create an image of your signature with a scanner. Then, insert the image in your letter.

Formatting Envelopes

Once you have finished writing your letter, you will need to put it in an envelope. Fortunately, with computers, putting addresses on envelopes is not difficult. Your word-processing program can capture the addresses from your letter (Figure 16.12).

Formatting for an Envelope

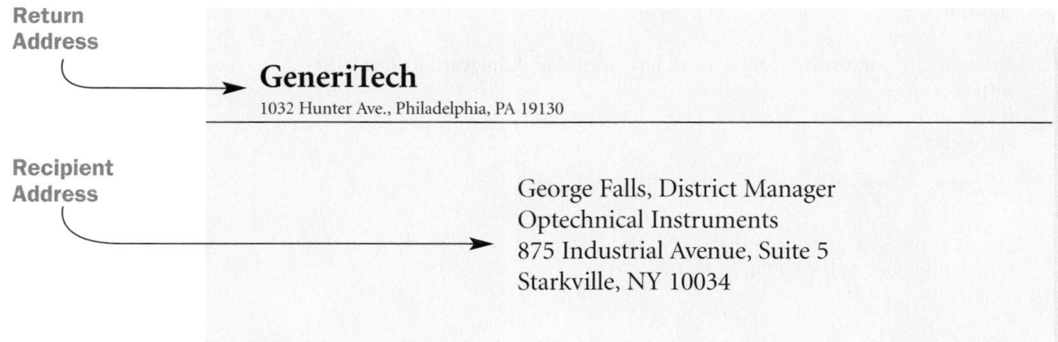

Figure 16.12: An envelope includes a return address and a recipient address.

Want more help with formatting your letter or memo? Go to **www.ablongman.com/johnsonweb/16.11**

Then, with the "Envelopes and Labels" function (or equivalent), you can have the word processor put the address on an envelope or label. Most printers can print envelopes.

An envelope should have two addresses, the *return address* and the *recipient address.* The return address is printed in the upper left-hand corner of the envelope, usually a couple lines from the top edge of the envelope. The recipient address is printed in the center of the envelope, usually about halfway down from the top edge of the envelope.

If your company has premade envelopes with the return address already printed on them, printing an envelope will be easier. You will only need to add the recipient address.

Formatting Memos

Memos are easier to format than letters because they include only a header and message.

HEADER Most companies have stationery available that follows a standard memo format (Figure 16.13). If memo stationery is not available, you can make your own by typing the following list:

Date:
To:
cc:
From:
Subject:

The *Date, To*, and *From* lines should include the date, who the memo is addressed to (the reader), and who it is from (you).

The *Subject* line should offer a descriptive and specific phrase that describes the content of the memo. Most readers will look at the subject line first to determine if they want to read the memo. If it is too general (e.g. "Project" or "FYI"), they may not read the memo. Instead, give them a more specific phrase like "Update on the TruFit Project" or "Accidental Spill on 2/2/04."

TAKE NOTE Instead of *Subject,* some writers and companies prefer the abbreviation *Re* (Regarding). Either *Subject* or *Re* is appropriate in memos.

The *cc* line (optional) includes names of any people who will receive copies of the memo. Often copies of memos are automatically sent to supervisors to keep them informed.

If possible, sign your initials next to your name on the *From* line. Since memos are not signed, these initials serve as your signature on the document.

MESSAGE Memos do not include a *Dear* line or any other kind of greeting. They just start out with the message. The block style (all lines set against the left margin and spaces between paragraphs) is preferred, though some writers indent the first line of each paragraph.

Longer memos should include headings to help readers identify the structure of the text. In some cases, you might choose to include graphics to support the written text.

It is important to note that memos do *not* include a closing or signature. When your conclusion is complete, the memo is complete. No closing or signature is needed.

LINK For more ideas about designing documents, go to Chapter 8, page 192.

Sample Memo

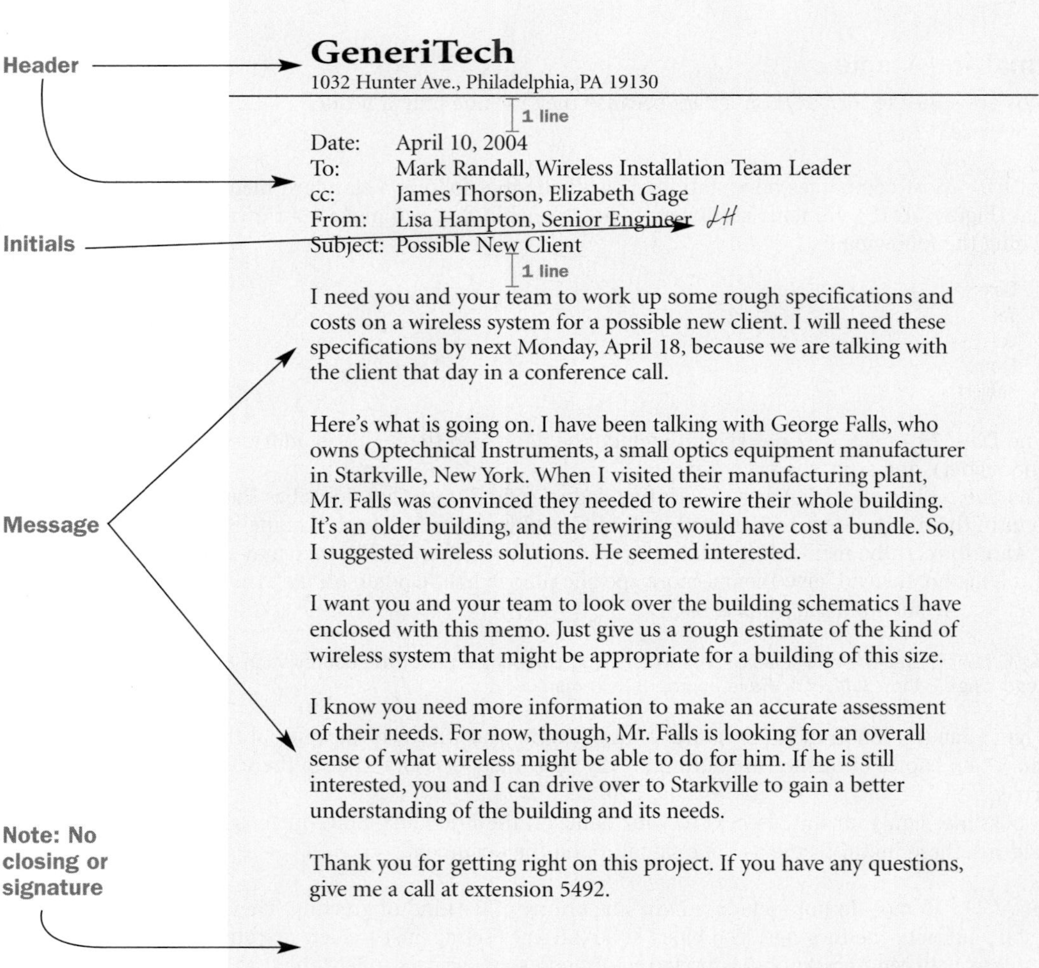

Header
Initials
Message
Note: No closing or signature
1 line
1 line
GeneriTech
1032 Hunter Ave., Philadelphia, PA 19130
Date: April 10, 2004
To: Mark Randall, Wireless Installation Team Leader
cc: James Thorson, Elizabeth Gage
From: Lisa Hampton, Senior Engineer LH
Subject: Possible New Client
I need you and your team to work up some rough specifications and costs on a wireless system for a possible new client. I will need these specifications by next Monday, April 18, because we are talking with the client that day in a conference call.
Here's what is going on. I have been talking with George Falls, who owns Optechnical Instruments, a small optics equipment manufacturer in Starkville, New York. When I visited their manufacturing plant, Mr. Falls was convinced they needed to rewire their whole building. It's an older building, and the rewiring would have cost a bundle. So, I suggested wireless solutions. He seemed interested.
I want you and your team to look over the building schematics I have enclosed with this memo. Just give us a rough estimate of the kind of wireless system that might be appropriate for a building of this size.
I know you need more information to make an accurate assessment of their needs. For now, though, Mr. Falls is looking for an overall sense of what wireless might be able to do for him. If he is still interested, you and I can drive over to Starkville to gain a better understanding of the building and its needs.
Thank you for getting right on this project. If you have any questions, give me a call at extension 5492.

Figure 16.13: The formatting of a memo uses a head that differs from a letterhead. Notice that there is no closing or signature.

HELP

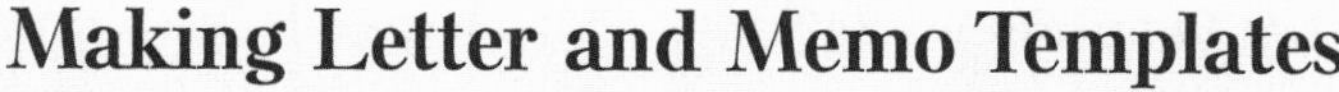

Making Letter and Memo Templates

Personal letters and memos looked rather bland before computers were available, because they used only minimal design features. Only companies could afford letterhead or memo headers.

Today, you can use the templates feature in your word-processing program to make your own headers for letters and memos. These templates come in especially handy when you are applying for jobs.

Using Existing Templates

Almost all word-processing software, like MS Word and Corel WordPerfect, includes a variety of letter and memo templates that you can immediately adapt to your own needs. When you select "New" or "New Document" in your word-processing program, you will be given a choice of different templates you can use (Figure A).

Using a Premade Template

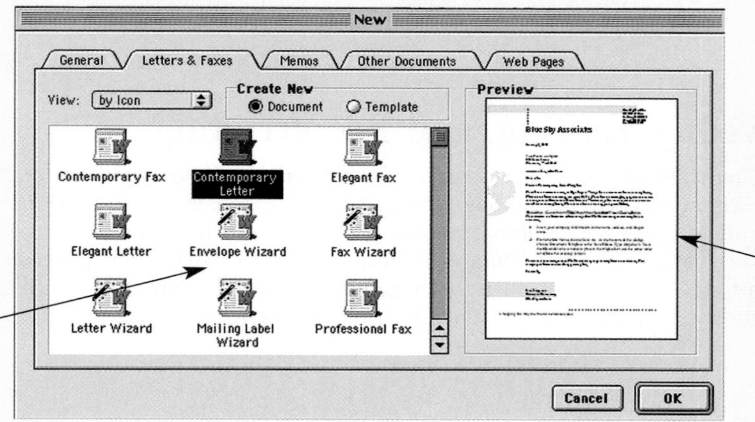

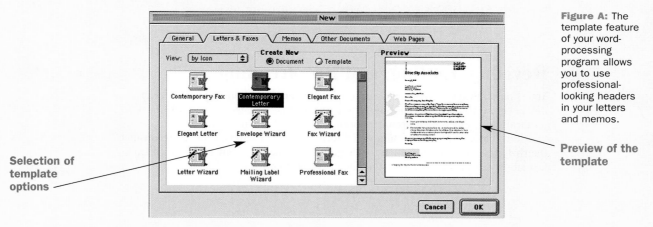

Selection of template options

Preview of the template

Figure A: The template feature of your word-processing program allows you to use professional-looking headers in your letters and memos.

At this point, you can select the template that most closely reflects the kind of document you need to write. The template will allow you to enter information such as an inside address or greeting. The computer will take care of the formatting of the document, though you should always check the format to make sure it looks appropriate after you have added your text.

Making Your Own Template

You can also make your own templates for letters and memos. In a blank document on your word processor, design a header that reflects your personality. You can add an icon or logo, while including your address (Figure B).

LINK To learn how to download graphics off the Internet, go to Chapter 9, page 234.

TAKE NOTE If you are using a graphic, even your college's logo, be sure to ask for permission to use it.

Want to make your own templates? Go to **www.ablongman.com/johnsonweb/16.12**

Personalized Letter Template

James Chosuke
1923 W. Lincolnway, Ames, IA 50010
515-555-2315

Figure B: You can quickly make up personalized templates for your letters and memos for work. For your personal use you can design a header that fits your personality.

When your template is complete, select "Save As." In the box that appears on your screen, you will find an option to save the file as a "template." Once you save the file as a template, the word-processing program will add it to the other templates, so you can use it in the future.

From then on, you can use your template whenever you want. These templates are especially handy when you are doing a job search and need to make numerous form letters.

Revising, Editing, and Proofreading

You should always leave some time for revising and proofreading your letter or memo before you send it out. Since these documents are one-on-one forms of communication, readers often become especially annoyed at smaller mistakes.

After you have finished drafting and formatting your letter or memo, reconsider the rhetorical situation in which the document will be used. Pay attention to any information or wording that might annoy or offend the readers.

Have I included only need-to-know information?

Is the purpose of the document stated or obvious in the introduction?

Does the document have a point?

Is it clear in the introduction what I want my readers to do?

How might secondary or tertiary readers use this information?

In what other contexts might the document be used?

Is the style appropriately formal or informal? Plain or persuasive?

Have I properly formatted the document?

If the message is especially important, let others look it over before you send it out. Your immediate supervisor would be a good person to ask for a critical review. In some cases, you may want to leave the document alone for a couple of hours. That way, you can revise it with a more critical eye.

LINK For more advice about proofreading, go to Chapter 10, page 274.

Revising and proofreading your work is very important. Since these letters and memos tend to be formal, you want them to represent your best work. Also, keep in mind that letters and memos are often filed, and they have a nasty habit of reappearing at unexpected moments (like performance reviews, court cases, or important meetings). You want them to reflect your true thoughts and your best work.

CHAPTER REVIEW

- Writing letters and memos can be a drain on your time. You should learn how to write them quickly and efficiently.
- Letters and memos are essentially the same kind of document, called a *correspondence*. They use different formats, but they tend to achieve the same purposes. Letters are used for messages that go outside the company. Memos are for internal messages.
- Letters and memos are always personal. They are intended to be a one-on-one communication to a reader, even if they are sent to more than one person.
- Letters and memos share the same basic features: header, introduction, informative body, and conclusion. They differ mostly in *format,* not content or purpose.
- The introduction should identify the subject, purpose, and point of the correspondence. If you want the readers to take action, put that action up front.
- The body of the correspondence should give the readers the information they need to take action or make a decision.
- The conclusion should thank the readers, restate your main point, and look to the future.
- To develop an appropriate style, use "you" style, create a tone, and avoid bureaucratic phrasing.
- Use standard formats for letters and memos. These formats will make the nature of your message easier to recognize.
- Be sure to revise your work as necessary and proofread it when you are finished.

EXERCISES AND PROJECTS

Individual or Team Projects

1. Write a memo to an instructor in which you request a letter of reference. Your memo should be polite, and it should offer suggestions about what the instructor should include in the letter of reference.

2. Find a sample letter or memo on the Internet. In a memo to your instructor, discuss why you believe the letter is effective or ineffective. Discuss how the content, organization, style, and design are effective/ineffective. Then, make suggestions for improvement. Sample letters and memos are available at www.ablongman.com/johnsonweb/16.2.

3. Think of something that bothers you about your college campus. To a named authority, write a letter in which you complain about this problem and discuss how it has inconvenienced you in some way. Offer some suggestions about how the problem might be remedied. Be sure to be tactful.

4. Imagine you are the college administrator who received the complaint letter in Exercise 3 above. Write a response letter to the complainant in which you offer a reasonable response to the writer's concerns. If the writer is asking for a change that requires a great amount of money, you may need to write a letter that refuses the request.

5. Create a letter template that you can use for your personal correspondence. It should include your name and address, as well as a symbol or logo that adds a sense of style and character to your correspondence.

Collaborative Project

With a group, choose three significantly different cultures that interest you. Then, research these cultures' different conventions, traditions, and expectations concerning letters and memos. You will find that correspondence conventions in countries such as Japan or Saudi Arabia are very different from those in the United States. The Japanese often find American correspondence to be blunt and rude. Arabs often find American correspondence to be bland (and rude, too).

Write a brief report, in memo form, to your class in which you compare and contrast these three different cultures' expectations in correspondence. In your memo, discuss some of the problems with not being aware of correspondence conventions in other countries. Then, offer some solutions that might help the others in your class become better intercultural communicators.

Present your findings to the class.

Revision Challenge

The memo shown in Figure A below needs to be revised before it is sent to its primary readers. Using the concepts and strategies discussed in this chapter, analyze the weaknesses of this document. Then, identify some ways it could be improved through revision.

Figure A: This memo needs some revision. How could it be improved?

ChemConcepts, LLC

Memorandum

Date: November 14, 2004
To: Laboratory Supervisors
cc: George Castillo, VP of Research and Development
From: Vicki Hampton, Safety Task Force
Re: FYI

It is the policy of the ChemConcepts to ensure the safety of its employees at all times. We are obligated to adhere to the policies of the State of Illinois Fire and Life Safety Codes as adopted by the Illinois State Fire Marshal's Office (ISFMO). The intent of these policies is to foster safe practices and work habits throughout companies in Illinois, thus reducing the risk of fire and the severity of fire if one should occur. The importance of chemical safety at our company does not need to be stated. Last year, we had four incidents of accidental chemical combustion in our laboratories. We needed to send three employees to the hospital due to the accidental combustion of chemicals stored or used in our laboratories. The injuries were minor and these employees have recovered; but without clear policies it is only a matter of time before a major accident occurs. If such an accident happens, we want to feel assured that all precautions were taken to avoid it, and that its effects were minimized through proper procedures to handle the situation.

In the laboratories of ChemConcepts, our employees work with various chemical compounds that are can cause fire or explosions if mishandled. For example, when stored near reducing materials, oxidizing agents such as peroxides, hydroperoxides and peroxyesters can react at ambient temperatures. These unstable oxidizing agents may initiate or promote combustion in materials around them. Of special concern are organic peroxides, the most hazardous chemicals handled in our laboratories. These compounds form extremely dangerous peroxides that can be highly combustible. We need to have clear policies that describe how these

Continued on following page

To download an electronic version of this revision challenge, go to **www.ablongman.com/johnsonweb/16.13**

Figure A:
(*continued.*)

kinds of chemicals should be stored and handled. We need policies regarding other chemicals, too. The problem in the past is that we have not had a consistent, comprehensive safety policy for storing and handling chemicals in our laboratories. The reasons for the lack of such a comprehensive policy are not clear. In the past, laboratories have been asked to develop their own policies, but our review of laboratory safety procedures shows that only four of our nine laboratories have written safety policies that specifically address chemicals. It is clear that we need a consistent safety policy that governs storage and handling of chemicals at all of our laboratories.

So, at a meeting on November 1, it was decided that ChemConcepts needs a consistent policy regarding the handling of chemical compounds, especially ones that are flammable or prone to combustion. Such a policy would describe in depth how chemicals should be stored and handled in the company's laboratories. It should also describe procedures for handling any hazardous spills, fires, or other emergencies due to chemicals. We are calling a mandatory meeting for November 30 from 1:00-5:00 in which issues of chemical safety will be discussed. The meeting will be attended by the various safety officers in the company, as well as George Castillo, VP of Research and Development. Before the meeting, please develop a draft policy for chemical safety for your laboratory. Make fifteen copies of your draft policy for distribution to others in the meeting. We will go over the policies from each laboratory, looking for consistencies. Then, merging these policies, we will draft a comprehensive policy that will be applicable throughout the corporation.

- What information in the memo goes beyond readers' need to know?
- How can the memo be reorganized to highlight its purpose and main point?
- What is the "action item" in the memo, and where should it appear?
- How can the style of the memo be improved to make the text easier to understand?
- How might design be used to improve the readers' understanding?

An electronic version of this memo are available at www.ablongman.com/johnsonweb/16.13. There, you can also download the file for revision.

CASE STUDY

The Nastygram

Shannon Phillips is the Testing Laboratory Supervisor at the Rosewood Medical Center. It is her laboratory's job to test samples, including blood samples, drawn from patients. After the samples are tested, one of her technicians returns a report on the sample to the doctor.

Last week, Shannon's laboratory was in the process of transferring over to a new facility. It was a complex move because the samples needed to be walked over on rolling carts to the new laboratory. There was no other way to make the transfer.

Then, the accident happened. One of the technicians who was pushing a cart of blood samples left it alone for a moment. The cart rolled down an incline, into the lobby, and turned over. Samples were strewn all over the floor.

The hazardous materials team cleaned up the mess, but the samples were all destroyed. In an e-mail, Shannon notified the doctors who ordered the tests that they would need to draw new samples from their patients. With her apologies, she tried her best to explain the situation and its remedy.

Most doctors were not pleased, but they understood the situation. However, Dr. Alice Keenan, the Director of the AIDS Center at the hospital, was quite angry about the accident. She wrote the memo shown in Figure A (next page) to Shannon.

Needless to say, Shannon was quite upset about the memo. She had done everything she could to avoid the accident, and now a doctor was going to ask the Board of Directors to fire her. At a minimum, Shannon might need to fire her technician, who had never made a mistake up to this point. She was also quite angry about the tone in the letter. It seemed so unprofessional.

If you were Shannon, how might you respond to this situation?

The Nastygram

Rosewood Medical Center

712 Hospital Drive, Omaha, Nebraska 68183

Memorandum

Date February 13, 2004
To: Shannon Phillips, Laboratory Supervisor
From: Dr. Alice Keenan, MD, Director of AIDS Center
cc: George Jones, CEO Rose Medical Center
Subject: Your Idiotic Blunder

I cannot believe you morons in the lab ruined the blood samples we sent from our patients. Do you realize how difficult it is to draw these samples in the first place? Our patients are already in great discomfort and in a weakened state. So, asking them for a second blood draw within only a few days is a big problem.

Not to mention the hazardous situation created by this idiotic blunder. Do you realize that someone in the lobby could have come into contact with these samples and perhaps contracted AIDS? Are you too obtuse to realize what kind of horrible consequences your mistake might have caused? You would have made us liable for huge damages if we were sued (which we would be).

Someone down there should be fired over this mistake! In my opinion, Shannon, you should be the first to go. I'm very concerned about your leadership abilities if you could allow something like this to happen. If I don't hear about a firing or your resignation in the next week, I'm going to bring a formal complaint to the Board of Directors at their next meeting.

Are you people all idiots down there? I've never heard of something this stupid happening at this hospital.

Figure A: Sometimes it is difficult to respond professionally to a nasty memo or letter like this one.

CHAPTER

17

Technical Definitions

CHAPTER CONTENTS

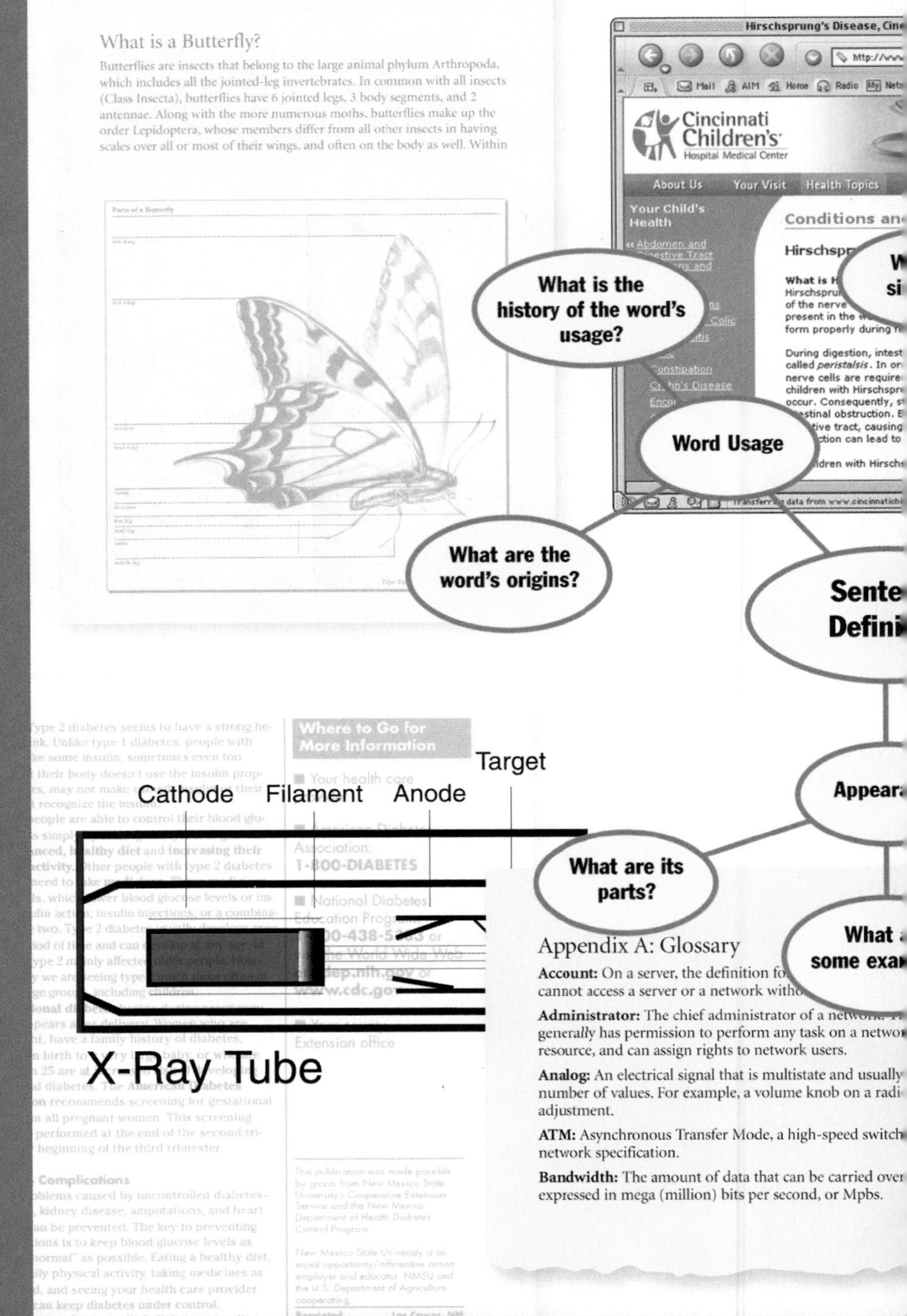

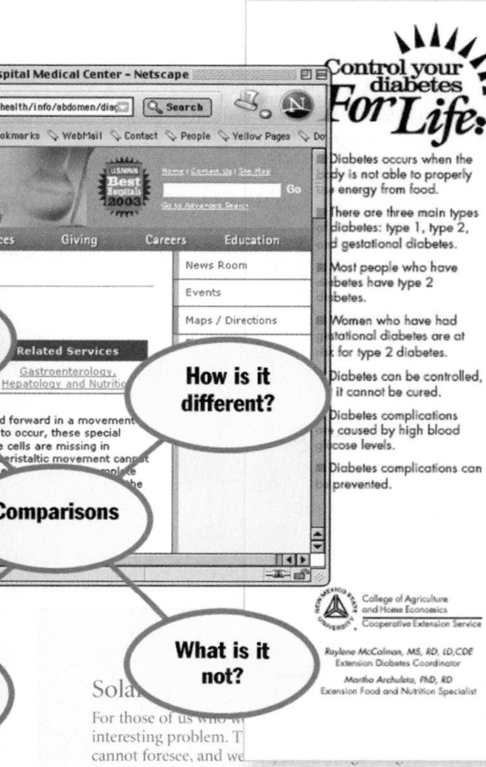

Control your diabetes For Life:

Circular 569 A

What is Diabetes?

Diabetes occurs when the body is not able to properly use energy from food.

There are three main types of diabetes: type 1, type 2, and gestational diabetes.

Most people who have diabetes have type 2 diabetes.

Women who have had gestational diabetes are at risk for type 2 diabetes.

Diabetes can be controlled, but it cannot be cured.

Diabetes complications are caused by high blood glucose levels.

Diabetes complications can be prevented.

Diabetes is a disease that occurs when the body is not able to properly use energy from food. The body needs **glucose** for growth and energy. Glucose is a kind of sugar we get from many of the foods we eat and is absorbed into the blood through the digestive system. Once in the blood, glucose needs to move into cells where it can be used for energy. A hormone called **insulin** acts like a key to open doors to cells, allowing glucose to enter. Diabetes occurs when either the **pancreas** doesn't make enough insulin, or the body isn't using insulin effectively. As a result, glucose stays in the blood. Over time, too much glucose in the blood can cause damage to tissues and result in diabetes complications, such as amputations, blindness, kidney disease, heart disease, and nerve and blood vessel damage.

Types of Diabetes

Type 1 diabetes occurs when the pancreas can no longer make insulin. Type 1 usually comes on abruptly and generally occurs in children and adults who are under age 30. Since the body is not capable of making insulin, people with type 1 diabetes need to take daily injections of insulin to stay alive. The causes of type 1 diabetes are not known. Scientists think that type 1 may have many causes: hereditary causes (when someone has parents or other family members with the disease) and causes both inside and outside the body, including viruses.

Type 2 diabetes, the most common form, affects about 90 to 95 percent of the people who have diabetes and seems to be related to being overweight and an inactive

College of Agriculture and Home Economics
Cooperative Extension Service

Raylene McCalman, MS, RD, LD, CDE
Extension Diabetes Coordinator

Martha Archuleta, PhD, RD
Extension Food and Nutrition Specialist

Sola...

For those of us who w... interesting problem. T... cannot foresee, and we... the delicate hardware that we send up into space.

2003/10/28 06:24 UT

Credit: SOHO, ESA/NASA

How does it look?

Thesaurus: English (US)

Looked Up: field

Meanings: province (noun), track (noun), acreage (noun), battlefield (noun), candidates (noun), occupation (noun)

Replace with Synonym: acreage

acreage, tillage, pasture, meadow, plot, garden, mead, grassland, moor

Replace | Look Up | Cancel | Previo

CHAPTER OBJECTIVES

In this chapter, you will learn:

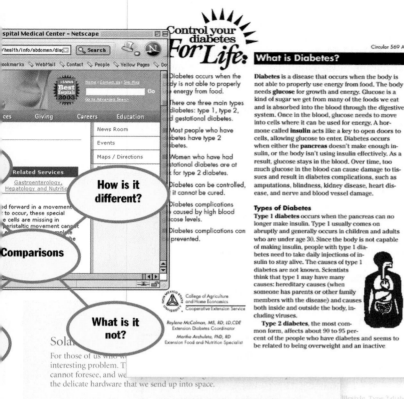

- How definitions are used in technical communication.
- How to write three kinds of definitions: parenthetical, sentence, and extended definitions.
- How to organize and draft each kind of definition.
- How to compose an extended definition.
- To use plain style to make descriptions clear and reliable.
- How to revise and edit definitions to ensure consistency.

Accurate technical definitions are essential in technical communication. Any new term or concept needs to be clearly defined before it can be used properly in technical documents. For example, think about all the new words that have come into the English language since the invention of the computer, like ***hypertext, software, Internet, information technology, bits, bytes, monitor, server,*** among numerous other new terms. Some of these words are completely new, like ***hypertext*** and ***bytes,*** taking on their own unique meanings. Other words, like ***server*** and ***hardware,*** have meanings adapted to the world of computers.

Each technical discipline has a vocabulary of specialized words that it uses.

- A mechanical engineer must be able to define words like *velocity, torque,* and *viscosity.*
- A medical professional needs to define words like *cerebellum, hepatitis, West Nile virus,* and *ankylosing spondyliti.*
- An anthropologist needs to understand the definitions of words like *flaking station, reflexivity,* and *karyotype.*

Knowing how to define the terms in your discipline is an important part of being a member of that discipline.

It is also important to recognize that different disciplines have their own distinct meanings for words. For example, consider the meanings of the word "field" in four different disciplines.

Discipline	Definition	Sample Usage
Physics	A region where a given effect (e.g., electricity, magnetism, or gravitation) exists	"The magnetic field influences the direction of the compass."
Medicine	An area of professional expertise	"He is an expert in the skin cancer field."
Geology	An area of study outside the office or laboratory	"She confirmed her theory with observations in the field."
Agronomy	An open area used for farming or pasture	"They let the cattle graze in the field."

Fortunately, computers and the Internet have made defining terms easier. At your fingertips are countless resources, like on-line dictionaries and glossaries, which can help you find precise definitions for the terms you use in your documents and presentations.

For websites that offer definitions of computer terms, go to **www.ablongman.com/johnsonweb/17.1**

But you will also occasionally need to write your own definitions for new terms or technical concepts. You will need to write accurate and precise definitions in various documents like specifications, reports, and any kind of description. You may also need to define the specialized terms of your field for nonexperts, who need to understand what you are talking about. Technical words are a natural part of a discipline's vocabulary, so they are unavoidable. Defining these terms is crucial to effective technical communication.

Types of Technical Definitions

Technical definitions are usually embedded within larger documents, but some are meant to stand alone. There are three types of definitions: *parenthetical, sentence,* and *extended* definitions.

Parenthetical definitions—To clarify the meaning of a term, you can use a parenthetical definition to provide an additional defining word or phrase.

Parenthetical definitions

A butterfly's thorax (its body) has three segments that bear four wings and six legs.

In 1929, Earnest Lawrence developed the first workable design for a cyclotron, a device that accelerates protons to high energies, helping scientists better explore the strange universe of subatomic particles.

As you can see in the first example above, sometimes a parenthetical definition appears in parentheses. Sometimes, as shown in the second example, the definition is set off with commas or dashes.

Sentence definitions—These definitions are whole sentences in which a term is defined by naming its "category" and the "distinguishing features" that differentiate it from its category.

Category

An ion is an atom that has a positive charge because one or more electrons have been stripped from it.

Distinguishing characteristics

Sentence definitions are most commonly found in dictionaries and glossaries. However, they are often also embedded in documents when a word or concept needs to be defined in exact terms.

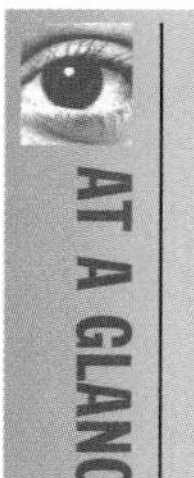

Three Types of Definitions

- Parenthetical definition—a word or phrase
- Sentence definition—a full sentence
- Extended definition—more than a sentence

Extended definitions—Extended definitions can fill a small paragraph or even run as long as several pages. In extended definitions, you can define complex terms very precisely by using techniques such as analogies, comparisons, contrasts, examples, negation, and graphics. Figure 17.1, for example, shows an extended definition of a butterfly from the *National Audubon Society Field Guide to North American Butterflies.*

As you will notice in this extended definition of butterflies, many words might be further defined. In this way, a definition can be extended indefinitely.

For websites with quick sentence definitions, go to **www.ablongman.com/johnsonweb/17.2**
To see samples of extended definitions, go to **www.ablongman.com/johnsonweb/17.3**

An Extended Definition

Figure 17.1: An extended definition starts with a sentence definition. Then it uses a variety of rhetorical tools to expand and sharpen the meaning of the term.

What Is a Butterfly?

Sentence definition → Butterflies are insects that belong to the large animal phylum Arthropoda, which includes all the jointed-leg invertebrates. In common with all insects **Comparisons** → (Class Insecta), butterflies have 6 jointed legs, 3 body segments, and 2 antennae. Along with the more numerous moths, butterflies make up the order Lepidoptera, whose members differ from all other insects in having scales over all or most of their wings, and often on the body as well. Within

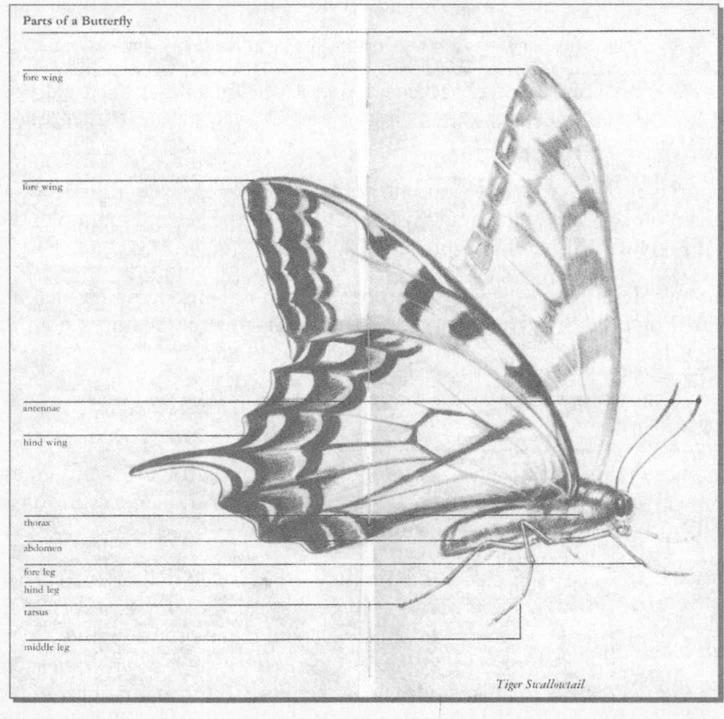

Source: M. Pyle, National Audubon Society Field Guide to North American Butterflies, *1981, pp. 14–16.*

Figure 17.1: *(continued.)*

Contrasts

Analogies

Division

Examples

the order, butterflies and moths are separated from each other by their wing venation, body structure, and habits. Butterflies fly during the day, while most moths are nocturnal. Butterflies at rest tend to hold the wings vertically over the back; in contrast, moths may either fold the wings tentlike over the back, or wrap them around the body, or extend them to the sides. Virtually all butterflies have knoblike clubs at the tips of their antennae; moths lack antennal clubs.

Butterflies are divided into two superfamilies: the true butterflies (Papilionoidea) and the skippers (Hesperioidea). True butterflies tend to have narrow bodies, long antennae, and brightly colored, full wings. Skippers are stocky, compact, and hairy: their short triangular wings are most often a shade of tawny-orange, brown, black, or gray. Butterflies are further broken down into families, genera, and species. The families comprise large groups of butterflies with a number of common characteristics. For example, both hairstreaks and elfins belong to the family of Gossamer Wings. In North America, there are 10 families. Genera are smaller groups of structurally related butterflies; the cloudywing skippers thus all belong to the genus Thorybes. Within each genus are various species, such as the Tiger Swallowtail or the Mourning Cloak. Members of a species breed with one another to produce similar offspring: there is very little hybridizing between species in nature. The species are further broken down into numerous subspecies, which are more or less distinct geographic populations. For example, the coloring of the Callippe Fritillary varies considerably among populations: the typical butterfly is yellow-brown, but the population in northern California and southern Oregon is orange-brown, while that found in the California Coast Ranges is dark brown with pale yellow spots. In this guide, we have included only those subspecies that are exceptionally distinctive.

Planning and Researching Technical Definitions

Usually, writing parenthetical and sentence definitions does not require a great amount of preparation. However, if you need to write an extended definition you will need to spend some time doing research and collecting information.

Planning

Whatever kind of definition you are writing, you should start out by gaining a full understanding of the term you are defining and the context in which that term is being used. A good way to start defining a term is to ask the Five-W and How Questions.

Who needs this definition, and what is their familiarity with terms in this field?

What amount of detail will be needed to accurately define this term for these readers?

Where will this definition be used in the document and elsewhere?

When will this definition be used?

Why is this definition needed?

How might this definition be used in the document and elsewhere?

Your answers to these questions will help you decide what kinds of information you should include in your definition. Expert readers in your own discipline probably don't need an extensive definition of a term. They simply want to know how you are using the term in your document. Nonexpert readers will probably need a more detailed definition.

Of course, there are exceptions. Nonexpert readers would quickly grow tired of a highly detailed definition of a common word like "butterfly." They may not need all the technical details. Expert readers, on the other hand, may need a highly detailed definition of "butterfly," because there are many insects that are called butterflies (skippers, longtails, moths) that are not true butterflies. An expert might want to know exactly what you mean by "butterfly."

Once you have briefly answered the Five-W and How questions, you are ready to start thinking about the subject, purpose, readers, and context of use for your definition.

TAKE NOTE If your definition will be part of a larger document, you can simply use your decisions about that document's purpose, readers, and context to guide your writing of the definition.

If your definition needs to stand alone, you should probably write down some notes that clarify the rhetorical situation in which your definition will be used.

SUBJECT Think about what kinds of information are needed to write an accurate definition of this term. What information does your reader need to know to understand the term? What information is not within the scope of your subject?

Need help defining the rhetorical situation? Go to
www.ablongman.com/johnsonweb/17.4

PURPOSE Write down *why* you are defining this term. Are you offering a new definition of this term? Are you trying to distinguish your definition of the term from other definitions? Are you striving for a certain level of accuracy? Your purpose statement might say something like

> My purpose is to define the word "butterfly" in this report, so that my readers will understand the differences between true butterflies (Papilionoidea) and other winged insects that are mistakenly called butterflies.

LINK For more information on defining your purpose, go to Chapter 2, page 23.

READERS When considering your readers, your first task should be to assess their level of expertise and what kinds of information they will need. After all, experts and nonexperts require different kinds of information to take action or make a decision.

Primary readers (action takers)—These readers need to know enough information to make informed choices about your subject. You need to gauge how much information they require, giving them enough to understand the term without overwhelming them with unnecessary details.

Secondary readers (advisors)—These readers may be experts in your field who are advising the primary readers. They will mainly be looking for accuracy in your definition.

Tertiary readers (evaluators)—Your tertiary readers might be just about anyone else who is interested in your document. They might include lawyers, auditors, accountants, or reporters who will be paying close attention to how you define your terms.

Gatekeepers (supervisors)—Your supervisors will be concerned about how you define any terms. They, too, will be looking for accuracy and clarity. Of course, your readers will all be concerned about the accuracy of your definitions. The challenge is to give your primary readers the right amount of information while also keeping the needs of these other readers in mind.

LINK For more reader analysis strategies, go to Chapter 3, page 55.

CONTEXT OF USE Definitions are used in a variety of different places and times. As you consider the context of use for your definition, think about the places where your readers might use or need your definition. Think about where it should appear in the text. Should it be early in the text? Should it appear in the appendix?

Definitions might seem rather cut-and-dried, but there are instances where economics, politics, and ethics can shape or influence how terms are defined. For example, politicians may define a word like "taxes" very differently. A "tax" to one politician may be a "user fee" to another.

LINK For more ideas about defining the context of use, go to Chapter 3, page 53.

Again, defining the rhetorical situation in great depth might seem like overkill for parenthetical and sentence definitions. But there are situations, especially with extended definitions, where this kind of thorough analysis will help you anticipate the

To find a dictionary specific to your field, go to **www.ablongman.com/johnsonweb/17.5**

needs of your readers and the contexts in which your definitions will be used. The additional effort is often worth the time spent.

Researching

When researching your definition, you should draw information from on-line, print, and empirical sources. Chapter 5 offers a full discussion of research methods. Here are a few strategies that are especially applicable to writing definitions.

Do background research—Start out by using print and on-line dictionaries to gather existing definitions. Locate information about the origins and history of the word. The "Help" section in this chapter describes how to use on-line sources to find information about words.

Find examples of usage—From a variety of sources, gather sentences in which the word is used. Also, take notes on how people use the word in everyday usage.

Compare and contrast—Identify similarities and differences between your subject and other things. You can make direct comparisons and note contrasts between similar things (e.g., "Keller Hall is larger and more modern than Ortega Hall."). Or, you can compare and contrast dissimilar things (e.g., "The upright granite rocks in this valley stand tall, unlike the smaller sandstone rocks found further down the river.").

Collect visuals—When possible, collect pictures and illustrations of your subject. You can also make your own visuals, using a camera, scanner, or drawings you made yourself.

LINK For more information about doing research, turn to Chapter 5, page 96.

The key to researching a definition is not to solely rely on the definitions you find in dictionaries. Dictionary definitions are rather generic and static. They rarely capture the full evolution and usage of a word, especially in technical disciplines. Instead, do research of your own to gain a fuller understanding of the word you are trying to define.

Organizing and Drafting Technical Definitions

When defining a term, you first need to gain a thorough understanding of it and the contexts in which it is used. In some situations, you may already have a rather firm understanding of the concept but cannot offer a clear definition. For example, almost anyone knows what an *atom* is, but few people would be able to offer a clear definition. So, as you begin organizing and drafting, think about the kinds of information you need to properly define your term.

Parenthetical Definitions

Parenthetical definitions use a word or phrase to define a term. The easiest way to come up with a parenthetical definition is to look in a dictionary, glossary, or thesaurus for a synonym (a different word that has almost the same meaning).

Traditional paper-based dictionaries are helpful at this point, but there are also many dictionaries and glossaries on the Internet, like Merriam Webster on-line (www.m-w.com), Dictionary.com, the Oxford English Dictionary (www.oed.com), and Onelook.com. Moreover, each field or discipline usually has an on-line dictionary or glossary that you can use. You can find them with an Internet search engine.

In these dictionaries, locate the word you are seeking to define. Then, use a word or phrase from that definition to define your term:

> If the circuit board is heated beyond 500°C, delamination (layer separation) will likely occur.

> In disbelief, we realized Kudzu (*Pueraria lobata*) was growing in this arid environment.

> We determined that the patient had gastritis, an inflammation of the mucous membrane of the stomach.

> The ketch (i.e., a multimasted sailboat) moved slowly out to sea.

TAKE NOTE This last example uses the Latin abbreviation, "i.e.," which means "that is." Sometimes using i.e. is a good way to signal that you are defining the term.

Avoid using words or phrases that are equally unfamiliar to your readers to define a term.

> Kenaf (East Indian hibiscus) is commonly used here for a variety of purposes.

> We will need to develop gateways—application-specific interfaces that link all seven layers of the OSI Model—when they are dissimilar at any or all levels.

Nonexperts might not know the meaning of "hibiscus" or "application-specific interface." As a result, these parenthetical definitions would not be helpful.

Sentence Definitions

Sentence definitions can appear in a variety of places in a document.

Body text—Occasionally, you will need to use a word that might be unfamiliar to your readers. In these situations, the sentence that follows the use of the word should be a definition.

> J. J. Thompson, using cathode rays, developed the concept of an electron. Today, an electron is defined as an elementary particle with a negative charge of about 1.6×10^{-19} coulomb.

Glossary—In highly technical documents, especially ones written for nonexperts, a glossary of terms can be placed toward the end of the document, usually in an appendix (Figure 17.2). A glossary contains terms that readers may want to refer to as they are reading the document.

Margin notes—In some documents, you might choose to put sentence definitions in the margins to help readers understand the meanings of technical terms.

Sentence definitions follow a rather standard three-part pattern, which includes (1) the item being defined, (2) the category of similar things to which the item

For samples of glossaries, go to **www.ablongman.com/johnsonweb/17.6**

belongs, and (3) any distinguishing features that separate this item from other items in its category.

A Glossary

Figure 17.2: A glossary is a helpful tool for looking up the words used in a document. They usually appear in an appendix.

Appendix A: Glossary

Account: On a server, the definition for a user of the server's services. A user cannot access a server or a network without a valid account.

Administrator: The chief administrator of a network. The administrator generally has permission to perform any task on a network and access any resource, and can assign rights to network users.

Analog: An electrical signal that is multistate and usually has an infinite number of values. For example, a volume knob on a radio is usually an analog adjustment.

ATM: Asynchronous Transfer Mode, a high-speed switched an multiplexed network specification.

Bandwidth: The amount of data that can be carried over a network, usually expressed in mega (million) bits per second, or Mpbs.

Source: Excerpted from Hallberg, Networking: A Beginner's Guide, *2003.*

Term	Category	Distinguishing Features
The chromosphere	is a region of the sun's atmosphere	that is close to the sun, thousands of miles thick, and composed chiefly of hydrogen gas.
Oxacillin is	a semisynthetic penicillin	that is used to control infections caused by penicillin-resistant staphylococci.
A solstice	is a time of the year	when the sun, having reached the tropical points, is farthest from the equator and appears to stand still; i.e., about 21st June (the summer solstice) and 22nd December (the winter solstice).

GO TO THE NET

Need more help writing a sentence definition? Go to **www.ablongman.com/johnsonweb/17.7**

Often the category is the hardest to name. The category should be broad enough to contain at least a few different items, but it should not be too broad, making it unhelpful to the readers.

You might want to use a diagram like the one in Figure 17.3 to help you sort out these three parts of the sentence definition. Put the name of the item in the smallest quarter circle. Then, in the middle area, note the possible categories into which the item might be classified. In the largest area, write down the features that distinguish the item from other similar items in these categories.

Diagram for Sentence Definitions

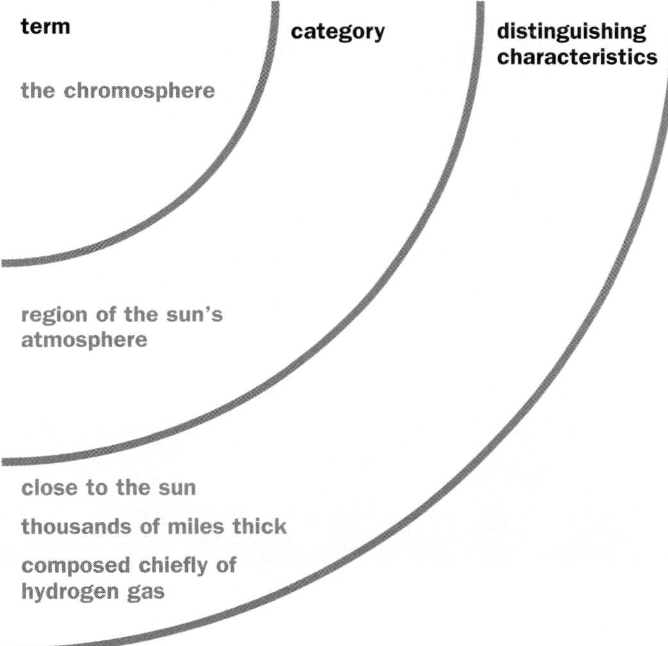

Figure 17.3: A diagram like this one can help you write a good sentence definition.

Again, dictionaries, available both on-line and in print, are a good place to start developing a sentence definition. Using dictionaries, you can compare a few different definitions and then devise your own. Remember, though, if you copy a definition directly from a dictionary, you should cite the dictionary as a source. For example, the following is a cited sentence definition used in a text:

> According to *Webster's Ninth New Collegiate Dictionary,* a gene is "an element of the germ plasm having a specific function in inheritance that is determined by a specific sequence of purine and pyrimidine bases in DNA or RNA and that serves to control the transmission of a hereditary character by specifying the structure of a particular protein or by controlling the function of other genetic material" (510).

Here, the page number is included so readers can check the dictionary as a source.

This diagram can be downloaded at **www.ablongman.com/johnsonweb/17.8**

Using On-line Sources for Definitions

The Internet is an amazing source for definitions. Your word processor also includes a thesaurus that you can use to help you write definitions.

On-line Dictionaries

Numerous dictionaries are available on the Internet. A favorite on-line dictionary is the Oxford English Dictionary (OED), which has been an ongoing project in print form since 1857. This dictionary identifies the original usage of English words and keeps track of new words. The OED also keeps tabs on English words used throughout the world, including North America, Australia, South Africa, India, and New Zealand, among other countries. Currently, it defines over half a million words.

The print version of the OED is a massive 20 volumes, but you can now look up definitions at the OED online through the Internet if your university or company is a subscriber. To check if you have access, go to the website (www.oed.com) and see if it allows you to enter the site. If you can't, you might want to ask your librarian how you can access the on-line version.

If you can access the OED, go to the website and click on the link "Enter OED Online." You will then be able to enter any English word in the website's search engine. Hit Return, and the definition will appear on your screen, along with a wealth of historical information on how the word has been used (Figure A).

OED Online

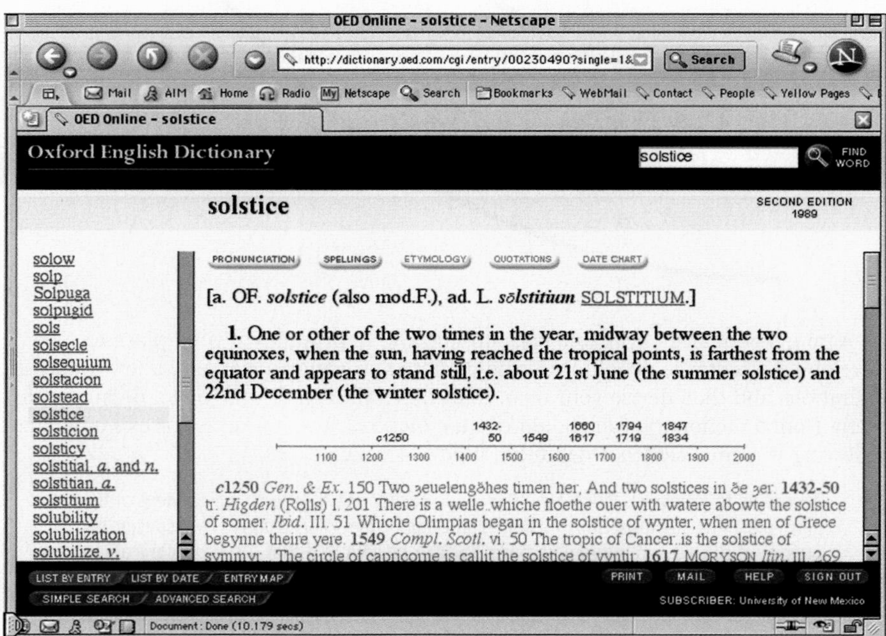

Figure A: It is simple and fun to use the on-line OED. Type the word in the box in the upper right-hand corner, and the dictionary will find the definition for you.

Source: Oxford English Dictionary Online, http://www.oed.com.

Want to see some on-line dictionaries, including the OED? Go to **www.ablongman.com/johnsonweb/17.10**

In Figure A, for example, the word "solstice" was entered. Initially, as shown, the dictionary will offer you the definition and the history of the word. The buttons at the top of the page (Pronunciation, Spellings, Etymology, Quotations, and Date Chart) allow you to gather further information.

One limitation to the OED is its conservative nature. You probably won't find the latest technical terms or slang in this dictionary. Before adding a new word, the OED typically waits for a word to become a regular part of the English language. Other on-line dictionaries are available if you want definitions of current or trendy words.

On-line Technical Dictionaries and Glossaries

As discussed earlier in this chapter, technical disciplines often use words in specific and unique ways. To help you define these words, you might locate an on-line *technical* dictionary or glossary. Most major disciplines and many subdisciplines have created dictionaries and glossaries to regulate the use of specialized words.

To find a technical dictionary or glossary in your discipline, you can use an Internet search engine. Type in "dictionary" and the name of the discipline (e.g., "computer science," "physics," "chemical engineering"). More than likely, the search engine will find one or more technical dictionaries and glossaries. At that point, you can find the definition you are looking for.

On-line Thesauruses

When defining a word, don't forget the thesaurus that comes with your word processor (Figure B). This thesaurus can offer a cluster of synonyms that are associated with the word you are trying to define.

A Thesaurus in a Word Processor

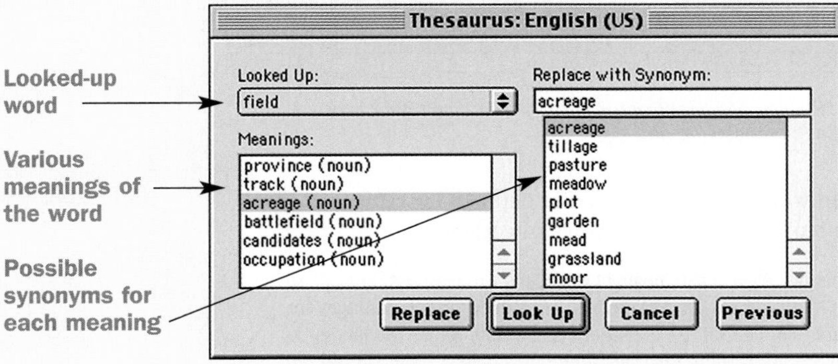

Figure B: The thesaurus in your word processor is a great source for synonyms to help define a word.

To use your word processor's thesaurus, type in the word you need to define. Then, select the **Thesaurus** command. A window like the one in Figure B will appear, displaying a cluster of synonyms that are most closely associated with the word you are defining.

Extended Definitions

Extended definitions are usually used for terms that need to be explained with utmost precision. These kinds of definitions are often found in larger documents like reports and websites, where they are used to explain a term thoroughly. They are also commonly found in guides and handbooks that are devoted to specific disciplines.

For example, Figure 17.4 shows an extended definition of a rare medical condition.

Extended Definition

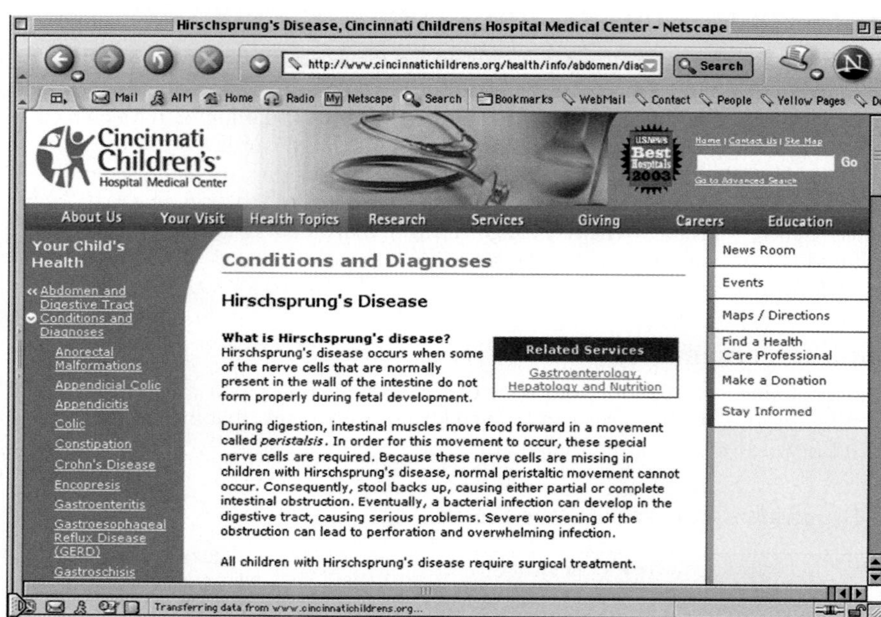

Figure 17.4: Extended definitions are commonly found on professional websites.

Source: Cincinnati Children's Hospital Medical Center, http://www.cincinnatichildrens.org/health/info/abdomen/diagnose/hirschsprung.htm.

In other cases, a common word needs to be exactly defined. Lawyers and insurance agents, for example, often need exact definitions of common terms.

> What is "negligence"? In its broadest sense, negligence means carelessness. But not all carelessness will result in a legal claim for negligence. To recover money damages for negligence, the plaintiff must prove that the defendant had a duty under the law to be careful, that the defendant wasn't careful, that the defendant's carelessness caused the plaintiff's injury, and that the plaintiff suffered damages, like medical expenses and pain and suffering, because of the defendant's carelessness.
>
> *Source:* U.S. Public Health Service, http://oep.osophs.dhhs.gov/dmat/training/phsfr/cha02/12cha02.htm.

As these examples show, most extended definitions start out with a sentence definition. Then, you can use a variety of techniques to extend the definition.

TAKE NOTE Using an Internet search engine, pick almost any subject. Then, type in "What is X?" replacing X with your subject (don't forget to use the quotation marks). More than likely, the search engine will find several webpages with extended definitions of your subject.

Extending a Definition

Drafting an extended definition is like drafting any other document. You are directly or indirectly claiming that a term can be defined a particular way. Then, you will need to offer support to back up that claim.

To help you start extending a definition, you might try using logical mapping to explore the many ways in which the word could be used or defined. As shown in Figure 17.5, start out with your sentence definition in the middle of the screen or a sheet of paper. Then, use logical mapping to explore all the different ways you might define the term.

Extending a Sentence Definition

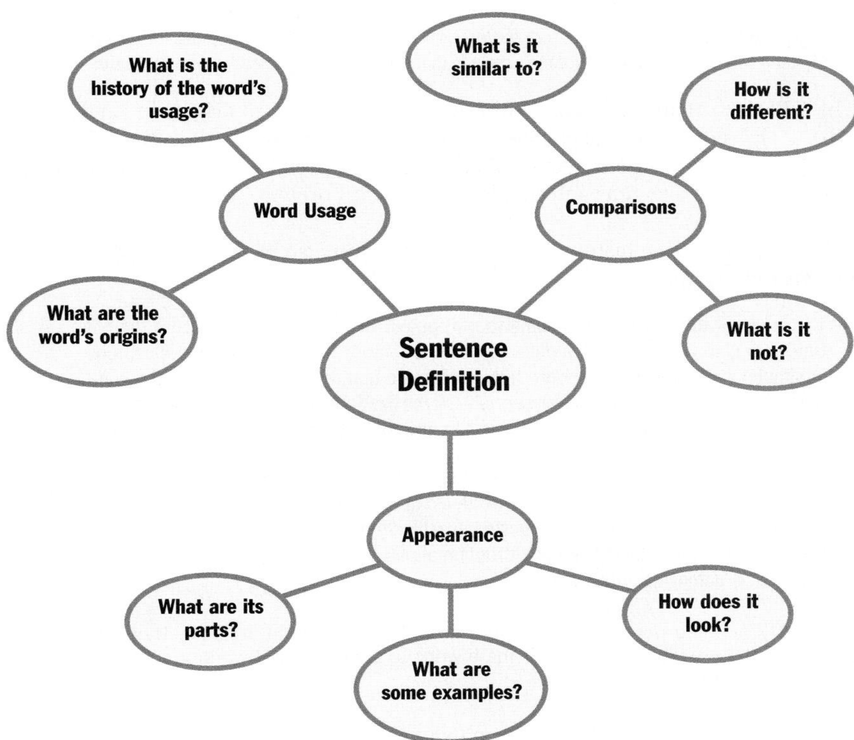

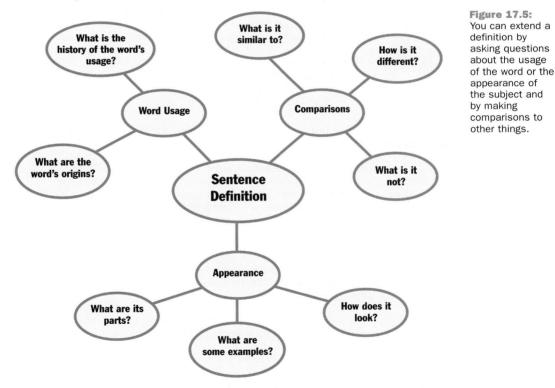

Figure 17.5: You can extend a definition by asking questions about the usage of the word or the appearance of the subject and by making comparisons to other things.

LINK For more help using logical mapping to develop new ideas, go to Chapter 5, page 103.

As your logical map will show, there are many different perspectives from which you might define a word.

Would you like to have a word of the day e-mailed to you automatically? Go to **www.ablongman.com/johnsonweb/17.11**

Once you have used your logical map to identify the possible perspective on the word, you can extend a definition discussing or by using some of the following techniques:

- Word origin
- Word history
- Examples
- Negation
- Division into parts
- Similarities and differences
- Analogy
- Graphics

WORD ORIGIN (ETYMOLOGY) A term's origin can often provide some interesting insights into its meaning.

> According to the *Oxford English Dictionary,* the word *cricket* comes from variations of the French word *criquer,* "to creak, rattle, crackle."
>
> The word *hysterical* includes the root *hyster-,* which comes from the Greek word for womb. Hysteria was believed to be a women's malady that was related to a disorder with the uterus.

The *OED* is probably the most authoritative source of word origins in English (see the "Help" sidebar in this chapter). The *OED* traces words back to their first print usage. But you can find technical dictionaries, especially ones specific to your discipline, that will also offer information about word origins.

WORD HISTORY The history of a word's usage can also offer interesting insights into its current meaning.

> The origin of the word *geek* is somewhat disputed. Geek actually comes from the old Dutch/German word, *geck,* that dates to the 16th century. It originally meant dupe, fool, or simpleton. However, some word historians argue that the American use of the word originally referred to carnival performers who bit the heads off animals, like snakes or chickens. This meaning gives a whole new sense to what it means to be a geek or geeked up.
>
> The word *sabotage* first meant making noise or destroying things with shoes (*sabot* means shoe in French). For example, an audience would disrupt musical performances by making noise with *sabots,* or workers would destroy machinery by striking it with their shoes. Since then, the word has evolved to mean disruption of military operations or manufacturing by deliberately damaging equipment.

TAKE NOTE Some word origins located on the Internet are not reliable. They are more urban myth than reality. So, before using a website's version of a word origin, try to confirm it with other sources.

EXAMPLES You might also include an example to clarify a term's usage. The example usually offers a specific usage of the word in a context familiar to readers. The following uses an example to define computer virus:

> Perhaps playing on instinctive fears of illness, the concept of a computer "virus" sends shudders through our society. For instance, in 1992 the media whipped up a frenzy about the Michelangelo virus, which was supposed to wipe out up to a quarter of American com-

For websites that offer word origins, go to **www.ablongman.com/johnsonweb/17.12**
For technical dictionaries that offer word histories, go to **www.ablongman.com/johnsonweb/17.13**

puter hard drives on March 6, the famed artist's birthday. Companies spent millions of dollars and valuable time preparing for this attack. In the end only about 10,000 computers were affected, a relatively small number.

NEGATION Often, you can define something by showing what it is not. Health officials, especially, seem to focus on defining terms by telling people what a disease is not.

According to the Saskatchewan Government Health website, HIV (human immunodeficiency virus) cannot be transmitted through casual contact with other people, like hugging, shaking hands, or eating meals prepared by people infected with HIV. It is not transmitted through sweat, saliva, or tears, so you cannot get the virus from computer keyboards, public phones, swimming pools, water fountains or other similar places (for more details, see http://www.health.gov.sk.ca/rr_aids_hiv_anonytest.html).

St John's wort is not a stimulant, and it won't cure all kinds of depression. Rather, it is a mild sedative. This is why St. John's wort has been found useful for treating patients suffering mild depression with high levels of anxiety.

DIVISION INTO PARTS There is a fine line between defining something and describing it. In your extended definition, you might describe your subject by breaking it down into parts.

According to the American Orchid Society, the orchid family is a group of flowering plants that has many different species and hybrids. Nevertheless, all orchids have some characteristics in common. Let's consider the common characteristics in the orchid family's flowers, leaves, and roots.

Flowers: The flowers of all orchids follow a similar pattern. There is a drum-shaped container, called the whorl, with three leaf-like sepals. Within the sepals is another whorl that contains three petals. One of these petals, called the labellum or lip, is different from the other two in color or size. The center of the flower includes the column, where the stamens and pistil are fused together.

Leaves: The leaves of an orchid can be folded or pleated, and they can be pliant or tough. The major veins of an orchid are usually parallel to each other.

Roots: The roots are typically adapted to the environment in which the plant grows. Most orchids have roots that protect interior tissues from water loss.

Source: The American Orchid Society, http://www.theaos.org.

LINK For more help dividing things into parts, go to Chapter 18, page 526.

SIMILARITIES AND DIFFERENCES You can compare your subject to things that are similar, showing their common characteristics and their differences.

Orangutans are similar to the other great apes (gorillas, chimpanzees, and bonobos) in several ways. Like other apes they primarily eat fruit and plants, though they will eat meat occasionally. They have prehensile hands and feet. They can stand but are most comfortable walking on all fours. Orangutans are different from other apes because they tend to be solitary animals. Also, their distinctive reddish color makes them stand out. Other apes tend to be black in color.

LINK To learn more about using comparison, go to Chapter 6, page 145.

For more examples of these techniques, go to **www.ablongman.com/johnsonweb/17.14**

ANALOGY It is helpful to compare your subject to something completely different but with some similar qualities.

> Imagine your body is a city. It has a circular system of arteries and veins that are like roads in a city. These arteries and veins contain white blood cells, which act like police patrolling for viruses and bacteria. The brain is the government, directing the city where it needs to go. And yes, your body has a mouth for importing the things it needs and its own sewage system for releasing things when it has used them.

Just about anything can be compared to anything else, some things with more success than others.

LINK To use analogies effectively, go to Chapter 7, page 182.

GRAPHICS A drawing, picture, diagram, or other kind of graphic is especially useful for describing something. The graphic should not stand alone. Rather, it should be used to support the written text. For example, the drawing of an X-ray tube in Figure 17.6 should not stand alone. Instead, it should support written text that offers further information on all these parts.

Using a Graphic in a Definition

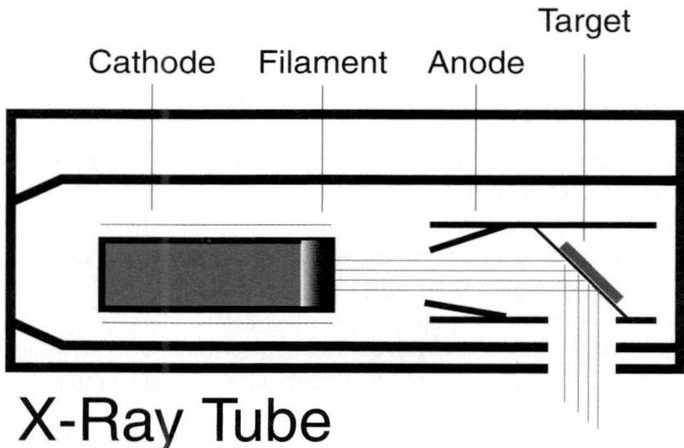

Figure 17.6: A graphic can be used to support the written text, but a graphic alone cannot replace the written text.

Any graphic used to define a term should label the parts of the subject. Make sure the labels on the graphic correspond directly to the terms you use in the written text.

LINK For more ideas on how to use graphics, go to Chapter 9, page 236.

In an extended definition, you can use a variety of techniques to define a term. For example, the definition of methamphetamine in Figure 17.7 uses several of the strategies discussed in this chapter.

GO TO THE NET

Interested in analogies? Go to
www.ablongman.com/johnsonweb/17.15

An Extended Definition

Figure 17.7: An extended definition uses a variety of techniques to explain the meaning of a term. In this definition of the word 'methamphetamine,' the author uses examples, shows similarities and differences, and includes an analogy.

Sentence definition →

Similarities and differences →

Analogy →

What Is Meth?

Meth (known on the street as "speed," "meth," "crank," "crystal meth," and "glass") is a central nervous system stimulant of the amphetamine family and has immense abuse potential. Like cocaine, methamphetamine is a powerful "upper" that produces alertness and elation, along with a variety of adverse reactions. The effects of methamphetamine, however, are much longer lasting than the effects of cocaine, yet the cost is much the same. For that reason, methamphetamine is sometimes called the "poor man's cocaine."

Methamphetamine is generally cheaper than cocaine and—because the body metabolizes it more slowly—much longer lasting. Methamphetamine's effects may last as much as 10 times longer than a cocaine user's high. With its long-lasting effects, methamphetamine binges may last up to a week, while cocaine binges rarely continue for more than 72 hours. When heavy cocaine users experience paranoia, it almost always disappears once the binge ends. For methamphetamine users, however, severe disturbance of mood and thought may be sustained well beyond the binge. Not infrequently, these effects persist for days, sometimes weeks. Similarly, the methamphetamine crash is more prolonged, and drug-related depression that users may experience upon awakening can be more severe than any experienced by cocaine users.

As with many drugs, methamphetamine, if prescribed by a physician, is legally available in the United States for the treatment of attention deficit disorders and obesity. Unfortunately, much of the methamphetamine available on the street is illicit methamphetamine from clandestine laboratories in the United States.

Methamphetamine can be ingested, inhaled, or injected. It is sold as a powder or in small chunks which resemble rock candy. It can be mixed with water for injection or sprinkled on tobacco or marijuana and smoked. Chunks of clear, high-purity methamphetamine ("ice," "crystal," "glass") are smoked in a small pipe, much as "crack" cocaine is smoked. Since methamphetamine will vaporize rapidly, some heat the drug and inhale the fumes that are released.

Source: Utah Attorney General's Office, http://attorneygeneral.utah.gov/Meth/Whatismeth.htm.

AT A GLANCE

Techniques for Extending a Definition

- Word Origin—What is the etymology of the word?
- Word History—How has this word been used in the past?
- Example—What is an example of the word's usage?
- Negation—What does not define this word?
- Division into Parts—How can my subject be divided?
- Similarities and Differences—What things can be compared and contrasted with my subject?
- Analogy—What is something completely different than my subject and yet similar in some important ways?
- Graphic—How can my subject be illustrated?

Using Style and Design in Technical Definitions

Technical definitions are almost always written in plain style, and their design is typically not flashy.

Keeping the Style Plain and Simple

When writing and revising a definition, you want to provide your readers with the straightforward meaning of the word. So, there is little need to be overly persuasive. Here are some suggestions for style in definitions:

> **Use only words that will be familiar to the readers**—If your readers are experts, a specialized vocabulary is fine, perhaps even preferred. If your readers are not experts, use common words and define any terms being used in specialized ways.
>
> **Keep sentences short**—Parenthetical and sentence definitions should use the least amount of words possible to provide an accurate definition. Extended definitions should avoid long sentences that go beyond breathing length
>
> **Use definitions within definitions**—When you use a specialized word within a definition, also include a parenthetical definition to define it.
>
> **Keep it visual**—Where possible, use words and phrases that allow readers to visualize the subject. Use color, texture, and shapes to define it. Add a graphic. Use analogies and similes. Show examples.

In almost all cases, the style in a definition should not stand out. Using plain style is probably the best approach whether you are writing a parenthetical, sentence, or extended definition.

LINK To learn more about using plain style, go to Chapter 7, page 162.

For more advice on writing plainly, go to
www.ablongman.com/johnsonweb/17.16

Designing for Clarity

Since most definitions are embedded within a larger document, they tend to adopt the design of the larger document. However, there are situations such as white papers or specifications, where extended definitions need to take on their own design. (Figure 17.8).

As you consider the definition's design, think about its readers, purpose, and context of use. Will the extended definition be used to define a new product or concept to experts? Will it be sent out to people around the country to clarify a subtle change in the meaning of a word? Will it be put on a website to help the public understand your company's products or services?

Good Style and Design in an Extended Definition

Figure 17.8: In definitions, clarity and accuracy are most important. Appropriate style and design can make the document easy to read and visually accessible.

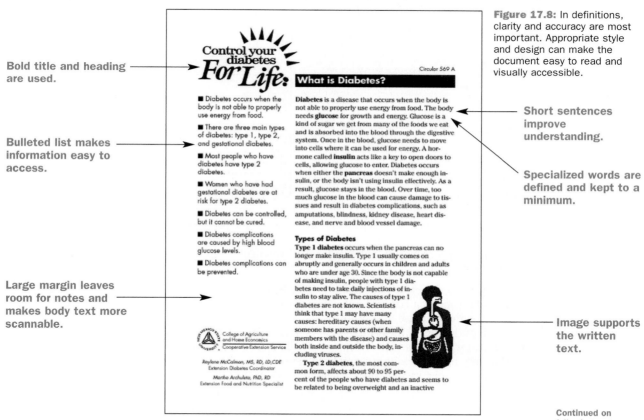

Bold title and heading are used.

Bulleted list makes information easy to access.

Large margin leaves room for notes and makes body text more scannable.

Short sentences improve understanding.

Specialized words are defined and kept to a minimum.

Image supports the written text.

Control your diabetes *For Life:*

Circular 569 A

What is Diabetes?

- Diabetes occurs when the body is not able to properly use energy from food.
- There are three main types of diabetes: type 1, type 2, and gestational diabetes.
- Most people who have diabetes have type 2 diabetes.
- Women who have had gestational diabetes are at risk for type 2 diabetes.
- Diabetes can be controlled, but it cannot be cured.
- Diabetes complications are caused by high blood glucose levels.
- Diabetes complications can be prevented.

Diabetes is a disease that occurs when the body is not able to properly use energy from food. The body needs **glucose** for growth and energy. Glucose is a kind of sugar we get from many of the foods we eat and is absorbed into the blood through the digestive system. Once in the blood, glucose needs to move into cells where it can be used for energy. A hormone called **insulin** acts like a key to open doors to cells, allowing glucose to enter. Diabetes occurs when either the **pancreas** doesn't make enough insulin, or the body isn't using insulin effectively. As a result, glucose stays in the blood. Over time, too much glucose in the blood can cause damage to tissues and result in diabetes complications, such as amputations, blindness, kidney disease, heart disease, and nerve and blood vessel damage.

Types of Diabetes

Type 1 diabetes occurs when the pancreas can no longer make insulin. Type 1 usually comes on abruptly and generally occurs in children and adults who are under age 30. Since the body is not capable of making insulin, people with type 1 diabetes need to take daily injections of insulin to stay alive. The causes of type 1 diabetes are not known. Scientists think that type 1 may have many causes: hereditary causes (when someone has parents or other family members with the disease) and causes both inside and outside the body, including viruses.

Type 2 diabetes, the most common form, affects about 90 to 95 percent of the people who have diabetes and seems to be related to being overweight and an inactive

College of Agriculture and Home Economics
Cooperative Extension Service

Raylene McCalman, MS, RD, LD,CDE
Extension Diabetes Coordinator

Martha Archuleta, PhD, RD
Extension Food and Nutrition Specialist

Source: New Mexico State University Extension Office.

Continued on following page

To see well- and badly designed definitions, go to **www.ablongman.com/johnsonweb/17.17**

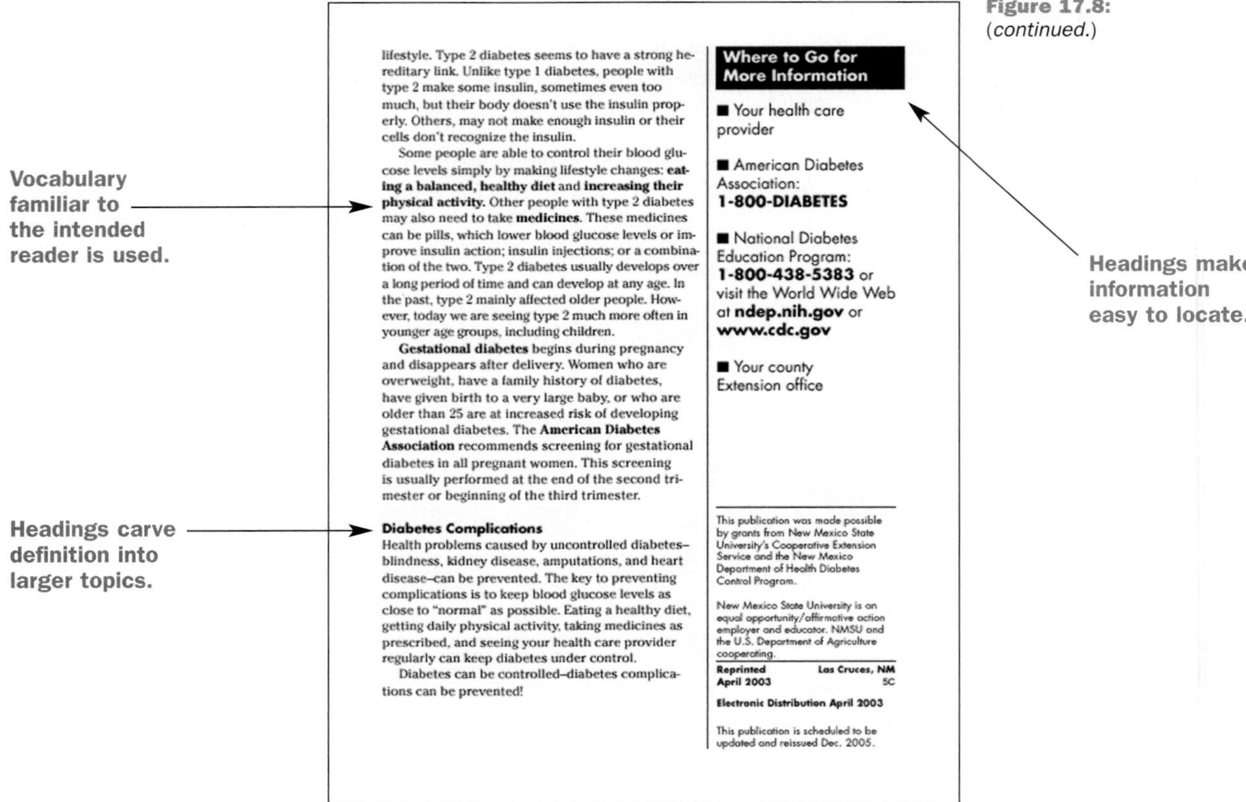

Vocabulary familiar to the intended reader is used.
Headings carve definition into larger topics.
Headings make information easy to locate.

lifestyle. Type 2 diabetes seems to have a strong hereditary link. Unlike type 1 diabetes, people with type 2 make some insulin, sometimes even too much, but their body doesn't use the insulin properly. Others, may not make enough insulin or their cells don't recognize the insulin.

Some people are able to control their blood glucose levels simply by making lifestyle changes: **eating a balanced, healthy diet** and **increasing their physical activity.** Other people with type 2 diabetes may also need to take **medicines.** These medicines can be pills, which lower blood glucose levels or improve insulin action; insulin injections; or a combination of the two. Type 2 diabetes usually develops over a long period of time and can develop at any age. In the past, type 2 mainly affected older people. However, today we are seeing type 2 much more often in younger age groups, including children.

Gestational diabetes begins during pregnancy and disappears after delivery. Women who are overweight, have a family history of diabetes, have given birth to a very large baby, or who are older than 25 are at increased risk of developing gestational diabetes. The **American Diabetes Association** recommends screening for gestational diabetes in all pregnant women. This screening is usually performed at the end of the second trimester or beginning of the third trimester.

Diabetes Complications

Health problems caused by uncontrolled diabetes–blindness, kidney disease, amputations, and heart disease–can be prevented. The key to preventing complications is to keep blood glucose levels as close to "normal" as possible. Eating a healthy diet, getting daily physical activity, taking medicines as prescribed, and seeing your health care provider regularly can keep diabetes under control.

Diabetes can be controlled–diabetes complications can be prevented!

Where to Go for More Information

- Your health care provider
- American Diabetes Association: **1-800-DIABETES**
- National Diabetes Education Program: **1-800-438-5383** or visit the World Wide Web at **ndep.nih.gov** or **www.cdc.gov**
- Your county Extension office

This publication was made possible by grants from New Mexico State University's Cooperative Extension Service and the New Mexico Department of Health Diabetes Control Program.

New Mexico State University is an equal opportunity/affirmative action employer and educator. NMSU and the U.S. Department of Agriculture cooperating.

Reprinted April 2003 **Las Cruces, NM** 5C

Electronic Distribution April 2003

This publication is scheduled to be updated and reissued Dec. 2005.

Figure 17.8: (*continued.*)

If you are given a free hand to design the text, you might consider some of the following design features:

- a larger title that stands out from the body text
- a typeface and font size that reflect your readers' characteristics and contexts in which they will use the document
- headings that break the definition into larger blocks of information
- a format with two or more columns that allows you to put graphics or explanatory information in a separate column
- a graphic or picture that offers a helpful image of the subject.

Technical definitions are rarely known for their innovative design. So, you should look for examples of previous definitions written at your company for guidance on design issues. Also, if you are writing a definition that will be embedded in a larger document, pay close attention to how that document is designed. Your definition should blend in seamlessly with the document's overall design.

LINK For more about designing documents, go to Chapter 8, page 192.

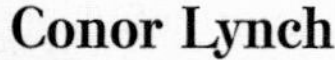

AT WORK

Conor Lynch
PROJECT CONTROLLER, RESIDENTIAL MANAGEMENT GROUP
The Residential Management Group manages large rental properties in New York City.

When are definitions important in the technical workplace?

Our company is currently renovating 10,000 apartments in a massive undertaking that involves every department in the company. Coordination among these departments is crucial to the project's success.

When the project began, however, simple vocabulary caused significant problems. When writing reports, proposals, or memoranda, each of our departments used simple words to describe the renovation, including nontechnical words such as "vacant,""completed," or "leased."

But to each department, those simple words carried different meanings. Our Leasing Unit, for example, considered an apartment *vacant* as soon as the old tenant moved out, but our lawyers did not consider an apartment *vacant* until we regained legal possession of the keys. And our Finance Department would not call an apartment *vacant* until the tenant's paperwork had gone through our accounting system. Because each of these processes might take weeks or months, we couldn't accurately count the number of vacant apartments.

The term "completed renovation" caused similar problems. Our Construction Management team designated an apartment as *complete* as soon as the punchlist was finished. Legal, on the other hand, would not use *complete* until the apartment had passed city inspection. And in the eyes of the Finance Department, an apartment was not *complete* until the contractor had been paid. Again, these differences in simple definitions led to disparities in monthly reporting.

The solution lay in more explicit language in our writing. Once the departments agreed upon one set of definitions, we could more accurately analyze our company and forecast its future success.

Revising, Editing, and Proofreading

Accuracy is very important in definitions, so you should leave plenty of time to revise and edit your work. Make sure others, especially experts, have a chance to offer commentary on the definition.

One thing to keep in mind as you revise is your readers' level of expertise in your field. After all, if you use words to define your subject that are unfamiliar to readers, you are only making the subject less clear.

Revising for Conciseness and Visual Detail

While revising, look for places where you can cut out any information that goes beyond your readers' need to know. Definitions can be extended endlessly, so you need to scale the amount of information to your readers' needs.

For worksheets to help you revise your definitions, go to **www.ablongman.com/johnsonweb/17.18**

Then, look for ways to use the senses, especially the visual, to bring your definition to life. Look for places where you can add color or texture to your writing. Add one or more graphics to help readers understand what you are defining. With these visual techniques, you will help readers gain a full three-dimensional understanding of the subject.

Always look for sentences that are too long or complex. It is tempting to use long sentences to just "get it right." That's fine in a draft, but the final version should use simple, plain sentences that readers can grasp quickly.

Editing for Accuracy and Consistency

Often, sentence definitions end up in a glossary, while extended definitions can be placed in an appendix. When readers take the time to look up these definitions, they will expect them to be absolutely accurate and consistent. So, as you are revising and proofreading, pay close attention to the preciseness and predictability of the definitions.

Accuracy is essential. If you have some doubts about whether a word is being defined properly, it is worth your time to double-check its meaning. Inaccuracies in definitions can lead to large problems down the line.

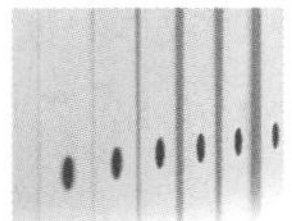

- Reliable, accurate technical definitions are essential in technical communication. They ensure clarity in technical documents.
- There are three types of definitions: parenthetical, sentence, and extended definitions.
- To determine the rhetorical situation for a definition, you can use the Five-W and How questions. Then, you should review your subject, purpose, readers, and context of use.
- Organizing and drafting a parenthetical definition involves finding a word or phrase to clarify the meaning of a term.
- Organizing and drafting a sentence definition involves determining the term, category, and distinguishing features of the object being defined.
- Drafting an extended definition may involve research into the background and meaning of a term, and it could involve using different techniques such as analogies or graphics to provide an extra dimension to your definition.
- A plain and simple style will make your definition clear.
- Revise and edit your definitions to make sure they are accurate, concise, and consistent. Use visual language and graphics where appropriate to illustrate the definition.

Individual or Team Projects

1. By yourself or with a team, list 10 technical words that are commonly used in your major or field. Then, using dictionaries, glossaries, and Internet search engines, find sentence definitions for each of these words. Finally, turn this list of definitions into a "glossary" that you can give your class or hand in to your instructor.

2. Pick a word that interests you. Then, go to the *Oxford English Dictionary* or another dictionary and find the history of this word. Pay special attention to its etymology and when it was first used in the English language. Then, write a memo to your instructor in which you discuss the word, its usage, and its history.

3. Here is a list of words and phrases with some interesting word histories. Pick a few and hunt down their history on the Internet. To your class, present your favorite word, showing how its history reflects (or doesn't) its usage today.

   America
   blackbox
   blurb
   boondoggle
   bug (as in computer bug)
   catch-22
   cybernetics
   doughnut
   glitch
   ground zero
   hermetic seal
   jeep
   jinx
   Murphy's Law
   on the fritz
   quark
   scapegoat
   spam (as in e-mail spam)
   the full monty
   the whole nine yards

4. Choose a word that is common in your major or field. Then, write an extended technical definition of that word for a nonexpert reader. Start out by defining your subject, purpose, readers, and context of use. Then, using various sources, research the meaning of the word. Write an extended definition in which you use examples, analogies, and similarities and differences among other techniques to expand the definition of the word.

5. Rewrite/redesign the sample definition "What Is Meth?" in this chapter for an audience of 12-year-old readers who are learning about illegal drugs. How can you change the content, organization, style, and design to better fit the needs of these younger readers?

Websites that discuss word histories are available at **www.ablongman.com/johnsonweb/17.19**

6. Of some alarm to language purists are the abbreviated words used in instant messaging and e-mails. Terms like "IMO" (in my opinion), "brb" (be right back), "lol" (laughing out loud), "rtfm" (read the * manual), "ttfn" (ta ta for now), and "ICQ" (I seek you) are commonly used codes in these electronic formats. With your team, produce a small glossary of these words, explaining what they mean and how they are used. Your glossary should be written for people who are new to instant messaging or e-mail.

Collaborative Projects

Your group has been assigned to develop a glossary of terms that are commonly used on your college campus. The glossary is being written for new students and their parents, as well as hapless college administrators and faculty who often don't know what their students are talking about. Your glossary will be put on the college's website, so people will have quick access to the words.

When your group is finished with its glossary, combine it with the glossaries from other groups in the class. Certainly, some of the words will overlap with the words defined by other groups. Probably other groups' definitions for these words will be somewhat different from yours. Negotiating these differences, develop a master list of definitions that the whole class can agree to. Pay special attention to where people disagree about the meanings of words. Discuss why these disagreements come about.

Revision Challenge

The following technical definition in Figure A was written for a communication company's website. Its purpose was to help explain why solar flares occasionally affect the company's operations. The definition was aimed primarily at the company's clients. However, the company also wanted to provide information about communication technology to schoolchildren and other people interested in satellites.

This draft is probably a couple revisions from complete. Using the concepts discussed in this chapter, identify places where this document could be strengthened. Then, develop a revision strategy for improving this website text.

- What content would you like to see added to this definition?
- Is there any content that goes beyond the readers' need to know?
- Does the definition have a clear introduction, body, and conclusion?
- Where and how could the style be improved to make it more readable?
- Do you think other graphics might be helpful?

An electronic version of this definition is available at www.ablongman.com/johnsonweb/17.20. There, you can also download the file for revision.

For an electronic version of the Revision Challenge text, go to **www.ablongman.com/johnsonweb/17.20**

Figure A: This document is a draft of some information that will appear on a website. Can you make this text more readable to a broader group of readers, including children?

Solar Flares

For those of us who work with satellites, solar flares are a particularly interesting problem. They are mostly unpredictable phenomena that we cannot foresee, and we are just now beginning to understand how they affect the delicate hardware that we send up into space.

Source: SOHO, ESA/NASA.

Continued on following page

Figure A:
(*continued.*)

Solar flares happen when a large amount of magnetic energy is discharged from the sun's atmosphere. When a solar flare occurs, observers on Earth and those using observation tools like the SOHO satellite witness a huge burst coming from the sun. In some cases, the sun also emits an incredible burst of x-rays, which are a form of light. Other particles like heavy nuclei, electrons, and protons are released too. This is called a coronal mass ejection. The picture included here shows one of the larger bursts that we have seen recently, which occurred on October 21, 2003. This flare disturbed many of the earth's communication systems and power systems when its emitted electromagnetic waves slammed into the earth's atmosphere and magnetosphere. Fortunately, it takes awhile for the solar storm to reach the earth, so we had time to put our satellites on standby.

The effects of a solar flare are significant. Here on earth, the effects are not immediately noticeable. Your radio or mobile phone may pick up some static. False alarms might be triggered on alarms, like your car. But up in space, the effects can be pretty serious. The emitted x-rays can charge up components on satellites, causing damage. Probably the best known effect of solar flares are the colorful auroras which are visible in the sky, especially in northern parts of the country.

Even though the effects are usually not immediately noticeable here on earth, they are still important. The high-energy particles that are released during a solar radiation storm can expose airplane passengers to radiation that is equivalent to ten chest x-rays. Also, navigational systems, especially ones close to earth or on earth can be bothered, creating orientation problems. Weather is effected by solar flares, too.

Another interesting thing about solar flares is that they tend to follow eleven year cycles. The last "solar maximum" was in 1989. During those periods there are many sunspots and coronal mass ejections. Satellites often need to be shut down to avoid damage. Even the SOHO satellite, which is used to observe the sun for the NOAA, cannot continue to operate during the largest of solar storms. It needs to be put on standby during the height of a solar storm.

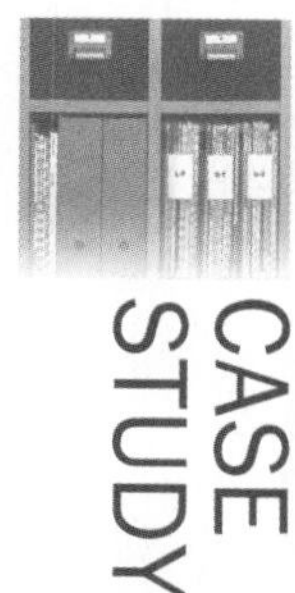

Fast-Food High School

The people of Edwardsville, like those in many communities, were growing concerned about their children's obesity problems. A recent study, commissioned by the Edwardsville School Board, reported that 34 percent of the children in the school district were overweight or obese. High school students, especially, were carrying extra pounds. Nearly 41 percent of the students at Edwardsville High School were overweight.

At a meeting in March, the School Board members decided to do something about this problem. They voted to remove the candy and soda machines from the schools, and they decided that only "nutritious food" would be served in school cafeterias.

The School Board then hired Sally Mayer, a nutritionist, to assess the high school cafeteria's current food offerings and make recommendations for changes to the menu. Her report to the School Board was due in two months.

Her first visit to the school confirmed that the food served in the cafeteria was typical fast food. She noted that students had a selection of pizza, burgers, hot dogs, and fried chicken strips. Overwhelmingly, students ate french fries and potato chips as sides, even though fruits and vegetables like apples, oranges, bananas, carrots, and celery were available. The cafeteria also had a salad bar, but only a few students were using it.

After observing lunch at the cafeteria, Sally met with the principal, Tom Young. He seemed like a person who genuinely cared for the students, but he was skeptical about changing their eating habits. He said, "We provide a choice. I don't think it's the school's job to regulate what kids eat. That's their parent's job. If we offer healthy food, students just won't eat it."

But Tom had another reason for resisting the move to nutritional foods—the expense. The cafeteria was a revenue source for the school. He told Sally that the profits from the cafeteria funded the after-school athletic intramural programs. The additional costs of nutritious foods would wipe out those revenues and possibly the intramural programs.

"Ironically," Tom pointed out, "the change to so-called 'nutritious food' would eliminate the kinds of physical activities and exercise these kids need to stay fit."

Nevertheless, Tom knew he had to make changes to satisfy the School Board. He said, "Much of the solution depends on how you define 'nutritious food.' We can make some small changes, like buying leaner meats and adding to the salad bar. Maybe we won't offer french fries a couple days a week. But, Sally, you're going to need to meet us halfway. We need you to define nutritious food in a way that's realistic."

Sally drove back to her office. Tom was right—her whole report hinged on her definition of nutritious food. If she followed a strict definition, there was a possibility that (1) the students would not eat the food, (2) the food would be expensive, and (3) the after-school intramural athletic programs would be cut. However, a loose definition, leading only to the minor changes Tom described, would not address the current nutrition problems in the school.

If you were Sally, how might you define nutritious food in your report? How would this definition influence your assessment of the school cafeteria's current offerings and your recommendations?

Want more information on fast food and school cafeterias? Go to **www.ablongman.com/johnsonweb/17.21**

CHAPTER

18

Technical Descriptions

CHAPTER CONTENTS

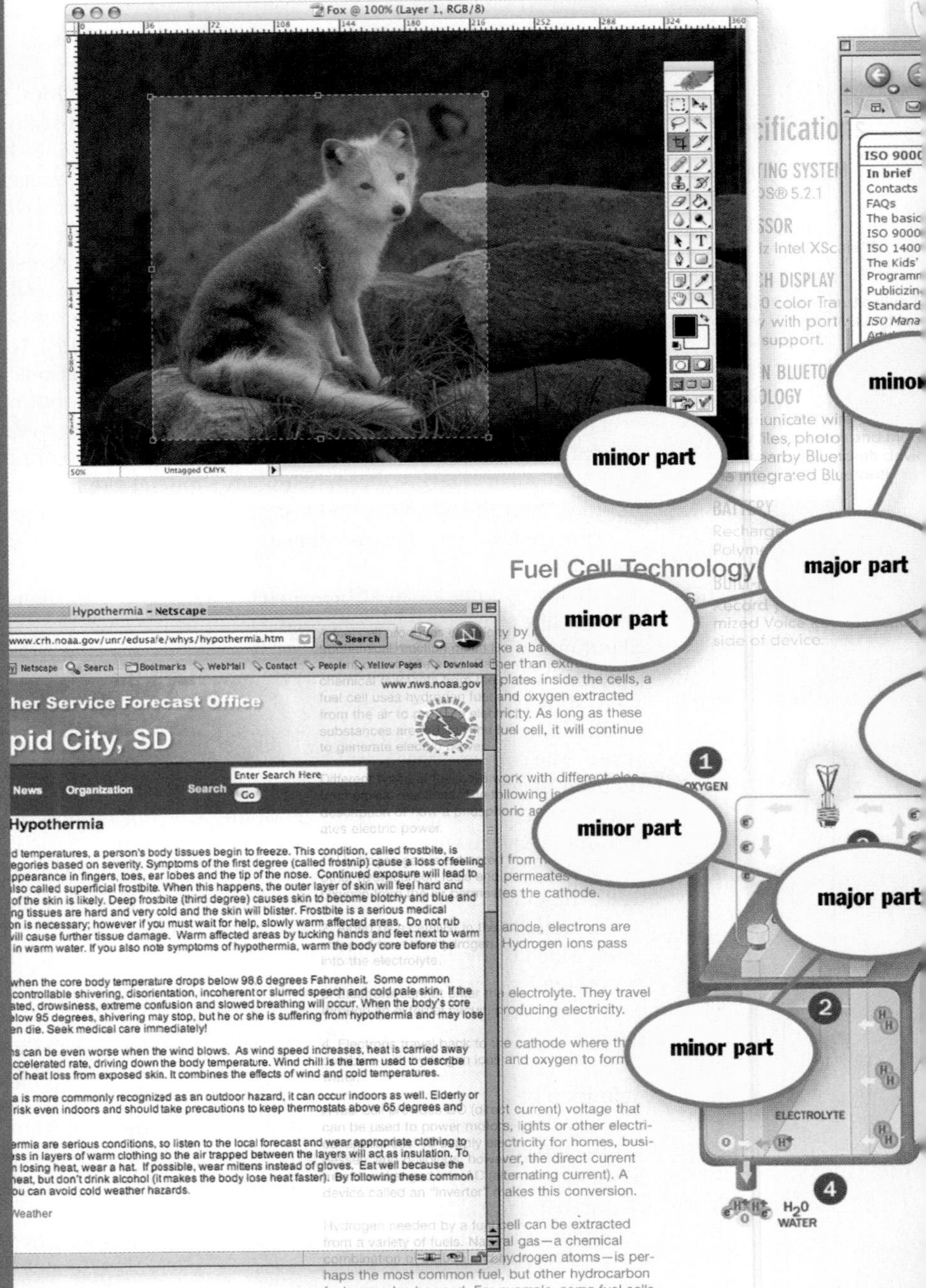

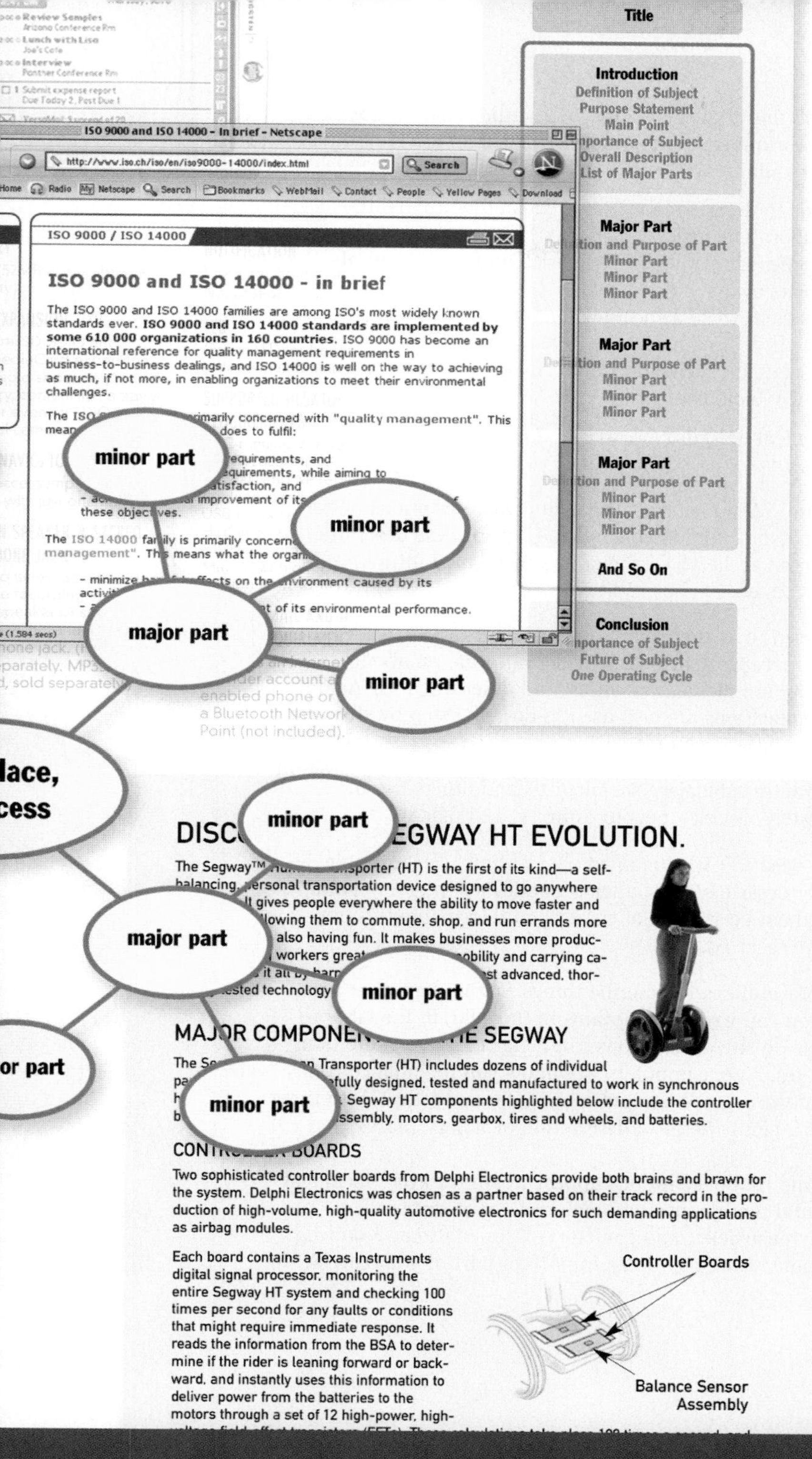

CHAPTER OBJECTIVES

In this chapter, you will learn:

- How descriptions are used in technical workplaces.
- Common features of descriptions.
- How to determine the rhetorical situation for a description.
- Strategies for partitioning objects, places, or processes into major and minor parts.
- Techniques for organizing and drafting descriptions.
- The use of plain style to make descriptions understandable.
- How to use page layout and graphics to highlight and illustrate important concepts.
- The importance of revising and editing descriptions for accuracy.

In any technical career, you will find that the ability to describe things, places, or processes is essential. In the workplace, descriptions are called a variety of different names, but they are used in similar ways:

- A medical researcher needs to describe how a new chemotherapy treatment attacks cancerous cells in laboratory mice.
- An astrophysicist needs to describe a newly discovered quasar in a spiral galaxy that is 300 light-years away.
- A mechanical engineer needs to write a specification for a new microprocessor that her team has developed.
- An architect needs to illustrate his ideas for a butterfly pavilion that will be the centerpiece of his town's new biological park.

There are a few types of technical descriptions, written for many different purposes in the technical workplace:

Basic technical description—Manufacturers use technical descriptions to describe their products for patents, quality control, and sales. These descriptions are often used to establish an archetype, or ideal, against which future products can be measured and tested.

Specifications (often referred to as the "specs")—Engineers write specifications to describe a product in great detail, providing exact information about features, dimensions, power requirements, capacities, and other qualities. As discussed in Chapter 17, specifications are also used to describe step by step how a product is assembled or a task is completed.

Field notes—Naturalists, anthropologists, sociologists, and others use field notes to help them accurately describe people, animals, and places.

Observations—Scientists need to make detailed observations. For example, medical personnel need to describe their patients' conditions in great detail. These observations help them keep track of changes in their patients' symptoms and the medical care received.

Computers have enhanced our abilities to describe things. With a laptop or personal digital assistant (PDA), you can write descriptions in the field, in the laboratory, or at your worktable. These on-site descriptions can be far more accurate than ones re-created from written notes. Also, computers will allow you to enhance your description with page layout software, graphics, and even audio and video. With these enhancements, you can present a highly detailed description of things, places, and processes.

Look around and you will find that descriptions appear in almost every technical document, including experimental reports, user manuals, reference materials, proposals, marketing literature, magazine articles, and conference presentations. Knowing how to write technical descriptions will likely be an important part of your career.

To see examples of technical descriptions, go to **www.ablongman.com/johnsonweb/18.1**

Basic Features of Technical Descriptions

A technical description can stand alone or be part of a larger document. A stand-alone technical description is a separate document with its own introduction, body, and conclusion. This kind of description will generally have the following features:

- specific and precise title
- introduction with an overall description
- description by features, functions, or stages of a process
- use of senses, similes, analogies, and metaphors
- use of graphics
- conclusion that shows the thing, place, or process in action.

When a technical description needs to stand alone as a separate document, it tends to follow a pattern like the one shown in Figure 18.1.

Technical Description

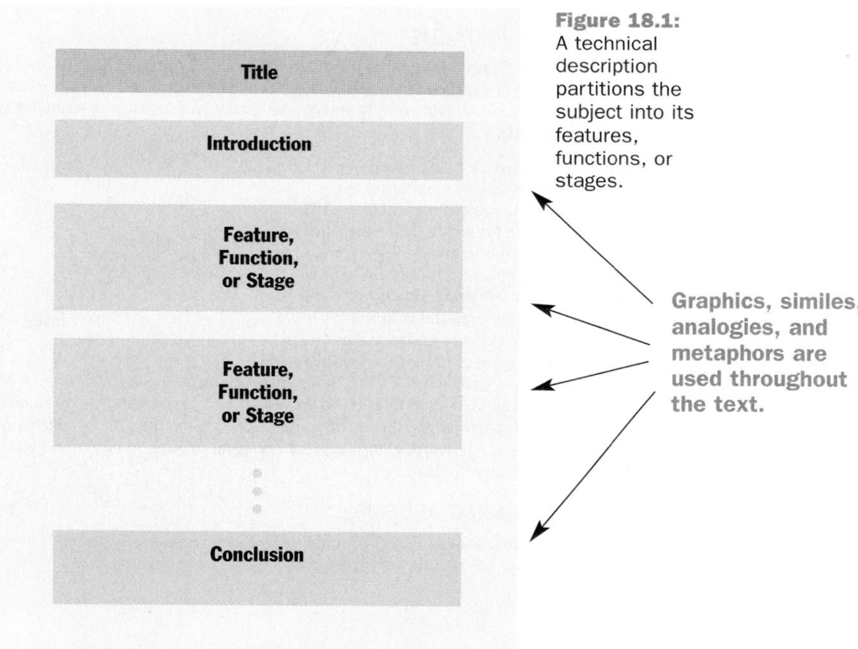

Figure 18.1: A technical description partitions the subject into its features, functions, or stages.

If your description is embedded in a larger document, like a report or proposal, you will likely need to adjust the content, organization, style, and design in ways that suit the larger document. The technical description in Figure 18.2, for instance, was placed on the Segway website. It needed to fit the look and style of the overall website.

An Extended Technical Description

Figure 18.2: This extended description of the Segway effectively partitions the subject by features and describes each feature in depth. In this well-written example, a conclusion was not included. This absence saves space but leaves the description lacking closure.

Definition of the product

Overall description

Importance stressed.

DISCOVER THE SEGWAY HT EVOLUTION.

The Segway™ Human Transporter (HT) is the first of its kind—a self-balancing, personal transportation device designed to go anywhere people do. It gives people everywhere the ability to move faster and carry more, allowing them to commute, shop, and run errands more efficiently while also having fun. It makes businesses more productive by allowing workers greater versatility, mobility and carrying capacity. It does it all by harnessing some of the most advanced, thoroughly tested technology ever created.

Forecasting of the structure of the description

MAJOR COMPONENTS OF THE SEGWAY

The Segway™ Human Transporter (HT) includes dozens of individual parts that have been carefully designed, tested and manufactured to work in synchronous harmony with one another. Segway HT components highlighted below include the controller boards, balance sensory assembly, motors, gearbox, tires and wheels, and batteries.

Major feature

CONTROLLER BOARDS

Metaphor is used to personify the product.

Two sophisticated controller boards from Delphi Electronics provide both brains and brawn for the system. Delphi Electronics was chosen as a partner based on their track record in the production of high-volume, high-quality automotive electronics for such demanding applications as airbag modules.

Minor features

Illustration shows minor parts.

Each board contains a Texas Instruments digital signal processor, monitoring the entire Segway HT system and checking 100 times per second for any faults or conditions that might require immediate response. It reads the information from the BSA to determine if the rider is leaning forward or backward, and instantly uses this information to deliver power from the batteries to the motors through a set of 12 high-power high-voltage field-effect transistors (FETs). These calculations take place 100 times a second, and the motors are adjusted at up to 1,000 times per second, responding far more quickly than the human body is capable of perceiving. Although each board is capable of operating the Segway HT after a failure (each board, in fact, powers electrical circuits on both motors), under normal conditions they share the load.

BALANCE SENSOR ASSEMBLY

The balance sensor assembly (BSA), supplied by Silicon Sensing Systems, is an elegantly designed, extremely robust, and yet incredibly sensitive piece of equipment.

Source: Segway, LLC, http://www.segway.com/segway/component_details.html.

Figure 18.2: (*continued.*)

This small cube, 3 inches on a side, is packed with five solid-state, vibrating-ring, angular-rate sensors ("gyroscopes") that use the Coriolis effect to measure rotation speed. These tiny rings are electromechanically vibrated in such a way that when they are rotated, a small force is

Clearly labeled diagram

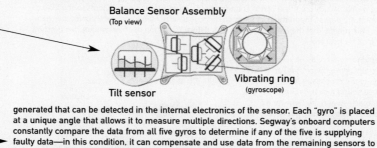

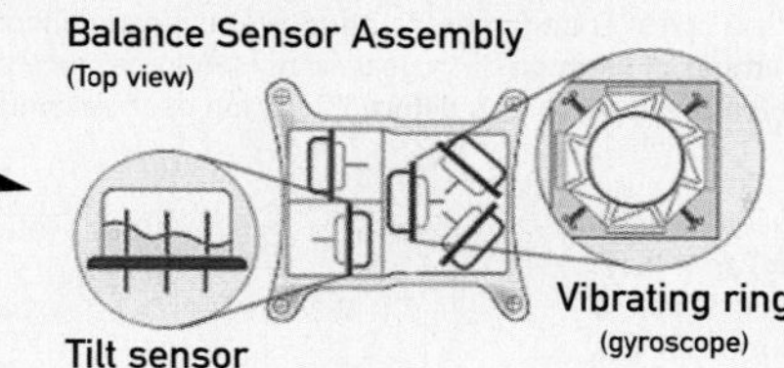

generated that can be detected in the internal electronics of the sensor. Each "gyro" is placed at a unique angle that allows it to measure multiple directions. Segway's onboard computers constantly compare the data from all five gyros to determine if any of the five is supplying faulty data—in this condition, it can compensate and use data from the remaining sensors to continue balancing through a controlled safety shutdown. Two tilt sensors filled with an electrolyte fluid provide a gravity reference in the same way your inner ear does for your own sense of balance. The BSA is monitored by two independent microprocessors and is split into two independent halves for redundancy. Even the communication between sides is performed optically to avoid electrical faults on one side propagating to the other.

Details add realism to the text.

Analogy is used to clarify a complex concept.

Major feature

MOTORS

The Segway HT's motors are unique in a number of respects. Produced by Pacific Scientific, a division of Danaher, they are the highest-power motors for their size and weight ever put into mass production. Each motor is capable of maintaining a power output of 1.5 kilowatts—that's 2 horsepower!

Definitions are used throughout the text to explain concepts.

The motors use brushless servo technology, meaning there are no contacts to wear, arc, and reduce performance. The magnets are constructed of an incredibly powerful rare-earth material: neodymium-iron-boron. Each motor is constructed with two independent sets of windings, each driven by a separate board and motor. Under normal conditions, both sets of windings work in parallel, sharing the load. In the event of a failure, the motor is designed to instantly disable the faulty side and use the remaining winding to maintain control of the Segway HT until it can be brought to a stop. The motor is carefully balanced to operate up to 8,000 rpm, allowing it to produce very high power levels in a small package. Feedback from the motor back to the Segway HT is provided by redundant, noncontact analog hall sensors that sense the positions of magnets with no moving parts other than the motor shaft itself.

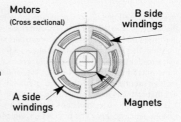

Motors
(Cross sectional)
B side windings
A side windings
Magnets

Continued on following page

Figure 18.2: *(continued.)*

Major feature →

GEARBOX

The Segway HT's gearbox, a joint effort between Axicon Technologies and Segway, is constructed more like a precision Swiss watch than a traditional gear drive from an automobile. A two-stage reduction system provides a 24:1 reduction, allowing the motor to operate at powerful, efficient speeds throughout the full range of speeds of the Segway HT. Each gear is cut to a helical profile, which creates a spiral engagement to minimize noise and increase the load capability of the gears. The number of teeth on each gear is chosen to produce noninteger gear ratios. This means that the teeth will mesh in a different location each revolution, maximizing the life of the gearbox.

← **A simile makes a comparison to something familiar.**

Definition explains technical term. →

Our engineers were so obsessed with the details on the Segway HT that they designed the meshes in the gearbox to produce sound exactly two musical octaves apart—when the Segway HT moves, it makes music, not noise.

Metaphor adds depth. →

The gearbox is pre-assembled and lubricated, and is designed to require no maintenance over the life of the Segway HT. It has been tested for thousands of miles under severe conditions.

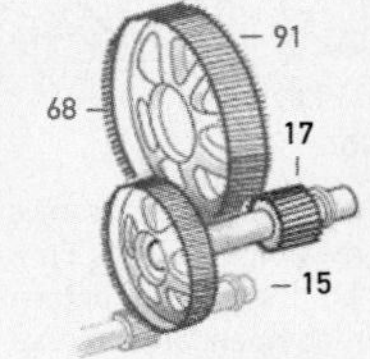

Gear Ratio:

$\frac{15}{68} \times \frac{17}{91} = \frac{1}{24}$

0.1861318813...

0.22058823529...

Gear Harmonics:

Input mesh:
15X motor speed

Interm. mesh:
$\frac{15}{68} \times 17 = 3.75\text{X}$ motor speed

Major feature →

TIRES AND WHEELS

Michelin was one of the earliest partners in the Segway HT design project. The tires on the Segway HT have been designed specifically for this product, using a unique tread compound (a silica-based compound instead of traditional carbon-based materials), giving enhanced traction and minimized marking on indoor floors, and a specially engineered tubeless construction that allows low pressure for comfort and traction while minimizing rolling resistance for long range.

Minor features

The tire is mounted on an equally unique wheel design: the wheel is constructed of a sophisticated engineering-grade thermoplastic chosen in partnership with GE Plastics. This composite material allows light weight and excellent durability, and actually reduces noise transmitted from the drive system. The wheel is molded around a forged steel hub, eliminating fasteners that can loosen over time. The attachment between this hub and the transmissions incorporates a unique (patent pending) taper and hex design, which allows a single nut to attach or remove the wheel while retaining the security of multiple-bolt systems more typically seen in conventional applications.

Tires & Wheels

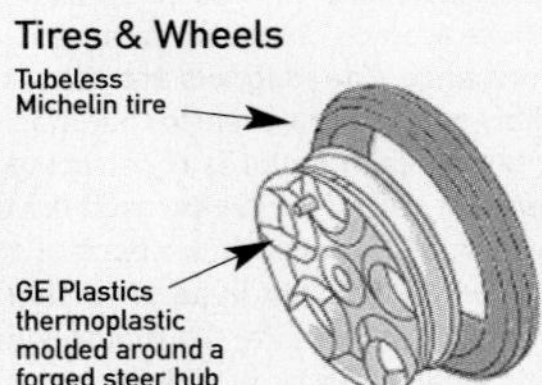

Figure 18.2: (*continued.*)

Major feature

BATTERIES

The Segway HT uses twin NiMH battery packs, designed in partnership with SAFT (a division of Alcatel), running at a nominal 72 volts. These nickel-metal hydride cells deliver the highest power of any currently available chemistry, optimized to maintain the Segway HT's balance under severe conditions. These are not your cell phone batteries!

Minor features

Each pack consists of an array of high-capacity cells and a custom-designed circuit board that constantly monitors the temperature and voltage of the pack in multiple locations. This assembly is enclosed in another unique application of GE thermoplastics—the battery box is sealed using a vibration welding technique that makes the outside of the pack a single, continuous structure—sealed from moisture and strong enough to survive the most extreme tests our durability engineers could throw at it.

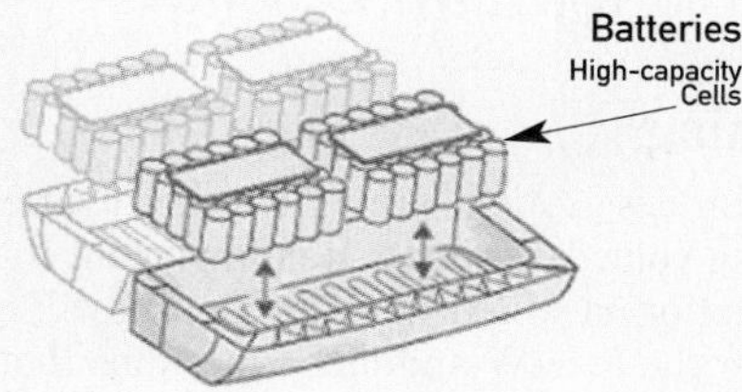

The internal electronics in the battery incorporate "smart" charging—the customer need only plug the Segway HT into the wall and the battery will choose the appropriate charge rate based on temperature, voltage, and level of charge. The batteries will quick charge, then automatically transition into a balance and maintenance charge mode. The Segway HT customer does not need to worry about memory or timing their charges—just plug it in.

Under normal operation, the Segway HT carefully monitors both batteries and automatically adjusts to drain the batteries evenly. In the unlikely event of a battery failure, the system is designed to use the second battery to operate the machine and allow it to continue balancing until it is brought to a safe stop.

Planning and Researching Technical Descriptions

During the planning and researching phase, you should identify what kinds of information your readers need, how they will use that information, and the contexts in which they will use it. Then, you need to research the subject to collect the content for your description.

Planning

Technical descriptions are written for a variety of people and uses. So, as you begin planning your description, it is important that you first have a good understanding of the situation in which your description will be used. A good way to start is to first consider the Five-W and How questions that will be important in your description.

Who might need this description?

Why is this description needed?

What details and facts should the description include?

Where will the description be used?

When will the description be used?

How will this description be used?

Once you have briefly answered these questions, you are ready to start defining the subject, purpose, readers, and context for your description.

SUBJECT Technical descriptions tend to be written about three types of subjects: objects, places, or processes. As you look over your subject, define its boundaries and major characteristics.

LINK For more help defining your subject, see Chapter 5, page 100.

PURPOSE Ask yourself what your description should achieve. Do you want it to provide exact detail, or are you just trying to familiarize your readers with the subject?

In one sentence, write down the purpose of your description. Here are some verbs that might help you write that sentence!

to describe	to represent
to illustrate	to clarify
to show	to reveal
to depict	to explain
to characterize	to portray

Your purpose statement might say something like the following:

The purpose of this description is to show how a fuel cell generates power.

In this description, I will explain the basic features of the International Space Station.

Need help defining the rhetorical situation? Go to
www.ablongman.com/johnsonweb/18.2

You should be able to write your purpose statement in one sentence. If it goes beyond one sentence, you likely need to be more specific about what you are trying to achieve.

LINK For more information on defining a document's purpose, see Chapter 2, page 23.

READERS Technical descriptions tend to be written for readers who are unfamiliar with the subject. So, your job is to help them understand your subject by describing it in terms and images that they will find familiar. You will also need to adjust the detail and complexity of your description to suit their specific interests and needs.

Primary readers (action takers) are readers who most need to understand your description. What information do they need to know to make a decision about your product, place, or process?

Secondary readers (advisors) will likely be experts in your area. They may be engineers, technicians, or scientists who are advising the primary readers about the strengths and weaknesses of the product, place, or process. How much technical detail and accuracy will these readers require to feel satisfied with your description?

Tertiary readers (evaluators) could include just about anyone who has an interest in the product, place, or process you are describing. Your technical description may be used by auditors, lawyers, reporters, or concerned citizens.

Gatekeeper readers (supervisors) may want to check your description for accuracy. Descriptions, especially specifications and observations, need to be exact. Your supervisors may want to review your materials for exactness and correctness.

LINK For more information on analyzing readers, see Chapter 3, page 42.

CONTEXT OF USE Imagine the places where your description might be used. Will the description be embedded in a report or proposal? Will it be placed in a larger document's appendix to provide additional details? Will salespeople use the description to promote the product or service? Will the description be published in a magazine? Will it be used as a specification to establish the ideal measurements of the product or service?

TAKE NOTE Visualize the kinds of situations in which the document might be used. You will need to adjust the content, organization, style, and design to fit those various situations.

Also, imagine your primary readers in a likely context, using your description. What physical, economic, ethical, and political factors will influence how they interpret the text?

LINK For more ideas about analyzing the context of use, see Chapter 3, page 55.

Worksheets are available to help you define readers and contexts of use. Go to **www.ablongman.com/johnsonweb/18.3**

Addressing ISO 9000/ISO 14000 Issues

One important issue involving context of use is whether your description needs to conform to ISO 9000 or ISO 14000 standards. These standards are accepted internationally and monitored by the International Organization for Standardization (ISO). Most high-tech companies, especially ones working for the government, follow these product quality guidelines. Figure 18.3 shows an introduction to these regulations drawn from the International Organization for Standardization website (www.iso.ch).

ISO 9000 and ISO 14000

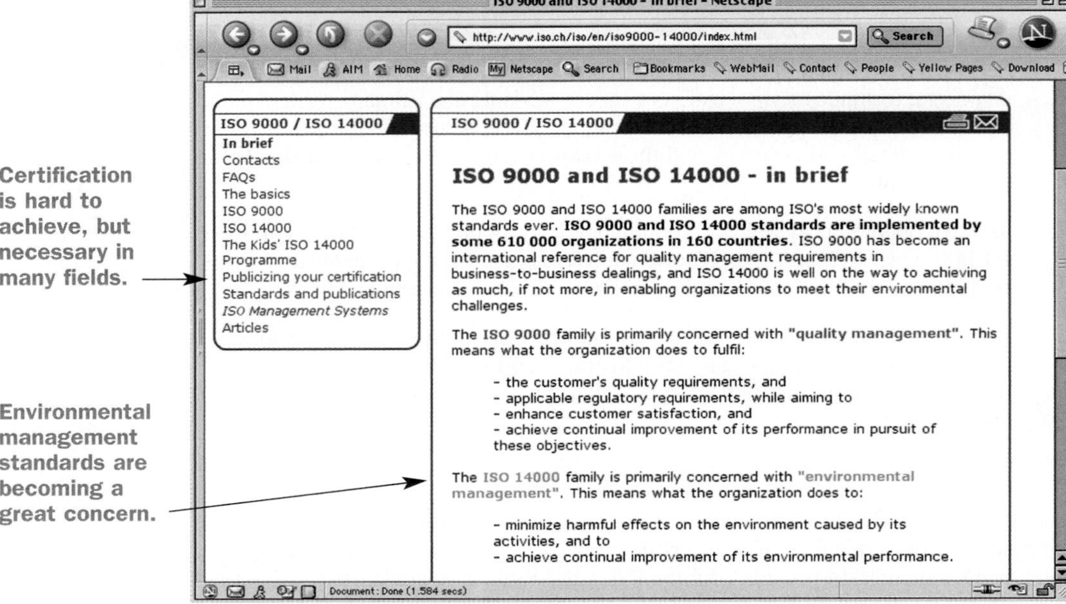

Figure 18.3: The ISO standards are crucial to maintaining quality and consistency in national and international manufacturing.

Source: International Organization for Standardization, http://www.iso.ch/iso/en/iso9000-14000/index.html.

The ISO regulations are too complex and too industry specific to be handled in depth here, but you should be aware that they exist. If your company follows ISO guidelines, any descriptions you write will need to reflect and conform to these standards.

Researching

In most cases, doing research for a technical description is primarily experiential. In other words, you will likely need to personally observe the object, thing, or process you are describing. Chapter 5 discusses research methods in depth. Here are some strategies that are especially applicable to writing descriptions.

To learn more about ISO guidelines, go to **www.ablongman.com/johnsonweb/18.4**

AT WORK

Chris Peterson

CAD TECHNICIAN, LARON, INC., KINGMAN, AZ

Laron, Inc., is an engineering company that specializes in heavy equipment.

How does computer-aided drafting (CAD) help write descriptions?

"A picture is worth a thousand words." It just doesn't get any better than that. A drawing can show a potential client your expertise for his or her particular needs. With the newer 3D CAD programs, such as Autodesk's® Inventor®, or SolidWorks®, you can produce awesome pictures. For video presentations, you can use CAD to walk your clients through anything, from their new home or new plant, to a tour through a new machine. When new video games come out, like PS2®, or Xbox®, what I hear kids (including big kids) say is "the graphics are awesome." CAD is a way to make your descriptions "awesome."

In order to impress your clients, you must master the CAD programs as well as your product. You can have the greatest talent to produce a "mousetrap," but if you lack the skills to present it, you will look like all the others, or worse! You may also be the greatest illustrator, but lack of understanding of the product you are promoting will shine out like a sore thumb.

Another place CAD is very helpful is for "internal clients" such as machinists, welders, carpenters, and pipe fitters. Here again you must have a good knowledge of your discipline as well as the CAD program. There is no substitute for knowledge of what you are trying to convey to the client.

DO BACKGROUND RESEARCH Before making direct observations, you should know as much as possible about your subject. On the Internet, use search engines to find as much information as you can. Then, collect print sources, like books, documents, and other literature. These materials should help you understand your subject better, so you can make more informed observations.

USE YOUR SENSES While making direct observations, use all of your available senses. As much as possible, take notes about how something looks, sounds, smells, feels, and tastes. Pay special attention to colors and textures, because they will add depth and vividness to your description.

TAKE MEASUREMENTS When possible, measure qualities like height, width, depth, and weight. If you cannot gain exact measurements, make estimates. In some cases, you can compare the subject to something of similar size (e.g., "The rough-legged hawk we observed was about the size of a large crow.").

DESCRIBE MOTION AND CHANGE Pay attention to your subject's movements. Look for patterns and note places where your subject deviates from those patterns. Also, take special note of any changes in your subject. Over time, you may notice that it evolves or transforms in some way.

Want to learn how to do close observations of things? Go to **www.ablongman.com/johnsonweb/18.5**

DESCRIBE THE CONTEXT Take copious notes about the surroundings of the subject. Pay attention to how your subject acts or interacts with the things and people around it.

COLLECT VISUALS If available, collect graphics that illustrate your subject, or create them yourself. You can make drawings or take pictures of your subject.

ASK SUBJECT MATTER EXPERTS (SMEs) If possible, find experts who can answer your questions. Your observations will give you material to work with, but there still may be gaps that an expert can fill. Also, an expert can often help you use small observations to develop a whole understanding of the subject.

LINK For more information on doing research, go to Chapter 5, page 96.

While researching, collect as much information as you can. When you are finished researching, you should determine how much your readers already know about the subject and how much they need to know. You can then prioritize your notes to suit their needs.

Partitioning the Subject

To describe something, you need to fully understand it. You need to go beneath the visible surface to discover how it works and how it affects the people who use it.

- If you are describing a product or artifact, take it apart if you can. Then, look at the object from a variety of different angles.
- If you are describing a place, go there and look around. Pay attention to what makes this place different or unique. Spend some time observing how people interact with the place.
- If you are describing a process, note its larger stages. Pay close attention to how each stage leads to and affects the following stage.

Once you are familiar with the subject, you can start describing it in words and images. The first question you should ask is, "How can I partition it?" Or, to put the question more simply, "How can I break it down into its major features, functions, or stages?" Your answer to this question will determine your *partitioning strategy* for describing your subject.

By features—You might partition the subject by separately describing its parts or features. For example, a description of a computer might describe it part by part by first partitioning it into a monitor, keyboard, external disk drives, and a central processing unit (CPU).

By functions—You might partition the subject by noting how its different parts function. A description of the International Space Station, for example, might partition it function by function into *Research, Power Generation, Infrastructure, Habitation,* and *Docking* sections.

By stages of a process—You could partition the subject chronologically by showing how it is assembled or how it works. A description of Hodgkin's dis-

For model descriptions that use each partitioning strategy, go to **www.ablongman.com/johnsonweb/18.6**

ease, for example, might walk readers step by step through *Detection, Diagnosis, Staging,* and *Remission* sections.

At this point, logical mapping can help you break down (partition) your subject into major and minor parts (Figure 18.4).

Partitioning with Logical Mapping

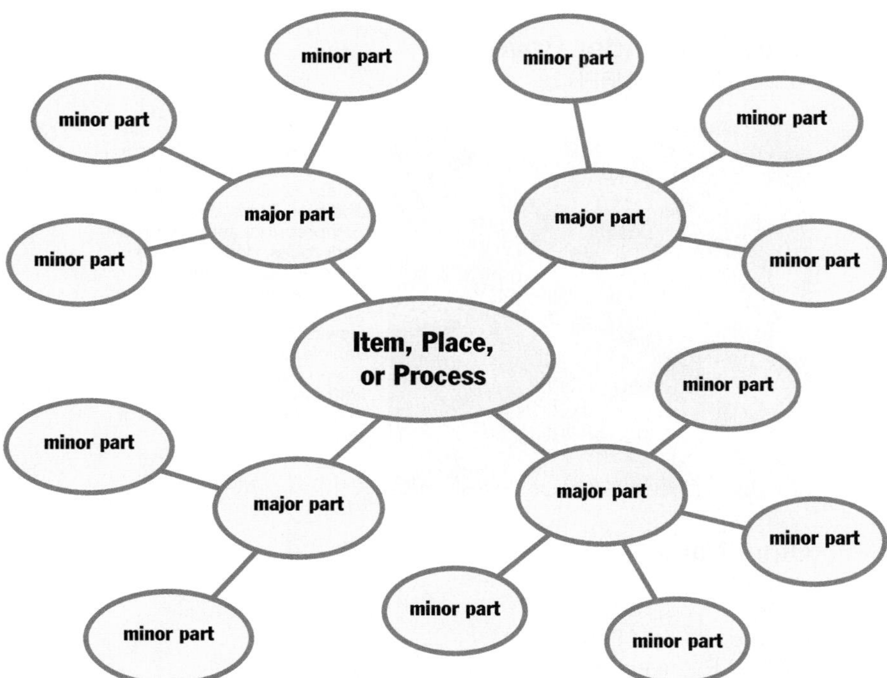

Figure 18.4: Partitioning means dividing the whole into its major and minor parts.

To use logical mapping to help describe something, follow these steps:

1. Put the name of your subject in the middle of your screen or a sheet of paper.
2. Write down the two to five major parts in the space around it.
3. Circle each major part.
4. Partition each major part into two to five minor parts.

LINK For more help using logical mapping, see Chapter 5, page 103.

For example, in Figure 18.5, NASA's description of the Mars Pathfinder Rover partitions the subject into

Chassis and suspension

On-board control system

Photographic system

Performance-monitoring system.

In this example, notice how the description carves the subject into its major parts. Then, each of these major parts is described in detail by paying attention to its minor parts.

Technical Description

Figure 18.5: This description of the Mars Pathfinder Rover shows how a subject can be partitioned into major and minor parts.

Mars Pathfinder Rover

NSSDC ID: MESURPR

Lander image of rover near The Dice (three small rocks behind the rover) and Yogi on sol 22. Color (red, green, and blue filters at 6:1 compression) image shows dark rocks, bright red dust, dark red soil exposed in rover tracks, and dark (black) soil. The APXS is in view at the rear of the vehicle, and the forward stereo cameras and laser light stripers are in shadow just below the front edge of the solar panel.

Other Name(s)

Rocky IV
MFEX
Microrover Flight Experiment
MESUR Pathfinder Rover
Sojourner

Launch Date/Time: 1996-12-04 at 06:58 UTC
On-orbit dry mass: 10.5 kg
Nominal Power Output: 13 W

Background information on subject

Description

The Mars Pathfinder was the second of NASA's low-cost planetary Discovery missions to be launched. The mission consists of a stationary lander and a surface rover. The mission had the primary

Source: National Space Science Data Center, http://nssdc.gsfc.nasa.gov/database/MasterCatalog?sc=MESURPR.

Figure 18.5: *(continued.)*

objective of demonstrating the feasibility of low-cost landings on and exploration of the Martian surface. This objective was met by tests of communications between the rover and lander, and the lander and Earth, tests of the imaging devices and sensors, and tests of the maneuverability and systems of the rover on the surface. The scientific objectives include atmospheric entry science, long-range and close-up surface imaging, rock and soil composition and material properties experiments, and meteorology, with the general objective being to characterize the Martian environment for further exploration. (Mars Pathfinder was formerly known as the Mars Environmental Survey (MESUR) Pathfinder.)

Major part: chassis and suspension ⟶

The rover, which has been named "Sojourner," is a six-wheeled vehicle, 280 mm high, 630 mm long, and 480 mm wide with a ground clearance of 130 mm, mounted on a "rocker-bogie" suspension. The rover was stowed on the lander at a height of 180 mm. At deployment, the rover extended to its full height and rolled down a deployment ramp at about 05:40 UT on 6 July 1997 (1:40 a.m. EDT). The rover was controlled by an Earth-based operator who used images obtained by both the rover and lander systems. Note that the time delay was between 10 and 15 minutes depending on the relative position of Earth and Mars over the course of the mission, requiring some autonomous control, provided by a hazard avoidance system on the rover.

Major part: control system ⟶

The on-board control system is an Intel 80C85 8-bit processor which runs about 100,000 instructions per second. The computer is capable of compressing and storing a single image on-board.

Minor parts ⟶

The rover is powered by 0.2 square meters of solar cells, which will provide energy for several hours of operations per sol (1 Martian day = 24.6 Earth hours). Non-rechargeable lithium thionyl chloride (LiSOCl2) D-cell batteries provide backup. All rover communications were done through the lander.

Continued on following page

Figure 18.5: (continued.)

Major part: photographic system

The rover is equipped with black and white and color imaging systems which were used to image the lander in order to assess its condition after touchdown. The goal was to acquire three black and white images spaced 120 degrees apart of the lander. Images of the surrounding terrain were also acquired to study size and distribution of soils and rocks, as well as locations of larger features. Imaging of the rover wheel tracks will be used to estimate soil properties. Imaging of the rover by the lander was also done to assess rover performance and soil and site properties.

Major part: performance monitoring system

The rover's performance was monitored to determine tracking capabilities, drive performance, thermal behavior, and sensor performance. UHF communications between the rover and lander were studied to determine the effectiveness of the link between the rover and lander. Assessments of rock and soil mechanics will be made based on abrasion of the wheels and adherence of dust.

Minor parts

An alpha-proton-X-ray spectrometer (APXS) is on-board the rover to assess the composition of rocks and soil. Images of all samples tested are transmitted to Earth.

Description of subject in use

The primary objectives were scheduled for the first seven sols, all within about 10 meters of the lander. The extended mission included slightly longer trips away from the lander, and even longer journeys were planned. Images were taken and experiments performed by the lander and rover until 27 September 1997 when communications were lost for unknown reasons.

Organizing and Drafting Technical Descriptions

With your subject partitioned into major and minor features, you are ready to start organizing and drafting your description. There are many different ways to describe your subject, but it is best to follow an organizational pattern that demonstrates an obvious logic that readers will immediately recognize. Figure 18.6 shows a basic pattern that might be followed.

Possible Outline for a Description or Specification

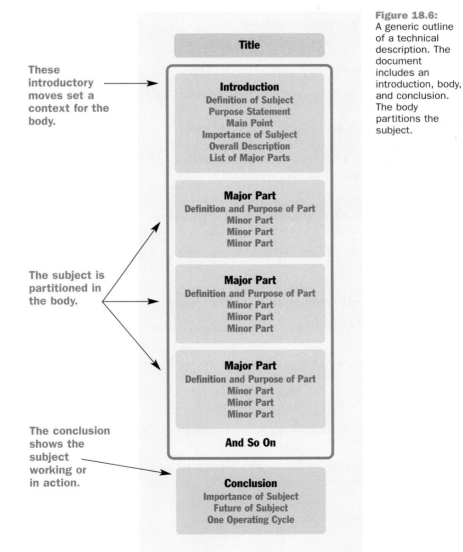

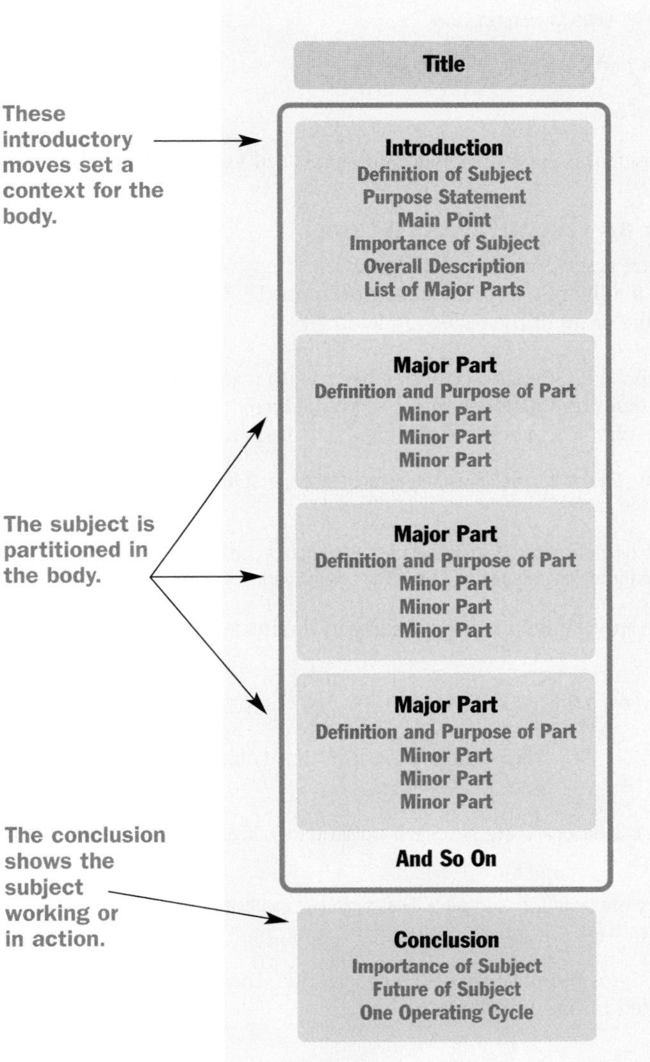

Figure 18.6: A generic outline of a technical description. The document includes an introduction, body, and conclusion. The body partitions the subject.

Specific and Precise Title

The title for your description can be long or short. In most cases, a concise title like "Mars Pathfinder Rover" would be fine. However, this title could suit a variety of documents about the Rover, including a report, a proposal, or a magazine article. If you want your title to clearly identify the purpose of the document, you might write something more exact, such as

Description of the Mars Pathfinder Rover

Here are some other sample titles:

Specifications for the XC-9000 Microprocessor

How Does a Fuel Cell Work?

Lung Cancer: Profile of a Killer

A good title for a description is a title that cannot be used for something else.

Introduction with an Overall Description

Like any document, your description should begin with an introduction that sets a framework, or context, for the rest of the document (Figure 18.6). Typically, the introduction will include some or all of the following features:

DEFINITION OF SUBJECT Often, descriptions begin with a sentence definition of the subject. A sentence definition includes three parts: the term, the class in which the subject belongs, and the characteristics that distinguish it from its class.

The International Space Station is a multinational research facility that will house six state-of-the-art laboratories in orbit.

Hodgkin's disease is a type of cancer that starts in the lymph nodes and other organs that are the body's system for making blood and protecting against germs.

The definition of the subject should appear early in the introduction, preferably in the first sentence.

LINK For more information on writing definitions, see Chapter 17, page 492.

PURPOSE STATEMENT Directly or indirectly state that you are describing something.

This description of the International Space Station will explain its major features, highlighting its research capabilities.

In this article, we will try to demystify Hodgkin's disease, so you can better understand its diagnosis and treatment.

If your title already states the purpose of the description, then a full purpose statement might not be needed in the introduction.

For examples of good introductions, go to **www.ablongman.com/johnsonweb/18.7**

MAIN POINT Give your readers an overall claim that your description will support or prove.

> Building the International Space Station is an incredible engineering feat that will challenge the best scientists and engineers in many different nations.
>
> To fight Hodgkin's disease, you first need to understand it.

IMPORTANCE OF THE SUBJECT Your readers may or may not be familiar with the object, place, or process you are describing. For unfamiliar readers, you might include a sentence or paragraph that stresses the importance of your subject.

> The ISS, when completed, will provide scientists and other researchers an excellent platform from which to study space.
>
> Hodgkin's disease is one of the most acute forms of cancer, and it needs to be aggressively treated.

OVERALL DESCRIPTION OF THE SUBJECT Usually, descriptions start by offering an overall look at the item being described.

> From a distance, the International Space Station looks like a large collection of white tubes with two rectangular solar panels jutting out like ears from its side.
>
> Hodgkin's disease spreads through the lymphatic vessels to other lymph nodes. It enlarges the lymphatic tissue, often putting pressure on vital organs and other important parts of the body.

An overall description should give your readers an image of the subject as a whole. When you begin describing the smaller features of the subject, this whole image will help your readers visualize how the parts fit together.

LIST OF THE MAJOR FEATURES, FUNCTIONS, OR STAGES In many descriptions, especially longer descriptions, the introduction will list the major features, functions, or stages of the subject.

> The International Space Station includes five main features: modules, nodes, trusses, solar power arrays, and thermal radiators.
>
> Once Hodgkin's has been detected, doctors will usually (1) determine the stage of the cancer, (2) offer treatment options, and (3) make a plan for remission.

You can then use this partitioning scheme to organize the body of your description.

LINK For more information on writing introductions, see Chapter 6, page 141.

Description by Features, Functions, or Stages in a Process

The body of your description will be devoted to describing the subject's features, functions, or stages.

Address each major part separately, defining it and describing it in detail. Within your description of each major part, identify and describe the minor parts.

To see some interesting descriptions from NASA, go to **www.ablongman.com/johnsonweb/18.8**

Definition of major part → **Modules** are pressurized cylinders of the habitable space on board the Station. They may contain research facilities, living quarters, and any vehicle operational systems and equipment the astronauts may need to access.

Minor parts

If needed, each of these minor parts could then be described separately. In fact, you could extend your description endlessly, teasing out the smaller and smaller features of the subject.

Figure 18.7 shows a description of a subject by "stages in a process." In this description of a fuel cell, the author walks readers through the energy generation process, showing them step by step how the fuel cell works.

Description by Senses, Similes, Analogies, and Metaphors

The key to a successful technical description is the use of detail to bring your subject to life—make it seem real. Therefore, a well-written description helps the readers visualize the thing, place, or process you are illustrating for them. To add this level of detail, you might consider using some of the following techniques:

DESCRIPTION THROUGH SENSES Humans experience the world through their five senses: sight, hearing, smell, touch, and taste. Consider each of these senses separately, asking yourself, "How does it look?" "How does it sound?" "How does it smell?" "How does it feel?" and "How does it taste?"

> A visit to a car manufacturing plant can be an overwhelming experience. Workers in blue jumpsuits seem to be in constant motion. Cars of every color—green, yellow, red—are moving down the assembly line with workers hopping in and out. The smell of welding is in the air, and you can hear the whining hum of robots at work somewhere else in the plant.

SIMILES A simile describes something by comparing it to something familiar to the readers ("A is like B").

> The mixed-waste landfill at Sandia Labs is like a football field with tons of toxic chemical and nuclear waste buried underneath it.

Similes are especially helpful for nonexpert readers, because they make the unfamiliar seem familiar.

ANALOGIES Analogies are like similes, but they work on two parallel levels ("A is to B as C is to D").

> Circuits on a semiconductor wafer are like tiny interconnected roads crisscrossing a city's downtown.

METAPHORS Metaphors are used to present an image of the subject by equating two different things ("A is B"). For example, consider these two common metaphors:

> The heart is a pump: It has valves and chambers, and it pushes fluids through a circulation system of pipes called arteries and veins.

> Ants live in colonies: A colony will have a queen, soldiers, workers, and slaves.

For more advice on using the senses, go to **www.ablongman.com/johnsonweb/18.9**
To see examples of similes, analogies, and metaphors in descriptions, go to **www.ablongman.com/johnsonweb/18.10**

A Technical Description: Stages in a Process

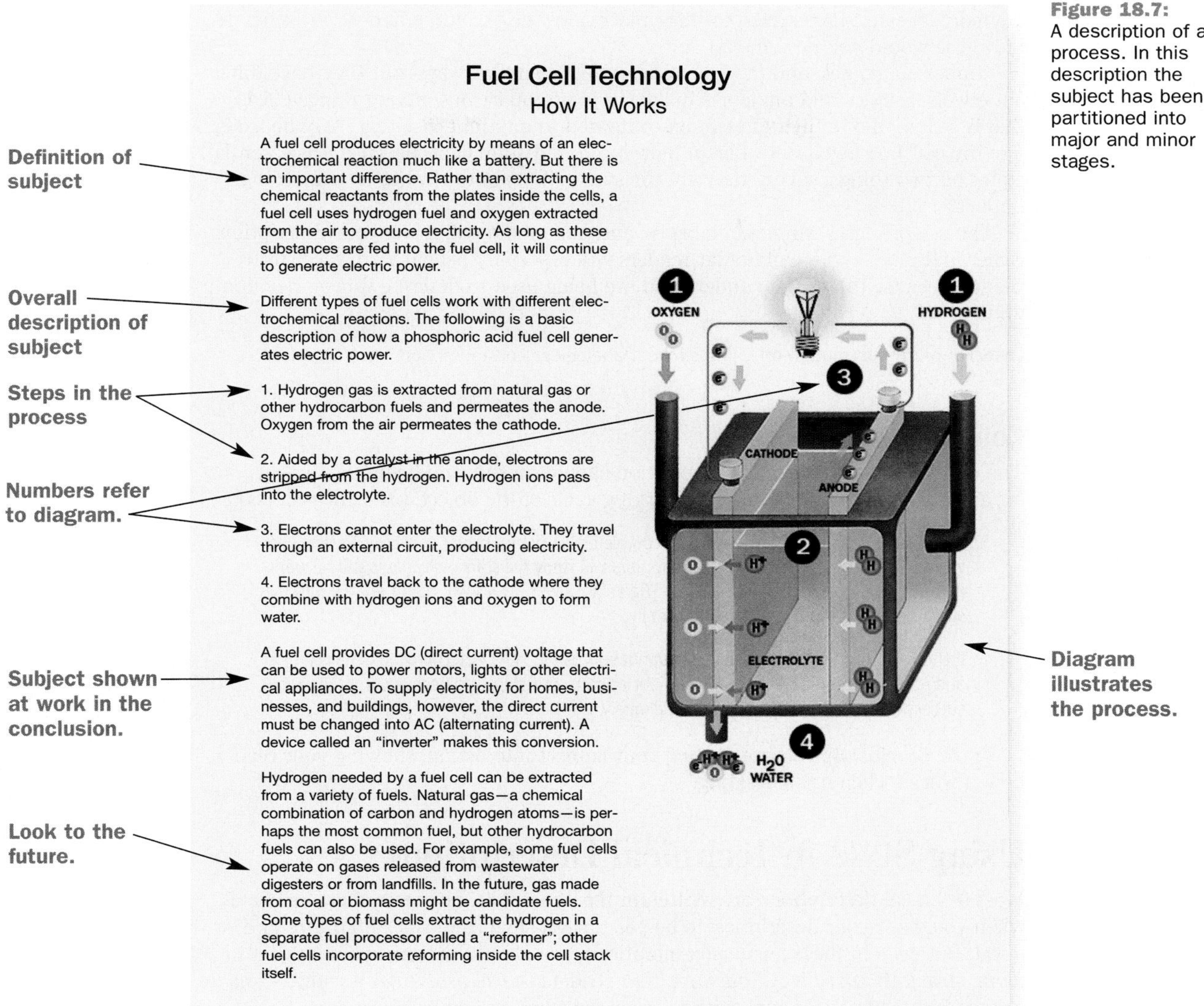

Fuel Cell Technology

How It Works

A fuel cell produces electricity by means of an electrochemical reaction much like a battery. But there is an important difference. Rather than extracting the chemical reactants from the plates inside the cells, a fuel cell uses hydrogen fuel and oxygen extracted from the air to produce electricity. As long as these substances are fed into the fuel cell, it will continue to generate electric power.

Different types of fuel cells work with different electrochemical reactions. The following is a basic description of how a phosphoric acid fuel cell generates electric power.

1. Hydrogen gas is extracted from natural gas or other hydrocarbon fuels and permeates the anode. Oxygen from the air permeates the cathode.

2. Aided by a catalyst in the anode, electrons are stripped from the hydrogen. Hydrogen ions pass into the electrolyte.

3. Electrons cannot enter the electrolyte. They travel through an external circuit, producing electricity.

4. Electrons travel back to the cathode where they combine with hydrogen ions and oxygen to form water.

A fuel cell provides DC (direct current) voltage that can be used to power motors, lights or other electrical appliances. To supply electricity for homes, businesses, and buildings, however, the direct current must be changed into AC (alternating current). A device called an "inverter" makes this conversion.

Hydrogen needed by a fuel cell can be extracted from a variety of fuels. Natural gas—a chemical combination of carbon and hydrogen atoms—is perhaps the most common fuel, but other hydrocarbon fuels can also be used. For example, some fuel cells operate on gases released from wastewater digesters or from landfills. In the future, gas made from coal or biomass might be candidate fuels. Some types of fuel cells extract the hydrogen in a separate fuel processor called a "reformer"; other fuel cells incorporate reforming inside the cell stack itself.

Figure 18.7: A description of a process. In this description the subject has been partitioned into major and minor stages.

Source: U.S. Department of Energy, http://www.fe.doe.gov/coal_power/fuelcells/fuelcells_howitworks.shtml.

Metaphors often become so accepted in technical writing that we no longer notice them. We hardly notice these "heart is a pump" or "ants live in colonies" metaphors anymore, because they are so commonplace. How else would you describe a heart? How else would you describe an ant colony?

Similes, analogies, and metaphors all work in similar ways, but they have different effects. Similes and analogies make simple comparisons among things ("A is like B"). Just about anything can be compared to anything else (e.g., "My house is like my car"). A metaphor offers a more profound statement, because it is essentially equating two things, saying they are the same. Metaphors tend to have a stronger impact.

The use of senses, similes, analogies, and metaphors will make your description richer and more vivid. Unfamiliar readers will especially benefit from these techniques, because things they understand are being used to describe things they don't understand.

LINK For more information on using similes, analogies, and metaphors, see Chapter 7, page 182.

Conclusion

The conclusion for a technical description should be short and concise. Conclusions for a description often portray one working cycle of the object, place, or process.

> When the International Space Station is completed, it will be a center of activity. Researchers will be conducting experiments. Astronomers will study the stars. Astronauts will be working, sleeping, exercising, and relaxing. The solar power arrays will pump energy into the station, keeping it powered up and running.

> Fighting Hodgkin's disease is difficult but not impossible. After detection and diagnosis, you and your doctors will work out treatment options and staging objectives. If treatment is successful, remission can continue indefinitely.

In the conclusion, you are putting your subject into action, showing your readers how it works when it is operating.

Using Style in Technical Descriptions

Most technical descriptions are written in the plain style. However, there are times when you want your description to be persuasive, such as in sales literature or a proposal. But even in these persuasive situations, you should describe your subject in a plain, straightforward way. You want your readers to feel as though the object, place, or process stands on its own, without need for any extra embellishments.

Here are some suggestions for style in descriptions:

Keep your words simple—The words used should be familiar to your readers, or you should define them. Jargon should be avoided with nonexpert readers.

Keep sentences short, within breathing length—Sentences that are too long only cloud your readers' abilities to visualize the subject. Your readers should be able to comfortably read each sentence in one breath.

To learn more about improving your style, go to **www.ablongman.com/johnsonweb/18.11**

Here, subject alignment is being used to keep a consistent perspective.

The given/new technique chains sentences together by using something from the previous sentence early in the next sentence.

Use subject alignment and given/new techniques to weave sentences together—Aligning your subjects and the given/new will make your text smoother and easier to follow.

Subject Alignment

The rover, which has been named "Sojourner," is a six-wheeled vehicle, 280 mm high, 630 mm long, and 480 mm wide with a ground clearance of 130 mm, mounted on a "rocker-bogie" suspension. The rover was stowed on the lander at a height of 180 mm. At deployment, the rover extended to its full height and rolled down a deployment ramp at about 05:40 UT on 6 July 1997 (1:40 a.m. EDT). The rover was controlled by an Earth-based operator who used images obtained by both the rover and lander systems.

Given/New

The golden eagle is one of the larger raptors found in the midwestern United States. Its large size makes this eagle almost unmistakable, and you will often see it perched on fence posts and telephone poles. From its perch, the eagle will swoop down to seize an unsuspecting rodent or small mammal.

Use the senses to add color, texture, taste, sound, and smell—Description doesn't just mean a visual description. Don't forget to use the senses as well. The style of your document will improve if you add other senses besides sight alone.

Use similes, analogies, and metaphors to add a visual dimension—These techniques will help you give your readers an overall image of the subject.

The best style for descriptions is an unnoticed style. So, as you are drafting and revising, find ways to use plain style techniques to improve your readers' ability to understand your meaning.

Designing Technical Descriptions

In most cases, the design of your technical description will depend on the context in which it will be used. A description used in your company's sales literature, for example, will usually look more colorful than a technical specification kept in your company's files. So, as you consider the design of your description, think carefully about where and how the document will be used.

Designing a Page Layout

The page layout of your description likely depends on your company's previous documentation or established style. If you are given a free hand to design the text, you might consider

- using a two-, three-, or four-column format.
- using lists to highlight minor parts of an object, place, or process.
- using a sidebar to focus on a particular part or function of the subject.
- using headings to show the organization of the content.
- placing data or measurements in a table.

To see well-designed descriptions, go to **www.ablongman.com/johnsonweb/18.12**

A Specification

Palm Tungsten T3

Specifications

OPERATING SYSTEM
Palm OS® 5.2.1

PROCESSOR
400MHz Intel XScale™

STRETCH DISPLAY
320×480 color Transflective TFT display with portrait and landscape support.

BUILT-IN BLUETOOTH® TECHNOLOGY
Communicate wirelessly and share files, photos and more with nearby Bluetooth devices via integrated Bluetooth.

BATTERY
Rechargeable 900mAh Lithium Polymer battery.

BUILT-IN MICROPHONE
Record your thoughts with optimized Voice Record button on side of device.

MEMORY
64MB (52MB actual storage capacity)

PALM EXPANSION SLOT
Supports SD, SDIO and MultiMediaCard expansion cards (sold separately) to add memory, content like a travel card, or even an SDIO card like a digital camera.

5-WAY NAVIGATOR BUTTON
Easily access important information with just one hand.

BUILT-IN SPEAKER & STEREO HEADPHONE JACK
Listen to tunes, games, videos or voice recordings with built-in mono speaker or stereo headphones with standard 3.5 mm headphone jack. (Headphones sold separately. MP3s require SD card, sold separately.)

NOTIFICATION
Vibration, Sound and LED Notification.

LED
Green LED for battery charging indication and alert notification.

SUPPORTED DESKTOP SYSTEM REQUIREMENTS
- PC running Windows 98/NT/2000/ME/XP (Windows NT, 2000 and XP require admin rights to install. Systems without USB ports, require the Palm HotSync® Cradle-Serial, SKU p10828U, sold separately) or Mac 9.1 or higher/Mac OS X, version 10.1.2 to 10.2.6

REMOTE EMAIL AND INTERNET ACCESS REQUIREMENTS
Requires an Internet Service Provider account and a data enabled phone or modem, or a Bluetooth Network Access Point (not included).

SIZE AND WEIGHT	
Height	4.3 in. (closed)
Width	3.0 in.
Thickness	.66 in.
Weight	(Handheld + Stylus) 5.5 oz

Source: Palm, Inc., http://www.palmone.com/us/products/handhelds/tungsten-t3/specs.html.

Figure 18.8: The design of this specification allows readers to quickly gain access to the information they need. Notice how information is presented in easy-to-access blocks.

Three-column format

Parts are described concisely.

Table is used to present data.

Image illustrates the product.

You don't need to restrict yourself to an uninteresting one-column format. Use your imagination. Figure 18.8, for example, shows how columns and tables can be used to pack a solid amount of information into one place, while still presenting the information in an attractive way.

LINK For more information on designing documents, see Chapter 8, page 192.

Using Graphics

Graphics are helpful in technical descriptions. Pictures, illustrations, and diagrams help the readers visualize your subject and its parts.

Using your computer, you can collect or create a wide range of graphics. Many free-use graphics are widely available on the Internet. (Reminder: Unless the site specifies that the graphics are free to use, you must ask permission to use them). If you cannot find graphics on the Internet, you can use a digital camera to take photographs that can be downloaded into your text. You can also use a scanner to digitize pictures, illustrations, and diagrams.

TAKE NOTE Drawing software packages, including the drawing features on your word-processing program, can help you make simple illustrations and diagrams.

Here are some guidelines for using graphics in a technical description:

- If possible, use a title and number with each graphic, so it can be referred to in the written text.
- Include captions that explain what the graphic shows.
- Label specific features in the graphic.
- Reference the graphic in the written text; e.g., "As shown in Figure 3. . ." or "(See Figure 3)."
- Place the graphic on the page where it is referenced or soon afterward.

It is not always possible to include titles, numbers, and captions with your graphics. In these situations, all graphics should appear immediately next to or immediately after the places where they are discussed.

Even simple graphics are helpful in technical descriptions. In almost all cases, you should plan to include some kind of picture, illustration, or diagram to support your written text.

LINK For more information on using graphics, see Chapter 9, page 236.

Revising, Editing, and Proofreading

You should save plenty of time to revise and edit your description. Often, what seems obvious or plain to you in a description is not obvious to your readers. Why? Because you probably understand the object, place, or process very well, while your readers are not familiar with it.

Need graphics? Go to
www.ablongman.com/johnsonweb/18.13
For websites that discuss photography, go to
www.ablongman.com/johnsonweb/18.14

HELP

Using Digital Photography in Descriptions

Digital cameras and scanners offer an easy way to insert visuals into your descriptions. These digitized pictures are inexpensive, alterable, and easily added to a text. Moreover, they work well in print and on-screen texts. Here are some photography basics to help you use a camera more effectively.

Resolution—Digital cameras will usually allow you to set the resolution at which you shoot pictures. If you are using your camera to put pictures on the web, you should use the 640 × 480 pixel setting. This setting will allow the picture to be downloaded quickly, because the file is smaller. If you want your picture to be a printed photograph, a minimum 1280 × 1024 pixel setting is probably needed. On-line pictures are usually best saved in jpeg format, while print photos should be saved as tiff files.

Shutter speed—Some digital cameras and better conventional cameras will allow you to adjust your shutter speed. The shutter speed determines how much light is allowed into the camera when you push the button. Shutter speeds are usually listed from 1/1000 of a second to 1 second. Slower speeds (like 1 second) capture more detail but risk blurring the image, especially if your subject is moving. Faster speeds (like 1/1000 second) will not capture as much detail, but they are good for moving subjects.

Cropping—Once you have downloaded your picture to your computer, you can "crop" the picture to remove things you don't want. Do you want to remove an old roommate from your college pictures? You can use the cropping tool to cut him or her out of the picture. Most word processors have a cropping tool that you can use to carve away unwanted parts of your pictures (Figure A).

Cropping a Digital Photograph

Fox @ 100% (Layer 1, RGB/8)
50%
Untagged CMYK
The cropping tool lets you frame the part of the picture you want.
The toolbar offers a variety of options for altering the picture.

Figure A: Using the cropping tool, you can focus the photograph on the subject. Or, you can eliminate things or people you don't want in the picture.

Retouching—One of the main advantages of digital photographs is the ease with which they can be touched up. Professional photographers make ample use of programs like Adobe Photoshop to manipulate their photographs. If the picture is too dark, you can lighten it up. If the people in the photo have "red-eye," you can remove that unwanted demonic stare.

One of the nice things about digital photography is that you can make any photograph look professional. With a digital camera, since the "bullets are free," as photographers say, you can experiment freely.

Revising for Conciseness

At this point, you should look for ways to make your description more concise. You should—

- comb through the text, looking for places where you go beyond need-to-know information.
- find places where you have included too many details for your readers. Trim out this excessive information.
- shorten any long sentences to make them easier to read.

When revising, try to boil your subject down to the fewest words possible, while still meeting the purpose of the description.

Editing and Proofreading for Accuracy

Above all, descriptions need to be accurate. As you are editing your description,

- check the accuracy of any figures.
- double-check measurements for exactness.
- confirm units of measurement.
- look up any words that you are not completely sure about.

TAKE NOTE There are several helpful dictionaries on the Internet. Don't hesitate to look up a word.

Proofreading is always important, but it is even more important if your description will be used as a sales tool or in a proposal. A typo or misspelling might be forgivable in your company's in-house materials, but your company's reputation is at stake when written materials are being shown to customers or clients. You want these materials to be error free.

You should show your description to your supervisor or colleagues before you finish it. Your familiarity with the subject might cause you to overlook problems in your text. So, letting someone with a fresh perspective look over your materials might help you locate trouble spots or errors.

Worksheets to help you revise your descriptions are available at **www.ablongman.com/johnsonweb/18.15**

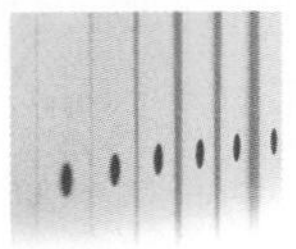

CHAPTER REVIEW

- Technical descriptions are written to describe objects, places, phenomena, and processes. They are important documents in all technical workplaces.
- Basic features of a technical description include a title, introduction, body, graphics, and conclusion.
- An object, place, or process can be partitioned according to its features, functions, or stages.
- Technical descriptions tend to be written in the "plain style," meaning words and sentences are kept simple, direct, and concise.
- To add a visual element to the description, use your senses, similes, analogies, and metaphors to describe the subject.
- Graphics are crucial in technical descriptions, because the purpose of the document is to allow the readers to "see" the object, place, or process. You can use pictures, illustrations, and diagrams to add graphics to your text.
- The design of the description will depend on how it is being used. In sales literature, the design will probably be colorful or ornate. A specification for the company's files might be rather plainly designed.

EXERCISES AND PROJECTS

Individual or Team Projects

1. Find a technical description on the Internet or in your workplace or home. First, determine the rhetorical situation (subject, purpose, readers, context) for which the description was written. Then, study its content, organization, style, and design. Write a two-page memo to your instructor in which you offer a critique of the description. What do you find effective about the description? What could be improved?

2. Your company sells a variety of different products, listed below. Choose one of these items and write a one-page technical description. Your description should be aimed at a potential customer who might purchase one of these products:

   plasma-screen television
   DVD player
   MP3 player
   bicycle
   clock radio
   telescope
   washing machine
   baby stroller
   toaster
   coffeemaker
   video camera

Sample descriptions are available at
www.ablongman.com/johnsonweb/18.16

3. People in your field of study (major) likely use tools, machines, or processes that are unfamiliar to nonexperts. Find one of these items and write a technical description for an audience of nonexperts. In your introduction, make sure you define the subject you are describing and stress its importance to your field.

4. Pick a building on your campus and write a technical description of it. Before drafting your description, though, identify some readers (e.g., new students, visitors, alumni) who might actually have a use for your description. Then, write and design your description in a way that suits their needs.

5. Find a common process that you can describe. Then, describe that process, walking your readers through its stages. In your description, you should define any jargon or technical terms that may be unfamiliar to your readers.

6. Study the description for the Segway or Mars Pathfinder Rover in this chapter. Write a two-page memo critiquing the description you choose. Do you think the description includes enough information? How might its content, organization, style, and design be improved?

Collaborative Project

Your group has been assigned to describe a variety of renewable energy sources that might be used in your state. These energy sources might include solar, wind, geothermal, and biomass power and fuel cells. Keeping the energy needs and limitations of your region in mind, offer a brief description of each of these renewable energy sources, showing how it works, its advantages, and its disadvantages.

In a report to your state's energy commissioner, describe these energy sources and discuss whether you think they offer possible alternatives to nonrenewable energy sources.

Revision Challenge

Figure A shows a description of hypothermia and frostbite that appears on the Rapid City, South Dakota National Weather Service Forecast Office website. The description presented on this webpage is adequate, but the organization, style, and design of the text are not accessible to some readers, especially children.

Imagine you are helping the National Weather Service create a website for children that discusses safety issues related to weather. How could you revise the description on this webpage to be suitable for children who are ten to fifteen years old?

Using the principles discussed in this chapter, analyze the content, organization, style, and design of this webpage. Then, revise this description of hypothermia and frostbite to improve its organization, style, and design. Make the text suitable for children.

To help you revise, you can download a version of this document at www.ablongman.com/johnsonweb/18.17.

A Description that Could Be Revised

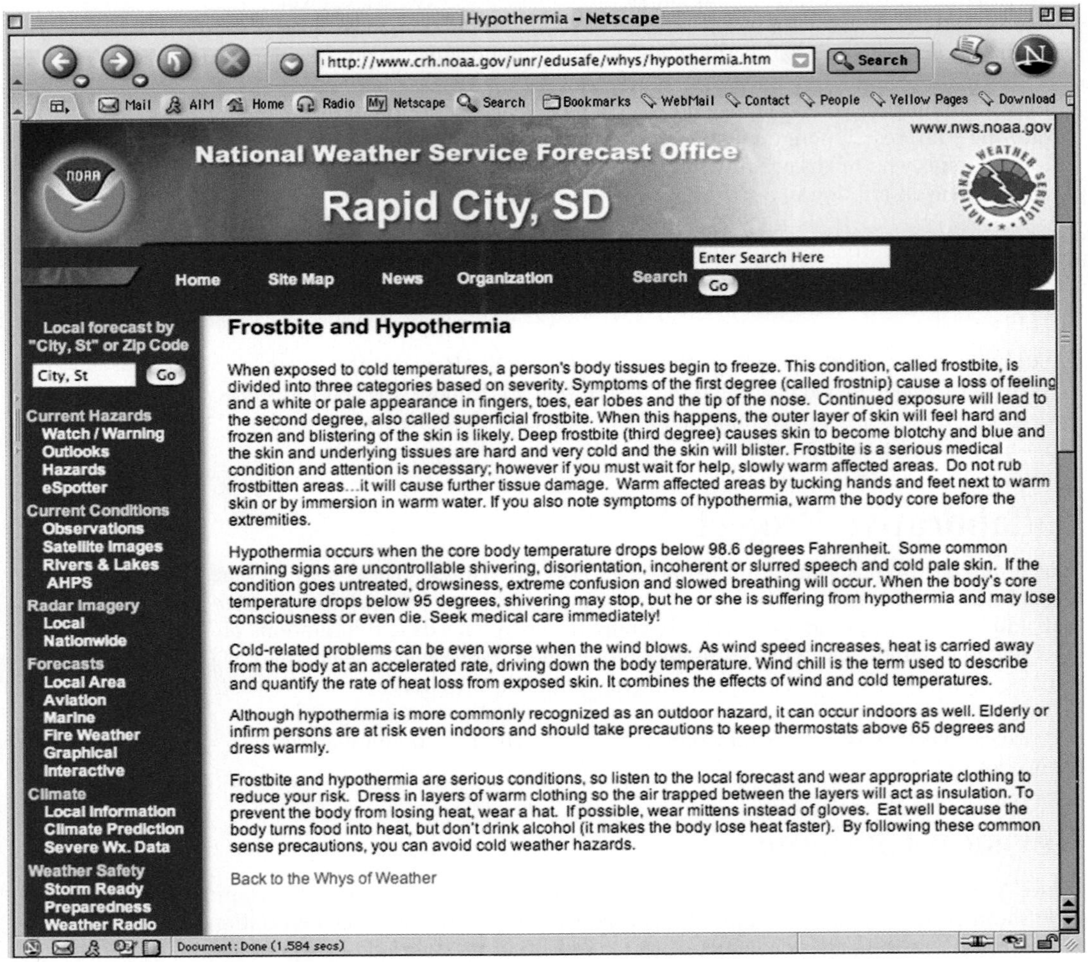

Hypothermia – **Netscape**

http://www.crh.noaa.gov/unr/edusafe/whys/hypothermia.htm Search

Mail AIM Home Radio Netscape Search Bookmarks WebMail Contact People Yellow Pages Download

www.nws.noaa.gov

National Weather Service Forecast Office

Rapid City, SD

Home Site Map News Organization Search Enter Search Here Go

Local forecast by "City, St" or Zip Code
City, St Go

Current Hazards
- Watch / Warning
- Outlooks
- Hazards
- eSpotter

Current Conditions
- Observations
- Satellite Images
- Rivers & Lakes
- AHPS

Radar Imagery
- Local
- Nationwide

Forecasts
- Local Area
- Aviation
- Marine
- Fire Weather
- Graphical
- Interactive

Climate
- Local Information
- Climate Prediction
- Severe Wx. Data

Weather Safety
- Storm Ready
- Preparedness
- Weather Radio

Frostbite and Hypothermia

When exposed to cold temperatures, a person's body tissues begin to freeze. This condition, called frostbite, is divided into three categories based on severity. Symptoms of the first degree (called frostnip) cause a loss of feeling and a white or pale appearance in fingers, toes, ear lobes and the tip of the nose. Continued exposure will lead to the second degree, also called superficial frostbite. When this happens, the outer layer of skin will feel hard and frozen and blistering of the skin is likely. Deep frostbite (third degree) causes skin to become blotchy and blue and the skin and underlying tissues are hard and very cold and the skin will blister. Frostbite is a serious medical condition and attention is necessary; however if you must wait for help, slowly warm affected areas. Do not rub frostbitten areas…it will cause further tissue damage. Warm affected areas by tucking hands and feet next to warm skin or by immersion in warm water. If you also note symptoms of hypothermia, warm the body core before the extremities.

Hypothermia occurs when the core body temperature drops below 98.6 degrees Fahrenheit. Some common warning signs are uncontrollable shivering, disorientation, incoherent or slurred speech and cold pale skin. If the condition goes untreated, drowsiness, extreme confusion and slowed breathing will occur. When the body's core temperature drops below 95 degrees, shivering may stop, but he or she is suffering from hypothermia and may lose consciousness or even die. Seek medical care immediately!

Cold-related problems can be even worse when the wind blows. As wind speed increases, heat is carried away from the body at an accelerated rate, driving down the body temperature. Wind chill is the term used to describe and quantify the rate of heat loss from exposed skin. It combines the effects of wind and cold temperatures.

Although hypothermia is more commonly recognized as an outdoor hazard, it can occur indoors as well. Elderly or infirm persons are at risk even indoors and should take precautions to keep thermostats above 65 degrees and dress warmly.

Frostbite and hypothermia are serious conditions, so listen to the local forecast and wear appropriate clothing to reduce your risk. Dress in layers of warm clothing so the air trapped between the layers will act as insulation. To prevent the body from losing heat, wear a hat. If possible, wear mittens instead of gloves. Eat well because the body turns food into heat, but don't drink alcohol (it makes the body lose heat faster). By following these common sense precautions, you can avoid cold weather hazards.

Back to the Whys of Weather

Document: Done (1.584 secs)

Source: National Weather Service, http://www.crh.noaa.gov/unr/edusafe/whys/hypothermia.htm.

Figure A: This description is somewhat unorganized. The style is stiff and the design is intimidating. How could this text be revised to make it more readable and interesting?

To download the text for revision, go to **www.ablongman.com/johnsonweb/18.17**

CASE STUDY

Memories of a Crane

Catherine Smith, an ornithology student from Illinois, was invited to participate in a bird count in southern Texas along the Gulf of Mexico coast. It sounded like fun, and she needed the experience. She and a couple of others were assigned to identify and count the birds in a farm field about a mile from the coast. They noticed that this field would soon be developed into housing.

The first day was difficult. There were so many different kinds of birds. Catherine and the others struggled to keep up. As they were eating lunch, Catherine spotted a large white crane standing in a distant field. Then, it flew away. "Better remember that one, folks," she said. "I haven't seen one of those before."

"None of us got a good look, so you will need to remember it for the rest of us," said one of the other team members.

After lunch, Catherine found the bird in her bird guide. It was a whooping crane. She marked it down in her field book.

A week later, she was back at school in Illinois. She received a call from the principal investigator in charge of the Texas bird count. He asked her if she was absolutely sure she had seen a whooping crane.

"I think I did," she replied.

"No," said the principal investigator, "you need to be certain. Either you did or you didn't. It's that important. The whooping crane is a very, very rare bird. If you saw one, the Endangered Species Act could be used to protect that area from development. I need you to be sure because we could go to court to protect the area if you can testify you saw this bird."

The principal investigator asked her to write down a description of exactly what she saw and send it to him. She turned to her bird guide to refresh her memory on just what she had seen.

Unfortunately, by the time she turned to write her description, she wasn't certain anymore whether she had seen that exact bird or not. It could have been a sandhill crane. She began to wonder whether she had made a mistake. But then, she was pretty sure the bird was like the one in her bird guide. As she wrote her description, she found herself turning to her bird guide to fill in visual details about the bird. Soon she was more confused than ever. Was she describing what she saw, or was she describing the bird in the book?

How should Catherine handle this situation?

To see websites about cranes, go to **www.ablongman.com/johnsonweb/18.18**

CHAPTER

19

Instructions

CHAPTER CONTENTS

	EMT-Paramedic Treatment Protocol 4501
Allergic Reaction/Anaphylaxis	**Page 1 of 2**

Anaphylaxis is an acute allergic reaction characterized by varying deg... distress, hypotension, wheezing, hives, non-traumatic edema, and tac... precipitated by a bite or sting or from exposure to certain drugs or all...

A. Perform **MAMP Protocol 4201**.

B. If from sting, remove injection mechanism, if present.

C. If patient is in mild distress with hives or itching but no or minimal respiratory distress (no wheezing or stridor):

1. Consider diphenhydramine (*Benadryl*) per **MCP order**.
   a. Adult: 25-50 mg, IM or slow IV.
   b. Pediatric: 1 mg/kg, IM or sl... 25 mg.
2. Maintain normal saline IV...
3. Reassess for improveme...
4. Transport and notify Medical Co...

D. If patient is in moderate distress with severe hives and/or m... distress (wheezing):

1. Immediately administer epinephrine, 1:1000:
   a. Adult: 0.3 mg SQ.
   b. Pediatric: 0.01 mg/kg SQ (maximum single dose of 0.3 mg).
   c. If age >50, **per MCP order**.

Task
Step 1
Step 2
Step 3
Step 4
minor steps
notes
minor steps
notes
minor steps
notes
minor steps
notes
Worst–Case Scenarios
http://www.worstcasescenarios.com/mainpage.htm
WORST-CASE SCENARIOS: online
STORE
FUN AND GAMES
THE TV SERIES
SURVIVAL COLUMN
ABOUT
CONTACT US
TRAVEL SURVIVAL
DATING & SEX SURVIVAL
GOLF SURVIVAL
HOLIDAY SURVIVAL
WORK SURVIVAL
SCENARIO ARCHIVE
Extreme Survival
How to Jump from a Building into a Dumpster
How to Survive If Your Parachute Fails to Open
How to Make Fire Without Matches
How to Avoid Being Struck by Lightning
Extreme Survival
How to Escape from a Sinking Car
1. As soon as you hit the water, open your window. This is your best chance of escape, because opening the door will be very difficult given the outside water pressure. (To be safe, you should drive with the windows and doors slightly open whenever you are near water or driving on ice.) Opening the window allows water to come in and equalize the pressure. Once the water pressure inside and outside the car is equal, you'll be able to open the door.
2. If your power windows won't work or you cannot roll your windows down all the way, attempt to break the glass with your foot,
Click image to enlarge and view captions

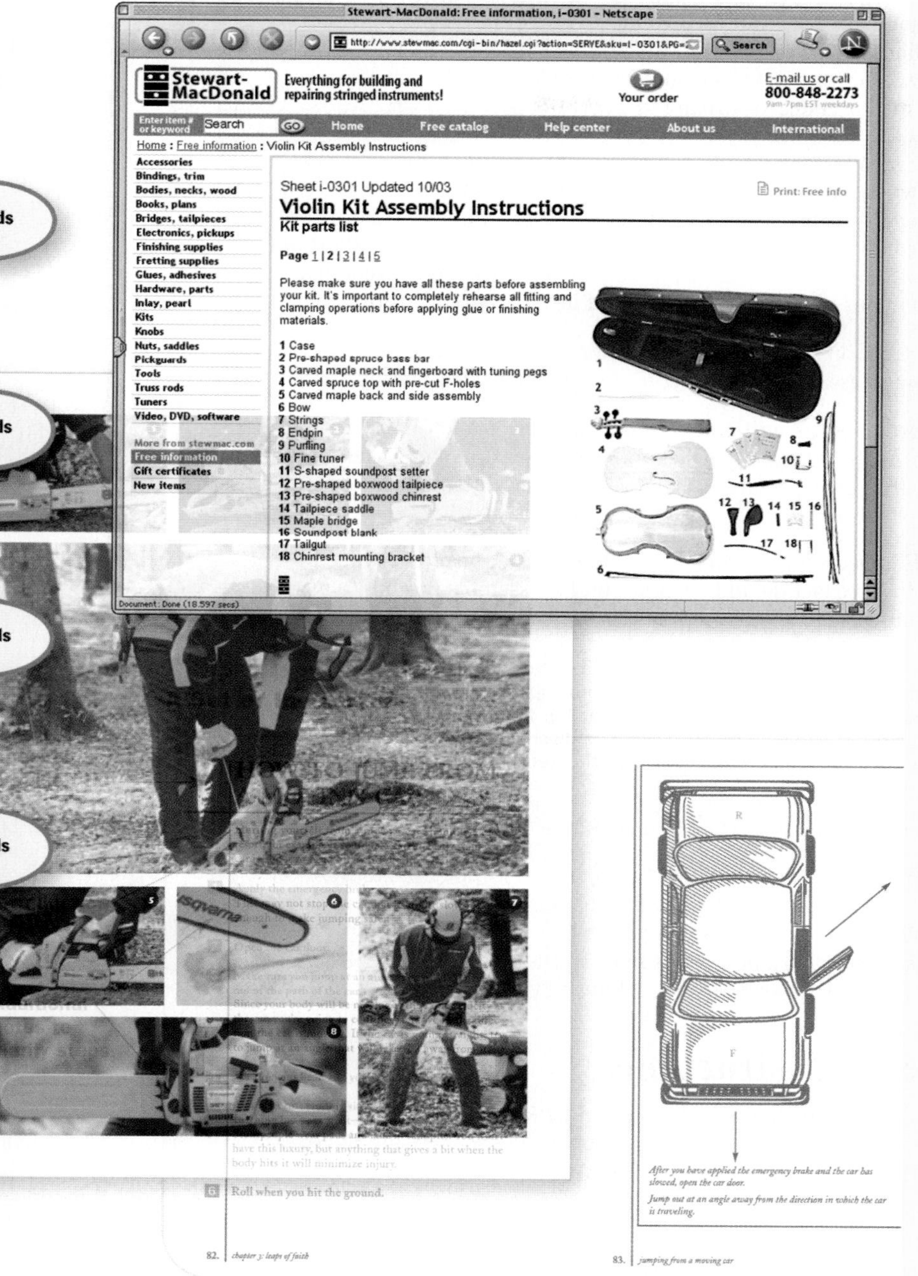

CHAPTER OBJECTIVES

In this chapter, you will learn:

- The importance of instructions, procedures, and specifications in the technical workplace.
- The basic features of instructions and procedures.
- How to plan and research instructions and procedures.
- How to organize and draft instructions and procedures.
- How style and design can be used to highlight and reinforce written text.
- Strategies for revising, editing, and proofreading instructions and procedures.

More than likely, you have read and used countless sets of instructions in your lifetime. Instructions are packaged with the products we buy, like phones, cameras, and televisions. They also help us complete everyday activities, like cooking dinner, setting up an e-mail account, or driving to a friend's house.

Instructions describe step by step how to complete a task.

Instructions describe step by step how to complete a task. In the technical workplace, instructions are written to describe a variety of tasks and activities. Some tasks are routine, like downloading software or drawing blood from a patient. Other tasks can be quite complicated, like assembling a satellite or piloting an airplane. Short or long, the instructions for doing these tasks follow similar patterns and conventions.

Three types of instructions are commonly produced in the technical workplace:

Instructions—Instructions describe how to perform a specific task. They typically describe how to assemble a product or do something step by step.

Procedures—Procedures are written to ensure consistency and quality in a workplace. In hospitals, for example, doctors and nurses are often asked to write procedures that describe how to handle emergency situations or care for a specific injury or illness.

Specifications—Engineers and technicians write specifications (often called the "specs") to describe in exact detail how a product is assembled or how a routine process is completed.

TAKE NOTE Documents called "specifications" are also written to describe products in exact detail (see Chapter 18). So, the dual use of the word "specifications" in technical workplaces might be confusing. In the end, though, remember that specifications are written to describe products or processes. The specifications discussed in Chapter 18 typically describe products. The specifications discussed in this chapter describe instructions or procedures.

LINK For more information on writing descriptions, go to Chapter 18, page 531.

To avoid confusion in this chapter, the word "instructions" will be used to mean instructions, procedures, and specifications. When the chapter discusses issues specific to procedures or specifications, those terms will be used.

Writing instructions is an important responsibility in the technical workplace. In this chapter, you will learn how to write them quickly and efficiently.

Basic Features of Instructions

Instructions tend to follow a consistent step-by-step pattern, whether you are describing how to make coffee or how to assemble an automobile engine. Here are the basic features of instructions:

- Specific and precise title.
- Introduction with background information.
- List of parts, tools, and conditions required.
- Sequentially ordered steps.

To see samples of various kinds of instructions, go to **www.ablongman.com/johnsonweb/19.1**

- Graphics.
- Safety information.
- Conclusion that signals completion of task.

Sequentially ordered steps are the centerpiece of a set of instructions, and they typically take up much of the space in the document (Figure 19.1).

Basic Organization of Instructions

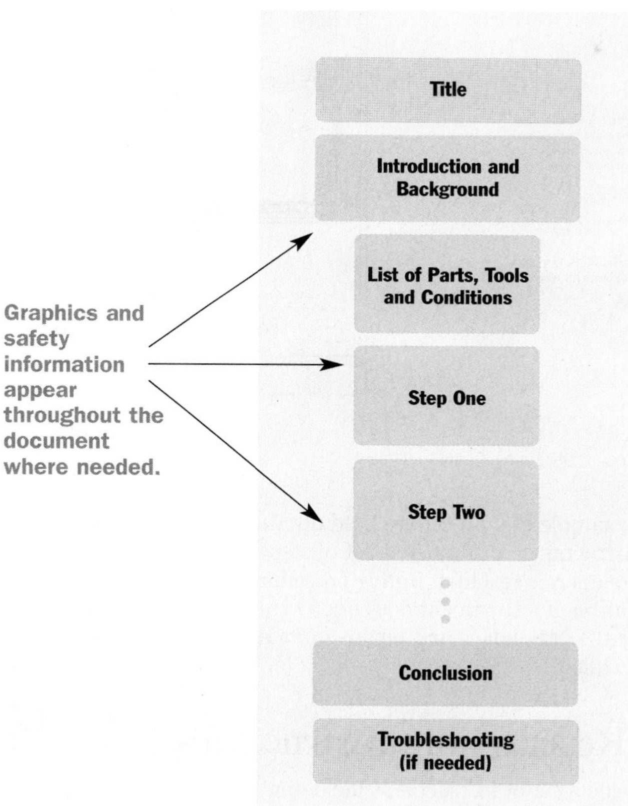

Figure 19.1: Instructions, including procedures and specifications use some or all of these parts.

The content, style, and design of your instructions will change to suit the readers and the contexts in which the document will be used. For example, Figure 19.2 shows survival instructions from an entertaining and useful book called *Worst-Case Scenarios.* These instructions use simple text and visuals to show how to jump out of a moving car—in case you ever need to. This set of instructions is designed to be easy to remember.

A Set of Instructions

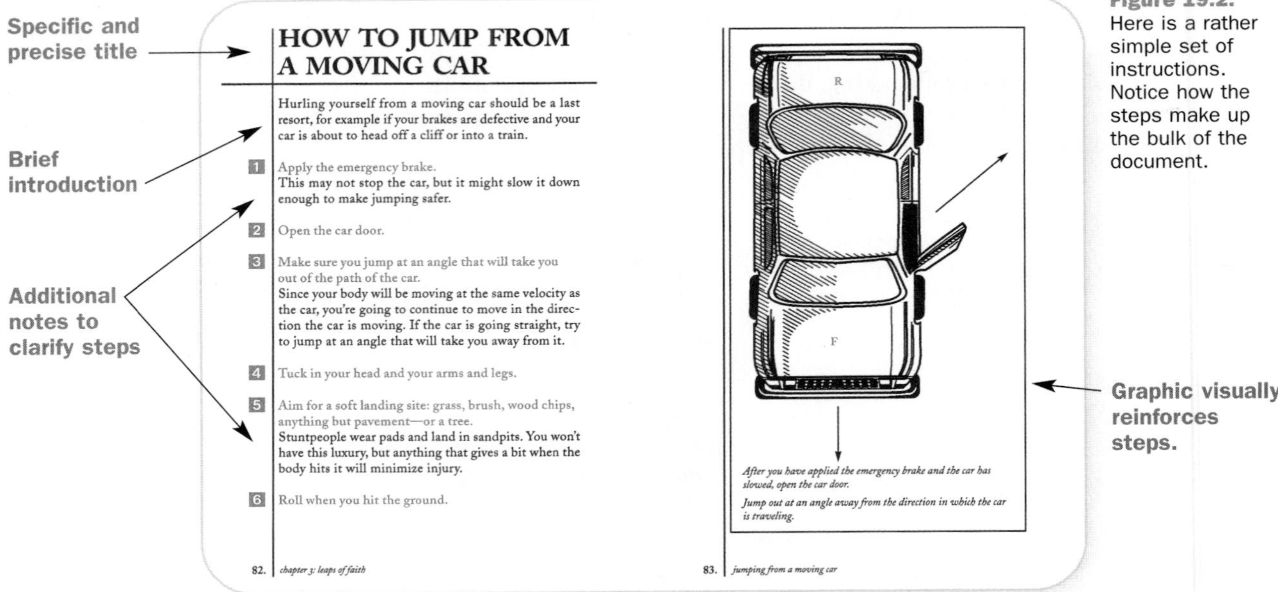

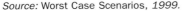

HOW TO JUMP FROM A MOVING CAR

Hurling yourself from a moving car should be a last resort, for example if your brakes are defective and your car is about to head off a cliff or into a train.

1. Apply the emergency brake.
   This may not stop the car, but it might slow it down enough to make jumping safer.
2. Open the car door.
3. Make sure you jump at an angle that will take you out of the path of the car.
   Since your body will be moving at the same velocity as the car, you're going to continue to move in the direction the car is moving. If the car is going straight, try to jump at an angle that will take you away from it.
4. Tuck in your head and your arms and legs.
5. Aim for a soft landing site: grass, brush, wood chips, anything but pavement—or a tree.
   Stuntpeople wear pads and land in sandpits. You won't have this luxury, but anything that gives a bit when the body hits it will minimize injury.
6. Roll when you hit the ground.

After you have applied the emergency brake and the car has slowed, open the car door.

Jump out at an angle away from the direction in which the car is traveling.

82. *chapter 3: leaps of faith*

83. *jumping from a moving car*

Figure 19.2: Here is a rather simple set of instructions. Notice how the steps make up the bulk of the document.

Source: Worst Case Scenarios, *1999.*

Figure 19.3 shows an example of a procedure used at a hospital. This procedure works like most sets of instructions. A numbered list of steps explains what to do if someone is having a severe allergic reaction. In this procedure, though, notice how the directions offer alternatives if a treatment does not work. Also, the procedure includes helpful visuals, like the gray boxes and pictures of a doctor, to signal places where a doctor should be called.

Planning and Researching Instructions

When you are asked to write a set of instructions, including procedures and specifications, you should first consider the situations in which they might be used. You also need to research the process you are describing, so you fully understand it and can describe it in detail.

Planning

When planning the document, you should first gain a thorough grasp of your readers and their needs. A good way to start is to answer the Five-W and How questions that define the rhetorical situation:

Who might use these instructions?

Why are these instructions needed?

A Procedure

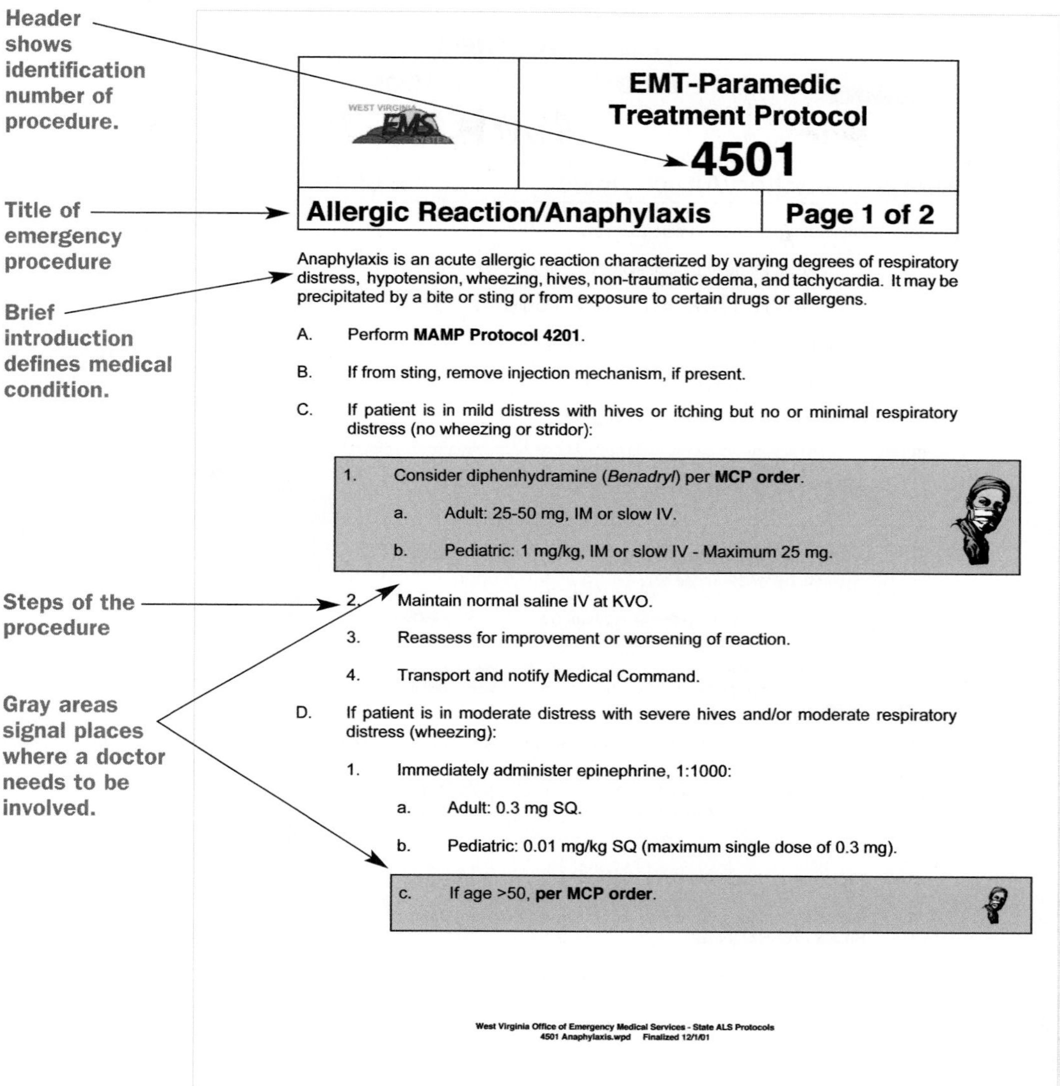

Header shows identification number of procedure.
Title of emergency procedure
Brief introduction defines medical condition.
Steps of the procedure
Gray areas signal places where a doctor needs to be involved.

	EMT-Paramedic Treatment Protocol **4501**
Allergic Reaction/Anaphylaxis	**Page 1 of 2**

Anaphylaxis is an acute allergic reaction characterized by varying degrees of respiratory distress, hypotension, wheezing, hives, non-traumatic edema, and tachycardia. It may be precipitated by a bite or sting or from exposure to certain drugs or allergens.

A. Perform **MAMP Protocol 4201**.

B. If from sting, remove injection mechanism, if present.

C. If patient is in mild distress with hives or itching but no or minimal respiratory distress (no wheezing or stridor):

1. Consider diphenhydramine (*Benadryl*) per **MCP order**.
   a. Adult: 25-50 mg, IM or slow IV.
   b. Pediatric: 1 mg/kg, IM or slow IV - Maximum 25 mg.
2. Maintain normal saline IV at KVO.
3. Reassess for improvement or worsening of reaction.
4. Transport and notify Medical Command.

D. If patient is in moderate distress with severe hives and/or moderate respiratory distress (wheezing):

1. Immediately administer epinephrine, 1:1000:
   a. Adult: 0.3 mg SQ.
   b. Pediatric: 0.01 mg/kg SQ (maximum single dose of 0.3 mg).
   c. If age >50, **per MCP order**.

West Virginia Office of Emergency Medical Services - State ALS Protocols
4501 Anaphylaxis.wpd Finalized 12/1/01

Source: *West Virginia EMS System, 2001.*

Figure 19.3: Procedures like this one are used for training. They also standardize care.

Continued on following page

Figure 19.3: (*continued.*)

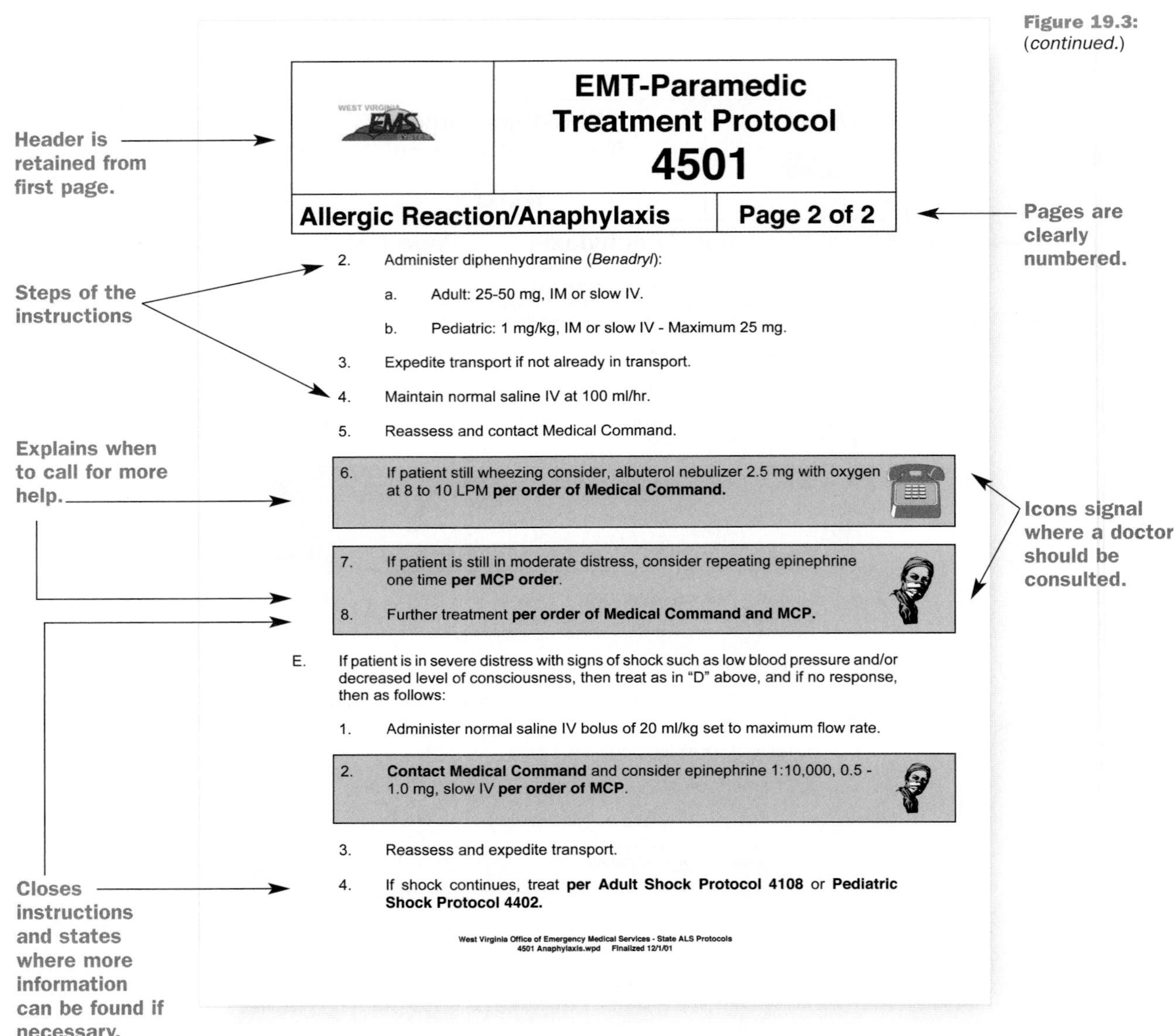

EMT-Paramedic Treatment Protocol 4501	
Allergic Reaction/Anaphylaxis	Page 2 of 2

2. Administer diphenhydramine (*Benadryl*):
   a. Adult: 25-50 mg, IM or slow IV.
   b. Pediatric: 1 mg/kg, IM or slow IV - Maximum 25 mg.
3. Expedite transport if not already in transport.
4. Maintain normal saline IV at 100 ml/hr.
5. Reassess and contact Medical Command.
6. If patient still wheezing consider, albuterol nebulizer 2.5 mg with oxygen at 8 to 10 LPM **per order of Medical Command.**
7. If patient is still in moderate distress, consider repeating epinephrine one time **per MCP order**.
8. Further treatment **per order of Medical Command and MCP.**

E. If patient is in severe distress with signs of shock such as low blood pressure and/or decreased level of consciousness, then treat as in "D" above, and if no response, then as follows:

1. Administer normal saline IV bolus of 20 ml/kg set to maximum flow rate.
2. **Contact Medical Command** and consider epinephrine 1:10,000, 0.5 - 1.0 mg, slow IV **per order of MCP**.
3. Reassess and expedite transport.
4. If shock continues, treat **per Adult Shock Protocol 4108** or **Pediatric Shock Protocol 4402.**

West Virginia Office of Emergency Medical Services - State ALS Protocols
4501 Anaphylaxis.wpd Finalized 12/1/01

WEST VIRGINIA
EMS
Header is retained from first page.
Pages are clearly numbered.
Steps of the instructions
Explains when to call for more help.
Icons signal where a doctor should be consulted.
Closes instructions and states where more information can be found if necessary.

What should the instructions include?

Where will the instructions be used?

When will the instructions be used?

How will these instructions be used?

Once you have answered these questions, you are ready to define the rhetorical situation that will shape how you write your instructions.

SUBJECT Make sure you are completely familiar with the task you are trying to describe. Give yourself time to use the product or follow the process. Try to identify any unexpected difficulties or dangers. Play with the product or process, looking for places where your readers might have trouble or make mistakes.

TAKE NOTE Don't make the expert's mistake. Experts have great familiarity with the product or process, so they often overlook potential problems for nonexperts. You need to anticipate these problems.

PURPOSE In most cases, the purpose of your instructions will seem obvious. Your instructions will show someone how to complete a task. But even if your purpose seems obvious, take a moment to consider and compose your purpose statement, limiting yourself to one sentence. That way, you will be very clear about what you are trying to accomplish.

Some key verbs for the purpose statement might include the following:

to instruct	to guide
to show	to lead
to illustrate	to direct
to explain	to train
to teach	to tutor

For example, here are a few purpose statements that might be used in a set of instructions:

> The purpose of these instructions is to show you how to use your new QuickTake 1031 digital video camera.
>
> These procedures will demonstrate the suturing required to complete and close up a knee operation.
>
> These specifications illustrate the proper use of the Series 3000 Router to trim printed circuit boards.

A version of your purpose statement will likely appear in the introduction of your instructions.

LINK For more information on defining a document's purpose, see Chapter 2, page 23.

READERS Of course, it is difficult to anticipate all the people who might use your instructions. But, people who decide to use a set of instructions usually have common characteristics, backgrounds, and motivations that you can use to make your instructions more effective.

Primary readers (action takers) are people who will use your instructions to complete a task. What is their skill level? How well do they understand the product or process? What is their age and ability? Typically instructions are written for the primary readers who have minimal experience with and knowledge of the task.

TAKE NOTE Try not to overestimate your readers' skills and understanding. If your instructions are detailed and use simple language, novice users will appreciate the added help. Experienced users may find your instructions a bit too detailed, but they can just skim information they don't need. In most cases, you are better off giving your readers more information than they need.

Secondary readers (advisors) are people who might supervise or help the primary readers complete the task. What is the skill level of these secondary readers? Are they training/teaching the primary readers? Are they helping them assemble the product or complete the process? Keep in mind that procedures and specifications are often used by supervisors to train and mentor new employees. The documents need to be clear to both trainer and trainee.

Tertiary readers (evaluators) often use instructions to check quality and look for problems. Auditors and quality experts will review procedures and specifications closely when evaluating products or processes in a technical workplace. Also, they are frequently used as evidence in lawsuits, so these kinds of readers need to be considered and their needs anticipated.

Gatekeeper readers (supervisors) will need to look over your instructions before they are sent out with a product or approved for use in the workplace. These gatekeeper readers may or may not be experts in your subject. They will be checking the accuracy, safety, and quality of your instructions.

LINK For more information on analyzing readers, see Chapter 3, page 42.

CONTEXT OF USE Put yourself in your readers' place for a moment. When and where will readers use your documentation? In their living room? At a workbench? At a construction site? In an office cubicle? At night? Each of these different places and times will require you to adjust your instructions to their needs.

Depending on the context of use, instructions can follow a variety of formats. They may be included in a user manual, or they could be part of a poster explaining how to accomplish a task. Increasingly, instructions are being placed on websites for viewing or downloading. Figure 19.4, for example, shows how the instructions from *Worst-Case Scenarios* are portrayed on a website.

Context of use also involves the safety and liability issues that are especially important in sets of instructions. If users of the instructions are at risk for injury or an accident, you are ethically obligated to warn them about the danger and tell them how to avoid it. Try to anticipate all the ways users might injure themselves or experience a mishap while following the instructions you are providing. Then, use warning

For reader analysis worksheets, go to **www.ablongman.com/johnsonweb/19.3**
To read about blunders in which writers did not anticipate their readers' needs, go to **www.ablongman.com/johnsonweb/19.4**

Instructions on a Website

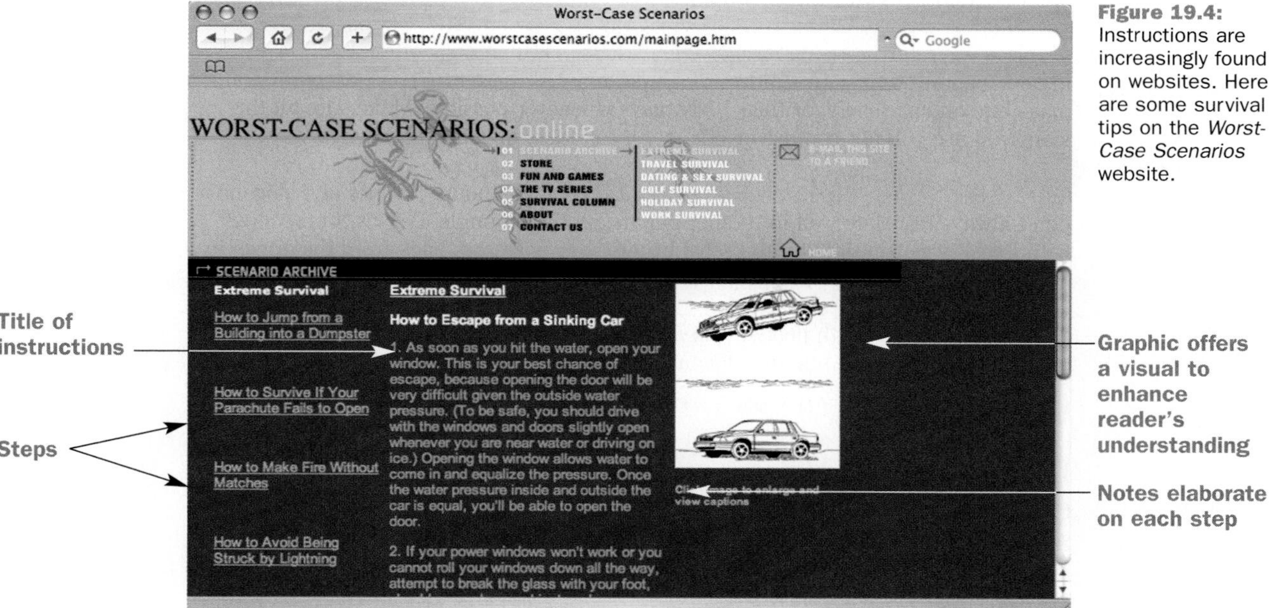

Worst-Case Scenarios
http://www.worstcasescenarios.com/mainpage.htm
WORST-CASE SCENARIOS: online
STORE
FUN AND GAMES
THE TV SERIES
SURVIVAL COLUMN
ABOUT
CONTACT US
TRAVEL SURVIVAL
DATING & SEX SURVIVAL
GOLF SURVIVAL
HOLIDAY SURVIVAL
WORK SURVIVAL
SCENARIO ARCHIVE
Extreme Survival
How to Jump from a Building into a Dumpster
How to Survive If Your Parachute Fails to Open
How to Make Fire Without Matches
How to Avoid Being Struck by Lightning
Extreme Survival
How to Escape from a Sinking Car
1. As soon as you hit the water, open your window. This is your best chance of escape, because opening the door will be very difficult given the outside water pressure. (To be safe, you should drive with the windows and doors slightly open whenever you are near water or driving on ice.) Opening the window allows water to come in and equalize the pressure. Once the water pressure inside and outside the car is equal, you'll be able to open the door.
2. If your power windows won't work or you cannot roll your windows down all the way, attempt to break the glass with your foot,
view captions
Title of instructions
Steps
Graphic offers a visual to enhance reader's understanding
Notes elaborate on each step

Figure 19.4: Instructions are increasingly found on websites. Here are some survival tips on the *Worst-Case Scenarios* website.

Source: Worst-Case Scenarios, http://www.worstcasescenarios.com.

statements and cautions (discussed later in this chapter) to help them avoid these problems.

Cross-Cultural Readers

As world trade continues to expand, instructions need to be written for cross-cultural readers. In fact, it is common today to include instructions written in two or more languages with products.

How should you address the needs of these cross-cultural readers? Here are a few suggestions.

TRANSLATE THE TEXT If the instructions are being sent to a place where people speak a non-English language, you should have them translated. For example, if your product is being sent to Japan, you should have the documentation translated for Japanese consumers. Or, if your product is being used by the Navajo nation in the southwestern United States, you should hire someone to do the translation.

USE BASIC ENGLISH If the instructions could be used by non-English speakers, you should use basic English. Avoid any jargon, idioms, and metaphors that will be understood only by North Americans (e.g., "senior citizens," "bottom line," or "back to square one"). Companies that do business with cross-cultural markets often maintain lists of basic words to use in documentation.

GO TO THE NET

For more advice on cross-cultural writing, go to **www.ablongman.com/johnsonweb/19.5**

USE ICONS CAREFULLY Some symbols commonly used in North America can be offensive to other cultures. A pointing hand, for example, is offensive in some Central and South American countries. An OK sign is offensive in Arab nations. Dogs are "unclean" animals in many parts of the world, so cartoon dogs often do not work well in documentation for multicultural readers. To avoid these problems, do not use animals, human characters, or their body parts whenever possible. Make sure all the icons you use will not confuse or offend your readers in some unintended way.

CHECK MEANINGS OF NAMES AND SLOGANS Famously, names and slogans don't always translate well into other languages. For example, Pepsi's "Come alive with Pepsi" was translated into "Pepsi brings your ancestors back from the grave" in Taiwan. In Spanish, Chevrolet's popular car, the "Nova" means "It doesn't go." Coors's slogan "Turn it loose" translated into "Suffer from diarrhea" in some Spanish-speaking countries. So, check with people who are familiar with the target culture and its language to see if your names and slogans have meanings other than those intended.

LINK For more information on working with cross-cultural readers, go to Chapter 3, page 60.

Researching

Once you have defined the rhetorical situation, you should spend some time doing research on the task you are describing. Research consists of gaining a thorough understanding of your subject. Here are a few research strategies that are especially useful when writing instructions.

DO BACKGROUND RESEARCH You should research the history and purpose of the product or process you are describing. If the product or process is new, find out why it was developed and study the documents that shaped its development. If the product or process is not new, determine whether it has evolved or changed. Also, collect any prior instructions, procedures, or specifications, that might help you write your own documentation.

MAKE OBSERVATIONS Observe people using the product. If, for example, you are writing instructions for using a coffeemaker, observe someone making coffee with it. Pay attention to his or her experiences, especially any mistakes. Your notes from these experiments will help you anticipate some of the situations and problems your readers will experience.

ASK SUBJECT MATTER EXPERTS (SMEs) Interview the experts who are very familiar with the product or have used the procedure. They may be able to give you some insight or pointers into how the product is actually used or the procedure is completed. They might also be able to point out trouble spots where other nonexperts might have problems.

USE YOUR SENSES As you work with the product or follow the process, pay attention to all your senses. Where appropriate, take notes about appearance, sounds, smells, textures, and tastes. These details will add depth to your instructions. They will also help your readers determine if they are following the directions properly.

To read about some other funny translation problems, go to **www.ablongman.com/johnsonweb/19.6**

DESCRIBE MOTION AND CHANGE Pay special attention to the motions of your subject and the way it changes as you complete the steps. Each step will lead to some kind of motion or change. By noting these motions and changes, you will be better able to describe them.

COLLECT VISUALS If available, collect graphics that illustrate the steps in your instructions. If necessary, you can create your own pictures or use drawings to illustrate your subject.

LINK For more information on doing research, go to Chapter 5, page 96.

Organizing and Drafting Instructions

Like other technical documents, instructions have an introduction, body, and conclusion. The introduction typically offers background on the task being described. The body describes the steps required to complete the task. The conclusion usually gives readers a chance to check their work.

Specific and Precise Title

The title of your instructions should clearly describe the specific task the reader will complete.

Not descriptive	RGS-90x Telescope
Descriptive	Setting Up Your RGS-90x Telescope
Not descriptive	Head Wound
Descriptive	Procedure for Treating a Head Wound

Introduction

The length of the introduction depends on the complexity of the task and your readers' familiarity with it. If the task is simple, your introduction might be only a sentence long. If the task is complex or your readers are unfamiliar with the product or process, your introduction may need to be a few paragraphs long.

Introductions for instructions should include some or all of the following moves.

STATE THE PURPOSE Simple or complex, all instructions should have some kind of statement of purpose.

> These instructions will help you set up your RGS-90x Telescope.
>
> The Remington Medical Center uses these procedures to treat head wounds in the emergency room.

STATE THE IMPORTANCE OF THE TASK You may want to stress the importance of the task or perhaps the importance of doing the task correctly.

Your RGS-90x Telescope is one of the most revolutionary telescope systems ever developed. You should read these instructions thoroughly so you can take full advantage of the telescope's numerous advanced features.

Head wounds of any kind should be taken seriously. The following procedure should be followed in all head wound cases, even the ones that do not seem serious.

DESCRIBE THE NECESSARY TECHNICAL ABILITY You may want to describe the necessary technical background that readers will need to complete the task. Issues like age, qualifications, education level, and prior training are often important considerations that should be mentioned in the introduction.

With advanced features similar to those found in larger and more specialized telescopes, the RGS-90x can be used by casual observers and serious astronomers alike. Some familiarity with microscopes is helpful but not needed.

Because head wounds are usually serious, a trained nurse should be asked to bandage them. Head wounds should never be bandaged by trainees without close supervision.

IDENTIFY THE TIME REQUIRED FOR COMPLETION If the instructions are complex, you may want estimate the time the readers will need for completion.

Initially, setting up your telescope should take about 15 to 20 minutes. As you grow more familiar with it, though, setup should take only 5 to 10 minutes.

Speed is important when treating head wounds. You may have only a few minutes before the patient goes into shock.

MOTIVATE THE READER An introduction is a good place to set a positive tone for your document. Add a sentence or two to motivate readers and make them feel positive about the task they are undertaking.

With push-button control, automatic tracking of celestial objects, and diffraction-limited imaging, an RGS-90x telescope may be all the telescope you will ever need. With this powerful telescope, you can study the rings of the planet Saturn or observe the feather structure of a bird from 50 yards away. This telescope will meet your growing interests in astronomy or terrestrial viewing.

Head wounds of any kind are serious injuries. Learn and follow these procedures, so you can effectively treat them without hesitation.

Procedures and specifications often also include motivational statements about the importance of doing the job right, as companies urge their employees to strive for the highest quality.

List of Parts, Tools, and Conditions Required

After the introduction, you should list the parts, tools, and conditions required for completing the task.

LIST THE PARTS REQUIRED This list should identify all the necessary items required to complete the instructions. In some cases, the parts may be included in the

Want to see some sample introductions for instructions? Go to **www.ablongman.com/johnsonweb/19.7**

product's package. If they are, your list of parts will allow the readers to check whether all the parts were included (Figure 19.5). Other items not included with the package, like an adhesive, batteries, and paint, should be mentioned at this point, so readers can collect these items before following the instructions.

A Parts List

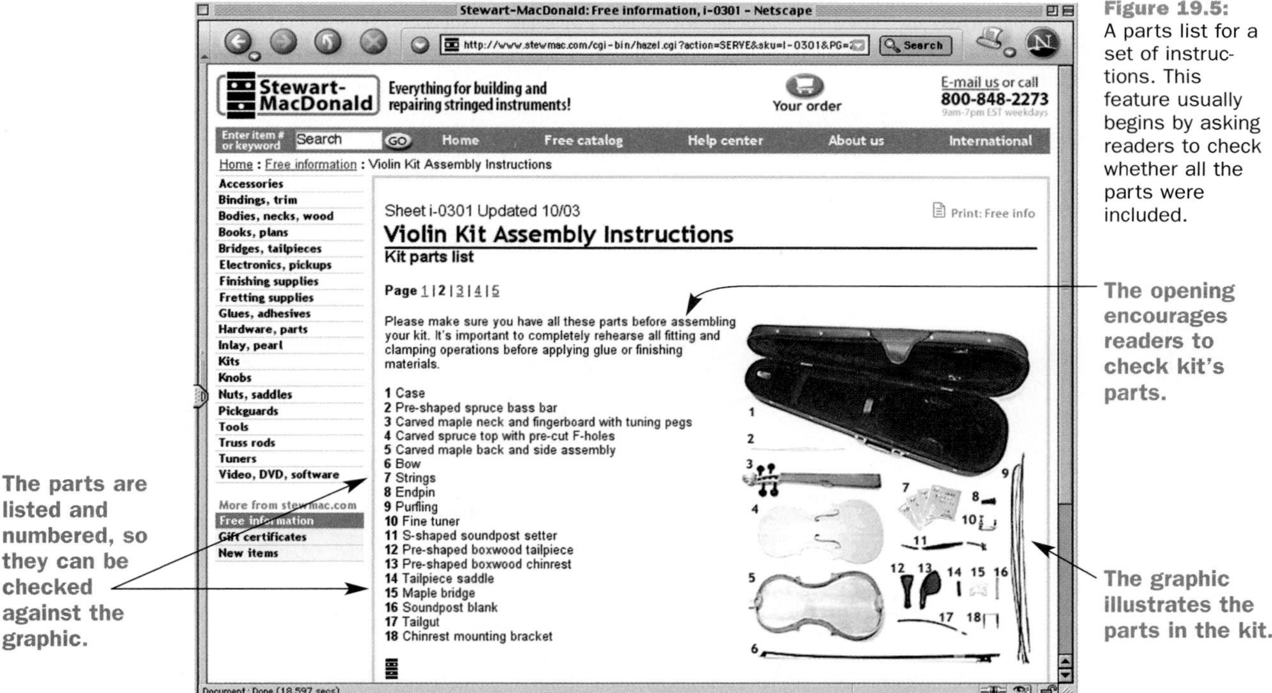

Figure 19.5: A parts list for a set of instructions. This feature usually begins by asking readers to check whether all the parts were included.

Source: Stewart-MacDonald, http://www.stewmac.com.

IDENTIFY TOOLS REQUIRED Nothing is more frustrating to readers than discovering midway through the instructions that they need a tool that was not previously mentioned. The required tools should be listed up front, so readers can gather them before starting.

SPECIFY SPECIAL CONDITIONS If any special conditions involving temperature, humidity, or light are required, mention them up front. Usually, instructions include a range for these items

> Paint is best applied when temperatures are between 50°F and 90°F.
>
> If the humidity is above 75 percent, do not solder the microchips onto the printed circuit board. High humidity may lead to a defective joint.

Sequentially Ordered Steps

The steps are the centerpiece of any set of instructions, and they will usually make up the bulk of the document. These steps need to be presented logically and concisely, allowing readers to easily understand them and complete the task.

As you carve the task you are describing into steps, you might use logical mapping to sort out the major and minor steps (Figure 19.6). First, put the overall task you are describing on the left-hand side of the screen or a sheet of paper. Then, break the task down into its major and minor steps. You might also, as shown in Figure 19.6, make note of any necessary hazard statements or additional notes that might be included.

Once you have organized the task into major and minor steps, you are ready to draft your instructions.

Using Logical Mapping to Identify Steps in Instructions

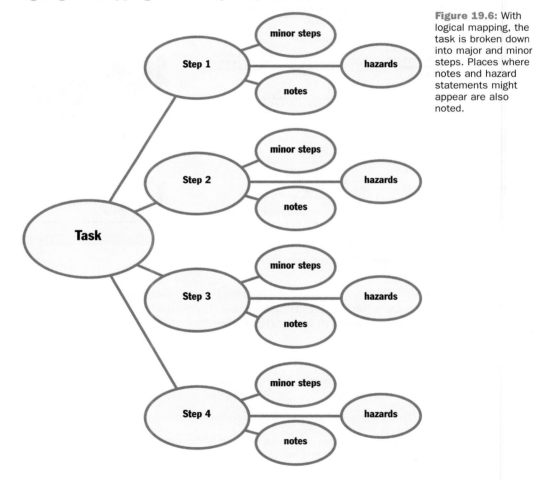

Figure 19.6: With logical mapping, the task is broken down into major and minor steps. Places where notes and hazard statements might appear are also noted.

USE COMMAND VOICE Steps should be written in *command voice,* or imperative mood. To use command voice, start each step with an action verb.

1. Place the telescope in an upright position on a flat surface.

2. Plug the coil cord for the Electronic Controller into the HBX port (see Figure 5).

In most steps, the verb should come first in the sentence. This puts the action up front, while keeping the pattern of the instructions consistent. The "you" in these sentences is not stated, but it is implied ("*You* plug the telescope in an upright position.").

STATE ONE ACTION PER STEP Each step should express only one action (Figure 19.7). You might be tempted to state two smaller actions in one step, but your readers will appreciate following each step separately.

Ineffective

2. Place the telescope securely on its side as shown in Figure 4 and open the battery compartment by simultaneously depressing the two release latches.

Revised

2. Place the telescope securely on its side as shown in Figure 4.

3. Open the battery compartment by simultaneously depressing the two release latches.

Instructions with Sequentially Ordered Steps

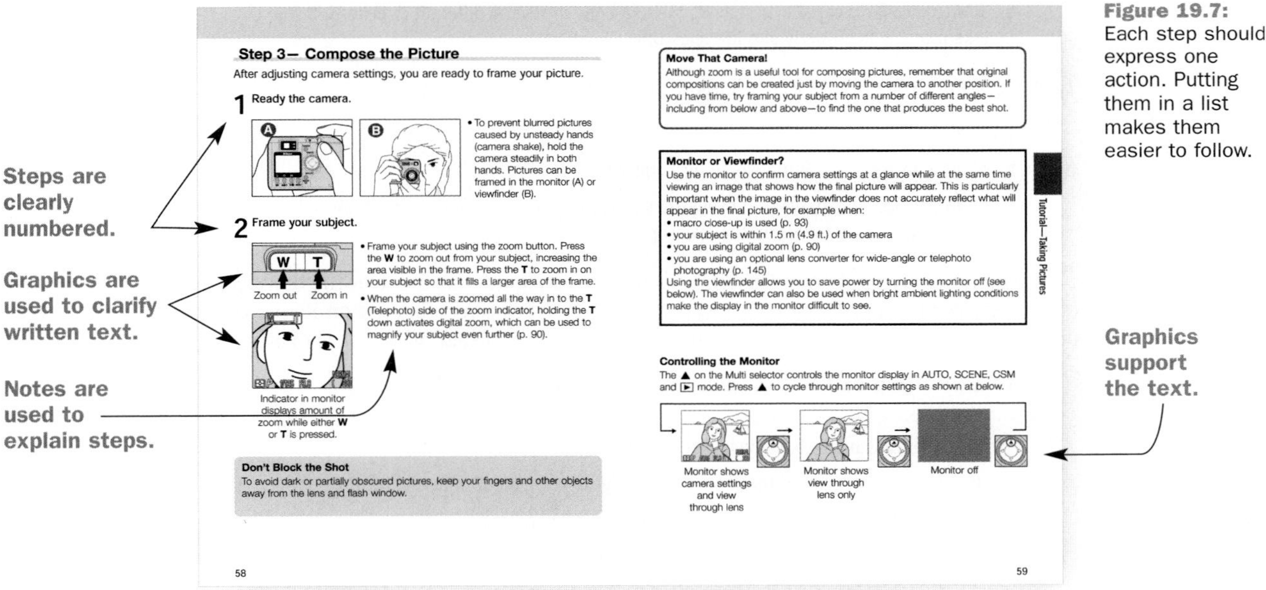

Steps are clearly numbered.
Graphics are used to clarify written text.
Notes are used to explain steps.
Graphics support the text.

Step 3— Compose the Picture

After adjusting camera settings, you are ready to frame your picture.

1 Ready the camera.

- To prevent blurred pictures caused by unsteady hands (camera shake), hold the camera steadily in both hands. Pictures can be framed in the monitor (A) or viewfinder (B).

2 Frame your subject.

- Frame your subject using the zoom button. Press the **W** to zoom out from your subject, increasing the area visible in the frame. Press the **T** to zoom in on your subject so that it fills a larger area of the frame.
- When the camera is zoomed all the way in to the **T** (Telephoto) side of the zoom indicator, holding the **T** down activates digital zoom, which can be used to magnify your subject even further (p. 90).

Indicator in monitor displays amount of zoom while either **W** or **T** is pressed.

Don't Block the Shot
To avoid dark or partially obscured pictures, keep your fingers and other objects away from the lens and flash window.

58

Move That Camera!
Although zoom is a useful tool for composing pictures, remember that original compositions can be created just by moving the camera to another position. If you have time, try framing your subject from a number of different angles—including from below and above—to find the one that produces the best shot.

Monitor or Viewfinder?
Use the monitor to confirm camera settings at a glance while at the same time viewing an image that shows how the final picture will appear. This is particularly important when the image in the viewfinder does not accurately reflect what will appear in the final picture, for example when:
- macro close-up is used (p. 93)
- your subject is within 1.5 m (4.9 ft.) of the camera
- you are using digital zoom (p. 90)
- you are using an optional lens converter for wide-angle or telephoto photography (p. 145)

Using the viewfinder allows you to save power by turning the monitor off (see below). The viewfinder can also be used when bright ambient lighting conditions make the display in the monitor difficult to see.

Tutorial—Taking Pictures

Controlling the Monitor
The ▲ on the Multi selector controls the monitor display in AUTO, SCENE, CSM and [▶] mode. Press ▲ to cycle through monitor settings as shown at below.

Monitor shows camera settings and view through lens → Monitor shows view through lens only → Monitor off

59

Figure 19.7: Each step should express one action. Putting them in a list makes them easier to follow.

Source: Nikon, Cool Pix 885 Guide.

However, when two actions must be completed at the same time, you should put them in the same sentence.

6. Insert a low-power eyepiece (e.g., 26mm) into the eyepiece holder and tighten the eyepiece thumbscrew.

You should state two actions in one step only when the two actions are dependent on each other. In other words, completion of one action should require the other to be handled at the same time.

KEEP THE STEPS CONCISE Use concise phrasing to describe each step. Short sentences are preferred, so readers can remember each step while they work. If your sentences are too long, consider moving some information into a follow-up "note" or "comment" that elaborates on the step.

7. Adjust the focus of the telescope with the focusing knob.

   The focusing knob is located at the rear of the telescope. Turning the knob clockwise will increase the focal length. Turning it counterclockwise will decrease the focal length.

NUMBER THE STEPS In most instructions, steps are presented in a numbered list. Start with the number 1 and mark each step sequentially with its own number. Notes or warnings should not be numbered, because they do not state steps to be followed.

Incorrect

8. Aim the telescope with the electronic controller.

9. Your controller is capable of moving the telescope in several different directions. It will take practice to properly aim the telescope.

There is no action in Step 9 above, so a number should not be used.

Correct

8. Aim the telescope with the electronic controller.

   Your controller is capable of moving the telescope in several different directions. It will take practice to properly aim the telescope.

An important exception to this "only number steps" guideline involves the numbering of procedures and specifications. Procedures and specifications often use an itemized numbering system in which lists of cautions or notes are "nested" within lists of steps.

10.2.1 Putting on Cleanroom Gloves, Hood, and Coveralls

10.2.1.1 Put on a pair of clean room gloves so they are fully extended over the arm and the coverall sleeves. Glove liners are optional.

10.2.1.2 Put a face/beard mask on, completely covering the mouth and nose.

10.2.1.2.1 *Caution: No exposed hair is allowed in the fab.*

10.2.1.2.2 *Caution: Keep your nose covered at all times while in the fab.*

10.2.1.2.3 *Note: Do not wear the beard cover as a face mask. A beard cover should be used with a face mask to cover facial hair.*

10.2.1.3 Put on coveralls.

Cautions and notes receive numbers in some procedures and specifications.

To see good and bad examples of instructions, go to **www.ablongman.com/johnsonweb/19.8**

In this more complex numbering scheme, comments and hazard statements receive a number. The purpose for this advanced numbering scheme is to make lines in the documentation easier to reference.

In some cases, you may also want to use *paragraph style* to write your instructions (Figure 19.8). In these situations, you can use headings or sequential transitions to highlight the steps. Numerical transitions ("first," "second," "third," "finally") are best in most cases. In shorter sets of steps, you might use transitions like "then," "next," "five minutes later," and "finally" to mark the actions. In Figure 19.8, headings are used to mark transitions among the major steps.

Paragraph style is usually harder to follow because readers cannot easily find their place again in the list of instructions. In some cases, though, paragraph style takes up less space and sounds more friendly and conversational.

ADD COMMENTS, NOTES, OR EXAMPLES After each step, you can include additional comments or examples that will help readers complete the action. Comments after steps might include additional advice or definitions for less experienced readers. Or, they might provide troubleshooting advice in case the step did not work out.

> **3. Locate a place to set your telescope.**
>
> Finding a suitable place to set up your telescope can be tricky. A paved area is optimal to keep the telescope steady. If a paved area is not available, find a level place where you can firmly set your telescope's tripod in the soil.

Comments and examples are often written in the "you" style to maintain a positive tone. Readers tend to respond more favorably to comments that address them directly and express information positively.

TAKE NOTE Comments and examples are usually not numbered, because they do not include actions. In some specifications and procedures, comments and examples are numbered for reference purposes.

PROVIDE FEEDBACK After a difficult step or group of steps, you might offer a paragraph of feedback to help readers assess their progress.

> When you finish these steps, the barrel of your telescope should be pointed straight up. The tripod should be stable, so it does not teeter when touched. The legs of the tripod should be planted firmly on the ground.

Writing Effective Steps

- Use command voice.
- State one action per step.
- Keep the steps concise.
- Number the steps.
- Add comments, notes, or examples.
- Provide feedback.
- Refer to the graphics.

REFER TO THE GRAPHICS In the steps, refer readers to any accompanying graphics. A simple statement like "See Figure 4" or "(Figure 4)" will notify them that a graphic is available that illustrates the step. After reading the step, they can look at the graphic to help them complete the step properly.

In some cases, graphics are not labeled in a set of instructions. In these situations, the graphic should appear immediately next to the step or below it so readers know which visual goes with each step.

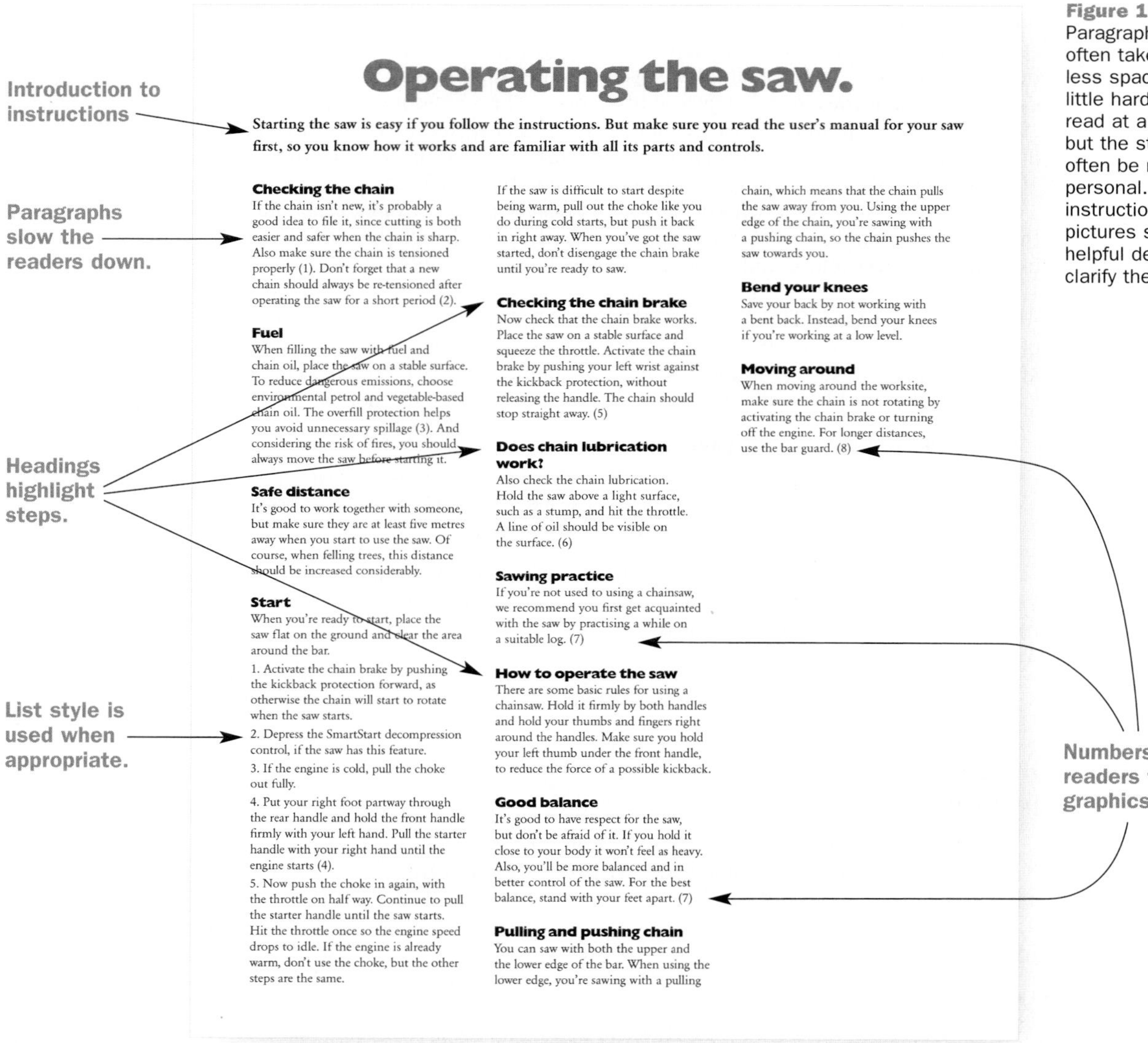

Figure 19.8: Paragraph style often takes up less space. It is a little harder to read at a glance, but the style can often be more personal. In these instructions the pictures show helpful detail to clarify the steps.

Introduction to instructions
Paragraphs slow the readers down.
Headings highlight steps.
List style is used when appropriate.
Numbers refer readers to the graphics.

Operating the saw.

Starting the saw is easy if you follow the instructions. But make sure you read the user's manual for your saw first, so you know how it works and are familiar with all its parts and controls.

Checking the chain
If the chain isn't new, it's probably a good idea to file it, since cutting is both easier and safer when the chain is sharp. Also make sure the chain is tensioned properly (1). Don't forget that a new chain should always be re-tensioned after operating the saw for a short period (2).

Fuel
When filling the saw with fuel and chain oil, place the saw on a stable surface. To reduce dangerous emissions, choose environmental petrol and vegetable-based chain oil. The overfill protection helps you avoid unnecessary spillage (3). And considering the risk of fires, you should always move the saw before starting it.

Safe distance
It's good to work together with someone, but make sure they are at least five metres away when you start to use the saw. Of course, when felling trees, this distance should be increased considerably.

Start
When you're ready to start, place the saw flat on the ground and clear the area around the bar.

1. Activate the chain brake by pushing the kickback protection forward, as otherwise the chain will start to rotate when the saw starts.
2. Depress the SmartStart decompression control, if the saw has this feature.
3. If the engine is cold, pull the choke out fully.
4. Put your right foot partway through the rear handle and hold the front handle firmly with your left hand. Pull the starter handle with your right hand until the engine starts (4).
5. Now push the choke in again, with the throttle on half way. Continue to pull the starter handle until the saw starts. Hit the throttle once so the engine speed drops to idle. If the engine is already warm, don't use the choke, but the other steps are the same.

If the saw is difficult to start despite being warm, pull out the choke like you do during cold starts, but push it back in right away. When you've got the saw started, don't disengage the chain brake until you're ready to saw.

Checking the chain brake
Now check that the chain brake works. Place the saw on a stable surface and squeeze the throttle. Activate the chain brake by pushing your left wrist against the kickback protection, without releasing the handle. The chain should stop straight away. (5)

Does chain lubrication work?
Also check the chain lubrication. Hold the saw above a light surface, such as a stump, and hit the throttle. A line of oil should be visible on the surface. (6)

Sawing practice
If you're not used to using a chainsaw, we recommend you first get acquainted with the saw by practising a while on a suitable log. (7)

How to operate the saw
There are some basic rules for using a chainsaw. Hold it firmly by both handles and hold your thumbs and fingers right around the handles. Make sure you hold your left thumb under the front handle, to reduce the force of a possible kickback.

Good balance
It's good to have respect for the saw, but don't be afraid of it. If you hold it close to your body it won't feel as heavy. Also, you'll be more balanced and in better control of the saw. For the best balance, stand with your feet apart. (7)

Pulling and pushing chain
You can saw with both the upper and the lower edge of the bar. When using the lower edge, you're sawing with a pulling chain, which means that the chain pulls the saw away from you. Using the upper edge of the chain, you're sawing with a pushing chain, so the chain pushes the saw towards you.

Bend your knees
Save your back by not working with a bent back. Instead, bend your knees if you're working at a low level.

Moving around
When moving around the worksite, make sure the chain is not rotating by activating the chain brake or turning off the engine. For longer distances, use the bar guard. (8)

Source: Husqvarna.

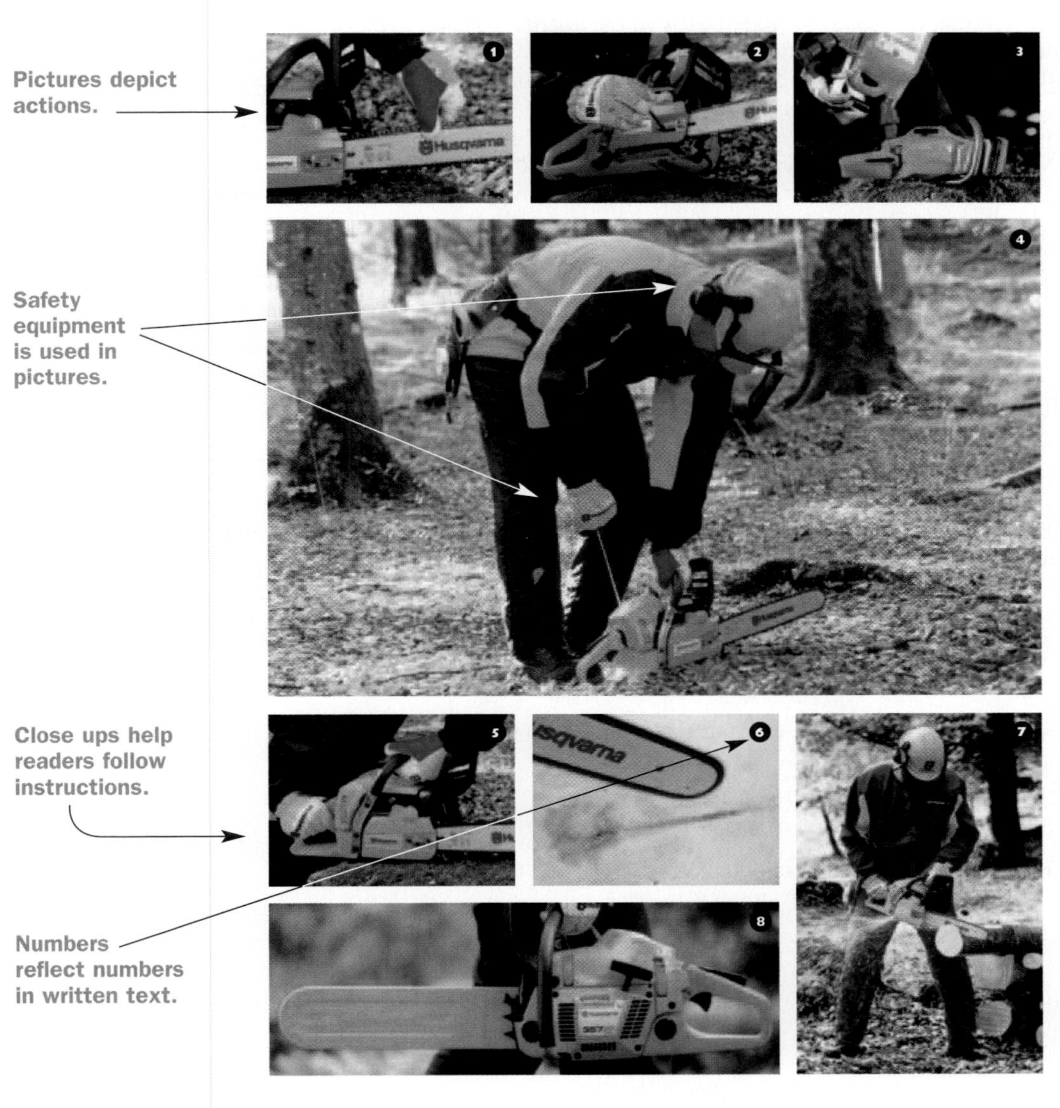
Pictures depict actions.
Safety equipment is used in pictures.
Close ups help readers follow instructions.
Numbers reflect numbers in written text.
1
2
3
4
5
6
7
8
Husqvarna

Figure 19.8: (*continued.*)

Safety Information

Safety information should be placed early in the instructions and in places where the reader will be completing difficult or dangerous steps. A common convention in technical writing is to use a three-level rating for safety information and warnings: *Danger, Warning,* and *Caution.*

DANGER Signals that readers may be at risk for serious injury or even death. This level of warning is the highest, and it should be used only when the situation involves real danger to readers.

> *Danger:* Do not remove grass from beneath your lawn mower while the engine is running (even if the blade is stopped). The blade can cause severe injury. To clear out grass, turn off the lawn mower and disconnect the spark plug before working near the blade.

WARNING Signals that the reader may be injured if the step is done improperly. To help readers avoid injury, warnings are used frequently in instructions.

> *Warning:* When heated, your solder iron will cause burns if it touches your skin. To avoid injury, always return the solder iron to its holder between uses.

CAUTION Alerts readers that mistakes may cause damage to the product or equipment. Cautions should be used to raise readers' awareness of difficult steps.

> *Caution:* The new oil filter should be tightened by hand only. Do not use an oil filter wrench for tightening, because it will cause the filter to seal improperly. If the filter is too tight, oil will leak through the filter's rubber gasket, potentially leading to major damage to your car's engine.

Safety information should tell your readers the following three things:

(1) the hazard,
(2) the seriousness of the hazard, and
(3) how to avoid injury or damage.

As shown in Figure 19.9, safety information should appear in two places:

- If a hazard is present throughout the procedure, readers should be warned before they begin following the steps. In these cases, danger and warning statements should appear between the introduction and the steps.
- If a hazard relates to a specific step, a statement should appear prominently before that step. It is important for readers to see the hazard statement before the step, so they can avoid damage or injury.

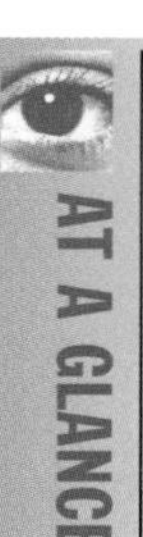

Labeling Hazards

- Danger—Risk of serious injury or death.
- Warning—Injury likely if step is handled improperly.
- Caution—Damage to the product or equipment is possible.

You can use symbols to highlight safety information. Icons are available to reinforce and highlight special hazards, like radioactive materials, electricity, or chemicals. Figure 19.10 shows a few examples of icons commonly used in safety information.

Want to learn more about using safety symbols? Go to **www.ablongman.com/johnsonweb/19.9**

Placement of Hazard Statements

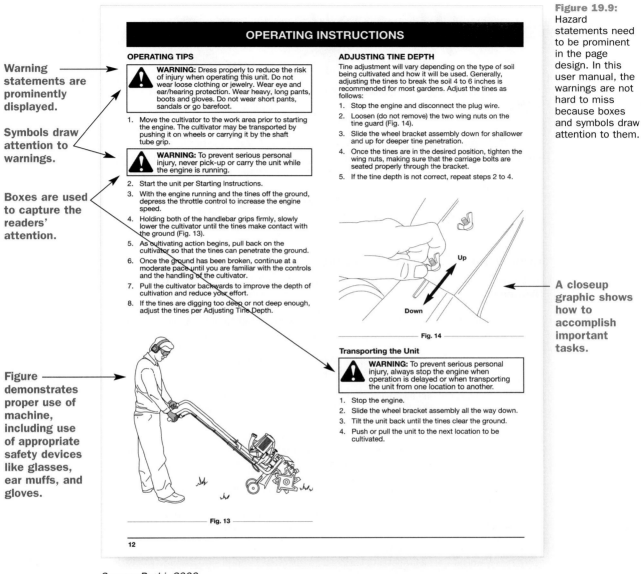

Warning statements are prominently displayed.

Symbols draw attention to warnings.

Boxes are used to capture the readers' attention.

Figure demonstrates proper use of machine, including use of appropriate safety devices like glasses, ear muffs, and gloves.

OPERATING INSTRUCTIONS

OPERATING TIPS

WARNING: Dress properly to reduce the risk of injury when operating this unit. Do not wear loose clothing or jewelry. Wear eye and ear/hearing protection. Wear heavy, long pants, boots and gloves. Do not wear short pants, sandals or go barefoot.

1. Move the cultivator to the work area prior to starting the engine. The cultivator may be transported by pushing it on wheels or carrying it by the shaft tube grip.

WARNING: To prevent serious personal injury, never pick-up or carry the unit while the engine is running.

2. Start the unit per Starting Instructions.
3. With the engine running and the tines off the ground, depress the throttle control to increase the engine speed.
4. Holding both of the handlebar grips firmly, slowly lower the cultivator until the tines make contact with the ground (Fig. 13).
5. As cultivating action begins, pull back on the cultivator so that the tines can penetrate the ground.
6. Once the ground has been broken, continue at a moderate pace until you are familiar with the controls and the handling of the cultivator.
7. Pull the cultivator backwards to improve the depth of cultivation and reduce your effort.
8. If the tines are digging too deep or not deep enough, adjust the tines per Adjusting Tine Depth.

Fig. 13

ADJUSTING TINE DEPTH

Tine adjustment will vary depending on the type of soil being cultivated and how it will be used. Generally, adjusting the tines to break the soil 4 to 6 inches is recommended for most gardens. Adjust the tines as follows:

1. Stop the engine and disconnect the plug wire.
2. Loosen (do not remove) the two wing nuts on the tine guard (Fig. 14).
3. Slide the wheel bracket assembly down for shallower and up for deeper tine penetration.
4. Once the tines are in the desired position, tighten the wing nuts, making sure that the carriage bolts are seated properly through the bracket.
5. If the tine depth is not correct, repeat steps 2 to 4.

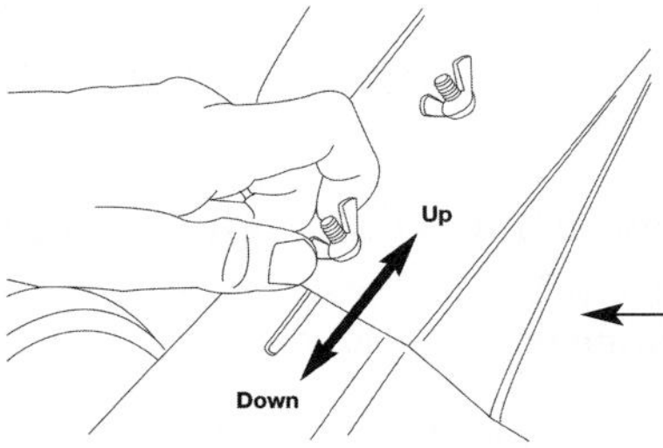

Up

Down

Fig. 14

A closeup graphic shows how to accomplish important tasks.

Transporting the Unit

WARNING: To prevent serious personal injury, always stop the engine when operation is delayed or when transporting the unit from one location to another.

1. Stop the engine.
2. Slide the wheel bracket assembly all the way down.
3. Tilt the unit back until the tines clear the ground.
4. Push or pull the unit to the next location to be cultivated.

12

Figure 19.9: Hazard statements need to be prominent in the page design. In this user manual, the warnings are not hard to miss because boxes and symbols draw attention to them.

Source: Ryobi, 2000.

Safety Symbols

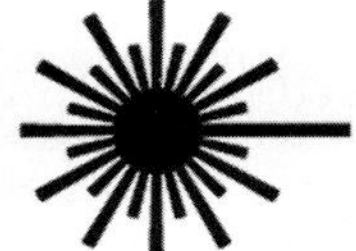

Figure 19.10: Here are a few examples of ISO 7000 safety symbols (hot surface, laser, radiation).

Source: Compliance Engineering Magazine, http://www.ce-mag.com/archive/02/03/peckham.html.

In our litigious culture, the importance of safety information should not be underestimated. Danger, warning, and caution notices will not completely immunize your company against lawsuits, but they will give your company some defense against legal action.

Lawsuits aside, though, you have an ethical obligation to look out for the well-being of your readers. You should look for places where they or their property are at risk.

Conclusion That Signals Completion of Task

When you have listed all the steps, you should offer a closing that tells readers that they are finished with the task. Closings can be handled a few different ways.

SIGNAL COMPLETION OF THE TASK Tell readers that they are finished with the instructions. Perhaps you might offer a few comments about the future.

> Congratulations! You are finished setting up your RGS-90x Telescope. You will now be able to spend many nights exploring the night skies.

> When completed, the bandaging of the head wound should be firm but not too tight. Bleeding should stop within a minute. If bleeding does not stop, call an emergency room doctor immediately.

DESCRIBE THE FINISHED PRODUCT You might describe the finished product or provide a graphic that shows how it should look.

> When you have completed setting up your telescope, it should be firmly set on the ground and the eyepiece should be just below the level of your eyes. With the Electronic Controller, you should be able to move the telescope horizontally and vertically with the push of a button. Figure 5 shows how a properly set-up telescope should look.

> Your bandaging of the patient's head should look like Figure B. The bandaging should be neatly wound around the patient's head with a slight overlap in the bandage strips.

OFFER TROUBLESHOOTING ADVICE Depending on the complexity of the task, you may end your instructions by anticipating some of the common problems that may occur. Simple tasks may require only a sentence or two of troubleshooting advice. More complex tasks may require a table that lists potential problems and

To download safety symbols, go to **www.ablongman.com/johnsonweb/19.10**

their remedies (Figure 19.11). Depending on your company's ability to provide customer service, you may also include a web address or a phone number where readers can obtain additional help.

> If you have any questions or need more information, please visit our website at www.yournewtelescope.com, or you can call our Customer Service Desk at 1-800-555-7865.

Closing statements provide closure for a set of instructions, procedures, or specifications. They signal completion of the task, so readers know they are finished.

Troubleshooting Guide

Troubleshooting Guide	
Problem	**Solution or Explanation**
Nothing is visible through the eyepiece.	Confirm that the dust cap is removed from the front lens of the telescope. Confirm that the flip mirror is set on viewing mode, not camera mode.
Object appears in the viewfinder, but not in telescope.	Check the alignment of the viewfinder. To realign the viewfinder, see page 8.
Images appear unfocused or distorted.	The magnification used is too high. Use a lower-power eyepiece. Warm conditions may cause heat waves that will distort images. The optics within the telescope may need time to adjust to the outside temperature. Wait a few minutes for the telescope to reach a stable temperature.
Telescope does not rotate when controller is used.	The batteries may be low. Replace them with fresh batteries. The telescope may have reached its full rotation limit. Loosen rotator lever and rotate telescope counterclockwise to the center position.

Figure 19.11: Troubleshooting guides are often provided in a table format with problems on the left and solutions on the right. Note the positive, constructive tone in this table.

Using Style in Instructions

People often assume that technical documentation should be dry and boring. But instructions can be—and sometimes should be—written in a more interesting style. There are ways, especially with consumer products, that you can use style to reflect readers' attitudes as they use your instructions.

To see examples of troubleshooting guides, go to **www.ablongman.com/johnsonweb/19.11**

HELP

On-line Documentation

One of the major changes brought about by computers is the availability of on-line documentation. Today, many user manuals and instructions for products are available through a website, provided on a CD-ROM, or included with the "Help" feature in a software package. On their websites, many companies offer electronic versions of their user manuals so customers can look up information or replace lost manuals.

The advantages of on-line documentation are numerous, and the disadvantages are few. The greatest advantage is reduced cost. After all, with some products, like software, the accompanying user manual costs more to print than the software itself. By putting the manual on a website, CD-ROM, or on-line "Help," a company can save thousands of dollars almost immediately. Also, on-line documentation can be updated regularly to reflect changes in the product or revisions to the documentation.

Companies are also putting specifications and procedures on-line, usually on the company's intranet. That way, all employees can use their computer to easily call up the document they need.

Several options are available for on-line documentation:

CD-ROM—Increasingly, computers have a CD burner as a standard feature. To make a CD, you can use web development software like MacroMedia Dreamweaver, Frontpage, or Adobe GoLive to create the files. Then, burn the files to a CD the same way you would save to a disk. Browser programs like

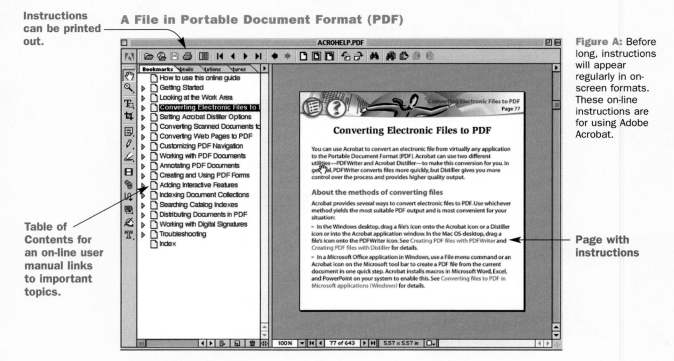

A File in Portable Document Format (PDF)

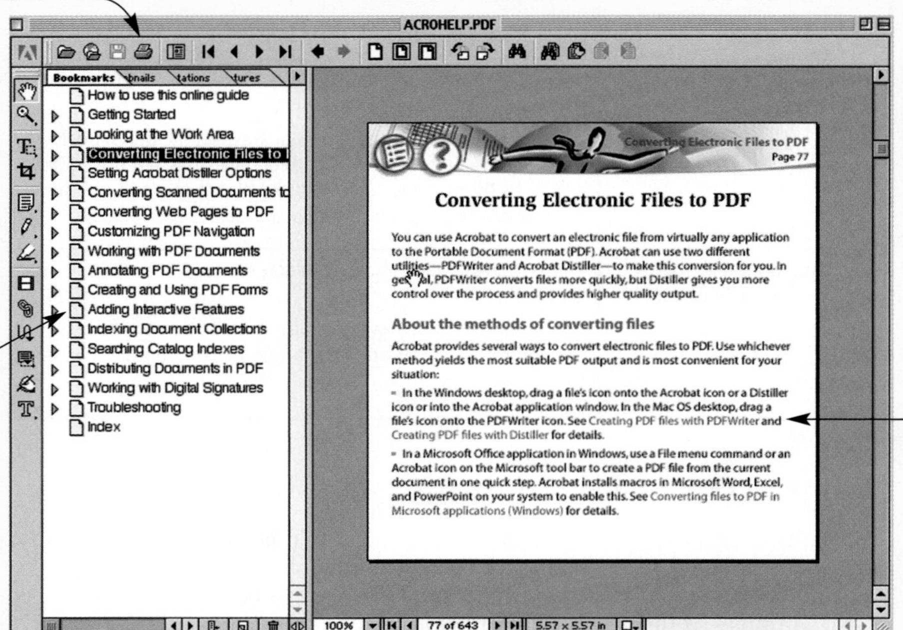

Figure A: Before long, instructions will appear regularly in on-screen formats. These on-line instructions are for using Adobe Acrobat.

Source: Adobe Acrobat Help.

To learn more about on-line documentation, go to **www.ablongman.com/johnsonweb/19.12**

Netscape Communicator, Safari, or Microsoft Explorer will be able to read these multimedia documents.

Website—You can put your documentation on a website for use through the Internet or a company intranet. These files can be created with web development software like Dreamweaver or Adobe GoLive.

Portable document format (PDF)—Software programs like Adobe Acrobat can turn your word-processing files into PDFs, which retain the formatting and color of the original document (Figure A). PDFs can be read by almost any computer, and they store information efficiently. They can also be password protected, so readers cannot tamper with the text. PDFs can be placed on a website for easy downloading.

On-line help—Increasingly, the on-line "Help" features that come with software packages are being used to present instructions (Figure B). On-line Help features allow readers to access the instructions more quickly, because they do not need to hunt around for the user manual. As more documents move on-line, it is likely instructions will increasingly be offered as on-line Help.

To write instructions as on-line Help, you will need Help-authoring software like RoboHelp and DoctoHelp. These programs simplify the writing of Help features.

On-line Help

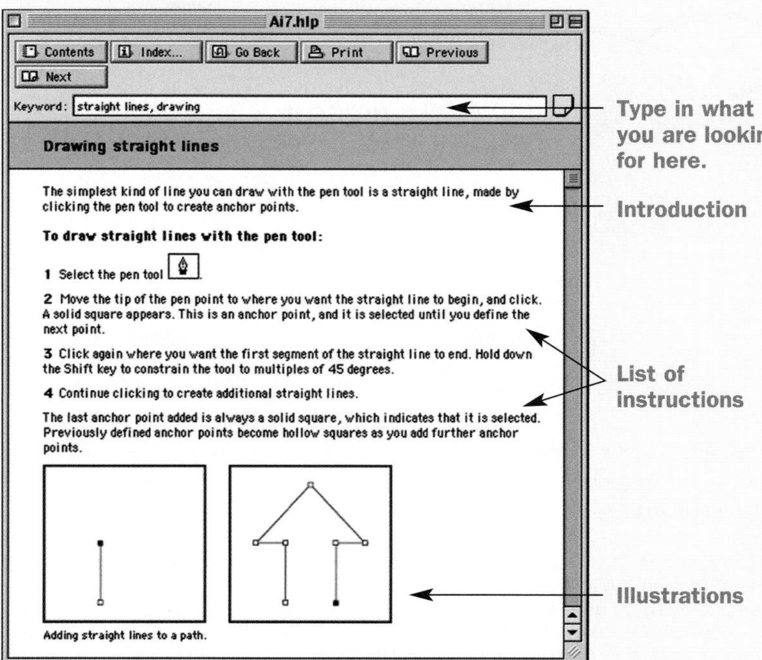

Figure B: On-line "Help" features are another place where instructions are commonly found, especially for software programs. These instructions are from the on-line Help feature in Adobe Illustrator.

Source: Adobe Illustrator 7.0 Help.

Want to know more about Help-authoring software? Go to **www.ablongman.com/johnsonweb/19.13**

For example, instructions for using a telescope should reflect readers' enthusiasm for their new ability to see into space. A procedure for bandaging a head wound should set a reassuring tone for a nurse who is learning the procedure.

How can you improve the style of your instructions? First, look at your original analysis of your readers and contexts in which your document will be used. Pay attention to your readers' needs, motives, and values. Try to identify the emotions and attitudes that shape how they are reading and using the instructions. Are they enthusiastic, frustrated, happy, apprehensive, or excited?

Identify a word that best reflects readers' feelings as they are using your instructions. Then, use logical mapping to come up with some words that are associated with that word (Figure 19.12).

TAKE NOTE You can use the "Thesaurus" function on your word-processing program to help you come up with synonyms for the word you choose.

Mapping a Tone for Instructions

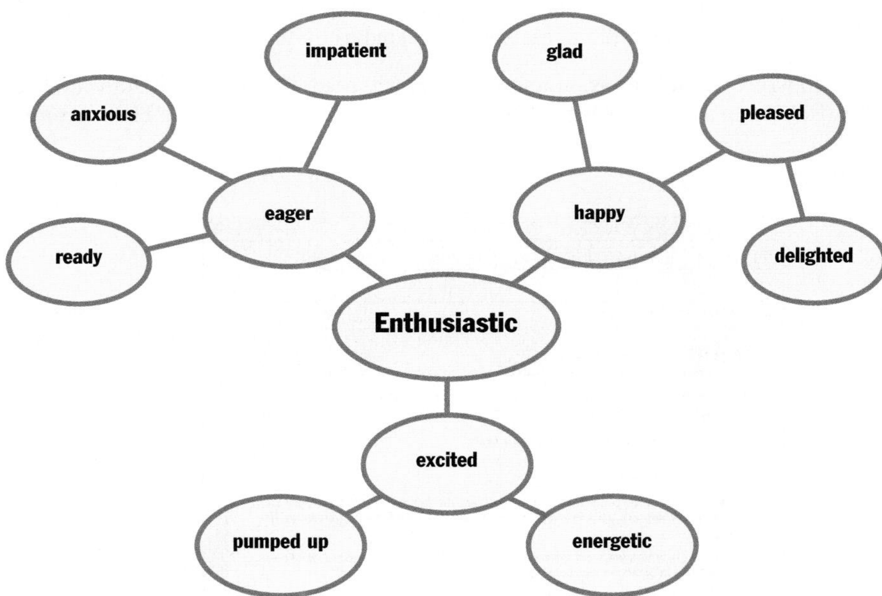

Figure 19.12: An enthusiastic tone can be created by finding synonyms associated with that word.

After you have found words that are associated with the appropriate tone, use them in the introduction, notes, and conclusion of your instructions. These words, when used strategically, will reflect your readers' attitudes or emotions.

TAKE NOTE If your readers have a negative attitude (perhaps they are annoyed that they need to read the instructions), you can use antonyms to counteract their feelings. For example, to sooth annoyed readers, words like satisfy, pleasure, please, delight, and fulfill can be used to counteract their negativity. Don't overuse them, though, because angry readers may detect your attempt to soothe them.

Increasingly, companies are seeing documentation as their main contact with customers or clients. So, they are using style to make their documentation more responsive to readers' needs, motives, and values. Ideally, the instructions will leave customers with a positive image of the company.

LINK For more ideas about improving style, go to Chapter 7, page 162.

Designing Instructions

With the content of the text composed, you can start designing the text and developing some graphics.

Page Layout

Instructions can be designed a variety of different ways. Sometimes a two-column format will allow you to put text on the left and graphics on the right (refer again to Figures 19.8 and 19.9 on pages 564–565 and 567). You can also use boxes and lines to highlight important information. At a minimum, headings should clearly show the levels of information in the text.

As you design your documentation, keep in mind your readers' specific needs and where they will be using the materials. Design should not be used to simply decorate the text. Instead, it should enhance readers' ability to access the information. Design should make the text more understandable.

TAKE NOTE If you are not sure how to design your instructions, study sample texts from your home or workplace. You can use these examples as models for designing your own documentation.

LINK For more help on document design, see Chapter 8, page 192.

Graphics

With computers, you have a variety of ways to add graphics to instructions. You can draw illustrations and diagrams. Or, you can use a digital camera or scanner to add graphics to your text.

When including graphics, it is best to number and title them. Then, refer to the graphics by number in the written text. That way, readers will be able to quickly locate them. Each graphic should appear next to or below the step that refers to it. Graphics should also appear on the same page as the step that refers to them.

TAKE NOTE When using instructions, readers should never need to turn the page to see a graphic. Their hands might be busy, not allowing them to turn the page.

In some cases, graphics do not need to be numbered and titled. In these situations, the graphics should be tightly grouped with the text, so readers can clearly see the relationship between the graphics and the text. For example, Figure 19.7 on page 561 showed an example of graphics tightly grouped with text.

LINK For more information on using graphics, see Chapter 9, page 234.

For tips on laying out pages for documentation, go to **www.ablongman.com/johnsonweb/19.14**

AT WORK

Amanda Jervis

MANAGER OF TRAINING, VIDIOM SYSTEMS CORPORATION, BOULDER, COLORADO

Vidiom Systems makes software and other products for interactive television.

What is the best way to prepare to write instructions for a product?

Writing documentation often means writing about something that has just been conceived, has not yet been created, and certainly has never been written about. No library or website can help you with this assignment. You must rely on your own research skills to create your own story about this product or service.

How do you invent content for a set of instructions when you don't have sources to rely on?

- Talk to developers, marketing specialists, and everyone in the company to learn their vision of the product.
- Interview the programmers in depth, asking them to explain the product at a simple level; copy down any diagrams they draw.
- Request all marketing and sales materials—Powerpoint presentations, flyers—and ask to attend a sales pitch.
- Attend all company meetings about the product, even developer reviews, and listen, listen, listen.

Finally, write up your information and give it to a project manager or developer for review. You'll always get great information when you give people an opportunity to point out that you're wrong.

Revising, Editing, and Proofreading

In many ways, instructions reflect the quality of the product, service, and manufacturer. If the documentation looks unprofessional or has errors, readers will doubt the quality of the company behind it. So, you should leave yourself plenty of time for revising, editing, and proofreading. Also, leave yourself some time to user-test your documentation on some real readers.

Revising for Content, Style, and Design

When you are finished drafting the instructions, spend time critically studying its content, style, and design.

- The content needs to be complete, including all the information readers need to complete the task. Look for places where steps are missing or unclear. Identify steps that would be clearer with a follow-up note or comment. Identify places where hazard statements are needed.
- The style should be concise and clear. Keep sentences, especially commands, short and to the point. Where possible, replace complex words with simpler, plainer terms. Meanwhile, use words that reflect readers' attitudes or emotions as they are using your documentation.

- The design should enhance readers' ability to follow the written instructions. Make sure you have used headings, notes, and graphics consistently. The page layout should also be consistent from the first page to the last.

As you revise, make sure the content, style, and design are appropriate for the readers you defined in your reader analysis. Your instructions should be usable by the least experienced, least knowledgeable person among your readers.

LINK For more advice on revision and editing, see Chapter 10, page 264.

User-Testing with Sample Readers

Testing your instructions on actual users can be handled formally or informally. Formal user testing involves finding real users of the documentation and observing them as they try to follow the instructions. Often, formal user testing includes videotaping the subjects as they use the documentation. You can then identify places where they stumble. Then, you can revise the text to fix those problems.

LINK For more information on user-testing documents, see Chapter 10, page 282.

Informal user testing can be used for in-house or less important forms of documentation. You should at least ask your supervisor or a few coworkers to look your document over before making it available to others. Often, your colleagues can help you highlight missing content, ambiguous sentences, or ineffective graphics.

Editing and Proofreading

Finally, you should carefully edit and proofread your instructions. Whether the document is being sent with a product, included in a book of procedures, or placed in the specifications file, it should be clearly written and mistake free.

You should actively question the content, organization, style, and design of your document. Be your own harshest critic. Then, while proofreading, pay close attention to grammar and spelling. Flawless documentation will strengthen your readers' faith in you and your company.

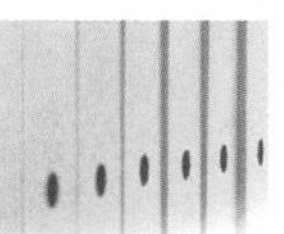

CHAPTER REVIEW

- Instructions describe step by step how to complete a task.
- Basic features of instructions include a specific and precise title; an introduction; a list of parts, tools, and conditions required; sequentially ordered steps; graphics; safety information; a conclusion; and troubleshooting information.
- Determine the rhetorical situation by asking the Five-W and one How questions and analyzing the document's subject, purpose, readers, and context of use.
- Organize and draft your instructions step by step, breaking them down into their major and minor actions.
- Safety information should (a) identify the hazard, (b) state the level of risk (Danger, Warning, or Caution), and (c) offer suggestions for avoiding injury or damage.

Worksheets to help you revise your instructions are available at **www.ablongman.com/johnsonweb/19.15**

- The style of your instructions does not need to be dry and boring. Try to use a style that reflects or counters readers' attitudes as they are following the instructions.
- Graphics offer important support for the written text. They should be properly labeled by number and inserted in the page where they are referenced in the written text.
- While revising your instructions, user-testing your instructions with sample readers is an important way to work out any bugs and locate places for improvement. Your observations of these sample readers should help you revise the document.

EXERCISES AND PROJECTS

Individual or Team Projects

1. Find a set of instructions in your home or workplace. Using concepts discussed in this chapter, develop a set of criteria to evaluate the content, organization, style, and design of these instructions. Then, write a two-page memo to your instructor in which you analyze the instructions. Highlight any strengths in the instructions and make suggestions for improvements.

2. In your home or workplace, find an ineffective set of instructions. First, identify its weaknesses in content, organization, style, and design. Then, revise the instructions to make them easier to use. Write a cover memo to your instructor in which you discuss the ways you revised and improved the set of instructions.

3. Turn the instructions for playing Klondike at the end of this chapter into a numbered list and use graphics to support the text. You can make minor changes to the wording of the text; however, try to keep the written text intact as much as possible.

4. On the Internet or at your home, find information on first aid (handling choking, treating injuries, using CPR, handling drowning, treating shock, dealing with alcohol or drug overdoses). Then, turn this information into a text that is specifically aimed at college students living on campus. You should keep in mind that these readers will be reluctant to actually read this text—until it is needed. So, write and design it in a way that will be both appealing before injuries occur and highly usable when an injury has occurred.

5. The case study at the end of this chapter presents a difficult ethical decision. Pretend you are Mark Harris in this situation. As Mark, write a memo to your supervisor in which you express your concerns about the product. Tell the supervisor what you think the company should do about the problem.

Collaborative Project

Have someone in your group bring to class an everyday household appliance (toaster, blender, hot air popcorn popper, clock radio, portable CD-player, etc). With your

For sample sets of effective and ineffective instructions, go to **www.ablongman.com/johnsonweb/19.16**
For first aid sites on the web, go to **www.ablongman.com/johnsonweb/19.17**

group, write and design a set of instructions for this appliance that would be appropriate for 8-year old children. Your instructions should keep the special needs of these readers in mind. The document should also be readable and interesting to these readers, so they will actually use it.

Revision Challenge

These instructions for playing Klondike are technically correct; however, they are hard to follow. Can you use visual design to revise these instructions to make them more readable?

Playing Klondike (Solitaire)

Many people know Klondike simply as Solitaire, because it is such a widely played solitaire game. Klondike is not the most challenging form of solitaire, but it is very enjoyable and known worldwide.

To play Klondike, use one regular pack of cards. Dealing left to right, make seven piles from twenty-eight cards. Place one card on each pile, dealing one fewer pile each round. When you are finished dealing, the pile on the left will have one card, the next pile on the left will have two cards, and so on. The pile farthest to the right will have seven cards. When you are finished dealing the cards, flip the top card in each pile face up.

You are now ready to play. You may move cards among the piles by stacking cards in decreasing numerical order (king to ace). Black cards are placed on red cards and red cards are placed on black cards. For example, a red four can be placed on a black five. If you would like to move an entire stack of face-up cards, the bottom card being moved must be placed on a successive card of the opposite color. For example, a face-up stack with the jack of hearts as the bottom card can only be moved to a pile with a black queen showing on top. You can also move partial stacks from one pile to another as long as the bottom card you are moving can be placed on the top face-up card on the pile to which you are moving it. If a face-down card is ever revealed on top of a pile, it should be turned face up. You can now use this card. If the cards in a pile are ever completely removed, you can replace the pile by putting a king (or a stack with a king as bottom face-up card) in its place.

The rest of the deck is called the "stock." Turn up cards in the stock one by one. If you can play a turned-up card on your piles, place it. If you cannot play the card, put it in the discard pile. As you turn up cards from the stock, you can also play the top card off the discard pile. For example, let us say you have an eight of hearts on top of the discard pile. You turn up a nine of spades from the stock, which you find can be played on a ten of diamonds on top of one of your piles. You can then play the eight of hearts on your discard pile on the newly placed nine of spades.

When an ace is uncovered, you may move it to a scoring pile separate from the seven piles. From then on, cards of the same suit may be placed on the ace in successive order. For example, if the two of hearts is the top face-up card in one of your piles, you can place it on the ace of hearts. As successive cards in the suit are the top cards in piles or revealed in the stock, you can place them on your scoring piles.

When playing Klondike properly, you may only go through the stock once (variations of Klondike allow you to go through the stock as many times as you like, three cards at a time). When you are finished going through the stock, count up the cards placed in your scoring piles. The total cards in these piles make up your score for the game.

Check your solitaire instructions against ones on the net. Go to **www.ablongman.com/johnsonweb/19.18**

CASE STUDY

The Flame

Mark Harris was recently hired as a design engineer at Fun Times Toy Manufacturers, Inc. In a way, the job was a dream come true. Mark liked designing toys, and he had two small children at home.

When he began working at the company, he inherited a project from the previous engineer, who had taken a job with another company. The project was an indoor campfire made out of a gas stove. The stove produced a flame about a foot tall in a pile of fake logs. The campfire was designed to be used outside, but it could be used inside with special precautions.

The prior engineer on the project had already built a prototype. All Mark needed to do was write the instructions for the product's user manual. So, he took the campfire home to try it out with his kids.

At home, the campfire started up easily in his living room, but Mark immediately noticed that the flame was somewhat dangerous. His kids tried to roast marshmallows over the fire, but he started to become a little nervous about the flame. He wondered if the couch and rug were at risk of catching on fire. Certainly, he could imagine situations where the campfire could be used incorrectly, causing burns or fire.

The next day, he stopped by his supervisor's office to express his concerns. His supervisor, though, waved off the problems. "Yeah, the previous guy mentioned some problems, too. All you need to do is fill the user manual with warnings and warning symbols. That should protect us from lawsuits."

As Mark walked out the door, his supervisor said, "Listen, we need that thing out the door in a couple weeks. Our manufacturers in Mexico have already retooled their plants, and we're ready to go. You need to get that manual done right now."

Mark went back to his office a bit concerned. Wouldn't his company still be negligent if people or their property were harmed? Would warnings and warning symbols really protect them from negligence? And even if the company couldn't be sued, would it be ethical to produce a potentially harmful product? What would you do if you were in Mark's place?

To find information on product liability, go to **www.ablongman.com/johnsonweb/19.19**

CHAPTER

20

Proposals

CHAPTER CONTENTS

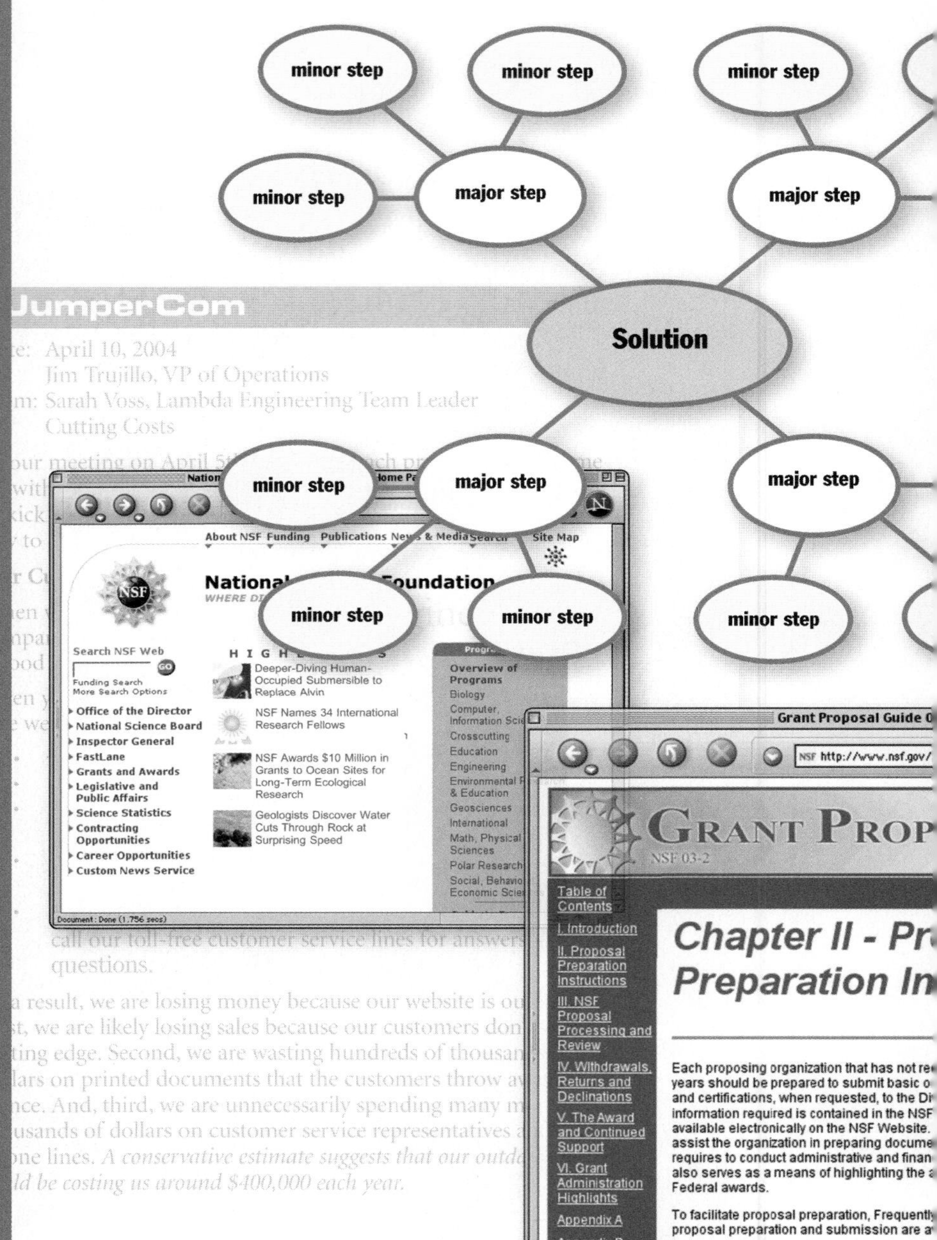

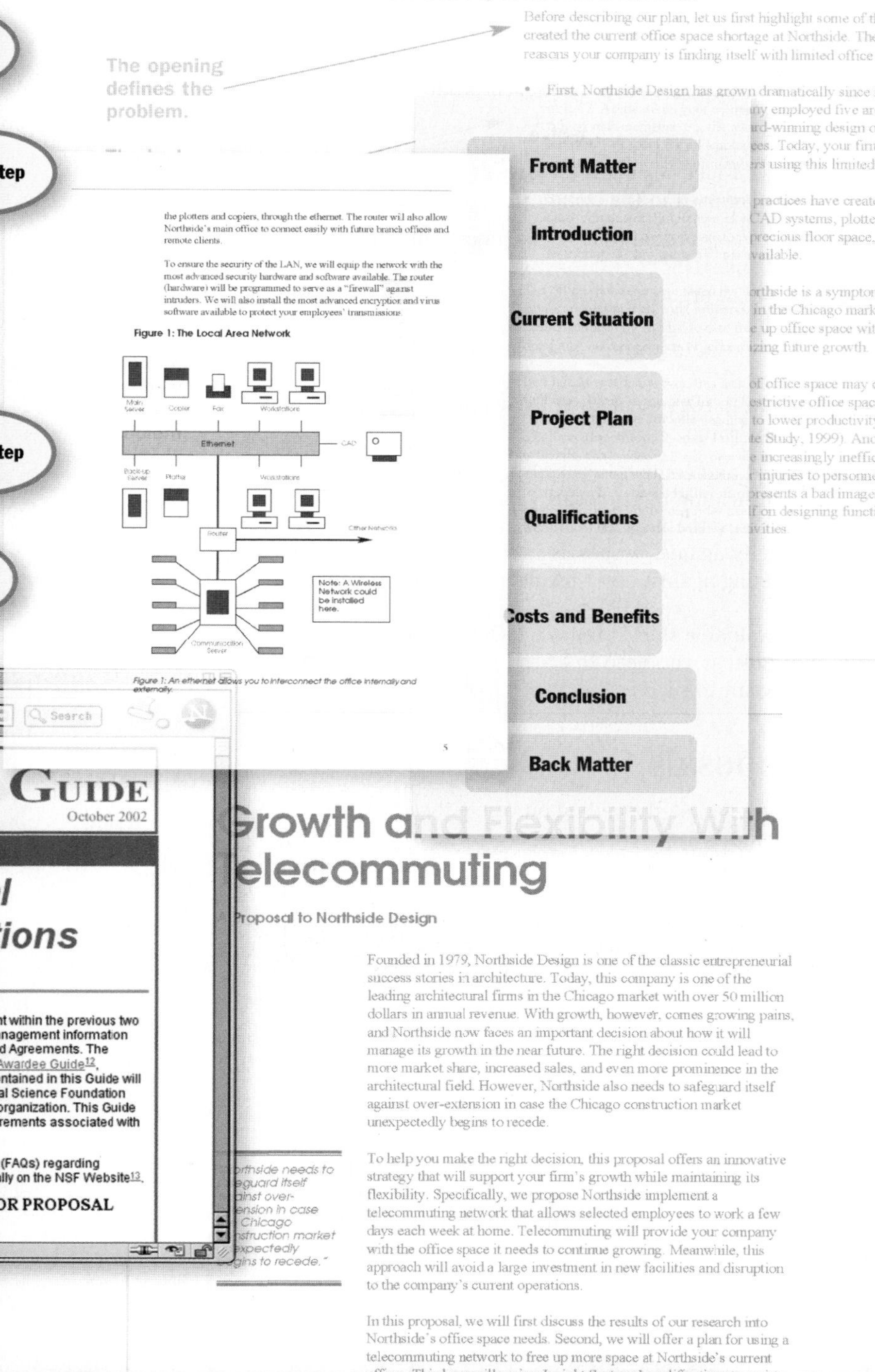

CHAPTER OBJECTIVES

In this chapter, you will learn:

- The purpose of proposals and their uses in the workplace.
- The basic features and types of proposals.
- How to plan and do research for a proposal.
- How to organize and draft the major sections in a proposal.
- Strategies for using plain and persuasive style to make a proposal influential.
- How document design and graphics can enhance a proposal.
- Techniques for revising, editing, and proofreading proposals.

Proposals are the lifeblood of the technical workplace. No matter what your field is, you will be asked to write proposals that describe projects, present ideas, and offer new strategies. Proposals are used to plan new projects, sell products and services, and suggest improvements.

Proposals are documents that present ideas or plans for consideration.

Proposals are documents that present ideas or plans for consideration. Here are just a few examples of how proposals are used in the technical workplace:

- An electronic engineer would use a proposal to describe a new kind of plasma screen television that he wants to develop.
- A manager would use a proposal to argue for the use of robots to automate the assembly line at her factory.
- A civil engineer would use a proposal to propose a monorail system in the downtown of a city.
- A biologist would use a proposal to request funding for her study of the effects of poaching on the African black rhino.

Computers have increased the speed and competitiveness of proposal development. They have heightened the sophistication of proposals, allowing writers to use graphics, color, and even video to enhance the persuasiveness of their ideas. Your proposals will usually be competing directly against proposals from other companies. In some cases, you might even be competing against other teams or divisions in your own company.

Effective proposal writing is an invaluable skill in today's technical workplaces. Almost all projects begin with proposals, so you need to learn how to write these important documents effectively if you want to succeed.

Basic Features of Proposals

Proposals are classified as internal or external, depending on where they are used and who will read them. *Internal* proposals are used within companies to plan or propose new projects or products. *External* proposals are used by companies to offer services or products to their clients.

Whether they are internal or external, proposals tend to have the following features:

- Introduction
- Description of the current situation
- Description of the project plan
- Review of qualifications
- Discussion of costs and benefits
- Graphics
- Budget

Figure 20.1 shows how these features are usually arranged in the document. As with other technical documents, though, you should not mechanically follow the pattern described here. The proposal genre is not a formula. It is only a model that offers a guideline for writing. You should alter this pattern to suit the needs of your proposal's subject, purpose, readers, and context of use.

Proposals are also classified as *solicited* or *unsolicited*, depending on whether they were requested.

To see sample proposals, go to **www.ablongman.com/johnsonweb/20.1**

Basic Pattern for a Proposal

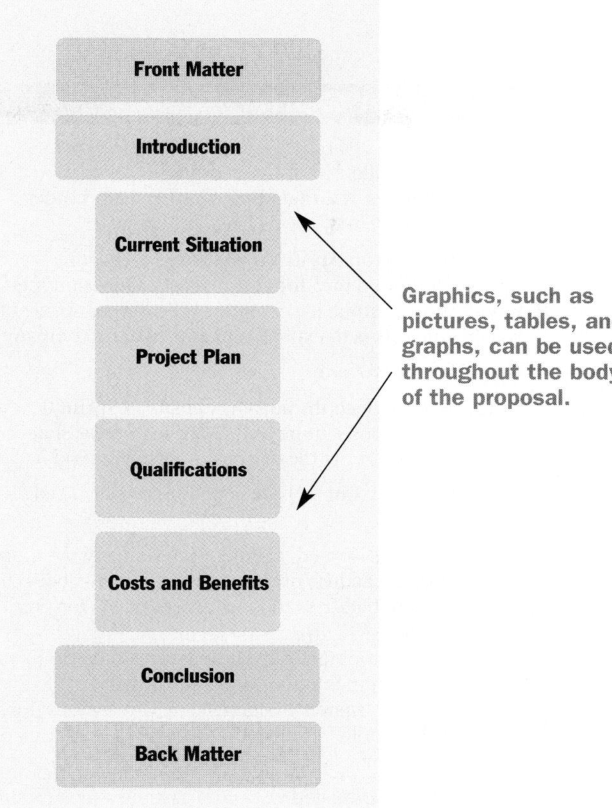

Figure 20.1: Proposals tend to address these areas, though not always in this order.

Solicited proposals are proposals requested by the readers. For example, the management of a company might request that teams or divisions submit internal proposals for new projects in the upcoming year. Supplier companies write solicited external proposals in response to advertisements called Requests for Proposals (RFP), which are prepared by companies seeking to have work done.

Unsolicited proposals are proposals that were not requested by the readers. For example, an employee or team might prepare an unsolicited internal proposal to pitch an innovative new idea to the company's management. Similarly, supplier companies might use unsolicited external proposals as sales tools to offer clients products or services they may need but have not requested.

Figure 20.2 shows a solicited proposal written for internal purposes. In this example, a team within the company is pitching its plan to overhaul the company's website. Notice how the proposal is used to persuade management to agree to the team's ideas.

To learn more about different kinds of proposals, go to **www.ablongman.com/johnsonweb/20.2**

An Internal Solicited Proposal

Figure 20.2: This smaller proposal is an internal proposal that is pitching a new idea to a manager. After a brief introduction, it describes the current situation and offers a plan for solving a problem. It concludes by highlighting the benefits of the plan.

Internal proposals are often written in memo format.

Background information signals that the proposal was solicited.

The main point of the proposal is stated up front.

This section describes the current situation.

JumperCom

Date: April 10, 2004
To: Jim Trujillo, VP of Operations
From: Sarah Voss, Lambda Engineering Team Leader
Re: Cutting Costs

At our meeting on April 5th, you asked each project team to come up with one good idea for cutting costs. Our team met on April 6th to kick around some ideas. At this meeting we decided that the best way to cut costs is to expand and enhance the company's website.

Our Current Website

When we developed the current website in Spring 1997, it served our company's purposes quite well. Websites were still new, and ours did a good job of serving the few customers who used it.

Seven years later, our website is no longer cutting edge—it's obsolete. The website

- looks antiquated, making our company seem out of touch
- does not address our customers' questions about current products
- does not address our customers' needs for product documentation
- is not a tool that our salespeople can use to provide answers and documentation to the customers
- does not answer frequently asked questions, forcing clients to call our toll-free customer service lines for answers to simple questions.

As a result, our outdated website is causing a few important problems. First, we are likely losing sales because our customers don't see us as cutting edge. Second, we are wasting hundreds of thousands of dollars on printed documents that the customers throw away after a glance. And, third, we are unnecessarily spending many more thousands of dollars on customer service representatives and tollfree phone lines. *A conservative estimate suggests that our outdated website could be costing us around $400,000 each year.*

Figure 20.2: *(continued.)*

This section offers a plan for the readers' consideration.

The plan is described step by step.

Renovating the Website

We believe a good way to cut costs and improve customer relations is to renovate the website. We envision a fully interactive site that customers can use to find answers to their questions, check on prices, and communicate with our service personnel. Meanwhile, our sales staff can use the website to discuss our products with clients. Instead of lugging around printed documents, our salespeople would use their laptop computers to show products or make presentations.

Renovating the site will require four major steps:

Step One: Study the Potential Uses of Our Website

With a consultant, we should study how our website might be better used by customers and salespeople. The consultant would survey our clients and salespeople to determine what kind of website would be most useful to them. The consultant would then develop a design for the website.

Step Two: Hire a Professional Web Designer to Renovate the Site

We should hire a professional web designer to implement our design, because modern websites are rather complex. A professional would provide us with an efficient, well-organized website that would include all the functions we are seeking.

Step Three: Train One of Our Employees to Be a Webmaster

We should hire or retrain one of our employees to be the webmaster of the site. We need someone who is working on the site daily and making regular updates. Being the webmaster for the site should be this employee's job description.

Step Four: User-Test the New Website with Our Customers and Salespeople

Once we have created a new version of the website, we should user-test it with our customers and salespeople. Perhaps we could pay some of our customers to try out the site and show us where it could be improved. Our salespeople will certainly give us plenty of feedback.

Continued on following page

Figure 20.2: *(continued.)*

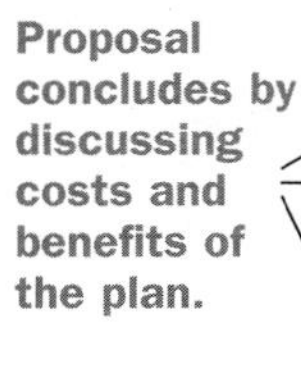

Proposal concludes by discussing costs and benefits of the plan.

At the end of this process, we would have a fully functioning website that would save us money almost immediately.

Costs and Benefits of Our Idea

Renovating the website would have many advantages:

- The new website will save us printing costs. We estimate that the printing costs at our company could be sliced in half—perhaps more—because our customers would be able to download our documents directly from the website, rather than ask us to send these documents to them. That's a potential savings of $300,000.
- The new website will provide better service to our customers. Currently, our customers go to the website first when they have questions. By providing more information in an interactive format, we can cut down dramatically on calls to our customer service center. We could save up to $120,000 in personnel costs and long-distance charges.
- Currently our sales staff will find the website a useful tool when they have questions. When products change, salespeople will immediately see those changes reflected on the website. As a result, more sales might be generated because product information will be immediately available on-line.

A quick estimate shows that a website renovation would cost us about $40,000. We would also need to shift the current webmaster's responsibilities from part time to full time, costing us about $20,000 per year more. The savings, though, are obvious. For an initial investment of $60,000 and a yearly investment of $20,000 thereafter, we will minimally save about $400,000 a year.

Thank you for giving us this opportunity to present our ideas. If you would like to talk with us about this proposal, please call me at 8-1204, or you can e-mail me at sarahv@jumpercom.net.

Another kind of proposal is the grant proposal. Researchers and nonprofit organizations prepare grant proposals to obtain funding for their projects. For example, one of the major funding sources for grants in science and technology is the National Science Foundation (NSF). Through its website, the NSF offers funding opportunities for scientific research (Figures 20.3 and 20.4).

The National Science Foundation Homepage

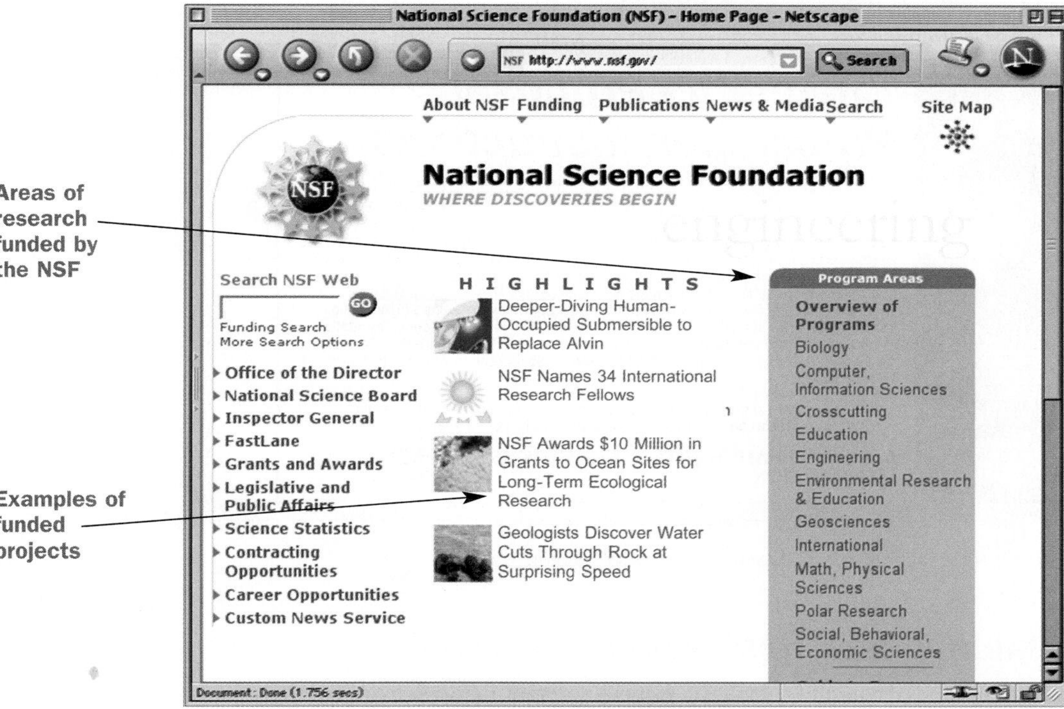

Figure 20.3: The National Science Foundation (NSF) website offers information on grant opportunities. The homepage, shown here, discusses some of the recent research projects that have received grants.

Source: National Science Foundation, http://www.nsf.gov.

GO TO THE NET

Want to learn more about the NSF? Go to **www.ablongman.com/johnsonweb/20.3**

NSF Proposal Preparation On-line Instructions

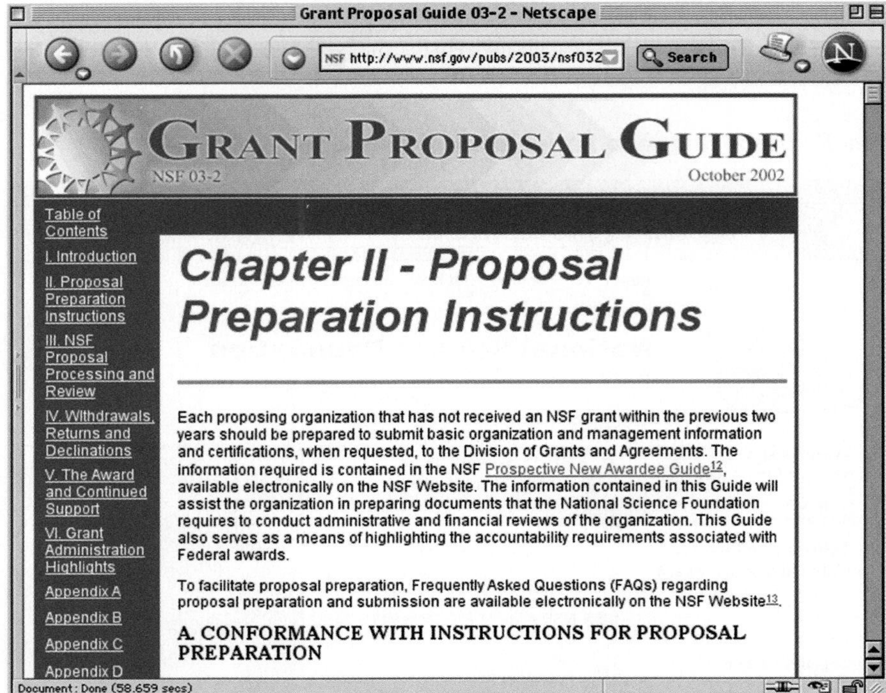

Figure 20.4: The NSF website also offers instructions on how to prepare proposals for grants.

Source: National Science Foundation, http://www.nsf.gov.

Planning and Researching Proposals

Because proposals are difficult to write, it is important that you follow a reliable writing process that will help you develop your proposal's content, organization, style, and design. An important first step in this process is to start with a planning and researching phase. During this phase, you will define the rhetorical situation and start collecting the content for the proposal.

Planning

A good way to start planning your proposal is to analyze the situations in which it will be used. Begin by answering the Five-W and How questions:

Who will be able to say "yes" to my ideas, and what are their characteristics?

Why is this proposal being written?

What information do the readers need to make a decision?

Where will the proposal be used?

When will the proposal be used?

How will the proposal be used?

Once you have answered these questions, you are ready to start thinking in depth about your proposal's subject, purpose, readers, and context of use. To begin this analysis, open a new document on your computer and jot down your understanding of the following issues:

SUBJECT Define exactly what your proposal is about. Where are the boundaries of the subject? What information do your readers expect you to include in the proposal? What need-to-know information must readers have if they are going to say "yes" to your ideas?

LINK For more help on defining need-to-know information, go to Chapter 5, page 115.

PURPOSE Clearly state the purpose of your proposal in one sentence. What should the proposal achieve? What do you want the proposal to do? By stating your purpose in one sentence, you will focus your writing efforts while making it easier for readers to understand what you are trying to accomplish.

Some key words for your purpose statement might include the following action verbs:

to persuade	to present
to convince	to propose
to provide	to offer
to describe	to suggest
to argue for	to recommend
to advocate	to support

A purpose statement might look something like this:

> The purpose of this proposal is to recommend that our company change its manufacturing process to include more automation.

LINK To read more about defining your purpose, go to Chapter 2, page 23.

READERS More than any other kind of document, proposals require you to fully understand your readers and anticipate their needs, values, and attitudes.

Primary readers (action takers) are the people who can say "yes" to your ideas. They need good reasons and solid evidence. They also hold values and attitudes that will shape how they interpret your ideas. Meanwhile, keep in mind that economic issues are always important to primary readers, so make sure you consider any money-related issues that might influence them.

Secondary readers (advisors) are usually experts in your field. They won't be the people who say "yes" to your proposal, but their opinions will be highly valued by your proposal's primary readers. You need to satisfy these advisors by offering enough technical information to demonstrate your understanding of the current situation and the soundness of your project.

Need help defining your purpose? Go to **www.ablongman.com/johnsonweb/20.4**

Tertiary readers (evaluators) can be just about anyone else who might have an interest in the project. They might include lawyers, journalists, and community activists, among others. You need to anticipate these readers' concerns, especially because tertiary readers can often undermine the project if you are not careful.

Gatekeepers (supervisors) are people at your own company who will need to look over your proposal before it is sent out. Your immediate supervisor is a gatekeeper, but you will likely also need to let other gatekeepers, like the company's accountants, lawyers, and technical advisors, look over the proposal before it is sent.

LINK For more strategies for analyzing your readers, see Chapter 3, page 42.

CONTEXT OF USE The document's context of use will also greatly influence how your readers will interpret the ideas in your proposal.

Physical context concerns the places they may read or use your proposal. Will readers look over your proposal at their desks, on their laptops, or in a meeting? Where will they discuss it?

Economic context involves the financial issues that will shape readers' responses to your ideas. How much money is available for the project? What economic trends will shape how your readers perceive the project? What are the financial limitations of the project?

Ethical context involves the ethical decisions that you and your readers will need to make. Where does the proposal touch on ethical issues? How might these ethical issues be resolved so they don't undermine the project? What are the legal issues involved with the proposal?

Political context concerns the people your proposal will affect. Who stands to gain or lose if your proposal is accepted? How will the proposal change relationships that are already in place? Would any larger political trends shape how the proposal is written or interpreted?

LINK For more help defining the context of use, turn to Chapter 3, page 53.

Something to keep in mind is that proposals, especially external proposals, are de facto legal contracts. They are legal documents that can be brought into court if a dispute occurs. So, you need to make sure that everything you say in the proposal is truthful and sincere, because the proposal may be used in a court case to prove (or disprove) that your company completed the promised work to the level expected.

If you are working with a team, it is especially important that all team members agree up front about the rhetorical situation. If all the team members start out with the same understanding of the rhetorical situation, you will avoid some of the all-too-typical misunderstandings that occur as proposals are written.

Worksheets are available to help you analyze your readers and contexts of use. Go to **www.ablongman.com/johnsonweb/20.5**

TAKE NOTE Save these comments about the subject, purpose, readers, and context of use. They will help you and your team stay on task as you begin drafting your proposal. In fact, a helpful practice is to print out your comments and tack them to the wall behind your computer. That way, your comments will keep you focused on your subject, purpose, readers, and context in which your proposal will be used.

Researching

After defining your proposal's rhetorical situation, you should start collecting and creating the content of your document. Chapter 5 of this book describes research methods in detail, so they won't be fully described here. However, here are some research strategies that are especially applicable to writing proposals:

DO BACKGROUND RESEARCH The key to writing a persuasive proposal is to first fully understand the problem you are trying to solve. First, you might go to the Internet to find as much information about your subject as you can. Second, locate print sources on your subject, like books, reports, news articles, and brochures. Third, interview, survey, and observe people who have a stake in the plan or project you are developing. Find out their views.

ASK SUBJECT MATTER EXPERTS (SMEs) Spend time interviewing experts who know a great amount about your subject. They can probably give you insight into the problem you are trying to solve and suggest some potential solutions. They might also tell you what has worked (and not worked) in the past.

PAY ATTENTION TO CAUSES AND EFFECTS All problems have causes, and all causes create effects. In your observations of the problem, try to identify the causes that are behind that problem. Then, try to identify some of the effects of the problem.

FIND SIMILAR PROPOSALS On the Internet or at your workplace, you can probably locate proposals that have dealt with similar problems in the past. These samples can help you frame the problem and better understand the causes and effects. They might also give you some insight into how similar problems have been solved in the past.

TAKE NOTE To avoid plagiarism or copyright violations, use sample proposals for guidance only. When you write your own proposal, you will need to develop your own words and ideas. The only exception to this rule is if your company owns the document you are using as a model.

COLLECT VISUALS Proposals are persuasive documents, so they often include plenty of graphics, like photographs, charts, illustrations, and graphs. Collect any materials, data, and information that will help you add a visual quality to your proposal. If appropriate, you might use a digital camera to take pictures to be added to the document.

LINK To learn more about doing research, turn to Chapter 5, page 96.

For more help on researching your subject, go to **www.ablongman.com/johnsonweb/20.6**

HELP

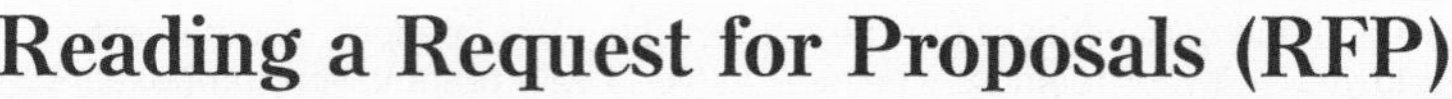

Reading a Request for Proposals (RFP)

A solicited proposal usually responds to a Request for Proposal (RFP)—an advertisement written by a client who has work available. These advertisements are typically written in highly technical language, so a first step in the proposal-writing process is to figure out what the client needs. RFPs are also called Information for Bids (IFBs), requests for quotes (RFQs), requests for applications (RFAs) and requests for offers (RFOs).

To interpret an RFP, use the Five-W and How questions to break it down into its basic elements. For example, consider the RFO in Figure A. What are some answers to the *who* question in this advertisement?

- Who is the client?
- Who is qualified to bid for the project?
- Who are the other people involved?
- Who is the point of contact (POC) who can answer your questions?

On your computer screen, jot down your answers to these kinds of *who*-related questions. Then, use the *what, where,* and *when* questions to help you sort out the other elements of the RFP. What does the client want? Where will the work be done? When does the project need to be completed? Your computer screen will soon fill up with information that was drawn from the RFP.

Sample RFO

NASA/Lyndon B. Johnson Space Center, Houston Texas, 77058-3696
A -- LONG-LIFE SPACE SHUTTLE VERNIER THRUSTER SOL 9-BH13-67-01-17P

POC Mary F. Thomas, Contract Specialist, Phone (281) 483-8828, Fax (281) 244-5337, Email mary.f.thomas1@jsc.nasa.gov -- Keith D. Hutto, Contracting Officer, Phone (281) 483-4165, Fax (281) 244-5337, Email keith.d.hutto1@jsc.nasa.gov WEB: Click here for the latest information about this notice, http://nais.msfc.nasa.gov/cgi- bin/EPS/bizops.cgi?gr=D&pin=73#9-BH13-67-0 01-17P. E-MAIL: Mary F. Thomas, mary.f.thomas1@jsc.nasa.gov. NASA/JSC plans to issue a Request for Offer (RFO) to develop technology for a long life thruster using mono-methyl hydrazine and nitrogen tetroxide that extends operating life to 300,000 seconds. The technology must eliminate coatings that can chip off and cause premature chamber failure and replacement. The deliverable shall be a report which details the development, analysis, and results of the testing. As an option in the contract, the vendor shall deliver a 25 lbf thrust chamber assembly to the NASA White Sands Test Facility in Las Cruces, New Mexico. The Government does not intend to acquire a commercial item using FAR Part 12. See Note 26. The NAICS Code and Size Standard are 541710 and 1,000 employees, respectively. The DPAS Rating for this procurement is DO-C9. The provisions and clauses in the RFO and model contract are those in effect through FAC 97-25. All qualified responsible sources may submit an offer which shall be considered by the agency. The anticipated release date of the RFO is on or about June 25, 2001 with an anticipated offer due date of on or about July 26, 2001. An ombudsman has been appointed -- See NASA Specific Note "B". The solicitation and any documents related to this procurement will be available over the Internet. These documents will be in Microsoft Office 97 format and will reside on a World Wide Web (WWW) server, which may be accessed using a WWW browser application. The Internet site, or URL, for the NASA/JSC Business Opportunities home page is http://nais.msfc.nasa.gov/cgi-bin/EPS/bizops.cgi?gr=C&pin=73 Prospective offerors shall notify this office of their intent to submit an offer. It is the offeror's responsibility to monitor the Internet site for the release of the solicitation and amendments (if any). Potential offerors will be responsible for downloading their own copy of the solicitation and amendments (if any). Any referenced notes can be viewed at the following URL: http://genesis.gsfc.nasa.gov/nasanote.html Posted 06/11/01 (D- SN50O564). (0162)

Figure A: Reading an RFO can be difficult. The 5-W and How questions can help you sort out the information, so you can better interpret it. This figure shows an advertisement for an upcoming RFO from NASA. Bidders can request the full RFO when it is available.

To see real RFPs, go to **www.ablongman.com/johnsonweb/20.7**

You will notice that the *why* and *how* questions are rarely answered in an RFP. The *why* question is especially difficult to answer, but you need to find an answer to this question before writing your proposal. Specifically, *why* does the client need someone to do the project? Initially, you can probably make some strategic guesses about why the client put the project out for bid. Then, using the Internet, the library, and your industry contacts, you should confirm whether your guesses are correct.

The *how* question is what your proposal is supposed to answer for the client. In your proposal, you will describe in detail *how* you, your team, or your company will do the work advertised in the RFP. Nevertheless, the client may give you hints in the RFP about what kinds of solutions or plans are likely acceptable. Read the RFP closely for any hint about how the client would like the *how* question answered.

Again, look at the RFO in Figure A. *Why* do you think the client wants someone to handle this project? *How* might you and your company help the client solve their problem?

When you have finished answering the Five-W and How questions, you should have a much better grasp of the RFP, including what the clients are looking for.

Organizing and Drafting Proposals

Writing the first draft of a proposal is always difficult. Why? Proposals describe the future—a future that you are trying to envision for your readers and yourself. Consequently, you will need to use your imagination to help create and describe the future you have in mind.

A good way to draft your proposal is to write it one section at a time. Think of the proposal as four or five separate mini-documents that could stand alone. When you finish drafting one section, move on to the next.

Writing the Introduction

As with all documents, the introduction of a proposal sets a context for the body of the document. A proposal's introduction will usually include up to six moves:

Move 1: Define the *subject.*

Move 2: State the *purpose.*

Move 3: State the *main point.*

Move 4: Stress the *importance of the subject.*

Move 5: Offer *background information* on the subject.

Move 6: *Forecast* the organization of the document.

These moves can be made in just about any order, depending on your proposal, and they are not all required. Minimally, your proposal's introduction should clearly identify your *subject, purpose,* and *main point.* The other three moves are helpful, but they are optional. Figure 20.5 shows a sample introduction that uses all six moves.

Introduction to a Proposal

Figure 20.5: This introduction makes all six "moves." As a result, it is somewhat lengthy. Nevertheless, this introduction prepares readers to understand the information in the body of the proposal.

Growth and Flexibility With Telecommuting

A Proposal to Northside Design

Offers background information. / **Defines subject.** / **Stresses importance.**

Founded in 1979, Northside Design is one of the classic entrepreneurial success stories in architecture. Today, this company is one of the leading architectural firms in the Chicago market with over 50 million dollars in annual revenue. With growth, however, comes growing pains, and Northside now faces an important decision about how it will manage its growth in the near future. The right decision could lead to more market share, increased sales, and even more prominence in the architectural field. However, Northside also needs to safeguard itself against over-extension in case the Chicago construction market unexpectedly begins to recede.

"Northside needs to safeguard itself against over-extension in case the Chicago construction market unexpectedly begins to recede."

States purpose. / **States main point.**

To help you make the right decision, this proposal offers an innovative strategy that will support your firm's growth while maintaining its flexibility. Specifically, we propose Northside implement a telecommuting network that allows selected employees to work a few days each week at home. Telecommuting will provide your company with the office space it needs to continue growing. Meanwhile, this approach will avoid a large investment in new facilities and disruption to the company's current operations.

The closing forecasts body of proposal.

In this proposal, we will first discuss the results of our research into Northside's office space needs. Second, we will offer a plan for using a telecommuting network to free up more space at Northside's current office. Third, we will review Insight Systems' qualifications to assist Northside with its move into the world of telecommuting. And finally, we will go over some of the costs and advantages of our plan.

1

To see sample proposal introductions, go to **www.ablongman.com/johnsonweb/20.8**

Your proposal's introduction should be concise. Usually introductions for proposals run about one to three paragraphs.

LINK For additional help on writing introductions, see Chapter 6, page 141.

Describing the Current Situation

The aim of the *current situation* section—sometimes called the *background* section—is to define the problem your plan will solve. In this section, you should offer readers an understanding of the current situation by clearly defining the problem, its causes, and its effects.

You should accomplish three things in this section of the proposal:

- Define and describe the problem.
- Discuss the causes of the problem.
- Discuss the effects of the problem if nothing is done.

For example, let us say you are writing a proposal to improve safety at your college or workplace. Your current situation section would first define the problem by proving there is a lack of safety and showing its seriousness. Then, it would discuss the causes and effects of that problem.

MAPPING OUT THE SITUATION Logical mapping is a helpful technique for developing your argument in this section. Here are some steps you can follow to map out the content:

1. Write the problem in the middle of your screen or piece of paper. Put a circle around it.
2. Write down the two to five major causes of that problem. Circle them, and connect them to the problem.
3. Write down some minor causes around each major cause, treating each major cause as a separate problem of its own. Circle the minor causes and connect them to the major causes.

Figure 20.6 illustrates how your logical map for the current situation section might look.

DRAFTING THE CURRENT SITUATION SECTION Your logical map of the current situation should help you define the basic content of this section. Now you are ready to turn your map into paragraphs and sentences. To begin, the current situation section should include an opening, a body, and a closing:

Opening—Identify and define the problem you will describe.

Body—Discuss the *causes* of the problem, showing how these causes brought about the problem.

Closing—Discuss the *effects* of not doing anything about the problem.

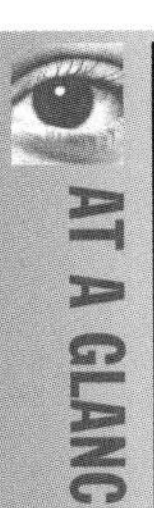

The Current Situation

- Define and describe the problem.
- Discuss the causes of the problem.
- Discuss the effects if nothing is done about the problem.

Mapping Out the Current Situation

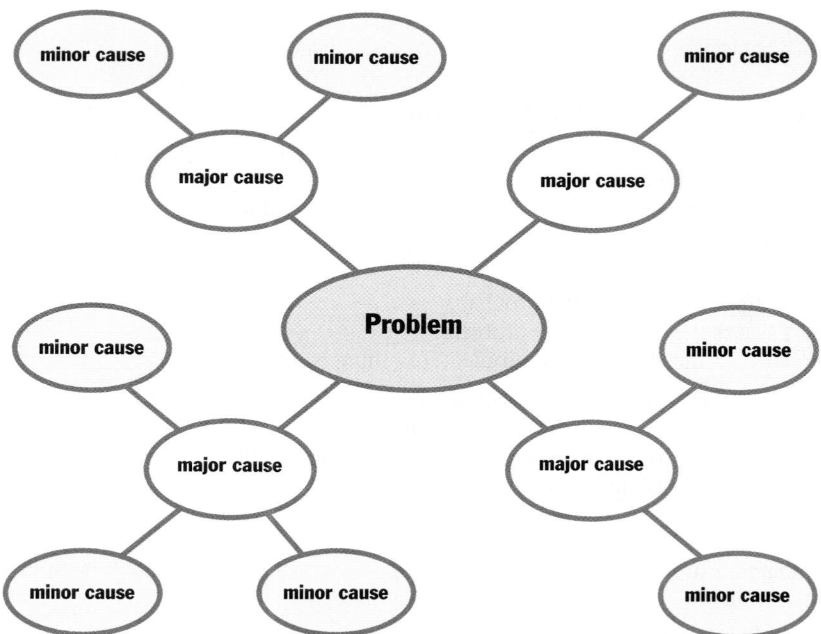

Figure 20.6: Logical mapping helps you figure out what caused the problem that you are trying to solve.

The length of the current situation section depends on your readers' familiarity with the problem. If readers are new to the subject, then several paragraphs or even pages might be required. If they fully understand the problem already, maybe only a paragraph or two are needed. The sample text in Figure 20.7 demonstrates how this section might be organized.

Figure 20.7 shows only one possible pattern. The current situation section can be organized many different ways. What is important is that the section define the problem and discuss its causes and effects.

Describing the Project Plan

A proposal's *project plan* section offers a step-by-step method for solving the problem. Your goal is to tell your readers *how* you would like to handle the problem and *why* you would handle it that way. In this section, you should—

- Identify the solution.
- State the objectives of the plan.
- Describe the plan's major and minor steps.
- Identify the deliverables or outcomes.

To see documents that analyze problems and causes, go to **www.ablongman.com/johnsonweb/20.9**

Example Current Situation Section

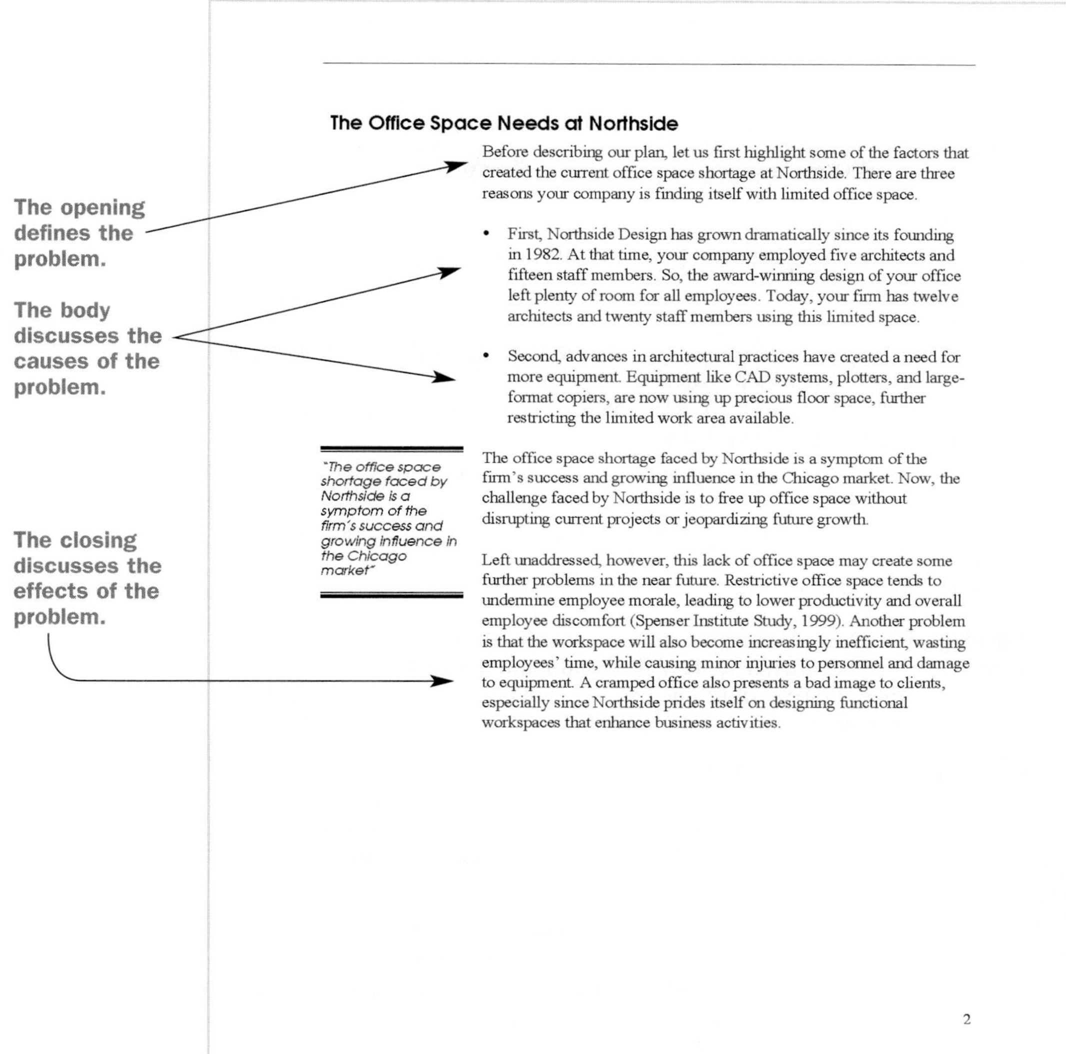

The Office Space Needs at Northside

Before describing our plan, let us first highlight some of the factors that created the current office space shortage at Northside. There are three reasons your company is finding itself with limited office space.

- First, Northside Design has grown dramatically since its founding in 1982. At that time, your company employed five architects and fifteen staff members. So, the award-winning design of your office left plenty of room for all employees. Today, your firm has twelve architects and twenty staff members using this limited space.
- Second, advances in architectural practices have created a need for more equipment. Equipment like CAD systems, plotters, and large-format copiers, are now using up precious floor space, further restricting the limited work area available.

"The office space shortage faced by Northside is a symptom of the firm's success and growing influence in the Chicago market"

The office space shortage faced by Northside is a symptom of the firm's success and growing influence in the Chicago market. Now, the challenge faced by Northside is to free up office space without disrupting current projects or jeopardizing future growth.

Left unaddressed, however, this lack of office space may create some further problems in the near future. Restrictive office space tends to undermine employee morale, leading to lower productivity and overall employee discomfort (Spenser Institute Study, 1999). Another problem is that the workspace will also become increasingly inefficient, wasting employees' time, while causing minor injuries to personnel and damage to equipment. A cramped office also presents a bad image to clients, especially since Northside prides itself on designing functional workspaces that enhance business activities.

2

The opening defines the problem.
The body discusses the causes of the problem.
The closing discusses the effects of the problem.

Figure 20.7: The current situation section includes an opening, a body, and a closing. The causes of the problem are discussed mainly in the body paragraphs, while the effects are usually discussed at the end of the section.

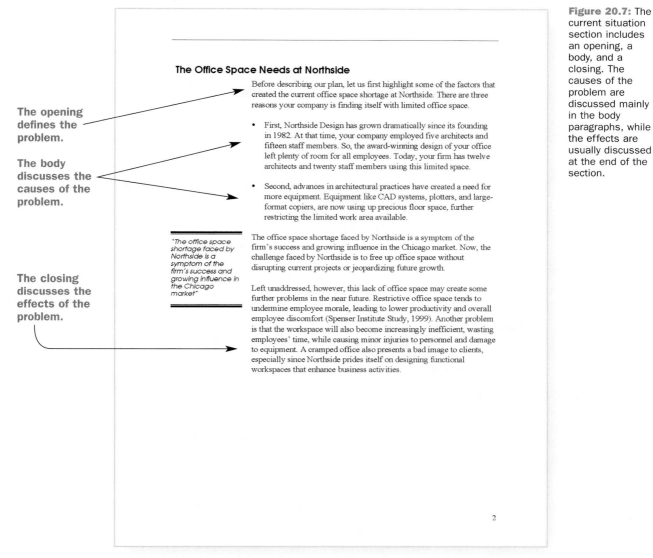

As you begin drafting this section, look back at your original purpose statement for the proposal, which you wrote during the planning phase. Now, imagine a solution that might achieve that purpose. Perhaps you already have a solution in mind. If not, brainstorm with your team members to determine what kinds of solutions might solve the problem you described in the current situation section.

TAKE NOTE Let yourself be creative and innovative while forming the project plan. Consider even the most far-fetched ideas. There are often several ways to solve any given problem. You never know the best solution until you have considered them all.

MAPPING OUT THE PROJECT PLAN When you have identified a possible solution, you can again use logical mapping to turn your idea into a plan:

1. Write your solution in the middle of your screen or a sheet of paper. Circle this solution.
2. Write down the two to five major steps needed to achieve that solution. Circle them and connect them to the solution.
3. Write down the minor steps required to achieve each major step. Circle them and connect them to the major steps.

As shown in Figure 20.8, your map should illustrate the basic steps in your plan.

DRAFTING THE PROJECT PLAN SECTION Like the current situation section before it, your project plan section will also have an opening, a body, and a closing. This section will describe step by step how you will achieve your project's purpose.

Opening—Identify your overall solution or plan to solve the problem. You can even give your plan a name to make it sound more real (e.g., the "Restore Central Campus Project"). Your opening might also include a list of project objectives so the readers can see what goals your plan is striving to achieve.

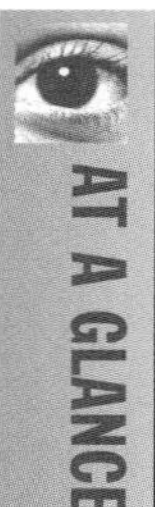

The Project Plan

- Identify the solution.
- State the objectives of the plan.
- Describe the plan's major and minor steps.
- Identify the deliverables or outcomes.

Body—Walk the readers through your plan step by step. Address each major step separately, discussing the minor steps needed to achieve that major step. It is also helpful to tell readers *why* each major and minor step is needed.

Closing—Summarize the final *deliverables,* or outcomes, of your plan. The deliverables are the goods and services that you will provide when the project is finished. Tell the readers what the end results of your plan will be.

TAKE NOTE Deliverables should be clearly identified so readers know what they are receiving for their investment in your ideas.

For more advice about writing project plans in proposals, go to **www.ablongman.com/johnsonweb/20.10**

Mapping Out a Project Plan

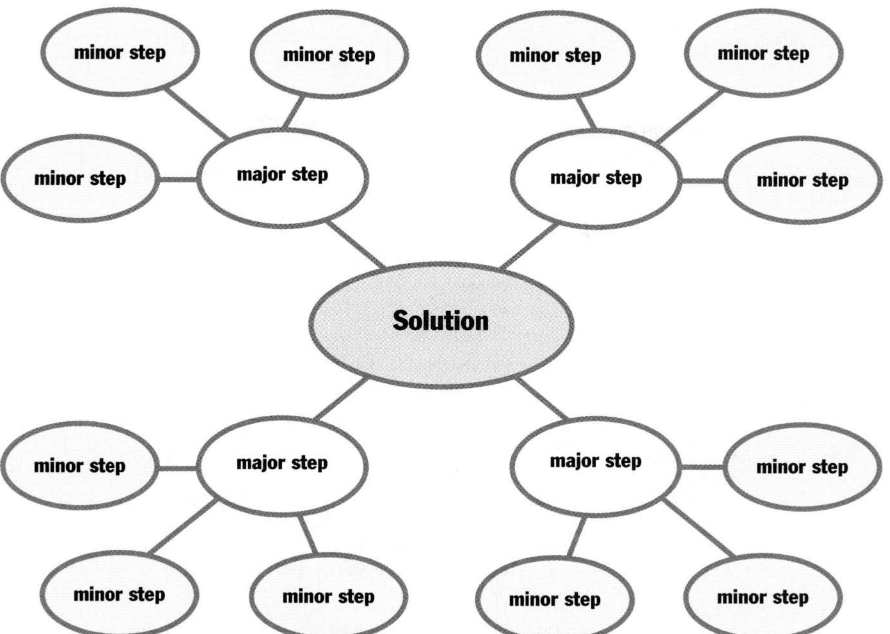

Figure 20.8: Logical mapping will help you figure out how to solve the problem. In a map, like the one shown here, you can visualize your entire plan by writing out the major and minor steps.

As shown in Figure 20.9, the project plan section balances the plan's steps with reasons why these steps are needed.

In most proposals, the project plan is the longest section of the document. This section needs to clearly describe your plan. Moreover, it needs to give your readers good reasons to believe your plan will work, while offering specific outcomes or results (deliverables).

Describing Qualifications

The qualifications section presents the credentials of your team or company, striving to prove that you are qualified to carry out the project plan. Minimally, the aim of the qualifications section is to show that your team or company is able to do the work. Ideally, however, you also want to prove that your team or company is *best qualified* to handle the project.

As you begin drafting this section, keep the following saying in mind: *What makes you different makes you attractive.* In other words, pay attention to the qualities that make your team or company different from your competitors. What are your company's strengths? What makes you better than the others?

Project Plan Section

Figure 20.9: An effective project plan section includes an opening, a body, and a closing. The opening states the solution and offers some objectives. The body walks the readers through the plan's steps. The closing identifies the major deliverables of the plan.

Objectives

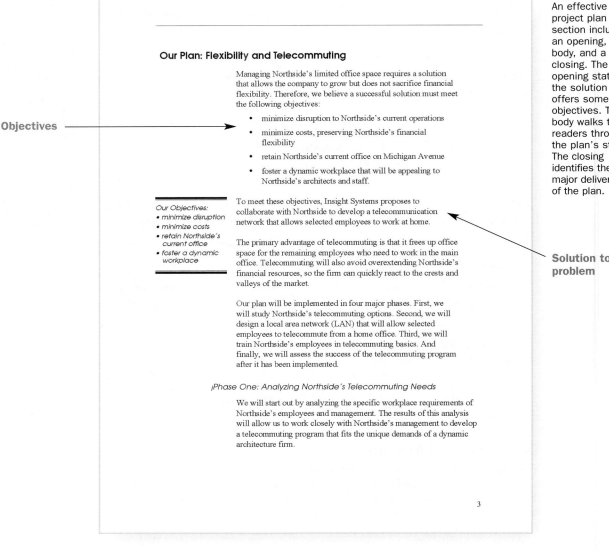

Solution to problem

Our Plan: Flexibility and Telecommuting

Managing Northside's limited office space requires a solution that allows the company to grow but does not sacrifice financial flexibility. Therefore, we believe a successful solution must meet the following objectives:

- minimize disruption to Northside's current operations
- minimize costs, preserving Northside's financial flexibility
- retain Northside's current office on Michigan Avenue
- foster a dynamic workplace that will be appealing to Northside's architects and staff.

Our Objectives:
- *minimize disruption*
- *minimize costs*
- *retain Northside's current office*
- *foster a dynamic workplace*

To meet these objectives, Insight Systems proposes to collaborate with Northside to develop a telecommunication network that allows selected employees to work at home.

The primary advantage of telecommuting is that it frees up office space for the remaining employees who need to work in the main office. Telecommuting will also avoid overextending Northside's financial resources, so the firm can quickly react to the crests and valleys of the market.

Our plan will be implemented in four major phases. First, we will study Northside's telecommuting options. Second, we will design a local area network (LAN) that will allow selected employees to telecommute from a home office. Third, we will train Northside's employees in telecommuting basics. And finally, we will assess the success of the telecommuting program after it has been implemented.

¡Phase One: Analyzing Northside's Telecommuting Needs

We will start out by analyzing the specific workplace requirements of Northside's employees and management. The results of this analysis will allow us to work closely with Northside's management to develop a telecommuting program that fits the unique demands of a dynamic architecture firm.

3

Figure 20.9: (*continued.*)

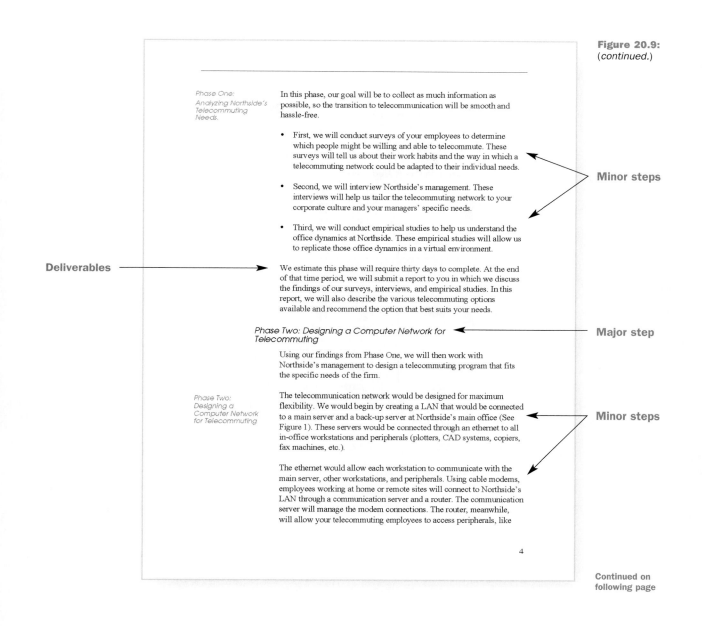

Phase One: Analyzing Northside's Telecommuting Needs.

In this phase, our goal will be to collect as much information as possible, so the transition to telecommunication will be smooth and hassle-free.

- First, we will conduct surveys of your employees to determine which people might be willing and able to telecommute. These surveys will tell us about their work habits and the way in which a telecommuting network could be adapted to their individual needs.
- Second, we will interview Northside's management. These interviews will help us tailor the telecommuting network to your corporate culture and your managers' specific needs.
- Third, we will conduct empirical studies to help us understand the office dynamics at Northside. These empirical studies will allow us to replicate those office dynamics in a virtual environment.

Minor steps

Deliverables

We estimate this phase will require thirty days to complete. At the end of that time period, we will submit a report to you in which we discuss the findings of our surveys, interviews, and empirical studies. In this report, we will also describe the various telecommuting options available and recommend the option that best suits your needs.

Phase Two: Designing a Computer Network for Telecommuting

Major step

Using our findings from Phase One, we will then work with Northside's management to design a telecommuting program that fits the specific needs of the firm.

Phase Two: Designing a Computer Network for Telecommuting

The telecommunication network would be designed for maximum flexibility. We would begin by creating a LAN that would be connected to a main server and a back-up server at Northside's main office (See Figure 1). These servers would be connected through an ethernet to all in-office workstations and peripherals (plotters, CAD systems, copiers, fax machines, etc.).

Minor steps

The ethernet would allow each workstation to communicate with the main server, other workstations, and peripherals. Using cable modems, employees working at home or remote sites will connect to Northside's LAN through a communication server and a router. The communication server will manage the modem connections. The router, meanwhile, will allow your telecommuting employees to access peripherals, like

4

Continued on following page

Figure 20.9: *(continued.)*

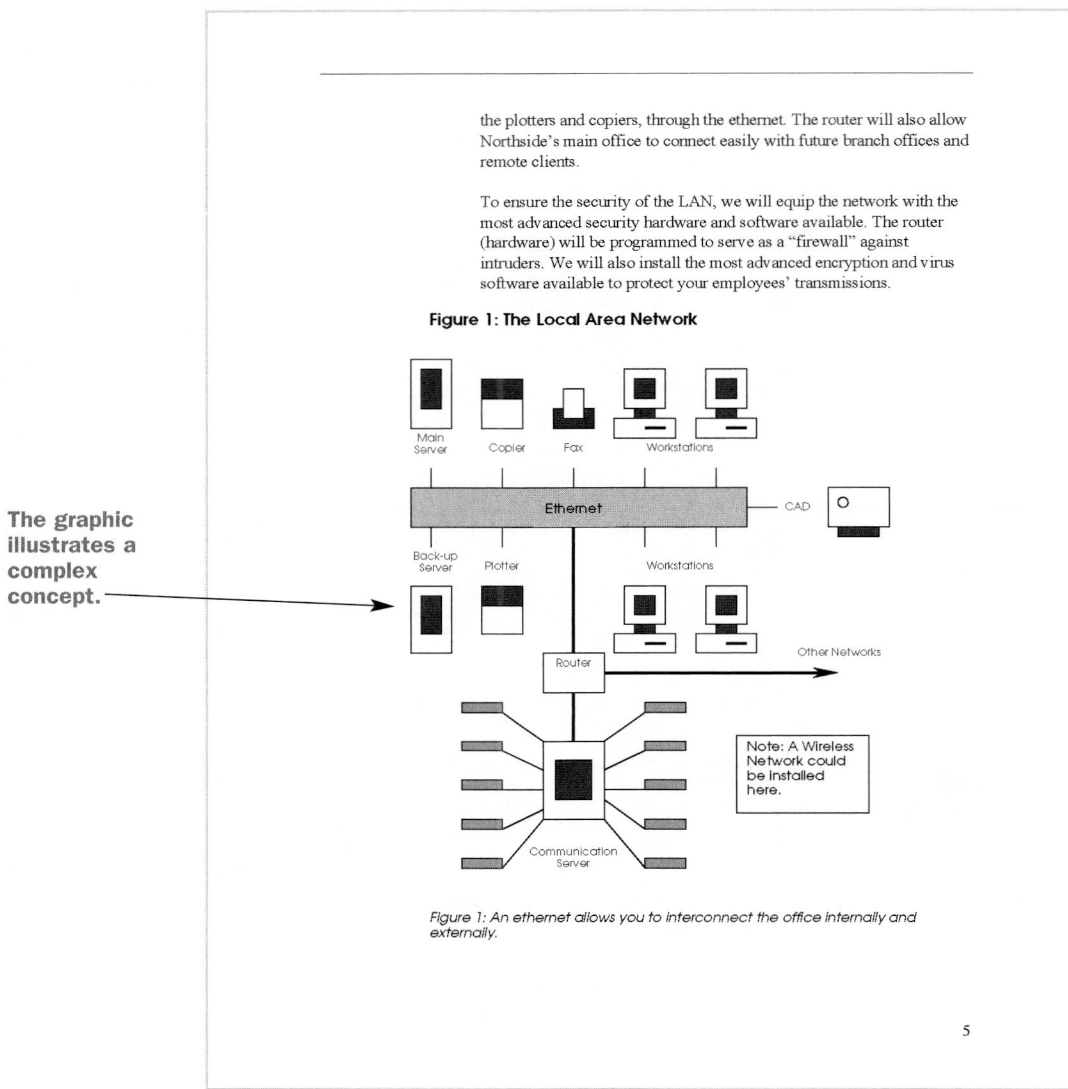

the plotters and copiers, through the ethernet. The router will also allow Northside's main office to connect easily with future branch offices and remote clients.

To ensure the security of the LAN, we will equip the network with the most advanced security hardware and software available. The router (hardware) will be programmed to serve as a "firewall" against intruders. We will also install the most advanced encryption and virus software available to protect your employees' transmissions.

Figure 1: The Local Area Network

Figure 1: An ethernet allows you to interconnect the office internally and externally.

5

The graphic illustrates a complex concept.

Figure 20.9: *(continued.)*

Overall, the advantage of this LAN design is that Northside's telecommuting employees will have access to all the equipment and services available in the main office. Meanwhile, even travelling employees who are visiting clients will be able to easily tap into the LAN from their laptop computers.

Deliverables

Phase Three: Training Northside's Employees

Experience has shown us that employees adapt quickly to telecommuting. Initially, though, we would need to train them how to access and use the LAN from workstations inside and outside the office.

Phase Three: Training Northside's Employees

To fully train your employees in telecommuting basics, we would need two afternoons (eight hours). We will show them how to communicate through the network and access peripherals at the main office. The training will also include time management strategies to help your employees adjust to working outside the office. We have found that time management training helps employees work more efficiently at home—often more efficiently than they work in the office.

Insight Systems maintains a 24-hour helpline that your employees can call if they have any questions about using the LAN. Also, our website contains helpful information on improving efficiency through telecommuting.

Phase Four: Assessing the Telecommuting Program

To ensure the effectiveness of the telecommuting network, we will regularly survey and interview your managers and employees to solicit their reactions and suggestions for improvements. These surveys and interviews will be conducted every three months for two years.

Phase Four: Assessing the Telecommuting Program

We will be particularly interested in measuring employee satisfaction with telecommuting, and we will measure whether they believe their efficiency has increased since they began working at home through the LAN. The results of these assessments will help us fine-tune the LAN to your employees' needs.

After each three-month survey, we will submit a progress report to Northside that discusses our findings. At the end of the two-year period, we will submit a full report that analyzes our overall findings and makes suggestions for improving the telecommuting program in the future.

6

In the qualifications section, you do not need to discuss every aspect of your team or company. Rather, you should offer just enough information to demonstrate that your team or your company is best qualified and able to handle the proposed project.

A typical qualifications section offers information on three aspects of your team or company:

> **Description of personnel**—Short biographies of managers who will be involved in the project; demographic information on the company's workforce; description of support staff.
>
> **Description of organization**—Corporate mission, philosophy, and history of the company; corporate facilities and equipment; organizational structure of the company.
>
> **Previous experience**—Past and current clients; a list of similar projects that have been completed; case studies that describe past projects.

Figure 20.10 shows a sample qualifications section that includes these three kinds of information about qualifications. Pay attention to how this section does more than describe the company—it makes an argument that the bidders are uniquely qualified to handle the project.

You should never underestimate the importance of the qualifications section in a proposal. In the end, your readers will not accept the proposal if they do not believe your team or company has the personnel, facilities, or experience to do the work. In this section, your job is to persuade them that you are uniquely or best qualified to handle the project.

Costs and Benefits

The costs and benefits section summarizes the advantages of saying "yes" to the proposal while also telling readers how much the project will cost.

Start out this section by making an obvious transition. Say something like, "Let us conclude by summarizing the costs and benefits of our plan." This kind of transition will wake up your readers, because they know that you are going to be telling them the most important points in the proposal.

Early in this section, tell them the costs. A good strategy for handling expenses is to simply state the costs up front without apology or a sales pitch.

> As shown in our budget, this renovation will cost $287,000.
>
> We anticipate the price for retooling your manufacturing plant will be $5,683,000.

Immediately after this statement of the costs, you should then describe the significant benefits of saying "yes" to the plan. Fortunately, you already identified many of these benefits earlier in the proposal. They appeared in two other sections: the project plan section and qualifications section.

> **Project plan section**—You identified some deliverables when you described the project plan. Summarize those deliverables for the readers and discuss the benefits of having them.

For more help writing a qualifications section, go to **www.ablongman.com/johnsonweb/20.11**

Qualifications Section

Figure 20.10: An effective qualifications section shows why your team or company is qualified, or the best qualified, to handle a project. In this sample, notice how the section makes a clean argument by appealing to experience.

When we complete this plan, Northside Design will have a fully functional telecommuting network that will allow selected employees to work from home. You should see an immediate improvement in productivity and morale. Meanwhile, you will be able to stay financially flexible to compete in the Chicago architectural market.

Qualifications at Insight Systems

The opening paragraph makes a claim that the section will support.

At Insight Systems, we know this moment is a pivotal one for Northside Design. To preserve and expand its market share, Northside needs to grow, but it cannot risk overextending itself financially. For these reasons, Insight Systems is uniquely qualified to handle this project, because we provide flexible, low-cost telecommuting networks that help growing companies stay responsive to shifts in their industry.

Management and Labor

With over seventy combined years in the industry, our management team offers the insight and responsiveness required to handle your complex growth needs. (The resumes of our management team are included in Appendix B).

"With over seventy combined years in the industry, our management team offers insight and responsiveness required to handle your complex growth needs."

Hanna Gibbons, our CEO, has been working in the telecommuting industry for over 20 years. After she graduated from MIT with a Ph.D. in computer science, she worked at Krayson International as a systems designer. Ten years later, she had worked her way up to Vice President in charge of Krayson's Telecommuting Division. In 1993, Dr. Gibbons took over as CEO of Insight Systems. Since then, Dr. Gibbons has built this company into a major industry leader with gross sales of $15 million per year.

Description of personnel

Frank Roberts, Chief Engineer at Insight Systems, has 30 years of experience in the networked computer field. He began his career at Brindle Labs, where he worked on artificial intelligence systems using analog computer networks. In 1985, he joined the Insight Systems team, bringing his unique understanding of networking to our team. Frank is very detail oriented, often working long hours to ensure that each computer network meets each client's exact specifications and needs.

7

Continued on following page

Figure 20.10: *(continued.)*

Lisa Miller, Insight System's Senior Computer Engineer, has successfully led the implementation of thirty-three telecommuting systems in companies throughout the United States. Earning her computer science degree at Iowa State, Lisa has won numerous awards for her innovative approach to computer networking. She believes that clear communication is the best way to meet her clients' needs.

Our management is supported by one of most advanced teams of high technology employees. Insight Systems employs twenty of the brightest engineers and technicians in the telecommunications industry. We have aggressively recruited our employees from the most advanced universities in the United States, including Stanford, MIT, Illinois, Iowa State, New Mexico, and Syracuse. Several of our engineers have been with Insight Systems since it was founded.

Corporate History and Facilities

Insight Systems has been a leader in the telecommuting industry from the beginning. In 1975, the company was founded by John Temple, a pioneer in the networking field. Since then, Insight Systems has followed Dr. Temple's simple belief that computer-age workplaces should give people the freedom to be creative.

"Insight Systems earned the coveted '100 Companies to Watch' designation from Business Outlook Magazine.*"*

Recently, Insight Systems earned the coveted "100 Companies to Watch" designation from *Business Outlook Magazine* (May 2004). The company has worked with large and small companies, from Vedder Aerospace to the Cedar Rapids Museum of Fine Arts, to create telecommuting options for companies that want to keep costs down and productivity high.

Description of organization

Insight Systems' Naperville office has been called "a prototype workspace for the information age." (*Gibson's Computer Weekly*, May 2000). With advanced LAN systems in place, only ten of Insight System's fifty employees actually work in the office. Most of Insight Systems' employees telecommute from home or on the road.

Experience You Can Trust

Our background and experience gives us the ability to help Northside manage its needs for a more efficient, dynamic office space. Our key to success is innovation, flexibility, and efficiency.

8

Qualifications section—When describing your team's or company's qualifications, you also showed that your organization is uniquely or best qualified to handle the project. You can now briefly remind your readers of the benefits of working with your company (e.g., qualified people, high-quality products and services, superior customer service, excellent facilities).

In most cases, your discussion of the benefits should add no new content to the proposal. Instead, this section should summarize the important benefits you discussed in earlier sections of the proposal. In Figure 20.11, for example, you will notice that this proposal's costs and benefits section really doesn't mention anything new to the proposal, except the cost of the project.

Why should you discuss the benefits at all, especially if you already mentioned them in the proposal? There are two main reasons:

- If you summarize the benefits at this point, readers will know exactly what they are going to receive if they say "yes" to your ideas.
- The costs are usually a bitter pill for most readers. Just telling them the price without reminding them of the benefits would make the price of the project hard to swallow.

At this point in the proposal, your readers are measuring the costs of your plan against its benefits. By discussing costs and benefits at the same time, you can show them how the benefits outweigh these costs.

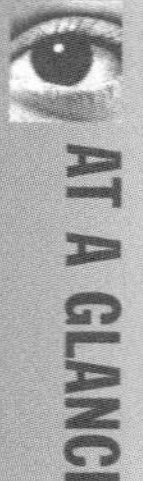

Concluding a Proposal

- Restate the proposal's main point (the solution).
- Say thank you.
- Describe the next step.
- Provide contact information.

Conclusion

The conclusion of a proposal should be concise, perhaps only one or two paragraphs. To draft an effective conclusion, you should include some or all of the following elements: a) restate the proposal's main point (the solution); b) say thank you; c) describe the next step; d) provide contact information.

Here at the end of the proposal is a good place to restate the main point of your proposal. Of course, you have already told readers your solution at least a couple of times. Tell them again. This last repetition will leave them with a clear statement of what you want to achieve.

Also, you might thank your readers for their consideration of your ideas. Much like a public speaker signaling the end of a speech by thanking listeners for their attention, you can end your proposal by thanking readers for considering your plan. Thanking your readers will end your proposal on a positive note.

Finally, when concluding, you should leave your readers with a clear sense of what they should do when they finish reading the proposal. Should they call you? Should they wait for you to call them? Should they set up a meeting with you? A well-written conclusion makes their next move obvious for them.

In Figure 20.11, for example, the conclusion has been added to the end of the costs and benefits section. The authors thank the readers for their consideration and ask them to call when they are finished looking over the proposal.

LINK For more information on writing conclusions, go to Chapter 6, page 155.

To see discussions of costs and benefits, go to **www.ablongman.com/johnsonweb/20.12**

Costs and Benefits Section

Figure 20.11: The costs and benefits section, as well as the proposal's conclusion, are shown here. To end on a strong note, the costs and benefits section summarizes the benefits of saying "yes." Afterward, the conclusion should thank the readers and tell them the next step.

The Plan's Benefits and Project Costs

Restate the solution.

To conclude, let us summarize the advantages of our plan and discuss the costs. Our preliminary research shows that Northside Design will continue to be a leader in the Chicago market. The strong economy coupled with Northside's award-winning designs will only increase the demand for your services. At Insight Systems, we believe the best way to manage Northside's growth is to implement a telecommuting network that will allow some of the company's employees to telecommute from home or on-site.

State the costs.

Cost is the most significant advantage of our plan. As shown in Appendix A, implementation of our plan would cost an estimated $230,640. We believe this investment in Northside's infrastructure will preserve your company's financial flexibility, allowing you to react quickly to the market's crests and valleys. But the advantages of our plan go beyond simple costs:

Our Plan's Benefits:
- *Limited costs*
- *It will not disrupt your current operations*
- *Employee morale will improve*

Summarize the benefits.

First, a telecommuting system will allow your current operations to continue without disruption. When the telecommuting network is ready to go on-line, your employees will simply need to attend two four-hour training sessions on using the LAN. Your management can then gradually convert selected employees into telecommuters.

Second, employee morale will benefit from using the telecommuting network. With fewer employees at the office, there will be more space available for the employees who need to be in the office each day. Also, studies have shown that telecommuting employees not only report more job satisfaction, they also increase their productivity.

When the telecommuting system is in place, Northside will be positioned for continued growth and leadership in the Chicago architectural market. Telecommuting will open up space at your current downtown office while maintaining the morale and productivity of your employees as your business continues to grow.

9

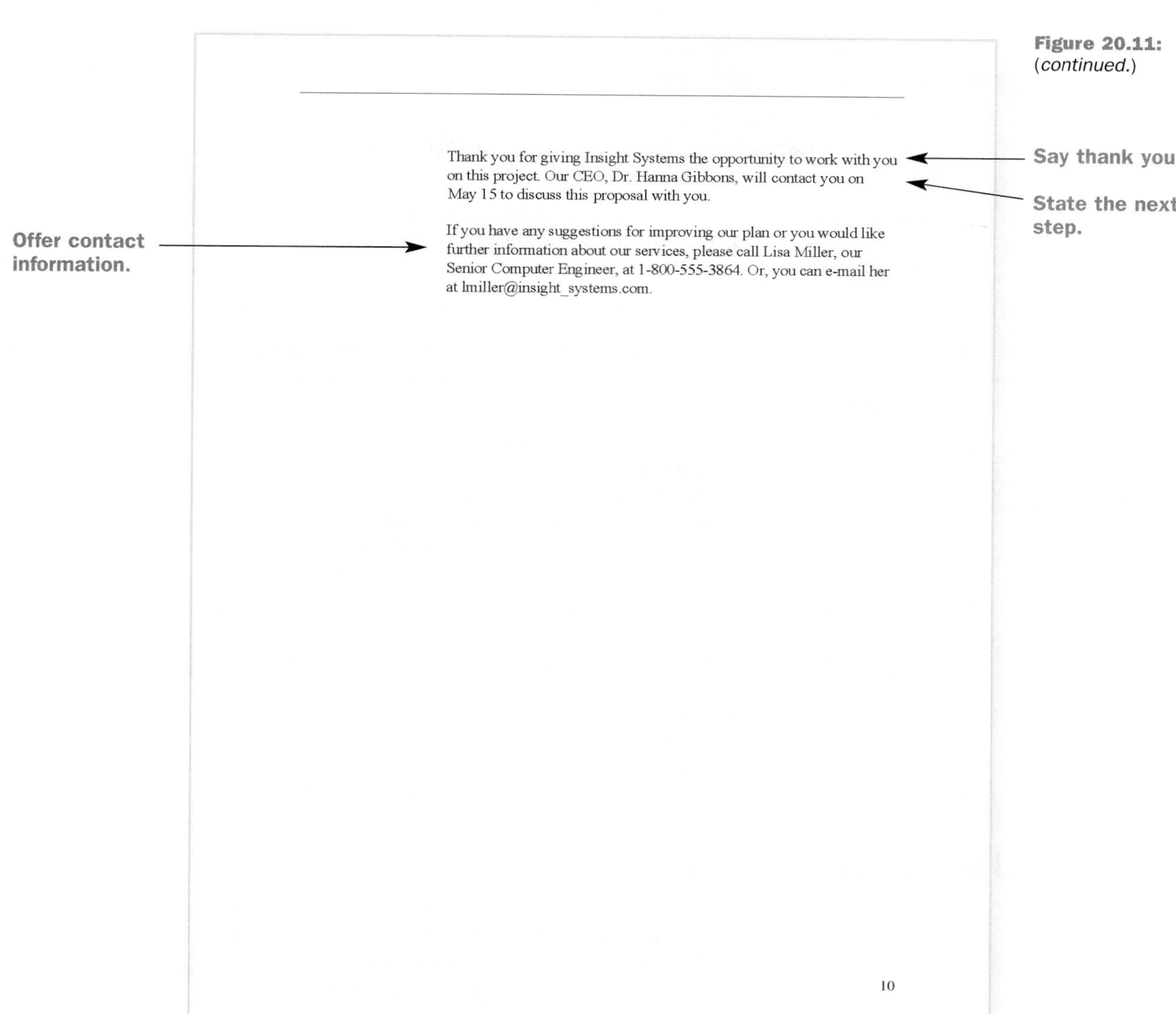

Thank you for giving Insight Systems the opportunity to work with you on this project. Our CEO, Dr. Hanna Gibbons, will contact you on May 15 to discuss this proposal with you.

If you have any suggestions for improving our plan or you would like further information about our services, please call Lisa Miller, our Senior Computer Engineer, at 1-800-555-3864. Or, you can e-mail her at lmiller@insight_systems.com.

10

Figure 20.11: (*continued.*)

Using Style in Proposals

Proposals are designed to both educate and persuade readers, so they tend to use a mixture of plain and persuasive style.

> **Plain style**—Use plain style in places where description is most important, like the current situation section, the project plan section, and the qualifications section.
>
> **Persuasive style**—Use persuasive style in places where readers are expected to make decisions, such as the proposal's introduction and the costs and benefits section.

LINK For more strategies for using plain and persuasive style, see Chapter 7, page 162.

One persuasive style technique, called "setting a tone," is particularly effective when writing a proposal. To use logical mapping to help you set a persuasive tone, follow these steps:

1. Determine how you want your proposal to sound. Do you want it to sound exciting, innovative, or progressive? Choose a word that best reflects the tone you want your readers to hear as they are looking over your proposal.
2. Put that word in the middle of your screen or sheet of paper. Circle it.
3. Write words associated with that word around it. Circle them also.
4. Keep mapping farther out until you have filled the page or screen.

Figure 20.12 shows a logical map of the word *progressive*. As shown in this map, you can quickly develop a set of words that are related to this word.

Once you have finished mapping out the tone, you can weave these words into your proposal to create a theme. As discussed in Chapter 7 of this book, if you use these words carefully and strategically in your proposal, your readers will hear this tone as they are reading your document. But, you do not want to overuse these words. Setting a tone is like adding spices in food. Used well, the spices will give the food an interesting flavor. When overused, they will make the taste too strong.

Designing Proposals

Computers have changed the way proposals are designed and delivered. Not long ago, it was common for proposals to include only minimal design with sparse graphics. Today, desktop publishing gives you a full range of page design options, including graphics, headers and footers, columns, and photographs, among other design elements. Increasingly, proposals are also being delivered in multimedia formats with video and sound.

Figure 20.13, for example, shows the first page of a well-designed proposal. Notice how the design of the proposal sets a professional, progressive tone for the whole document—even before you start reading.

When designing your proposal, you should consider three issues: graphics, page design, and medium:

Want more help developing your proposal's style? Go to **www.ablongman.com/johnsonweb/20.13**

Mapping to Set a Tone

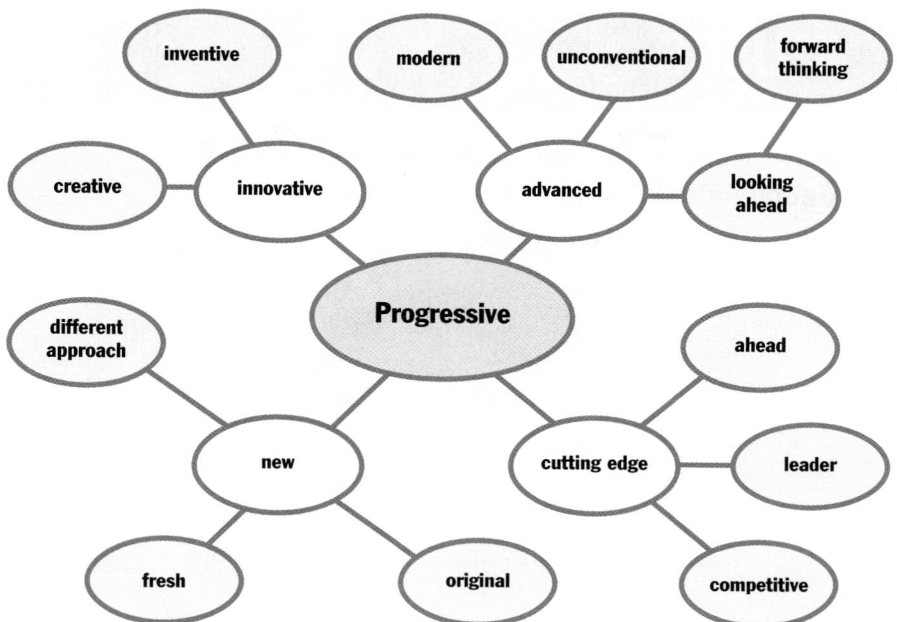

Figure 20.12: Logical mapping can help you develop a tone for your proposal. Weaving these words into a text would make the document sound "progressive."

Graphics—In proposals, it is common to include charts, graphs, maps, illustrations, photographs, and other kinds of graphics. You should look for places in your proposal where graphics can be used to reinforce important points.

LINK For more information on using graphics, see Chapter 9, page 234.

Page design—Page layouts for proposals vary from simple to elaborate. At a minimum, you should use headings, lists, and graphics. More elaborate page layouts might include multiple columns, margin comments, pull quotes, and sidebars. Choose a page design that suits your readers and the context in which they will use the proposal.

LINK For more information on page and screen layout, see Chapter 8, page 192.

Medium—The appropriate medium is also an important choice with the advent of computers. Paper is still the norm for most proposals, but increasingly companies are using CD-ROMs, websites, and presentation software to deliver their ideas.

LINK For more information on developing oral presentations with presentation software, see Chapter 11, page 302.

In the rush to get the proposal out the door, you might be tempted to skip the design phase of the writing process. You might even convince yourself that visual

To see samples of well-designed and badly designed proposals, go to **www.ablongman.com/johnsonweb/20.14**

Sample Proposal Design

Header adds color and anchors the top of the page.

Title of proposal is bold and easy to read.

Headings provide access points into text.

Two-column format makes text more scannable.

Figure 20.13: The page design of this proposal sets a professional tone while helping readers scan for important information. The tone is set with positive words and phrasings, while the graphics and headings make the information easy to access.

Graphic supports text while adding color.

USGS science for a changing world — NRCS Natural Resources Conservation Service

A Proposal for Upgrading the National-Scale Soil Geochemical Database for the United States

The most requested data from the U.S. Geological Survey's (USGS) National Geochemical Database is a set of 1,323 soil samples. Why? Consider the following examples:

Example 1—Imagine for a moment that you are employed by an environmental regulatory agency of either the Federal or a State Government. Your assignment is to establish a "remediation value" for arsenic in soil at a contaminated site where a wood preservative facility once operated. Current arsenic values in soils at the facility range from 15 to 95 ppm and you must decide the concentration of arsenic that is acceptable after remediation efforts are completed. Scientists refer to the natural or native concentration of an element in soils as the "background concentration." Given the fact that arsenic occurs naturally in all soils, how would you determine the background concentration of arsenic in soils for this particular area?

Example 2—Your environmental consulting firm has been assigned to work with a team of specialists conducting a risk-based assessment of land contaminated with lead, zinc, and cadmium from a metal foundry. The assessment would determine the likelihood of adverse health or ecological effects caused by the contaminants. Again, an important part of this determination is, "What is the background concentration of these elements in the soil?"

What data are available for persons responsible for making the determinations of background concentrations for soils contaminated with potentially toxic metals? The most-often-quoted data set for background concentrations of metals and other trace elements in soil of the conterminous United States consists of only 1,323 samples collected during the 1960s and 1970s by the U.S. Geological Survey (Boerngen and Shacklette, 1981; Shacklette and Boerngen, 1984). (There is a similar data set for Alaska (Gough and others, 1984, 1988)). Samples for the "Shacklette data" were collected from a depth of about 1 ft, primarily from noncultivated fields having native vegetation, and samples were analyzed for more than 40 elements. Data in this study represent about one sample per 2,300 mi^2, indicating that very few samples were collected in each State. For example, the State of Arizona is covered by only 47 samples, and Pennsylvania has only 16. Despite the low number of samples, this data set is still being used on a regular basis to determine background concentrations of metals in soil to aid in remediation or risk-based assessments of contaminated land.

The only other national-scale soil geochemical data set for the United States was generated by the Natural Resources Conservation Service (NRCS), formerly the Soil Conservation Service (Holmgren and others, 1993). This data set consists of 3,045 samples of agricultural soil collected from major crop-producing areas of the conterminous United States. The primary purpose of this study was to assess background levels of lead and cadmium in major food crops and in soils on which these crops grow. Thus, the samples were only analyzed for five metals—lead, cadmium, copper, zinc, and nickel.

ARSENIC

PPM

KILOMETERS

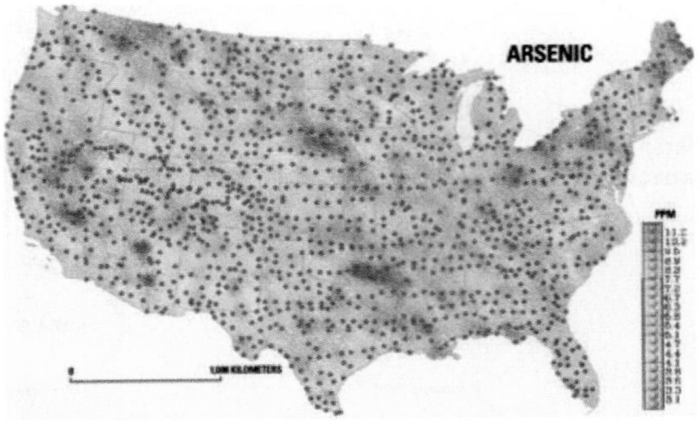

Figure 1. Map of arsenic distribution in soils and other surficial materials of the conterminous United States based on 1,323 sample localities as represented by the black dots.

The Shacklette data set allows us to produce geochemical maps for specific elements, such as that shown on figure 1 for arsenic (Gustavsson and others, 2001). A map produced from such sparse data points obviously carries a large degree of uncertainty with it and does not have the resolution needed to answer many of the questions raised by land-management and regulatory agencies, earth scientists, and soil scientists. An example of the poor data set resolution is illustrated for Pennsylvania (fig. 2). The State is divided into major soil taxonomic units referred to as Suborders (Soil Survey Staff, 1999). Suborders group similar soil types in any region. The dots represent the sample points from the Shacklette data set. The few sample points shown on figure 2 illustrate that this data set would be inadequate for someone who must define the arsenic content of a given soil. At this time, no data set exists that will allow us to make these kinds of determinations.

The USGS and NRCS are currently studying the feasibility of a national-scale soil geochemical survey that will increase the sample density of the Shacklette data set by at least a factor of 10. This project, called Geochemical Landscapes, began in October 2002. The first 3 years will be devoted to determining how such a survey should be conducted. Therefore, we are actively soliciting input from potential customers of the new data. Interested members of the private sector, government, or academic communities

U.S. Department of the Interior
U.S. Geological Survey

Printed on recycled paper

USGS Fact Sheet FS-015-03
March 2003

Source: U.S. Geological Survey, 2003.

design doesn't matter to readers. Don't fool yourself. Good design is very important in proposals. Design makes the proposal more attractive while helping readers locate important information in the document.

Revising, Editing, and Proofreading

You and your team should always leave plenty of time to revise, edit, and proofread your proposal. At this point in the writing process, you can dramatically strengthen the content, organization, style, and design of the document.

Your time revising and editing will be well spent. After all, in today's competitive technical workplace, a carefully revised and edited proposal is often the difference between success and failure.

As you revise, edit, and proofread your proposal, consider the following three areas:

Review the rhetorical situation—When you have completed a draft, look back at your original notes on the proposal's subject, purpose, readers, and context of use. In some cases your subject and purpose might have evolved as you drafted the proposal. Or, perhaps you now have a better understanding of your readers and the context in which the proposal will be used.

Edit for ethics—Proposals almost always involve ethical issues of some kind. Ask yourself whether the proposal stretches the truth in places. Does it hide important information? Are the graphics drawn ethically? Unethical proposals can lead to lawsuits or lost clients, so make sure your proposal anticipates any ethical issues that might be involved.

LINK For more information on ethics, see Chapter 4, page 70.

Proofread—You might be surprised how many proposals are rejected simply because of grammatical mistakes, typos, and misspellings. The importance of good proofreading makes sense. After all, why would readers entrust their money and faith with you if your proposal demonstrates carelessness and lack of quality? You need to proofread your text and graphics carefully for any errors.

LINK For more information on editing and proofreading, see Chapter 10, page 264.

Solid revision and editing are often the difference between proposals that succeed and those that fail. Make sure you reserve plenty of time for revising and editing.

Revision worksheets for proposals are available at **www.ablongman.com/johnsonweb/20.15**

Jane Perkins
COMMUNICATIONS CONSULTANT, A.T. KEARNEY, CHICAGO, ILLINOIS
A.T. Kearney is a consulting firm that works in high-tech industries.

What makes a proposal successful?

Writing proposals is critical work at A.T. Kearney, a global management consulting firm comprised primarily of MBAs with engineering backgrounds. It is how we obtain the majority of our business. Our success in writing proposals depends on key aspects of our proposal process:

- *Web-based processes for data and expertise.* We have formal processes in place to proactively gather important data that is necessary for providing potential clients with decision-making information. We also rely on an informal and collaborative network for tapping into expertise throughout the firm. These processes help us write proposals that include the necessary level of detail that persuades stakeholders.
- *A cohesive proposal and selling process.* For many opportunities, we develop a series of proposal documents, which are components of a negotiation/selling process, often extending over months. For example, an opportunity might begin with a formal RFP to down-select competitors. The RFP often specifies providing both on-line and print versions with strict requirements on content, length, and software application. For the next round, we would probably prepare a PowerPoint presentation for the client buyers. This round would open the door for refinements on the scope and initiate negotiations. The proposal process might then conclude with a proposal letter to summarize the aspects we had discussed.
- *Innovation and internal education.* Our proposal writing spurs much innovation and forward thinking, as we respond to evolving client needs and opportunities, and we build on recent project work. Proposal teams are not only educating clients through proposals but also important, internal stakeholders, such as those responsible for ensuring financials and quality delivery of the project. The use of eRooms and collaborative applications facilitates this learning process. Additionally, expert reviews at specified stages of the proposal process are part of our culture—leadership insights and approvals help the proposal team, while the latest thinking is spread across the firm.

As you can see, a well-written proposal is only part of a much larger process.

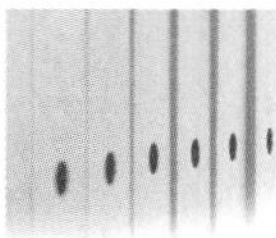

CHAPTER REVIEW

- Proposals are documents that present ideas or plans for consideration.
- Proposals can be internal or external and solicited or unsolicited.
- Proposals usually include five sections: (1) introduction, (2) current situation, (3) project plan, (4) qualifications, and (5) costs and benefits.
- When planning the proposal, start out by defining the Five-W and How questions. Then define the rhetorical situation (subject, purpose, readers, context of use).
- When organizing and drafting the proposal, each major section should be written separately with an opening, a body, and a closing.
- Proposals should use a combination of plain and persuasive style to motivate readers.
- Design is absolutely essential in professional proposals. You should look for places where page layout and graphics will improve the readability of your proposal.
- Don't skimp on the revision, editing, and proofreading of the proposal. A proposal that demonstrates poor quality in writing usually signals inattention to quality by the bidder (you).

EXERCISES AND PROJECTS

Individual or Team Projects

1. Analyze the proposal, "No anchoring areas for Flower Garden Banks in the Northwestern Gulf of Mexico," at the end of this chapter. Study the content, organization, style, and design of the proposal. Does it cover the four areas (current situation, project plan, qualifications, costs and benefits) that were discussed in this chapter? Where does it include more or less information than you would expect in a proposal?

   Write a two-page memo to your teacher in which you discuss whether you believe the proposal was effectively written and designed. Discuss the content, organization, style, and design in specific details. Highlight the proposal's strengths and suggest ways in which it could be improved.

2. Find an RFP at www.fedbizopps.gov or in the classifieds of your local newspaper. Analyze the RFP according to the Who, What, Where, When, Why, and How questions. Then, prepare a presentation for your class in which you (1) summarize the contents of the RFP, (2) discuss why you believe the RFP was sent out, and (3) explain what kinds of projects would be suitable for this RFP.

3. Find a proposal that demonstrates weak style and/or design. Using the style and design techniques discussed in this chapter and Chapters 8 through10, revise the document so it is more persuasive and more visual. Locate places where blocks of text could be turned into lists, and identify places where graphics would reinforce the text.

To find sample proposals for analysis, go to
www.ablongman.com/johnsonweb/20.16

Collaborative Project: Improving Campus

As students, you are aware—perhaps more than the administrators—of some of the problems on campus. Write a proposal that analyzes the causes of a particular problem on campus and then offers a solution that the administration might consider implementing. The proposal should be written to a *named* authority on campus who has the power to say "yes" to your proposal and put it into action.

Some campus-based proposals you might consider include the following:

- improving lighting on campus to improve safety at night.
- improving living conditions in your dorm or fraternity/sorority.
- creating a day-care center on campus.
- creating an adult commuter room in the student union.
- improving campus facilities for the handicapped.
- improving security in buildings on campus.
- creating a University Public Relations office run by students.
- increasing access to computers on campus.
- improving the parking situation.
- reducing graffiti and/or litter on campus.
- improving food service in the student union.
- helping new students make the transition into college.
- changing the grading system at the university.
- encouraging more recycling on campus.
- reducing the dependence on cars among faculty and students.

Use your imagination to come up with a problem to which you can offer a solution. You don't need to be the person who implements the program. Just offer some guidance for the administrators.

TAKE NOTE In the qualifications section of your proposal, you will likely need to recommend that someone else should do the work.

Revision Challenge

This proposal from the International Maritime Organization might be hard to read for non-experts. How would you revise the style to make it easier to understand?

Sample Proposal

Figure A: This proposal is good, but it could be improved. How could you make it more persuasive to non-expert readers?

INTERNATIONAL MARITIME ORGANIZATION

IMO

E

SUB-COMMITTEE ON SAFETY OF NAVIGATION
46th session
Agenda item 3

NAV 46/3/3
5 April 2000
Original: ENGLISH

ROUTEING OF SHIPS, SHIP REPORTING AND RELATED MATTERS

No anchoring areas for Flower Garden Banks in the Northwestern Gulf of Mexico

Submitted by the United States

SUMMARY	
Executive summary:	This document sets forth a proposal for mandatory no anchoring areas on three coral reef banks ("Flower Garden Banks") in the northwestern Gulf of Mexico by ships greater than 30.48 meters (100 feet in overall length), for consideration and approval by the Sub-Committee on Safety of Navigation and forwarding to the Maritime Safety Committee for adoption. This proposal should be considered in conjunction with NAV 46/3/2, which proposes amendment of the General Provisions on Ships' Routeing to provide for the measure proposed in this document.
Action to be taken:	Paragraph 14
Related documents:	NAV 46/3/2

Introduction

1 This proposal must be read in conjunction with NAV 46/3/2, which is a proposal by the United States to amend the General Provisions on Ships' Routeing (GPSR) to provide for a no anchoring area measure. Consistent with the manner in which the Sub-Committee handled the issue of amendment of the GPSR to provide for the designation of archipelagic sealanes and the Indonesian proposal to designate such sealanes, NAV 46 should first consider the United States proposal to amend the GPSR and then this proposal to establish three specific no anchoring areas.

I:\NAV\46\3-3.DOC

For reasons of economy, this document is printed in a limited number. Delegates are kindly asked to bring their copies to meetings and not to request additional copies.

Continued on following page

Source: International Maritime Organization, 2000.

Figure A: (*continued.*)

2 The United States proposes the establishment of mandatory no anchoring areas to be adopted by the IMO for ships greater than 30.48 metres (100 feet in overall length) on Flower Garden Banks, which are three coral reef banks in the northwestern Gulf of Mexico. The geographic co-ordinates of these three areas are set forth in the annex. The size of the three areas combined is approximately 42.34 square nautical miles.

3 This proposed measure can reasonably be expected to significantly prevent and reduce the risk of damage to the coral marine environment by ships, without restricting the sea area available for navigation. The sizes of the areas and the proposed measures are limited to what is essential for the interests of safe navigation and the protection of the marine environment.

Background

4 Damage to the coral banks by anchoring of large ships has been well demonstrated. Significant damage is caused by the direct impact of large, heavy anchors and from the dragging and swinging of large anchor cables and chains. These activities destroy coral heads and create gouges and scars that destabilize the reef structure. There may be a minimal adverse impact on the shipping industry from complying with this anchoring measure. In contrast, anchoring on these coral banks causes significant adverse environmental effects to these resources. Coral such as that in Flower Garden Banks takes thousands of years to build. The regeneration of the reef from anchor damage may never occur. If optimal conditions for regeneration exist, it would still take hundreds and perhaps thousands of years for the reef to return to its condition before the damage.

5 The proposed areas have been designated by the United States as Flower Garden Banks National Marine Sanctuary. These areas are unique even among the world's coral reefs. The banks contain the northernmost coral reefs on the North American continental shelf and support the most highly developed offshore hard-bank communities in the region. They are of considerable interest to scientists because they are isolated from other reef systems by over 550 kilometers and exist in areas where slight changes in hydrographic conditions could significantly reduce reef community health. There are also organisms otherwise unknown on the world's continental shelves. These organisms are generally associated with a hypersaline, anoxic brine seep having a chemosynthetic energy base analogous to that found at deep-sea hydrothermal vents.

6 The reefs in Flower Garden Banks crest at approximately 15 meters below the water surface and extend downward to 46 meters depth, where the hermatypic (reef building) corals are replaced by reefal communities dominated by coralline algae and sponges. This deeper "algal terrace" covers most surfaces down to a depth of 90 meters. The area has at least 20 species of hermatypic corals, 80 species of algae, 196 known macro-invertebrate species, and supports more than 200 fish species. The reef-building corals and coralline algae construct and maintain the substratum and, through a multitude of relationships, largely control the structure of benthic communities occupying the banks. As a primary building-block for the entire ecosystem of the banks, the coral and algae are by far the most important organisms in the Flower Garden Banks ecosystem.

Figure A: *(continued.)*

7 This area is transited by commercial ships, many of which are en route to and from the United States ports in Texas and Louisiana. A major east-west shipping fairway passes approximately 11 kilometers south of Flower Garden Banks. Other large ships in the area generally follow fairways approximately 65 kilometers to the west and 80 kilometers to the east of Flower Garden Banks.

8 In addition to transiting ships, the other human activities that take place around and within Flower Garden Banks National Marine Sanctuary are commercial fishing, recreational pursuits, and research. Oil and gas exploration and development activities also occur in the area surrounding the sanctuary, but they are strictly regulated so as to prevent adverse impacts on the sanctuary. These human activities–including the anchoring of vessels–are strictly regulated or prohibited.

9 There is clear evidence of anchoring damage to Flower Garden Banks from large ships. Scars or tracks of pulverized coral have been documented by studies conducted by submersibles or divers. The largest scar from anchoring found to date extends for approximately 1.7 kilometers and resembles a continuous, "roadcut-like" gouge into the bank. Another crater-like scar measures approximately 50 meters in diameter. Chain scars from the swinging of ships on their anchors are evident on many corals. There are hundreds of abraded, fractured, and toppled coral colonies that appear to be from the dragging of anchors or anchor cables and chains. Loose coral pieces act as agents of further injury to the living coral, particularly during heavy seas and storms as the pieces are repeatedly driven into and around the living coral.

Proposal

10 The United States proposes mandatory no anchoring areas for ships greater than 30.48 meters (100 feet in overall length) to protect coral in Flower Garden Banks National Marine Sanctuary from destruction. Hydrographic surveys of the areas have been conducted and appropriate aids to navigation exist. The proposal will not interfere with any existing patterns of ship traffic. Other areas are available for ships to anchor, in particular in recommended anchorage areas outside of area ports. Additionally, because this no anchoring area measure allows ships to continue to navigate through the area, it does not have the effect of limiting the sea area available for navigation.

11 The significant damage that has been, and continues to be, caused by anchoring in this unique area is sufficient justification for the establishment of a mandatory measure. The United States has attempted to address this problem through domestic anchoring regulations and by notifying mariners through various means, such as notices to mariners; however, ships continue to anchor in this area. Thus, it is appropriate at this time to pursue a measure through the IMO.

12 Safety considerations also support establishment of this measure. As set forth in NAV 46/3/2, the safety of a ship can depend on the ability of its anchor to hold and the character of the bottom is of prime importance. Coral provides an unstable anchoring bottom. The scars and gouges are clear evidence that anchors tend to drag along the bottom rather than hold in the coral. Because there are a number of platforms and pipelines in this area it is very important from a safety perspective for ships to anchor only in areas where the bottom will provide good holding ground.

Continued on following page

Figure A: *(continued.)*

- 4 -

Additional Considerations

13 The United States has taken several steps domestically to protect Flower Garden Banks National Marine Sanctuary from damage by anchoring. A sanctuary regulation was promulgated in 1992 prohibiting the anchoring of ships greater than 30.48 meters (100 feet). The United States has also adopted several regulations prohibiting and restricting the anchoring by smaller vessels. Notifications of these regulations are published on the U.S. nautical charts, in notices to mariners, the United States Code of Federal Regulations, and on the NOAA sanctuary web page (http://www.sanctuaries.nos.noaa.gov).

Conclusion

14 The Sub-committee is asked to approve this proposal for a measure establishing mandatory no anchoring areas for ships which are greater than 30.48 meters (100 feet in overall length) in the area set forth in the annex and forward the proposal to the Maritime Safety Committee for its adoption. The United States requests that the effective date of implementation be six months after adoption.

***

I:\NAV\46\3-3.DOC

Figure A: *(continued.)*

ANNEX

East Flower Garden Bank
(NAD 83)(Chart No. 11340)

Point Number	Latitude (N)	Longitude (W)
E-1	27°52'.91	93°37'.70
E-2	27°53'.60	93°38'.40
E-3	27°55'.24	93°38'.68
E-4	27°57'.53	93°38'.56
E-5	27°58'.48	93°37'.78
E-6	27°59'.04	93°35'.54
E-7	27°59'.03	93°35'.17
E-8	27°55'.39	93°34'.26
E-9	27°54'.08	93°34'.32
E-10	27°53'.46	93°35'.09
E-11	27°52'.88	93°36'.96

West Flower Garden Bank
(NAD 83)(Chart No. 11340)

Point Number	Latitude (N)	Longitude (W)
W-1	27°49'.19	93°50'.76
W-2	27°50'.22	93°52'.18
W-3	27°51'.23	93°52'.87
W-4	27°51'.56	93°52'.85
W-5	27°52'.85	93°52'.42
W-6	27°55'.03	93°49'.74
W-7	27°54'.99	93°48'.64
W-8	27°54'.60	93°47'.18
W-9	27°54'.26	93°46'.83
W-10	27°53'.61	93°46'.86
W-11	27°52'.97	93°47'.26
W-12	27°50'.69	93°47'.38
W-13	27°49'.20	93°48'.72

Stetson Bank
(NAD 83)(Chart Nos. 11300, 11330, 11340)

Point Number	Latitude (N)	Longitude (W)
S-1	28° 09'.52	94° 18'.53
S-2	28° 10'.17	94° 18'.50
S-3	28° 10'.13	94° 17'.40
S-4	28° 09'.48	94° 17'.43

I:\NAV\46\3-3.DOC

CASE STUDY

The Mole

Henry Espinoza knew this proposal was important to his company, Valen Industries. If the proposal was successful, it could be worth millions of dollars in short-term and long-term projects for his employer. For Henry, winning the project meant a likely promotion to vice president and a huge raise in salary. The company's CEO was calling him daily to check up and see if it was going well. That meant she was getting stressed out about the project.

Through the grapevine, Henry knew his company really had only one major competitor for the project. So, the odds were good that Valen Industries would win the contract. Of course, Henry and his team needed to put together an innovative and flawless proposal to win, because their major competitor was going all out to get this project.

Then, one day something interesting happened. One of Henry's team members, Vera Houser, came to him with an e-mail from one of her friends, who worked for their main competitor. In the e-mail, this person offered to give Valen Industries a draft of its competitor's proposal. Moreover, this friend would pass along any future drafts as the proposal evolved. In the e-mail, Vera's friend said he was very frustrated working at his "slimy" company, and he was looking for a way out. He also revealed that his company was getting "inside help" on the proposal, but Henry wasn't sure what that meant.

Essentially, this person was hinting that he would give Henry's team a copy of the competitor's proposal and work as a mole if they considered hiring him at Valen Industries. Henry knew he would gain a considerable advantage over his only competitor if he had a draft of its proposal to look over. The ability to look at future drafts would almost ensure that he could beat his competitor. Certainly, he thought, with this kind of money on the line, a few rules could be bent. And it seemed like his competitor was already cheating by receiving inside help.

Henry went home to think it over. What would be the ethical choice in this situation? How do you think Henry should handle this interesting opportunity?

Discussions of proprietary information are available at **www.ablongman.com/johnsonweb/20.18**

CHAPTER

21

Activity Reports

CHAPTER CONTENTS

MERCURY WHITE PAPER

The Utility Hazardous Air Pollutants Regulatory Determination
To reduce the risk mercury poses to people's health, EPA Administrator Carol M. Browner announced that the Environmental Protection Agency will regulate emissions of mercury and other air toxics from coal- and oil-fired electric utility steam generating units (power plants).
EPA plans to propose a regulation to control air toxics emissions, including mercury, from coal- and oil-fired power plants by the end of 2003. This regulation will be one more important piece of an Agency-wide effort to protect people and wildlife from exposure to the toxic pollutant mercury.

Sources and Fate of Mercury in the Environment
Like all elements, the same amount of mercury has existed on the planet since the Earth was formed. However, the amount of mercury mobilized and released into the environment has increased since the beginning of the industrial age. Mercury moves through the environment as a result of both natural and human activities. The human activities that are most responsible for causing mercury to enter the environment are burning materials (such as batteries), fuels (such as coal) that contain mercury, and certain industrial processes. These activities produce air pollution containing mercury.

Based on EPA's National Toxics Inventory, the highest emitters of mercury to the air include coal-burning power plants, municipal waste combustors, medical waste incinerators and hazardous waste combustors. Mercury emissions from these and other sources are transported through the air and eventually deposit to water and land, where humans and wildlife are exposed.

Most of the mercury entering the environ... directly contaminate land and water as a ... disposal of waste-containing batteries an... either directly or through air deposition, it... form, methylmercury. Bioaccumulation... of the food web (for example, predatory... even millions of times greater than the co...

Mercury deposition can occur very close... emitted, it can be transported great dista... deposition rates in the U.S. occur in the ... scattered areas in the Southeast. Approx... the U.S. comes from domestic human-m... from human-made sources located outsi... and natural sources. While the U.S. cont... mercury, it still contributes more than it ... transported outside our borders.

More information on the sources, fate, a... 1997 Mercury Report to Congress and 1...

Red Hills Health Sciences Center *RH*

Testing and Research Division
201 Hospital Drive, Suite A92
Red Hills, CA 92698

March 10, 2004

To: Brian Jenkins, Safety Assurance Officer
From: Hal Chavez, Testing Laboratory Supervisor
Subject: Incident Report: Fire in Laboratory

I am reporting a fire in Testing Laboratory 5, which occurred yesterday, March 9, 2004, at 3:34 p.m.

The fire began when a sample was being warmed with a bunsen burner. A laboratory notebook was left too close to the burner, and it caught fire. One of our laboratory assistants, Vera Cather, grabbed the notebook and threw it into a medical waste container. The contents of the waste container then lit on fire, filling the room with black smoke. At that point, another laboratory assistant, Robert Jackson, grabbed the fire extinguisher and emptied its contents into the waste container, putting out the fire. The overhead sprinklers went off, dousing the entire room.

Even though everyone seemed fine, we decided to send all lab personnel down to the emergency room for an examination. While we were in the waiting room, Vera Cather developed a cough and her eyes became red. She was held for observation and released that evening when her condition was stable. The rest of us were looked over by the emergency room doctors, and they suggested that we stay out of the laboratory until it was thoroughly cleaned.

I asked the hospital's HazMat team to clean up the mess that resulted from the fire. We had been working with samples of *Borrelia burgdorferi* bacteria, which causes Lyme disease. I was not sure if the

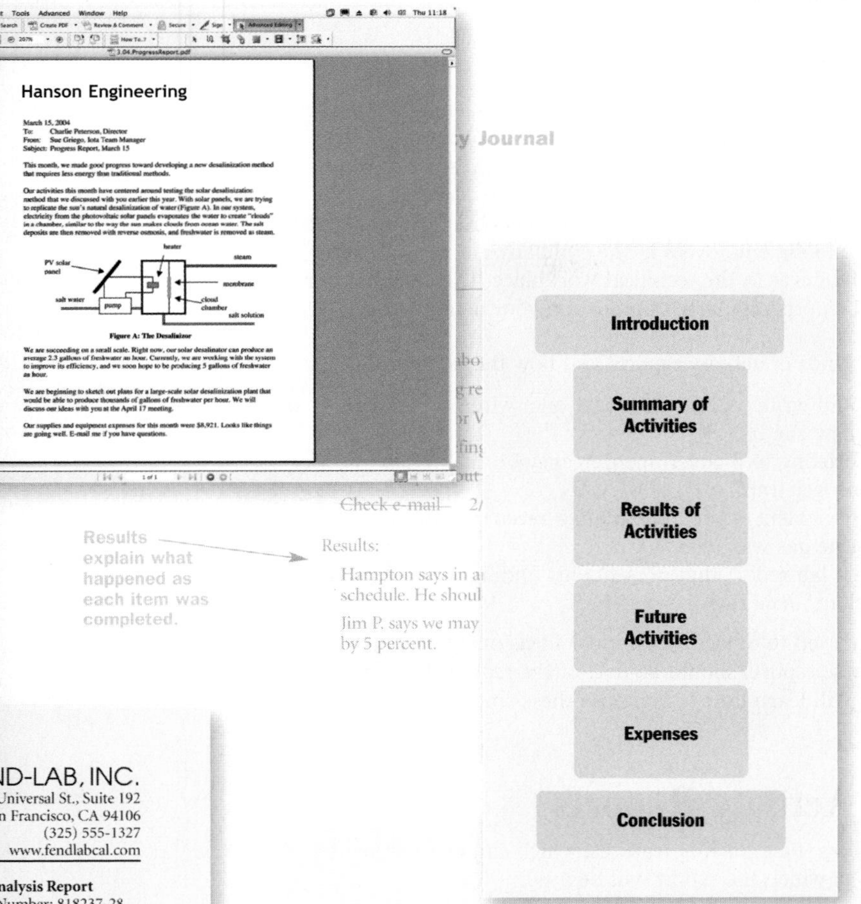

ND-LAB, INC.
Universal St., Suite 192
n Francisco, CA 94106
(325) 555-1327
www.fendlabcal.com

nalysis Report
Number: 818237-28
Sampling: 091204
Date: 091404
s Date: 091904
ian: Alice Valles

ting for mold at the
3910 S. Randolph
ormal amounts of
ctions among the

the test site with two
lk Physical Sampling.
d the building where
water stains, dusty
transparent tape, we
ent agar in petri
laboratory, where it

TIGER INDUSTRIE

Automation Development Divisio

Date: 29 February 2004
To: Hal Roberts, Division Head
From: Sally Fenker, Green Robot Project Manager
Subject: Progress Report on Green Robot Project

We are pleased to report that we have made significant progr
coding for the X53 Manufacturing Robot.

In September, we coded for the lateral movement of the robo
We developed several usable subroutines that should improv
placement accuracy.

So far, the results of our work are looking good:

- placement accuracies of .009 mm
- speed of 20 placements per minute
- frozen arm only 3 times in 100 trials.

In March, we hope to complete development of the software.

CHAPTER OBJECTIVES

In this chapter, you will learn:

- About the different kinds of activity reports and how they are used in the workplace.
- The basic features of activity reports.
- How to determine the rhetorical situation for an activity report.
- How to organize and draft an activity report.
- Strategies for using an appropriate style.
- How to design and format activity reports.
- The importance of editing and proofreading activity reports.

Many companies are using computer networks to develop management structures that are less hierarchical. These companies require fewer levels of managers than before, because computer networks help top corporate executives better communicate with employees throughout the company.

These "flatter" management structures require more communication, quicker feedback, and better accountability among employees in the company. As a result, activity reports are more common than ever in the technical workplace. Throughout your career, you will use these small reports regularly to report on your activities each week or month.

Here are a few examples of kinds of activity reports and how they might be used:

- An electrical engineer would write a progress report on a wire harness her team is developing for a new car.
- A scientist would give a briefing to a government agency on the possibilities for a hydrogen-powered passenger train.
- A chemist would write an incident report to explain a recent accident in the laboratory in which chlorine gas was released.
- A technician might write a lab report that presents his finding of malaria in mosquitoes captured in South America.

Activity reports are typically used to objectively present ideas or information within the company. Writing these reports should be part of the regular flow of your workweek. In this chapter, you will learn how to compose these smaller reports quickly and efficiently.

Basic Features of Activity Reports

An activity report usually includes the following features, which can be modified to suit the needs of the situations in which the report will be used:

- Introduction
- Summary of activities
- Results of activities or research
- Future activities or research
- Incurred or future expenses
- Graphics

Keep in mind that activity reports are flexible and informal, so you should adjust this pattern to suit your needs. The content, organization, style, and design of activity reports depend on the situation in which they are used.

Figure 21.1 shows the basic pattern for an activity report. Sections can be added and subtracted from this basic pattern. For example, in some activity reports, you might need to include a section that offers recommendations. Or, you might decide not to include a section discussing expenses.

Types of Activity Reports

Even though they are similar in most ways, activity reports are called a variety of names that reflect their different purposes. Despite their different names, the goal of

attern for an Activity Report

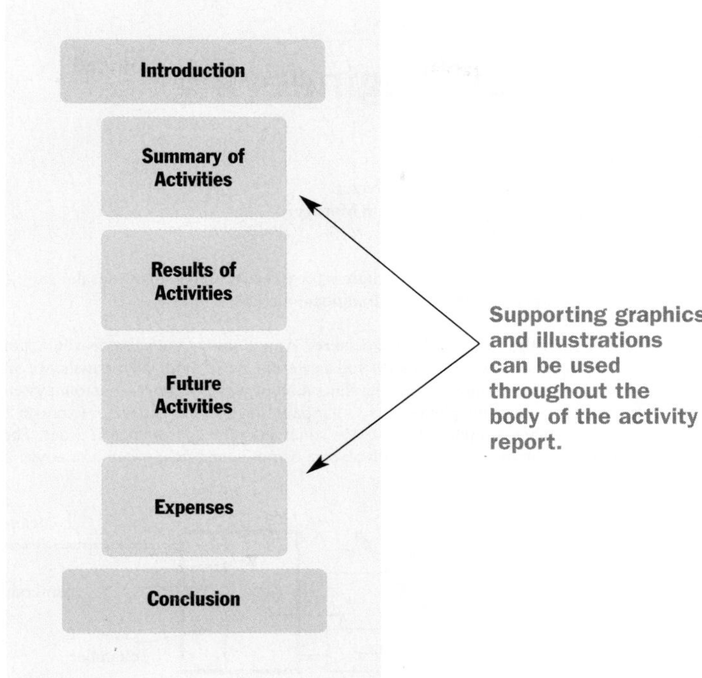

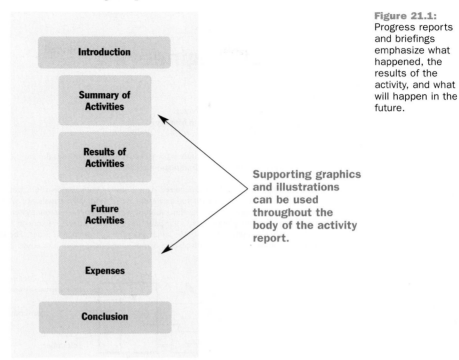

Figure 21.1: Progress reports and briefings emphasize what happened, the results of the activity, and what will happen in the future.

all activity reports is to objectively inform readers about (a) what happened, (b) what is happening, and (c) what will happen in the near future.

Progress Reports

A progress report, also called a status report, is written to inform management about the progress or status of a specific project. These reports are usually written at regular intervals—weekly, biweekly, or monthly—to update management on what has happened since the last progress report was submitted. Management may also periodically request a progress report to stay informed about your or your team's activities.

A typical progress report will provide the following information:

- a summary of completed activities.
- a discussion of ongoing activities.
- a forecast of future activities.

Figure 21.2, for example, shows a progress report that is designed to update management on a project.

To see other sample progress reports, go to **www.ablongman.com/johnsonweb/21.1**

A Progress Report

Figure 21.2: A progress report describes the team's activities and discusses future activities.

Hanson Engineering

March 15, 2004
To: Charlie Peterson, Director
From: Sue Griego, Iota Team Manager
Subject: Progress Report, March 15

Subject, purpose and main point are identified up front.

This month, we made good progress toward developing a new desalinization method that requires less energy than traditional methods.

Ongoing activities are described objectively.

Our activities this month have centered around testing the solar desalinization method that we discussed with you earlier this year. With solar panels, we are trying to replicate the sun's natural desalinization of water (Figure A). In our system, electricity from the photovoltaic solar panels evaporates the water to create "clouds" in a chamber, similar to the way the sun makes clouds from ocean water. The salt deposits are then removed with reverse osmosis, and freshwater is removed as steam.

The graphic supports the text.

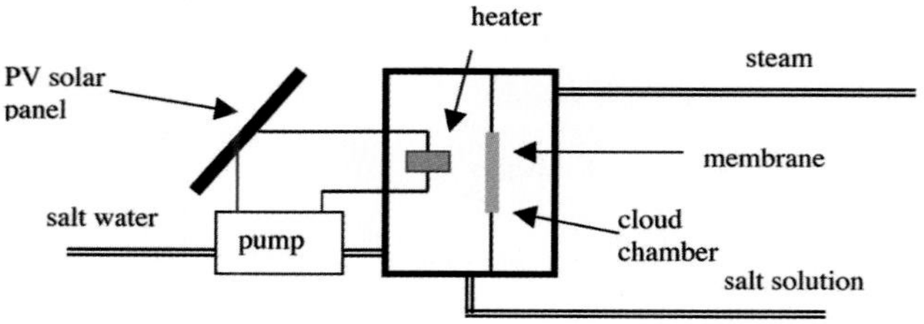

Figure A: The Desalinizor

Results are presented.

We are succeeding on a small scale. Right now, our solar desalinator can produce an average 2.3 gallons of freshwater an hour. Currently, we are working with the system to improve its efficiency, and we soon hope to be producing 5 gallons of freshwater an hour.

Report ends with a look to the future and a brief conclusion.

We are beginning to sketch out plans for a large-scale solar desalinization plant that would be able to produce thousands of gallons of freshwater per hour. We will discuss our ideas with you at the April 17 meeting.

Our supplies and equipment expenses for this month were $8,921. Looks like things are going well. E-mail me if you have questions.

Briefings and White Papers

Briefings and white papers are used to inform management or clients about an important issue. Typically, briefings are presented orally, while white papers are provided in print. Occasionally briefings will appear as "briefs" in written form.

Briefings and white papers typically present gathered facts in a straightforward and impartial way. They include the following kinds of information:

- a summary of the facts.
- a discussion of the importance of these facts.
- a forecast about the importance of these facts in the future.

A good briefing presents the facts as concisely as possible, leaving time for questions and answers. When you brief an audience on the facts, try to do so as objectively as possible. Then, interpret the importance of those facts based on evidence, not on speculation.

Similarly, white papers typically do not advocate for any specific side or course of action. Instead, they present the facts in a straightforward way and offer an objective assessment of what those facts mean. It is up to the readers to decide what actions are appropriate after they have finished reading the document.

Figure 21.3 shows the first page of a six-page white paper from the U.S. Environmental Protection Agency. The white paper discusses the effects of mercury in our environment.

Incident Reports

Incident reports describe an event, usually an accident or irregular occurrence, and identify what corrective actions have been taken. As with other kinds of activity reports, incident reports present the facts as objectively as possible. They provide the following information:

- a summary of what happened (the facts).
- a discussion of why it happened.
- a description of how the situation was handled.
- a discussion of how the problem will be avoided in the future.

It is tempting, especially when an accident was your fault, to make excuses or offer apologies, but the incident report is not the place. As with other activity reports, you should concentrate on the facts. Describe what happened as honestly and clearly as possible. You can make excuses or apologize later.

Figure 21.4 shows a typical incident report in which management is notified of an accident in a laboratory.

To see white papers, go to **www.ablongman.com/johnsonweb/21.2**
Want to see other incident reports? Go to **www.ablongman.com/johnsonweb/21.3**

A White Paper

Figure 21.3: A white paper presents technical information objectively, allowing readers to make decisions based on the facts.

Subject and main point are identified up front.

Background on the subject is offered.

Facts are presented objectively.

Shows that more information is available elsewhere.

MERCURY WHITE PAPER

The Utility Hazardous Air Pollutants Regulatory Determination
To reduce the risk mercury poses to people's health, EPA Administrator Carol M. Browner announced that the Environmental Protection Agency will regulate emissions of mercury and other air toxics from coal- and oil-fired electric utility steam generating units (power plants).
EPA plans to propose a regulation to control air toxics emissions, including mercury, from coal- and oil-fired power plants by the end of 2003. This regulation will be one more important piece of an Agency-wide effort to protect people and wildlife from exposure to the toxic pollutant mercury.

Sources and Fate of Mercury in the Environment
Like all elements, the same amount of mercury has existed on the planet since the Earth was formed. However, the amount of mercury mobilized and released into the environment has increased since the beginning of the industrial age. Mercury moves through the environment as a result of both natural and human activities. The human activities that are most responsible for causing mercury to enter the environment are burning materials (such as batteries), fuels (such as coal) that contain mercury, and certain industrial processes. These activities produce air pollution containing mercury.

Based on EPA's National Toxics Inventory, the highest emitters of mercury to the air include coal-burning power plants, municipal waste combustors, medical waste incinerators and hazardous waste combustors. Mercury emissions from these and other sources are transported through the air and eventually deposit to water and land, where humans and wildlife are exposed.

Most of the mercury entering the environment is the result of air emissions; however, mercury also can directly contaminate land and water as a result of the release of industrial wastewater or from the disposal of waste-containing batteries and other sources of mercury. Once mercury enters waters, either directly or through air deposition, it can bioaccumulate in fish and animal tissue in its most toxic form, methylmercury. Bioaccumulation means that the concentration of mercury in predators at the top of the food web (for example, predatory fish and fish-eating birds and mammals) can be thousands or even millions of times greater than the concentrations of mercury found in the water.

Mercury deposition can occur very close to the source or, depending on the chemical form in which it is emitted, it can be transported great distances – even crossing international borders. The highest deposition rates in the U.S. occur in the southern Great Lakes, the Ohio Valley, the Northeast and scattered areas in the Southeast. Approximately 60 percent of the mercury deposition that occurs in the U.S. comes from domestic human-made sources of pollution. The remaining 40 percent comes from human-made sources located outside of the U.S., re-emitted mercury from historic U.S. sources, and natural sources. While the U.S. contributes only about 3 percent of the global atmospheric pool of mercury, it still contributes more than it receives. Approximately two-thirds of U.S. emissions are transported outside our borders.

More information on the sources, fate, and risks of mercury in the environment can be found in EPA's 1997 Mercury Report to Congress and 1998 Utility Air Toxics Report to Congress.

Source: U.S. Environmental Protection Agency, 2003.

An Incident Report

Figure 21.4: An incident report is not the place to make apologies or place blame. You should state the facts as objectively as possible.

Red Hills Health Sciences Center *RH*

Testing and Research Division
201 Hospital Drive, Suite A92
Red Hills, CA 92698

March 10, 2004

To: Brian Jenkins, Safety Assurance Officer

From: Hal Chavez, Testing Laboratory Supervisor

Subject: Incident Report: Fire in Laboratory

Subject and purpose are stated up front.

I am reporting a fire in Testing Laboratory 5, which occurred yesterday, March 9, 2004, at 3:34 p.m.

What happened is described objectively.

The fire began when a sample was being warmed with a bunsen burner. A laboratory notebook was left too close to the burner, and it caught fire. One of our laboratory assistants, Vera Cather, grabbed the notebook and threw it into a medical waste container. The contents of the waste container then lit on fire, filling the room with black smoke. At that point, another laboratory assistant, Robert Jackson, grabbed the fire extinguisher and emptied its contents into the waste container, putting out the fire. The overhead sprinklers went off, dousing the entire room.

What was done about it is noted.

Even though everyone seemed fine, we decided to send all lab personnel down to the emergency room for an examination. While we were in the waiting room, Vera Cather developed a cough and her eyes became red. She was held for observation and released that evening when her condition was stable. The rest of us were looked over by the emergency room doctors, and they suggested that we stay out of the laboratory until it was thoroughly cleaned.

I asked the hospital's HazMat team to clean up the mess that resulted from the fire. We had been working with samples of *Borrelia burgdorferi* bacteria, which causes Lyme disease. I was not sure if the

Continued on following page

Figure 21.4: *(continued.)*

waste container held any of our discarded samples. So, I thought it appropriate to clean up the laboratory with the utmost care. Even if the samples were in the waste container, it would be unlikely that the bacteria survived the fire, but I asked the HazMat team to do a Type 3 cleaning anyway.

The costs are specified.

The Hazmat team will be charging us $2,405 for the cleaning. The water damage to the laboratory was about $3,529. We will pay these costs out of our operating budget for now. We will file a claim with the Center's insurance company.

What will happen in the future is described.

In the future, we will be more careful about fire hazards in the laboratory. We are currently developing policies to avoid these kinds of situations in the future. We will also develop an action and evacuation plan to handle these sorts of situations if they occur again.

Contact information concludes the memo.

Thank you for your attention. If you have any questions or would like to talk further about this incident, please call me at 5-9124.

Laboratory Reports

Laboratory reports are written in response to experiments, tests, or inspections. Students in science lab classes are no doubt familiar with lab reports. They tend to describe the experiment, present the results, and discuss the results. The same is true of laboratory reports that present the results of testing or inspections. Lab reports typically include the following kinds of information:

- a summary of the experiment (methods).
- a presentation of the results.
- a discussion of the results.

Lab reports, like other activity reports, emphasize the facts. Here is not the place to speculate or develop a new theory. Instead, your lab report should present the results as objectively as possible and use those results to support the reasoned discussion that follows.

Figure 21.5 shows an example of a laboratory report. In this report, the writer describes the results of the testing as objectively as possible.

Other lab reports can be viewed at **www.ablongman.com/johnsonweb/21.4**

A Lab Report

Figure 21.5: A lab report walks readers through the methods, results, and discussion. Then, it offers any conclusions, based on the facts.

FEND-LAB, INC.
2314 Universal St., Suite 192
San Francisco, CA 94106
(325) 555-1327
www.fendlabcal.com

Test Address
NewGen Information Technology, LLC
3910 S. Randolph
Slater, CA 93492

Client
Brian Wilson
Phone: 650-555-1182
Fax: 650-555-2319
e-mail: brian_wilson@cssf.edu

Mold Analysis Report
Report Number: 818237-28
Date of Sampling: 091204
Arrival Date: 091404
Analysis Date: 091904
Technician: Alice Valles

Lab Report: Mold Test

The introduction states the subject, purpose, and main point.

In this report, we present the results of our testing for mold at the offices of NewGen Information Technology, at 3910 S. Randolph in Slater, California. Our results show above-normal amounts of allergenic mold, which may lead to allergic reactions among the residents.

Testing Methods

On 12 September 2004, we took samples from the test site with two common methods: Lift Tape Sampling and Bulk Physical Sampling.

Methods are described, explaining how the study was done.

Lift Tape Sampling. We located 10 areas around the building where we suspected mold or spores might exist (e.g., water stains, dusty areas, damp areas). Using 8-cm-wide strips of transparent tape, we lifted samples and pressed them into the nutrient agar in petri dishes. Each sample was sealed and sent to our laboratory, where it was allowed to grow for one week.

Continued on following page

Figure 21.5: *(continued.)*

Bulk Physical Sampling. We located 5 additional areas where we observed significant mold growth in ducts or on walls. Using a sterilized scraper, we removed samples from these areas and preserved them in plastic bags. In one place, we cut a 1-inch-square sample from carpet padding because it was damp and contained mold. This sample was saved in a plastic bag. All the samples were sent to our laboratory.

At the laboratory, the samples were examined through a microscope. We also collected spores in a vacuum chamber. Mold species and spores were identified.

Results of Microscopic Examination

The following chart lists the results of the microscope examination:

Mold Found	**Location**	**Amount**
Trichoderma	Break room counter	Normal growth
Geotrichum	Corner, second floor	Normal growth
Cladosporium	Air ducts	Heavy growth
Penicillium spores	Corkboard in bathroom	Normal growth

Descriptions of molds found:

Trichoderma: Trichoderma is typically found in moistened paper and unglazed ceramics. This mold is mildly allergenic in some humans, and it can create antibiotics that are harmful to plants.

Geotrichum: Geotrichum is a natural part of our environment, but it can be mildly allergenic. It is usually found in soil in potted plants and on wet textiles.

Cladosporium: Cladosporium can cause serious asthma and it can lead to edema and bronchiospasms. In chronic cases, this mold can lead to pulmonary emphysema.

Penicillium: Penicillium is not toxic to most humans in normal amounts. It is regularly found in buildings and likely poses no threat.

Results are presented objectively, without interpretation.

2

Figure 21.5: *(continued.)*

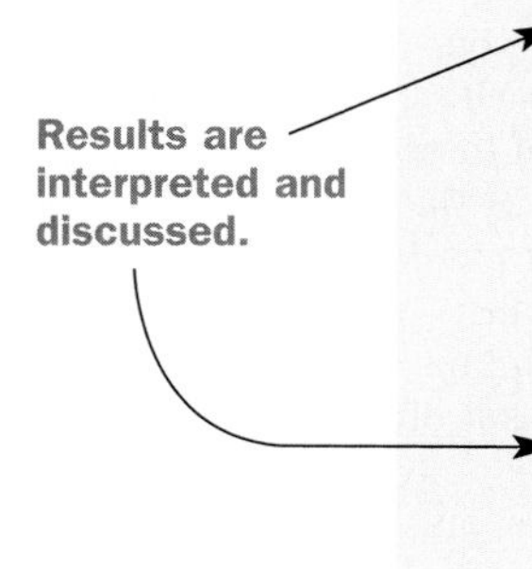

Results are interpreted and discussed.

Discussion of Results

It does not surprise us that the client and her employees are experiencing mild asthma attacks in their office, as well as allergic reactions. The amount of Cladosporium, a common culprit behind mold-caused asthma, is well above average. More than likely, this mold has spread throughout the duct system of the building, meaning there are probably no places where employees can avoid coming into contact with this mold and its spores.

The other molds found in the building could be causing some of the employees' allergic reactions, but it is less likely. Even at normal amounts, Geotrichum can cause irritation to people prone to mold allergies. Likewise, Trichoderma could cause problems, but it would not cause the kinds of allergic reactions the client reports. Penicillium in the amounts found would not be a problem.

The results of our analysis lead us to believe that the Cladosporium is the main problem in the building.

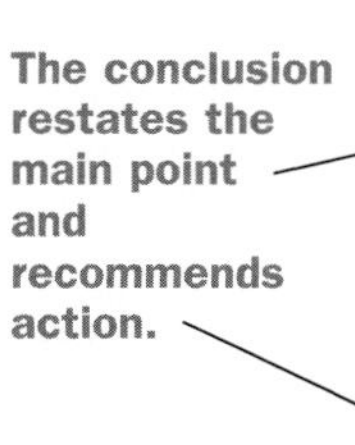

The conclusion restates the main point and recommends action.

Conclusions

The mold problem in this building will not go away over time. Cladosporium has obviously found a comfortable place in the air ducts of the building. It will continue to live there and send out spores until it is removed.

We suggest further testing to confirm our findings and measure the extent of the mold problem in the building. If our findings are confirmed, the building will not be safely habitable until a professional mold remover is hired to eradicate the mold.

Ignoring the problem would not be wise. At this point, the residents are experiencing mild asthma attacks and occasional allergic reactions. These symptoms will only grow worse over time, leading to potentially life-threatening situations.

Contact us at (325) 555-1327 if you would like us to further explain our methods and/or results.

3

Planning and Researching Activity Reports

One of the nice things about writing activity reports is that you have already developed most of the content and you are probably familiar with your readers. So, minimal planning is required, and the research has been mostly completed. These internal reports, after all, simply describe your activities over the past week or month.

A good workplace practice you might adopt is keeping an *activity journal* on your computer hard drive or in a notebook at your office. In your journal, start out each day by jotting down the things you need to accomplish. As you complete each of these activities, note the dates and times they were completed and the results (Figure 21.6).

Computers can also help you keep track of your activities. Groupware programs like Lotus Notes and MS Outlook can maintain "To Do" lists while keeping a calendar of your activities. Similarly, personal digital assistants (PDAs) can also maintain a "To Do" list as well as your calendar. When you complete a task, you can cross it off your list. The PDA will keep track of the time and date.

At first, keeping an activity journal will seem like extra work. But, you will soon realize that your journal keeps you on task and saves you time in the long run. Moreover, when you need to report on your activities for the week or month, you will have a record of all the things you accomplished.

Page from an Activity Journal

February 2, 2004

To Do List:

~~Call Hampton about Quality Report~~

~~Finish collecting results in Laboratory~~ 2/2, 9:23

Prepare slides for Wednesday's presentation

Prepare for briefing at 1:00 on project

~~Call Phillips about new productivity standards~~

~~Check e-mail~~ 2/2, 9:50

Results:

Hampton says in an e-mail his part of the project is on schedule. He should be ready to demo next Thursday.

Jim P. says we may be asked to increase productivity by 5 percent.

The "To Do" list keeps track of what needs to be handled today.

Results explain what happened as each item was completed.

Figure 21.6: An activity journal will help you keep track of your daily activities and when they were completed.

For links to software packages that help you keep activity journals, go to **www.ablongman.com/johnsonweb/21.5**

Analyzing the Rhetorical Situation

With your notes in front of you, you are ready to plan out your activity report. You should begin by briefly answering the Five-W and How questions:

Who might read or use this activity report?

Why do they want the report?

What information do they need to know?

Where will the report be used?

When will the report be used?

How might the report be used?

After considering these questions, you can begin thinking about the rhetorical situation that will shape how you write the activity report or present your briefing.

SUBJECT The subject of your report includes your recent activities. Include only information your readers need to know, and remove any unneeded want-to-tell information. After all, you accomplished many things that *could* be mentioned in the report, but your readers don't need to hear about all of them.

PURPOSE The purpose of your report is to describe what happened and what will happen in the future. In your introduction, state your purpose directly:

> In this memo, I will summarize our progress on the Hollings project during the month of August 2004.
>
> The purpose of this briefing is to update you on our research into railroad safety in northwestern Ohio.

You might use some of the following action verbs to describe your purpose:

to explain	to show	to demonstrate
to illustrate	to present	to exhibit
to justify	to account for	to display
to outline	to summarize	to inform

LINK To learn more about defining a purpose, go to Chapter 2, page 23.

READERS Think about the people who will need to use your report. The readers of activity reports tend to be your supervisors. Occasionally, though, reports are read by clients (lab reports or briefings) or used to support testimony (white papers). An incident report, especially when it concerns an accident, may have a range of readers who plan to use the document in a variety of ways. An insurance adjuster, for example, may use the incident report very differently than the quality control officer at your company. You need to compose your report to suit both of these readers' needs.

LINK For more ideas about reader analysis, turn to Chapter 3, page 42.

To learn more about groupware and PDAs, go to **www.ablongman.com/johnsonweb/21.6**

CONTEXT OF USE The context of use for your activity report will vary. In reality, most of your readers will simply scan and file your report. Similarly, oral briefings are not all that exciting. Your listeners will perk up for the information that interests them, but they are mostly checking to see if you are making progress.

Nevertheless, take a moment to check whether your activity report discusses any topics that involve troublesome ethical or political issues. Throughout your career, you will find that these small reports seem to surface when things go wrong at your company. When mistakes happen, auditors and lawyers will go through your activity reports, looking for careless statements or admissions of fault. So, make sure your statements reflect your actual actions and the results of your work.

Moreover, if you are reporting expenses in your activity report, make sure they are accurate. Auditors and accountants will look at these numbers closely. If your numbers don't add up, you may have some explaining to do.

LINK For more help defining the context, go to Chapter 3, page 42.

AT WORK

Tom Weber

ELECTRICAL ENGINEER, U.S. DEPARTMENT OF DEFENSE, WASHINGTON, D.C.

Tom Weber works in Washington, D.C., as a technical advisor for two government agencies within the U.S. Department of Defense.

What is a white paper and how are they used?

In Washington, D.C., engineers and scientists often act as advisors to the executive and legislative branches and play an important role in briefing policymakers and congressional staff on technical issues. Because time for meetings is limited, these briefings are usually accompanied by a short technical document, or *white paper,* that explains the issue in question.

Although meant to be an objective technical appraisal, white papers will often influence a policy decision or promote a specific program or research agenda. These papers are short, usually four pages or fewer, and cover the technical goals and benefits of a new policy or program. For example, a white paper on power system reliability, with a technical explanation of recent problems and proposed solutions, might be given to administrators at the Department of Energy considering new regulatory policies for the electric power industry. A white paper on the benefits of nanotechnology might be given to congressional staff working on next year's budget priorities for federally funded research and development programs.

In a white paper, the writer must clearly explain, to a mostly nontechnical audience, the relevant technical issues, and come to a logical conclusion as to the best decision and path forward within current political realities. All this must be accomplished in just a few pages. With the incredible information flow in our nation's capital, most government decision makers will have only a few minutes to review any one document.

Worksheets to help you define your purpose, readers, and context of use can be found at **www.ablongman.com/johnsonweb/21.7**

Organizing and Drafting Activity Reports

Remember, drafting activity reports should not take too much time. If you find yourself taking more than an hour to write one, you are probably spending too much time on this routine task.

To streamline your efforts, remember that all technical documents have an introduction, a body, and a conclusion. Each of these parts of the document makes predictable moves that you can use to guide your drafting of the report.

Writing the Introduction

Readers of your activity report are mostly interested in the facts. So, your introduction needs only to give them a brief framework for understanding those facts. To provide this framework, you should concisely

- define your subject.
- state your purpose.
- state your main point.

Figure 21.7 shows an example of a concise introduction. In this example, the first sentence identifies the subject ("retooling the Wichita Falls plant") and states the main point ("we made significant progress"). The second sentence states the purpose of the document.

Introduction for on Activity Report (Minimal)

Figure 21.7: An introduction should state the subject, purpose, and main point of the activity report or briefing.

Blue Ribbon Manufacturing 

Date: September 30, 2004
To: Alice Freedman, District Manager
From: Jim Fredrickson, Manufacturing
Subject: 9/31 Briefing on Retooling Progress

In September 2004, we made significant progress toward retooling the Wichita Falls plant. In this briefing, we will describe our progress since the end of August, and we will outline our activities for the next month.

Subject line clearly defines subject of the briefing.

Introduction is concise, stating the purpose and main point.

To see more examples of activity report memos, go to **www.ablongman.com/johnsonweb/21.8**

If your readers are not familiar with your project (e.g., you are giving a demonstration to clients), you might want to expand the introduction by also offering background information, stressing the importance of the subject, and forecasting the body of the report. Figure 21.8 shows the introduction to a document that would accompany a demonstration.

This example includes all the typical moves found in an introduction. The first paragraph offers background information. The second paragraph states the purpose and defines the subject. Then, the author stresses the importance of the subject and states the main point of the demonstration. The last three sentences forecast what will happen during the demonstration.

Full Introduction for an Activity Report

Wilson National Laboratory
Always Moving Forward

Nanotech Micromachines Demonstration for Senators Laura Geertz and Brian Hanson
Presented by Gina Gould, Head Engineer

Background information is offered for unfamiliar readers. → Nanotechnology is the creation and utilization of functional materials, devices, and systems with novel properties and functions that are achieved through the control of matter, atom by atom, molecule by molecule, or at the macromolecular level. A revolution has begun in science, engineering, and technology, based on the ability to organize, characterize, and manipulate matter systematically at the nanoscale.

Purpose and main point are mentioned here. → In this demonstration, we will show you how the 5492 Group at Wilson National Laboratory is applying breakthroughs in nanotechnology science toward the development of revolutionary new micromachines. Our work since 2002 has yielded some amazing results that might dramatically expand the capacity of these tiny devices.

Forecasting shows the structure of the briefing. → Today, we will first show you a few of the prototype micromachines we have developed with nanotechnology principles. Then, we will present data gathered from testing these prototypes. And finally, we will discuss future uses of nanotechnology in micromachine engineering.

Figure 21.8: When readers are less familiar with the subject, you might add background information, stress the importance of the subject, and forecast the rest of the document.

To see more examples of briefs, go to **www.ablongman.com/johnsonweb/21.9**

TAKE NOTE The length of your introduction depends on the readers' familiarity with your work and project. A familiar reader, like your supervisor, needs only the subject, purpose, and main point stated. An unfamiliar reader might also want background information, a statement about the importance of the subject, and forecasting.

LINK For more advice on writing introductions, turn to Chapter 6, page 141.

Writing the Body

In the body of the activity report, provide up to four things:

Summary of activities—In chronological order, summarize the project's two to five major events since your previous activity report. Don't dwell too much on the details. Instead, tell your readers briefly what you did and highlight any advances or setbacks in the project.

Results of activities or research—In order of importance, list the two to five most significant results or outcomes of your project. To aid a scanning reader, you might even use bullets to highlight these results.

Future activities or research—Tell readers what you will be doing during the next work period.

Expenses—If asked, you should state the costs incurred over the previous week or month. Make sure you highlight any places where costs are deviating from the project's budget.

The body of the activity report shown in Figure 21.9 includes these four items.

Writing the Conclusion

The conclusion should be as brief as possible. You should

- restate your main point.
- restate your purpose.
- make any recommendations, if appropriate.
- look to the future.

These concluding moves should be made in a maximum of two to four sentences.

> To conclude, in this demonstration our goal was to update you on our progress toward developing nanotechnology micromachines. Overall, it looks like we are making solid progress toward our objectives, and we seem to be on schedule. Over the next couple of months, we will be facing some tough technical challenges. At that point, we will know if micromachines are feasible with current technology.

The conclusion shown in Figure 21.9 is probably more common than this example. It concisely restates the memo's main point, ending the memo.

Keep in mind that most readers do not read the conclusion closely, if at all. So, your conclusion should not bring up any new information.

LINK For more ideas about writing conclusions, go to Chapter 6, page 155.

Progress Report

Figure 21.9: A progress report keeps things brief, discussing what happened, the results, and what will happen in the future.

TIGER INDUSTRIES
Automation Development Division

Memo header identifies primary readers and subject of report.

Date: 29 February 2004
To: Hal Roberts, Division Head
From: Sally Fenker, Green Robot Project Manager
Subject: Progress Report on Green Robot Project

The introduction is concise.

We are pleased to report that we have made significant progress in coding for the X53 Manufacturing Robot.

The month's activities are summarized without interpretation.

In September, we coded for the lateral movement of the robot's arm. We developed several usable subroutines that should improve its placement accuracy.

The results are clearly stated.

So far, the results of our work are looking good:

- placement accuracies of .009 mm
- speed of 20 placements per minute
- frozen arm only 3 times in 100 trials.

Future activities are discussed.

In March, we hope to complete development of the software. One member of our team is currently debugging. Right now, it looks like we're on schedule for the May rollout of the robot. Of course, that depends on the folks over in manufacturing.

Update on costs is given.

At this point, we're running a bit over budget, due to higher-than-expected contractor costs. We're perhaps $2,000 over.

The conclusion is brief and looks ahead.

Otherwise, we're on schedule and looking forward to finishing. Call me or e-mail me if you have any questions.

Making and Using PDFs

Increasingly, companies are asking for activity reports to be submitted in portable document format (PDF). Documents in PDF retain the exact look of the original, including the page layout and graphics. However, they do not require the same amount of memory space as other kinds of documents, so they are easily sent through e-mail, stored on a computer, or placed on a website for downloading.

PDFs also have some other helpful features:

- They have an almost identical appearance on all computers, so your file won't look different if someone is using a different kind of computer (PC, Macintosh, PDA).
- They include password protection to keep a file from being altered.
- They can include an electronic signature to signal your ownership or support of the document.

Also, if you are scanning documents, PDF is probably the most efficient way to save them in a usable format (Figure A).

Scanned Document in Portable Document Format

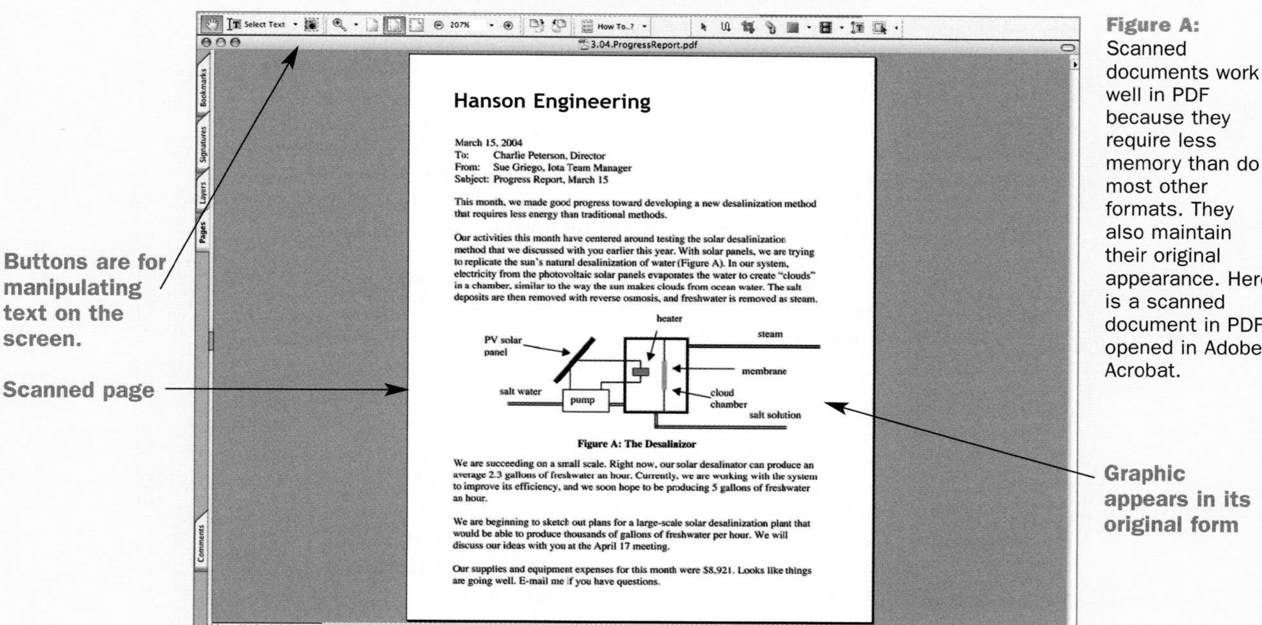

Hanson Engineering

March 15, 2004
To: Charlie Peterson, Director
From: Sue Griego, Iota Team Manager
Subject: Progress Report, March 15

This month, we made good progress toward developing a new desalinization method that requires less energy than traditional methods.

Our activities this month have centered around testing the solar desalinization method that we discussed with you earlier this year. With solar panels, we are trying to replicate the sun's natural desalinization of water (Figure A). In our system, electricity from the photovoltaic solar panels evaporates the water to create "clouds" in a chamber, similar to the way the sun makes clouds from ocean water. The salt deposits are then removed with reverse osmosis, and freshwater is removed as steam.

heater
PV solar panel
steam
membrane
salt water
pump
cloud chamber
salt solution

Figure A: The Desalinizor

We are succeeding on a small scale. Right now, our solar desalinator can produce an average 2.3 gallons of freshwater an hour. Currently, we are working with the system to improve its efficiency, and we soon hope to be producing 5 gallons of freshwater an hour.

We are beginning to sketch out plans for a large-scale solar desalinization plant that would be able to produce thousands of gallons of freshwater per hour. We will discuss our ideas with you at the April 17 meeting.

Our supplies and equipment expenses for this month were $8,921. Looks like things are going well. E-mail me if you have questions.

Figure A: Scanned documents work well in PDF because they require less memory than do most other formats. They also maintain their original appearance. Here is a scanned document in PDF opened in Adobe Acrobat.

Until recently, you would usually need a program called Adobe Acrobat to create and read PDF documents (Adobe Acrobat Reader can be downloaded free of charge from www.adobe.com). Today, word-processing programs like MS Word and Corel WordPerfect will allow you to convert text files into PDF files. With Acrobat, you can also convert other kinds of texts, like PowerPoint or Presentation files, into PDF format.

Want to learn more about using PDFs? Go to **www.ablongman.com/johnsonweb/21.10**

To save a file as a PDF:

1. Select "Print" in your word processor or presentation program.
2. Select "Print as PDF" in the Print Box that pops up.
   Instead of making a hard copy on your printer, your computer will convert your document or presentation into a PDF file.
3. Save this new PDF document on your hard drive.

Now you can store the PDF file, send it as an attachment, or put it on a website for downloading.

As technical workplaces continue evolving toward the goal of the "paperless office" it is likely that PDFs will become a typical format for sending documents of all kinds, including activity reports.

Using Style and Design in Activity Reports

Generally, activity reports follow a plain style and use simple design. These documents are mostly informative, not overly persuasive, so writers try to keep them rather straightforward.

Using a Plain Style

As you revise your document with style in mind, pay attention to the following items:

Sentences—Using plain style techniques, make sure that (1) the subject is the "doer" of most sentences and (2) the verb expresses the action in most sentences. Where appropriate, eliminate any prepositional chains.

Paragraphs—Each paragraph should begin with a topic sentence that makes a direct statement or claim that the rest of the paragraph will support. This topic sentence will usually appear as the first or second sentence of each paragraph.

LINK For more information on using plain style, see Chapter 7, page 162.

Tone—Since activity reports are often written quickly, you should make sure you are projecting an appropriate tone. It might be tempting to be sarcastic or humorous, but here is not the place. After all, you never know how the activity report might be used in the future.

For briefings, you want to project a professional tone. If you have negative information to convey, state it candidly with no apologies.

LINK For more information on using an appropriate tone, go to Chapter 7, page 181.

Using Design and Graphics

The design of your activity report should be straightforward also. Usually, the design of these documents is governed by a standard format, like a memo format or

For more advice on plain style, go to **www.ablongman.com/johnsonweb/21.11**

perhaps a standardized form for lab reports. Your company probably will specify the format for activity reports. Otherwise, you might use the templates available with your word-processing program.

LINK For help using templates, go to Chapter 16, page 475.

If you want to add any visuals, you should center them in the text and place them after the point where you referred to them. Even though activity reports tend to be short, you should still label the graphic and refer to it by number in the text.

If you are presenting a briefing orally, you should look for ways to include graphics to support your presentation. Graphs are always helpful to show trends in any data. If appropriate, you might also use a digital camera to snap some pictures to be included in the report.

Revising, Editing, and Proofreading

Even though activity reports and briefings are usually considered "informal" documents, you should devote some time to revising, editing, and proofreading them. Your reports, after all, might turn up unexpectedly in the future. Perhaps they might be evaluated during your performance review. Or, perhaps they might be studied by auditors who are trying to figure out where mistakes were made. An activity report that has not been revised, edited, or proofread will make you look unprofessional in these situations.

Revising and Editing Your Activity Reports

While revising and editing, look back at your original notes about the rhetorical situation in which the activity report might be used:

SUBJECT Does your report provide all the information that readers need to know? Have you included any additional want-to-tell information that can be removed?

PURPOSE Does the report achieve your intended purpose and the purpose you stated in the introduction of the document? If you want readers to do something, is it stated up front or clearly in the body?

READERS Does the report satisfy the needs of the primary readers (action takers) and secondary readers (advisors)? If an auditor, lawyer, or reporter read the document, would they be satisfied that it is accurate and truthful?

CONTEXT OF USE Are the format and design appropriate for the places where the document will be used? Can readers scan the document quickly to locate the most important information? If the report was taken out of context, would it still project a sense of professionalism?

LINK For more information on editing documents, turn to Chapter 10, page 264.

Need templates for letters and memos? Go to **www.ablongman.com/johnsonweb/21.12**
For worksheets to help you revise, go to **www.ablongman.com/johnsonweb/21.13**

Proofreading Your Document

Give your activity report a close proofreading before printing it out or hitting the Send button. These documents are typically written quickly, so they often contain small typos, misspellings, and grammar errors. Allow yourself a few minutes to read the text sentence by sentence to locate any annoying errors.

You may even print out a copy of the document and look it over on paper. Often, errors you did not catch on the screen are more noticeable in the paper version.

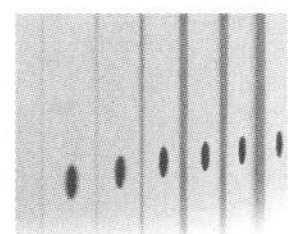

CHAPTER REVIEW

- Activity reports include progress reports, briefings, white papers, reports that accompany demonstrations, incident reports, and lab reports.
- An activity report typically includes the following sections: introduction, summary of activities, results of activities, future activities, expenses, and conclusion.
- While preparing to write an activity report, analyze the rhetorical situation by anticipating the readers and the context in which the report will be used.
- The style and design of activity reports should be plain and straightforward.
- Edit and proofread activity reports carefully. In the future, they may be used in an unexpected context, so you want them to reflect a sense of professionalism.

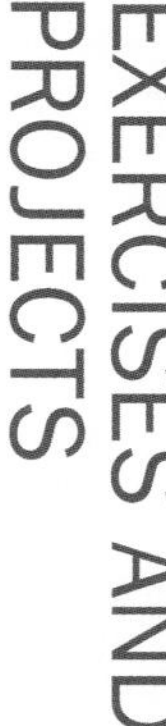

EXERCISES AND PROJECTS

Individual or Team Projects

1. For a week, keep a journal that tracks your activities related to school or work. Each day, make up a "to do" list. Then, as you complete each task, cross it off and write down the results of the task. In a memo to your instructor, summarize your activities for the week and discuss whether the activity journal was a helpful tool or not.

2. While you are completing another large project in this class or another, write a progress report to your instructor in which you summarize what has been accomplished and what still needs to be completed. The progress report should be submitted in memo format.

3. Think back to an accident that occurred in your life. Write an incident report in which you explain what happened, the actions you took, and the results of those actions. Then, discuss how you made changes to avoid that kind of accident in the future.

4. While completing a larger project for your class (like a proposal or report), give a two- to three-minute briefing on that project to your class. Summarize your activities, discuss the results, and discuss what will happen in the future.

To view information worksheets like this one, go to **www.ablongman.com/johnsonweb/21.14**

Collaborative Project

Your group has been asked to develop a standardized information sheet that helps students report accidents on your campus. Think of all the different kinds of accidents that might happen on your campus. Your information sheet should explain how to report an accident to the proper authorities on campus. Encourage the users of the information sheet to summarize the incident in detail, discuss the results, and make recommendations for avoiding similar accidents in the future.

Of course, numerous potential accidents might occur on campus. Your group may need to categorize them so readers can find the right authorities.

Revision Challenge

This activity report in Figure A is intended to update students on the recent changes in computer use policies at a small college. How would you revise this report to help it achieve its purpose?

Figure A: This activity report needs help. How might you revise it?

Smith College

Office of the Provost

Date: August 4, 2004
To: Smith College students
From: Provost George Richards
Subject: File-Sharing

Smith College has worked hard to develop a top-notch computer network on campus to provide access to the Internet for a variety of purposes. As a Smith student, you are welcome to use these computers for legal purposes, You will find the college's computer usage policies explained in the *Computer Policy for Smith College,* version 03.28.03, which is in effect until eclipsed by a revision.

Illegal downloading and sharing of copyrighted materials is a problem, especially music files, over our network. Violating our computer usage policies puts the college at risk for copyright infringement lawsuits. Our information technology experts also tell me that these activities slow down our network because these files require large amounts of bandwidth.

Recently, we have installed network tools and filters that allow us to detect and block illegal file sharing. We are already warning students about illegal use. After October 1, 2004, we will begin disciplining people who share files illegally by suspending their computer privileges. Repeated violations will be referred to the college's Academic Integrity Review Board.

You should know that if a complaint is filed by copyright owners, Smith College must provide your name and address to prosecutors. Our computer usage policy explicitly states that using college computers for illegal activities is forbidden and those who do will be prosecuted. Thank you for giving us this opportunity to stress the seriousness of this situation. We expect your compliance.

GO TO THE NET

To download an electronic copy of this Revision Challenge file, go to **www.ablongman.com/johnsonweb/21.15**

CASE STUDY

Bad Chemistry

Amanda Jones worked as a chemical engineer at BrimChem, one of the top plastics companies in the country. Recently, her division had hired a bright new chemical engineer named Paul Hanson. Paul was tall and good-looking, and he was always polite. At lunch during Paul's first week, Amanda and a coworker teased him about being a "Chippendales guy." Paul laughed a little, but it was apparent that the comment offended him. So, Amanda was careful from then on about her comments regarding his appearance.

A few months after he started, Paul went to a convention and came back somewhat agitated. Amanda asked him what was wrong. After a pause, Paul told her that one of the managers, Linda Juno, had made a pass at him one evening at the convention, suggesting he come up to her room "for a drink." When he declined, she became angry and said, "Paul, you need to decide whether you want to make it in this company." She didn't speak to him for the rest of the convention.

Paul told Amanda he was a bit worried about keeping his job with the company, since he was still on "probationary" status for his first year. Being on probation meant Linda or anyone else could have him fired for the slightest reason.

Later that day, though, Linda came down to Paul's office and seemed to be patching things up. After Linda left his office, Paul flashed Amanda a "thumbs up" signal to show that things were all right.

The next week, Amanda was working late and passed by Paul's office. Linda was in his office giving him a backrub. He was clearly not enjoying it. He seemed to be making the best of it, though, and he said, "OK, thank you. I better finish up this report."

Linda was clearly annoyed and said, "Paul, I let you off once. You better not disappoint me again." A minute later, Linda stormed out of Paul's office.

The next day, Paul stopped Amanda in the parking lot. "Amanda, I know you saw what happened last night. I'm going to file a harassment complaint against Linda. If necessary, I'm going to sue the company if I'm fired. I'm tired of being harassed by her and other women in this company."

Amanda nodded. Then Paul asked, "Would you write an incident report about what you saw last night? I want to put some materials on file." Amanda said she would.

A couple of days later, Paul was fired for a minor mistake. Amanda hadn't finished writing up the incident report.

If you were Amanda, what would you do at this point? If you wrote the report, what would you say and how would you say it?

To find links on sexual harassment in the workplace, go to **www.ablongman.com/johnsonweb/21.16**

CHAPTER

22

Analytical Reports

CHAPTER CONTENTS

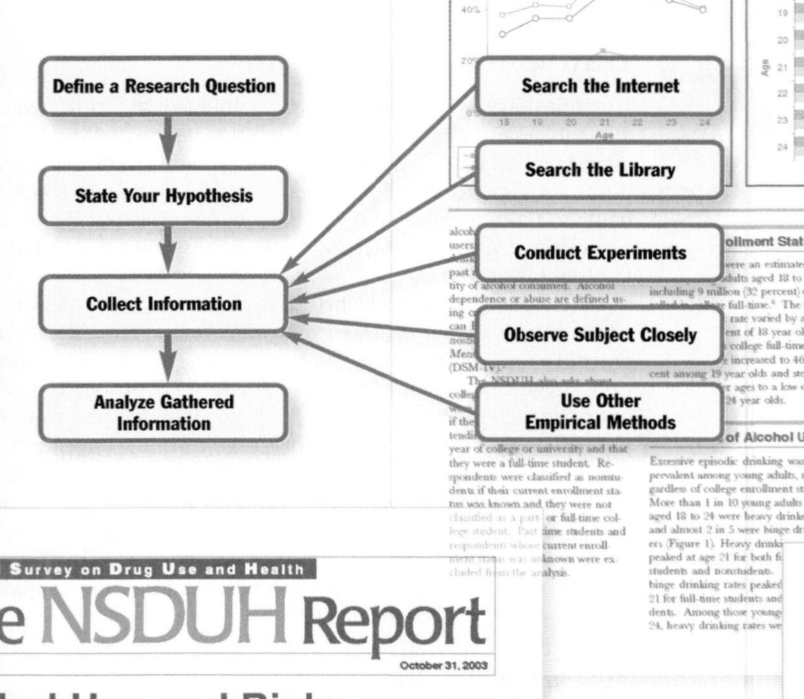

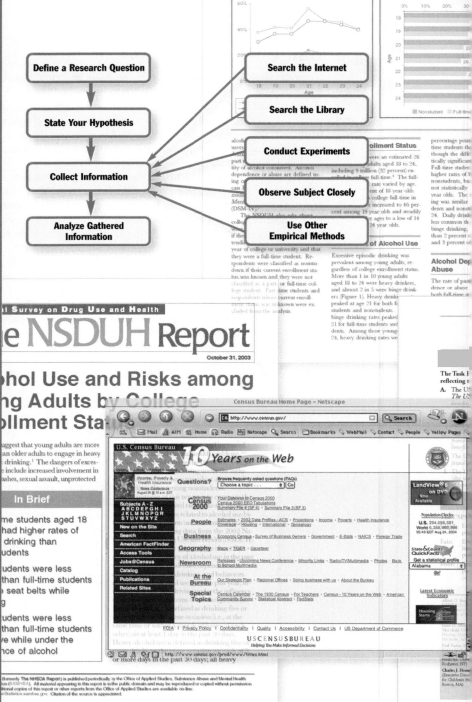

Can solar power be used to heat Engineering Hall?

gather information

library

internet

E Hall heating costs

develop an evaluation criteria

analyze E Hall

measure space (m^3)

calculate R values

analyze models

passive solar

active solar

conduct interviews

administrators

solar experts

physical plant engineers

AHRQ Pub. No. 04-526
November 2003

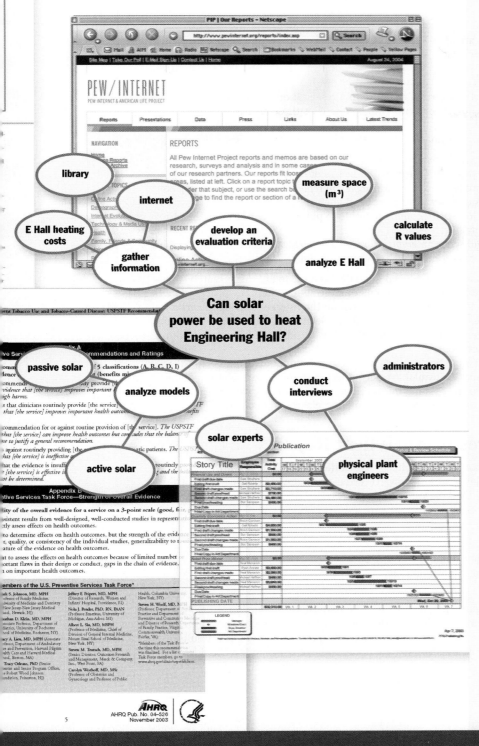

CHAPTER OBJECTIVES

In this chapter, you will learn:

- About the various kinds of analytical reports used in technical workplaces.
- How to use the IMRaD pattern for organizing reports.
- How to determine the rhetorical situation for your reports.
- How to develop a methodology for collecting and analyzing information.
- Strategies for organizing and drafting an analytical report.
- How to use style and design to highlight important information and make it understandable.
- Strategies for revising, editing, and proofreading analytical reports.

Analytical reports are among the most common formal documents produced in the technical workplace. The purpose of analytical reports is to present the results of research.

An analytical report is a formal response to a research question.

An analytical report is a formal response to a research question. The report defines a research methodology, presents results, discusses those results, and makes recommendations.

Here are just a few ways analytical reports are used in the technical workplace:

- A civil engineer would use a *recommendation report* to suggest ways to strengthen the structural integrity of a bridge after an earthquake.
- A medical researcher would use a *completion report* to present her findings on the effectiveness of a new treatment for diabetes.
- A manager would use a *feasibility report* to explore the use of robots to automate an assembly line.
- A biologist would write an *empirical research report* to present the results of her research on the migration patterns of monarch butterflies.

Research is the foundation of a good analytical report. To aid this research, computers allow us to access incredible amounts of information, and they process data in highly sophisticated ways. Even a modest search on the Internet quickly unearths a small library of information on any given subject. Meanwhile, the number-crunching capabilities of computers allow us to highlight subtle trends in data collected from experiments and observations.

As a result, writing analytical reports involves a combination of managing the large amount of information available on a subject and doing empirical research that clarifies or adds to the existing information. In this chapter, you will learn how to write analytical reports that present information clearly and concisely.

Basic Features of Analytical Reports

Because analytical reports are used in so many different ways, it is hard to pin down a basic pattern that they tend to follow. Nevertheless, you will find that analytical reports typically include the following basic features:

- Introduction
- Methodology or research plan
- Results
- Discussion of the results
- Conclusions or recommendations
- Graphics

The diagram in Figure 22.1 shows a general pattern for organizing an analytical report. To help you remember this generic report pattern, you might do what many researchers do—memorize the acronym IMRaD (Introduction, Methods, Results, and Discussion). This acronym outlines the main areas that most reports will address.

LINK For more strategies on organizing information in large documents, see Chapter 6, page 132.

To read more about the IMRaD pattern, go to **www.ablongman.com/johnsonweb/22.1**

The IMRaD pattern for reports is flexible and should be adapted to the specific situation in which you are writing. For example, in some reports, the methodology section can be moved into an appendix at the end of the report, especially if readers do not need a step-by-step description of the research approach. In other reports, you may find it helpful to combine the results and discussion sections into one larger section.

Like patterns for other documents, the IMRaD pattern is not a formula to be followed strictly. Rather, it is a guide to help you organize the information you have collected.

TAKE NOTE The IMRaD pattern actually reflects the steps in the scientific method: (1) identify a research question or hypothesis, (2) create a methodology to study the question, (3) generate results, (4) discuss those results, showing how they answer the research question or support the hypothesis.

LINK For more information on the scientific method, turn to Chapter 5, page 114.

Basic Pattern for an Analytical Report

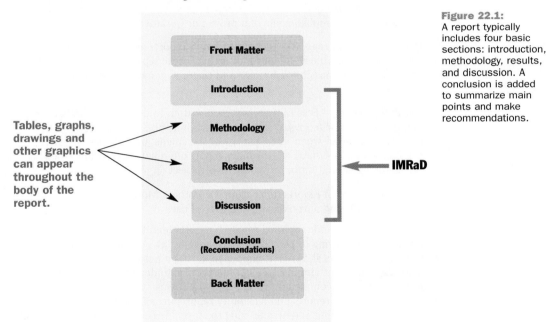

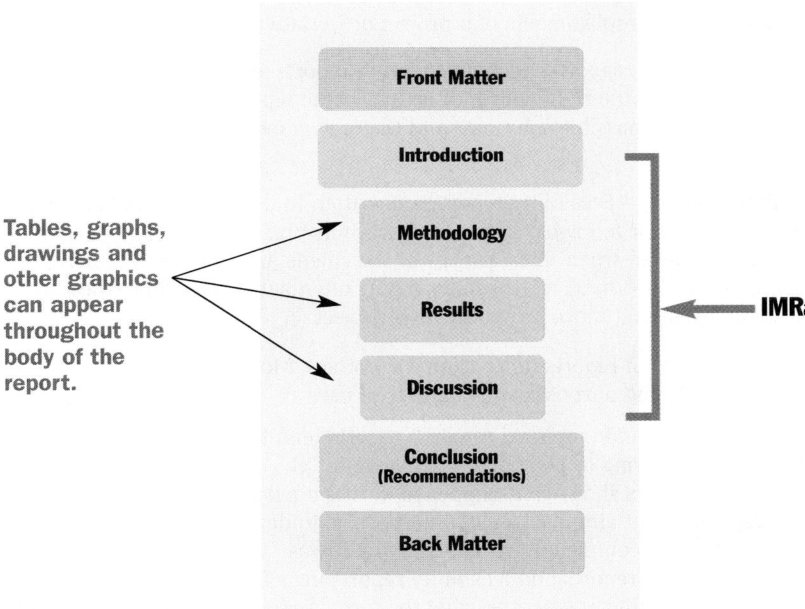

Figure 22.1: A report typically includes four basic sections: introduction, methodology, results, and discussion. A conclusion is added to summarize main points and make recommendations.

Types of Analytical Reports

Reports can be brief or long. Brief analytical reports might take up only a few pages. Longer reports can grow to hundreds of pages, especially when the project is highly technical or complex.

The main qualities that distinguish analytical reports from activity reports—which you learned about in the previous chapter—are their (1) level of formality and (2) argumentative nature. Analytical reports are formal documents that present findings and make recommendations. They also typically make an argument, using the results of research to answer an important question, prove a point, or recommend an appropriate course of action.

Analytical reports can be divided into five basic categories:

Research reports—The purpose of a research report is to present the findings of a study. Research reports often stress the causes and effects of problems or trends, showing how factors in the past have created the current situation.

Empirical research reports—Empirical research reports are written when a scientific project is completed. They first define a research question and hypothesis. Then, they describe the methods of study and the results of the study. And finally, they discuss these results and draw conclusions about what the study found. These reports often start out as laboratory reports and can evolve into published scientific articles.

Completion reports—Most technical projects conclude with a completion report. These documents are used to report back to management or the client, assessing the accomplishments of a project or initiative.

Recommendation reports—Recommendation reports are often used to make suggestions about the right course of action. These reports are used to study a problem, present possible solutions, and then recommend what actions should be taken.

Feasibility reports—Feasibility reports are written to determine whether developing a product or following a course of action is possible. Usually, these reports are produced when management or the clients are not sure whether something can be done. The feasibility report often helps determine whether the company should move forward with a project.

How do these types of reports differ from each other? Mostly, their differences result from their differing purposes and contexts of use.

- Research reports and empirical research reports tend to present and discuss results without trying to persuade the readers to take a course of action.
- Completion reports show what happened during a project and discuss the results. An important aim in completion reports is to demonstrate that the project was properly completed—or explain why it wasn't.
- Recommendation reports and feasibility reports usually endorse a course of action. By presenting and synthesizing the facts, these reports actively attempt to persuade the readers to follow a specific path.

For example, Figure 22.2 shows a research report on alcohol use among young adults. Like any research report, it defines a research question and offers a methodology for studying that research question. Then, it presents the results of the study and discusses those results. The report does not advocate a particular course of action. If this report were a recommendation report, it would make recommendations for ways to address alcohol use in young adults. (Later in this chapter, Figure 22.7 starting on page 670 is an example of a feasibility report. Figure 22.10 starting on page 690 is an example of a recommendation report.)

For examples of analytical reports, go to **www.ablongman.com/johnsonweb/22.2**

A Research Report

Figure 22.2: A research report does not typically recommend a specific course of action. It concentrates on presenting and discussing the facts.

National Survey on Drug Use and Health

The NSDUH Report

October 31, 2003

A clear title for the report is placed up front.

Alcohol Use and Risks among Young Adults by College Enrollment Status

Background information stresses the importance of the subject.

Studies suggest that young adults are more likely than older adults to engage in heavy episodic drinking.[1] The dangers of excessive alcohol use include increased involvement in fatal vehicle crashes, sexual assault, unprotected sex, violence, property damage, and increased risk of physical and mental health problems.[2,3] Studies conducted during the 1990s suggest that excessive drinking and driving under the influence of alcohol are more prevalent among college students than among nonstudents.[4]

Main points are placed up front in an easy-to-access box.

In Brief

- Full-time students aged 18 to 21 had higher rates of binge drinking than nonstudents
- Nonstudents were less likely than full-time students to use seat belts while driving
- Nonstudents were less likely than full-time students to drive while under the influence of alcohol

The purpose is clearly stated here.

The purpose of this report is to compare the use of alcohol, perceived risks of heavy alcohol use, and risk behaviors related to alcohol use by full-time college students and nonstudents at each age from 18 to 24 using data from the 2002 National Survey on Drug Use and Health (NSDUH).

The subject is defined.

The NSDUH asks respondents about the quantity and frequency of alcohol use in the past month, symptoms of alcohol dependence or abuse, perceptions of drinking, and behaviors associated with alcohol use over the past year, including seat belt use, driving under the influence of alcohol, and symptoms of alcohol problems. Binge alcohol use is defined as drinking five or more drinks on the same occasion (i.e., at the same time or within a couple of hours of each other) on at least 1 day in the past 30 days. Heavy alcohol use is defined as drinking five or more drinks on the same occasion on each of 5 or more days in the past 30 days; all heavy

The NSDUH Report (formerly The NHSDA Report) is published periodically by the Office of Applied Studies, Substance Abuse and Mental Health Services Administration (SAMHSA). All material appearing in this report is in the public domain and may be reproduced or copied without permission from SAMHSA. Additional copies of this report or other reports from the Office of Applied Studies are available on-line: http://www.DrugAbuseStatistics.samhsa.gov. Citation of the source is appreciated.

Continued on following page

Source: Office of Applied Studies Substance Abuse and Mental Health Services Administration [SAMUSA], 2003.

Figure 22.2: (*continued.*)

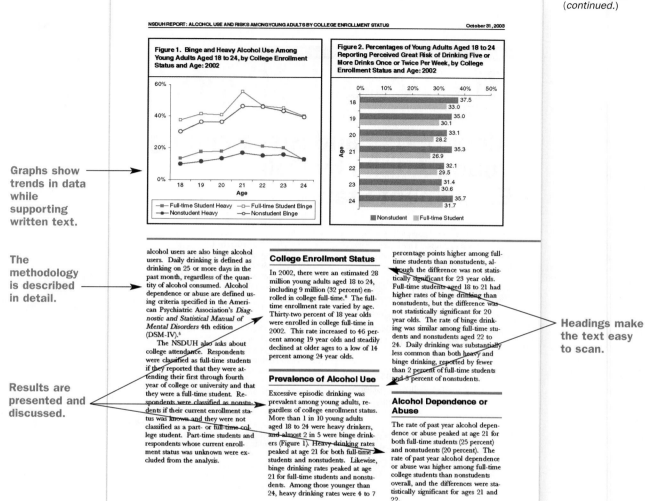

NSDUH REPORT: ALCOHOL USE AND RISKS AMONG YOUNG ADULTS BY COLLEGE ENROLLMENT STATUS October 31, 2003

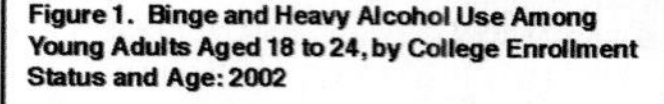

Figure 1. Binge and Heavy Alcohol Use Among Young Adults Aged 18 to 24, by College Enrollment Status and Age: 2002

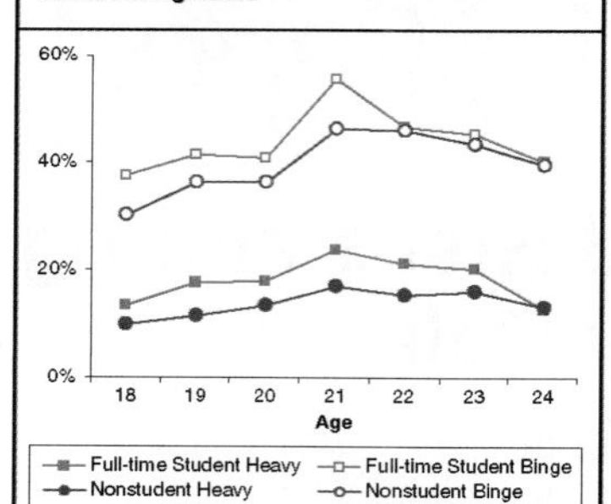

Figure 2. Percentages of Young Adults Aged 18 to 24 Reporting Perceived Great Risk of Drinking Five or More Drinks Once or Twice Per Week, by College Enrollment Status and Age: 2002

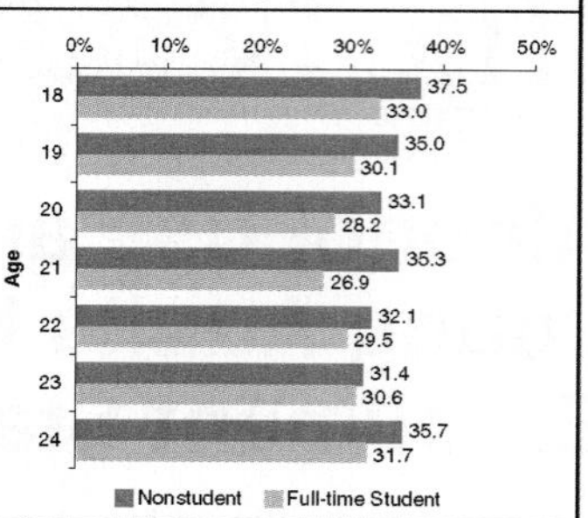

Graphs show trends in data while supporting written text.

The methodology is described in detail.

Results are presented and discussed.

Headings make the text easy to scan.

alcohol users are also binge alcohol users. Daily drinking is defined as drinking on 25 or more days in the past month, regardless of the quantity of alcohol consumed. Alcohol dependence or abuse are defined using criteria specified in the American Psychiatric Association's *Diagnostic and Statistical Manual of Mental Disorders* 4th edition (DSM-IV).[5]

The NSDUH also asks about college attendance. Respondents were classified as full-time students if they reported that they were attending their first through fourth year of college or university and that they were a full-time student. Respondents were classified as nonstudents if their current enrollment status was known and they were not classified as a part- or full-time college student. Part-time students and respondents whose current enrollment status was unknown were excluded from the analysis.

College Enrollment Status

In 2002, there were an estimated 28 million young adults aged 18 to 24, including 9 million (32 percent) enrolled in college full-time.[6] The full-time enrollment rate varied by age. Thirty-two percent of 18 year olds were enrolled in college full-time in 2002. This rate increased to 46 percent among 19 year olds and steadily declined at older ages to a low of 14 percent among 24 year olds.

Prevalence of Alcohol Use

Excessive episodic drinking was prevalent among young adults, regardless of college enrollment status. More than 1 in 10 young adults aged 18 to 24 were heavy drinkers, and almost 2 in 5 were binge drinkers (Figure 1). Heavy drinking rates peaked at age 21 for both full-time students and nonstudents. Likewise, binge drinking rates peaked at age 21 for full-time students and nonstudents. Among those younger than 24, heavy drinking rates were 4 to 7 percentage points higher among full-time students than nonstudents, although the difference was not statistically significant for 23 year olds. Full-time students aged 18 to 21 had higher rates of binge drinking than nonstudents, but the difference was not statistically significant for 20 year olds. The rate of binge drinking was similar among full-time students and nonstudents aged 22 to 24. Daily drinking was substantially less common than both heavy and binge drinking, reported by fewer than 2 percent of full-time students and 3 percent of nonstudents.

Alcohol Dependence or Abuse

The rate of past year alcohol dependence or abuse peaked at age 21 for both full-time students (25 percent) and nonstudents (20 percent). The rate of past year alcohol dependence or abuse was higher among full-time college students than nonstudents overall, and the differences were statistically significant for ages 21 and 22.

Figure 22.2:
(continued.)

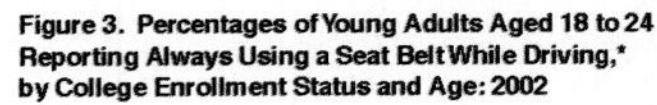

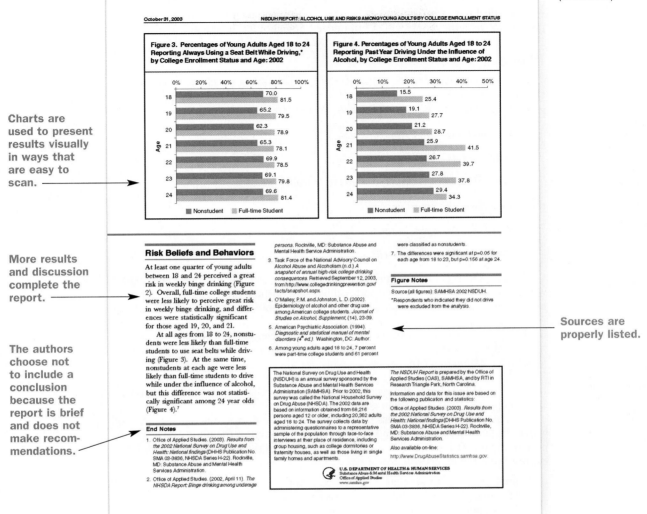

Figure 3. Percentages of Young Adults Aged 18 to 24 Reporting Always Using a Seat Belt While Driving,* by College Enrollment Status and Age: 2002

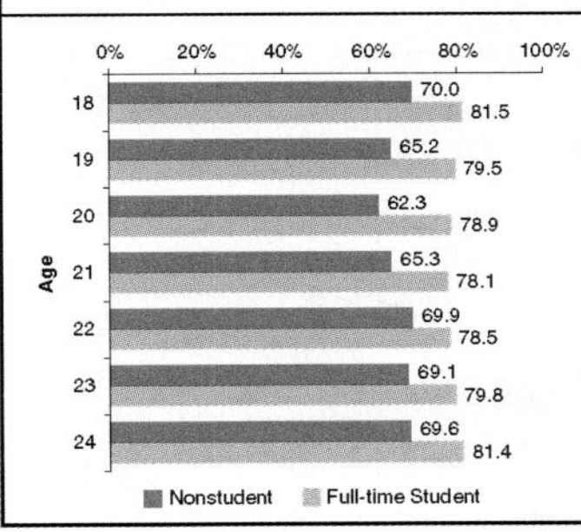

Figure 4. Percentages of Young Adults Aged 18 to 24 Reporting Past Year Driving Under the Influence of Alcohol, by College Enrollment Status and Age: 2002

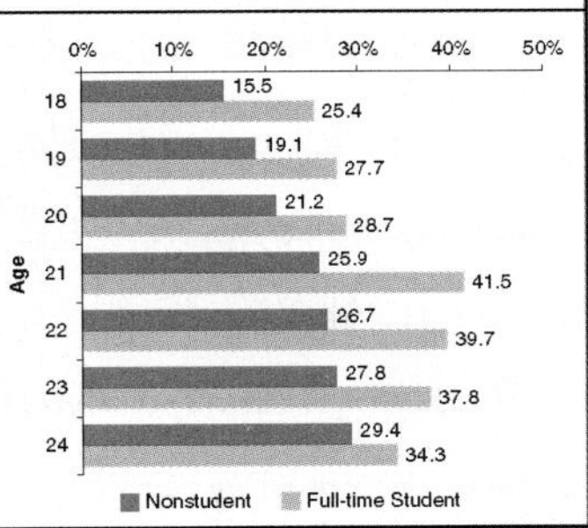

Charts are used to present results visually in ways that are easy to scan.

Risk Beliefs and Behaviors

At least one quarter of young adults between 18 and 24 perceived a great risk in weekly binge drinking (Figure 2). Overall, full-time college students were less likely to perceive great risk in weekly binge drinking, and differences were statistically significant for those aged 19, 20, and 21.

At all ages from 18 to 24, nonstudents were less likely than full-time students to use seat belts while driving (Figure 3). At the same time, nonstudents at each age were less likely than full-time students to drive while under the influence of alcohol, but this difference was not statistically significant among 24 year olds (Figure 4).[7]

More results and discussion complete the report.

The authors choose not to include a conclusion because the report is brief and does not make recommendations.

End Notes

1. Office of Applied Studies. (2003). *Results from the 2002 National Survey on Drug Use and Health: National findings* (DHHS Publication No. SMA 03-3836, NHSDA Series H-22). Rockville, MD: Substance Abuse and Mental Health Services Administration.
2. Office of Applied Studies. (2002, April 11). *The NHSDA Report: Binge drinking among underage persons.* Rockville, MD: Substance Abuse and Mental Health Service Administration.
3. Task Force of the National Advisory Council on Alcohol Abuse and Alcoholism (n.d.) *A snapshot of annual high-risk college drinking consequences.* Retrieved September 12, 2003, from http://www.collegedrinkingprevention.gov/facts/snapshot.aspx
4. O'Malley, P. M. and Johnston, L. D. (2002). Epidemiology of alcohol and other drug use among American college students. *Journal of Studies on Alcohol, Supplement,* (14), 23-39.
5. American Psychiatric Association. (1994). *Diagnostic and statistical manual of mental disorders (4th ed.)* Washington, DC: Author.
6. Among young adults aged 18 to 24, 7 percent were part-time college students and 61 percent were classified as nonstudents.
7. The differences were significant at $p=0.05$ for each age from 18 to 23, but $p=0.156$ at age 24.

Sources are properly listed.

Figure Notes

Source (all figures): SAMHSA 2002 NSDUH.

*Respondents who indicated they did not drive were excluded from the analysis.

The National Survey on Drug Use and Health (NSDUH) is an annual survey sponsored by the Substance Abuse and Mental Health Services Administration (SAMHSA). Prior to 2002, this survey was called the National Household Survey on Drug Abuse (NHSDA). The 2002 data are based on information obtained from 68,216 persons aged 12 or older, including 20,362 adults aged 18 to 24. The survey collects data by administering questionnaires to a representative sample of the population through face-to-face interviews at their place of residence, including group housing, such as college dormitories or fraternity houses, as well as those living in single family homes and apartments.

The NSDUH Report is prepared by the Office of Applied Studies (OAS), SAMHSA, and by RTI in Research Triangle Park, North Carolina.

Information and data for this issue are based on the following publication and statistics:

Office of Applied Studies. (2003). *Results from the 2002 National Survey on Drug Use and Health: National findings* (DHHS Publication No. SMA 03-3836, NHSDA Series H-22). Rockville, MD: Substance Abuse and Mental Health Services Administration.

Also available on-line:
http://www.DrugAbuseStatistics.samhsa.gov

U.S. DEPARTMENT OF HEALTH & HUMAN SERVICES
Substance Abuse & Mental Health Services Administration
Office of Applied Studies
www.samhsa.gov

Planning and Researching Analytical Reports

An analytical report can be a large, complex document, so it is important that you prepare to write the report with a full understanding of its rhetorical situation.

Planning

You should start planning the document by first identifying the elements of the rhetorical situation.

Begin planning by answering the Five-W and How questions:

Who might read this report?

Why was this report requested?

What kinds of information or content do readers need?

Where will this report be read?

When will the report be used?

How will this report be used?

With the answers to these questions fresh in your mind, you can begin defining the writing situation in which your report will be used. Spend a little time writing one or more sentences that describe the report's subject, purpose, readers, and context of use:

SUBJECT What exactly will the report cover, and what are the boundaries of its subject? What information and facts do readers need to know to make a decision? What information don't they need?

PURPOSE What should the report accomplish, and what do my readers expect it to accomplish? What is its main goal or objective?

You should be able to express the purpose of your report in one sentence. A good way to begin forming your purpose statement is to complete the phrase, "The purpose of my report is to . . . ." You can then use some of the following action verbs to express what the report will do:

to analyze	to develop	to determine
to examine	to formulate	to recommend
to investigate	to devise	to decide
to study	to create	to conclude
to inspect	to generate	to recommend
to assess	to originate	to resolve
to explore	to produce	to select

READERS Who are the primary readers (action takers), secondary readers (advisors), and tertiary readers (evaluators)? Who are the gatekeeper readers for this report?

Primary readers (action takers) are the people who need the report's information to make some kind of decision. Anticipate the decision they need to make

To find worksheets to help you analyze readers and contexts of use, go to **www.ablongman.com/johnsonweb/22.3**

and provide them the information they require. If they need recommendations, provide them in a direct, obvious way.

Secondary readers (advisors) are usually experts or other specialists who will advise the primary readers. They will probably be most concerned about the accuracy of your facts and the validity of your reasoning. Provide them enough technical detail and data to feel confident about your conclusions and recommendations.

Tertiary readers (evaluators) might be people you didn't expect to read the report, like reporters, lawyers, auditors, and perhaps even historians. Anticipate these kinds of audiences to avoid making unfounded statements that might be used to harm you or your company.

Gatekeeper readers (supervisors) will probably include your immediate supervisor. However, other gatekeepers, like your company's legal or technical experts, might need to look over your report before it is sent to the primary readers.

LINK For more help defining your readers and their characteristics, go to Chapter 3, page 42.

CONTEXT OF USE Where, when, and how will the report be used? What are the economic, political, and ethical factors that will influence the writing of the report and how readers will interpret it?

Physical context—Consider the various places where your report might be used, such as meetings or at a conference. What adjustments will be needed to make it more readable/usable in these situations?

Economic context—Anticipate the monetary issues that may influence how your readers will interpret the results and recommendations in your report, especially if you are making recommendations for change.

Political context—Think about the politics involved with your report. On a micropolitical level (i.e., office politics), determine who stands to gain or lose from the information in your report. On a macropolitical level, consider the larger political trends that will shape the reception of your report.

Ethical context—Consider any legal or ethical issues that might affect your report and the methods you will use to collect information. For example, if you are doing a study that involves people or animals, you may need to secure permission release forms. If you need access to sensitive information of any kind, you may find there are legal roadblocks.

LINK For more information on defining the context of use, go to Chapter 3, page 53.

If you are writing a report with a team or group, collaborate on your answers to these questions about the rhetorical situation. Then, print the answers out and give each member a copy. If your team begins the project with a clear understanding of the subject, purpose, readers, and context of use, you will likely avoid unnecessary conflict and wasted time.

For more information about legal issues in research, go to **www.ablongman.com/johnsonweb/22.4**

Researching

With the rhetorical situation fresh in your mind, you can start collecting information for your report. It is important that you define a research question and develop a plan (a methodology) for conducting research on your subject.

Research in technical fields typically follows a predictable process:

1. Define a research question.
2. State a hypothesis.
3. Develop a research methodology.
4. Collect information following the research methodology.
5. Analyze gathered information.

The most effective methodologies are ones that collect information from a variety of sources (Figure 22.3).

Researching a Subject

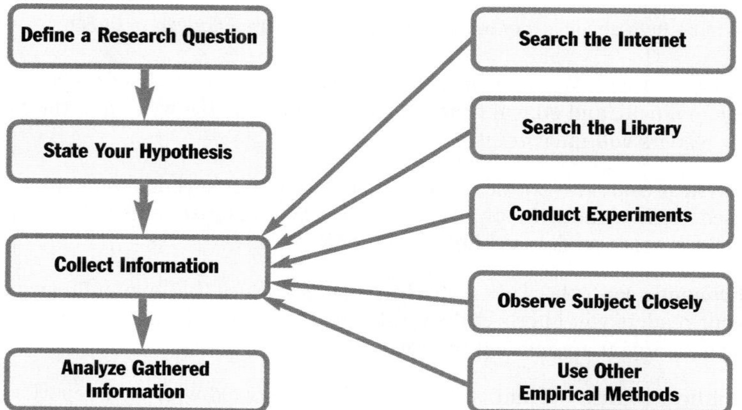

Figure 22.3: A research methodology is a plan for gathering information, preferably from a variety of sources.

DEFINE A RESEARCH QUESTION AND HYPOTHESIS Reports are usually written to answer a specific *research question* or test a *hypothesis*. So, you should begin by defining the question you are trying to answer. Write down the question in one sentence.

> Could we convert one of our campus buildings to a renewable heating source, like solar?
>
> Why are the liver cancer rates in Horn, Nevada, higher than the national average?
>
> How much would it cost to automate our factory in Racine, Wisconsin?
>
> Is it feasible to reintroduce wolves into the Gila Wilderness Area?

At this point, you should also write down a one-sentence hypothesis. A hypothesis is essentially an educated guess or tentative explanation that answers your research question. You don't need a hypothesis at this point; however, some people like to begin with their best try at answering the research question. With your hypothesis

Want help defining a hypothesis? Go to **www.ablongman.com/johnsonweb/22.5**

stated, you will have a better idea about what direction your research will take and what you are trying to prove or disprove.

> We believe we could convert a building like Engineering Hall to a solar heating source.
>
> Our hypothesis is that liver cancer rates in Horn, Nevada, are high because of excessive levels of arsenic in the town's drinking water.
>
> Automating our Racine plant could cost $2 million, but the savings will offset that figure in the long run.
>
> Reintroducing wolves to the Gila Wilderness Area is feasible, but there are numerous political obstacles and community fears to be overcome.

Remember, though, that a hypothesis is really just a possible answer to your research question. As you move forward with your research, you will likely find yourself modifying this hypothesis to fit the facts. In some cases, you might even need to abandon your original hypothesis completely and come up with a new one.

DEVELOP A METHODOLOGY Once you have defined your research question and stated a hypothesis, you are ready to develop your research *methodology.* A methodology is the series of steps you will take to answer your research question or test your hypothesis.

A good way to invent your methodology is to use logical mapping (Figure 22.4).

1. Write your research question in the middle of a sheet of paper or your computer screen.
2. Identify the two to five major steps you would need to take to answer that question. For example, if you are researching the use of solar power to generate heat for a building, one major step in your methodology might be to use the Internet to collect information on solar power and its potential applications.
3. Identify the two to five minor steps required to achieve each major step. In other words, determine which smaller steps are needed to achieve each major step.
4. Keep filling out and revising your map until you have fully described the major and minor steps in your methodology.

Your logical map will help you answer the How questions about your methodology. By writing down your major and minor steps, you are describing how you would go about studying your subject.

Allow yourself to be creative at this point. As you keep mapping out farther, your methodology will likely evolve in front of you. You can cross out some steps and combine others.

When you have finished mapping, you may or may not want to immediately turn your methodology into sentences and paragraphs. Some writers of reports prefer to keep their methodology in outline form until they are ready to draft the report, because they know the methodology will be modified before the final report is written. Scientific research projects, however, often require the methodology to be written out completely before the project begins.

TAKE NOTE A credible methodology is one that is "replicable," meaning the readers can obtain the same results if they redo your research. As you are inventing and writing your methodology, ask yourself whether someone in your research area could duplicate your work.

For websites that discuss methodologies, go to **www.ablongman.com/johnsonweb/22.6**

Using Logical Mapping to Develop a Methodology

Figure 22.4: When mapping a methodology, ask yourself how you might answer the research question. Then decide on the major and minor steps in your methodology.

COLLECT INFORMATION Your methodology is your road map to collecting information. Once you have written your methodology down, you can use it to guide your research. The information you collect will become the results section of your report.

There are many places to find information:

Internet searches—With some well-chosen keywords, you can use your favorite search engine to start collecting information on the subject. Cut and paste the materials you may need and bookmark helpful websites. Websites like the ones from the U.S. Census Bureau and the Pew Charitable Trust offer a wealth of information (Figures 22.5 and 22.6). Make sure you properly cite any sources when you enter them into your notes.

LINK For more strategies for using search engines, see Chapter 5, page 106.

Library research—Using your library's catalogs and indexes, start finding articles, reports, books, and other print documents that discuss your subject. You should make copies of the materials you find or use a scanner to save them on your computer. Again, keep track of where you found these materials, so you can cite them in your report.

For help with Internet research, go to **www.ablongman.com/johnsonweb/22.7**

Experiments and observations—Your report might require experiments and measurable observations to test your hypothesis. Each field follows different methods for conducting experiments and observations. Learn about research methodologies used in your field and use them to generate empirical data.

Other empirical methods—You might conduct interviews, pass out surveys, or do case studies to generate empirical information. It is preferable that a report include some empirical material that you generated yourself, because self-generated results will help strengthen the credibility of your report.

To avoid any problems with plagiarism or copyright, you need to carefully cite your sources in your notes. It's fine to cut and paste from the Internet or duplicate passages from print sources, as long as you quote or paraphrase them properly and cite them accurately when you write your report. Chapter 5 offers note-taking strategies that will help you avoid any issues of plagiarism or copyright. Chapter 6 discusses copyright in more depth.

LINK For more information on plagiarism and copyright, go to Chapter 5, page 127.

U.S. Census Bureau Website

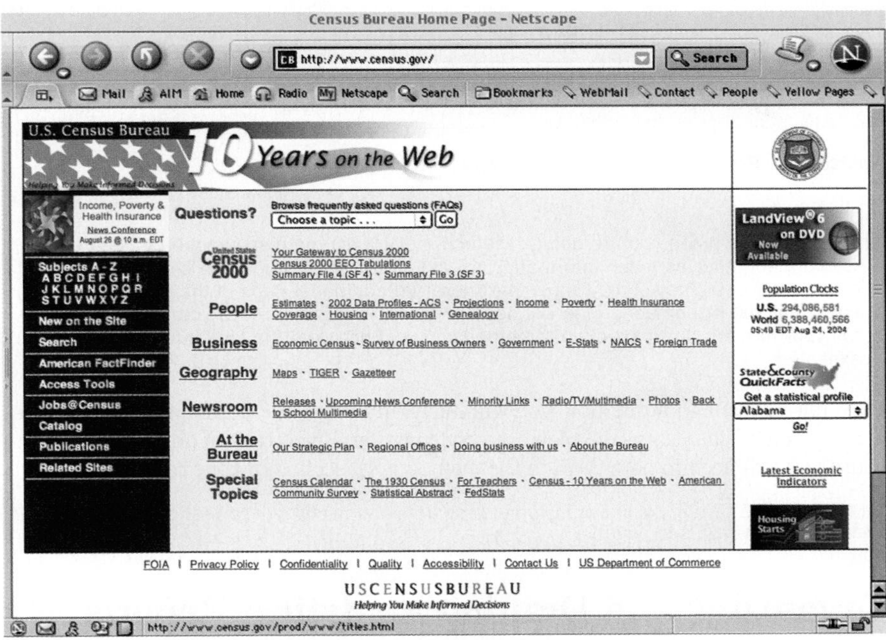

Figure 22.5: The U.S. Census Bureau website is an excellent resource for statistics. The website offers statistics on income, race, trade, and lifestyle issues.

Source: United States Census Bureau, http://www.census.gov.

To find other great sources of information, go to **www.ablongman.com/johnsonweb/22.8**

An Archive on the Internet

Figure 22.6: The Pew Research Center's archive offers reports on the Internet and American life. Pew collects data on religion, consumption, lifestyle issues, and many other topics.

Source: Pew Research Center, http://www.pewinternet.org/reports/index.asp.

TAKE NOTE When you are doing research, never become discouraged by dead ends. Just pull together as much information as you can, wherever you can find it. Dead ends are not failures. They simply show that there are additional gaps in the available knowledge that you or someone else will need to fill. In your final report, you may need to tell your readers that part of your methodology did not yield the kind information you expected.

While collecting information, you will likely find that your methodology will evolve. These kinds of changes in the plan are very common. Keep track of any changes, because you will need to note them in the methodology section of your report.

LINK Chapter 5 offers a full discussion of research. You might turn to that chapter for more research strategies, see page 96.

Organizing and Drafting Analytical Reports

You can begin organizing and drafting your report as soon as you have enough information to start writing. In other words, you don't need to collect *all* your information before you begin drafting your report.

Organizing your information and drafting a report are not difficult if you stay focused on your purpose. Make a concerted effort to include only need-to-know information in the report. It might be tempting to include everything you collected, but

AT WORK

Janice Starzyk

SENIOR ANALYST, FUTRON CORPORATION, BETHESDA, MARYLAND

Futron is an aerospace consulting firm.

How has the writing of reports changed in the last 10 years?

My career is in industry analysis, so writing reports has been a central part of my daily job functions. Working for consulting firms in multiple industries, I have written reports on projects ranging from short newsletter briefs on the transportation of nuclear waste to 300-page studies on the commercial satellite industry.

The style of writing always conforms to the audience for which the document is intended, be it government agencies, foreign companies without a grasp of American idioms, trendy high-tech magazines, or the low-tech executive trying to learn what his or her company really does. The level of technical detail, the complexity of language, the tone, and the subjectivity of the content must be carefully crafted to satisfy the needs of the customer.

Recently, much of my writing has been focused on a newer medium—PowerPoint. Even consulting projects that include an in-depth written report must be accompanied by summary slides for high-level managers who do not have time to read the full report. The art form of the industry analyst has become the craft of writing bullet points that convey chapters of information in a quick presentation format. This craft I will likely spend the rest of my career trying to perfect.

don't do this. Anything beyond need-to-know information will only muddle your document, making the most important ideas harder to find.

Writing the Introduction

Let's be honest. Reports are not the most interesting documents to read. Given the slightest excuse, your readers will start thumbing through your report's pages, looking for the main points. That's why your introduction needs to grab their attention and give them good reasons to read your report closely.

Your report should begin with an introduction that sets a framework, or context, for the rest of the document (Figure 22.7 on pages 670–681). Typically, the introduction will include some or all of the following moves:

DEFINE THE SUBJECT In the introduction, clearly state what your report is about—and perhaps what it is not about. Reports have a tendency to grow far too large when writers are not clear about the scope of the subject.

STATE THE PURPOSE Reports are usually rather blunt about their purpose. In the introduction, you can make the following kinds of statements:

> The purpose of this report is to show how Engineering Hall can be converted to solar heating with minimal cost.

Have writer's block? Go to **www.ablongman.com/johnsonweb/22.9**

In this report, we present the results of our research into higher liver cancer rates in Horn, Nevada.

In most cases, the more straightforward your purpose statement, the better. After all, you want your readers to know immediately what you are trying to do. If your readers have a clear idea of your purpose up front, they will be able to better understand your methods, results, conclusions, and recommendations.

LINK For more information on writing a purpose statement, see Chapter 2, page 23.

STATE THE MAIN POINT Your readers will be most attentive at the beginning of the report. So, it is a good idea to put your main point (i.e., your overall conclusion) in the introduction. That way, as they read the report, they can retrace your steps as you came to that conclusion.

We argue that the best way to add solar heating to Engineering Hall is to use a combination of direct gain methods and hydrosolar tubes. We recommend that the university begin working on this project this summer, using faculty and students from the Engineering Department as designers and builders of the system.

Our conclusion is that Horn's high liver cancer rates are primarily due to excessive levels of arsenic in the town's drinking water.

Of course, you are giving away the ending by putting your main conclusion in the introduction. But then, reports are not mystery stories. By telling readers your main point up front, you will focus their attention and make their reading of the report more efficient.

STRESS THE IMPORTANCE OF THE SUBJECT Your readers may not immediately recognize the importance of a report's subject. To avoid this problem, tell your readers in the introduction *why* they should pay attention to your report.

In the near future, the United States will need to wean itself off nonrenewable energy sources like oil and natural gas. The conversion of Engineering Hall to solar heating will give us a starting point for studying how the conversion to renewable energy could be made on our campus.

Critics of our findings might point out that Horn is just one small town with high levels of liver cancer—it is not, they might say, the norm. We think they are wrong. Many Western towns have high levels of arsenic in their tap water. If our findings are true, removal of arsenic from tap water could be an important step toward improving health in the region.

OFFER BACKGROUND INFORMATION Reports often start out with some historical information to familiarize readers with the subject. Background information can be simple, such as reminding readers that they requested the report on a particular date.

On March 14, the President of Kellen College, Dr. Sharon Holton, asked us to explore ways to begin converting the campus to renewable energy sources. We decided to use one building, Engineering Hall, as a test case for studying solar heating options.

To see sample reports with good introductions, go to **www.ablongman.com/johnsonweb/22.10**

More extensive background information might tell the history of the subject, giving readers some context for understanding the contents of the report.

> Horn, Nevada, was settled in 1893 when a rich vein of silver was discovered in the Madres mountains nearby. For decades, mining was the primary source of income, though legalized gambling later became an alternative source of revenue. Today, Horn is becoming a haven for retirees who live here in the winter.
>
> The people of Horn, especially longtime residents, have always recognized that their cancer rates were especially high. They call it the "Horn Curse." In 1998 medical researchers began to suspect arsenic. They found that levels of arsenic in the drinking water were consistently higher than recommended limits set by the federal government. Local politicians, however, have resisted recommendations to remove arsenic from the Horn water supply, mostly due to cost and perhaps a resistance to government intrusion.

Usually, background information will tell your readers information they already know or won't find too contentious. The purpose of background information is to give your readers a familiar place from which to begin reading the report.

FORECAST THE REMAINDER OF THE REPORT Your readers will find forecasting in the introduction helpful. Even a quick sentence can help them develop a framework for understanding your document:

> This report includes four sections: (a) our methodology for gathering information, (b) the results of our research, (c) a discussion of those results, and (d) our conclusions and recommendations.

AT A GLANCE

Moves in an Introduction

- Define the subject.
- State the purpose.
- State the main point.
- Stress the importance of the subject.
- Offer background information.
- Forecast the remainder of the report.

Forecasting is also a nice way to signal the end of the introduction, while making the transition into the body of the report.

These moves do not need to be made in this order, nor are they all required. Minimally, your report should tell readers your subject, purpose, and main point in the introduction. You can use the other moves to fill out the introduction where appropriate.

LINK For more information on writing introductions, see Chapter 6, page 141.

Describing Your Methodology

Following the introduction, reports typically include a methodology section that describes step by step how the study was conducted. The aim of the methodology section is to describe the steps you went through to collect information on the subject. This section should include an opening, body, and closing.

TAKE NOTE This section could also be called the "Methods," "Research Plan," or "Information-Gathering Phase," among other titles.

OPENING In the opening paragraph, start out by describing your overall approach to collecting information (Figure 22.7). If you are following an established methodology, you might mention where it has been used before and who used it.

LINK For help organizing sections in larger documents, see Chapter 6, page 145.

BODY In the body of the methods section, walk your readers step by step through the major steps of your study. As you highlight each major step, you should also discuss the minor steps that were part of it.

CLOSING To close the methodology section, you might discuss some of the limitations of the study. For example, your study may have been conducted with a limited sample (e.g., college students in a midwestern university). Perhaps time limitations restricted your ability to collect comprehensive data. All methodologies have their limitations. By identifying your study's limitations, you will show your readers that you are aware that other approaches may yield different results.

Summarizing the Results of the Study

In the results section, you should summarize the major *findings* of your study. A helpful guideline is to discuss only the two to five major results. That way, you can avoid overwhelming your readers with a long list of findings, especially ones that are less significant.

TAKE NOTE Limiting yourself to two to five major results is only a guideline. If you need to include more than five results, then do so. But check whether all of the included results are important to support your argument in the report.

This section should include an opening, body, and perhaps a closing.

- In your opening paragraph for the results section, briefly summarize your major results (Figure 22.7).
- In the body of this section, devote at least one paragraph to each of these major results, using data to support each finding.
- In the closing (if needed), you can again summarize your major results.

Your aim in this section is to present your findings as objectively and clearly as possible. Therefore, state your results with minimal interpretation. Wait until the discussion section to offer your interpretation of the results.

If your study generated numerical data, you should use tables, graphs, and charts to present your data in this section. As discussed in Chapter 14, these graphics should support the written text, not replace it. You should include a verbal description of your findings as well as tables, graphs, and charts that present the data.

LINK For help making tables, charts, and graphs, see Chapter 9, starting on page 234.

Discussing Your Results

The discussion section is where you will analyze the results of your research. As you look over the results of your study, what are the two to five major conclusions you

To see samples of well-written methodologies, go to **www.ablongman.com/johnsonweb/22.11**

might draw from your information or data? What are your results telling you about the subject?

The discussion section should start out with an opening paragraph that briefly states your overall conclusions about the results of your study (Figure 22.7). Then, in the body of this section, you should devote a paragraph or more to each of your conclusions. Discuss the results of your study, showing what you think your results show.

Stating Your Overall Conclusions and Recommendations

The conclusion of a report should be concise. To conclude the report, you should make the following six moves, which are typical in a larger document:

MAKE AN OBVIOUS TRANSITION A heading like "Our Recommendations" or "Summary" will cue readers that you are concluding. Or, you can use phrases like "In conclusion" or "To sum up" to signal that the report is coming to an end.

RESTATE THE MAIN POINTS Boiled down to its essence, what did your study show or demonstrate? Tell your readers what you proved, disproved, or did not prove.

STATE YOUR RECOMMENDATIONS If you have been asked to make recommendations, your conclusion should identify two to five actions that your readers should consider. A good way to handle recommendations is to list them in a bulleted list. Tell your readers exactly what you believe should be done (Figure 22.7).

RESTRESS THE IMPORTANCE OF THE STUDY Tell your readers why you believe your study was important. Tell them why you think the results and recommendations should be taken seriously.

LOOK TO THE FUTURE You might discuss future research paths that could be pursued. Or, you could describe the future you envision if readers follow your recommendations.

Moves in the Conclusion

- Make an obvious transition.
- Restate the main points.
- State your recommendations.
- Restress the importance of the study.
- Look to the future.
- Say thank you and offer contact information.

SAY THANK YOU AND OFFER CONTACT INFORMATION After thanking your readers for their interest, you might also provide contact information, such as a phone number and e-mail address. Readers who have questions or want to comment on the report will then be able to contact you.

Like the introduction, your conclusion doesn't need to make these moves in this order, nor are they all necessary. Minimally, you should restate your main point and state any recommendations. You can use the other moves to help end your report on a positive note.

LINK For more ideas on writing conclusions, see Chapter 6, page 155.

To see reports with well-written conclusions, go to **www.ablongman.com/johnsonweb/22.12**

A Feasibility Report

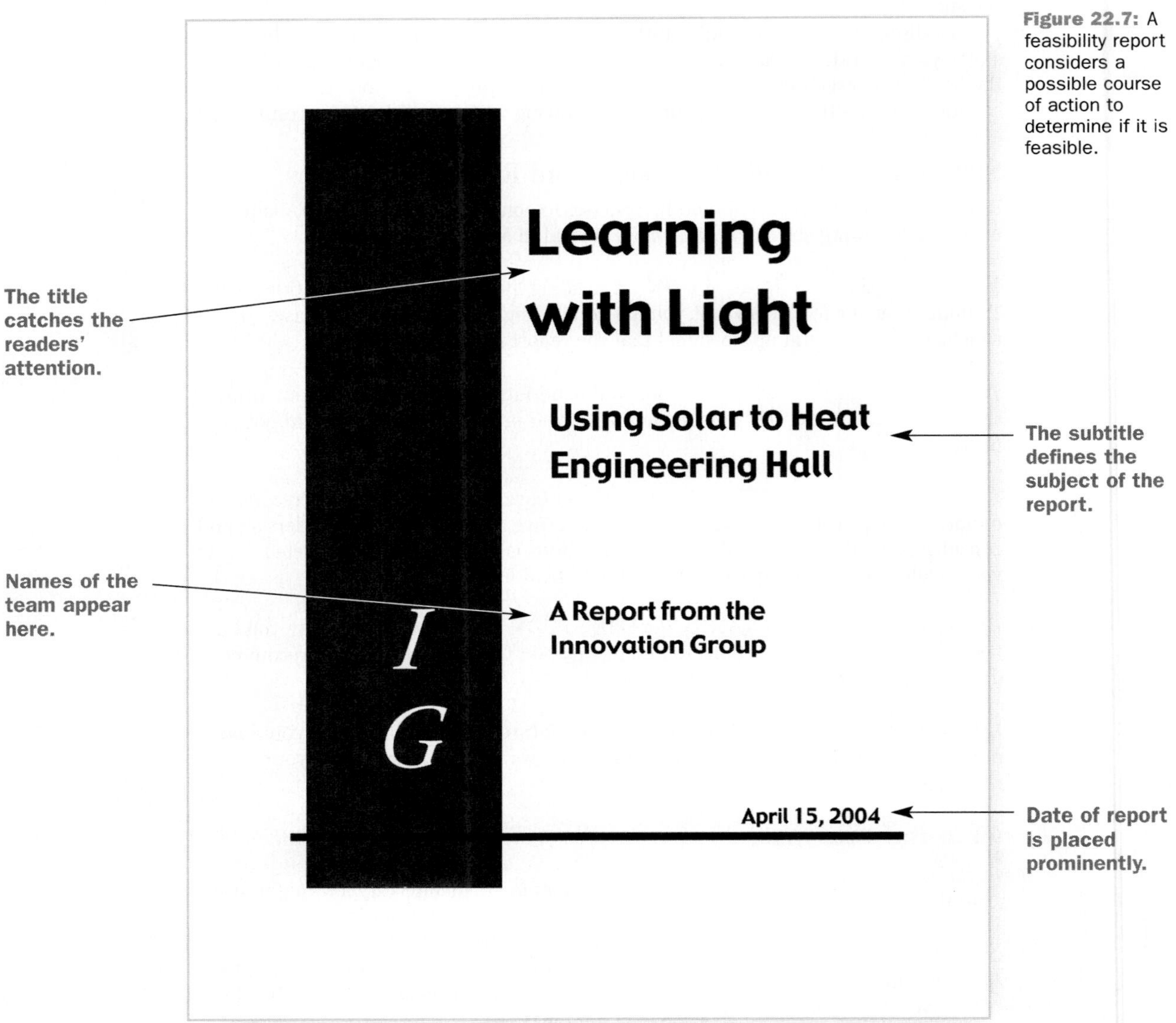

Learning
with Light
Using Solar to Heat
Engineering Hall
A Report from the
Innovation Group
I
G
April 15, 2004
The title catches the readers' attention.
Names of the team appear here.
The subtitle defines the subject of the report.
Date of report is placed prominently.

Figure 22.7: A feasibility report considers a possible course of action to determine if it is feasible.

Figure 22.7: (*continued.*)

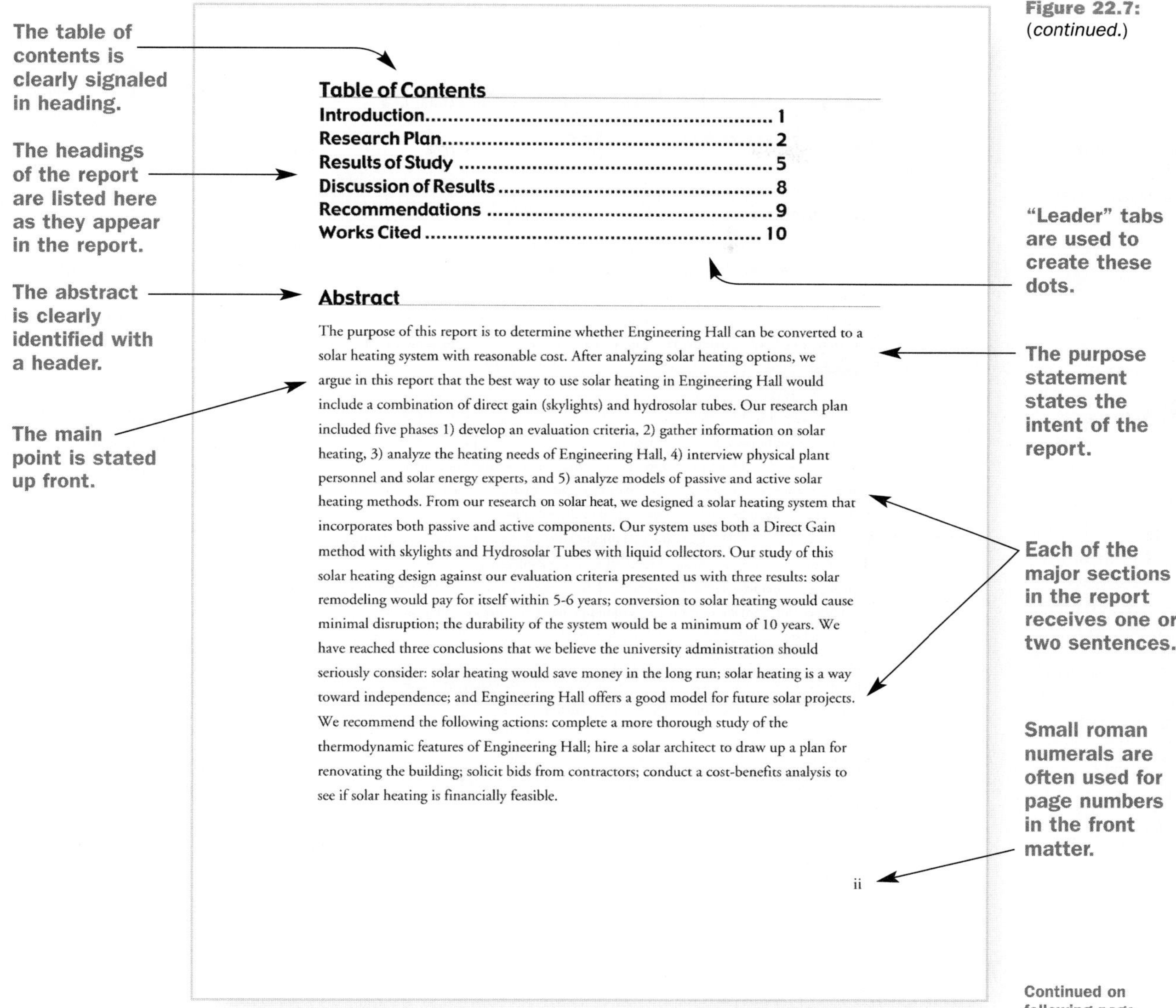

The table of contents is clearly signaled in heading.
The headings of the report are listed here as they appear in the report.
The abstract is clearly identified with a header.
The main point is stated up front.
"Leader" tabs are used to create these dots.
The purpose statement states the intent of the report.
Each of the major sections in the report receives one or two sentences.
Small roman numerals are often used for page numbers in the front matter.

Table of Contents

Abstract

The purpose of this report is to determine whether Engineering Hall can be converted to a solar heating system with reasonable cost. After analyzing solar heating options, we argue in this report that the best way to use solar heating in Engineering Hall would include a combination of direct gain (skylights) and hydrosolar tubes. Our research plan included five phases 1) develop an evaluation criteria, 2) gather information on solar heating, 3) analyze the heating needs of Engineering Hall, 4) interview physical plant personnel and solar energy experts, and 5) analyze models of passive and active solar heating methods. From our research on solar heat, we designed a solar heating system that incorporates both passive and active components. Our system uses both a Direct Gain method with skylights and Hydrosolar Tubes with liquid collectors. Our study of this solar heating design against our evaluation criteria presented us with three results: solar remodeling would pay for itself within 5-6 years; conversion to solar heating would cause minimal disruption; the durability of the system would be a minimum of 10 years. We have reached three conclusions that we believe the university administration should seriously consider: solar heating would save money in the long run; solar heating is a way toward independence; and Engineering Hall offers a good model for future solar projects. We recommend the following actions: complete a more thorough study of the thermodynamic features of Engineering Hall; hire a solar architect to draw up a plan for renovating the building; solicit bids from contractors; conduct a cost-benefits analysis to see if solar heating is financially feasible.

ii

Continued on following page

Figure 22.7: (*continued.*)

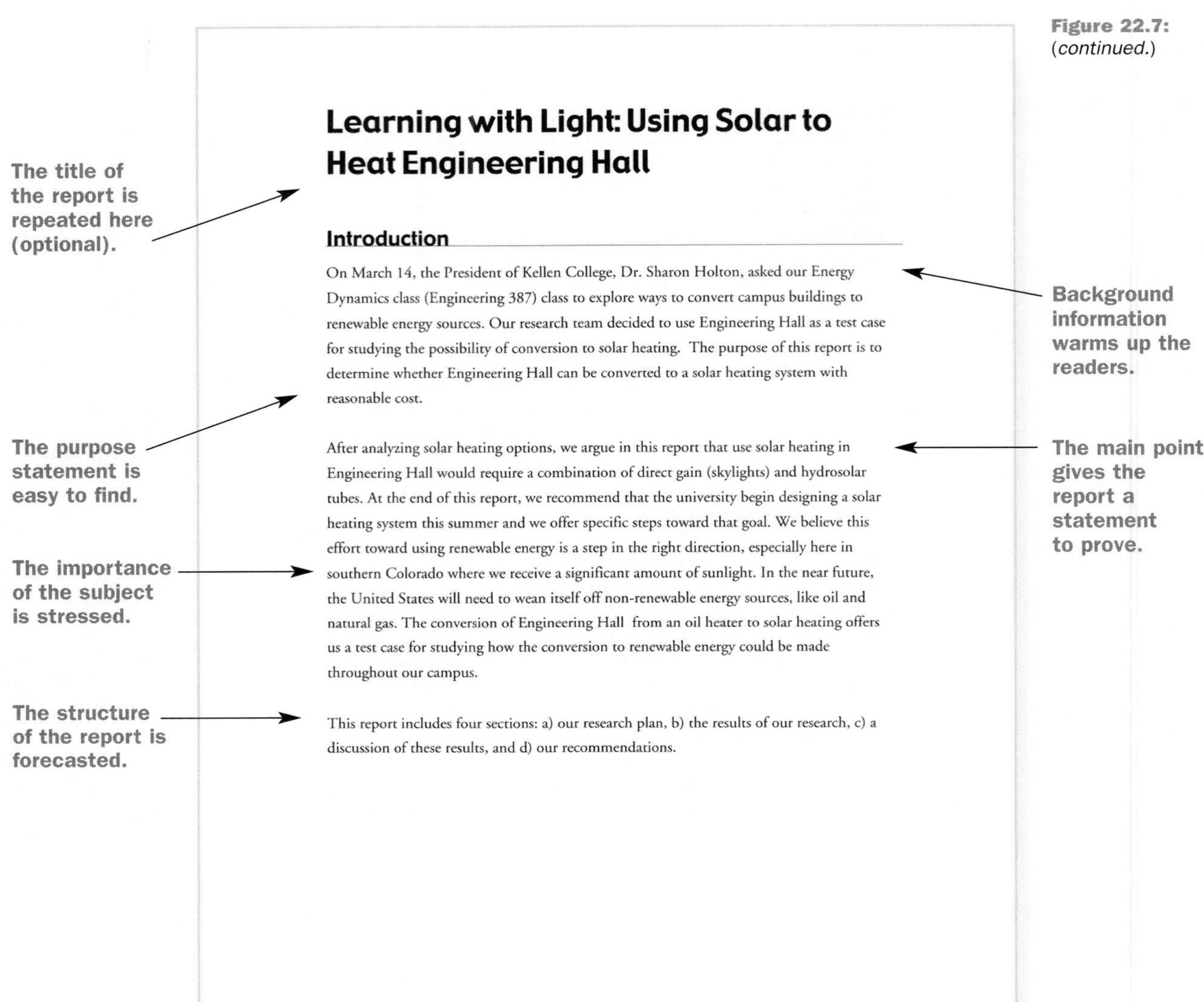

The title of the report is repeated here (optional).
Background information warms up the readers.
The purpose statement is easy to find.
The main point gives the report a statement to prove.
The importance of the subject is stressed.
The structure of the report is forecasted.
Learning with Light: Using Solar to Heat Engineering Hall
Introduction
On March 14, the President of Kellen College, Dr. Sharon Holton, asked our Energy Dynamics class (Engineering 387) class to explore ways to convert campus buildings to renewable energy sources. Our research team decided to use Engineering Hall as a test case for studying the possibility of conversion to solar heating. The purpose of this report is to determine whether Engineering Hall can be converted to a solar heating system with reasonable cost.
After analyzing solar heating options, we argue in this report that use solar heating in Engineering Hall would require a combination of direct gain (skylights) and hydrosolar tubes. At the end of this report, we recommend that the university begin designing a solar heating system this summer and we offer specific steps toward that goal. We believe this effort toward using renewable energy is a step in the right direction, especially here in southern Colorado where we receive a significant amount of sunlight. In the near future, the United States will need to wean itself off non-renewable energy sources, like oil and natural gas. The conversion of Engineering Hall from an oil heater to solar heating offers us a test case for studying how the conversion to renewable energy could be made throughout our campus.
This report includes four sections: a) our research plan, b) the results of our research, c) a discussion of these results, and d) our recommendations.

Figure 22.7: (*continued.*)

The opening of the research section includes a summary of the methodology.

Research Plan

To study whether heating Engineering Hall with solar is feasible, we followed a five-part research plan:

Phase 1: Develop an evaluation criteria
Phase 2: Gather information on solar heating
Phase 3: Analyze the heating needs of Engineering Hall
Phase 4: Interview physical plant personnel and solar energy experts
Phase 5: Analyze models of passive and active solar heating that would be appropriate for this building.

The heading states a major step.

Phase 1: Develop Evaluation Criteria

In consultation with President Holton and Carlos Riley, Director of Campus Planning, we determined that a successful conversion to solar heating would need to meet three criteria:

The evaluation criteria are listed and defined.

Cost-Effectiveness—the solar heating system would need to be cost-neutral in the long run. In other words, any costs of converting to solar heating would need to be offset by the savings of the system.

Minimal Disruption—conversion needs to cause only minimal disruption to the use of the building. Any construction would need to occur mostly over the summer or in a way that allowed the building to be used.

Durability—the heating system would need to be reliable over time. It would need to last at least 20 years with routine maintenance.

These criteria were the basis of our evaluation of solar heating systems in the Results and Discussion sections later in this report.

2

Continued on following page

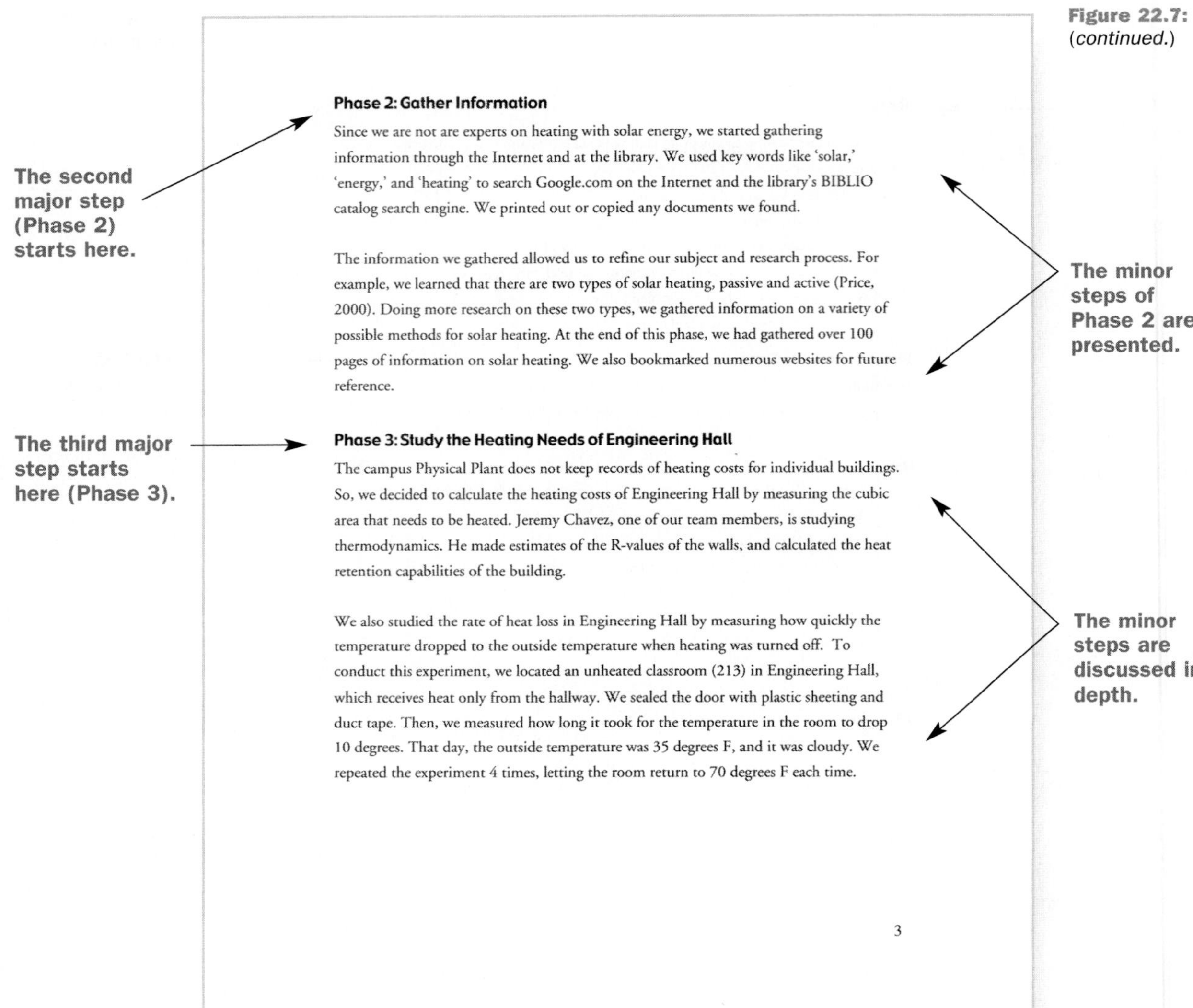

The second major step (Phase 2) starts here.
The minor steps of Phase 2 are presented.
The third major step starts here (Phase 3).
The minor steps are discussed in depth.

Phase 2: Gather Information

Since we are not are experts on heating with solar energy, we started gathering information through the Internet and at the library. We used key words like 'solar,' 'energy,' and 'heating' to search Google.com on the Internet and the library's BIBLIO catalog search engine. We printed out or copied any documents we found.

The information we gathered allowed us to refine our subject and research process. For example, we learned that there are two types of solar heating, passive and active (Price, 2000). Doing more research on these two types, we gathered information on a variety of possible methods for solar heating. At the end of this phase, we had gathered over 100 pages of information on solar heating. We also bookmarked numerous websites for future reference.

Phase 3: Study the Heating Needs of Engineering Hall

The campus Physical Plant does not keep records of heating costs for individual buildings. So, we decided to calculate the heating costs of Engineering Hall by measuring the cubic area that needs to be heated. Jeremy Chavez, one of our team members, is studying thermodynamics. He made estimates of the R-values of the walls, and calculated the heat retention capabilities of the building.

We also studied the rate of heat loss in Engineering Hall by measuring how quickly the temperature dropped to the outside temperature when heating was turned off. To conduct this experiment, we located an unheated classroom (213) in Engineering Hall, which receives heat only from the hallway. We sealed the door with plastic sheeting and duct tape. Then, we measured how long it took for the temperature in the room to drop 10 degrees. That day, the outside temperature was 35 degrees F, and it was cloudy. We repeated the experiment 4 times, letting the room return to 70 degrees F each time.

3

Figure 22.7: (*continued.*)

Figure 22.7: *(continued.)*

It is fine to acknowledge the limitations of your study.

Admittedly, our experiment was crude and probably didn't yield us completely reliable results, but it did give us some data that we could use to calculate how much heat was needed to maintain room temperature in the building. We used these results to supplement Jeremy's calculations of heat retention capabilities.

Fourth major step begins.

Phase 4: Interviewing Physical Plant Personnel and Solar Heating Experts

We then set up a four interviews with experts in heating issues—two interviews with employees of the campus Physical Plant and two with solar heating experts.

Minor steps are discussed.

At the physical plant, we interviewed Jerry Madden and Gina Brin on March 21st. Mr. Madden is Director of the Physical Plant, where he has worked for 20 years. Gina Brin is the Physical Plant's energy efficiency expert, and she has worked at the plant for 2 years. We asked them a series of questions about the feasibility of using solar heating in Engineering Hall and their willingness to participate in such a program. During the interviews, both Mr. Madden and Ms. Brin offered us brochures on solar heating. Mr. Brin showed us a plan he had sketched out for using solar heating in Henson Hall, a smaller, more energy-efficient building on campus. Mr. Madden also directed us to some websites about solar heating that we were not aware of. Meanwhile, Ms. Brin looked over our calculations regarding heating needs of Engineering Hall and said we were probably not too far off the correct measurements.

Our interviews with the solar heating experts were conducted on April 21 and 23. We met separately with Jane Karros, who is a solar energy consultant, and Jim Smothers, who is president of the Southern Colorad Solar Energy Institute. They answered our questions about the feasibility of heating an aging building with solar. From these two experts, we learned a couple of new methods for active solar heating that we had not found to that point. Mr. Smothers gave us photographs and schematic drawings of heating systems used to heat college campus buildings in Japan.

4

Continued on following page

Figure 22.7:
(*continued.*)

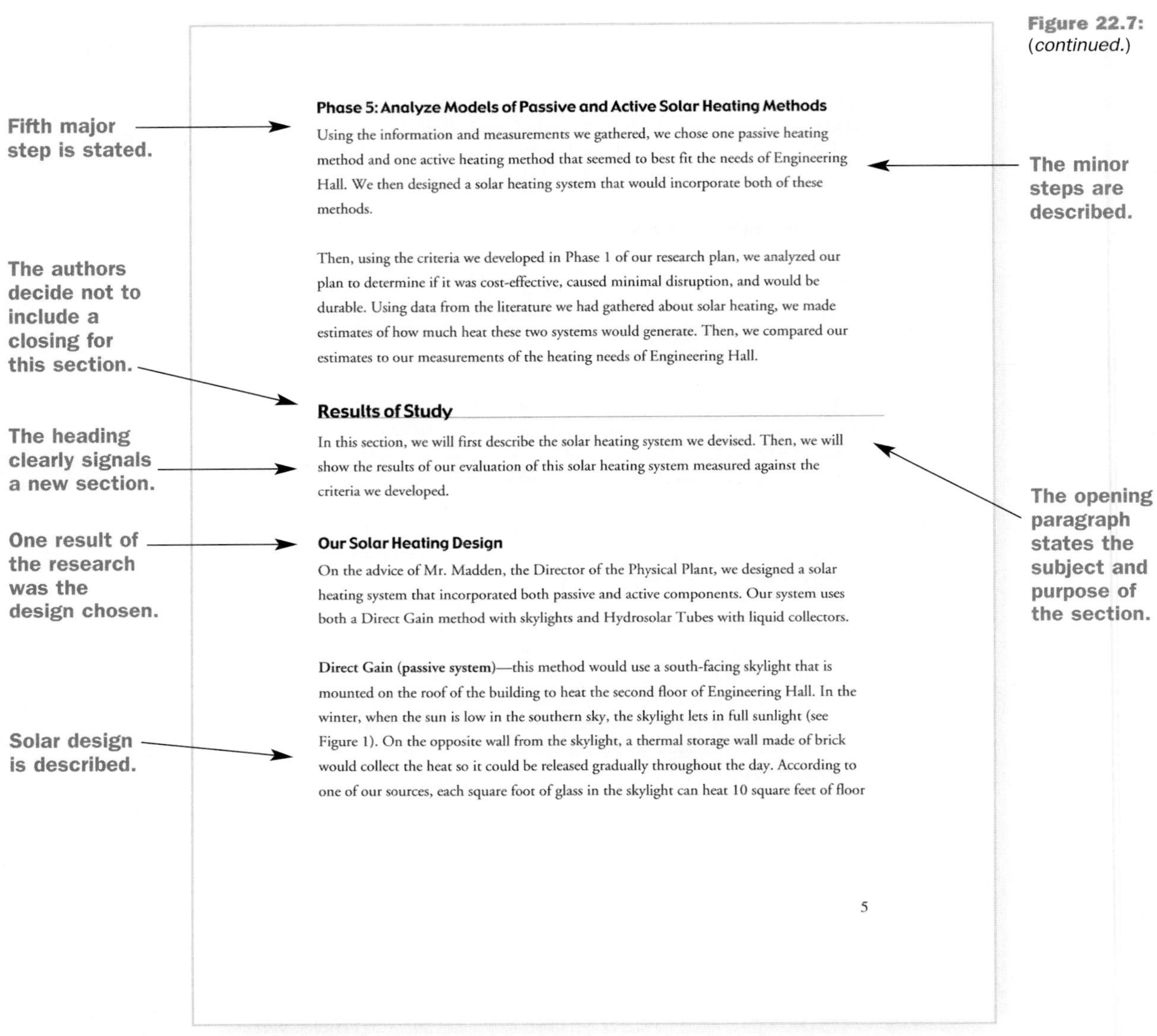

Fifth major step is stated.
The minor steps are described.
The authors decide not to include a closing for this section.
The heading clearly signals a new section.
The opening paragraph states the subject and purpose of the section.
One result of the research was the design chosen.
Solar design is described.

Phase 5: Analyze Models of Passive and Active Solar Heating Methods

Using the information and measurements we gathered, we chose one passive heating method and one active heating method that seemed to best fit the needs of Engineering Hall. We then designed a solar heating system that would incorporate both of these methods.

Then, using the criteria we developed in Phase 1 of our research plan, we analyzed our plan to determine if it was cost-effective, caused minimal disruption, and would be durable. Using data from the literature we had gathered about solar heating, we made estimates of how much heat these two systems would generate. Then, we compared our estimates to our measurements of the heating needs of Engineering Hall.

Results of Study

In this section, we will first describe the solar heating system we devised. Then, we will show the results of our evaluation of this solar heating system measured against the criteria we developed.

Our Solar Heating Design

On the advice of Mr. Madden, the Director of the Physical Plant, we designed a solar heating system that incorporated both passive and active components. Our system uses both a Direct Gain method with skylights and Hydrosolar Tubes with liquid collectors.

Direct Gain (passive system)—this method would use a south-facing skylight that is mounted on the roof of the building to heat the second floor of Engineering Hall. In the winter, when the sun is low in the southern sky, the skylight lets in full sunlight (see Figure 1). On the opposite wall from the skylight, a thermal storage wall made of brick would collect the heat so it could be released gradually throughout the day. According to one of our sources, each square foot of glass in the skylight can heat 10 square feet of floor

5

Figure 22.7: *(continued.)*

space (Solar Thermal Energy Group 2003). If so, we would need 200 square feet of glass to heat the 2000 sq. ft. on the second floor of Engineering Hall. In other words, we would need the equivalent of a 5 ft. by 40 ft. skylight across the roof of the building.

Graphics can help illustrate complex concepts.

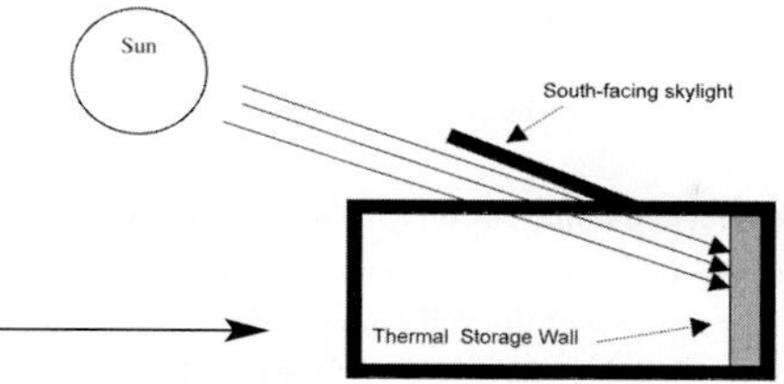

Figure 1: Direct Gain Method with Skylights and a Thermal Storage Wall

Hydrosolar Tubes (active system) —this method would place liquid solar collectors by the southern-facing base of the building to gather heat from the sun (Langa, 1981). The heated water would then be piped into hot water registers inside the first-floor rooms, with a pump that runs on solar electricity (Figure 2). The circulating water would heat each room through registers placed along the walls (Meeker & Boyd, 1983, Eklund, et al., 1979). We decided we would need six of these systems—one for each south-facing room in Engineering Hall. Each system can heat a room of 600 square feet on a sunny day, allowing us to heat the 3000 sq. ft. of space on the first floor of the building.

A graphic helps the readers visualize the project.

Figure 2: Hydrosolar Tubes with Liquid Collector and a Register

We also determined that the solar heating system would not be able to stand alone. A small backup electric heating system would need to be installed for the occasional cloudy days (which are rare here in south-central Colorado).

6

Continued on following page

Figure 22.7: *(continued.)*

Here begins a section that describes the specific results generated by the researchers.

The first result is described objectively.

Note how numbers are used to support the presentation of results.

The second result is presented with support.

Evaluation of Solar Heating Design

Our evaluation of this solar heating design against our criteria presented us with three results:

- Solar remodeling would pay for itself within 5-6 years.
- Conversion to solar heating would cause minimal disruption.
- The durability of the system would be a minimum of 10 years.

Result: Solar remodeling would pay for itself within 5–6 years

Our measurements show that a combination of passive and active systems like the ones we designed would pay for itself in 5-6 years. With help from Mr. Madden, we estimate that installing the skylights and adding a thermal wall would cost about $24,000. Installing the liquid collectors with tanks, pumps, and registers would cost $23,000. The backup electric heating system would cost about $5,000 and cost about $500 a year to run. The total cost of the system would be $57,000 to heat the building for ten years. This compares favorably with our estimate that the current heating system costs about $4,500 per year for a total of $45,000 over ten years. Renovating the oil heater, Ms Brin from the Physical Plant estimates, would cost around $20,000. Thus, converting to solar saves us $8000 over ten years.

Result: Conversion to solar heating would cause minimal disruption

Jane Karros, one of the solar consultants we interviewed, estimated that installation of a solar heating system would take approximately 3 months. If the work was begun late in the spring semester and completed by the end of the summer, the disruption to the operations of the building would be minimal. Certainly, the disruption would be no more than a typical renovation of a building's heating system.

7

Figure 22.7: (*continued.*)

The third result is presented.

Result: The durability of the system would be a minimum of 10 years

The literature we gathered estimates that the durability of this solar heating system would be a minimum of 10 years (Price 2000). The direct gain system, using the skylight and thermal storage wall, could be used indefinitely with minor upkeep. The hydrosolar tube system, using liquid solar collectors and a solar pump, would likely need to be renovated in 10 years. These systems have been known to go 20 years without major renovation.

Discussion of Results

Based on our research and calculations, we have reached three conclusions, which suggest that a solar heating system would be a viable option for Engineering Hall.

The discussion section starts out with an opening paragraph to redirect the discussion.

Solar heating would save money in the long run

Engineering Hall is one of the oldest, least efficient buildings on campus. Yet, our calculations show that putting in skylights and liquid collectors would be sufficient to heat the building on most days. According to our estimates, the solar remodeling would more than pay for itself in 10 years. Moreover, we would eliminate the need for the current oil heating system in the building, which will likely need to be replaced in that time period.

The results of the research are discussed.

Solar heating is a way toward independence

Our dependence on imported oil and natural gas puts our society at risk a few different ways (US DOE 2004). Engineering Hall's use of oil for heating pollutes our air, and it contributes to our nation's dependence on other countries for fuel. By switching over to solar heating now, we start the process of making ourselves energy independent. Engineering Hall's heating system will need to be replaced soon anyway. Right now would be a good time to think seriously about remodeling to use solar.

8

Continued on following page

Figure 22.7: (continued.)

Note how the authors are offering their opinions in this section.

Engineering Hall offers a good model for future projects

Engineering Hall is one of the older buildings on campus. Consequently, it presents some additional renovation challenges that we would not face with the newer buildings. We believe that Engineering Hall provides an excellent model for conversion to solar energy because it is probably one of the more difficult buildings to convert. If we can make the conversion with this building, we can almost certainly make the conversion with other buildings.

It might also be noteworthy that our discussions with the Mr. Madden and Ms. Brin at the Physical Plant showed us that there is great enthusiasm for making this kind of renovation to buildings like Engineering Hall.

The recommendations section is clearly signaled here.

Recommendations

In conclusion, we think the benefits of remodeling Engineering Hall to use solar heating clearly outweigh the costs. We recommend the following actions:

- Complete a more thorough study of the heating needs of Engineering Hall.
- Hire a solar architect to draw up a plan for renovating the building.
- Solicit bids from contractors.
- Conduct a formal cost-benefits analysis to see if solar heating is financially feasible for Engineering Hall.

The recommendations are put in bullet form to make them easy to read.

Here is the main point and a look to the future.

If all goes well, remodeling of Engineering Hall could be completed in the summer of 2006. When the remodeling is complete, the college should begin saving money within ten years. Moreover, our campus would house a model building that could be studied to determine whether solar heating is a possibility for other buildings on campus.

9

Figure 22.7: (*continued.*)

Offering contact information is a good way to end the report.

Thank you for your time and consideration. After you have looked over this report, we would like to meet with you to discuss our findings and recommendations. Please call Dan Garnish at 555-9294.

Appendix: Works Cited

The materials cited in the report are listed here in APA format.

Eklund, K. (1979). *The solar water heater workshop manual* (2nd ed.). Seattle, WA: Ecotape Group.

Langa, F. (1981). *Integral passive solar water heating book.* Davis, CA: Passive Solar Institute.

Meeker, J. & Boyd, L. (1983). Domestic hot water installations: The great, the good, and the unacceptable. *Solar Age 6*, 28-36.

Price, G. (2000). *Solar remodeling in southern New Mexico.* Las Cruces, NM: NMSU Energy Institute.

Solar Thermal Energy Group (2003). *Solar home and solar collector plans.* Retrieved March 30 from the World Wide Web: http://www.jc-solarhomes.com.

US Department of Energy (2004). Residential solar heating retrofits. Retrieved March 30 from the World Wide Web: http://www.eere.energy.gov/consumerinfo/factsheets/ac6.html.

10

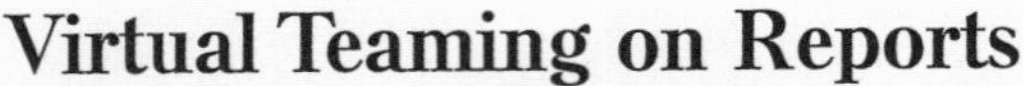

HELP

Virtual Teaming on Reports

Reports tend to be written in teams, because they often represent the research or work of a team of people. In technical workplaces, people rarely have time to meet in person to work on a report. So they collaborate virtually through their computers. Here are some strategies for improving your work in virtual teams.

Plan virtual meeting times—E-mail is useful for staying in touch, but you might also use instant messaging or phone conferencing to talk about the progress of the project. To hold a virtual meeting, set regular times when everyone can participate in an instant messaging conversation or phone conference.

LINK For more information on using e-mail and instant messaging, see Chapter 12, page 330.

Assign responsibilities to members of the team—One reason why team projects fail is because people are not sure what they are supposed to be doing. Some responsibilities could include *document coordinator, researchers, document designer,* and *editor.* Everyone on the team should help draft the document, but people with specific responsibilities will lead the project at various times.

LINK For more advice about working in teams, see Chapter 13, page 352.

Set a project schedule—Reports need to be written in stages, so setting milestones toward completion is a good way to keep the team on task. To establish a project schedule, work backward from the due date. Chart when each part of the report should be completed. As you are writing the report, these dates on the schedule will become milestones to signal whether you are on schedule. Groupware programs like Microsoft's Project, Corel's iGrafx Process, and AEC Software's Fasttrack Schedule can help you plan whole projects electronically (Figure A).

Don't let slackers ruin your efforts—There are people out there who will let you do all the work. You will find them in school and in the workplace. The secret is to not let them sabotage your project. When your team starts the project, talk among yourselves about how to handle slackers. Often, if the team asks a slacker to get to work, he or she will.

In some cases, you may need to call in a supervisor or instructor to mediate a slacker situation. If all else fails, write a memo or e-mail to your supervisor or instructor to make him or her aware of the situation. Your instructor may suggest you "fire" your slacker team member.

Be honest—Virtual teaming relies on the honesty of team members, because checking up on each other is difficult. As you form your team, stress the importance of being honest about the progress of work. In other words, agree among yourselves not to tell team members something is finished if it isn't. If anyone is having trouble completing his or her part of the project, that person should be honest and tell the others. That way, someone else can help out. Honesty ensures that one person won't let the others down.

Want to know more about virtual teaming? Go to
www.ablongman.com/johnsonweb/22.13

A Groupware Calendar

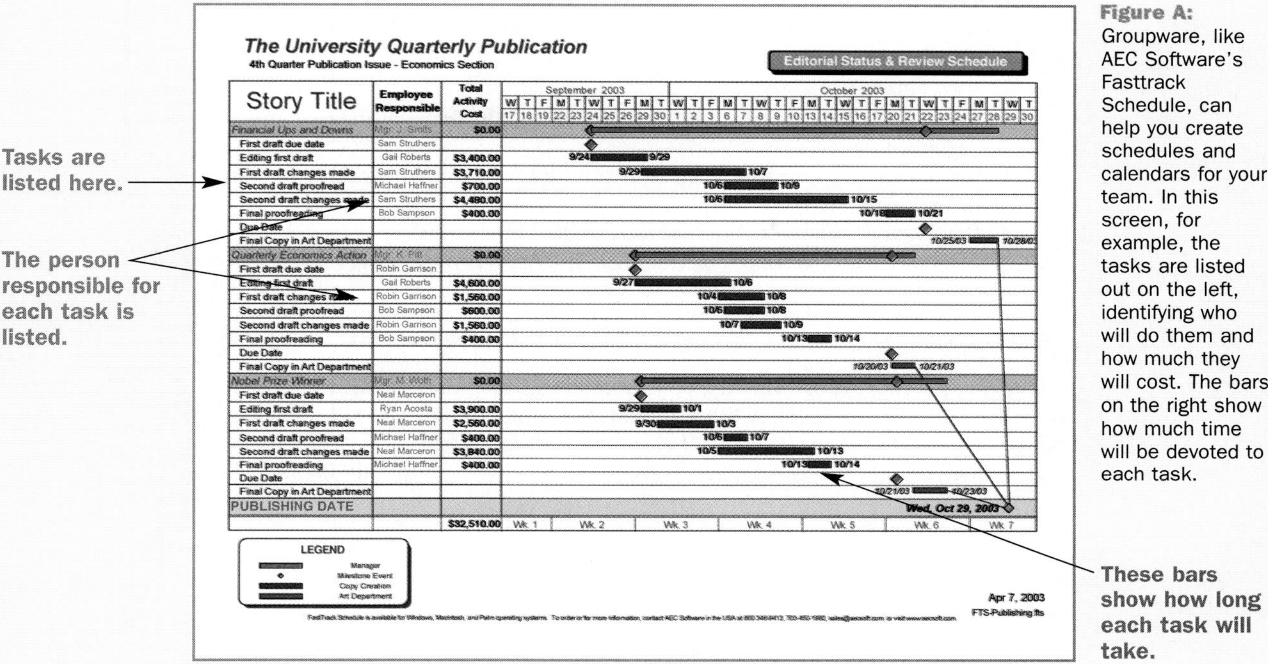

Figure A: Groupware, like AEC Software's Fasttrack Schedule, can help you create schedules and calendars for your team. In this screen, for example, the tasks are listed out on the left, identifying who will do them and how much they will cost. The bars on the right show how much time will be devoted to each task.

Source: AEC Software, 2003.

Drafting Front Matter and Back Matter

Most reports also include front matter and back matter. Front matter includes the letter of transmittal, cover page, table of contents, and other items that are placed before the first page of the main report. Back matter includes appendixes, glossaries, and indexes that are placed after the main report.

Developing Front Matter

Front matter may include some or all of the following items:

LETTER OR MEMO OF TRANSMITTAL Typically, reports are accompanied by a letter or memo of transmittal. A well-written letter or memo gives you an opportunity to make positive personal contact with your readers before they look through your report.

LINK For more information on writing letters and memos of transmittal, see Chapter 16, page 459.

To learn more about groupware, go to **www.ablongman.com/johnsonweb/22.14**

TITLE PAGE Title pages are an increasingly common feature in analytical reports. A well-designed title page sets a professional tone, while introducing readers to the subject of the report. The title page should include all or some of the following features:

- a specific title for the report.
- the names of primary readers, their titles, and the name of their company or organization.
- the names of the writers, their titles, and the name of the writers' company or organization.
- the date on which the report was submitted.
- company logos, graphics, or rules (lines) to enhance the design.

TAKE NOTE Before computers, it was common to center all the items on the title page. With computers, you can move items around on the cover to create a more dynamic look.

The title for your report should give your readers a clear idea about what the report contains (Figure 22.8). A title like "Solar Heating" probably is not specific enough. A more descriptive title like "Learning with Light: Using Solar Heating in Engineering Hall" gives readers a solid idea about what the report will discuss.

LINK For graphic design strategies, see Chapter 8, page 192.

Designing the Title Page

Learning with Light:

Using Solar Heating in Engineering Hall

A Report from the Innovation Group

October 25, 2004

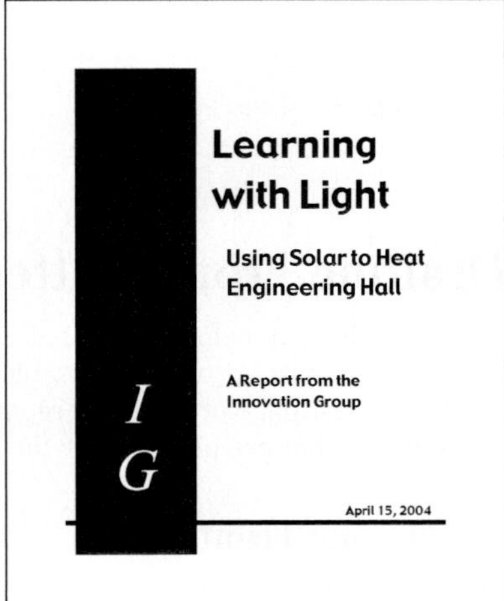

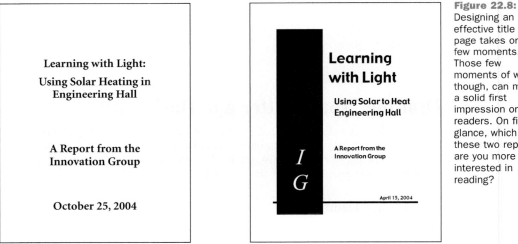

Figure 22.8: Designing an effective title page takes only a few moments. Those few moments of work, though, can make a solid first impression on the readers. On first glance, which of these two reports are you more interested in reading?

ABSTRACT OR EXECUTIVE SUMMARY Your readers usually won't have time to read all the information in your report. So, if your report is longer than 10 pages, you should consider including an abstract or executive summary.

An abstract is a summary of the report that uses the phrasing in the report and follows its organizational structure. When writing an abstract, you should draw key sentences directly from the report itself. Start out with the purpose statement of the

Want to see some sample texts that use good and bad design? Go to **www.ablongman.com/johnsonweb/22.15**

report, and then state the main point. From there, draw one or two key sentences from each major section. In an abstract for a report, for example, you would probably include these items in the following order:

- Purpose statement (one sentence)
- Main point (one sentence)
- Methodology (one or two sentences)
- Results (one or two sentences)
- Discussion (one or two sentences)
- Recommendations (one or two sentences)

You should modify the sentences in places to make the abstract readable, but try to retain the phrasing of the original report as much as possible.

Purpose of report leads the abstract.

The main point comes second.

The remainder of the abstract mirrors the structure of the report (methodology, results, discussion, recommendations).

Abstract

The purpose of this report is to determine whether Engineering Hall can be converted to a solar heating system with reasonable cost. After analyzing solar heating options, we argue in this report that the best way to add solar heating to Engineering Hall would include a combination of direct gain (skylights) and hydrosolar tubes. Our research plan included five phases: (1) develop evaluation criteria, (2) gather information on solar heating, (3) analyze the heating needs of Engineering Hall, (4) interview physical plant personnel and solar energy experts, and (5) analyze models of passive and active solar heating methods. From our research, we designed a solar heating system that incorporates both passive and active components. Our system uses both a direct gain method with skylights and hydrosolar tubes with liquid collectors. Our evaluation of this solar heating design against our evaluation criteria presented us with three results: Solar remodeling would pay for itself within 5–6 years; conversion to solar heating would cause minimal disruption; the durability of the system would be a minimum of 10 years. We have reached three conclusions that we believe the university administration should seriously consider: Solar heating would save money in the long run; solar heating is a way toward independence; and Engineering Hall offers a good model for future solar projects. We recommend the following actions: complete a more thorough study of the thermodynamic features of Engineering Hall; hire a solar architect to draw up a plan for renovating the building; solicit bids from contractors; and conduct a cost-benefits analysis to see if solar heating is financially feasible.

An executive summary is a concise, *paraphrased* version of your report (usually one page) that highlights the key points in the text. The two main differences between an abstract and an executive summary are that (a) the summary does not follow the organization of the report, and (b) the summary does not use the exact phrasing of the report. In other words, a summary paraphrases the report and organizes the information to highlight the key points.

The purpose of the report is placed early in the summary.

The main point is also placed up front.

Report Summary

This report was written in response to a challenge to our Energy Dynamics class (Engineering 387) from Dr. Sharon Holton, President of Kellen College. She asked us to develop options for converting campus buildings to renewable energy sources. In this report, we discuss the possibility of converting Engineering Hall to a solar heating system. We conclude that heating Engineering Hall with solar sources would require a combination of direct gain (skylights) and hydrosolar tubes. The combination of these two solar technologies would ensure adequate heating for almost all the building's heating needs. A backup heater could be retained for sustained cold spells.

To see sample abstracts and summaries, go to **www.ablongman.com/johnsonweb/22.16**

The remainder of the summary organizes information in order of importance.

To develop the information for this report, we followed a five step research plan: (1) develop an evaluation criteria, (2) gather information on solar heating, (3) analyze the heating needs of Engineering Hall, (4) interview physical plant personnel and solar energy experts, and (5) analyze models of passive and active solar heating that would be appropriate for this building.

The results of our research are mostly anecdotal, but they show that solar heating is possible, even for an older building on campus. We believe that our results show that Engineering Hall can be a model for developing solar heating systems around campus, because it is truly one of the more difficult buildings at Kellen to convert to solar heating. Newer buildings on campus would almost certainly be easier to convert. We conclude by pointing out that solar heating would save money in the long run. In the case of Engineering Hall, solar remodeling would pay for itself in 5–6 years.

We appreciate your time reading this report. If you have any questions or would like to meet with us, please call Dan Garnish at 555-9294.

TAKE NOTE The executive summary will often duplicate the contents of the introduction, but it should not replace the introduction. Instead, it should be written so that it can stand alone, apart from the rest of the report.

TABLE OF CONTENTS If your report runs over 10 pages, you should consider adding a table of contents. A table of contents is helpful to readers in two ways. First, it helps them quickly access the information they need in the report. Second, it offers an overall outline of the contents of the report. Since reports tend to be larger documents, your readers will appreciate a quick summary of the report's contents.

In the table of contents, the headings should be the same as the ones used in your report. Then, use tabs or leader tabs to line up the page numbers on the right side. Leader tabs are used to insert a line of dots or dashes from the heading to the page number.

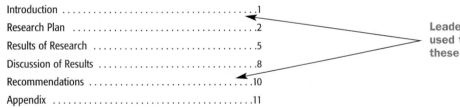

Table of Contents

Introduction 1

Research Plan 2

Results of Research 5

Discussion of Results 8

Recommendations 10

Appendix 11

Leader tabs are used to put in these dots.

Developing Back Matter

Back matter in a report might include appendixes, a glossary of terms, and calculations. Keep in mind, though, that most readers will never look at the back matter. So, if you have something important to say, do not say it here.

APPENDIXES Appendixes are storage areas for information that may or may not be useful. For example, additional data tables or charts not needed in the body of the report might be displayed in an appendix. You might include clippings from newspapers or magazines.

TAKE NOTE Do not put all your graphics in the appendix, because readers will not look for them. If you want readers to actually look at your charts, tables, and other graphics, put them in the body of the report.

GLOSSARY OF TERMS Depending on the complexity of your report and the familiarity of your readers with the subject, you may want to include a short glossary of terms. When creating a glossary, look back over your report and highlight words that may not be familiar to your nonexpert readers. Then, in a glossary at the back of the report, list these terms and write sentence definitions for each.

LINK For more information on writing sentence definitions, see Chapter 17, page 493.

CALCULATIONS In highly technical reports, you may want to include your calculations in the back matter. Here is where you can demonstrate how you arrived at your figures in the report.

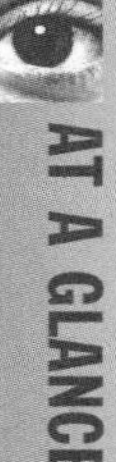

Front Matter and Back Matter

Front Matter—items that appear before the main report

- Letter or Memo of Transmittal
- Title Page
- Abstract or Executive Summary
- Table of Contents

Back Matter—items that appear after the main report

- Appendixes
- Glossary of Terms
- Calculations

Using Style in Analytical Reports

Reports have a reputation for being difficult to read. They are often crammed full of facts, data, information, and jargon that are meaningful to experts but not clear to nonexperts.

For this reason, the most readable reports are written in the plain style, with limited use of persuasion. Reports are persuasive in an unstated way, putting the emphasis on the soundness of the methodology, the integrity of the results, and the reasonableness of the discussion. Plain style will make your report sound straightforward and clear to the readers.

LINK For more help on using the plain style, see Chapter 7, page 187.

While revising your report, you might pay close attention to the following plain style techniques:

Make "doers" the subjects of sentences—Reports tend to overuse the passive voice, making them harder to read than necessary. If the passive voice is required in your field, use it. But if you want your writing to be more effective, make your sentences active by moving the "doers" into the subjects of the sentences.

Passive: Saplings had been eaten during the winter by the deer we monitored, because they were desperate for food.

Active: The deer we monitored ate saplings to survive the winter because they were desperate for food.

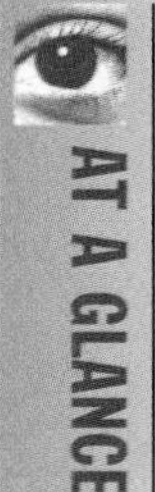

AT A GLANCE

Improving Style in Analytical Reports

- Make "doers" the subjects of sentences.
- Use breathing-length sentences.
- Eliminate nominalizations.
- Define jargon.

Use breathing-length sentences—Reports are notorious for using sentences that are far too long to be understood. As you write and revise the report, look for places where your sentences are too long to be stated in one breath. These sentences should be shortened to breathing length or divided into two breathing-length sentences.

Eliminate nominalizations—Reports are often littered with nominalizations that make the meaning of sentences cloudy. You can revise these sentences for clarity:

Nominalization: This report offers <u>a presentation</u> of our findings.

Revised: This report <u>presents</u> our findings.

Nominalization: We made <u>a decision</u> to initiate <u>a replacement</u> of the Collings CAD software.

Revised: We decided to replace the Collings CAD software.

Revised Further: We replaced the Collings CAD software.

Nominalizations may make the report sound more formal, but your readers will appreciate their elimination for the sake of clarity.

Define jargon and specialized terms—Jargon should not be completely eliminated in reports. Instead, these words should be defined for nonexperts. When you need to use a specialized term, use a sentence definition or parenthetical definition to clarify what the word means.

Sentence Definition: A gyrocompass is a directional finding device that uses a gyroscope to compensate for the earth's rotation and thus points to true north.

Parenthetical Definition: Spotting a blue grouse, <u>a plump, medium-sized bird with feathered legs and bluish gray plumage,</u> is especially difficult because they are well camouflaged and live in higher mountain areas.

LINK For more information on writing sentence and parenthetical definitions, see Chapter 17, pages 492–493.

Want to learn about passive voice? Go to **www.ablongman.com/johnsonweb/22.17**
Need a definition for a jargon term? Go to **www.ablongman.com/johnsonweb/22.18**

Designing Analytical Reports

People rarely read reports word for word. Instead, they scan these documents. Therefore, you should use document design and graphics to highlight your main points and offer readers access points to start reading.

Choosing a Document Design

Your report's design should reflect the subject of the report and the preferences of your readers. An exciting subject or progressive readers might allow you to experiment with page layout, design of headings, and use of graphics. You might look for ways to use multicolumn formats, which allow you to use pullout quotes, sidebars, and other page layout enhancements (Figure 22.9). A report written to more conservative readers might call for a more traditional one-column format with limited graphics.

LINK For more help on document design, see Chapter 8, page 192.

Page Layouts for Reports

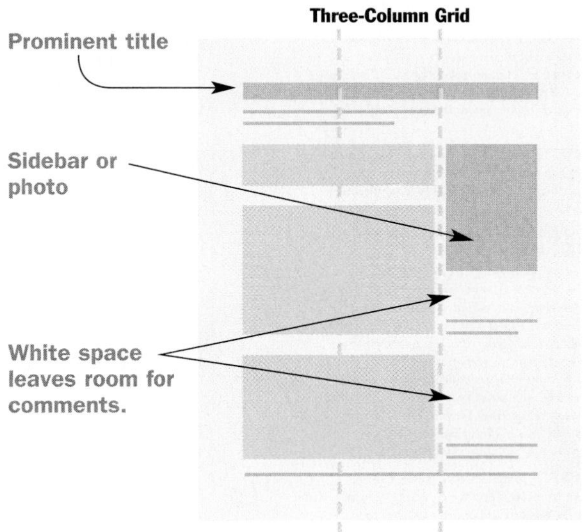

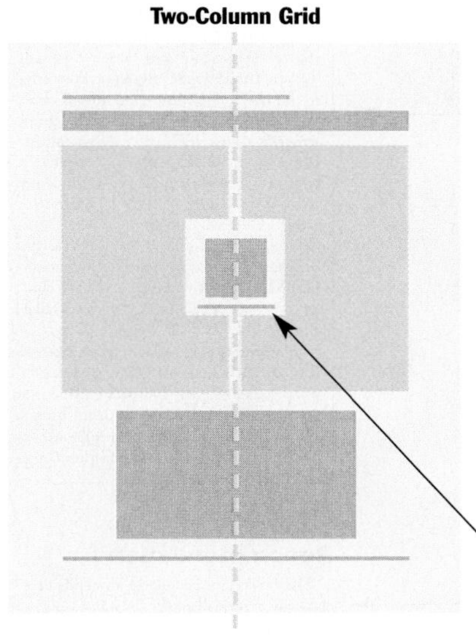

Figure 22.9: Reports don't need to look boring. A little attention to design will make your report inviting and easier to read. In these page layouts, grids have been used to balance the texts, leaving room for margin text and other access points.

Whatever you do, make sure you choose a document design that fits your subject, readers, and the document's context of use. Often, as a project comes to a close, writers do not put enough time into the design of their report. As a result, the report makes a bad first impression on readers. With a little work, you can create a design that makes a positive first impression, while making the information readers need easier to find.

Want more tips about designing page layouts? Go to **www.ablongman.com/johnsonweb/22.19**

Figure 22.10, for example, shows a well-designed report. The writers have designed this document to make it highly scannable. In reports, you should also strategically use lists to make the text more scannable. Look for places in your report where sentences can be turned into bulleted or numbered lists.

LINK For more information on using lists, see Chapter 8, page 215.

Designing a Report

The introduction also serves as a summary of the report.

Recommendations are made up front in this report so they are easy to locate.

Counseling to Prevent Tobacco Use and Tobacco-Caused Disease

Recommendation Statement

U.S. Preventive Services Task Force

This statement summarizes the U.S. Preventive Services Task Force (USPSTF) recommendations on counseling to prevent tobacco use and tobacco-caused disease. It updates the 1996 recommendations contained in the *Guide to Clinical Preventive Services,* second edition. Explanations of the ratings and strength of overall evidence are given in Appendix A and Appendix B, respectively. The complete information on which this statement is based, including evidence tables and references, is available in the Public Health Service (PHS) "Clinical Practice Guideline: Treating Tobacco Use and Dependence." The USPSTF recommendation and PHS clinical practice guideline on this topic are available through the USPSTF Web site (http://www.preventiveservices.ahrq.gov) and through the National Guideline Clearinghouse™ (http://www.guideline.gov). This recommendation statement is also available in print through the AHRQ Publications Clearinghouse (call 1-800-358-9295 or e-mail ahrqpubs@ahrq.gov).

Summary of Recommendations

The USPSTF strongly recommends that clinicians screen all adults for tobacco use and provide tobacco cessation interventions for those who use tobacco products. **A recommendation.**

The USPSTF found good evidence that brief smoking cessation interventions, including screening, brief behavioral counseling (less than 3 minutes), and pharmacotherapy delivered in primary care settings, are effective in increasing the proportion of smokers who successfully quit smoking and remain abstinent after 1 year. Although most smoking cessation trials do not provide direct evidence of health benefits, the USPSTF found good evidence that smoking cessation lowers the risk for heart disease, stroke, and lung disease. The USPSTF concluded that there is good indirect evidence that even small increases in the quit rates from tobacco cessation counseling would produce important health benefits, and that the benefits of counseling interventions substantially outweigh any potential harms.

The USPSTF strongly recommends that clinicians screen all pregnant women for tobacco use and provide augmented pregnancy-tailored counseling to those who smoke. **A recommendation.**

The USPSTF found good evidence that extended or augmented smoking cessation counseling (5–15 minutes) using messages and self-help materials tailored for pregnant smokers, compared with brief generic counseling interventions alone, substantially increases abstinence rates during pregnancy, and leads to increased birth weights. Although relapse rates are high in the post-partum period, the USPSTF concluded that reducing smoking during pregnancy is likely to have substantial health benefits for both the baby and the expectant mother. The USPSTF concluded that the benefits of smoking cessation counseling outweigh any potential harms.

The USPSTF concludes that the evidence is insufficient to recommend for or against routine screening for tobacco use or interventions to prevent and treat tobacco use and dependence among children or adolescents. I **recommendation.**

The USPSTF found limited evidence that screening and counseling children and adolescents in the primary

Corresponding Author: Alfred O. Berg, MD, MPH, Chair, U.S. Preventive Services Task Force, c/o Project Director, USPSTF, Agency for Healthcare Research and Quality, 540 Gaither Road, Rockville, MD 20850, e-mail: uspstf@ahrq.gov.

1

Source: U.S. Preventive Services Task Force, 2003.

Figure 22.10: This recommendation report puts its recommendations up front where readers can easily find them. Also, notice how the report uses such visual design features as a double-column format, headings, and lists to make the information easy to scan.

Figure 22.10: *(continued.)*

The two-column format makes the report easy to scan.

These bold headings help readers find the information they need.

Here are the results of the study.

care setting are effective in either preventing initiation or promoting cessation of tobacco use. As a result, the USPSTF could not determine the balance of benefits and harms of tobacco prevention or cessation interventions in the clinical setting for children or adolescents.

Clinical Considerations

- Brief tobacco cessation counseling interventions, including screening, brief counseling (3 minutes or less), and/or pharmacotherapy, have proven to increase tobacco abstinence rates, although there is a dose-response relationship between quit rates and the intensity of counseling. Effective interventions may be delivered by a variety of primary care clinicians.
- The "5-A" behavioral counseling framework provides a useful strategy for engaging patients in smoking cessation discussions: (1) *Ask* about tobacco use; (2) *Advise* to quit through clear personalized messages; (3) *Assess* willingness to quit; (4) *Assist* to quit; and (5) *Arrange* follow-up and support. Helpful aspects of counseling include providing problem-solving guidance for smokers to develop a plan to quit and to overcome common barriers to quitting and providing social support within and outside of treatment. Common practices that complement this framework include motivational interviewing, the 5 R's used to treat tobacco use (*relevance, risks, rewards, roadblocks, repetition*), assessing readiness to change, and more intensive counseling and/or referrals for quitters needing extra help.[1–3] Telephone "quit lines" have also been found to be an effective adjunct to counseling or medical therapy.[4]
- Clinics that implement screening systems designed to regularly identify and document a patient's tobacco use status increased their rates of clinician intervention, although there is limited evidence for the impact of screening systems on tobacco cessation rates.[5]
- FDA-approved pharmacotherapy that has been identified as safe and effective for treating tobacco dependence includes several forms of nicotine replacement therapy (ie, nicotine gum, nicotine transdermal patches, nicotine inhaler, and nicotine nasal spray) and sustained-release bupropion. Other medications, including clonidine and nortriptyline, have been found to be efficacious and may be considered.
- Augmented pregnancy-tailored counseling (eg, 5–15 minutes) and self-help materials are recommended for pregnant smokers, as brief interventions are less effective in this population. There is limited evidence to evaluate the safety or efficacy of pharmacotherapy during pregnancy. Tobacco cessation at any point during pregnancy can yield important health benefits for the mother and the baby, but there are limited data about the optimal timing or frequency of counseling interventions during pregnancy.
- There is little evidence addressing the effectiveness of screening and counseling children or adolescents to prevent the initiation of tobacco use and to promote its cessation in a primary care setting, but clinicians may use their discretion in conducting tobacco-related discussions with this population, since the majority of adult smokers begin tobacco use as children or adolescents.

Discussion

Tobacco use, cigarette smoking in particular, is the leading preventable cause of death in the United States, resulting in 440,000 deaths annually.[6] Smoking has been attributed to over 155,000 deaths annually from neoplasms, 80,000 deaths annually from ischemic heart disease, and over 17,000 deaths annually from cerebrovascular disease. Smoking also affects health outcomes of people other than the smokers, with smoking during pregnancy resulting in the deaths of about 1,000 infants annually. Significant risks associated with smoking during pregnancy include premature births, spontaneous abortions, stillbirths, and intrauterine growth retardation. Additionally, environmental tobacco smoke contributes to the deaths of an estimated 38,000 people annually from lung cancer and heart disease.[7]

There is good quality evidence that smoking cessation lowers the risk for heart disease, stroke, and lung disease.[8,9] However, despite smoking's established risks and the health benefits of quitting, 23% of adults in the United States continue to

2

Continued on following page

A header is used as a consistent feature in the report.

Figure 22.10: (*continued.*)

smoke[10] and more than 2,000 adolescents become regular tobacco users daily.[11] Nearly 90% of smokers start by age 18, and 25% of teen smokers remain addicted as adults.[12] Because 70% of smokers see a physician each year,[5] clinicians have a unique opportunity to intervene.

The discussion interprets the results of the study.

The USPSTF found good quality evidence examining the efficacy of various levels of intensity of tobacco cessation counseling by clinicians based on a meta-analysis of 43 studies.[5] Compared with no intervention, minimal counseling, lasting less than 3 minutes, has been shown to increase overall tobacco abstinence rates. Increasing session length and frequency increased efficacy in a dose-response manner.[5] There is limited evidence to determine the optimal duration and periodicity of tobacco counseling interventions.

A meta-analysis of 7 studies found that abstinence rates were higher (16.8% vs 6.6%) for pregnant smokers receiving pregnancy-tailored counseling and self-help materials compared with pregnant smokers receiving brief counseling or "usual care."[5]

The USPSTF found limited evidence of the efficacy of counseling children or adolescents in the clinical primary care setting, but found that school- and classroom-based smoking cessation programs may be more effective than no intervention among tobacco users who attend these programs.[13] As with tobacco cessation programs for adults in the community setting, programs with a greater number of counseling sessions and increasing intensity of follow-up had higher quit rates.[14]

Several FDA-approved pharmacotherapies have been identified as safe and effective in helping adults to quit smoking. Nicotine products, including nicotine gum, transdermal patch, nicotine nasal spray, and nicotine inhaler, have all been studied in comparison with placebo. There are good quality studies to support the abstinence rates among people who use these products compared with those who do not: 18% to 31% versus 10% to 17%.[5] (Although nicotine lozenges are currently available, at the time of this review they were not FDA-approved and therefore not included in this recommendation statement.) There are fair quality studies showing that combining the nicotine patch with either gum or nasal spray is more efficacious than using a single form of nicotine replacement therapy alone.[15,16] Sustained-release bupropion has been shown to be efficacious compared with placebo, with an estimated cessation rate of 23% to 38% compared with 17%.[17,18] Other pharmacotherapies, including clonidine and nortriptyline, have been shown to result in higher smoking cessation rates when compared with placebo, although their use may be limited by side effects.[5] There is little evidence on the safety and efficacy of tobacco cessation pharmacotherapy for the pregnant woman, the fetus, or the nursing mother and child.[19] Therefore, pharmacotherapy for pregnant women may be considered when the likelihood of quitting and its potential benefits outweighs the risks of the therapy and continued smoking. Likewise, there is little evidence on the safety and efficacy of tobacco cessation pharmacotherapy in children or adolescents.

There is good evidence supporting the effectiveness of community and population-based approaches to reducing environmental tobacco smoke (eg, tobacco price increases, clean indoor air laws, anti-tobacco media campaigns); many of these interventions are especially effective among adolescent populations.[4] However, in the clinical setting, limited studies with mixed results address the effect of parental counseling on reducing secondhand smoke exposure of children and reducing parental smoking rates.[20-22] Future research is required to define the effectiveness of screening, counseling, and pharmacotherapy for children, adolescents, and their parents in the primary care setting.

Notice how the authors are making an argument here, based on the facts.

Recommendations of Others

The Public Health Service guideline can be accessed at: http://www.surgeongeneral.gov/tobacco/smokesum.htm. Tobacco-related recommendations from the CDC *Guide to Community Preventive Services* can be accessed at: http://www.thecommunityguide.org/tobacco.

Policies of the American Academy of Family Physicians can be accessed at: http://www.aafp.org/x7112.xml. Recommendations of the American Academy of Pediatrics relating to tobacco can be accessed at: http://www.aap.org/policy/re0041.html, http://www.aap.org/policy/re9716.html, and http://www.aap.org/policy/re9801.html.

The report ends with more recommendations from other sources.

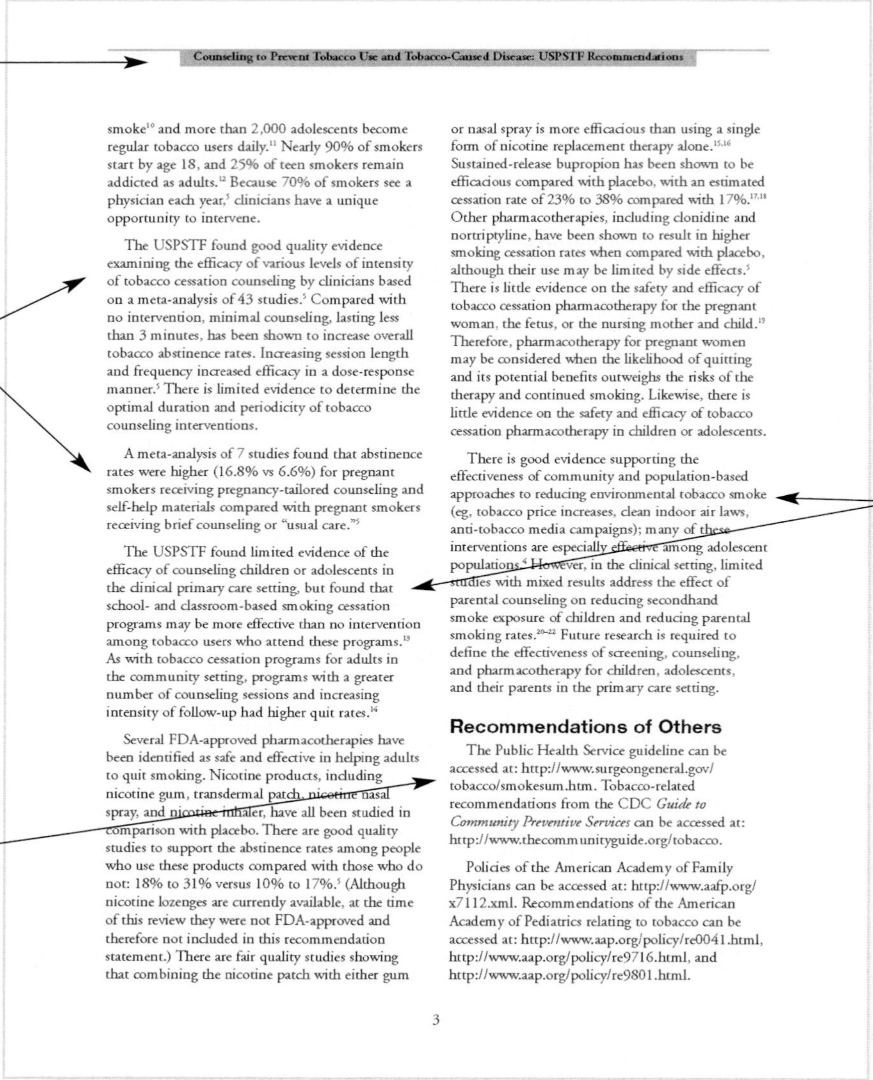

3

Figure 22.10: *(continued.)*

Recommendations of the Canadian Task Force can be accessed at: http://www.ctfphc.org.

References

In this report, the authors have chosen to list references sequentially (by number) instead of alphabetically.

1. Miller W, Rolnick S. Motivational interviewing: preparing people to change addictive behavior. New York: Guilford, 1991.
2. Anderson JE, Jorenby DE, Scott WJ, Fiore MC. Treating tobacco use and dependence: an evidence-based clinical practice guideline for tobacco cessation. *Chest.* 2002;121(3):932–941.
3. Prochaska JO, Velicer WF. The transtheoretical model of health behavior change. *Am J Health Promot.* 1997;12(1):38–48.
4. CDC. Strategies for reducing exposure to environmental tobacco smoke, tobacco-use cessation, and reducing initiation in communities and health-care systems. A report on recommendations of the Task Force on Community Preventive Services. *MMWR.* 2000;49(No. RR-12);1–11.
5. Fiore MC, Bailey WC, Cohen SJ, et al. Treating tobacco use and dependence. Rockville MD: Department of Health and Human Services, Public Health Service, 2000.
6. CDC. Annual smoking-attributable mortality, years of potential life lost, and economic costs—United States, 1995–1999. *MMWR.* 2002;51:300–303.
7. National Cancer Institute. Health effects of exposure to environmental tobacco smoke: the report of the California Environmental Protection Agency. Smoking and tobacco control monograph 10. Bethesda, Maryland: U.S. Department of Health and Human Services, National Institutes of Health, National Cancer Institute, 1999.
8. Department of Health and Human Services. The health benefits of smoking cessation: a report of the Surgeon General. Rockville, MD: Department of Health and Human Services, 1990.
9. U.S. Preventive Services Task Force. *Guide to Clinical Preventive Services.* 2nd ed. Washington, DC: Office of Disease Prevention and Health Promotion; 1996.
10. CDC. Prevalence of Current Cigarette Smoking Among Adults and Changes in Prevalence of Current and Some Day Smoking—United States, 1996–2001. *MMWR.* 2003;52:303–307.
11. Substance Abuse and Mental Health Services Administration. Office of Applied Studies, National Household Survey on Drug Abuse, 1999–2001.
12. U.S. Department of Health and Human Services. Preventing tobacco use among young people. A report to the Surgeon General. Atlanta, GA: U.S. Department of Health and Human Services, Public Health Service, CDC, NCCDPH, Office on Smoking and Health. 1994.
13. Sussman S, Lichtman K, Ritt A, Pallonen UE. Effects of thirty-four adolescent tobacco use cessation and prevention trials on regular users of tobacco products. *Subst Use Misuse.* 1999;34(11):1469–1503.
14. Stanton WR, Smith KM. A Critique of Evaluated Adolescent Smoking Cessation Programs. *Adolesc.* 2002;25:427–438.
15. Fagerstrom KO. Combined use of nicotine replacement products. *Health Values.* 1994;18(3):15–20.
16. Fagerstrom KO. Effectiveness of nicotine patch and nicotine gum as individual versus combined treatments for tobacco withdrawal symptoms. *Psychopharmacology.* 1993;111:271–277.
17. Hurt RD, Sachs DP, Glover ED, et al. A comparison of sustained-release bupropion and placebo for smoking cessation. *N Engl J Med.* 1997;337(17):1195–1202.
18. Jorenby DE, Leischow S, Nides M, et al. A controlled trial of sustained-release bupropion, a nicotine patch, or both for smoking cessation. *N Engl J Med.* 1999;340(9):685–691.
19. Wisborg K, Henriksen TB, Jespersen LB, Secher NJ. Nicotine patches for pregnant smokers: a randomized controlled trial. *Obstet Gynecol.* 2000;96(6):967–971.
20. Wall MA, Severson HH, Andrews JA, Lichtenstein E, Zoref L. Pediatric office-based smoking intervention: impact on maternal smoking and relapse. *Pediatrics.* 1995;96(4Pt1):622–628.
21. Groner JA, Ahijevych K, Grossman LK, Rich LN. The impact of a brief intervention on maternal smoking behavior. *Pediatrics.* 2000;105(1Pt3):267–271.
22. Severson HH, Andrews JA, Lichtenstein E, Wall M, Akers L. Reducing maternal smoking and relapse: long-term evaluation of a pediatric intervention. *Prev Med.* 1997;26(1):120–130.

Continued on following page

Figure 22.10: *(continued.)*

The appendixes are used to define concepts and ratings in the report.

Appendix A
U.S. Preventive Services Task Force—Recommendations and Ratings

The Task Force grades its recommendations according to one of 5 classifications (A, B, C, D, I) reflecting the strength of evidence and magnitude of net benefit (benefits minus harms):

A. The USPSTF strongly recommends that clinicians routinely provide [the service] to eligible patients. *The USPSTF found good evidence that [the service] improves important health outcomes and concludes that benefits substantially outweigh harms.*

B. The USPSTF recommends that clinicians routinely provide [the service] to eligible patients. *The USPSTF found at least fair evidence that [the service] improves important health outcomes and concludes that benefits outweigh harms.*

C. The USPSTF makes no recommendation for or against routine provision of [the service]. *The USPSTF found at least fair evidence that [the service] can improve health outcomes but concludes that the balance of benefits and harms is too close to justify a general recommendation.*

D. The USPSTF recommends against routinely providing [the service] to asymptomatic patients. *The USPSTF found at least fair evidence that [the service] is ineffective or that harms outweigh benefits.*

I. The USPSTF concludes that the evidence is insufficient to recommend for or against routinely providing [the service]. *Evidence that [the service] is effective is lacking, of poor quality, or conflicting and the balance of benefits and harms cannot be determined.*

Appendix B
U.S. Preventive Services Task Force—Strength of Overall Evidence

The USPSTF grades the quality of the overall evidence for a service on a 3-point scale (good, fair, poor):

Good: Evidence includes consistent results from well-designed, well-conducted studies in representative populations that directly assess effects on health outcomes.

Fair: Evidence is sufficient to determine effects on health outcomes, but the strength of the evidence is limited by the number, quality, or consistency of the individual studies, generalizability to routine practice, or indirect nature of the evidence on health outcomes.

Poor: Evidence is insufficient to assess the effects on health outcomes because of limited number or power of studies, important flaws in their design or conduct, gaps in the chain of evidence, or lack of information on important health outcomes.

Members of the U.S. Preventive Services Task Force*

Alfred O. Berg, MD, MPH, Chair, USPSTF (Professor and Chair, Department of Family Medicine, University of Washington, Seattle, WA)

Janet D. Allan, PhD, RN, CS, FAAN, Vice-chair, USPSTF (Dean, School of Nursing, University of Maryland Baltimore, Baltimore, MD)

Paul Frame, MD (Tri-County Family Medicine, Cohocton, NY, and Clinical Professor of Family Medicine, University of Rochester, Rochester, NY)

Charles J. Homer, MD, MPH (Executive Director, National Initiative for Children's Healthcare Quality, Boston, MA)

Mark S. Johnson, MD, MPH (Professor of Family Medicine, University of Medicine and Dentistry of New Jersey-New Jersey Medical School, Newark, NJ)

Jonathan D. Klein, MD, MPH (Associate Professor, Department of Pediatrics, University of Rochester School of Medicine, Rochester, NY)

Tracy A. Lieu, MD, MPH (Associate Professor, Department of Ambulatory Care and Prevention, Harvard Pilgrim Health Care and Harvard Medical School, Boston, MA)

C. Tracy Orleans, PhD (Senior Scientist and Senior Program Officer, The Robert Wood Johnson Foundation, Princeton, NJ)

Jeffrey F. Peipert, MD, MPH (Director of Research, Women and Infants' Hospital, Providence, RI)

Nola J. Pender, PhD, RN, FAAN (Professor Emeritus, University of Michigan, Ann Arbor, MI)

Albert L. Siu, MD, MSPH (Professor of Medicine, Chief of Division of General Internal Medicine, Mount Sinai School of Medicine, New York, NY)

Steven M. Teutsch, MD, MPH (Senior Director, Outcomes Research and Management, Merck & Company, Inc., West Point, PA)

Carolyn Westhoff, MD, MSc (Professor of Obstetrics and Gynecology and Professor of Public Health, Columbia University, New York, NY)

Steven H. Woolf, MD, MPH (Professor, Department of Family Practice and Department of Preventive and Community Medicine and Director of Research Department of Family Practice, Virginia Commonwealth University, Fairfax, VA).

*Members of the Task Force at the time this recommendation was finalized. For a list of current Task Force members, go to www.ahrq.gov/clinic/uspstfab.htm.

AHRQ Pub. No. 04-526
November 2003

Using Graphics to Clarify and Enhance Meaning

Increasingly, graphics are essential features of reports. Tables and graphs are especially helpful tools for displaying data. As you write the report, you should actively look for places where tables might be used to better display your data or information. Also, look for places where a graph could show trends in the data.

LINK For more information on making and using graphs, see Chapter 9, page 244.

In a report, graphics should *not* be used to embellish the text. Instead, make sure that each graphic you include is worth the space that it will take up. Graphics should be titled and numbered so they can be referred to in the text (e.g., "See Table 2"). Place each graphic on the page where it is referenced or very soon afterward.

Graphics have some other benefits. They tend to break up large blocks of text, making the report more readable. They also offer access points that stop readers from scanning by encouraging them to look into the text for an explanation of the graphic.

Revising, Editing, and Proofreading

Always leave at least a couple of days to revise and edit your report. Since these documents are usually rather large, it often takes time to edit them into their final form.

Revising the Content, Style, and Design

Revise the report from three different perspectives—content, style, and design:

Content—When revising for content, look for places where there is too much information or too little. Working paragraph by paragraph, ask yourself whether the information included is need-to-know information or need-to-tell information. Then, cut out all the information that does not contribute to the overall argument in the report. Also, recheck any calculations to make sure they are accurate.

Style—Since reports are often written in teams, the style can be uneven or inconsistent. Even reports that you write alone will often sound uneven, because your style may have evolved while you were drafting the text. To avoid these uneven spots in your writing, use the "plain style" strategies discussed in Chapter 7 to smooth out your writing. Then, use the "persuasive style" strategies to add color and emphasis to your writing where needed.

LINK For more information on improving style, see Chapter 7, page 162.

Design—Make sure the design is consistent across all the pages. Then, check whether your graphics are properly labeled and referenced in the text. Finally, make sure you have included page numbers on each sheet, except the cover.

For worksheets to help you revise and edit your report, go to **www.ablongman.com/johnsonweb/22.20**

Editing and Proofreading the Draft

You and your team should carefully proofread the report before submitting it.

If you're like most people, you are probably up against a deadline at this point. So, it might be tempting to simply skip the editing and proofreading phases of the writing process. Don't do it. After all, nothing sabotages a report faster than typos, spelling errors, and grammar errors. Your readers will certainly notice your sloppy work, and they will immediately assume that you were equally careless in your research.

LINK For more tips on editing and proofreading, see Chapter 10, page 264.

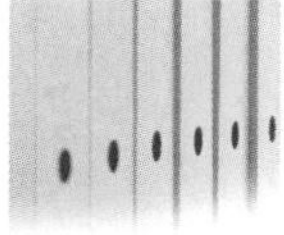

CHAPTER REVIEW

- Various types of analytical reports are used in the workplace, including research reports, completion reports, recommendation reports, feasibility reports, and empirical research reports.
- Use the IMRaD acronym to remember the generic report pattern: Introduction, Methods, Results, and Discussion.
- Determine the rhetorical situation for your report by considering the subject, purpose, readers, and context of use for your report.
- Define your research question, and formulate a hypothesis to be tested. Then, develop a methodology and identify the two to five major steps you will need to take to establish or refute your hypothesis.
- Collect information using library, Internet, and empirical sources.
- Organize and draft your report following the separate "moves" for an introduction, body, and conclusion.
- Both the style and the design of analytical reports should aim to clarify the contents of the report.
- It is vital to leave enough time to revise, edit, and proofread an analytical report to ensure professional quality.

Individual or Team Projects

1. On the Internet or elsewhere, find a report on a topic you care about. Analyze the report by paying close attention to its content, organization, style, and design. Does the report include only need-to-know information? Is the report organized according to the IMRaD pattern? Is the style plain and appropriate to the subject of the report? Does the design enhance the reading of the report?

   Write a two-page memo to your instructor in which you critique the report. In your memo, highlight the strengths of the report and offer suggestions for improvements.

2. Write a small report (two or three pages) in which you conduct a preliminary study of a scientific or technical topic that interests you. In your report, create a short methodology that will allow you to collect some overall information about the subject. Then, present the results of your study. In the conclusion, talk about how you might conduct a larger study on the subject and the limitations you might face as you enlarge the project.

3. Devise a research question of interest to you. Then, using logical mapping, sketch out a methodology that would allow you to study that research question in depth. When you have finished outlining the methodology, write a one-page memo to your instructor in which you describe the methodology and discuss the kinds of results your research would produce.

4. Write an executive summary of an article from a scientific or technical journal in your field. You should be able to find journals in your campus library. Your summary should paraphrase the article, highlighting its main points, methodology, results, discussion, and conclusions.

Collaborative Project: Problems in the Community

With a team of others in your class, choose a local problem in your community on which you would like to write an analytical report. The readers of your report should be people who can take action on your findings, like the mayor or the city council.

After defining the rhetorical situation in which your report will be used, develop a step-by-step methodology for studying the problem. Then, collect information on the topic. Here are some ideas for topics that might interest you:

- Driving under the influence of alcohol or drugs
- Violence
- Underage drinking
- Illegal drug use
- Homelessness
- Teen pregnancy
- Water usage

Sample reports for analysis are available at
www.ablongman.com/johnsonweb/22.21

- Pollution
- Graffiti
- Environmental health

In your report, present the results of your study and discuss those results. In the conclusion of your report, make recommendations about what local authorities can do to correct the problem or improve the situation.

Revision Challenge

The introduction to a report shown here is somewhat slow and clumsy. How could you revise this introduction to make it sharper and more interesting?

Introduction

Let us introduce ourselves. We are the Putnam Consulting Firm, LLC, based out of Kansas City, KS. We specialize in the detection and mitigation of lead-related problems. We have been in business since 1995, and we have done many studies much like the one we did for you.

Our CEO is Lisa Vasquez. She has been working with lead-related issues for nearly two decades. In her opinion, lead poisoning is one of the greatest epidemics facing the United States today. The effects of lead on children, especially poor children, is acute. Dr. Vasquez has devoted her life to addressing the lead problem so today and tomorrow's children won't be damaged by continued neglect of the issue.

One of today's silent villains is lead poisoning. It is one of the most prevalent sources of childhood health problems. Each year, several hundred children in Yount County are treated for serious cases of lead poisoning. There are countless others who are suffering silently because their symptoms are either too minor or they go unnoticed by parents, teachers, and other authorities. For example, consider the case of Janice Brown in northwestern Yount County. She showed the effects of lead poisoning in her cognitive development, leading to a lower IQ rating. When her house was tested for lead poisoning, it was found that her parents had sandblasted the sides of the home to remove the old paint. Lead had saturated the ground around Janice's home and there was residual lead dust everywhere in the home. The levels of lead in Janice's blood were nearly 10 times the amount that causes problems.

You will find other troubling stories like this one in this report. We were hired by the Yount County Board of Directors to complete this study on the amounts of lead in the county. Unfortunately, we have found that Yount County is in particular trouble. Much of the county's infrastructure and housing was built when lead use was at its peak. Housing in the county relies heavily on lead pipes. Most of the houses were painted inside and out with lead-based paint.

Ironically, lead poisoning is also one of the most preventable problems that face children today. The continual persistence of the problem is a stain on our nation, and it is a particular problem that Yount County must face. In this report, we offer some recommendations.

To see a revision of this introduction, go to **www.ablongman.com/johnsonweb/22.22**

The X-File

George Franklin worked as an environmental engineer for Outdoor Compliance Associates. He and his team were responsible for writing research reports called environmental impact studies (EISs) on sites that would be developed for housing or businesses. An EIS was required by the government before any work could begin.

George's responsibility on the team was to track down any historical uses for the site. For example, if a gas station was once on the site, his job was to make sure the old underground holding tanks had been removed. If the site once held a factory, his job was to make sure that chemicals had not been dumped in the area. George loved being an environmental detective.

A year ago, George's team had written an EIS for a site where a new apartment complex was planned. The site was about a mile from a major research university. While researching the site, George discovered that it had housed part of the city's waste treatment center until the early 1950s. George's discovery wasn't a problem, though, because any contaminants would have disappeared long ago. So, George's team wrote a favorable EIS, clearing the site for development. The building plan was approved by the city, and construction started soon after.

Then, yesterday, an old, yellowed file mysteriously appeared in George's office mailbox. The file was marked "Confidential" and had no return address. George looked inside.

In the file, there was a report that nuclear weapons research had once been done at the university during the 1940s. Not recognizing the potential harm of the nuclear waste, the scientists had sent tons of radioactive waste down the drain. The nuclear waste ended up at the city's old waste treatment center. The waste, including the nuclear waste, was then spread around the grounds of the waste treatment plant.

George grabbed his Geiger counter and went out to the building site. The apartment development was now half built. When he pulled out his Geiger counter, it immediately began detecting significant levels of radiation.

The radiation wasn't high enough to violate government standards, but it was close. George knew that the building permit would have been granted, even if knowledge of this site's nuclear past was known. However, he also knew that as soon as people heard about the radiation, no one would want to live in the apartments. That would be a disaster for the developer building them.

If he reported the radiation, there was a good chance George's company would be sued by the builder for missing this important problem with the property. Moreover, other builders would likely never again hire his company to write an EIS. George would almost certainly lose his job, and the company he worked for might be forced out of business. But then, he could keep it quiet. The radiation, after all, was not above government standards.

If you were George, how would you handle this situation?

More information on nuclear waste disposal is available at **www.ablongman.com/johnsonweb/22.23**

Appendixes

Appendix A: Grammar and Punctuation Guide

Appendix B: English as a Second Language Guide

Appendix C: Documentation Guide

APPENDIX

A Grammar and Punctuation Guide

This guide is a reference tool to help you handle mechanical issues in writing. The guide has two parts: (1) the Top 12 Grammar Mistakes, and (2) a Punctuation Refresher. The Top 12 Grammar Mistakes will show you how to solve the twelve most common grammar mistakes. The Punctuation Refresher will update you on punctuation rules, showing you how to properly use these marks.

Perhaps you have heard writing teachers say, "Writing isn't about grammatical correctness. Writing is about forming your ideas and expressing yourself clearly." In a limited sense, they are right. Good writing *is* more than simple correctness.

In another sense, they are quite wrong. Reliable punctuation and grammatical correctness are essential if you are going to clearly express yourself, especially in technical documents and presentations. Any grammar and punctuation mistakes in your writing will undermine even your best ideas. Moreover, readers will make important judgments about you and your work based on grammar. If you send a document littered with grammatical errors to your supervisor or your company's clients, they will question your attention to quality, your commitment to the project, and even your intelligence. That's bad.

Mastering basic grammar and punctuation does not take long. If you haven't worked on improving your grammar since high school, take some time to refresh yourself on these rules.

The Top 12 Grammar Mistakes

Don't be one of those people who is always apologizing with statements like "I'm not good at grammar." These kinds of statements do not excuse your grammatical problems; they only indicate that you have not taken the time to learn basic grammar rules. If you are one of these grammar apologists, it is time to spend the little time necessary to master the rules. An hour or two of study will make those grammatical problems disappear.

In this section, we will go over the Top 12 Grammar Mistakes. Plenty of other grammar mistakes exist, but you will find that the mistakes discussed here account for the vast majority of grammatical problems. If you learn how to avoid/correct these twelve mistakes, your writing will be nearly free of error.

Comma Splice

A significant percentage of grammar errors are comma splices. A comma splice occurs when two complete sentences are spliced together with a comma.

Incorrect

The machine kept running, we pulled the plug.

We moved the telescope just a little to the left, the new nova immediately came into view.

Want more information on comma splices? Go to **www.ablongman.com/johnsonweb/A.1**

In these examples, notice how the parts before and after the commas could stand alone as sentences. The comma is *splicing* these sentences together.

How can you fix these comma splices? There are a few options.

Correct The machine kept running, so we pulled the plug.

The machine kept running; we pulled the plug.

The machine kept running. We pulled the plug.

These sentences use conjunctions (*so, and, but, yet*), semicolons, and periods to turn the comma splices into grammatically correct sentences.

Correct We moved the telescope just a little to the left, and the new nova immediately came into focus.

We moved the telescope just a little to the left; then, the new nova immediately came into focus.

We moved the telescope just a little to the left. The new nova immediately came into focus.

Avoiding the stealth comma splice

A sneaky comma splice error that might not look like one is when a comma is used before a conjunctive adverb (*therefore, however, moreover, furthermore, consequently, accordingly*). In these situations, a semicolon is needed.

Incorrect The computer software program worked flawlessly, however we were very worried that there were some bugs yet to be found.

Correct The computer software program worked flawlessly; however, we were very worried that there were some bugs yet to be found.

To avoid this kind of comma splice, just remember that a semicolon is needed in sentences where conjunctive adverbs like *however* and *therefore* are used to join two sentences.

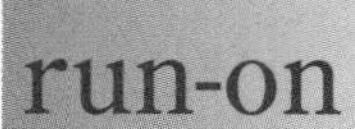

Run-on Sentence

The run-on sentence error is a close cousin of the comma splice. In a run-on sentence, two or more sentences have been crammed into one.

Incorrect The computer suddenly crashed it had a virus.

The Orion nebula lies about 1,500 light-years from the sun the nebula is a blister on the side of the Orion molecular cloud that is closest to us.

Run-on sentences are corrected the same way as comma splices. You can use conjunctions (*and, but, or, nor, for, because, yet, however, furthermore, hence, moreover, therefore*) to fix them. Or, you can divide the sentences with a semicolon or period.

Correct The computer suddenly crashed, because it had a virus.

The computer suddenly crashed; we guessed it had a virus.

The computer suddenly crashed. It had a virus.

The Orion nebula lies about 1,500 light-years from the sun. The nebula is a blister on the side of the Orion molecular cloud that is closest to us.

The Orion nebula lies about 1,500 light-years from the sun; moreover, the nebula is a blister on the side of the Orion molecular cloud that is closest to us.

In most cases, run-on sentences are best fixed by adding a period, thus separating the two sentences completely.

frag

Fragment

A fragment, like the name suggests, is an incomplete sentence. A fragment typically occurs when the sentence is missing a subject or a verb.

Incorrect

After walking through the forest for an hour. We decided to leave.

The new motherboard was shipped back to the manufacturer. Because it was not working.

Our observations of the child convinced us. That he was learning disabled.

As shown in these examples, fragments almost always come about because a sentence has been mistakenly divided into two parts. One of the two parts is a fragment, missing a subject or verb.

To fix these fragment errors, fuse the sentences back together.

Correct

After walking through the forest for an hour, we decided to leave.

The new motherboard was shipped back to the manufacturer, because it was not working.

Our observations of the child convinced us that he was learning disabled.

Sometimes writers, especially creative writers, will use fragments on purpose to have a jarring effect on readers. In technical writing, you don't want to jar your readers. Leave the creative uses of fragments to the creative writers.

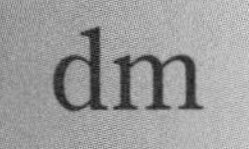

Dangling Modifier

Ah, yes, the grammar error with the funny name. Who said grammar isn't fun? A dangling modifier occurs when a phrase does not properly modify the subject.

Incorrect

While eating lunch, the acid boiled over and destroyed the testing apparatus.
(The acid is apparently eating lunch while it does damage to the testing apparatus. That's some acid!)

After driving to Cleveland, our faithful cat was a welcome sight.
(The cat apparently drove the car to Cleveland. Bad kitty!)

These kinds of errors are common—and often funny. To avoid them (and your readers' grins), you need to make sure that the introductory phrase is modifying the subject of the sentence. Here are corrections of these sentences:

Worksheets on run-ons, fragments, and comma splices are available at **www.ablongman.com/johnsonweb/A.2**

Correct While we were eating lunch, the acid boiled over and destroyed the testing apparatus.

After driving to Cleveland, we were glad to see our faithful cat.

Notice how the information before the comma modifies the subject of the sentence, which immediately follows the comma.

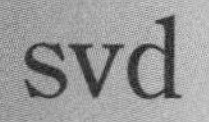

Subject-Verb Disagreement

Subject-verb disagreements occur when the subject of the sentence does not match the verb. Singular subjects should go with singular verbs, while plural subjects should have plural verbs. Here are a few sentences with subject-verb disagreement problems:

Incorrect The windows, we discovered after some investigation, was the reason for heat loss in the house.

(The singular verb *was* does not match the plural subject *windows*.)

The robin, unlike sparrows and cardinals, do not like sunflower seeds.

(The subject *robin* is singular, while the verb phrase *do not like* is plural.)

Either my DVD player or my stereo were blowing the fuse.

(The subject *DVD player or my stereo* is singular, while the verb *were* is plural.)

In these examples, the writer forgot that the subject was singular. Here are the correct sentences:

Correct The windows, we discovered after some investigation, were the reason for heat loss in the house.

Robins, unlike sparrows and cardinals, do not like sunflower seeds.

Either my DVD player or my stereo was blowing the fuse.

When *or* is used in the subject, as in this third example, the verb needs to agree with the noun that follows *or*. In this sentence, the use of *or* means we are treating the DVD player and stereo separately. So, a singular verb is needed.

This third example brings up an interesting question. What if one or both of the words flanking the *or* is plural? Again, the answer is that the verb will agree with the noun that comes after the *or*.

The speeding cars or the reckless motorcyclist was responsible for the accident.

Either the falling branches or the high winds were responsible for the damage.

But, you say, the first sentence just doesn't seem right, even though it is grammatically correct. To make the sentence sound right, you might swap the two items in the subject, so you can use a plural verb.

The reckless motorcyclist or the speeding cars were responsible for the accident.

Misused Apostrophe

Usually, when writers misuse apostrophes, they are trying to make a possessive out of a word that already shows possession.

Some great grammar handbooks are on-line. Go to **www.ablongman.com/johnsonweb/A.3**

Incorrect The dog is their's.

The best score was her's.

Who's calculator is this?

In these situations, the writer assumes that the possession needs to be signaled by an *'s*. But the words themselves signal possession, so the apostrophe is not necessary.

Correct The dog is theirs.

The best score was hers.

Whose calculator is this?

The most common (and annoying) apostrophe error is the confusion between *its* and *it's*. To avoid this problem, remember that *it's* always means *it is*—period. The word *its* always shows possession. Meanwhile the word *its'* is never correct in the English language.

Incorrect Its clear that there are many bugs in the software.

The dog growled when we tried to take away it's bone.

Correct It's clear that there are many bugs in the software.

The dog growled when we tried to take away its bone.

Usually, *it's* and *its* are misused when a writer adds an *'s* to *it* to signal possession. A simple way to avoid this problem completely is to check all your uses of *it's*. If *it's* does not mean *it is*, you need to eliminate the apostrophe.

mc

Misused Comma

Commas are designed to mark pauses in a sentence. Some people, though, feel the need to add a comma whenever they sense a shift in the sentence.

Incorrect Can we all agree, that the project was a success?
(The comma marks a pause that should not exist.)

The car was a red, Mercedes sedan.
(The adjectives do not need to be separated by a comma.)

He ate the new asparagus pizza, but didn't like it.
(The phrase after *but* cannot stand alone, so the comma should not divide the sentence.)

We should also consider, whether the reservoir has enough water to last the summer.
(The comma marks a pause that should not exist.)

The commas in these sentences should all be removed to make them grammatically correct.

The second and third of these examples of misused commas might be debated. An author might point out that he or she wants readers to pause at these points; consequently, the comma was included. Again, though, technical writing is not creative writing. Simpler and clearer is better. You should remove these unnecessary commas.

Want to see people rage about the *it's/its* problem? Go to **www.ablongman.com/johnsonweb/A.4**

Pronoun-Antecedent Disagreement

A pronoun-antecedent disagreement error usually occurs when a writer forgets whether the subject is plural or singular.

Incorrect Anyone who thinks Nintendo is better than Sony Playstation should have their head examined.

Like the scientists, we were sure the rocket was going to blast off, but it wasn't long before you knew it was a dud.

In these cases, words later in the sentence do not agree with pronouns earlier in the sentence. In the first sentence, *anyone* is singular while *their* is plural. In the second sentence, *we* is a first-person noun while *you* is a second-person pronoun. Here are a couple of ways to correct these sentences:

Correct People who think Nintendo is better than Sony Playstation should have their heads examined.

(The subject is made plural (*people*) and an *s* is added to *head* to make it plural.)

Anyone who thinks Nintendo is better than Sony Playstation should have his or her head examined.

(The subject (*anyone*) is kept singular, and the plural *their* is changed to the singular *his or her*.)

Like the scientists, we were sure the rocket was going to blast off, but it wasn't long before I knew it was a dud.

(The *you* was changed to *I*, keeping the whole sentence in first person.)

Like the scientists, we were sure the rocket was going to blast off, but it wasn't long before we knew it was a dud.

(The *you* was changed to *we*, again keeping the whole sentence in first person.)

In most cases, the secret to avoiding these subtle errors is to check whether the pronouns later in the sentence match the subject.

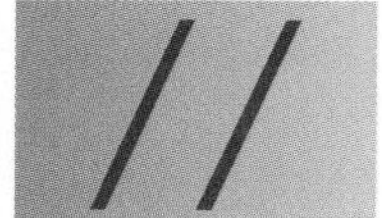

Faulty Parallelism

Lists can be difficult to manage in sentences. A good rule of thumb is to remember that each part of the list needs to be parallel in structure to other parts of the list.

Incorrect After the interview, we went out for dinner, had a few drinks, and a few jokes were told.

(In this sentence, the third part of the list, *jokes were told*, is not parallel with the first two parts, *went out for dinner* and *had a few drinks*.)

Our survey shows that people want peace, most want to own a home, and that many are worried about their jobs.

(In this sentence, the second item in the list, *most want to own a home*, is not parallel with the first and third items.)

To correct the sentence, make the items in the list parallel.

Correct After the interview, we went out for dinner, had a few drinks, and told a few jokes.

Our survey shows that people want peace, most want to own a home, and many are worried about their jobs.

For practice avoiding parallelism problems, go to **www.ablongman.com/johnsonweb/A.5**

To avoid faulty parallelism errors, pay special attention to any lists you write. Check whether the items are parallel in phrasing.

Pronoun Case Error (*I* and *Me, We* and *Us*)

Often, people are confused about when to use *I* or *me* and *we* or *us*. Here is a simple way to make the right decision about which one to use. If you are using the word as the subject of the sentence or phrase, use *I* or *we*. Anywhere else, use *me* or *us*.

Incorrect

Jones's team and me went down to the factory floor to see how things were going.
(The word *me* is misused here, because it is part of the subject of the sentence. In this case, *I* should have been used.)

When the roof fell in, the manager asked Fred and I to start developing a plan for cleaning the mess up.
(The word *I* is misused in this sentence, because the phrase *Fred and I* is not the subject. In this case, the phrase *Fred and me* should have been used.)

Things were getting pretty ugly in there, so us unimportant people slipped out the back door.
(The phrase *us unimportant people* is the subject of the phrase that follows the comma. Therefore, the phrase *we unimportant people* should have been used.)

Remember, if the word is being used as the subject of a phrase or sentence, use *I* or *we*. Anywhere else, use *me* or *us*. Here are the correct versions of these sentences:

Correct

Jones's team and I went down to the factory floor to see how things were going.

When the roof fell in, the manager asked Fred and me to start developing a plan for cleaning the mess up.

Things were getting pretty ugly in there, so we unimportant people slipped out the back door.

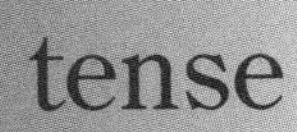

Shifted Tense

Sentences can be written in past, present, or future tense. In most cases, neighboring sentences should reflect the same tense. Shifting tenses can make readers feel like they are hopping back and forth in time.

Incorrect

Few countries possess nuclear weapons, but many countries tried to build them.
(Here, *possess* is in present tense, while *tried* is in past tense.)

Parts flew everywhere on the factory floor as the robot finally breaks down.
(Now *flew* is in past tense, while *breaks* is in present tense.)

The advances in microchip technology allowed electronics to become much smaller, which leads to today's tiny electronic devices.
(The word *allowed* is in past tense, while *leads* is in present tense.)

We found ourselves staring in disbelief. The excavation site is vandalized by people who wanted to have a campfire.
(Here, *found* is in past tense, while *is vandalized* is in present tense.)

To revise these sentences, make the tenses consistent.

Correct

Few countries possess nuclear weapons, but many countries are trying to build them.

Parts flew everywhere on the factory floor when the robot finally broke down.

The advances in microchip technology allowed electronics to become much smaller, which led to today's tiny electronic devices.

We found ourselves staring in disbelief. The excavation site had been vandalized by people who wanted to have a campfire.

Tense shifts are not always wrong. Sometimes a sentence or phrase needs to be in a different tense than those around it. When checking sentences for unnecessary tense shifts, look for places where tense shifts cause more confusion than clarity.

Vague Pronoun

Occasionally, a writer uses a pronoun, seeming to know exactly who or what the pronoun refers to, while readers are left scratching their heads, trying to figure out what the writer means.

Incorrect

Fred and Javier went to the store, and then he went home.
(Who does *he* refer to, Fred or Javier?)

They realized that the inspection of the building was not going well. It was fundamentally unsound.
(In this sentence, *it* could refer to the inspection or the building.)

We really had a great week. Our program review went well, and we made huge strides toward finishing the project. This is why we are taking all of you to lunch.
(What does *this* refer to? The great week? The program review? The huge strides? All of them?)

The camera captured the explosion as it ripped apart the car. It was an amazing experience.
(In these sentences, the multiple uses of *it* are confusing. In the first sentence, *it* might refer to the camera or the explosion. In the second sentence, *it* might refer to the taking of the picture or the explosion. Or, the final *it* might just be a weak subject for the second sentence that doesn't refer directly to anything in the previous sentence.)

Correcting these sentences mostly involves rewording them to avoid the vague pronoun.

Correct

Fred and Javier went to the store, and then Javier went home.

They realized that the inspection of the building was not going well. The inspection process was fundamentally unsound.

We really had a great week. Our program review went well, and we made huge strides toward finishing the project. For these reasons, we are taking all of you to lunch.

The camera captured the explosion as it ripped apart the car. The explosion was an amazing experience.

A common cause of vague pronouns is the overuse of "It is . . ." and "This is . . ." to begin sentences. These kinds of sentences force readers to look back at the previous sentence to figure out what *it* or *this* refers to. In some cases, two or three possibilities might exist.

To avoid these problems, train yourself to avoid using "It is . . ." and "This is . . ." sentences. Occasionally, these sentences are fine, but some writers rely on them too much. You are better off minimizing their use in your writing.

Punctuation Refresher

Punctuation rules are as important to writing as the rules of the road are to driving a car. If you don't know the rules, accidents will happen.

More than likely, you probably already have a good sense of the punctuation rules. However, there are quirks and exceptions that you need to learn as your writing skills advance to a new level. So, spend a little time here refreshing your memory on punctuation rules.

Period, Exclamation Point, Question Mark

Let's start with the most basic marks, the period, exclamation point, and question mark. These punctuation marks signal the end of a sentence, or a full stop.

> We need to test the T6 Robot on the assembly line.
>
> The acid leaked and burned through the metal plate beneath it!
>
> Where can we cut costs to bring this project under budget?

The period signals the end of a standard sentence. The exclamation point signals surprise or strong feelings. The question mark signals a query.

Periods can also be used with abbreviations and numbers.

> A.D.
>
> Feb.
>
> Fig. 8
>
> Dr. Valerie Hanks
>
> 56.21 cm

Question marks can also be used in a series.

> How will global warming affect this country? Flood coastal cities? Create drought in the Southwest? Damage fragile ecosystems?

Using periods, exclamation points, and question marks with quotes

A common mistake with periods, exclamation points, and question marks is their misuse with quotation marks. In almost all cases, these punctuation marks are placed inside the quotation marks.

Incorrect

> He said, "The audit team reported that we are in compliance".
>
> We asked him, "Do you really think we are going to finish the project on time"?

For websites on punctuation, go to **www.ablongman.com/johnsonweb/A.6**

Correct He said, "The audit team reported that we are in compliance."

We asked him, "Do you really think we are going to finish the project on time?"

The one exception to this rule is if a quoted statement is placed within a question.

Did he really say, "The company will be closing the Chicago office"?

Commas

Now let's move on to commas. In the English language, commas are the most flexible, useful, and therefore problematic punctuation mark. In most cases, a comma signals a pause in the flow of a sentence.

When he hiked in the mountains east of Ft. Collins, he always took along his compass and map to avoid getting lost.
(This comma signals that an introductory phrase is finished.)

My PDA is a helpful organizing tool, and it makes a great paperweight too.
(This comma signals that two independent clauses are joined with a conjunction [*and, yet, but, so, because*].)

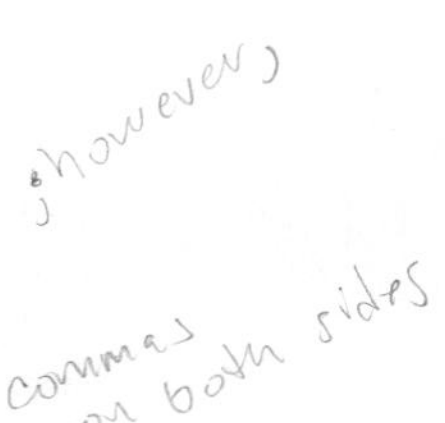

Our company's CEO, the engineering genius, is scheduled to meet with us tomorrow.
(These commas set off information that is not essential to understanding the sentence.)

We are, however, having some luck locating new sources of silicon on the open market.
(These commas set off a conjunctive adverb [*therefore, however, furthermore, nevertheless, moreover*].)

The archaeological dig yielded pottery shards, scraping tools, and a corn meal grinding stone.
(These commas separate items in a series.)

Using commas with quotation marks

Using commas with quotation marks can be problematic. Just remember that commas almost always go before, not after, the quotation mark.

Albert Einstein said, "God does not play dice."
(Here the comma sets off a speaker tag. Notice that the comma is placed before the quotation mark.)

"I'm having trouble hearing you," she said, "so I'm switching over to a new phone."
(Again, the commas in this sentence offset the speaker tag. Note that both commas come before the quotation marks.)

Using commas in numbers, dates, and place names

Commas are also used with numbers, dates, and place names.

Reporters estimated the rally drew nearly 10,000 people.

We first noticed the problem on January 10, 2004.

For great pizza you need to go to Uncle Pete's in Naperville, Illinois.

Need more help with commas? Go to **www.ablongman.com/johnsonweb/A.7**

Removing excess commas

There is some flexibility in the use of commas. Many editors recommend an "open" punctuation style that eliminates commas where they do not help comprehension. For example, the following sentences are both correct:

> In minutes, she whipped up the most amazing apple pie.
>
> In minutes she whipped up the most amazing apple pie.

Here, the comma after *minutes* is not aiding understanding, so it can be removed. Be careful, though. Sometimes the lack of a comma can cause some confusion.

Confusing Soon after leaving the airplane needed to turn back for mechanical reasons.

Not Confusing Soon after leaving, the airplane needed to turn back for mechanical reasons.

Keep in mind that misused commas are responsible for many grammatical errors, as we discussed in the Top 12 Grammar Mistakes earlier in this guide.

Semicolon and Colon

Semicolons and colons are less common than periods and commas, but they can be helpful in some situations. In a sentence, semicolons and colons signal a partial stop.

The trick to properly using these marks is to remember this simple rule: In most cases, the phrase on either side of a semicolon or colon should be able to stand alone as a separate sentence (an independent clause).

> We were not pleased with the results of the study; however, we did find some interesting results that gave us ideas for new research.
>
> (The semicolon joins two independent clauses. The second clause [the part starting with *however*] supports the first clause.)
>
> Commuting to work by car requires nerves of steel: Each mile brings you into contact with people who have no respect for the rules of the road.
>
> (Here, the colon also divides two independent clauses. In this example, though, the colon signals that the two parts of the sentence are equivalent to each other.)

How do you know when to use a semicolon or a colon? It depends on whether the part of the sentence following the mark is *lesser* or *equal* to the first part. If lesser, use the semicolon. If equal, use the colon.

TAKE NOTE It helps to keep in mind that the semicolon is like a comma, while the colon is like a period. The semicolon signals a strong pause, and the colon signals a stop.

If you want to avoid problems, use semicolons and colons only when necessary. When considering these punctuation marks in a sentence, you should ask yourself whether a period or a conjunction (*and, but, so, yet*) would make the sentence easier to understand. After all, joining sentences together with semicolons and colons can often create long, difficult-to-read sentences.

Let's turn to the exceptions. As with most punctuation rules, there are exceptions to the rule that semicolons and colons are used with independent clauses.

Websites to help you master the semicolon and colon are available at **www.ablongman.com/johnsonweb/A.8**

Using semicolons in a series

Semicolons can be used to punctuate complicated lists in a sentence.

We have offices in Boston, Massachusetts; Freetown, New York; and Sedona, Arizona.

Using colons to lead off lists

Colons can be used to signal a list.

Four steps are required to complete the process: (1) preparing the workspace, (2) assembling the model, (3) painting the model, and (4) checking quality.

Keep in mind the following issues when searching for a job:

- You can't get the job if you don't apply.
- Jobs don't always go to the people with the most experience.
- What makes you different makes you interesting.

Using colons in titles, numbers, and greetings

Colons are commonly used in titles of books and articles, numbers, and greetings in letters and memos.

The Awakening: The Irish Renaissance in Nineteenth-Century Boston

Genesis 2:18

11:45 A.M.

They won by a 3:1 ratio.

Dear Mr. Franklin:

Using semicolons and colons with quotation marks

Unlike commas or periods, semicolons and colons should appear after the quotation mark in a sentence.

Land Commissioner George Hampton claimed, "The frogs will survive the draining of the lake"; but he was clearly wrong.

One of my favorite chapters in Leopold's *A Sand County Almanac* is "Thinking Like a Mountain": This essay is his best work.

Whenever possible, though, you should avoid these kinds of situations. In both of these examples, the sentences could be rearranged or repunctuated to avoid these awkward, though correct, uses of the semicolon or colon.

Misusing the colon

A common misuse of the colon is to use it with an incomplete sentence.

The reasons for our dissatisfaction are: low quality, late work, and slow response.
(The colon is misused, because the phrase before the colon cannot stand alone as a separate sentence.)

For example:

(Again, the phrase before the colon cannot stand alone as a sentence. In this case, a dash or comma should be used. Or, turn the phrase "For example" into a complete sentence.)

In his report, Bill Trimble claims: "We have a golden opportunity to enter the Japanese market."

(Yet again, the information before the colon cannot exist as a separate sentence. In this case, a comma should have been used instead of the colon.)

As a rule, the information before the colon should *always* be able to stand alone as a complete sentence. Here are the correct versions of these sentences:

The reasons for our dissatisfaction are the following: low quality, late work, and slow response.

For example, consider these interesting situations:

In his report, Bill Trimble makes this important statement: "We have a golden opportunity to enter the Japanese market."

Notice how all three of these examples have independent clauses (full sentences) before the colon.

’ Apostrophe

The apostrophe has two important jobs in the English language: (1) to signal contractions and (2) to signal possession.

Using an apostrophe to signal a contraction

An apostrophe that signals a contraction identifies the place where two words have been fused and letters removed.

They're going to the store today.

He really isn't interested in the project.

They shouldn't have taken that road.

Contractions should be used only in informal writing. They signal a familiarity with the readers that could seem too informal for some situations. Some other common contractions include *won't, shouldn't, it's, I'm, you've, wouldn't,* and *couldn't.*

Using an apostrophe to signal possession

An apostrophe is also used to signal possession. With a singular noun, an *'s* is added to signal possession. Joint possession is usually signaled with an *s'*.

We have decided to take Anna's car to the convention.

She stepped right into a badger's den.

The players' bats were missing before the game.

When plural nouns do not end in an *s,* you should use an *'s* to create the plural.

We rode the children's bikes.

The men's briefcases were left near the door.

When singular nouns end in an *s*, you should add an *'s* to show possession.

They met in Mary Jones's office.

Charles's computer was shorting out.

Using apostrophes to show possession with two or more nouns

When you are showing possession with multiple nouns, your use of the apostrophe depends on your meaning. If two nouns are acting as one unit, only the last noun needs an apostrophe to signal possession.

We decided to accept Grim and Nether's proposal.

But if you are signaling possession for several separate nouns, each needs an apostrophe.

I found it difficult to buy meaningful gifts for Jane's, Valerie's, and Charles's birthdays.

Using apostrophes to signal plurals with numbers, acronyms, and symbols

You should use apostrophes to signal plurals of numbers, acronyms, and symbols, but do so sparingly. Here are a couple of situations where apostrophes would be appropriate:

The *a*'s just kept appearing when I typed *x*'s.

Is it necessary to put ©'s on all copyrighted documents?

In most cases, though, do not include apostrophes to show a plural if they do not aid the meaning of the text.

The police discovered a warehouse full of stolen TVs.

The 1870s were a tough time for immigrants.

The war in Afghanistan demonstrated the importance of NGOs.

In the basement, a crate of dead CPUs sat unnoticed.

Quotation Marks

Quotation marks are used to signal when you are using someone else's words. Quotation marks should not be used to highlight words. If you need to highlight words, use italics.

Using quotation marks to signal a quote

Quotation marks are used to frame an exact quotation from another person.

In *The Panda's Thumb,* Gould states, "The world, unfortunately, rarely matches our hopes and consistently refuses to behave in a reasonable manner."

"Not true" was her only response to my comment.

He asked me, "Are you really working on that project?"

Only use quotation marks when you are copying someone else's exact words. If you are only paraphrasing what someone else said, do not use quotation marks.

In *The Panda's Thumb,* Gould argues that nature often does not meet our expectations, nor does it operate in predictable ways.

She rejected my comment as untrue.

He asked me whether I was working on the project.

Also, when paraphrasing, avoid the temptation to highlight words with quotation marks.

Using quotation marks to signal titles

Titles of works that are part of larger works, like articles, songs, or documents, should be set off with quotation marks.

Time published an article called "The Silicon Valley Reborn."

A classic blues tune covered by the Yardbirds was "I'm a Man."

The report, "Locating Evidence of Ancient Nomads in Egypt," is available on-line.

Titles of books and other full works should not set in quotation marks. They should be italicized.

Taking the Quantum Leap, by Fred Wolf, is a very helpful book.

Revolver is one of the Beatles' best albums.

Using quotation marks to signal a quote or title within another quote

When quoting something within another quote, you should use single quote marks to set it off.

Tim Berra shows the weakness of the creationist argument by quoting one of its strongest advocates: "Morris wrote 'the only way we can determine the true age of the earth is for God to tell us what it is."'

One of the physicists at the conference remarked, "I cannot believe that Einstein's 1905 paper on special relativity, 'On the Electrodynamics of Moving Bodies,' is already a century old."

Using quotation marks to signal irony

Quotation marks are often used incorrectly to highlight words and slang terms.

Incorrect One problem with "free-trade policies" is that the laborers who work for Third World countries almost work for free.

I found working with her to be "wonderful," because she is so "attentive" and "understanding."

More help with quotation marks is available at **www.ablongman.com/johnsonweb/A.9**

The sentences above do not need quotation marks to set off these words. If you want to highlight words that are not direct quotes, use italics.

You can, however, use quotation marks to signal irony by quoting another person's misuse of a term or phrase.

> Conservation is more than Vice President Cheney's notion of a "personal virtue."

> The *Matrix* is an entertaining film, but it's hard to accept the "Biblical significance" that Clarke and others claim for this highly violent movie.

Using quotation marks with in-text citations

One of the exceptions to placing periods inside the quotation marks is when they are used with in-text citations.

> In his article on ancient dams, Abbas points out that "water-driven power systems have been around for thousands of years" (p. 67).

Here, note that the period comes after the in-text citation, not within the closing quotation mark.

Dashes and Hyphens

The uses of dashes and hyphens follow some rather specific rules. There are actually two types of dashes, the "em dash" and the "en dash." The em dash is the longer of the two (the width of an *m*), and it is the more widely used dash. The en dash is a bit shorter (the width of an *n*), and it is less widely used. Hyphens are shorter than the two dashes.

Using em dashes to highlight asides from the author or the continuation of a thought.

An em dash is typically used to insert comments from the author that are asides to the readers.

> At the meeting, Hammons and Jenkins—this is the ironic part—ended up yelling at each other, even though they both intended to be peacemakers.

> We must recognize the continuing influence of Lamarckism in order to understand much social theory of the recent past—ideas that become incomprehensible if forced into the Darwinian framework we often assume for them.

An em dash can be made with two hyphens (--). Most word processors will automatically change two dashes into an em dash. Otherwise, a series of keystrokes (usually shift-command-hyphen) will create this longer dash.

Using en dashes in numbers and dates

It might seem trivial, but there is a difference between en dashes and em dashes. An en dash is almost always used with numbers and dates.

> Copernicus (1473–1543) was the first European to make a cogent argument that the earth went around the sun rather than the sun going round the earth.

Young and Chavez argue conclusively that Valles Bonita is really a dormant sunken volcano, called a *caldera* (pp. 543–67).

As you can see in these examples, the en dash is slightly shorter than the em dash.

Using the hyphen to connect prefixes and make compound words

The hyphen is mainly used to connect prefixes with words or connect two or more words into compound words.

neo-Platonists

one-to-one relationship

trisomy-21

four-volume set of books

3.4-billion-year-old life

One thing to notice is how hyphens are used to create compound adjectives but not compound nouns. You can write "four-volume set of books," where *four-volume* is an adjective. But, you would need to write, "the four volumes of books," because the word *volumes* is being used as a noun. Hyphens are usually used to make compound adjectives, but not compound nouns.

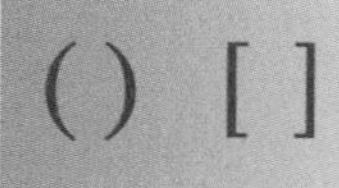

Parentheses and Brackets

Parentheses and brackets are handy for setting off additional information, like examples, definitions, references, lists, and asides to the readers.

Using parentheses to include additional information

Parentheses are often used to include additional information or refer readers to a graphic.

When hiking through the Blanca Mountains, you will be surprised by the wide range of animals you will see (e.g., elk, deer, hawks, eagles, and the occasional coyote).

The data we collected show a sharp decline in alcohol use when teens become involved in constructive, nontelevision activities (see Figure 3).

These unicellular organisms show some plantlike features (many are photosynthetic) and others show more animal-like features.

Using parentheses to clarify a list

Parentheses can be used to clarify the elements of a long list.

When meeting up with a bear in the wild, (1) do not run, (2) raise your arms to make yourself look bigger, (3) make loud noises, and (4) do not approach the animal.

Only three things could explain the mechanical failure: (a) the piston cracked, (b) one of the push rods came loose, or (c) the head gasket blew.

For more help with dashes, hyphens, parentheses, and brackets, go to **www.ablongman.com/johnsonweb/A.10**

Using brackets to include editorial comments or replace a pronoun

Brackets are less common than parentheses, but they can be helpful for inserting editorial comments or replacing a pronoun in a quote.

> Though pictures of the moon are often spectacular, *any view of the moon from earth is slightly blurred* [emphasis mine].

> Shea points out, "Whether he intended it or not, [Planck] was the originator of the quantum theory."

In this second example, the word *he* was replaced with *Planck* to make the meaning of the quote clearer.

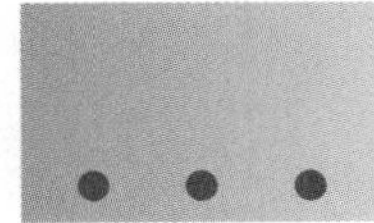

Ellipses

Ellipses are used to show that information in a quote was removed or to indicate the trailing off of a thought. An ellipse is made with three dots, with spaces between each dot (. . .), not (...).

Using an ellipse to signal that information in a quote has been removed

Sometimes a passage, especially a longer one, includes more information than you want to quote. In these cases, ellipses can be used to trim out the excess.

> As historian Holton writes, "What Bohr had done in 1927 . . . was to develop a point of view that allowed him to accept the wave-particle duality as an irreducible fact" (117).

Using an ellipse to show that a thought is trailing off

At the end of a sentence, you might use an ellipse to urge the reader to continue the thought.

> For those who don't want to attend the orientation, we can find much less pleasant ways to spend your day. . . .

When an ellipse ends the sentence, use an additional dot to make four (. . . .). The extra dot is a period that signals the end of the sentence.

APPENDIX B

English as a Second Language Guide

The English language is a composite of many languages. As a result, English has many maddening exceptions and inconsistencies in spelling, syntax, and usage. If English is not your native language, you should pay special attention to these irregularities, so your writing and speaking will be consistent with the writing and speaking of fluent speakers.

This English as a Second Language (ESL) Guide will not help you learn English. Instead, this guide concentrates on three major sources of irregularities in technical English—use of articles, word order of adjectives and adverbs, and verb tenses. As you master the English language, you should first concentrate on handling these three sources of irregularities. Then, you can turn to other ESL resources to help you refine your use of English.

Using Articles Properly

Perhaps the most significant source of ESL problems is the use of articles (*a, an, the*) in English. For example, the sentences "The computer broke down" and "A computer broke down" have significantly different meanings in English. The use of *the* suggests that a specific computer broke down. The use of *a* suggests that a computer—among many computers—broke down.

Use *the* to refer to specific items

If you are referring to a specific item, use *the* to signal the noun.

The planet Saturn has been bright during the last week.

The car stalled, so we started walking to the nearest service station.

The professor asked us to work harder on the next assignment.

Use *a* to refer to nonspecific items

If you are referring to a nonspecific item, use *a* to signal the noun.

A planet was found circling a star in the Orion system.

A car stalled in the road, so we needed to drive around it.

A professor asked us to attend the party.

Do not use an article with uncountable things

Articles should be used only with things that can be counted, like *the eight cars, the five bikes,* or *an orange.* When items are not counted or cannot be counted, do not use an article.

He surfaced for air.

Links to ESL sources on the Internet are available at **www.ablongman.com/johnsonweb/B.1**

They decided to have tea with dinner.

Rice grown in Asia tastes better than rice grown in North America.

Putting Adjectives and Adverbs in the Correct Order

Compared to some languages, English is flexible in its syntax. Nevertheless, word order in sentences is important for expressing the meaning you intend. In this section, we will go over the two major sources of word-order problems for ESL writers—adjectives and adverbs.

Using adjectives in the proper order

In English, adjectives should be placed in the proper order.

Improper The red beautiful sailboat came into the bay.

Proper The beautiful red sailboat came into the bay.

Fluent speakers of English will still understand the improper sentence, but it will sound odd to them. To properly order adjectives, you can use the following hierarchy of adjectives:

1. article, determiner, or possessive (*a, an, the, this, that, those, my, our, their, Lisa's*)
2. ordinal (*first, second, third, final, next*)
3. quantity (*one, two, three, more, some, many*)
4. size and shape (*big, tiny, large, circular, square, round*)
5. appearance (*beautiful, filthy, clean, damaged, old, young, ancient*)
6. color (*red, yellow, black, green*)
7. substance (*wool, copper, wood, plastic*).

TAKE NOTE This list was modified from Hult and Huckin (2001, pp. 395–396). Hult and Huckin offer a more comprehensive approach.

When properly ordered, adjectives can be strung together indefinitely in a sentence.

A third, large, beautiful, ancient, red, wooden sailboat came into the bay.

Keep in mind, though, that you should not string together too many adjectives. More than three or four adjectives strung together can be difficult to understand.

Using adverbs in proper places

Adverbs usually modify verbs in clauses. Adverbs can be used in a variety of places in the clause. Three guidelines are especially helpful for placing adverbs.

GUIDELINE 1: *Adverbs involving time and place usually go after the verb.*

The cat went *outside* when the children came over.

She arrived *late* to the lecture.

He went *promptly* to his professor's office for help.

Want to practice adjective order? Go to **www.ablongman.com/johnsonweb/B.2**

GUIDELINE 2: *Adverbs that show frequency usually go before the verb.*

The cat *typically* runs outside when the children come over.

She *usually* arrives late to the lecture.

He *always* went to his professor's office for help.

GUIDELINE 3: *Do not put an adverb between the verb and object of a clause*

Improper Maria drives *recklessly* her car on the interstate.

Kim eats *quickly* his breakfast before he goes to class.

Proper Maria drives her car *recklessly* on the interstate.

Kim eats his breakfast *quickly* before he goes to class.

Using Verb Tenses Appropriately

Proper use of verbs and verb phrases can be difficult in English, even for fluent speakers. So, learn them as best you can, and be patient while you are mastering the numerous English tenses.

To use tenses appropriately, remember that English, like most languages, has *past, present,* and *future* tenses. Each of these tenses, then, has four verbal aspects, which are called *simple, progressive, perfect,* and *perfect progressive.*

Simple—indicates whether the event happened, is happening, or will happen.

Progressive—indicates that the event was, is, or will be in progress at a specific time.

Perfect—indicates that the event was, is, or will be completed by a specific point in time.

Perfect progressive—indicates whether the event was, is, or will be progressing until a specific point in time.

Altogether, English has twelve tenses that need to be learned. Let us consider them separately.

Past Tense

Past tense refers to events that have already happened.

Simple Past Tense

Victor walked to the store yesterday.

Past Progressive Tense

Victor was walking to the store yesterday, when he was nearly hit by a car.

Past Perfect Tense

Before going to class, Victor had walked to the store.

Past Perfect Progressive Tense

Victor had been walking to the store before class.

More information on adverbs is available at **www.ablongman.com/johnsonweb/B.3**

Present Tense

Present tense refers to events that are happening at the moment.

Simple Present Tense
I enjoy talking with my friends.

Present Progressive Tense
I am enjoying talking with my friends this evening.

Present Perfect Tense
I have enjoyed talking with my friends this evening.

Present Perfect Progressive Tense
I have been enjoying talking with my friends this evening.

Future Tense

Future tense refers to events that will happen.

Simple Future Tense
The movie will start at 6:00 this evening.

Future Progressive Tense
The movie will be starting at 6:00 this evening.

Future Perfect Tense
By 6:10, the movie will have started already.

Future Perfect Progressive Tense
By 6:10, the movie will have been playing for nearly 10 minutes.

When learning English, you should begin by mastering the simple past, present, and future forms of the verbs. Then, as you grow more comfortable with the language, you can begin using the progressive, perfect, and perfect progressive aspects.

When you are unsure about whether to use progressive, perfect, or perfect progressive, just revert to the simple form. In most cases, these simple sentences will be correct, though perhaps a bit awkward sounding to fluent speakers of English.

Still not sure about verb tenses? Go to **www.ablongman.com/johnsonweb/B.4**

APPENDIX C

Documentation Guide

Documenting sources is an important part of doing research. As you collect information on your subject, you should keep track of the sources from which you drew quotes and ideas. Then, cite these sources inside your text and use them to create a list of references at the end of your document.

When should you cite and document a source? The answer to this question depends on the kind of document you are writing. A scientific report, for example, requires more citation than a technical description or a set of instructions. The best way to determine the level of documentation is to consider your readers' needs. How much citing and documenting do they need to feel confident in your work?

Some commonly documented materials include the following:

Quotes or ideas taken from someone else's work—If others wrote it or thought it before you did, you should cite them as the owners of their words and ideas. Otherwise, you might be accused of lifting their work. In important cases, you or your company might be sued for using someone else's ideas.

Materials that support your ideas—You can build the credibility of your work by showing that others have discussed this topic before. Readers are going to be highly skeptical of your work if they think you are pulling your ideas out of thin air.

Sources of any data or facts—Any numbers or facts that you did not generate yourself need to be carefully cited and documented. That way, readers can check your sources for accuracy.

Materials that refer to your subject—You want to demonstrate that you are aware of the broader conversation on your subject. By citing sources, including those with which you disagree, you show that you have a complete understanding of the issues involved.

Historical sources on your subject—To build a background for understanding your subject, you should include any sources that might help readers understand its history.

Graphics taken from on-line or print sources—On the Internet, it is easy to download graphs, tables, images, and photographs. Sometimes you will need permission to use these items. Minimally, though, you should cite the sources from which they were taken.

Let us review the three most common documentation styles for technical communication:

- The **APA documentation style** from the American Psychological Association is widely used in engineering and the sciences.
- The **CBE documentation style** from the Council of Science Editors (previously the Council of Biology Editors) is used primarily in biological and medical sciences, though it is gaining popularity in other scientific fields.
- The **MLA documentation style** from the Modern Language Association is used in the humanities. This style is not commonly used in engineering and science;

Links to sources on APA, MLA, and CBE styels can be found at
www.ablongman.com/johnsonweb/C.1

however, it is sometimes used when scholars approach technical issues from cultural, historical, rhetorical, or philosophical perspectives.

Literally hundreds of other documentation styles are available. So, find out which documentation style is used in the organization or company for which you work.

Citing and documenting a source requires two items, an *in-text citation* and a *full entry* in the list of References, Works Cited, or Bibliography at the end of the document.

TAKE NOTE Footnotes and endnotes are not common in APA or CBE style, and they are increasingly rare in MLA style. If you want to use footnotes or endnotes, consult the style guide you are using for instructions. Footnotes and endnotes will not be covered here because they are rarely used in technical documents.

We will discuss the most common in-text citations and full-entry patterns. If one of the following models does not fit your needs, you should consult the style guide (APA, CBE, or MLA) that you are following.

APA Documentation Style

The APA documentation style is most common in the natural and human sciences, except in fields related to biology and medicine. The official source for this style is the *Publication Manual of the American Psychological Association,* 5th ed. (2001).

When using the APA style, you will need to include in-text citations and a list of alphabetically arranged references at the end of your document.

APA In-Text Citations

The APA style follows an author-year system for in-text citations, meaning the author and year are usually cited inside the text.

Individual Authors

Individual authors are cited using their last name and the date of the article.

One study reports a significant rise in HIV cases in South Africa in one year (Brindle, 2000).

One study reported a 12.2% rise in HIV cases in only one year (Brindle, 2000, p. 843).

Brindle (2000) reports a significant rise in HIV cases in South Africa in one year.

Brindle (2000) reports a 12.2% rise in HIV cases in South Africa in one year (p. 843).

In most cases, only the author and year need to be noted, as shown in the first example. If you are reporting a specific fact or number, however, you should cite the page from which it was taken.

Multiple Authors

If an article has two or more authors, you should use the ampersand symbol (&) to replace the word *and* in the in-text citation. The word *and* should be used in the sentence itself.

To find on-line sources that discuss APA style, go to **www.ablongman.com/johnsonweb/C.2**

(Thomas & Linter, 2001)

According to Thomas and Linter (2001) . . .

Technical documents often have more than two authors. In these cases, cite all the names the first time the work is referenced. Afterward, use the last name of the first author followed by "et al."

First Citation of Work

(Wu, Gyno, Young & Reims, 2003)

As reported by Wu, Gyno, Young, and Reims (2003) . . .

Subsequent Citations of Work

(Wu et al., 2003)

As reported by Wu et al. (2003) . . .

Corporate or Unknown Authors

When the author of the document is a corporation or the author is unknown, the in-text citation uses the name of the corporation or the first prominent word in the title of the document.

First Citation of Work

(National Science Foundation [NSF], 2004)

("Results," 2002)

(*Silent*, 2002)

Subsequent Citations of Work

(NSF, 2004)

("Results")

(*Silent*)

Notice in these examples that the first word of a journal article title should be put in quotation marks, while the first word of a book title should be put in italics.

Paraphrased Materials

When citing paraphrased materials, only the year and page number are needed because the authors' names are typically mentioned in the sentence. In many cases, only the year is needed.

Franks and Roberts report that aptitude for visual thinking runs in families (2003, p. 76).

The instinct for survival, according to Ramos (2004), is strong in the Mexican wolf.

Jones (2001) argues that finding a stand of dead trees near an industrial plant is a good indicator that something is seriously wrong (pp. 87–88).

Two or More Works in Same Parentheses

In some cases, several documents will state similar information. If so, you should cite them all.

> Studies have shown remarkable progress toward reviving the penguin population on Vostov Island (Johnson & Smith, 2001; Tamili, 2002; Hinson & Kim, 2004).

Personal Communication and Correspondence

The APA style discourages putting any forms of personal communication in the references list. Personal communication includes conversations, e-mail, letters, and even interviews. So, in-text citations are the only citations for these sources in a document.

> Bathers (personal communication, December 5, 2003) pointed out to me that . . .

These sources are not listed in the references list because they do not provide information that is retrievable by readers.

The Reference List for APA Style

When using the APA style, your references should be listed in alphabetical order at the back of the document. In your references, you should list only items that were actually cited in your document. Leave any uncited documents out of the list of references. After all, if a document is important enough to list in your references, it should be important enough to cite in the text itself.

In most cases, the reference list is identified by the centered heading "References." Entries can be single spaced or double spaced. Also, in most cases, each reference should use a hanging indent style (i.e., the second line and subsequent lines should be indented). The first line should be flush with the left margin.

The following list includes examples of the APA style. This list is not comprehensive. If you do not find a model here for a document you are adding to your references, you should check the *Publication Manual of the American Psychological Association,* 5th ed. Or, check one of several websites that offer examples of APA style.

1. **Website or Webpage, Author Known**

   Jaspers, F. (2001). *Einstein online.* Retrieved March 9, 2003, from the World Wide Web: http://www.einsteinonlinetoo.com

2. **Website or Webpage, Corporate Author**

   National Wildlife Service. (2002). *Managing forest on your land.* Retrieved September 8, 2004, from the World Wide Web: http://www.nws.gov/manageyourforest.htm

3. **Webpage, Author Unknown**

   Skin cancer treatments debated. (2004, January 1). *CNN.com.* Retrieved January 19, 2004, from the World Wide Web: http://www.cnn.com/2004/HEALTH/conditions/01/19/skincancer.treatment.ap/index.html

4. **Webpage, On-line Periodical**

   Grinspoon, D. (2004, January 7). Is Mars ours? *Slate Magazine.* Retrieved January 19, 2004, from the World Wide Web: http://slate.msn.com/id/2093579

Having trouble finding an APA model for a source? Go to **www.ablongman.com/johnsonweb/C.3**

5. **Book, One Author**

Jones, S. (2001). *Darwin's ghost: The origin of species updated.* New York, NY: Balantine Books.

6. **Book, More than One Author**

Pauling, L., & Wilson, E. B. (1935). *Introduction to quantum mechanics.* New York, NY: Dover Publications.

7. **Book, Corporate or Organization Author**

American Phychiatric Association. (1994). *Diagnostic and statistical manual of mental disorders* (4th ed.). Washington, DC: American Psychiatric Association.

8. **Book, Edited Collection**

Mueller-Vollmer, K. (Ed.). (1990). *The hermeneutics reader.* New York, NY: Continuum.

9. **Book, Translated**

Habermas, J. (1979). *Communication and the evolution of society.* (T. McCarthy, Trans.). Boston, MA: Beacon Press.

10. **Book, Author Unknown**

Handbook for the WorkPad c3 PC Companion. (2000). Thornwood, NY: IBM.

11. **Book, Second Edition or Beyond**

Williams, R., & Tollet, J. (2000). *The non-designer's web book* (2nd ed.). Berkeley, CA: Peachpit.

12. **Book, Dissertation or Thesis**

Simms, L. (2002). *The hampton effect in fringe desert environments: An ecosystem under stress.* Unpublished doctoral dissertation, University of New Mexico.

13. **Book, Electronic**

Darwin, C. (1862). *On the various contrivances by which British and foreign orchids are fertilised by insects.* London, UK: John Murray. Retrieved January 1, 2004, from the World Wide Web: http://pages.britishlibrary.net/charles.darwin3/orchids/orchids_fm.htm

14. **Document, Government Publication**

Greene, L. W. (1985). *Exile in paradise: The isolation of Hawai'i's leprosy victims and development of Kalaupapa settlement, 1865 to present.* Washington, DC: U.S. Department of the Interior, National Park Service.

15. **Document, Pamphlet**

The Colorado Health Network. (2002). *Exploring high altitude areas.* Denver, CO: TCHN.

16. **Film or Video Recording**

Jackson, P. (Director), Osborne, B., Walsh, F., & Sanders, T. (Producers). (2002). *The lord of the rings: The fellowship of the ring* [Film]. Hollywood, CA: New Line Productions.

17. **CD-ROM**

Masters of renaissance art [CD-ROM]. (2000). Chicago, IL: Revival Productions.

Need more help with on-line sources? Go to **www.ablongman.com/johnsonweb/C.4**

18. **Article, Journal with Continuous Pagination**

Boren, M. T., & Ramey, J. (1996). Thinking aloud: Reconciling theory and practice. *IEEE Transactions on Professional Communication, 39,* 49–57.

19. **Article, Journal without Continuous Pagination**

Kadlecek, M. (1991). Global climate change could threaten U.S. wildlife. *Conservationist 46*(1), 54–55.

20. **Article, Edited Book**

Katz, S. B., & Miller, C. R. (1996). The low-level radioactive waste siting controversy in North Carolina: Toward a rhetorical model of risk communication. In G. Herndl & S. C. Brown (Eds.), *Green culture: Environmental rhetoric in contemporary America* (pp. 111–140). Madison, WI: U of Wisconsin P.

21. **Article, Magazine**

Appenzeller, T. (2004, February). The case of the missing carbon. *National Geographic,* 88–118.

22. **Article, Newspaper**

Hall, C. (2002, November 18). Shortage of human capital envisioned, Monster's Taylor sees worker need. *Chicago Tribune,* p. E7.

23. **Article, Author Unknown**

The big chill leaves bruises. (2004, January 17). *Albuquerque Tribune,* p. A4.

24. **Article, CD-ROM**

Hanford, P. (2001). Locating the right job for you. *The electronic job finder* [CD-ROM]. San Francisco, CA: Career Masters.

25. **Song or Recording**

Myer, L. (1993). Sometimes alone. *Flatlands* [CD]. Ames, IA: People's Productions.

26. **Television or Radio Program**

Harris, R. (2003, January 6). Destination: The south pole. Washington, DC: National Public Radio. Retrieved February 4, 2004, from the World Wide Web: http://discover.npr.org/features/feature.jhtml?wfId=904848

27. **Personal Correspondence, E-Mail, or Interview**

In the APA style, a personal correspondence is not listed in the reference list. Instead, the information from the correspondence should be written in the in-text citation:

> This result was confirmed by J. Baca (personal communication, March 4, 2004).

Creating the APA Reference List

In the APA style, the reference list is placed at the end of the document or in an appendix. The sources referenced in the document should be listed alphabetically by author's last name.

References

Assel, R., Cronk, K., & Norton, D. (2003). Recent trends in Laurentian Great Lakes ice cover. *Climatic Change 57,* 185–204.

Hoffmann, A., & Blows, M. (1993). Evolutionary genetics and climate change: Will animals adapt to global warming? In P. M. Kareiva, J. G. Kingsolver & R. B. Huey (Eds.), *Biotic interactions and global change* (pp. 13–29). Sunderland, MA: Sinauer.

Houghton, J. (1997). Global warming: The complete briefing (2nd ed). Cambridge, MA: Cambridge UP.

Kadlecek, M. (1991). Global climate change could threaten U.S. wildlife. *Conservationist 46* (1), 54–55.

Sherwood, K., & Idso, C. (2003). Is the global warming bubble about to burst? Retrieved March 4, 2004, from the World Wide Web: http://www.co2science.org/edit/v6_edit/v6n37edit.htm

CBE Documentation Style (Citation-Sequence)

The CBE documentation style is most commonly used in biological and medical fields, though it is gaining popularity in other scientific fields. The official source for this style is *Scientific Style and Format: The CBE Manual for Authors, Editors, and Publishers,* 6th ed. (1994).

TAKE NOTE Unfortunately, the CBE Manual has become noticeably dated, especially where electronic texts are involved. So, other organizations have offered semiofficial updates to the style. You can find links to these updates on this book's website at www.ablongman.com/johnsonweb/C.6

The *CBE Manual* describes two citation methods. The first method, called the *author-year* system, is very similar to the APA style, so it will not be discussed here. The second method, called the *citation-sequence* system, will be discussed here. If you need to use the CBE author-year system, you can consult the *CBE Manual* or websites that offer examples of this system.

In the citation-sequence system, sources are referred to by number in the text, usually with a superscript number similar to a footnote.

> This bacteria has been shown[1] to grow at a significant rate when exposed to black light.

When referring to multiple sources, a dash is used to signal the range of sources.

> Several studies[3–8, 10] have illustrated this relationship.

In some situations, editors will ask for the citations to use numbers in parentheses (3, 12–15) or brackets [2–9, 11] instead of superscript numbers.

Need to use the CBE author-year system? Go to **www.ablongman.com/johnsonweb/C.5**

In the references list at the end of the document, the sources are numbered and listed in the order they were cited in the text

The advantage of the citation-sequence system is that readers feel less disruption than with the author-year system, because the superscript numbers are less intrusive. However, a disadvantage is that readers need to flip back to the list of references to see author names for any sources of information.

The Reference List for CBE Citation-Sequence Style

The following list includes examples of the CBE citation-sequence style. This list is not comprehensive. If you do not find a model here for a document you are adding to your references, you should check *Scientific Style and Format: The CBE Manual for Authors, Editors, and Publishers,* 6th ed.

One of the more frustrating aspects of the CBE style is its lack of clarity about citing on-line sources. According to the Council of Science Editors website (http://www.councilscienceeditors.org), the 7th edition of the *CBE Manual* will include information on electronic sources, closely following a CBE style update created by National Library of Medicine (NLM). The NLM update, the *National Library of Medicine Recommended Formats for Bibliographic Citation, Supplement: Internet Formats,* is available at http://www.nlm.nih.gov/pubs/formats/recommendedformats.html.

Consequently, the models below reflect examples drawn from two sources. For print sources, the *CBE Manual,* 6th ed., was used. For on-line sources, the NLM update of the CBE style was used.

The format of the list of references for the CBE citation-sequence style is somewhat different from that of reference lists following the APA or MLA styles:

- Sources are numbered (1, 2, 3, and so on) to reflect the order in which they were cited.
- The items in the references list are all flush left against the margin (no hanging indent).
- When a citation refers to a specific page or set of pages in a stand-alone document, the full text reference includes the number(s) after a *p.* or a *pp.* (e.g., p. 23 or pp. 123–36). When the citation is referring to the whole book, the reference includes the whole page count of the book followed by a *p.* (e.g., 231 p.).

Items in the reference list can be single spaced or double spaced.

1. **Website or Webpage, Author Known**

   12. Jaspers F. Einstein online [Internet]. Downers Grove (IL): Einstein Inc.; 2001 [cited 2003 Mar. 9]. Available from: http://www.einsteinonlinetoo.com

2. **Website or Webpage, Corporate Author**

   34. National Wildlife Service [NWS]. Managing forest on your land [Internet]. Washington (DC): NWS; 2003 [cited 2004 Sep 8]. Available from: http://www.nws.gov/manageyourforest.htm

For more information on CBE style updates, go to **www.ablongman.com/johnsonweb/C.6**

3. **Webpage, Author Unknown**

   3. [Anonymous]. Skin cancer treatments debated. CNN.com [Internet]. 2004 [cited 2004 Jan 1]. Available from: http://www.cnn.com/2004/HEALTH/conditions/01/19/skincancer.treatment.ap/index.html

4. **Webpage, On-line Periodical**

   7. Grinspoon D. Is Mars ours? Slate Magazine [Internet] 2004 Jan 7 [cited 2004 Jan 19]. Available from: http://slate.msn.com/id/2093579

5. **Book, One Author**

   23. Jones S. Darwin's ghost: the origin of species updated. New York (NY): Balantine Books. 377 p.

6. **Book, More than One Author**

   2. Pauling L, Wilson EB. Introduction to quantum mechanics. New York (NY): Dover Publications; 1935. p. 38.

7. **Book, Corporate or Organization Author**

   11. American Phychiatric Association. Diagnostic and statistical manual of mental disorders. 4th ed. Washington (DC): American Psychiatric Association; 1994. 886 p.

8. **Book, Edited Collection**

   22. Mueller-Vollmer K, editor. The hermeneutics reader. New York (NY): Continuum; 1990. pp. 203–12.

9. **Book, Translated**

   14. Habermas J. Communication and the evolution of society. McCarthy T, translator. Boston (MA): Beacon Press; 1979. p. 156.

10. **Book, Author Unknown**

   13. [Anonymous]. Handbook for the WorkPad c3 PC Companion. Thornwood (NY): IBM; 2000. 253 p.

11. **Book, Second Edition or Beyond**

   21. Williams R, Tollet J. The non-designer's web book. 2nd ed. Berkeley (CA): Peachpit; 2000. pp. 123–7.

12. **Book, Dissertation or Thesis**

   18. Simms L. The hampton effect in fringe desert environments: an ecosystem under stress [dissertation]. Albuquerque (NM): University of New Mexico; 2002. 113 p. Available from: University Microfilms, Ann Arbor (MI); AAD02-56.

13. **Book, Electronic**

   13. Darwin C. On the various contrivances by which British and foreign orchids are fertilised by insects. London (UK): John Murray; 1862 [cited 2002 Sep 5]. 245 p. Available from: http://pages.britishlibrary.net/charles.darwin3/orchids/orchids_fm.htm

14. **Document, Government Publication**

   6. Greene LW. Exile in paradise: the isolation of Hawai'i's leprosy victims and development of Kalaupapa settlement, 1865 to present. Available from: U.S. Department of the Interior, National Park Service, Washington (DC); 1985. 231 p.

15. Document, Pamphlet

23. The Colorado Health Network [TCHN]. Exploring high altitude areas. Available from: TCHN, Denver (CO); 2002. 4 p.

16. Film or Video Recording

16. The lord of the rings: the fellowship of the ring [DVD]. Jackson P, director. Osborne B, Walsh F, Sanders T, producers. Hollywood (CA): New Line Productions; 2002. 1 DVD: 210 min, sound, color.

17. CD-ROM

7. Masters of renaissance art [CD-ROM]. Chicago (IL): Revival Productions; 2000.

18. Article, Journal with Continuous Pagination

34. Boren MT, Ramey J. Thinking aloud: reconciling theory and practice. IEEE Transactions on Professional Communication 1996; 39: 49–57.

19. Article, Journal without Continuous Pagination

32. Lenhoff R, Huber L. Young children make maps! Young Children 2000; 55(5): 6–12.

20. Article, Edited Book

1. Katz SB, Miller CR. The low-level radioactive waste siting controversy in North Carolina: toward a rhetorical model of risk communication. In: Herndl G, Brown SC, editors. Green culture: environmental rhetoric in contemporary America. Madison (WI): U of Wisconsin P; 1996. pp. 111–40.

21. Article, Magazine

12. Appenzeller T. The case of the missing carbon. National Geographic 2004 Feb: 88–118.

22. Article, Newspaper

6. Hall C. Shortage of human capital envisioned, monster's Taylor sees worker need. Chicago Tribune 2002 Nov 18; Sect E; 7 (col 2).

23. Article, Author Unknown

3. [Anonymous]. The big chill leaves bruises. Albuquerque Tribune 2004 Jan 17; Sect A; 4 (col 1).

24. Article, CD-ROM

21. Hanford P. Locating the right job for you. The electronic job finder [CD-ROM]. San Francisco (CA): Career Masters; 2001.

25. Song or Recording

12. Myer L. Sometimes alone. Flatlands [CD]. Ames (IA): People's Productions; 1993.

26. Television or Radio Program

4. Harris R. Destination: the south pole [recording]. Washington (DC): National Public Radio. 2003 Jan 6 [cited 2004 Jan 19]. Available from: http://discover.npr.org/features/feature.jhtml?wfId=904848

27. Personal Correspondence, E-Mail, or Interview

10. Baca J. Personal interview. 2004 Mar 4.

Can't find a CBE model for your source? Go to **www.ablongman.com/johnsonweb/C.7**

Creating the CBE Reference List (Citation-Sequence Style)

In the CBE style, the reference list is placed at the end of the document or in an appendix. The sources are listed by number in the order in which they were referenced in the text.

References

1. Hoffmann A, Blows M. Evolutionary genetics and climate change: will animals adapt to global warming? In: Kareiva P, Kingsolver J, Huey R, editors. Biotic interactions and global change. Sunderland (MA): Sinauerl; 1993. pp. 13–29.

2. Sherwood K, Idso C. Is the global warming bubble about to burst? [Internet]. 2003 [cited 2004 Mar 4]. Available from: http://www.co2science.org/edit/v6_edit/v6n37edit.htm

3. Assel R, Cronk K, Norton D. Recent trends in Laurentian Great Lakes ice cover. Climatic Change 2003; 57: 185–204.

4. Kadlecek M. Global climate change could threaten U.S. wildlife. Conservationist 1991; 46(1): 54–55.

5. Houghton J. Global warming: the complete briefing. 2nd ed. Cambridge (MA): Cambridge UP; 1997. p. 12.

MLA Documentation Style

The MLA Documentation Style is not commonly used in technical or scientific fields. It is most commonly used in the arts and humanities. Nevertheless, there are occasions where the MLA style is requested, because it is a widely used documentation style. The official source for this style is the *MLA Handbook for Writers of Research Papers,* 6th ed. (2003).

When using the MLA style, you will need to use in-text citations and a list of alphabetically arranged references at the end of your document.

MLA In-Text Citations

The MLA style follows an *author-page number* system for in-text citations, meaning the author and page number are usually cited inside the text.

Individual Authors

Individual authors are cited using their last name and page number(s) from which the information was drawn. If the year is significant, put it after the author's name in parentheses.

One study reports a significant rise in HIV cases in South Africa in one year (Brindle 834).

One study reported a 12.2 percent rise in HIV cases in only one year (Brindle 843).

Brindle (2000) reports a significant rise in HIV cases in South Africa in one year.

Brindle (2000) reports a 12.2 percent rise in HIV cases in South Africa in one year (834).

In most cases, only the author and page number need to be noted, as shown in the first example above. In the MLA style, the year of publication is not usually a large concern, so include the year only if it is necessary.

Multiple Authors

If an article has two or more authors, you should use the word *and* to connect the authors.

(Thomas and Linter 130)

According to Thomas and Linter (2001) the number of mammals in this area were dramatically reduced during the Ice Age (130).

Technical documents often have more than two authors. In these cases, cite all the names the first time the work is referenced. Afterward, use the last name of the first author followed by "et al."

First Citation of Work

(Wu, Gyno, Young, and Reims 924)

Subsequent Citations of Work

(Wu et al. 924)

As reported by Wu et al., the Permian Age . . .

Corporate or Unknown Authors

When the author of the document is a corporation or the author is unknown, the in-text citation uses the name of the corporation or the first prominent word in the title of the document.

(National Science Foundation 76)

("Results" 91)

(*Silent* 239)

As shown here, if the source is an article, put the first prominent word in quotes. If it is a book, put it in italics.

Paraphrased Materials

Because the authors' names are typically mentioned in the sentence, citing paraphrased materials requires only a mention of the page number.

Franks and Roberts report that aptitude for visual thinking runs in families (76).

The instinct for survival, according to Ramos, is strong in the Mexican wolf (198–201).

Jones argues that finding a stand of dead trees near an industrial plant is a good indicator that something is seriously wrong (87–88).

More information on the MLA style is available at **www.ablongman.com/johnsonweb/C.8**

Two or More Works in Same Parentheses

In some cases, several documents will state similar information. In these cases, you should cite them all.

> Studies have shown remarkable progress toward reviving the penguin population on Vostov Island (Hinson and Kim 330; Johnson and Smith 87; Tamili 102).

The Reference List for MLA Style

When using the MLA style, your references should be listed in alphabetical order at the back of the document. In your references list, you should include only items that were actually cited in your document. Leave any uncited documents out of the list of references. After all, if a document is important enough to list in your references, it should be important enough to cite in the text itself.

In most cases, the reference list is identified by the centered heading "References." Entries can be single spaced or double spaced, but leave a space between each entry. Also, in most cases, each reference should use a hanging indent style (i.e., the second line and subsequent lines should be indented). The first line should be flush with the left margin.

The following list includes examples of the MLA style. This list is not comprehensive. If you do not find a model here for a source you are adding to your references, you should check the *MLA Handbook for Writers of Research Papers,* 6th ed. Or, check the several websites available that offer examples of MLA style.

1. **Website or Webpage, Author Known**

   Jaspers, Fred. *Einstein Online.* 9 Mar. 2003 <http://www.einsteinonlinetoo.html>.

2. **Website or Webpage, Corporate Author**

   National Wildlife Service. *Managing Forest on Your Land.* 8 Sep. 2002. 10 Oct. 2004 <http://www.nws.gov/manageyourforest.htm>.

3. **Webpage, Author Unknown**

   "Skin Cancer Treatments Debated." *CNN.com* 1 Jan. 2004 <http://www.cnn.com/2004/HEALTH/conditions/01/19/skincancer.treatment.ap/index.html>.

4. **Webpage, On-line Periodical**

   Grinspoon, David. "Is Mars Ours?" *Slate Magazine* 7 Jan. 2004, 19 Jan. 2004. <http://slate.msn.com/id/2093579>.

5. **Book, One Author**

   Jones, Steve. *Darwin's Ghost: The Origin of Species Updated.* New York, NY: Balantine Books, 2001.

6. **Book, More than One Author**

   Pauling, Linus, and E. Bright Wilson. *Introduction to Quantum Mechanics.* New York, NY: Dover Publications, 1935.

7. **Book, Corporate or Organization**

   American Psychiatric Association. *Diagnostic and Statistical Manual of Mental Disorders.* 4th ed. Washington, DC: American Psychiatric Association, 1994.

Can't find an MLA model for your source? Go to **www.ablongman.com/johnsonweb/C.9**

8. **Book, Edited Collection**

Mueller-Vollmer, Kurt, ed. *The Hermeneutics Reader.* New York, NY: Continuum, 1990.

9. **Book, Translated**

Habermas, Jurgen. *Communication and the Evolution of Society.* Trans. Thomas McCarthy. Boston, MA: Beacon Press, 1979.

10. **Book, Author Unknown**

Handbook for the WorkPad c3 PC Companion. Thornwood, NY: IBM, 2000.

11. **Book, Second Edition or Beyond**

Williams, Robin, and John Tollet. *The Non-Designer's Web Book.* 2nd ed. Berkeley, CA: Peachpit, 2000.

12. **Book, Dissertation or Thesis**

Simms, Laura. "The Hampton Effect in Fringe Desert Environments: An Ecosystem Under Stress." Diss. U of New Mexico, 2002.

13. **Book, Electronic**

Darwin, Charles. *On the Various Contrivances by which British and Foreign Orchids are Fertilised by Insects.* London, UK: John Murray, 1862. 1 Jan. 2004 <http://pages.britishlibrary.net/charles.darwin3/orchids/orchids_fm.htm>.

14. **Document, Government Publication**

Greene, Linda W. *Exile in Paradise: The Isolation of Hawai'i's Leprosy Victims and Development of Kalaupapa Settlement, 1865 to Present.* Washington, DC: US Department of the Interior, National Park Service, 1985.

15. **Document, Pamphlet**

Exploring High Altitude Areas. Denver, CO: TCHN, 2002.

16. **Film or Video Recording**

The Lord of the Rings: The Fellowship of the Ring. Dir. Peter Jackson. Prod. Barrie Osborne, Peter Jackson, Fran Walsh, and Tim Sanders. New Line Productions, 2002.

17. **CD-ROM**

Masters of Renaissance Art. CD-ROM. Chicago, IL: Revival Productions, 2000.

18. **Article, Journal with Continuous Pagination**

Boren, M. Ted, and Judith Ramey. "Thinking Aloud: Reconciling Theory and Practice." *IEEE Transactions on Professional Communication* 39 (1996): 49–57.

19. **Article, Journal without Continuous Pagination**

Lenhoff, Rosalyn, and Lynn Huber. "Young Children Make Maps!" *Young Children* 55.5 (2000): 6–12.

20. **Article, Edited Book**

Katz, Steven B., and Carolyn R. Miller. "The Low-Level Radioactive Waste Siting Controversy in North Carolina: Toward a Rhetorical Model of Risk Communication." *Green Culture: Environmental Rhetoric in Contemporary America.* Ed. Carl G. Herndl and Stuart C. Brown. Madison, WI: U of Wisconsin P, 1996. 111–40.

21. **Article, Magazine**

Appenzeller, Tim. "The Case of the Missing Carbon." *National Geographic* Feb. 2004: 88–118.

22. **Article, Newspaper**

Hall, Cheryl. "Shortage of Human Capital Envisioned, Monster's Taylor Sees Worker Need." *Chicago Tribune* 18 Nov. 2002: E7.

23. **Article, Author Unknown**

"The Big Chill Leaves Bruises." *Albuquerque Tribune* 17 Jan. 2004: A4.

24. **Article, CD-ROM**

Hanford, Peter. "Locating the Right Job For You." *The Electronic Job Finder.* CD-ROM. San Francisco, CA: Career Masters, 2001.

25. **Song or Recording**

Myer, Larry. "Sometimes Alone." *Flatlands.* Ames, IA: People's Productions, 1993.

26. **Television or Radio Program**

Harris, Richard. "Destination: The South Pole." All Things Considered. National Public Radio. 6 Jan. 2003. 4 Feb. 2004 <http://discover.npr.org/features/feature.jhtml?wfId=904848>.

27. **Personal Correspondence, E-Mail, or Interview**

Baca, James. Personal interview. 4 Mar. 2004.

Creating the MLA Reference List

In the MLA style, the reference list is placed at the end of the document or in an appendix. The sources referenced in the document should be listed alphabetically.

References

Assel, Robert, Kevin Cronk, and David Norton. "Recent Trends in Laurentian Great Lakes Ice Cover." *Climatic Change* 57 (2003): 185–204.

Hoffmann, Amber, and Marlin Blows. "Evolutionary Genetics and Climate Change: Will Animals Adapt to Global Warming?" *Biotic interactions and Global Change.* Ed. Paul M. Kareiva, John G. Kingsolver and Renee B. Huey. Sunderland, MA: Sinauer, 1993. 13–29.

Houghton, James. *Global Warming: The Complete Briefing.* 2nd ed. Cambridge, MA: Cambridge UP, 1997.

Kadlecek, Mary. "Global Climate Change Could Threaten U.S. Wildlife." *Conservationist* 46.1 (1991): 54–55.

Sherwood, Kevin and Craig Idso. "Is the Global Warming Bubble About to Burst?" 4 Mar. 2004 <http://www.co2science.org/edit/v6_edit/v6n37edit.htm>.

References

AEC Software. (2003). The university quarterly publication. Retrieved August 8, 2003, from the World Wide Web: http://www.aecsoftware.com/downloads/documents/pdf/examples/fasttrack/Publishing-res.pdf

Allen, L., & Voss, D. (1997). *Ethics in technical communication: Shades of gray.* New York, NY: Wiley.

American Institute of Aeronautics and Astronautics. (2003). View survey of how design engineers evaluate their education. Retrieved November 2, 2003, from the World Wide Web: http://www.aiaa.org/tc/de

American Lung Association. (2003). Top stories. Retrieved August 1, 2003, from the World Wide Web: http://www.lungusa.org

American Psychological Association. (2001). *Publication manual of the American Psychological Association* (5th ed.). Washington, DC: American Psychological Association.

American Red Cross. (2003). In the news. Retrieved August 2, 2003, from the World Wide Web: http://www.redcross.org

Americans for Tax Reform. (2003). *Americans For Tax Reform cost of government day chart room.* Retrieved May 10, 2004, from the World Wide Web: http://www.atr.org/pdffiles/cogdcharts2003.pdf

Aristotle. (1991). *On rhetoric.* (G. Kennedy, Trans.). New York, NY: Oxford UP.

Arnheim, R. (1969). *Visual thinking.* Berkeley, CA: U of California P.

Australian Resuscitation Council. (2002). *Basic life support flowchart.* Retrieved January 4, 2004, from the World Wide Web: http://www.resus.org.au/public/bls_flow_chart.pdf

Aversa, A. (2003). *Galaxy simulations.* Retrieved February 3, 2004, from the World Wide Web: http://www.u.arizona.edu/~aversa/galaxysims.pdf

Avery, D., & Avery, A. (2001, August 31). Hope for sustainable farming in gene-altered crops. *The Christian Science Monitor.* Retrieved May 5, 2003, from the World Wide Web: http://www.csmonitor.com/2001/0830/p9s1-coop.html

Belbin, R. (1981). Management teams. New York, NY: Wiley.

Benson, P. (1985). Writing visually: Design considerations in technical publications. *Technical Communication, 32,* 35–39.

Berg, A. (2003). *Counseling to prevent tobacco use and tobacco-caused disease.* Washington, DC: U.S. Preventive Services Task Force.

Bernhardt, S. (1986). Seeing the text. *College Composition and Communication, 30,* 66–78.

Bledsoe, L., & Bibhuti, S. (2001). *Campus survey report: Safety perception and experiences of violence.* Retrieved July 18, 2004, from the World Wide Web: http://www.louisville.edu/provost/womenctr/peacc/executive%20summary.pdf

Blum, D., & Knudson, M. (1997). *Field guide for science writers.* Oxford, UK: Oxford UP.

Boisjoly, R. (1985). SRM o-ring erosion/potential failure criticality. In Report of the Presidential Commission on the Space Shuttle *Challenger* Accident. Retrieved March 10, 2004, from the World Wide Web: http://science.ksc.nasa.gov/shuttle/missions/51-l/docs/rogers-commission/table-of-contents.html

Booth, W., Columb, G.C., & Williams, J. (1995). *The craft of research.* Chicago, IL: U of Chicago P.

Bosley, D. (2001). *Global contexts.* Boston, MA: Allyn & Bacon.

Brody, M. (1998). *Speaking your way to the top.* Boston, MA: Allyn & Bacon.

Brusaw, C., Alred, G., & Oliu, W. (1993). *Handbook of technical writing* (4th ed.). New York, NY: St. Martin's Press.

Buckingham County Council. (2003). Flooding—cause and effect. Retrieved January 14, 2004, from the World Wide Web: http://www.buckscc.gov.uk/emergency_planning/ff_cause_and_effect_2003.htm

Buzan, T. (1994). *The mind map book: How to use radiant thinking to maximize your brain's untapped potential.* New York, NY: Dutton.

California Energy Commission. (2001). *Costs and benefits of a biomass-to-ethanol production industry in California: Commission report.* Sacramento, CA: California State Government.

Campbell, G. M. (2002). *Bulletproof presentations.* Franklin Lake, NJ: Career Press.

Canadian Rockhound. (2003). Welcome to Canada's free online earth science magazine. Retrieved January 17, 2003, from the World Wide Web: http://www.canadianrockhound.com

Carey, S. S. (1998). *A beginner's guide to the scientific method.* (2nd ed.). Belmont, CA: Wadworth Publishing.

Carson, R. (2002). *Silent spring.* New York, NY: Mariner.

Chaney, L., & Martin, J. (2004). *Intercultural business communication* (3rd ed.). Upper Saddle River, NJ: Pearson Prentice Hall.

Cincinnati Children's Hospital Medical Center. (2003). Hirschsprung's Disease. Retrieved September 1, 2003, from the World Wide Web: http://www.cincinnatichildrens.org/health/info/abdomen/diagnose/hirschsprung.htm

Clark, H., & Haviland, S. (1977). Comprehension and the given-new contract. In Freedle, R. (Ed.). *Discourse production and comprehension* (pp. 1–40). Norwood, NJ: Ablex.

Clark, R. (1971). *Einstein: His life and times.* New York, NY: World Publishing.

Clement, D. E. (1987). Human factors, instructions, and warning, and product liability. *IEEE Transactions on Professional Communication, 30,* 149–56.

Council of Biology Editors (1994). *The CBE manual for authors, editors, and publishers.* Cambridge, UK: Cambridge UP.

Curlew Communications. (2003). *Hemyock Castle.* Retrieved May 21, 2003, from the World Wide Web: http://www.hemyockcastle.co.uk/bodiam.htm

De Vries, M. A. (1994). *Internationally yours: Writing and communicating successfully in today's global marketplace.* New York, NY: Houghton Mifflin.

Deming, W. E. (2000). *Out of crisis.* Cambridge, MA: MIT Press.

Dombrowski, P. (2000). *Ethics in technical communication.* Needham Heights, MA: Allyn & Bacon.

Dragga, S. (1997). A question of ethics: Lessons from technical communicators on the job. *Technical Communication Quarterly, 6,* 161–78.

Dumas, J. S., & Redish, J. C. (1993). *A practical guide to usability testing.* Norwood, NJ: Ablex.

Einstein, A. (1939). *August 2, 1939 letter to Franklin Roosevelt.* Retrieved September 15, 2003, from the World Wide Web: http://www.anl.gov/OPA/frontiers96arch/aetofdr.html

Farkas, D., & Farkas, J. (2002). *Principles of web design.* New York, NY: Longman.

Fermi National Laboratory. (2003). Fermilab. Retrieved August 10, 2003, from the World Wide Web: http://www.fermilab.gov

Gibaldi, J. (2003). *MLA handbook for writers of research papers* (6th ed.). New York, NY: Modern Language Association.

Goddard Institute for Space Studies. (2004). Global temperature trends: 2002 summation. Retreived June 10, 2004, from the World Wide Web: http://www.giss.nasa.gov/research/observe/surftemp

Google. (2003). *Google.* Retrieved August 10, 2003, from the World Wide Web: www.google.com

Guilford, C. (1996). Paradigm online writing assistant. Retrieved October 5, 2003, from the World Wide Web: http://www.powa.org

Gurak, L. (2000). *Oral presentations for technical communication.* Boston, MA: Allyn & Bacon.

Hallberg, B. A. (2003). *Networking: A beginner's guide.* (3rd ed.). New York, NY: McGraw-Hill.

Haneda, S., & Shima, H. (1983). Japanese communication behavior as reflected in letter writing. *Journal of Business Communication, 19,* 19–32.

Haviland, S., & Herbert, C. (1974). What's new: Acquiring new information as a process in comprehension. *Journal of Verbal Learning and Verbal Behavior, 13,* 512–21.

Herrington, T. (2003). *A legal primer for the digital age.* New York, NY: Pearson Longman.

Hoft, N. (1995). *International technical communication.* New York, NY: Wiley.

Hult, C. & Huckin, T. (2001). *The new century handbook* (Brief ed.). Boston, MA: Allyn & Bacon.

Husqvarna. (2002). *Working with a chainsaw.* Åsbro, Sweden: Electrolux.

IBM. (2003). IBM Lotus Notes. Retrieved April 23, 2003, from the World Wide Web: http://www.lotus.com/products/product4.nsf/wdocs/notesnamepage

Indiana Prevention Resource Center at Indiana University. (2001). Marijuana use among Indiana 6th–12th graders, 2001. Retrieved January 10, 2004, from the World Wide Web: http://www.iprc.indiana.edu/drug_stats/youth2001/s2001-26.gif

InfoSpace (2003). *Metacrawler, Advanced Search.* Retrieved March 13, 2003, from the World Wide Web: http://www.metacrawler.com/info.metac/search/advance.htm

InfoSpace (2003). *Metacrawler.* Retrieved March 13, 2003, from the World Wide Web: http://www.metacrawler.com

Institute of Electrical and Electronics Engineers (1990). *IEEE code of ethics.* Retrieved May 10, 2003, from the World Wide Web: http://www.ieee.org/portal/site/mainsite/menuitem.818c0c39e85ef176fb2275875bac26c8/index.jsp?&pName=corp_level1&path=about/whatis&file=code.xml&xsl=generic.xsl

International Maritime Organization. (2000). *No anchoring areas for Flower Garden Banks in the northwestern Gulf of Mexico.* London, UK: International Maritime Organization.

International Organization for Standardization. (2003). ISO 9000 and ISO 14000—in brief. Retrieved December 19, 2003, from the World Wide Web: http://www.iso.ch/iso/iso9000-14000/index.html

Johnson-Sheehan, R., & Baehr, C. (2001). Visual-spatial thinking: Thinking differently about hypertexts. *Technical Communication, 48,* 37–57.

Johnson-Sheehan, R. (2002). *Writing proposals: A rhetoric for managing change.* Boston, MA: Longman.

Jones, D. (1998). *Technical writing style.* Boston, MA: Allyn & Bacon.

Kiester, E., & Kiester, W. (2003, July). Birdbrain breakthrough. *Smithsonian,* 36–37.

Koffka, K. (1935). *Principles of gestalt psychology.* New York, NY: Harcourt.

Kostelnick, C., & Roberts, D. (1998). *Designing visual language.* Boston, MA: Allyn & Bacon.

Laib, N. (1993). *Rhetoric and style.* Englewood Cliffs, NJ: Prentice Hall.

Lanham, R. (1999). *Revising business prose* (4th ed.). New York, NY: Pearson Longman.

Leopold, A. (1966). *A Sand county almanac.* New York, NY: Ballantine.

Lipnack, J., & Stamps, J. (2000). *Virtual teams: People working across boundaries with technology.* New York, NY: Wiley.

Markel, M. M. (1991). A basic unit on ethics for technical communicators. *Journal of Technical Writing and Communication, 21,* 327–50.

Mathes, J., & Stevenson, D. (1976). *Designing technical reports.* Indianapolis, IN: Bobbs-Merrill.

McColman, R., & Archuleta, M. (2003). *Control your diabetes for life: What is diabetes?* Las Cruces, NM: New Mexico State University.

Microsoft. (2003). Project Standard 2003 overview. Retrieved March 5, 2003, from the World Wide Web: http://www.microsoft.com/office/project/prodinfo/standard/overview.mspx

Miller, C. R., Larsen, J., & Gaitens, J. (1996). Communication in the workplace: What can NCSU students expect? Retrieved November 2, 2003, from http://www.chass.ncsu.edu/ccstm/pubs/no2/index.html

Monster. (2004). Get work. Network. Do it all day. Retrieved September 8, 2004, from the World Wide Web: http://www.monster.com

Moore, P., & Fitz, C. (1993). Using gestalt theory to teach document design and graphics. *Technical Communication Quarterly, 2,* 389–410.

NASA. (2003). Mars Pathfinder Rover. Retrieved June 15, 2003, from the World Wide Web: http://nssdc.gsfc.nasa.gov/database/MasterCatalog?sc=MESURPR

National Foundation for Cancer Research. (1998). *On your health: Cancer detection.* Bethesda, MD: NFCR.

National Human Genome Research Institute. (2003). *National Human Genome Research Institute.* Retrieved December 19, 2003, from the World Wide Web: http://www.genome.gov

National Library of Medicine (2001). *National Library of Medicine recommended formats for bibliographic citation: Supplement: Internet formats.* Bethesda, MD: NLM.

National Library of Medicine. (2003). *Medline Plus.* Retrieved November 6, 2004, from the World Wide Web: http://www.nlm.nih.gov/medlineplus/medlineplus.html

National Ocean Service. (2003). *National Ocean Service Accomplishments.* Washington, DC: National Ocean Service.

National Science Foundation. (2002). Grant proposal guide. Retrieved October 8, 2004, from the World Wide Web: http://www.nsf.gov/pubs/2003/nsf032/032_2.htm

National Science Foundation. (2004). Highlights. Retrieved October 8, 2004, from the World Wide Web: http://www.nsf.gov

National Survey on Drug Use and Health. (2003). *Alcohol use and risks among young adults by college enrollment status.* Washington, DC: Office of Applied Studies, Substance Abuse, and Mental Health Services Administration.

Nikon. (2002). *The Nikon guide to digital photograph with the Coolpix 885 digital camera.* Tokyo, Japan: Nikon.

Nokia. (2003). *User guide: Nokia 2100.* Finland: Nokia.

Nuclear Energy Institute. (2003). Benefits of nuclear energy. Retrieved April 18, 2003, from the World Wide Web: http://www.nei.org/doc.asp?catnum=2&catid=118

Ocean Alliance. (2002). Welcome to the Ocean Alliance. Retrieved August 2, 2004, from the World Wide Web: http://www.oceanalliance.org

Ong, W. (1982). *Orality and literacy.* London, UK: Routledge.

Online Ethics Center for Engineering and Science. (2003). Welcome to the online ethics center for engineering and science. Retrieved July 5, 2003, from the World Wide Web: http://www.onlineethics.org

Oxford English Dictionary. (2003). Solstice. Retrieved June 13, 2003, from the World Wide Web: http://www.oed.com

Pakiser, L., & Shedlock, K. (1997). *Earthquakes.* Retrieved February 10, 2003, from the World Wide Web: http://pubs.usgs.gov/gip/earthq1/earthqkgip.html

Palm. (2003). Palm Tungsten T3. Retrieved December 10, 2003, from the World Wide Web: http://www.palmone.com/us/products/handhelds/tungsten-t3/specs.html

Pearsall, T. (2001). *Elements of technical writing.* Needham Heights, MA: Allyn & Bacon.

Peckham, G. (2003). Safety symbols. *Compliance engineering* [website]. Retrieved March 13, 2004, from the World Wide Web: http://www.ce-mag.com/archive/02/03/peckham.html

Pew Research Center. (2003). Reports. Retrieved August 6, 2003, from the World Wide Web: http://www.pewinternet.org/reports/index.asp

Piven, J., & Borgenicht, D. (1999). *The worst-case scenario survival handbook.* San Francisco, CA: Chronicle.

Piven, J., & Borgenicht, D. (2003). How to escape from a sinking car. *Worst-case scenarios online.* Retrieved October 8, 2003, from the World Wide Web: http://www.worstcasescenarios.com/mainpage.htm

Plotnik, A. (1982). *The elements of editing.* New York, NY: Macmillan.

Population Action International. (2004). Condoms Count 2004 data update. Retrieved April 10, 2004, from the World Wide Web: http://www.populationaction.org/resources/publications/condomscount/graphs.htm

Price, J. (1999). *Outlining goes electronic.* Stamford, CT: Ablex.

Pyle, R. (1981). *National Audubon Society field guide to North American butterflies.* New York, NY: Knopf.

Reynolds, S., & Valentine. D. (2004). *Guide to cross-cultural communication.* Upper Saddle River, NJ: Pearson Prentice Hall.

Rooksby, E. (2002). *E-mail and ethics: Style and ethical relations in computer-mediated communication.* New York, NY: Routledge.

Rubin, J. (1994). *Handbook of usability testing.* New York, NY: Wiley.

Rude, C. (1998). *Technical editing* (2nd ed.). Needham Heights, MA: Allyn & Bacon.

Ryobi. (2000). *510r 4-cycle garden cultivator operator's manual.* Chandler, AZ: Ryobi.

Safetyline Institute. (1998). Gas laws. Retrieved June 30, 2004, from the World Wide Web: http://www.safetyline.wa.gov.au/institute/level2/course16/lecture47/l47_02.asp

Schriver, K. (1997). *Dynamics in document design.* New York, NY: Wiley.

Schultz, H. (2000). *The elements of electronic communication.* Needham Heights, MA: Allyn & Bacon.

Segway. (2004). A gyro is not just a sandwich. Retrieved August 5, 2004, from the World Wide Web: http://www.segway.com/segway/component_details.html

Shapiro, G. (1998, March). The ABCs of asthma. *Discover, 35,* 30–33.

Shenk, D. (1998). *Data smog: Surviving the information glut, revised and updated edition.* San Francisco, CA: HarperCollins.

Silyn-Roberts, H. (2000). *Writing for science and engineering: Papers, presentations, and reports.* Boston, MA: Butterworth-Heinemann.

Silyn-Roberts, H. (1998). Using engineers' characteristics to improve report writing instruction. *Journal of Professional Issues in Engineering Education and Practice, 124,* 12–16.

State of Utah Office of the Attorney General. (2004). What is meth? Retrieved October 10, 2004, from the World Wide Web: http://attorneygeneral.utah.gov/Meth/Whatismeth.htm

Stauber, J., & Rampton, S. (1995). *Toxic sludge is good for you! Lies, damn lies and the public relations industry.* Monroe, ME: Common Courage Press.

Stewart-MacDonald. (2003). Violin kit assembly instructions. Retrieved October 10, 2003, from the World Wide Web: http://www.stewmac.com/shop/Kits/Violin_Kits/1/Violin_Kits/Instructions/I-5342.html

Terminello, V., & Reed, M. (2003). *E-mail: Communicate effectively.* Upper Saddle River, NJ: Prentice Hall.

The Field Museum. (2001). *The Field Museum.* Retrieved June 10, 2003, from the World Wide Web: http://www.fieldmuseum.org

The Writing Center at Rensselaer Polytechnical Institute. (2003). Revising prose. Retrieved October 5, 2003, from the World Wide Web: http://www.rpi.edu/dept/llc/writecenter/web/revise.html

TiVo. (2002). *Start here.* Aviso, CA: TiVo.

Toastmasters International. (2004). Making effective communication a worldwide reality. Retrieved April 5, 2004, from the World Wide Web: http://www.toastmasters.org

Tuckman, B.W. (1965). Development sequence in small groups. *Psychological Bulletin, 63,* 384–99.

Tufte, E. (1983). *The visual display of quantitative information.* Cheshire, MA: Graphics Press.

U.S. Census Bureau. (2003). Homepage. Retrieved August 5, 2003, from the World Wide Web: http://www.census.gov

U.S. Centers for Disease Control. (2003). *Fight the bite.* Retrieved November 2, 2003, from the World Wide Web: http://www.cdc.gov/ncidod/dvbid/westnile/index.htm

U.S. Centers for Disease Control. (2003). The rabies virus. Retrieved April 18, 2004, from the World Wide Web: http://www.cdc.gov/ncidod/dvrd/rabies/the_virus/virus.htm

U.S. Centers for Disease Control. (2003). *West Nile Virus (WNV) infection: Information for clinicians.* Retrieved November 2, 2003, from the World Wide Web: http://www.cdc.gov/ncidod/dvbid/westnile/index.htm

U.S. Centers for Disease Control. (2004). Emergency department visit data. Retrieved August 4, 2004, from the World Wide Web: http://www.cdc.gov/nchs/about/major/ahcd/ercharts.htm

U.S. Copyright Office. (2003). About copyright. Retrieved March 2, 2003, from the World Wide Web: http://www.loc.gov/copyright

U.S. Department of Energy. (2003). Fuel cell technology: How it works. Retrieved from the World Wide Web: http://www.fe.doe.gov/coal_power/fuelcells/fuelcells_howitworks.shtml

U.S. Environmental Protection Agency. (2000). *National water quality inventory, 2000 report.* Washington, DC: U.S. EPA.

U.S. Environmental Protection Agency. (2002). Global warming impacts: Forests. Retrieved March 2, 2003, from the World Wide Web: http://yosemite.epa.gov/oar/globalwarming.nsf/content/ImpactsForests.html

U.S. Environmental Protection Agency. (2004). *IPM for schools: A how-to manual.* Washington, DC: U.S. EPA.

U.S. Federal Bureau of Investigations (2003). FBI history. Retrieved September 18, 2003, from the World Wide Web: http://www.fbi.gov/fbihistory.htm

U.S. Geological Survey. (2002). Eruption history of Kilauea. Retrieved July 10, 2004, from the World Wide Web: http://hvo.wr.usgs.gov/kilauea/history/main.html

U.S. Geological Survey. (2003). *A proposal for upgrading the national-scale soil geochemical database for the United States.* Washington, DC: U.S. Geological Survey.

U.S. Geological Survey. (2004). Geology. Retrieved February 10, 2004, from the World Wide Web: http://geology.usgs.gov/index.shtml

U.S. National Weather Service. (2004). Frostbite and hypothermia. Retrieved June 10, 2004, from the World Wide Web: http://www.crh.noaa.gov/unr/edusafe/whys/hypthermia.htm

U.S. Office of Energy Efficiency and Renewable Energy. (2004). Solar decathalon, 2005. Retrieved February 10, 2004, from the World Wide Web: http://www.eere.energy.gov/solar_decathalon

University of Chicago Press. (2003) *Chicago manual of style* (15th ed.). Chicago, IL: U of Chicago P.

University of New Mexico General Library. (2003). Homepage. Retrieved August 10, 2003, from the World Wide Web: http://elibrary.unm.edu

University of North Carolina Writing Center. (2003). Improving your writing style. Retrieved October 5, 2003, from the World Wide Web: http://www.unc.edu/depts/wcweb

U.S. Environmental Protection Agency. (2000). Global warming—impacts. Retrieved February 5, 2004, from the World Wide Web: http://yosemite.epa.gov/oar/globalwarming.nsf/content/ImpactsForests.html

Velasquez, M. G. (2002). *Business ethics: Concepts and cases* (5th ed.). Upper Saddle River, NJ: Prentice Hall.

Velotta, C. (1987). Safety labels: What to put in them, how to write them, and where to place them. *IEEE Transactions on Professional Communication, 30,* 121–26.

Vucetich, J. (2003). Population data from the wolves and moose of Isle Royale. Retrieved May 6, 2003, from the World Wide Web: http://www.isleroyalewolf.org/wolf-moose_data.htm

Washington State Department of Ecology. (2003). *North Creek cleanup plan.* Retrieved February 9, 2004, from the World Wide Web: http://www.ecy.wa.gov/programs/wq/tmdl/watershed/north-creek/solution.html

West Virginia Office of Emergency Medical Services. (2001). EMT-paramedic treatment protocol 4501. Charleston, WV: West Virginia EMS System.

White, J. (1988). *Graphic design for the electronic age.* New York, NY: Watson-Guptill.

Williams, J. (1990). *Style.* Chicago, IL: U of Chicago P.

Williams, R., & Tollett, J. (2000). *The non-designer's web book* (2nd. ed.). Berkeley, CA: Peachpit.

Williams, R. (2004). *The non-designer's design book* (2nd. ed.). Berkeley, CA: Peachpit.

Credits

Page 8 Reprinted with permission of National Human Genome Research Institute. **Page 10** PowerStock/Javier Larrea. **Page 12** Reprinted with permission of National Ocean Service. **Page 26** Reprinted with permission of TiVo, Inc. **Page 33** ImageState/Photosure.com. **Page 37** Reprinted with permission of The American National Red Cross. © Copyright 2003, The American National Red Cross. All rights reserved. **Pages 48–49** Reprinted with permission of The Smoking Gun. **Page 52** Reprinted with permission of Google. **Pages 56–57** Reprinted with permission of Centers for Disease Control. **Pages 58–59** Reprinted with permission of Centers for Disease Control. **Page 60** © Dorling Kindersley. For further information see www.dkimages.com. **Page 65** Reprinted with permission of Apple Computer, Inc. **Page 73** Getty Images/Time Life Pictures/Alfred Eisenstaedt. **Pages 74–75** Reprinted with permission of Argonne National Laboratory. **Page 82** Reprinted with permission of U.S. Copyright Office. **Page 85** Reprinted with permission of the Institute of Electrical and Electronics Engineers, Inc. **Page 87** Reprinted with permission of The Online Ethics Center for Engineering and Science. **Page 98** AGEFotostock/PowerStock **Page 107** Reprinted with Permission of InfoSpace, Inc., © 2004 InfoSpace, Inc. All rights reserved. **Page 108** Reprinted with permission of InfoSpace, Inc., © 2004 InfoSpace, Inc. All rights reserved. **Page 109** Reprinted with permission of The University of New Mexico. **Pages 112–113** Reprinted with permission of Linda Bledsoe and Bibhuti Sar. This work was supported under 1999-WA-VX-0012 from the U.S. Department of Justice "Grants to Combat Violent Crimes Against Women on Campuses." The opinions, findings, and conclusions expressed in this document are those of the authors and do not necessarily represent the official position of the U.S. Department of Justice. **Page 114** Superstock/PowerStock **Pages 120–121** Reprinted with permission of U.S. Geological Survey. **Page 145** Reprinted with permission of Canadian Rockhound. **Page 148** Reprinted with permission of Buckinghamshire County Council, UK. **Page 149** Reprinted with permission of Curlew Communications, Ltd. **Page 151** Reprinted with permission of U.S. Environmental Protection Agency. **Page 152** Reprinted with permission of Alan G. Aversa. **Pages 153–155** Reprinted with permission of Washington State Department of Ecology. **Page 167** Reprinted with permission of the University of North Carolina. **Page 168** Reprinted with permission of Paradigm Online Writing Assistant. **Page 169** Reprinted with permission of Rensselaer Polytechnic Institute. **Page 171** Reprinted with permission of Nuclear Energy Institute. **Pages 186–187** Reprinted with permission of Dennis Avery and Alex Avery and The Christian Science Monitor. **Page 196** Reprinted with permission of Patagonia; photo courtesy of Topher Donahue © 2001 Patagonia, Inc. **Page 197** Reprinted with permission of Smithsonian Magazine. **Page 198** Reprinted with permission of Fermilab National Accelerator Laboratory. **Page 202** Reprinted with permission of U.S. Environmental Protection Agency. **Page 204** Reprinted with permission of Federal Bureau of Investigation. **Page 208** Reprinted with permission of The Field Museum. Image of Eternal Egypt in with permission from The British Museum. Image of Baseball as America with permission by the Baseball Hall of Fame. **Page 213** Reprinted with permission of Nokia. **Page 219** Reprinted with permission of the National Library of Medicine. **Page 226** © Dorling Kindersley. For further information see www.dkimages.com. **Pages 230–232** Reprinted with permission of U.S. Environmental Agency. **Page 238** Reprinted with permission of Population Action International. **Page 239** Reprinted with permission of Corel Corporation © 2004 and its licensors. All rights reserved. **Page 239** Reprinted with permission of Americans for Tax Reform Foundation. **Page 241** Reprinted with permission of Indiana Prevention Resource Center. **Page 245** Reprinted with permission of The Safetyline Institute, Government of Western Australia. **Page 246** Reprinted with permission of Centers for Disease Control. **Page 248** Reprinted with permission of National Foundation for Cancer Research. **Page 249** "Fire Arms Accessible to Children," 2003, reprinted with permission of Rand Corporation. **Page 250** Reprinted with permission of Australian Resuscitation Council. **Page 253** Reprinted with permission of Corel Corporation © 2004 and its licensors. All rights reserved. **Page 254** Reprinted with permission of The Internet Public Library at the University of Michigan. **Page 256** Reprinted with permission of Centers for Disease Control. **Page 257** Reprinted with permission of the New Mexico Space Grant Consortium. **Page 257** Images of balance and eagle in Figure 9.22 reprinted with permission of Centers for Disease Control. Image of sneezing person in Figure 9.22 reprinted with permission of International Association for Food Protection. **Page 258** Reprinted with permission of Office of Energy Efficiency and Renewable Energy. **Page 259** Reprinted with permission of The Gallup Organization. **Page 279** Reprinted with permission of U.S. Geological Survey. **Page 285** © Dorling Kindersley. For further information see www.dkimages.com. **Page 299** Doug Martin/Photo Researchers, Inc. **Page 300** © Dorling Kindersley. For further information see www.dkimages.com. **Page 302** Getty Images/Stone/Frank Herholdt **Page 307** © Dorling Kindersley. For further information see www.dkimages.com. **Page 311** Getty Images/Image Bank/Ghislain & Marie David de Lossy. **Page 316** Graph shown in Figure 11.16 reprinted with permission of NASA Goddard Institute for Space Studies. **Page 318** Getty Images/Stone/Walter Hodges. **Page 322** Reprinted with permission of Toastmasters International. **Page 359** Reprinted with permission of Microsoft Corporation. **Page 371** Alamy/Dynamic Graphics Group/Creatas. **Page 373** Reprinted with permission of IBM. **Page 383** Reprinted with permission of American Lung Association. **Page 392** Reprinted with permission of Ocean Alliance. **Page 393** Reprinted with permission of Ocean Alliance. **Page 394** Reprinted with permission of Ocean Alliance. **Page 399** Reprinted with permission of U.S. Geological Survey. **Page 409** Reprinted with permission of Monster. **Page 432** © Dorling Kindersley. For further information see www.dkimages.com. **Page 437** © Dorling Kindersley. For further information see www.dkimages.com. **Pages 488–489** Reprinted with permission of Chanticleer Press. **Page 496** Reprinted with permission of the Oxford English Dictionary. **Page 498** Reprinted courtesy of Cincinnati Children's Hospital Medical Center. **Page 503** Reprinted with permission of Utah Attorney General's Office. **Pages 505–506** Reprinted with permission of New Mexico State University. **Pages 511–512** Reprinted with permission of SOHO, ESA, NASA. **Pages 518–521** Reprinted with permission of Segway, LLC. **Page 524** Reprinted with permission of International Organization for Standardization. The material taken from ISO publication "ISO 9000 and ISO 14000—in brief" is reproduced with the permission of the International Organization for Standardization, ISO. The complete text is posted on ISO's Web site, www.iso.org. Copyright remains with ISO. **Pages 528-530** Reprinted with permission of NASA. **Page 535** Reprinted with permission of U.S. Department of Energy. **Page 538** Reprinted with permission of Palm, Inc. **Page 544** Reprinted with permission of National Weather Service. **Page 550** Reprinted with permission of Chronicle Books. **Pages 551–552** Reprinted with permission of West Virginia EMS System. **Page 555** Reprinted with permission of Chronicle Books. **Page 559** Reprinted with permission of Stewart-MacDonald. **Page 561** Reprinted with permission of Nikon Corporation. **Pages 564–565** Reprinted with permission of Husqvarna. **Page 567** Reprinted with permission of Ryobi Outdoor Products. **Page 568** Reprinted with permission of Geoffrey Peckham. **Page 570** Reprinted with permission of Adobe Systems Incorporated. **Page 571** Reprinted with permission of Adobe Systems Incorporated. **Page 587** Reprinted with permission of National Science Foundation. **Page 588** Reprinted with permission of National Science Foundation. **Page 612** Reprinted with permission of U.S. Geological Survey. **Pages 617–621** Reprinted with permission of International Maritime Organization. **Pages 655–657** Reprinted with permission of National Survey on Drug Use and Health. **Page 663** Reprinted with permission of U.S. Census Bureau. **Page 664** Reprinted with permission of Pew Center. **Page 683** Reprinted with permission of AEC Software. **Pages 690–694** Reprinted with permission of U.S. Preventative Services Task Force.

Photos in HELP boxes: PhotoAlto/ImageState/Eric Audras.

Photos in AT WORK boxes: Alamy/Jim Pickerell.

Index

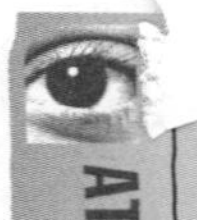

AT A GLANCE

…ies, which are interspersed throughout each chapter, serve as study aides and self-check reviews of major topic discussions.